Fundamental Aspects of Ultrathin Dielectrics on Si-based Devices

NATO Science Series

A Series presenting the results of activities sponsored by the NATO Science Committee. The Series is published by IOS Press and Kluwer Academic Publishers, in conjunction with the NATO Scientific Affairs Division.

General Sub-Series

A. Life Sciences   IOS Press
B. Physics   Kluwer Academic Publishers
C. Mathematical and Physical Sciences Kluwer Academic Publishers
D. Behavioural and Social Sciences   Kluwer Academic Publishers
E. Applied Sciences   Kluwer Academic Publishers
F. Computer and Systems Sciences   IOS Press

Partnership Sub-Series

1. Disarmament Technologies   Kluwer Academic Publishers
2. Environmental Security   Kluwer Academic Publishers
3. High Technology   Kluwer Academic Publishers
4. Science and Technology Policy   IOS Press
5. Computer Networking   IOS Press

The Partnership Sub-Series incorporates activities undertaken in collaboration with NATO's Partners in the Euro-Atlantic Partnership Council – countries of the CIS and Central and Eastern Europe – in Priority Areas of concern to those countries.

NATO-PCO-DATA BASE

The NATO Science Series continues the series of books published formerly in the NATO ASI Series. An electronic index to the NATO ASI Series provides full bibliographical references (with keywords and/or abstracts) to more than 50000 contributions from international scientists published in all sections of the NATO ASI Series.
Access to the NATO-PCO-DATA BASE is possible via CD-ROM "NATO-PCO-DATA BASE" with user-friendly retrieval software in English, French and German (© WTV GmbH and DATAWARE Technologies Inc. 1989).

The CD-ROM of the NATO ASI Series can be ordered from: PCO, Overijse, Belgium.

3. High Technology – Vol. 47

Fundamental Aspects of Ultrathin Dielectrics on Si-based Devices

edited by

Eric Garfunkel
Rutgers University,
Piscataway, New Jersey, U.S.A.

Evgeni Gusev
Rutgers University,
Piscataway, New Jersey, U.S.A.

and

Alexander Vul'
A.F. Ioffe Physico-Technical Institute of the Russian Academy of Science,
St. Petersburg, Russia

Kluwer Academic Publishers

Dordrecht / Boston / London

Published in cooperation with NATO Scientific Affairs Division

Proceedings of the NATO Advanced Research Workshop on
Fundamental Aspects of Ultrathin Dielectrics on Si-based Devices:
Towards an Atomic Scale Understanding
St. Petersburg, Russia
August 4–8, 1997

A C.I.P. Catalogue record for this book is available from the Library of Congress.

ISBN 0-7923-5007-3 (HB)

Published by Kluwer Academic Publishers,
P.O. Box 17, 3300 AA Dordrecht, The Netherlands.

Sold and distributed in the U.S.A. and Canada
by Kluwer Academic Publishers,
101 Philip Drive, Norwell, MA 02061, U.S.A.

In all other countries, sold and distributed
by Kluwer Academic Publishers,
P.O. Box 322, 3300 AH Dordrecht, The Netherlands.

Printed on acid-free paper

All Rights Reserved
© 1998 Kluwer Academic Publishers
No part of the material protected by this copyright notice may be reproduced or utilized in any form or by any means, electronic or mechanical, including photocopying, recording or by any information storage and retrieval system, without written permission from the copyright owner.

Printed in the Netherlands

TABLE OF CONTENTS

Preface   ix

Introduction

Ultrathin dielectrics in silicon microelectronics - an overview   1
    L.C. Feldman, E.P. Gusev and E. Garfunkel

Section 1. Recent advances in experimental studies of SiO$_2$ films on Si

Study of the Si/SiO$_2$ interface using positrons: present status and prospects   25
    J.M.M. de Nijs and M. Clement

Medium energy ion scattering studies of silicon oxidation and oxynitridation   39
    E. Garfunkel, E.P. Gusev, H.C. Lu, T. Gustafsson and M.L. Green

Synchrotron and conventional photoemission studies of oxides and
N$_2$O oxynitrides   49
    Z.-H. Lu

Stress in the SiO$_2$/Si structures formed by thermal oxidation   65
    A. Szekeres

Section 2. Theory of the SiO$_2$/Si and SiO$_x$N$_y$/Si systems

Modeling the oxide and the oxidation process: can silicon oxidation be solved?   79
    A.M. Stoneham and C.J. Sofield

Core-level shifts in Si(001)-SiO$_2$ systems: the value of first-principle
investigations   89
    A. Pasquarello, M.S. Hybertsen, G.-M. Rignanese and R. Car

A simple model of the chemical nature of bonds at the Si-SiO$_2$ interface and
its influence on the electronic properties of MOS devices   103
    H.Z. Massoud

Chemical perspectives on growth and properties of ultrathin SiO$_2$ layers   117
    G.F. Cerofolini and N. Re

A theoretical model of the Si/SiO$_2$ interface   131
    A. Markovits and C. Minot

Section 3: Growth mechanism, processing, and analysis of (oxy)nitridation

Spatially-selective incorporation of bonded-nitrogen into ultra-thin gate
dielectrics by low-temperature plasma-assisted processing   147
    G. Lucovsky

Isotopic labeling studies of oxynitridation in nitric oxide (NO) of Si and SiO$_2$ 165
    I. Trimaille, J.-J. Ganem, L.G. Gosset, S. Rigo, I.J.R. Baumvol,
    F.C. Stedile, F. Rochet, G. Dufour and F. Jolly

Thermal routes to ultrathin oxynitrides   181
    M.L. Green, D. Brasen, L.C. Feldman, E. Garfunkel, E.P. Gusev,
    T. Gustafsson, W.N. Lennard, H.C. Lu and T. Sorsch

Nitrogen in ultra thin dielectrics   191
    H.B. Harrison, H.-F. Li, S. Dimitrijev and P. Tanner

Endurance of EEPROM-cells using ultrathin NO and NH$_3$ nitrided tunnel oxides   217
    A. Mattheus, A. Gschwandtner, G. Innertsberger, A. Grassl and A. Talg

Effects of the surface deposition of nitrogen on the oxidation of silicon   227
    T.D.M. Salgado, I.J.R. Baumvol, C. Radtke, C. Krug and F.C. Stedile

Section 4: Initial oxidation and surface science issues

Surface, interface and valence band of ultra-thin silicon oxides   241
    T. Hattori

Low temperature ultrathin dielectrics on silicon and silicon carbide surfaces:
from the atomic scale to interface formation   257
    P.G. Soukiassian

Interaction of O$_2$ and N$_2$O with Si during the early stages of oxide formation   277
    A.A. Shklyaev

Scanning tunneling microscopy on oxide and oxynitride formation, growth
and etching of Si surfaces   289
    H. Neddermeyer, T. Doege, E. Harazim, R. Kliese, A. Kraus,
    R. Kulla, M. Mitte and B. Röttger

The interaction of oxygen with Si(100) in the vicinity of the oxide
nucleation treshold   309
    V.D. Borman, V.I. Troyan, Yu.Yu. Lebedinski

Section 5: Electrical properties and microscopic models of defects

Tunneling transport and reliability evaluation in extremely thin gate oxides 315
M. Hirose, Y. Mizubayashi, K. Morino, M. Fukuda and S. Miyazaki

Electrical defects at the SiO$_2$/Si interface studied by EPR 325
J.H. Stathis

Towards atomic scale understanding of defects and traps in oxide/nitride/oxide and oxynitride systems 335
V.A. Gritsenko

A new model of photoelectric phenomena in MOS structures: outline and applications 343
H.M. Przewlocki

Point defect generation during Si oxidation and oxynitridation 359
C. Tsamis and D. Tsoukalas

Optically induced switching in bistable structures: heavily doped n$^+$ - polysilicon - tunnel oxide layer - n - silicon 375
V.Yu. Osipov

Heterojunction Al/SiO$_2$/n-Si device as an Auger transistor 383
E.V. Ostroumova and A.A. Rogachev

Radiation induced behavior in MOS devices 391
V.V. Emelianov, G.I. Zebrev, O.V. Meshurov, A.V. Sogoyan and R.G. Useinov

Section 6: Hydrogen/Deuterium issues

Hydrogenous species and charge defects in the Si-SiO$_2$ system 397
E.H. Poindexter, C.F. Young and G.J. Gerardi

The role of hydrogen in the formation, reactivity and stability of silicon (oxy)nitride films 411
F.H.P.M. Habraken, E.H.C. Ullersma, W.M. Arnoldbik and A.E.T. Kuiper

Hydrogen-induced donor states in the MOS-system: hole traps, slow states and interface states 425
J.M.M. de Nijs, K.G. Druijf and V.V. Afanas'ev

Section 7: New substrates (SiC, SiGe) and SOI technologies

Future trends in SiC-based microelectronic devices   431
    A.A. Lebedev and V.E. Chelnokov

The initial phases of SiC-SiO$_2$ interface formation by low-temperature (300°C) remote plasma-assisted oxidation of Si and C faces on flat and vicinal 6H SiC   447
    G. Lucovsky and H. Niimi

Challenges in the oxidation of strained SiGe layers   461
    V. Craciun, J.-Y. Zhang and I.W. Boyd

The current status and future trends of SIMOX/SOI, new technological applications of the SiC/SOI system   477
    J. Stoemenos

Local tunnel emission assisted by inclusions contained in buried oxides   493
    L. Meda and G.F. Cerofolini

Appendix

Authors index   503

List of workshop participants   505

PREFACE

The goal of this NATO Advanced Research Workshop entitled "Fundamental Aspects of Ultrathin Dielectrics on Si-based Devices: Towards an Atomic-scale Understanding", which was held in St. Petersburg from August 4 to 8, 1997, was to: (i) report recent progress on the "**atomic-scale**" physical, chemical, and electrical understanding of ultrathin dielectric films on silicon-based materials, (ii) identify priority directions for future basic research in the field, and (iii) promote interdisciplinary scientific exchange between participants. Over 50 scientists from leading academic, governmental, and industrial laboratories in 18 countries and 15 scientists from local (St. Petersburg) Institutes and Universities participated in the Workshop.

The Workshop covered the following topics: ultrathin dielectrics (oxides, nitrides, and oxynitrides) in Si-based microelectronics; recent advances in experimental studies of SiO_2/Si; the theory of the SiO_2/Si system; oxynitride growth mechanism and processing; surface preparation, initial oxidation, and surface science issues; new ultrathin dielectric characterization methods; electrical properties and microscopic models of defects; H/D issues and devices reliability; and new substrates (SiC, SiGe) and technologies (SOI, etc.).

The first day of the Workshop began by reviewing recent advances in experimental studies of the SiO_2/Si system in the ultrathin regime with emphasis on (i) a microscopic understanding of the growth mechanism (as revealed by high depth resolution and isotopic labeling techniques), and (ii) the atomic-scale configuration of the SiO_2/Si interface as determined by photoemission. The afternoon session was devoted to the current status of the theory of the SiO_2/Si and SiO_xN_y/Si systems. In particular, phenomenological approaches of modeling oxidation kinetics and more sophisticated state-of-the-art ab-initio schemes to calculate interfacial microstructure were discussed.

New results on the timily topic of silicon oxynitridation were presented on the second day. Specifically, the participants discussed growth mechanism, various processing issues, analysis, and applications of (oxy)nitridation. A significant part of the discussion was devoted to the ideal nitrogen profile in the film (the current thinking is nitrogen should be placed near both dielectric/Si and poly-Si gate/dielectric interfaces) and reaction/processing pathways which can be used to achieve the ideal profile. Seventeen papers were presented at the evening poster session, which was followed by a panel discussion reviewing results of the first two days. Also discussed were: mobile species and elemental processes during oxidation, oxide microstructure, thermodynamics and transport of nitrogen in ultrathin films, the mechanism of nitrogen as a diffusion barrier and electronic band structure.

Surface science approaches, using STM, photoemission, optical and other techniques, certainly needed to understand sub-5 nm films, were reviewed in the morning session on the third day. This was followed by a session on electrical properties and microscopic models of defects which determine ULSI device performance and are, therefore, of particular concern for the semiconductor industry.

The fourth day of the Workshop brought discussions of novel techniques for ultrathin dielectric characterization (including positron annihilation, free electron lasers, photoelectric and nonlinear optical techniques), new substrates (SiC, SiGe), and new technologies (SOI). It was clear that these materials and methods represent new, relatively unexplored, and very exciting topics in the field of Si-based science and technology. In the evening, a second open discussion took place where the following issues were brought up: electrical defects at the interface and their identification, interface structure and the ways it can be understood and optimized, new techniques which complement conventional tools, new materials, the search for alternative dielectrics, and other issues.

The final day of the Workshop was devoted to hydrogen/deuterium issues and the role of H/D in device reliability. Progress in EPR and NRA studies was reviewed.

There was a general concensus that despite more than 30 years of intense research, some basic issues of the SiO_2/Si system still remain unclear, especially at the atomic scale. An "atomic-scale" understanding of growth mechanisms, microstructure, and defects in both "traditional" (SiO_2) and novel dielectrics (oxynitrides in the short term) and substrates is still critical and will be the focus of near-future research. New sub-nanometer resolution techniques and traditional surface science tools, on one hand, and state-of-the art calculational methods, on the other, form a powerful base to address many unresolved issues. Many participants suggested that a similar meeting/workshop be held in 2-3 years to review results of new experiments that result in part from the ideas generated at this Workshop.

The enjoyment of the week came not only from the excellent presentations and stimulating discussions, but from the beauty of St. Petersburg amplified by gorgeous summer weather. The foreigners among us were not only delighted in the "physical" beauty of the city (architecture, canals, palaces, parks, etc.) but enjoyed rich cultural experiences (the Hermitage, the Palaces, opera, ballet) and the hospitality of local people (at the opening reception in the St. Petersburg Science House, the banquet, and during walking tours in the city).

The editors would like to thank the members of the International Advisory Committee, Dr. G.F. Cerofolini, Prof. L.C. Feldman, Prof. V.G. Litovchenko, Prof. S. Rigo, and Prof. A.M. Stoneham for their valuable help in selecting participants from different countries and developing a scientific program. We would also like to thank Dr. M.L. Green, Dr. E.H. Poindexter, Prof. G. Lucovsky, Dr. F.R. McFeely, Prof. H. Massoud, and Prof. T. Tsakalakos for their encouragement during the early stages of the planning of the Workshop, Drs. S. Kidalov and Yu. Osipov for perfect local organization in St. Petersburg, Mrs. D. White for secretarial assistance, Dr. H.C. Lu for his help in developing the webpage (http://www.physics.rutgers.edu/NATO_dielectrics/), and all participants for their high-quality presentations, active participation (including peer-reviewing papers presented in this book) and fruitful discussions at the Workshop. The Workshop would not have been possible without financial support from High Technology Program of the NATO Scientific Division directed by Dr. J.A. Rausell-Colom. We also greatly appreciate financial contributions from our co-sponsors: the Russian Foundation for Basic Research, the U.S. Office of Naval Research (Dr. L. Cooper), the U.S. Army Research Office (Dr. J. Rowe), SGS-Thomson Microelectronics (Dr. G. Cerofolini and Dr. G. Ferla) and the Laboratory for Surface Modification at Rutgers University (Prof. T.E. Madey).

October 1997
Piscataway, New Jersey

St.Petersburg, Russia

Eric Garfunkel
Evgeni Gusev
Alexander Vul'

Ultrathin Dielectrics Workshop co-chairs:
Evgeni Gusev, Alexander Vul' and Eric Garfunkel

Ultrathin Dielectrics Workshop participants

ULTRATHIN DIELECTRICS IN SILICON MICROELECTRONICS

An overview

L.C. FELDMAN
Department of Physics and Astronomy, Vanderbilt University, Nashville, TN 37235, USA

E.P. GUSEV and E. GARFUNKEL
Department of Chemistry, and Laboratory for Surface Modification, Rutgers University, Piscataway, NJ 08854, USA

The paper reviews recent progress and current scientific issues ultrathin gate dielectrics on silicon based devices. We discuss microstructural aspects and electrical defects of the SiO_2/Si interface, oxidation mechanism, the very initial stages of the interaction of oxygen with silicon surfaces in the "surface science" limit, roughness at the SiO_2/Si interface and roughening that occurs during initial oxidation under certain conditions, hydrogen and deuterium in thin SiO_2 films, silicon oxynitridation and nitridation, "alternative" high-K dielectrics, and processing issues.

1. Historical perspective

The ages of civilization are often denoted by the dominant material of the time; the stone age, the bronze age and iron age. Many would say that we are now in the silicon age. Certainly the last forty years have given rise to a revolution in technology and, more significantly, a dramatic change in the culture of society. The silicon transistor and development of the integrated circuit have produced a massive change not only in technology and the economy but in culture and thinking commensurate with the great materials revolutions of time.

Those familiar with the field are aware that the underlying science that has enabled this revolution is the atomic scale understanding and manipulation of the materials and processes that make up silicon-based microelectronic devices.[1] This is particularly applicable to the silicon/silicon dioxide interface, the heart of the modern transistor and the subject of this book.

This year, 1997, represents a significant anniversary in the history of physics and the history of the transistor, perhaps the most far reaching "product" of quantum science. Exactly one hundred years ago marks the identification of the electron [2] and the beginning of the development of the quantitative atomistic description of matter. This discovery was followed by the rapid development of quantum mechanics, first applied at the atomic level and then later to the understanding of more complicated systems, including solids. Fifty years ago marks the first demonstration of the transistor - the first transistor was actually germanium, not silicon.[3] And almost forty years ago marks the first demonstration of the silicon MOSFET, metal-oxide semiconductor field effect transistor (Fig. 1) which has evolved into the key component of most ULSI devices.

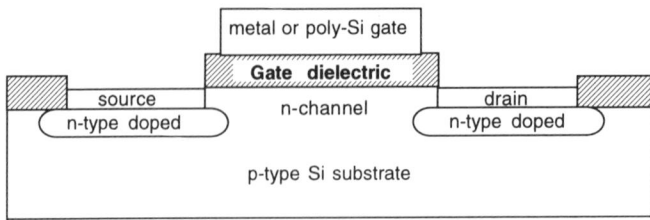

Fig. 1 Schematic of a simple n-channel MOSFET

Perhaps no one truly understood that this device, which lent itself to a unique planar geometry, would gave rise to the integrated circuit concept and all its ramifications. Furthermore the planar geometry lends itself to "scaling", the reduction in individual device size that leads to the characteristics of higher speed, greater density (more devices/area), and lower power - all parameters in the correct direction to revolutionize our world.[4-9] The scaling behavior is described by the well-known Moore's law[10, 11], i.e. exponential growth of chip complexity due to increasing wafer size and reducing minimum feature size accompanied by concurrent improvement of technology processes and circuit design.(Fig. 2) As an illustration, the minimum feature size has dropped from tens of microns in early 60s to the current value of 0.25 µm and is projected to be below 0.05 µm in about 15 years. (Table 1) For gate dielectrics, these numbers translate into a gate dielectric thickness of ~ 1 - 2 µm in the early 60s, ~ 4 - 5 nm in 1997, and an equivalent thickness of <1 nm in 2012. (Table 1).

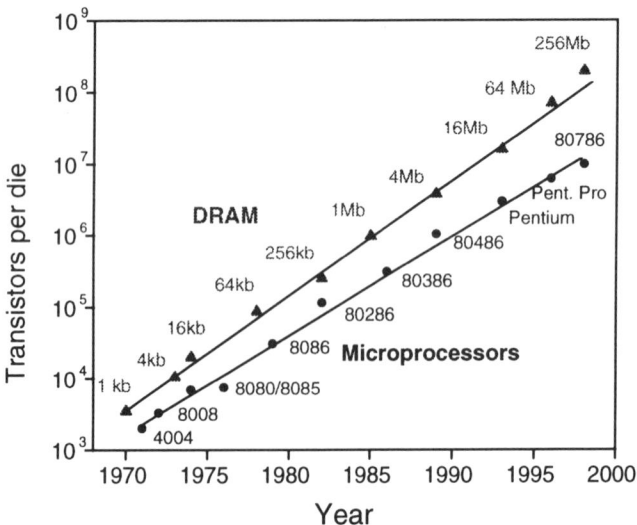

Fig. 2 Integrated circuit scaling (following a revised Moore's law, i.e. circuit complexity doubles every 18 months [11])

TABLE 1. Technology roadmap characteristics in the area of thermal/thin films[12]

First Year of IC Production	1997	1999	2001	2003	2006	2009	2112
DRAM generation	256M	1G		4G	16G	64G	256G
Minimum Feature Size, µm	0.25	0.18	0.15	0.13	0.10	0.07	0.05
Equivalent Oxide Thickness, nm	4 - 5	3 - 4	2 - 3	2 - 3	1.5 - 2	< 1.5	< 1.0

The success of the MOSFET hinged on a most important and fundamental detail; namely the limitations imposed by surface states. Even ideal surfaces can possess electronic states that represent sufficient charge to make the MOSFET inoperable. Hans Quiesser, in his exciting book, Conquest of the Microchip[13], notes: "what finally saved the day was that an incredibly stable oxide of silicon can be wrapped around the crystal to protect it." Another historical account, The History of Engineering & Science in the Bell System[14], describes "surface state problems were resolved by an unexpected discovery....Kahng and Atalla found that silicon and clean, thermally grown SiO_2 interfaces contain sufficiently small surface states to realize a true field effect transistor in silicon." And in keeping with the anniversary theme, forty years ago marks the 1957 Proceedings of the Electrochemical Society, where Frosch and Derick first reported the beneficial properties of the SiO_2/Si materials system. [15]

In the last forty years this materials interface has received an enormous amount of scientific attention.[16-35] The subject has given rise to a field unto itself with specialists, sub-specialties and a drive to understand and improve this remarkable structure. It would be frightening to count the number of scientific papers that have reported on this interface; suffice it to say that estimates put the number at 50,000!

Nevertheless, notwithstanding all this attention, outstanding scientific issues remain at the very forefront, limiting the further evolution of silicon science and technology. Indeed these issues have become even more critical. For as the device has scaled down, with the corresponding reduction of silicon dioxide thickness from microns to nanometers, the silicon/silicon dioxide interface becomes more prominent and more limiting. We are close to fundamental limits where the need for understanding and atomic control becomes ever more critical.

2. Recent Progress and Current Scientific Issues of Ultrathin Dielectrics on Si-based Devices

Despite an extensive history of exciting results and technological progress[16, 17, 20, 21, 23-27, 29-34, 36-40], there are still many unresolved fundamental problems concerning dielectric growth and microstructure, especially in the now technologically important ultrathin (<5 nm) film regime.

2.1 KEY ISSUES OF THE Si/SiO_2 INTERFACE: DEFECTS AND STRUCTURE

2.1.1 Electrical Defects at the Interface
It is now well established that ULSI reliability and electrical properties are strongly dependent on the quality of the silicon-gate oxide interface region. [24, 26, 30, 36, 38, 40-

49] For example, channel mobility, leakage current, time-dependent breakdown, and hot-electron induced effects have all been correlated with the oxide structure and defects at the SiO2/Si interface. Capacitance-voltage (C-V) measurements, deep-level transient spectroscopy (DLTS), electronic paramagnetic resonance (EPR), and charge pumping measurements have shown that ideal behavior is perturbed by several kinds of electrical defects. These include interface traps, fixed oxide charge, border traps near the interface, oxide trapped charge, and mobile ionic charge. (Comprehensive reviews of electrically active defects in the SiO_2/Si and SiO_xN_y systems can be found elsewhere[26, 27, 36, 43, 44, 50]). Although the electrical defects are controlled by fabrication conditions, oxidation ambient, etc.[51], relatively little is known about the atomic configuration of these defects, especially for ultrathin oxide films. Pioneering EPR experiments by E. Poindexter and colleagues argued that interface traps are related to dangling bonds at the interface, so-called P_b-centers.[26] Recently, it has been shown that P_b centers have different configurations for Si(111) and Si(100) surfaces[52], and that the general relationship between P_b centers and interface traps may be questionable. The mechanism of breakdown of ultrathin dielectrics is also not fully understood. [53-57]

Much of our knowledge of defects has come from relating electrical device parameters with electronic states of physically and chemically reasonable atomic structures in the SiO_2/Si system.[26, 27, 36, 40, 45] Unfortunately the situation becomes more complicated for ultrathin films, where it is unclear to what extent traditional electrical methods (C-V, I-V, etc.) can be used to characterize such thin films. For instance, low-frequency C-V measurements cannot be used because of high direct tunnel currents. Another issue which becomes increasingly important as gate oxide thickness shrinks below 5 nm is that "physical" measurements of oxide thickness (as monitored by techniques such as XTEM, XPS, ellipsometry) differ from each other (depending on the parameters used, e.g. photoelectron meanfree path in XPS and refractive index in ellipsometry) and are slightly different from the "electrical" thickness (as measured by C-V and another electrical tools).[58] The electrical thickness is generally larger due to poly-Si depletion effects and quantum effects in the silicon substrate. The difference can be as high as ~ 1.0 nm.

2.1.2 Microstructure of the Interface, Near-interfacial Oxide and Near-interfacial Silicon

Oxide structure, especially in the near interfacial region, appears to be one of the critical parameters that determines device performance. Recent experiments[59] have shown that the difference in the electrical reliability of the SiO_2/Si(100) system, as compared to SiO_2/Si(111), is caused by the difference in oxide quality or oxide structure (even when the interface microroughness is of the same level), indicating the importance of ultra-thin oxide microstructure studies.

It is generally agreed that there is a transition region (of altered structure and/or stoichiometry) between crystalline silicon and SiO_2.[16, 20, 24, 30, 32, 60-78] However, the thickness of this region has been reported to vary from 0.5 nm to 3 nm (even 7 nm!). Although such a large scatter can be attributed to differences in oxidation procedure and oxide thickness, more important factors are that the width of the transition region depend strongly on both the probing technique and the definition of the transition region. For instance, from ellipsometry measurements[23, 79], the transition region is understood as a layer with optical properties different from both bulk oxide and crystalline silicon, although how different is not very clear. Photoemission studies[16, 20, 32, 80-92] consider the transition region as a layer of Si(n+) suboxide states with local electronic configurations different from pure Si and SiO_2 (although the interpretation of the Si 2p photoemission is still under debate[93-98]). Ion scattering experiments suggest a small non-stoichiometric region, but a rather large strained region in the underlying silicon.[70, 71] One should also mention that a quantitative analysis of the transition region width (and the total thickness of thin oxides) by the above mentioned techniques is difficult because: for example, the refractive index of thin films[79, 99-102] (ellipsometry and other optical techniques) and photoelectron mean path[16, 20, 58, 92, 103, 104] (photoemission) are not universally agreed upon, and, moreover, both change as the composition of the transition region changes. Cross-sectional TEM yields beautiful images of gate oxides, some as thin as 1 - 1.5 nm; however, averaging information over the full thickness of the cross-section slice results in a limited thickness resolution. In addition, this technique is not sensitive to chemical composition. Scanning TEM (STEM) has recently been developed to combine the spatial abilities of TEM with chemical characterization from EELS.[64, 65, 105, 106] However, to date, preliminary studies of the SiO_2/Si system by this promising technique have been done only on poor-quality oxides.[64, 65] Thus, despite extensive work neither the atomic-scale structure nor the composition (or gradient) in the transition region are well understood - there has been no universally accepted "conventional" model.[107, 108]

Although there is no definitive evidence for epitaxial oxide growth on silicon during thermal oxidation, some HRTEM[90, 109] and x-ray diffraction/scattering experiments[78, 110-114] have argued that a thin ordered oxide region exists between crystalline Si(100) and amorphous SiO_2. Some reports indicate that only a small fraction of the oxide is ordered.[114, 115] The accurate atomic position of the atoms in this region is still not well defined; even the structure of this ordered phase (cristobalite[90, 110, 116], tridymite[78, 109], or other models[113, 117]) is not agreed upon. Other workers have explicitly stated that they do not observe oxide or interface periodicity using HRTEM [118-121]. Since a completely amorphous interface SiO_2 layer might leave more interfacial Si dangling bonds unsaturated, some order on the SiO_2 side of the interface could help explain the very low electrical defect density at the SiO_2/Si interface. Microvoids (of ~ 1 nm in size and with low concentration of ~ 10^9 cm^{-2}) have been observed in the oxide and interface region by positron annihilation technique.[122] Stress is also believed to occur in the near-interfacial oxide because of lattice mismatch between crystalline Si and the oxide and different thermal expansion coefficients.[23, 60, 75]

2.2 SILICON OXIDATION

In 1965, B. Deal and A. Grove published a seminal paper in which they presented results on silicon (dry and wet) oxidation and proposed a model to describe the kinetics of the growth of relatively thick oxide films.[123] According to this model, the oxide grows via molecular oxygen (or water) diffusion through the oxide film and reaction with silicon at the SiO_2/Si interface. Interestingly, Deal and Grove noted that the initial very thin oxidation regime was anomalous. It is precisely this "anomalous regime" that is now used and studied. It has been observed that the oxidation kinetics for ultrathin films differ from a simple Deal-Grove model extrapolation to thin layers, although this issue is still under debate.[19, 21, 23, 31, 77, 90, 102, 124-126] Several phenomenological models, such as the parallel oxidation model[127], Massoud's model[19, 128], space charge model[129], reactive layer model [37, 130], and others[90, 116, 124, 131-143], have been proposed to account for this "faster" initial growth. (A review of the models can be found elsewhere [21, 23, 25, 31, 37]). Some of the models fit the experimental data on oxidation kinetics quite satisfactorily, although often with a number of physically unjustified empirical fitting parameters. Furthermore, most of the models do not have direct experimental support; an analysis of kinetic results alone does not allow one to distinguish conclusively between models. Key questions are the nature of diffusing species, where oxidation takes place [72, 74, 90, 144-148], how oxidation effects interface structure [118, 121, 149], and how these phenomena depend on oxidation parameters, such as temperature, pressure, and post-oxidation treatment. It is also quite possible that a "steady-state" approximation employed in the Deal-Grove model breaks down in the limit of ultrathin films. In addition, some of the key assumptions of the model, such as a first order quasi-chemical reaction at a geometrical interface, are probably not valid for ultrathin films. Current thinking about silicon oxidation includes two additional spatial regions for the reaction of incident oxygen: (i) the reaction of oxygen with incompletely oxidized silicon in the near-interfacial (~ 1-2 nm) region, and (ii) a surface oxygen exchange reaction, as has became evident in isotopic labeling experiments [74, 90, 144-148, 150-153] Finally, we note that although the Deal-Grove model and its modifications are still used in process simulators (e.g. TMA/SUPREME code), there is no generally accepted atomistic-level model to be incorporated into the simulators for the ultrathin growth regime.[154] This is especially true for ultrathin oxynitride growth (section 2.6).

2.3 VERY INITIAL STAGES OF Si OXIDATION IN THE "SURFACE SCIENCE" LIMIT

Oxide phase formation during the initial 0-1 nm of growth is also not fully understood. [29, 81, 83, 85, 149, 151, 155-173] This initial oxidation regime plays a critical role in the formation of gate oxides because it may significantly effect the final structure of the interface and, therefore, device properties. In fact, two step oxidation (i.e. a very thin thermal oxide followed by deposition/growth of a thicker dielectric film) is used in many processes and may become even more important in the future for a stacked oxide(oxynitride)/high-K dielectric (section 2.7). There is an unresolved debate in the literature concerning what may be considered the lateral homogeneity of the oxide during the first few monolayers of growth: Does the initial oxidation of silicon proceed in a layer-by-layer fashion or does it grow non-uniformly through the formation of 3D oxide islands? There is a small but growing number of publications supporting layer-by-layer mode.[85, 118, 121, 149, 171, 172, 174, 175] Using a real-time HRTEM, Ross and Gibson observed that steps on the Si(111) surface were immobile during the initial oxidation, indicative of layer-by-layer growth.[121, 149] Some photoemission results reveal a stepwise increase in the amount of oxide or Si(n+) suboxide states on both Si(111) and Si(100) implying layer-by-layer growth.[85, 171] However, in other studies, this was not confirmed.[88] In addition to the above mentioned problem of the identification of the Si(n+) suboxide states[94-96], quantitative analysis in photoemission experiments is dependent on several important factors including background subtraction procedure[92] and the source of the primary radiation[89]; experiments[89] taken on the same samples with conventional (Al K_α) and synchrotron sources yielded concentrations of the suboxide states that differed by more than a factor of two. STM experiments[155, 156, 161, 162, 165-169, 173, 176-186] have given a new insight on the initial interaction of oxygen with Si, demonstrating multiple bonding configurations of oxygen on the surface and that, at submonolayer coverages, especially at elevated temperatures[161, 162, 166, 167], surface oxides grow laterally as 2D islands. Unfortunately, STM studies of silicon oxidation are mostly limited to the mono(double) layer regime because of the large SiO_2 band gap. Also, change of surface morphology can be confused in some cases with "chemical" effects due to the presence of oxygen on the surface.

2.4 ROUGHNESS

Roughness is one of the factors which is now recognized as a critical processing parameter.[59, 187-201] As interface roughness increases both the breakdown field strength[187, 193, 194, 202, 203] and channel mobility[188, 194, 203] decrease. CV measurements show that both the interface traps and the fixed oxide charge depend on interface roughness, with higher defect density for rougher interfaces.[187] Interface roughness depends on oxidation temperature, oxidation rate, post-oxidation annealing, and pre-oxidation cleaning.[120, 178, 187, 204] However the microscopic origin of interface roughness and the mechanisms causing the interface to roughen are not clear. The variation of the interface roughness during oxidation (e.g. if the interface gets rougher[90], smoother[118, 121, 204-207] or preserves the original roughness[119, 208]) is not agreed upon.

The invention and development of scanning probe microscopes, STM and AFM, over the past decade shed light on surface morphology on the atomic-scale and stimulated application of these powerful tools to study various systems, including the SiO_2/Si system.[184] Other techniques[209, 210], including spectroscopic ellipsometry[62, 66, 206], light scattering[209], Nomarski microscopy[211], and X-ray diffraction[204, 212] have also been used to characterize wafer roughness. Results obtained by these methods frequently differ from each other. A major source of this discrepancy is the different lateral resolution and analyzed areas afforded by each probe[210, 211]; roughness is now usually quoted as function of k. Roughness at a buried interface is more difficult to characterize than surface roughness.

Using HRTEM Gibson and colleagues[149] have observed that roughness increased when approaching the phase boundary[157, 213, 214] (in P-T space) for which a transition from oxide growth (passive oxidation) to oxide decomposition via SiO desorption (active oxidation) takes place. This was also seen by STM[162, 167] and medium energy ion scattering (MEIS) [215]. These observations imply the existence of a separate "roughening" regime in the (P-T) phase space between passive and active oxidation in O_2. All these studies were done in UHV chambers at oxygen pressures not higher than 10^{-5} Torr, far from the "commercial" process. It is unclear if the "roughening" regime exists at much higher pressures, closer to real device processing conditions, or for other reactive gases (e.g. N_2O). Understanding the reasons for surface roughening and the oxidation conditions under which the surface roughens would offer important new insights in device processing. The observation of roughening also demonstrates the strong correlation between processing parameters, growth mechanism(s) and the microstructure.

2.5 HYDROGEN/DEUTERIUM ISSUES

It has long been realized that hydrogen plays an important role in silicon based technology.[48, 216] The relatively high background level of hydrogen-containing molecules in processing ambients (CVD reactors, etc.) and the high diffusivity of hydrogen in silica led to a belief that hydrogen is ubiquitous in the SiO_2/Si system.[216-221] In fact, nuclear reaction analysis experiments have revealed that the hydrogen[222] (deuterium [223]) concentration in SiO_2 films can be as high as in the 10^{21} cm^{-3} range.

Hydrogen is not necessarily detrimental and some have argued that a number of beneficial effects arise from hydrogen in the film. For example, it is generally believed that hydrogen passivates dangling bonds at the interface which in turn further reduces the concentration of surface states.[26, 216] On the other hand, it has been demonstrated that hydrogen can be associated with instabilities and degradation of MOS-based devices, especially under irradiation.[48, 220, 224-226] In addition to the electrical effects, hydrogen may be involved in other chemical reactions in the SiO_2 films (e.g. oxidation reactions in nominally dry and wet ambients, nitridation in ammonia[227], etc.) though these reactivity aspects are still not well understood.[228]

Replacing hydrogen for its heavier isotope, deuterium, offers advantages both for (i) studying fundamental aspects of hydrogen/deuterium behavior in ultrathin SiO_2 films, and (ii) for practical applications. Several groups have successfully used deuterium (low background level) to mimic basic aspects of hydrogen reactivity/diffusivity in the oxide.[223, 229-234] It has been demonstrated recently that processing of the SiO_2/Si system in D_2 (instead of H_2) results in: (i) a retardation of the rate of interface trap built-up by a factor of 2.6 - 4.5 [235], (ii) transistor lifetime improvement by factors of 10 - 50 [236, 237]. It is clear that the difference in mass between hydrogen and deuterium is not enough to explain these significant isotopic effects. Electronic effects also need to be taken into account. [238]

2.6 ULTRATHIN OXYNITRIDE FILMS

While "pure" SiO_2 films have served as the gate dielectric since the birth of integrated circuits, the traditional SiO_2 dielectric faces a number of problems and challenges in the sub-5 nm thickness regime. Dopant (boron) diffusion into and through the ultrathin oxide from the poly-Si gate becomes a critical issue. Oxynitrides (SiO_xN_y) are leading candidates to replace conventional "pure" silicon gate oxides for sub-0.25µm devices.[49, 239] Nitrogen incorporated into the oxide has been shown to form a barrier against boron diffusion in p+ gate MOSFETs.[106, 240-251] In addition, a small concentration of nitrogen near the interface appears to reduce hot-electron degradation and improve breakdown properties.[44, 55, 225, 245, 246, 252-258] On the other hand, a high concentration of nitrogen near the interface may cause lower channel mobility and boron accumulation in the oxide[259].

Although it is clear that the nitrogen distribution and concentration in the film are important factors which determine reliability of ultrathin oxynitrides, there is still no consensus as how to best optimize these two factors. One current argument is that the majority of the nitrogen should be located closer to the dielectric/poly-Si gate interface (to prevent boron diffusion), while a smaller amount should be placed near the film/Si substrate interface (to improve hot-electron immunity). Accurate nitrogen depth profiles have been difficult to measure because of limitations of secondary ion mass spectroscopy (SIMS)[260] and etch-back profiling methods. Several groups have demonstrated different nitrogen distributions for RTO vs. furnace grown films; and NH_3 vs. NO vs. N_2O grown films; and following reoxidation in O_2/N_2O. [151, 243-247, 255, 261-265]

The mechanism(s) of silicon oxide nitridation and oxynitride growth, an issue directly related to the nitrogen distribution particularly to nitrogen incorporation near the interface and removal from the film, is also not fully understood. Initial experiments on oxide nitridation in NH_3 showed a significant amount of nitrogen incorporated into the film, though a drawback of processing in ammonia is that too much hydrogen penetrates into the film (section 2.5). [239, 266] The oxidation of silicon in N_2O is particularly attractive because of O_2 - N_2O processing similarities. However, among other factors, oxynitridation in N_2O is complicated by the fast gas-phase decomposition of the molecule into N_2, O_2, NO, and O at typical oxidation temperatures (800-1100°C) with the rate and branching ratio of the products strongly dependent on the processing conditions, temperature, residence time, gas flow, and oxidation reactor type (furnace vs. RTP).[245, 251, 267, 268] In addition, overall nitrogen incorporation into ultrathin dielectric from the N_2O source may not be enough to supress boron penetration.[248] NO is now believed to be responsible for nitrogen incorporation into the film when introducing N_2O (NO) in a furnace or RTP reactor.[245, 261, 262, 265, 269-271] Recent experiments demonstrated however a difference in the mechanism of silicon oxynitridation in NO vs. N_2O.[151, 247, 254, 255, 258, 265] Thermal oxynitridation in N_2O results in atomic oxygen in gas phase which, in turn, cause nitrogen removal from the film (occurring simultaneously with nitrogen incorporation).[262, 264] This is one of the reasons why oxynitridation in NO (during which there is no efficient mechanism of nitrogen removal from the film) results in a higher concentration of incorporated nitrogen, which makes NO processing attractive for ultrathin films. Another practical consequence of the low reactivity of NO towards nitrogen removal from the film is that it allows to layered nitrogen structures (with N both near the interface and near the dielectric surface), as has been demonstrated using sequential $NO/O_2/NO$ exposures.[264]

Local atomic and electronic configurations of nitrogen in ultrathin oxynitrides remain open questions. Photoemission experiments suggest the existence of Si-N bonds (similar to the Si_3N_4 phase), though this may not be the only local configuration for nitrogen.[151, 242, 245, 255, 263, 272, 273] EPR measurements on oxynitrides also show multiple electronic configurations, with signatures of a pure SiO_2 phase, a pure Si_3N_4 phase, and some mixed bonding states.[50]

There are a number of "non-thermal" techniques which allow one to form ultrathin oxynitride or nitride films on silicon. They include (remote) plasma deposition methods[249, 253, 274-276], low pressure chemical vapor deposition[228, 277], jet vapor deposition[278], low energy ion implantation[279] (followed by oxidation) and atomic layer deposition[280] methods. In all cases, the oxide phase is thermodynamically more favorable than the nitride, as follows from the (bulk) phase diagram of the Si-N-O system.[281, 282] For example, Si_2N_2O, the only known stable ternary phase, can be formed (under equilibrium conditions) at extremely low partial pressures of oxygen.[281, 282]

2.7 "ALTERNATIVE" DIELECTRICS FOR SUB-0.1 μm DEVICES - THE NECESSITY OF HIGH-K MATERIALS

One of the fundamental limits of ultrathin oxide/oxynitride films is direct tunneling (current) which grows exponentially with decreasing film thickness.[49, 198, 283, 284] For sub-2 nm films, the current becomes larger than 1 A/cm^2 which, in turn, requires a significant (and unacceptable) power dissipation in logic devices, especially for portable applications. To overcome the direct tunneling problem, the "physical" thickness of the dielectric should be kept large, much thicker than the direct tunneling limit. On the other hand, ULSI scaling is driving a reduction in thickness (increasing capacitance) for next generation fast switching devices. One way to ameliorate these conflicting needs is to replace conventional SiO_2 by a material with a higher dielectric constant, preferably a much higher one. This would allow one to make a film with a thicker "physical" thickness, but the "equivalent" electrical thickness (with respect to pure SiO_2) and the direct tunneling current would be much reduced.

Several high-K materials (Ta_2O_5 Refs.[33, 285-294], TiO_2 Refs.[295-298], Y_2O_3 Refs. [299-305], ferroelectrics Refs.[306-313], CeO_2 Refs. [314-317], ZrO_2 Refs.[318, 319], and HfO_2 Refs.[320, 321]) are now being explored as a gate dielectric. Some of our useful knowledge of these materials comes from research and development studies of their use as capacitors for DRAM and other applications. Though these materials (Table 2) have a dielectric constant larger than SiO_2 (as a reference, $\varepsilon(SiO_2)=3.8$, $\varepsilon(Si_3N_4)=7.8$, ε(BST ferroelectrics) ~ 200 - 300), the search for the best high-K material to be used as a future gate dielectric still continues. The material of choice must have a set of important properties including: (i) a high dielectric constant (preferably >25), (ii) high thermal stability (especially with respect to Si); (iii) perfect stoichiometry (which presumably minimizes intrinsic defects and traps in the film), (iv) a low concentration of interface states and stability of the interface during thermal treatments and external radiation; (v) resistance to dopant diffusion, (vi) low leakage currents; (vii) large band gap (> 4 eV) and an appropriate barrier height (> 1 eV relative to Si); (viii) a low thermal budget, defect-free processing; and (ix) manufacturability and integration with silicon technology. Ideally, this material should have properties as good as SiO_2 on Si, but with a much higher dielectric constant. This is a "tall order" and there is no clear favorite at the moment.

A very thin buffer layer of silicon oxide, oxynitride or nitride may still be required between the silicon substrate and the high-K dielectric in order to minimize interface states and to be a diffusion barrier between the layers. In this case, the equivalent thickness of the stacked dielectric will have contributions from both the thickness of the buffer layer and the equivalent thickness of the high-K layer. To keep the overall equivalent thickness less than 1.5 nm, the buffer layer should consist of not more than 1-2 atomic layers. Needless to say, this mono/double layer should be nearly perfect and the new interface between the buffer layer and the high-K material should be as good

Table 2. Some important properties of metal oxides (from G.V. Samsonov (ed.), The Oxide Handbook, IFI/Plenum, NY-Wash.-London, 1973).

Oxide	MgO	Al$_2$O$_3$	SiO$_2$	TiO$_2$	Cr$_2$O$_3$	Y$_2$O$_3$	ZrO$_2$	Nb$_2$O$_5$	La$_2$O$_3$	CeO$_2$	HfO$_2$	Ta$_2$O$_5$	WO$_3$
Dielectric constant	8 - 10.5	10.5 - 12	3.5 - 4.1	80[a] 167[b] [30][1]	9.2	14	[12-16][2]	50[c] 30[d] 35[e]	20.8	21.2	[16][3] [45][4]	[25][3,5]	20[f] 29[g]
Band gap, eV	7.3	8.8[6]	9	3.05 2.59[7]	4.8	2.1	2 3.95[7]	1 - 3.2	5.4	3.4	4.5[7]	~5[8]	2.2 - 2.5
Work function, eV	3.1-4.4	4.8	5	3.9 6.2		2 - 3.9	3.1 - 5.8		2.8 - 4.2	3.2	2.8 3.75	4.6	
Density, g/cm^3	3.65	3.97	2.2[j] 2.65[h] 2.32[i]	3.84[k] 4.17[l] 4.24[m]	5.21	4.84	5.56[n] 6.28[o]	4.95	6.51	7.13	10.01	8.73	6.47
Heat of formation, $\Delta H°_{298}$ 10^6 J/kg·mole	602	1676	858[j] 911[h]	942[m] 867[k]	1130	1758		1906	1799	1089	1114	2047	841
Entropy, $S°_{298}$ 10^3 J/kg·mole·deg	27	51	42	50	81	124	50	136	153	74	59	143	62
Free energy of formation ΔF_{298} 10^6 J/kg·mole	-570	-1580	-805	-882	-1047	-1796	-1037	-1771	-1703	-971	-1080	-1970	-763
Melting point, K	3073	2319	1993	2128	2573	2649	2963	1783	2573	2670	3063	2150	1743
Thermal conductivity at 373 K/773 K, W/m·deg.	27/ 8.8	30/ 9.1	1.6/ 0.5[r]	6.5/ 3.6			1.9/ 2.1						
Thermal expansion coeff., 10^{-6} deg.	14-15	8	3[p] 0.5[r]	7 - 8	9.6	9.3	5 - 8			3.6 8.5	5.86. 4		

Notes: dielectric constants shown in brackets [...] correspond to thin film values; others to bulk values.
a - parallel to optical axis (for rutile) ; b - perpendicular to optical axis (for rutile); c,d and e - in direction of axis a,b and c, respectively, f - density 2.9 g/cm^3; g - density 4.2 g/cm^3; h - quartz, i - trydimite, j - cristabolite; k - anatase, l - brokite, m - rutile; n - monoclinic system, o - cubic system; p - crystalline; r - silica glass.
Additional Refs:
[1] - H.S. Kim, D.C. Gilmer, S.A. Campbell. D.L. Polla. Appl. Phys. Lett. 69 (1996) 3860; [2] - T.S. Kalkur, Y.C. Lu, Thin Solid Films 207 (1992) 193;
[3] - K. Kukli, J. Ihanus, M. Ritala, M. Leskela, Appl. Phys. Lett. 68 (1996) 3737; [4] - C.T. Hsu, Y.K. Su, M. Yokoyama, Jpn. J. Appl. Phys. 31 (1992) 2501;
[5] - S.K. Tiku, IEEE Electron Devices ED31 (1984) 105; [6] - R.H. French, J. Am. Ceram. Soc. 73 (1990) 477:
[7] - N.I. Medvedeva, V.P. Zhukov, M.Ya. Khodos, V.A. Gubanov, A.L. Ivanovskii, Phys. Stat. Sol. B160 (1990) 517(theory); S.D. Khanin, A.L. Ivanovskii, Phys. Stat. Sol. B174 (1992) 449

as the SiO$_2$/Si interface. This fact motivates and justifies further fundamental studies of ultrathin silicon oxides, oxynitrides and nitrides, especially on the atomic scale. Finally we note that in parallel with the search for high-K dielectrics, new (metal-based) gate materials are beginning to be explored.

2.8 PROCESSING ISSUES

It is generally agreed that oxidation/annealing parameters play an important role in device manufacturing.[24, 30, 51] However, in most cases, these parameters are chosen empirically and not optimized. For example, although there is no doubt concerning the significance of post-oxidation annealing, a fundamental understanding behind the "processing window" for post-oxidation treatments is absent. Recent Auger and photoemission experiments show that a short anneal at 900°C (or higher) reduces the concentration of suboxide states at the interface.[68] On the other hand, EPR studies suggest that longer anneals at these temperatures results in an irreversible degradation of the interface due to formation of silicon monoxide.[322] More systematic work on characterization and understanding of rapid thermal processing (RTP) is required as the industry is moving towards a single wafer manufacturing concept.[248, 277, 323-326] Atomic layer "epitaxy"[327, 328] and other deposition methods[249, 274, 329] of ultrathin dielectrics are also under investigation, especially for low-temperature processing.

Other fundamental aspects of oxidation also affect the fabrication cycle. For example, oxidation is known to release silicon interstitials which cause a number of detrimental effects, such as stacking fault formation and oxidation enhanced diffusion of dopants.[17, 133, 134, 330-334] (In contrast, nitridation of silicon in ammonia generates vacancies[335]). Basic oxidation studies are needed to shed light on this process and possibly control the defect injection. Finally we stress the increasing importance of ultraclean processing.[202, 203, 336-339]

3. Summary

The papers in this book summarize the presentations, discussions and ideas from the NATO Advanced Research Workshop entitled "Fundamental Aspects of Ultrathin Dielectrics on Si-Based Devices: Towards an Atomic Level Understanding" and captures the spirit of our current understanding and the continuing urgency of this problem. The book is divided into sections, roughly equivalent to how the presentations were grouped during the Workshop. In addition to the topics discussed above, some important papers are also included which discuss new materials (SiC, SiGe) [340], new structures (SOI) [341-343], and novel experimental tools (second harmonic generation [344], MEIS, etc.).

References

1. Huff, H.R. (1997) Twentieth century silicon microelectronics, in H. Z. Massoud, H. Iwai, C. Claeys and R. B. Fair (eds.), *International ULSI Symposium at the 191st ECS meeting*.
2. Squires, G. (1997) J.J. Thomson and the discovery of the electron, *Physics World* **10(4)**, 33.
3. Hoddeson, L. (1994) Research on crystal rectifiers during World War II and the invention of the transistor, *History and Technology* **11**, 121.
4. Hu, C. (1993) Future CMOS scaling and reliability, *Proc. of the IEEE* **81**, 682.
5. Sugano, T. (1993) A perspective on next generation silicon devices, *Jpn. J. Appl. Phys.* **32**, 261.
6. Taur, Y., Mii, Y.J., Frank, D.J., Wong, H.S., Buchanan, D.A., Wind, S.J., Rishton, S.A., Sai-Halasz, G.A., and Nowak, E.J. (1995) CMOS scaling into the 21st century: 0.1 µm and beyond, *IBM J. Res. Develop.* **39**, 245.
7. Toriumi, A. (1996) 0.1 µm complementary MOS and beyond, *J. Vac. Sci. Technol. B* **14**, 4020.
8. Wann, C.H., Noda, K., Tanaka, T., Yoshida, M., and Hu, C. (1996) A comparative study of advanced MOSFET concepts, *IEEE Trans. Electron Devices* **43**, 1742.
9. Bohr, M.T. (1996) Technology development strategies for the 21st century, *Appl. Surf. Sci.* **100/101**, 534.
10. Moore, G.E. (1965) Cramming more components into integrated circuits, *Electronics Magazine* **38(8)**, 114.
11. Schaller, R.R. (1997) Moore's law: Past, present, and future, *IEEE Spectrum* **34(6)**, 52.
12. *The National Technology Roadmap for Semiconductors*, Semiconductor Industry Association, (1997).
13. Queisser, H. (1988) *The Conquest of the Microchip*, Harvard University Press, Cambridge.
14. Millman, S. (Ed.) (1983) *A History of Engineering and Science in the Bell System*, Bell Telephone Labs.
15. Frosch, C.J. and Derick, L. (1957) Surface protection and selective masking during diffusion in silicon, *Proc. of the Electrochem. Soc.*, 547.
16. Raider, S.I. and Flitsch, R. (1978) X-ray photoelectron spectroscopy of SiO_2-Si interfacial regions: Ultrathin oxide films, *IBM J. Res. Develop.* **22**, 294.
17. Fair, R.B. (1981) Oxidation, impurity diffusion, and defect growth in silicon - An overview, *J. Electrochem. Soc.* **128**, 1360.
18. Nicollian, E.H. and Brews, J.R. (1982) *MOS Physics & Technology*, Willey and Sons, .
19. Massoud, H.Z., Plummer, J.D., and Irene, E.A. (1985) Thermal oxidation of silicon in dry oxygen: Growth rate enhancement in the thin regime, *J. Electrochem. Soc.* **132**, 2693.
20. Grunthaner, F.J. and Grunthaner, P.J. (1986) Chemical and electronic structure of the SiO_2/Si interface, *Mat. Sci. Rep.* **1**, 65.
21. Mott, N.F. (1987) On the oxidation of silicon, *Phil. Mag. B* **55**, 117.
22. Helms, C.R. and Deal, B.E. (Eds.) (1988) The Physics and Chemistry of Si and the Si/SiO_2 Interface, Plenum Press, NY.
23. Irene, E. (1988) Models for the oxidation of silicon, *Crit. Rev. Sol. St. Mat. Sci.* **14**, 175.
24. Balk, P. (Ed.) (1988) *The $Si-SiO_2$ system*, Elsevier, Amsterdam.
25. Deal, B.E. (1988) Historic perspectives of silicon oxidation, in C. R. Helms and B. E. Deal (Eds.), *The Physics and Chemistry of SiO_2 and the $Si-SiO_2$ Interface* , Plenum Press, NY, p. 5.
26. Poindexter, E.H. (1989) MOS interface states: overview and physicochemical perspective, *Semic. Sci. Technol.* **4**, 961.
27. Fowler, W.B. (1991) Theory of defects and defect processes in silicon dioxide and the silicon-silicon dioxide interface, *Rev. Sol. St. Sci.* **5**, 435.
28. Helms, C.R. and Deal, B.E. (Eds.) (1993) The Physics and Chemistry of SiO_2 and the $Si-SiO_2$ Interface II , Plenum Press, NY.
29. Engel, T. (1993) The interaction of molecular and atomic oxygen with Si(100) and Si(111), *Surf. Sci. Rep.* **18**, 91.
30. Helms, C.R. and Poindexter, E.H. (1994) The $Si-SiO_2$ system: its microstructure and imperfections, *Rep. Prog. Phys.* **57**, 791.
31. Sofield, C.J. and Stoneham, A.M. (1995) Oxidation of silicon: the VLSI gate dielectric?, *Semic. Sci. Technol.* **10**, 215.
32. Hattori, T. (1995) Chemical structure of the SiO_2/Si interface, *CRC Crit. Rev. Solid State Mater. Sci.* **20**, 339.
33. Balk, P. (1995) Dielectrics for field effect technology, *Advanced Materials* **7**, 703.
34. Massoud, H.Z., Poindexter, E.H., and Helms, C.R. (eds) (1996) The Physics and Chemistry of SiO_2 and the $Si-SiO_2$ Interface - 3, , The Electrochemical Society, Pennington, NJ.
35. T. Hori, Gate Dielectrics and MOS ULSI, 1997, Springer, Berlin.
36. Conley, J.F. and Lenahan, P.M. (1996) A review of electron spin resonance spectroscopy of defects in thin films SiO_2 on Si, in H. Z. Massoud, E. H. Poindexter and C. R. Helms (Eds.), *The Physics and Chemistry of SiO_2 and the $Si-SiO_2$ Interface - 3* , The Electrochemical Soc., Pennington, NJ, p. 214.

37. Mott, N.F., Rigo, S., Rochet, F., and Stoneham, A.M. (1989) Oxidation of silicon, *Phil. Mag. B.* **60**, 189.
38. Oldham, T.R., McLean, F.B., Jr, H.E.B., and McGarrity, J.M. (1989) An overview of radiation-induced interface traps in MOS structures, *Semic. Sci. Technol.* **4**, 986.
39. Weber, W. and Brox, M. (1993) Physical properties of SiO_2 and its interface to silicon in microelectronic applications, *MRS Bulletin* **12**, 36.
40. Stesmans, A. (1989) The $Si=Si_3$ defect at various $(111)Si/SiO_2$ and $(111)Si/Si_3N_4$ interfaces, *Semicond. Sci. Technol.* **4**, 1000.
41. Anderson, W.R., Wheeler, R.G., and Ma, T.P. (1992) Observation of interface traps in the silicon conduction band at the $(100)Si/SiO_2$ interface at 4.2 K, *Appl. Phys. Lett.* **61**, 1107.
42. Stesmans, A. and van Gorp, G. (1990) Maximum density of P_b centers at the $(111)Si/SiO_2$ interface after vacuum anneal, *Appl. Phys. Lett.* **57**, 2663.
43. Fleetwood, D.M., Winokur, P.S., Reber, R.A., Meisenheimer, T.L., Schwank, J.R., Shaneyfelt, M.R., and Riewe, L.C. (1993) Effect of oxide traps, interface traps, and "border traps" on metal-oxide-semiconductor devices, *J. Appl. Phys.* **73**, 5058.
44. Fleetwood, D.M. and Saks, N.M. (1996) Oxide, interface, and border traps in thermal, N_2O, and N_2O nitrided oxides, *J. Appl. Phys.* **79**, 1583.
45. Edwards, A.H. (1987) Theory of the P_b center at the $<111>Si/SiO_2$ interface, *Phys. Rev. B.* **36**, 9638.
46. Stathis, J.H., Buchanan, D.A., Quinlan, D.L., Parsons, A.H., and Kotecki, D.E. (1993) Interface defects of ultrathin rapid-thermal oxide on silicon, *Appl. Phys. Lett.* **62**, 2682.
47. Cartier, E. and DiMaria, D.J. (1993) Hot-electron dynamics in SiO_2 and the degradation of the Si/SiO_2 interface, *Microelectronic Engineering* **22**, 207.
48. Stahlbush, R.E. (1996) Slow and fast state formation caused by hydrogen, in H. Z. Massoud, E. H. Poindexter and C. R. Helms (eds.), *The Physics and Chemistry of SiO_2 and the $Si-SiO_2$ Interface - 3*, The Electrochemical Society, Pennington, NJ p. 525.
49. Buchanan, D.A. and Lo, S.H. (1997) Reliability and integration of ultrathin gate dielectrics for advanced CMOS, *Microelectronics Engineering* **36**, 13.
50. Poindexter, E.H. and Warren, W.L. (1995) Paramagnetic point defects in amorphous thin films of SiO_2 and Si_3N_4: Updates and additions, *J. Electrochem. Soc.* **142**, 2508.
51. Razouk, R.R. and Deal, B.E. (1979) Dependence of interface state density on silicon thermal oxidation process variables, *J. Electrochem. Soc.* **126**, 1573.
52. Stathis, J.H. and Dori, L. (1994) Fundamental chemical differences among P_b defects on (111) and (100) silicon, *Appl. Phys. Lett.* **58**, 1641.
53. Degraeve, R., Roussel, P., Maes, H.E., and Groeseneken, G. (1996) A new analytic model for the description of the intrinsic oxide breakdown statistics of ultra-thin oxides, *Microelectronics and Reliability* **36**, 1651.
54. Ludeke, R. and Wen, H.J. (1997) Gate oxide characterization with ballistic electron emission microscopy, *Microelectronic Engineering* **36**, 255.
55. Matsuoka, T., Taguchi, S., Taniguchi, K., Hamaguchi, C., Kakimoto, S., and Takagi, J. (1995) Thickness dependence of N_2O-oxynitridation effects on breakdown of thermal oxides, *IEICE Trans. Electron.* **E78-C**, 248.
56. Silvestre, C. and Hauser, J.R. (1995) Time dependent dielectric breakdown measurements on RPECVD and thermal oxides, *J. Electrochem. Soc.* **142**, 3881.
57. Oka, S. and Katayama, M. (1997) Breakdown mechanism of oxide grown on czochralski silicon wafers, *Jpn. J. Appl. Phys.* **36**, 1995.
58. Lu, Z.H., McCaffrey, J.P., Brar, B., Wilk, G.D., Wallace, R.M., Feldman, L.C., and Tay, S.P. (1997) SiO_2 film thickness metrology by X-ray photoelectron spectroscopy, *Appl. Phys. Lett.* **71**, in press.
59. Nakamura, K., Ohmi, K., Yamamoto, K., Makihara, K., and Ohmi, T. (1994) Silicon wafer orientation dependence of MOS device reliability, *Jpn. J. Appl. Phys.* **33**, 500.
60. Brunner, K., Abstreiter, G., Kolbesen, B.O., and Meul, H.W. (1989) Strain at $Si-SiO_2$ interfaces studied by micro-raman spectroscopy, *Appl. Surf. Sci.* **39**, 116.
61. Taniguchi, K., Tanaka, M., Hamaguchi, C., and Imai, K. (1990) Density relaxation of silicon dioxide on (100) silicon during thermal annealing, *J. Appl. Phys.* **67**, 2195.
62. Yakovlev, V.A., Liu, Q., and Irene, E. (1992) Spectroscopic immersion ellipsometry study of the mechanism of Si/SiO_2 interface annealing, *J. Vac. Sci. Technol. A* **10**, 427.
63. Terada, N., Haga, T., Miyata, N., Moriki, K., Fujisawa, M., Morita, M., Ohmi, T., and Hattori, T. (1992) Optical absorption in ultrathin silicon oxide films near the SiO_2/Si interface, *Phys. Rev. B* **46**, 2312.
64. Batson, P.E. (1993) Simultaneous STEM imaging and electron energy loss spectroscopy with atomic-column sensitivity, *Nature* **366**, 727.

65. Batson, P.E., Browning, N.D., and Muller, D.A. (1994) EELS at buried interfaces: pushing towards atomic resolution, *Microsc. Soc. Amer. Bulletin* **24**, 371.
66. Nguyen, N.V., Chandler-Horowitz, D., Amirtharaj, P.M., and Pellegrino, J.G. (1994) Spectroscopic ellipsometry determination of the properties of the thin underlying strained Si layer and the roughness at the SiO_2/Si interface, *Appl. Phys. Lett.* **64**, 2688.
67. Awaji, N., Sugita, Y., Nakanishi, T., Ohkubo, S., Takasaki, K., and Komiya, S. (1996) High-precision X-ray reflectivity study of ultrathin SiO_2 on Si, *J. Vac. Sci. Technol.* A **14**, 971.
68. Lucovsky, G., Banerjee, A., Hinds, B., Claflin, B., Koh, K., and Yang, H. (1997) Minimization of sub-oxide transition regions at Si-SiO_2 interfaces by 900°C rapid thermal annealing, *Microelectr. Engineering* **36**, 207.
69. Hasegawa, E., Ishitani, A., Akimoto, K., Tsukiji, M., and Ohta, N. (1995) SiO_2/Si interface structures and reliability characteristics, *J. Electrochem. Soc.* **142**, 273.
70. Haight, R. and Feldman, L.C. (1982) Atomic structure of the Si-SiO_2 interface, *J. Appl. Phys.* **53**, 4884.
71. Feldman, L.C. (1988) The stoichiometry and structure of the SiO_2/Si Interface: Ion scattering studies, in C. R. Helms and B. E. Deal (Eds.), *The Physics and Chemistry of SiO_2 and the Si-SiO_2 Interface*, Plenum Press, NY p. 199.
72. Gusev, E.P., Lu, H.C., Gustafsson, T., and Garfunkel, E. (1996) Initial oxidation of silicon: New ion scattering results in the ultrathin regime, *Appl. Surf. Sci.* **104/105**, 329.
73. Gusev, E.P., Lu, H.C., Gustafsson, T., and Garfunkel, E. (1997) Silicon oxidation and oxynitridation in the ultrathin regime: Ion scattering studies, *Brazil. J. Phys.* **27**, 302.
74. Gusev, E.P., Lu, H.C., Gustafsson, T., and Garfunkel, E. (1995) The growth mechanism of thin silicon oxide films on Si(100) studied by medium energy ion scattering, *Phys. Rev. B* **52**, 1759.
75. Bjorkman, C.H., Fitch, J.T., and Lucovsky, G. (1990) Correlation between midgap interface state density and thickness-averaged oxide stress and strain at Si/SiO_2 interfaces formed by thermal oxidation of Si, *Appl. Phys. Lett.* **56**, 1983.
76. Kosowsky, S.D., Pershan, P.S., Krisch, K.S., Bevk, J., Green, M.L., Brasen, D., Feldman, L.C., and Roy, P.K. (1997) Evidence of annealing effects on a high-density Si/SiO_2 interfacial layer, *Appl. Phys. Lett.* **70**, 3119.
77. Vul', A.Y., Makarova, T.L., Osipov, V.Y., Zinchik, Y.S., and Boitsov, S.K. (1992) Kinetics of silicon oxidation and structure of oxide films of thickness less than 50 A, *Sov. Phys. Semicond.* **26**, 62.
78. Brugemann, L., Bloch, R., Press, W., and Gerlach, P. (1990) Surface and interface topography of amorphous SiO_2/crystalline Si(100) studied by X-ray diffraction, *J. Phys.: Condens. Matter* **2**, 8869.
79. Irene, E. (1993) Application of spectoscopic ellipsometry to microelectronics, *Thin Sol. Films* **233**, 96.
80. Grunthaner, F.J., Grunthaner, P.J., Vasquez, R.P., Lewis, B.F., Maserjian, J., and Madhukar, A. (1979) High resolution X-ray photoemission spectroscopy as a probe of local atomic structure: Application to amorphous SiO_2 and the Si-SiO_2 interface, *Phys. Rev. Lett.* **43**, 1683.
81. Tabe, M., Chiang, T.T., Lindau, I., and Spicer, W.E. (1986) Initial stages of thermal oxidation of the Si(111) 7x7 surface, *Phys. Rev. B* **34**, 2706.
82. Grunthaner, P.J., Hecht, M.H., Grunthaner, F.J., and Johnson, N.M. (1987) The localization and crystallographic dependence of Si suboxide species at the SiO_2/Si interface, *J. Appl. Phys.* **61**, 629.
83. D'Evelyn, M.P., Nelson, M.M., and Engel, T. (1987) Kinetics of the adsorption of O2 and of the desorption of SiO on Si(100): a molecular beam, XPS, and ISS study, *Surf. Sci.* **186**, 75.
84. Nakazawa, Z. and Sekiyama, H. (1990) Photoemission studies of Si surface oxidation using synchrotron radiation, *J. Appl. Phys.* **56**, 2108.
85. Borman, V.D., Gusev, E.P., Lebedinski, Y.Y., and Troyan, V.I. (1991) Direct observation of the layer-by-layer growth of initial oxide layers on Si(100) surface at the thermal oxidation, *Phys. Rev. Lett.* **67**, 2387.
86. Himpsel, F.J., McFeely, F.R., Taleb-Ibrahimi, A., Yarmoff, J.A., and Hollinger, G. (1988) Microscopic structure of the SiO_2/Si interface, *Phys. Rev. B* **38**, 6084.
87. Niwano, M., Katakura, H., Takeda, Y., Takakuwa, Y., Miyamoto, N., Hiraiwa, A., and Yagi, K. (1991) Photoemission study of the SiO_2/Si interface structure of thin oxide films on Si(100), (111) and (110) surfaces, *J. Vac. Sci. Technol.* A **9**, 195.
88. Himpsel, F.J., Lapiano-Smith, D.A., Morar, J.F., and Bevk, J. (1993) Local bonding at SiO_2/Si interface, in C. R. Helms and B. E. Deal (Eds.), *The Physics and Chemistry of SiO_2 and the Si-SiO_2 Interface, II*, Plenum Press, NY p. 237.
89. Lu, Z.H., Graham, M.J., Jiang, D.T., and Tan, K.H. (1993) SiO_2/Si(100) interface studied by Al K_α X-ray and synchrotron radiation XPS, *Appl. Phys. Lett.* **63**, 2941.
90. Rochet, F., Rigo, S., Froment, M., d'Anterroches, C., Maillot, C., Roulet, H., and Dufour, G. (1986) The thermal oxidation of silicon: The special case of the growth of very thin films, *Adv. Phys.* **35**, 339.

91. Seiger, M.T., Luh, D.A., Miller, T., and Chiang, T.C. (1996) Photoemission extended fine structure study of the $SiO_2/Si(111)$ interface, *Phys. Rev. Lett.* **77**, 2758.
92. Iwata, S. and Ishizaka, A. (1996) Electron spectroscopic analysis of the SiO_2/Si system and correlation with metal-oxide-semiconductor device characteristics, *J. Appl. Phys.* **79**, 6653.
93. Morgen, P., Hoffer, U., Wurth, W., and Umbach, E. (1989) Initial stages of oxygen adsorption on Si(111): The stable state, *Phys. Rev. B* **39**, 3720.
94. Banaszak-Holl, M.M. and McFeely, F.R. (1993) Si/SiO_2 interface : new structures and well-defined model systems, *Phys. Rev. Lett.* **71**, 2441.
95. Banaszak-Holl, M.M., Lee, S., and McFeely, F.R. (1994) Core-level photoemission and the structure of the Si/SiO_2 interface:, *Appl. Phys. Lett.* **65**, 1097.
96. McFeely, F.R., Zhang, K.Z., Banaszak-Hall, M.M., Lee, S., and Bender-IV, J.E. (1996) An inquiry concerning the principles of Si2p core-level photoemission shift assignments at the Si/SiO_2 interface, *J. Vac. Sci. Technol. B* **14**, 2824.
97. Pasquarello, A., Hybertsen, M.S., and Car, R. (1995) Si 2p core-level shifts at the $Si(001)$-SiO_2 interface: A First principles study, *Phys. Rev. Lett.* **74**, 1024.
98. Kageshima, H. and Tabe, M. (1996) Theoretical calculation of core-level shifts for O/Si(111) surfaces, *Surf. Sci.* **351**, 53.
99. Kim, S.Y. and Irene, E.A. (1995) An evaluation of errors in determining the refractive index and thickness of thin SiO_2 films using a rotating analyzer ellipsometer, *Rev. Sci. Instr.* **66**, 5277.
100. Jellison, G.E. (1991) Examination of thin SiO_2 films on Si using spectroscopic polarization modulation ellipsometry, *J. Appl. Phys.* **69**, 7627.
101. Hebert, K.J., Zafar, S., Irene, E.A., Kuehn, R., McCarthy, J.E., and Demirlioglu, E.K. (1996) Measurement of the refractive index of thin SiO_2 films using tunneling current oscillations and ellipsometry, *Appll. Phys. Lett.* **68**, 266.
102. Hebert, K.J., Labayen, T., and Irene, E.A. (1996) A measurement of the refractive index of ultrathin SiO2 films and a reevaluation of the thermal Si oxidation kinetics in the thin film regime, in H. Z. Massoud, E. H. Poindexter and C. R. Helms (Eds.), *The Physics and Chemistry of SiO_2 and the Si-SiO_2 Interface - 3* , The Electrochemical Society, Pennington, NJ pp. 81.
103. Hochella, M.F. and Carim, A.H. (1988) A reassessment of electron escape depth in silicon and thermally grown silicon dioxide thin films, *Surf. Sci.* **197**, L260.
104. Lennard, W.N., Massoumi, G.R., Mitchell, I.V., Tang, H.T., and Mitchell, D.F. (1994) Measurements of thin oxide films of $SiO_2/Si(100)$, *Nucl. Instr. Meth. B* **85**, 42.
105. Browning, N.D., Chisholm, M.F., and Pennycook, S.J. (1993) Atomic-resolution chemical analysis using a scanning transmission electron microscope, *Nature* **366**, 143.
106. Fukuda, H., M.Yasuda, Iwabuchi, T., and Ohno, S. (1992) Characterization of $SiO_2/Si(100)$ interface structure of N_2O-oxynitrided ultrathin SiO_2 films, *Appl. Surf. Sci.* **60/61**, 359.
107. Pasquarello, A., Hybertsen, M.S., and Car, R. (1996) Structurally relaxed models of the $Si(001)$-SiO_2 interface, *Appl. Phys. Lett.* **68**, 625.
108. Ohdomari, I., Akatsu, H., Yamakoshi, Y., and Kishimoto, K. (1987) Study of the interface structure between Si(100) and thermally grown SiO_2 using a ball-and-spoke model, *J. Appl. Phys.* **62**, 3751.
109. Ourmazd, A., Taylor, D.W., Rentscheir, J.A., and Bevk, J. (1987) Si to SiO2 transformation: Interfacial structure and mechanism, *Phys. Rev. Lett.* **53**, 743.
110. Fouss, P.H., Norton, H.J., Brennan, S., and Fisher-Colbrie, A. (1988) X-ray scattering studies of the Si-SiO_2 interface, *Phys. Rev. Lett.* **60**, 600.
111. Renaud, G., Fouss, P.H., Ourmazd, A., Bevk, J., Freer, B.S., and Hahn, P.O. (1991) Native oxidation of the Si(001) surface: Evidence for an interfacial phase, *Appl. Phys. Lett.* **58**, 1044.
112. Hirosawa, I., Akimoto, K., Tatsumi, T., Mizuki, J., and Matsui, J. (1990) SiO_2/Si interface study with synchrotron radiation X-ray diffraction, *J. Crystal Growth* **103**, 150.
113. Rabedeau, T.A., Tidswell, I.M., Pershan, P.S., Berk, J., and Freer, B.S. (1991) X-ray reflectivity studies of $SiO_2/Si(001)$, *Appl. Phys. Lett.* **59**, 3422.
114. Munkholm, A., Brennan, S., Comin, F., and Ortega, L. (1995) Observation of a distributed epitaxial oxide in thermally grown SiO_2 on Si(001), *Phys. Rev. Lett.* **75**, 4254.
115. Takahashi, I., Shimura, T., and Harada, J. (1993) X-ray diffraction evidence for epitaxial microcrystallinity in thermally oxidized SiO_2 thin films on the Si(001) surface, *J. Phys. Condensed Matter* **5**, 6525.
116. Tiller, W.A. (1981) On the kinetics of the thermal oxidation of Si, *J. Electrochem. Soc.* **128**, 689.
117. Harp, G.R., Saldin, D.K., and Tonner, B.P. (1993) Finite-size effects and short-range cristalline order in Si and SiO_2 studied by x-ray absorption fine structure spectroscopy, *J. Phys. Condens. Matt.* **5**, 5377.

118. Gibson, J.M. and Lanzerotti, M.Y. (1989) Observation of interfacial atomic steps during silicon oxidation, *Nature* **340**, 128.
119. Ravindra, N.M., Narayan, J., Fathy, D., Stivastava, J.K., and Irene, E.A. (1987) Silicon oxidation and Si-SiO$_2$ interface of thin oxide, *J. Mater. Res.* **2**, 216.
120. Akutsu, H., Sami, Y., and Ohdomari, I. (1991) Evalution of SiO$_2$/(001)Si interface roughness with HRTEM and simulation, *Phys. Rev. B* **44**, 1616.
121. Ross, F.M. and Gibson, J.M. (1992) Dynamic observation of interface propagation during silicon oxidation, *Phys. Rev. Lett.* **68**, 1782.
122. Peng, J.P., Lynn, K.G., Asoka-Kumar, P., Becker, D.P., and Harshman, D.R. (1996) Study of the SiO$_2$-Si interface using variable energy positron two-dimensional angular correlation of annihilation radiation, *Phys. Rev. Lett.* **76**, 2157.
123. Deal, B.E. and Grove, A.S. (1965) General relationship for the thermal oxidation of silicon, *J. Appl. Phys.* **36**, 3770.
124. Wolters, D.R. and Zegers van Duynhoven, A.T.A. (1989) Kinetics of dry oxidation of silicon, *Appl. Surf. Sci.* **39**, 81.
125. Dutta, T. and Ravindra, N.M. (1992) Silicon oxidation in the thin oxide regime, *Phys. Stat. Sol.* **134**, 447.
126. Kao, S.C. and Doremus, R.H. (1993) Oxidation of silicon in oxygen: measurement of film thickness and kinetics, in C. R. Helms and B. E. Deal (Eds.), *The Physics and Chemistry of SiO$_2$ and the Si-SiO$_2$ interface*, Plenum Press, NY p. 23.
127. Delarious, J.M., Helms, C.R., Kao, D.B., and Deal, B.E. (1989) Parallel oxidation model for Si including both molecular and atomic oxygen mechanisms, *Appl. Surf. Sci.* **39**, 89.
128. Massoud, H.Z., Plummer, J.D., and Irene, E.A. (1985) Thermal oxidation of silicon in dry oxygen: Accurate determination of the kinetic rate constant, *J. Electrochem. Soc.* **132**, 1745.
129. Schafer, S.A. and Lyon, S.A. (1985) New model of the rapid initial oxidation of Si, *Appl. Phys. Lett.* **47**, 154.
130. Stoneham, A.M., Grovenor, C.R.M., and Cerezo, A. (1987) Oxidation and the structure of the silicon/oxide interface, *Phil. Mag. B* **55**, 201.
131. Blank, J. (1978) A revised model for the oxidation of Si by oxygen, *Appl. Phys. Lett.* **33**, 424.
132. Cristy, S.S. and Condon, J.B. (1981) A model for oxidation of Si by oxygen, *J. Electrochem. Soc.* **128**, 2170.
133. Dunham, S.T. (1988) Interstitial fluxes during silicon oxidation, in C. R. Helms and B. E. Deal (eds.), *The Physics and Chemistry of SiO$_2$ and the Si-SiO$_2$ Interface*, Plenum Press, NY pp. 477.
134. Taniguchi, K., Shibata, Y., and Hamaguchi, C. (1989) Theoretical model for self-interstitial generation at the Si/SiO$_2$ interface during oxidation, *J. Appl. Phys.* **65**, 2723.
135. Moharir, S.S. and Chandorkar, A.N. (1989) An interface reaction mechanism for the dry oxidation of silicon, *J. Appl. Phys.* **65**, 2171.
136. Doremus, R.H. (1989) Oxidation of silicon: Is there a slow interface reaction?, *J. Appl. Phys.* **66**, 4441.
137. Takakuwa, Y., Nihei, M., Horie, T., and Miyamoto, N. (1994) Thermal oxidation mechanism based on formation and diffusion of volatile SiO molecules, *J. Non-Crystalline Solids* **179**, 345.
138. Cerofolini, G.F., LaBruna, G., and Meda, L. (1995) Enhanced Si Oxidation in O$_2$ and O$_2$:F$_2$, *Appl. Surf. Sci* **89**, 361.
139. Dimitrijev, S. and Harrison, H.B. (1996) Modeling of growth of thin silicon oxides, *J. Appl. Phys.* **80**, 2467.
140. Peng, K.Y., Wang, L.C., and Slattery, J.C. (1996) A new theory for silicon oxidation, *J. Vac. Sci. Technol. B* **14**, 3316.
141. Kamohara, S. and Kamigaki, Y. (1991) Activation energy enhancement during initial silicon oxide growth in dry oxygen, *J. Appl. Phys.* **69**, 7871.
142. Whidden, T.K., Thanikasalam, P., Rack, M.J., and Ferry, D.K. (1995) Initial oxidation of silicon (100): A unified chemical model for thin and thick growth rates and interface structure, *J. Vac. Sci. Technol. B* **13**, 1618.
143. Hu, S.M. (1984) Thermal oxidation of Si: Chemisorption and linear rate constant, *J. Appl. Phys.* **55**, 4095.
144. Han, C.J. and Helms, C.R. (1988) O^{18} tracer study of Si oxidation in dry O$_2$ using SIMS, *J. Electrochem. Soc.* **135**, 1824.
145. Gusev, E.P., Lu, H.C., Gustafsson, T., and Garfunkel, E. (1996) New features of silicon oxidation in the ultrathin regime: an ion scattering study, in H. Z. Massoud, E. H. Poindexter and C. R. Helms (Eds.), *The Physics and Chemistry of SiO$_2$ and the Si-SiO$_2$ Interface - 3*, The Electrochem. Soc., Pennington, NJ, p. 49.
146. Trimaille, I. and Rigo, S. (1989) Use of ^{18}O isotopic labelling to study thermal dry oxidation of silicon as a function of temperature and pressure, *Appl. Surf. Sci.* **39**, 65.
147. Ganem, J.J., Battistig, G., Rigo, S., and Trimaille, I. (1993) A study of the initial stages of the oxidation of silicon using O^{18} and RTP, *Appl. Surf. Sci.* **65/66**, 647.

148. Ganem, J.J., Trimaille, I., Andre, P., Rigo, S., Stedile, F.C., and Baumvol, I.J.R. (1997) Diffusion of near surface defects during the thermal oxidation of silicon, *J. Appl. Phys.* **81**, 8109.
149. Ross, F.M., Gibson, J.M., and Twesten, R.D. (1994) Direct observations of interface motion during the oxidation of silicon, *Surf. Sci.* **310**, 243.
150. Lu, H.C., Gustafsson, T., Gusev, E.P., and Garfunkel, E. (1995) An isotopic labeling study of the growth of thin oxide films on Si(100), *Appl. Phys. Lett.* **67**, 1742.
151. Lu, H.C., Gusev, E.P., Gustafsson, T., Garfunkel, E., Green, M.L., Brasen, D., and Feldman, L.C. (1996) High resolution ion scattering study of silicon oxynitridation, *Appl. Phys. Lett.* **69**, 2713.
152. Gustafsson, T., Garfunkel, E., Gusev, E.P., Haberle, P., Lu, H.C., and Zhou, J.B. (1996) Structural studies of oxide surfaces, *Surf. Rev. Lett.* **3**, 1561.
153. Hussey, R.J., Bisaillion, D.A., Sproule, G.I., and Graham, M.J. (1993) The growth and transport in thermal oxide films formed on silicon, *Corros. Sci.* **35**, 917.
154. Plummer, J.D. (1996) Silicon oxidation kinetics - from Deal-Grove to VLSI process models, in H. Z. Massoud, E. H. Poindexter and C. R. Helms (Eds.), *The Physics and Chemistry of SiO_2 and the Si-SiO_2 Interface - 3*, The Electrochemical Society, Pennington, NJ.
155. Avouris, P. and Lyo, I.-W. (1991) Probing and inducing surface chemistry with the STM: the reactions of Si(111)-7 x 7 with H_2O and O_2, *Surf. Sci.* **242**, 1.
156. Avouris, P. and Cahill, D. (1992) STM studies of Si(100)2x1 oxidation: defect chemistry and Si ejection, *Ultramicroscopy* **42/44**, 838.
157. Baklanov, M.R., Kruchinin, V.N., Repinsky, S.M., and Shklyaev, A.A. (1989) Initial stages of the interaction of nitrious oxide and oxygen with the Si(100) surface under low pressure, *React. Solids* **7**, 1.
158. Borman, V.D., Gusev, E.P., Lebedinski, Y.Y., and Troyan, V.I. (1994) The mechanism of submonolayer oxide phase formation on a silicon surface upon the thermal oxidation, *Phys. Rev. B* **49**, 5415.
159. Engstrom, J.R., Bonser, D.J., Nelson, M.M., Engel, T. (1991) The reaction of atomic oxygen with Si(100) and Si(111). I. Oxide decomposition, active oxidation and the transition to passive oxidation, *Surf. Sci.* **256**, 317.
160. Engstrom, J.R., Bonser, D.J., and Engel, T. (1992) The reaction of atomic oxygen with Si(100) and Si(111). II Adsorption, passive oxidation and the effect of coincident ion bombardment, *Surf. Sci.* **268**, 238.
161. Feltz, A., Memmert, U., and Behm, R.J. (1992) In situ STM imaging of high temperature oxygen etching of Si(111)7x7 surfaces, *Chem. Phys. Lett.* **192**, 271.
162. Feltz, A., Memmert, U., and Behm, R.J. (1994) High temperature STM studies on the interaction of O_2 with Si(111)-(7x7) surfaces, *Surf. Sci.* **314**, 34.
163. Horie, T., Takakuwa, Y., and Miyamoto, N. (1994) 2D growth and decomposition of initial thermal SiO_2 layers on Si(100), *Jpn. J. Appl. Phys.* **33**, 4684.
164. Kobayashi, Y., Prabhakaran, K., and Ogino, T. (1995) Thermal clustering of very thin oxide formed on Si surfaces by N_2O/O_2 adsorption, *Surf. Sci.* **329**, 167.
165. Pelz, J.P. and Koch, R.H. (1991) Successive oxidation stages and annealing behavior of the Si(111)7x7 surface observed with STM and STS, *J. Vac. Sci. Tech. B* **9**, 775.
166. Seiple, J., Pecquet, J., Meng, Z., and Pelz, J.P. (1993) Elevated temperature oxidation and etching of Si(111)7x7 surface observed with STM, *J. Vac. Sci. Technol. A* **11**, 1649.
167. Seiple, J. and Pelz, J.P. (1994) STM study of oxide nucleation and oxidation induced roughening at elevated temperatures on Si(100), *Phys. Rev. Lett.* **73**, 999.
168. Seiple, J. and Pelz, J.P. (1995) Evolution of atomic-scale roughening on Si(001)-(2x1) surfaces resulting from high temperature oxidation, *J. Vac. Sci. Technol. A* **13**, 772.
169. Udagawa, M., Niwa, M., and Sumita, I. (1993) The initial stages of the thermal oxidation of Si(100)2x1 surface studied by STM, *Jpn. J. Appl. Phys.* **32**, 282.
170. Westermann, J., Nienhaus, H., and Monch, W. (1994) Oxidation stages of clean and H-terminated Si(100) at room temperature, *Surf. Sci.* **311**, 101.
171. Onishi, K. and Hattori, T. (1994) Periodic changes in SiO_2/Si(111) interface structures during oxidation, *Jpn. J. Appl. Phys.* **33**, L676.
172. Morita, M. and Ohmi, T. (1994) Characterization and control of native oxide on silicon, *Jpn. J. Appl. Phys.* **33**, 370.
173. Dujardin, G., Mayne, A., Comtet, G., Hellner, L., Jamet, M., Goff, E.L., and Millet, P. (1996) New model of the initial stages of Si(111)-(7x7) oxidation, *Phys. Rev. Lett.* **76**, 3782.
174. Tsai, V., Wang, X.S., Williams, E.D., Schneir, J., and Dixson, R. (1997) Conformal oxides on Si surfaces, *Appl. Phys. Lett.* **71**, 1495.
175. Fujita, S., Watanabe, H., Maruno, S., Ichikawa, M., and Kawamura, T. (1997) Observation of oxide/Si(001) interface during layer-by-layer oxidation by scanning reflection electron microscopy, *Appl. Phys. Lett.* **71**, 885.

176. Leibsle, F.M., Samsavar, A., and Chiang, T.C. (1988) Oxidation of Si(111)7x7 as studied by STM, *Phys. Rev. B* **38**, 5780.
177. Avouris, P. (1990) Atom resolved surface chemistry using the STM, *J. Phys. Chem.* **94**, 2246.
178. Niwa, M., Iwasaki, H., Watanabe, Y., Sumita, I., Akutsu, N., and Y.Akutsu (1992) Statistical properties of atomic-scale Si/SiO$_2$ interface roughness studied by STM, *Appl. Surf. Sci.* **60/61**, 39.
179. Ono, Y., Tabe, M., and Kageshima, H. (1993) STM observation of thermal oxide growth on Si(111)7x7 surfaces, *Phys. Rev. B* **48**, 14291.
180. Wurm, K., Kliese, R., Hong, Y., Rottger, B., Wei, Y., Neddermeyer, H., and Tsong, I.S.T. (1994) Evolution of surface morphology of Si(100) during oxygen adsorption at elevated temperatures, *Phys. Rev. B* **50**, 1567.
181. Udagawa, M., Niwa, M., and Sumita, I. (1994) Local ordering and lateral growth of initial thermal oxide of Si(001), *Jpn. J. Appl. Phys.* **33**, 375.
182. Hasegawa, T., Kohno, M., Hosaka, S., and Hosoki, S. (1994) Initial stages of oxygen adsorption on Si(111)-7x7 by STM, *Surf. Sci.* **312**, L753.
183. Fujita, S., Maruno, S., Watanabe, H., and Ichikawa, M. (1996) Nanostructure fabrication using the selective thermal desorption of SiO$_2$ induced by electron beams, *Appl. Phys. Lett.* **69**, 638.
184. Neddermeyer, H. (1996) Scanning tunneling microscopy of semiconductor surfaces, *Rep. Prog. Phys.* **59**, 701.
185. Wilk, G.D., Wei, Y., Edwards, H., and Wallace, R.M. (1997) In situ Si flux cleaning technique for producing atomically flat Si(100) surfaces at low temperature, *Appl. Phys. Lett.* **70**, 2288.
186. Watanabe, H., Fujita, S., and Ichikawa, M. (1997) Thermal decomposition of ultrathin oxide layers on Si(111) surfaces mediated by surface Si transport, *Appl. Phys. Lett.* **70**, 1095.
187. Hahn, P.O. and Henzler, M. (1984) The Si-SiO$_2$ interface: correlation of atomic structure and electrical properties, *J. Vac. Sci. Technol.* **2**, 574.
188. Goodnick, S.M., Ferry, D.K., Wilmsen, C.M., Liliental, Z., Fathy, D., and Krivanek, O.L. (1985) Surface roughness at the Si(100)-SiO$_2$ interface, *Phys. Rev. B* **32**, 8171.
189. Carim, A.H. and Bhattacharyya, A. (1985) Si/SiO$_2$ interface roughness: structural observation and electrical consequences, *Appl. Phys. Lett.* **46**, 872.
190. Ross, A., Bergkvist, M., and Ribbing, C.G. (1988) Determination of Si/SiO$_2$ interface roughness by diffuse scattering measurement, *Appl. Optics* **27**, 4660.
191. Ogura, A. (1991) Improvement of SiO$_2$/Si interface flatness by post oxidation anneal, *J. Electrochem. Soc.* **138**, 807.
192. Ohmi, T., Kotani, K., Teramoto, A., and Miyashita, M. (1991) Dependence of electron channel mobility on Si-SiO$_2$ interface microroughness, *IEEE Electron. Device Lett.* **12**, 652.
193. Offenberg, M., Liehr, M., and Rubloff, G.W. (1991) Surface etching and roughening in integrated processing of thermal oxides, *J. Vac. Sci. Technol. A* **9**, 1058.
194. Ohmi, T., Miyashita, M., Itano, M., Imaoka, T., and Kawanabe, I. (1992) Dependence of thin-oxide films quality on surface microroughness, *IEEE Trans. Electr. Dev.* **39**, 537.
195. Niwa, M., Udagawa, M., Okada, K., Kouzazki, T., and Sinclair, R. (1993) Atomic-scale planarization of SiO$_2$/Si(001) interfaces, *Appl. Phys. Lett.* **63**, 675.
196. Anderson, W.R., Lombardi, D.R., and Wheller, R.G. (1993) Determination of Si/SiO$_2$ interface roughness using weak localization, *IEEE Electr. Dev. Lett.* **14**, 351.
197. Niwa, M., Kouzaki, T., Okada, K., Udagawa, M., and Sinclair, R. (1994) Atomic-order planarization of ultrathin SiO$_2$/Si(001) interfaces, *Jpn. J. Appl. Phys.* **33**, 388.
198. Hirose, M., Hiroshima, M., and Yasaka, T. (1994) Characterization of silicon surface microroughness and tunneling transport through ultrathin gate oxide, *J. Vac. Sci. Technol. A* **12**, 1864.
199. Ohmi, T., Matsumoto, K., Nakamura, K., Makihara, K., Takano, J., and Yamamoto, K. (1995) Influence of silicon wafer orientation on very thin oxide quality, *J. Appl. Phys.* **77**, 1159.
200. Dadap, J.I., Doris, B., Deng, Q., Downer, M.C., Lowell, J.K., and Diebold, A.C. (1994) Randomly oriented Angstrom-scale microroughness at the Si(100)/SiO$_2$ interface probed by optical second harmonic generation, *Appl. Phys. Lett.* **64**, 2139.
201. Yamanaka, T., Fang, S.J., and Helms, C.R. (1996) Correlation between inversion layer mobility and surface roughness measured by AFM, *IEEE Electron Device Lett.* **17**, 178.
202. Offenberg, M., Liehr, M., Rubloff, G.W., and Holloway, K. (1990) Ultraclean, integrated procesing of thermal oxide structures, *Appl. Phys. Lett.* **57**, 1254.
203. Ohmi, T. (1993) ULSI reliability through ultraclean processing, *Proceedings of the IEEE* **81**, 716.
204. Tang, M.T., Evans-Lutterodt, K.W., Higashi, G.S., and Boone, T. (1993) Roughness of the silicon(001)/SiO$_2$ interface, *Appl. Phys. Lett* **62**, 3144.

205. Suzuki, M., Homma, Y., Kudoh, Y., and Yabumoto, N. (1993) Roughness evaluation of thermally oxidized Si(111) surface by scanning force microscopy, *Jpn. J. Appl. Phys.* **32**, 1419.
206. Liu, Q., Wall, J.F., and Irene, E.A. (1994) Si/SiO$_2$ interface studies by spectroscopic immersion ellipsometry and AFM, *J. Vac. Sci. Technol. A* **12**, 2625.
207. Koga, J., Takagi, S.I., and Toriumi, A. (1996) Observation of oxide-thickness-dependent interface roughness in Si MOS structure, *Jpn. J. Appl. Phys.* **35**, 1440.
208. Homma, Y., Suzuki, M., and Yabumoto, N. (1992) Observation of atomic step morphology on silicon oxide surfaces, *J. Vac. Sci. Tech.* **A10**, 2055.
209. Bennett, J. (1992) Recent developments in surface roughness characterization, *Meas. Sci. and Tech.* **3**, 1119.
210. Teichert, C., MacKay, J.F., Savage, D.E., Legally, M.G., Brohl, M., and Wagner, P. (1995) Comparison of surface roughness of polished silicon wafers measured by light scattering topography, soft-X-ray scattering, and AFM, *J. Appl. Phys.* **66**, 2346.
211. Malik, I.J., Pirooz, S., Shive, L.W., Davenport, A.J., and Vitus, C.M. (1993) Surface roughness of silicon wafers on different lateral length scales, *J. Electrochem. Soc.* **140**, L75.
212. Tang, M.T., Evans-Lutterodt, K.W., Green, M.L., Feldman, L.C., Higashi, G.S., and Boone, T. (1994) Growth temperature dependence of the Si(001)/SiO$_2$ interface width, *Appl. Phys. Lett* **64**, 748.
213. Smith, F.W. and Ghidini, G. (1982) Reaction of oxygen with Si(111) and (100): Critical conditions for the growth of SiO$_2$, *J. Electrochem. Soc.* **129**, 1300.
214. Shklyaev, A.A. and Suzuki, T. (1995) Branching of critical conditions for Si(111)-(7x7) oxidation, *Phys. Rev. Lett.* **75**, 272.
215. Lu, H.C., Gusev, E.P., Garfunkel, E., and Gustafsson, T. (1996) An ion scattering study of the interaction of oxygen with Si(111): surface roughening and oxide growth, *Surf. Sci.* **341**, 111.
216. Poindexter, E.H., Gerardi, G.J., and Keeble, D.J. (1996) Hydrogen speciations in electronic silica, in H. Z. Massoud, E. H. Poindexter and C. R. Helms (Eds.), *The Physics and Chemistry of SiO$_2$ and the Si-SiO$_2$ Interface - 3*, The Electrochemical Society, Pennington, NJ p. 172.
217. Doremus (1973) *Glass Science*, John Wiley & Sons, New York.
218. Shelby, J.E. (1977) Molecular diffusion and solubility of hydrogen isotopes in vitreous silica, *J. Appl. Phys.* **48**, 3387.
219. Griscom, D.L. (1985) Diffusion of radiolytic molecular hydrogen, *J. Appl. Phys.* **58**, 2524.
220. Gale, R., Chew, H., Feigl, F.J., and Magee, C.W. (1988) Current-induced charges and hydrogen species distribution in MOS silicon dioxide films, in C. R. Helms and B. E. Deal (Eds.), *The Physics and Chemistry of SiO$_2$ and the Si-SiO$_2$ Interface*, Plenum Press, NY pp. 177.
221. George, M.A., Bohling, D.A., Wortman, J.J., Melzak, J.A., and Hames, G.A. (1993) Hydrogen content of silicon and thermal oxidation induced moisture generation in an integrated thermal processing reactor, *J. Vac.Sci. Technol. B* **11**, 86.
222. Krauser, J., Weidinger, A., and Braunig, D. (1996) Hydrogen distribution at the oxide/silicon interface reflecting the microscopic structure of the near-interface region, in H. Z. Massoud, E. H. Poindexter and C. R. Helms (Eds.), *The Physics and Chemistry of SiO2 and the Si-SiO2 Interface - 3*, The Electrochemical Society, Pennington, NJ p. 184.
223. Baumvol, I.J.R., Stedile, F.C., Radtke, C., Freire, F.L., Gusev, E.P., Green, M.L., and Brasen, D. (1998) Nuclear reaction analysis of deuterium near the SiO$_2$/Si interface, *Nucl. Instr. Meth.*, in press.
224. Brower, K.L. (1990) Dissociation kinetics of hydrogen-passivated (111) Si-SiO$_2$ interface defects, *Phys. Rev. B* **42**, 3444.
225. Cartier, E., Buchanan, D.A., and Dunn, G.J. (1994) Atomic hydrogen-induced interface degradation of reoxidized-nitrided silicon dioxide on silicon, *Appl. Phys. Lett.* **64**, 901.
226. Stathis, J.H. and Cartier, E. (1994) Atomic hydrogen reactions with P_b centers at the (100)Si/SiO$_2$ interface, *Phys. Rev. Lett.* **72**, 2745.
227. Baumvol, I.J.R., Stedile, F.C., Ganem, J.J., Trimaille, I., and Rigo, S. (1996) Thermal nitridation of SiO$_2$ films in ammonia: the role of hydrogen, *J. Electrochem. Soc.* **143**, 1426.
228. Habraken, F.H.P.M. and Kuiper, A.E.T. (1994) Silicon nitride and oxynitride films, *Materials Sci. and Eng. Rept.* **R12**, 123.
229. Johnson, N.M., Biegelsen, D.K., Moyer, M.D., Deline, V.R., and C.A. Evans, J. (1981) Deuterium at the Si-SiO$_2$ interface detected by SIMS, *Appl. Phys. Lett.* **38**, 995.
230. Myers, S.M. (1987) Interaction of deuterium gas with dry SiO$_2$: an ion-beam study, *J. Appl. Phys.* **61**, 5428.
231. Qiu, Q., Arai, E., and Ohji, Y. (1991) Study of deuterium diffusion behavior in SiO$_2$ by means of the D(^{3}He,p)^{4}He reaction, *Nucl. Instr. Meth. B* **56/57**, 816.

232. Park, H. and Helms, C.R. (1992) The effect of annealing treatment on the distribution of deuterium in silicon and in silicon/silicon oxide systems, *J. Electrochem. Soc.* **139**, 2042.
233. Fukuda, H., Ueno, T., Kawarada, H., and Ohdomari, I. (1993) Effect of deuterium anneal on $SiO_2/Si(100)$ interface traps and electron spin resonance signals of ultrathin SiO_2 films, *Jpn. J. Appl. Phys.* **32**, L569.
234. Muraoka, K., Takagi, S., and Toriumi, A. (1996) Evidence for assymmetrical hydrogen profile in thin D_2O oxidized SiO_2 by SIMS and modified TDS, *Extended Abstract of the 1996 International Conference on Solid State Devices and Materials*, 500.
235. Saks, N.S. and Rendell, R.W. (1992) The time-dependence of post-irradiation interface trap build-up in deuterium-annealed oxides, *IEEE Trans. Nucl. Sci.* **39**, 2220.
236. Lyding, J.M., Hess, K., and Kizilyalli, I.C. (1996) Reduction of hot electron degradation in metal oxide semiconductor transistors by deuterium processing, *Appl. Phys. Lett.* **68**, 2526.
237. Devine, R.A.B., Autran, J.-L., Warren, W.L., Vanheusdan, K.L., and Rostaing, J.-C. (1997) Interfacial hardness enhancement in deuterium annealed 0.25 µm channel MOSFET, *Appl. Phys. Lett.* **70**, 2999.
238. Van de Valle, C.G. and Jackson, W.B. (1996) Comment on "Reduction of hot electron degradation in metal oxide semiconductor transistors by deuterium processing", *Appl. Phys. Lett.* **69**, 2441.
239. Hori, T., Akamatsu, S., and Odake, Y. (1992) Deep-submicrometer technology with reoxized or annealed nitrided-oxide gate dielectrics prepared by rapid thermal processing, *IEEE Trans. Electron Devices* **39**, 118.
240. Fukuda, H., Arakawa, T., and Ohno, S. (1990) Highly reliable thin nitrided SiO_2 films formed by rapid thermal processing in an N_2O ambient, *Jpn. J. Appl. Phys.* **29**, L2333.
241. Hwang, H., Ting, W., Maiti, B., Kwong, D.L., and Lee, J. (1990) Electrical characteristics of ultrathin oxynitride gate dielectrics prepared by rapid thermal oxidation of silicon in N_2O, *Appl. Phys. Lett.* **57**, 1010.
242. Carr, E.C. and Buhrman, R.A. (1993) Role of interfacial nitrogen in improving thin silicon oxides grown in N_2O, *Appl. Phys. Lett.* **63**, 54.
243. Okada, Y., Tobin, P.J., and Lakhotia, V. (1993) Evaluation of interfacial nitrogen concentration of RTP silicon oxynitrides by reoxidation, *J. Electrochem. Soc.* **140**, L87.
244. Tang, H.T., Lennard, W.N., Zinke-Allmang, M., Mitchell, I.V., Feldman, L.C., Green, M.L., and Brasen, D. (1994) Nitrogen content of oxynitride films on Si(100), *Appl. Phys. Lett.* **64**, 64.
245. Tobin, P.J., Okada, Y., Ajuria, S.A., Lakhotia, V., Feil, W.A., and Hedge, R.I. (1994) Furnace formation of silicon oxynitride thin dielectrics in N_2O, *J. Appl. Phys.* **75**, 1811.
246. Okada, Y., Tobin, P.J., and Ajuria, S.A. (1994) Furnace grown silicon oxynitrides using NO, *IEEE Trans. Electr. Dev.* **41**, 1608.
247. Yao, Z.Q., Harrison, H.B., Dimitrijev, S., Yeow, Y.T., and Sweatman, D. (1994) High-quality ultrathin dielectric films grown on Si in NO ambient, *Appl. Phys. Lett.* **64**, 3584.
248. Green, M.L., Brasen, D., Evans-Lutterodt, K.W., Feldman, L.C., Krisch, K., Lennard, W., Tang, H.T., Manchanda, L., and Tang, M.T. (1994) RTO of silicon in N_2O between 800 and 1200°C: Incorporated nitrogen and roughness, *Appl. Phys. Lett.* **65**, 848.
249. Hattangady, S.V., Niimi, H., and Lucovsky, G. (1995) Controlled nitrogen incorporation at the gate oxide surface, *Appl. Phys. Lett.* **66**, 3495.
250. Green, M.L., Brasen, D., Feldman, L.C., Lennard, W., and Tang, H.T. (1995) Effect of incorporated nitrogen on the kinetics of thin rapid thermal N_2O oxides, *Appl. Phys. Lett.* **67**, 1600.
251. Ellis, K.A. and Buhrman, R.A. (1996) Furnace gas-phase chemistry of silicon oxynitridation in N_2O, *Appl. Phys. Lett.* **68**, 1696.
252. Hori, T. (1993) Nitrided gate oxide CMOS technology for improved hot-carrier reliability, *Microelectronic Engineering* **22**, 245.
253. Landheer, D., Tao, Y., Xu, D.X., Sproule, G.I., and Buchanan, D.A. (1995) Defects generated by Fowler-Nordheim injection in silicon dioxide films produced by plasma-enhanced chemical-vapour deposition with nitrous oxide and silane, *J. Appl. Phys.* **78**, 1818.
254. Bhat, M., Han, L.K., Wristers, D., Yan, J., Kwong, D.L., and Fulford, J. (1995) Effect of chemical composition on the electrical properties of NO-nitrided SiO_2, *Appl. Phys. Lett.* **66**, 1225.
255. Hedge, R.I., Tobin, P.J., Reid, K.G., Maiti, B., and Ajuria, S.A. (1995) Growth and surface chemistry of oxynitride gate dielectric using nitric oxide, *Appl. Phys. Lett.* **66**, 2882.
256. Matsuoka, T., Taguchi, S., Ohtsuka, H., Taniguchi, K., Hamaguchi, C., and Uda, K. (1996) Hot-carrier-induced degradation of N_2O-oxynitrided gate oxide NMOSFETs, *IEEE Trans. Electron. Dev.* **43**, 1364.
257. Chang-Liao, K.-S. and Chen, L.-C. (1997) Metal-Oxide-Si capacitors hot-electron and radiation hardness improvement by gate electrodes deposited using amorphous Si and gate oxides rapid thermal annealed in N_2O, *Jpn. J. Appl. Phys.* **36**, L604.

258. Alessandri, M., Clementi, C., Crivelli, B., Ghidini, G., Pellizzer, F., Martin, F., Imai, M., and Ikegawa, H. (1997) Nitridation impact on thin oxide charge trapping, *Microelectronic Engineering* **36**, 211.
259. Wristers, D., Han, L.K., Chen, T., Wang, H.H., and Kwong, D.L. (1996) Degradation of oxynitride gate dielectric reliability due to boron diffusion, *Appl. Phys. Lett.* **68**, 2094.
260. Frost, M.R. and Magee, C.W. (1996) Characterization of nitrided SiO_2 thin films using SIMS, *Appl. Surf. Sci.* **104/105**, 379.
261. Saks, N.S., Ma, D.I., and Fowler, W.B. (1995) Nitrogen depletion during Si oxidation in N_2O, *Appl. Phys. Lett.* **67**, 374.
262. Carr, E.C., Ellis, K.A., and Buhrman, R.A. (1995) Nitrogen profiles in thin SiO_2 in N_2O: the role of atomic oxygen, *Appl. Phys. Lett.* **66**, 1492.
263. Lu, Z.H., Tay, S.P., Cao, R., and Pianetta, P. (1995) The effect of rapid thermal N_2O oxynitridation on the oxide/Si(100) interface structure, *Appl. Phys. Lett.* **67**, 2836.
264. Lu, H.C., Gusev, E.P., Gustafsson, T., Green, M.L., Brasen, D., and Garfunkel, E. (1997) Compositional and mechanistic aspects of ultratrhin oxynitride film growth on Si(100), *Microelectronic Engineering* **36**, 29.
265. Gusev, E.P., Lu, H.C., Gustafsson, T., Garfunkel, E., Green, M.L., and Brasen, D. (1997) The composition of ultrathin oxynitrides thermally grown in NO, *J. Appl. Phys.* **82**, 896.
266. Baumvol, I.J.R., Stedile, F.C., Ganem, J.J., Trimaille, I., and Rigo, S. (1996) Thermal nitridation of SiO_2 with ammonia, *J. Electrochem. Soc.* **143**, 2938.
267. Hartig, M.J. and Tobin, P.J. (1996) A model for the gas-phase chemistry occurring in a furnace N_2O oxynitride process, *J. Electrochem. Soc.* **143**, 1753.
268. Gupta, A., Toby, S., Gusev, E.P., Lu, H.C., Li, Y., Green, M.L., Gustafsson, T., and Garfunkel, E. (1997) Nitrous oxide gas phase chemistry during silicon oxynitride growth, *to be published*
269. Baumvol, I.J.R., Stedile, F.C., Ganem, J.J., Trimaille, I., and Rigo, S. (1996) Nitrogen transport during rapid thermal growth of silicon oxynitride in N_2O, *Appl. Phys. Lett.* **69**, 2385.
270. Baumvol, I.J.R., Stedile, F.C., Ganem, J.J., Trimaille, I., and Rigo, S. (1997) Isotopic tracing during rapid thermal growth of silicon oxynitride films on Si in O_2, NH_3, and N_2O, *Appl. Phys. Lett.* **70**, 2007.
271. Ganem, J.J., Rigo, S., Trimaille, I., Baumvol, I.J.R., and Stedile, F.C. (1996) Dry oxidation mechanisms of thin dielectric films formed under N_2O using isotopic tracing methods, *Appl. Phys. Lett.* **68**, 2366.
272. Yao, Z.Q. (1995) The nature and distribution of nitrogen in silicon oxynitride grown on Si in a nitric oxide ambient, *J. Appl. Phys.* **78**, 2906.
273. Lu, Z.H., Hussey, R.J., Graham, M.J., Cao, R., and Tay, S.P. (1996) Rapid thermal N_2O oxynitride on Si(100), *J. Vac. Sci. Technol. B* **14**, 2882.
274. Hattangady, S.V., Niimi, H., and Lucovsky, G. (1996) Integrated processing of silicon oxynitride films by combined plasma and rapid-thermal processing, *J. Vac. Sci. Technol. A* **14**, 3017.
275. Kraft, R., Schneider, T.P., Dostalik, W.W., and Hattangady, S. (1997) Surface nitridation of silicon dioxide with a high density nitrogen plasma, *J. Vac. Sci. Technol. B* **15**, 967.
276. Kobayashi, H., Mizokuro, T., Nakato, Y., Yoneda, K., and Todokoro, Y. (1997) Nitridation of silicon oxide layers by nitrogen plasma generated by low energy electron impact, *Appl. Phys. Lett.* **71**, 1978.
277. Hill, W.L., Vogel, E.M., Misra, V., McLarty, P.K., and Wortman, J.J. (1996) Low pressure rapid thermal CVD of oxynitride gate dielectrics for N-channel and P-channel MOSFET's, *IEEE Trans. Electron Devices* **43**, 15.
278. Ma, T.P. (1997) Gate dielectric properties of silicon nitride films formed by jet vapor deposition, *Appl. Surf. Sci.* **117/118**, 259.
279. Pan, J.S., Wee, A.T.S., Huan, C.H.A., Tan, H.S., and Tan, K.L. (1997) AES analysis of nitridation of Si(100) by 2-10 keV N_2^+ ion beams, *Appl. Surf. Sci.* **115**, 166.
280. Goto, H., Shibahara, K., and Yokoyama, S. (1996) Atomic layer controlled deposition of silicon nitride with self-limiting mechanism, *Appl. Phys. Lett.* **68**, 3257.
281. Du, H., Tressler, R.E., and Spear, K.E. (1989) Thermodynamics of the Si-N-O system and kinetic modelling of oxidation of Si_3N_4, *J. Electrochem. Soc.* **136**, 3210.
282. Hillert, M., Jonsson, S., and Sundman, B. (1992) Thermodynamic calculation of the Si-N-O system, *Z. Metallkd.* **83**, 648.
283. Schenk, A. and Heiser, G. (1997) Modeling and simulation of tunneling through ultrathin gate dielectrics, *J. Appl. Phys.* **81**, 7900.
284. Yoshida, T., Imafuku, D., Alay, J.L., Miyazaki, S., and Hirose, M. (1995) Quantitative analysis of tunneling through ultrathin gate oxides, *Jpn. J. App. Phys.* **34**, L903.
285. Hwu, J.G. and Lin, S.T. (1990) Electrical characterization of the insulating property of Ta_2O_5 in Al-Ta_2O_5-SiO_2-Si capacitors by a low-frequency CV technique, *IEEE Proc.-G, Circuits, devices and systems.* **137**, 390.

286. Zaima, S., Furuta, T., and Koide, Y. (1990) Conduction mechanism of leakage current in Ta_2O_5 films on Si prepared by LPCVD, *J. Electrochem. Soc.* **137**, 2876.
287. Resetic, A. and Metikos-Hukovic, M. (1992) A.C. impedance investigation of Ta_2O_5 film for use as a storage capacitor dielectric, *Thin Solid Films* **30**, 176.
288. Sundaram, K., Choi, W.K., and Ling, C.H. (1993) Quasi-static and high-frequency C-V measurements on Al/Ta_2O_5/SiO_2/Si, *Thin Solid Films* **230**, 145.
289. Ishibashi, K., Patnaik, B.K., and Parikh, N.R. (1994) Thermal stability of thin poly-Si/Ta_2O_5/TiN capacitors for dynamic random access memory applications, *J. Vac. Sci. Technol.* **B 12**, 2822.
290. Kamiyama, S., Suzuki, H., and Ishitani, A. (1994) Highly reliable ultra-thin tantalum oxide capacitors for ULSI DRAMs, *IEICE transactions on electronic* **77**, 379.
291. Kwon, K.W., Kang, C.S., and Ahn, S.T. (1996) Thermally robust Ta_2O_5 capacitor for the 256-Mbit DRAM, *IEEE Trans. Electron Devices* **43**, 919.
292. Lo, G.Q., Kwong, D.L., and Lee, S. (1993) Reliability characteristics of MOS capacitors with chemical vapor deposited Ta_2O_5 gate dielectrics, *Appl. Phys. Lett.* **62**, 973.
293. An, C.H. and Sugimoto, K. (1994) Ellipsometric examination of structure and growth rate of metallorganic chemical vapor deposited Ta_2O_5 films on Si(100), *J. Electrochem. Soc.* **141**, 853.
294. Autran, J.L., Devine, R., Chaneliere, C., and Balland, B. (1997) Fabrication and characterization of Si-MOSFETs with PECVD amorphous Ta_2O_5 gate insulator, *IEEE Electron Device Lett.* **18**, 447.
295. Abe, Y. and Fukuda, T. (1994) TiO_2 thin films formed by electron cyclotron resonance plasma oxidation at high temperatures and their application to capacitor dielectrics, *Jpn. J. Appl. Phys.* **33**, 1248.
296. Kim, H.S., Gilmer, D.C., and Polla, D.L. (1996) Leakage current and electrical breakdown in metal-organic chemical vapor deposited TiO_2 dielectrics on silicon substrates, *Appl. Phys. Lett.* **69**, 3860.
297. Kim, H.S., Campbell, S.A., and Gilmer, D.C. (1997) Charge trapping and degradation in high-permittivity TiO_2 dielectric films, *IEEE Electron Device Letters* **18**, 465.
298. Campbell, S.A., Gilmer, D.C., and Yan, J. (1997) MOSFET transistors fabricated with high permitivity TiO_2 dielectrics., *IEEE Trans. Electron Devices* **44**, 104.
299. Gurvitch, M., Manchanda, L., and Gibson, J.M. (1987) Study of thermally oxidized yttrium films on silicon, *Appl. Phys. Lett.* **51**, 919.
300. Onisawa, K., Fuyama, M., and Tamura, K. (1990) Dielectric properties of rf-sputtered Y_2O_3 thin films, *J. Appl. Phys.* **68**, 719.
301. Sharma, R.N. and Rastogi, A.C. (1993) Compositional and electronic properties of chemical-vapor-deposited Y_2O_3 thin film-Si(100) interfaces, *J. Appl. Phys.* **74**, 6691.
302. Ling, C.H. (1993) Interfacial polarisation in Al-Y_2O_3-SiO_2-Si capacitor, *Electronics letters* **29**, 1676.
303. Sharma, R.N. and Rastogi, A.C. (1994) Structure and composition of interfacial silicon oxide layer in chemical vapor deposited Y_2O_3-SiO_2 bilayer dielectrics for MIS devices, *J. Appl. Phys.* **76**, 4215.
304. Choi, S.C., Cho, M.H., Whangbo, S.W., Whang, C.N., Kang, S.B., Lee, S.I., and Lee, M.Y. (1997) Epitaxial growth of Y_2O_3 films on Si(100) without an interfacial oxide layer, *Appl. Phys. Lett.* **71**, 903.
305. Agarwal, M., DeGuire, M.R., and Heuer, A.H. (1997) Synthesis of yttrium oxide thin films with and without the use of organic self-assembled monolayers, *Appl. Phys. Lett.* **71**, 891.
306. Kalkur, T.S., Kulkarni, J., and Lu, Y.C. (1991) Metal-Ferroelectric-Semiconductor characteristics of bismuth titanate films on silicon, *Ferroelectrics* **116**, 135.
307. Arita, K., Fujii, E., and de_Araujo, C. (1994) Application of ferroelectric thin films to Si devices, *IEICE Trans. Electronics* **77**, 392.
308. Arita, K., Fujii, E., and Suzuoka, N. (1994) Si LSI Process technology for integrating ferroelectric capacitors, *Jpn. J. Appl. Phys.* **33**, 5397.
309. Auciello, O., Gifford, K.D., and Kingon, A.I. (1994) Control of structure and electrical properties of lead-zirconium-titanate-based ferroelectric capacitors produced using a layer-by-layer ion beam sputter-deposition technique, *Appl. Phys. Lett.* **64**, 2873.
310. Chen, H.D., Udayakumar, K.R., and Niles, L.C. (1995) Dielectric, ferroelectric, and piezoelectric properties of lead zirconate titanate thick films on silicon substrates, *J. Appl. Phys.* **77**, 3349.
311. Tokumitsu, E., Nakamura, R., and Ishiwara, H. (1997) Nonvolatile memory operations of Metal-Ferroelectric-Insulator-Semiconductor (MFIS) FET's using PLZT/STO/Si(100) structures, *IEEE Electron Device Letters* **18**, 160.
312. Massoud, H. (1997) Device physics and simulation of Metal/Ferroelectric Film/p-type silicon capacitors, *Microelectronics Engineering* **36**, 95.
313. Kotecki, D.E. (1996) High-K dielectric materials for DRAM capacitors, *Semicond. International* **11**, 109.
314. Inoue, T., Yamamoto, Y., and Koyama, S. (1990) Epi-growth of CeO_2 layers on Si, *Appl. Phys. Lett.* **56**

315. Chikyow, T., Bedair, S.M., and El-Masry, N.A. (1994) Reaction and regrowth control of CeO_2 on Si(111) surface for the silicon-on-insulator structure, *Appl. Phys. Lett.* **65**, 1030.
316. Tye, L., El-Masry, N.A., and Bedair, S.M. (1994) Electrical characteristics of epitaxial CeO_2 on Si(111), *Appl. Phys. Lett.* **65**, 3081.
317. Yoshimoto, M., Shimozono, K., and Koinuma, H. (1995) Room-temperature epitaxial growth of CeO_2 thin films on Si(111) substrates for fabrication of sharp oxide/silicon interface, *Jpn. J. Appl. Phys.* **34**, L688.
318. Kalkur, T.S. and Lu, Y.C. (1992) Electrical characteristics of ZrO_2-based metal-insulator-semiconductor structures on p-Si, *Thin Solid Films* **207**, 193.
319. Hwang, C.S. and Kim, H.J. (1993) Deposition and characterization of ZrO_2 thin films on silicon substrate by MOCVD., *J. Materials Res.* **8**, 1361.
320. Hsu, C.T., Su, Y.K., and Yokoyama, M. (1992) High dielectric constant of RF-sputtered HfO_2 thin films, *Jpn. J. Appl. Phys.* **31**, 2501.
321. Kuo, C.T., Kwor, R., Jones, K. (1992) Study of sputtered HfO_2 thin films on Si, *Thin Solid Films* **213**, 257.
322. Stesmans, A. and Afanas'ev, V.V. (1996) Thermally induced interface degradation in (111)Si/SiO_2 traced by electron spin resonance, *Phys. Rev. B* **54**, R11129.
323. Moslehi, M.M., Shatas, S.C., and Saraswat, K.C. (1985) Thin SiO_2 insulators grown by rapid thermal oxidation of silicon, *Appl. Phys. Lett.* **47**, 1353.
324. Deaton, R.S. and Massoud, H.Z. (1992) Manufacturability of rapid thermal oxidation of silicon: Oxide thickness, oxide thickness variation, and system dependency, *IEEE Trans. Semicond. Manufacturing* **5**, 347.
325. Lucovsky, G., Yasuda, T., Ma, Y., Hattangady, S.V., Xu, X.L., Misra, V., Hornung, B., and Wortman, J.J. (1994) Control of Si-SiO_2 interface properties in MOS devices prepared by plasma-assisted and RTP, in S. P. Murarka, K. Rose, T. Ohmi and T. Seidel (Eds.), *Interface Control of Electrical, Chemical, and Mechanical Properties*, MRS, vol. 318, p. 81.
326. Roozeboom, F. (Ed.) (1995) *Advances in Rapid Thermal and Integrated Processing*, Kluwer Academic, Dordrecht - Boston - London.
327. George, S.M., Sneh, O., and Way, J.D. (1994) Atomic layer controlled deposition of SiO_2 and Al_2O_3 using ABAB... binary reaction sequence chemistry, *Appl. Surf. Sci.* **82/83**, 460.
328. Suntola, T. (1989) Atomic layer epitaxy, *Mater. Sci. Rept.* **4**, 261.
329. Ma, Y., Yasuda, T., and Lucovsky, G. (1994) Ultrathin device quality oxide-nitride-oxide heterostructure formed by remote plasma enhanced CVD, *Appl. Phys. Lett.* **64**, 2226.
330. Hu, S.M. (1974) Formation of stacking faults and enhanced diffusion in the oxidation of silicon, *J. Appl. Phys.* **45**, 1567.
331. Leroy, B. (1987) Stresses and silicon interstitials during the oxidation of silicon, *Phil. Mag. B* **55**, 159.
332. Dunham, S.T. and Plummer, J.D. (1986) Point defect generation during oxidation of silicon in dry oxygen. I. Theory, *J. Appl. Phys.* **59**, 2541.
333. Dunham, S.T. (1992) Interaction of silicon point defects with SiO_2 films, *J. Appl. Phys.* **71**, 685.
334. Tsamis, C., Kouvatsos, D.N., and Tsoukalas, D. (1996) Influence of N_2O oxidation of silicon on point defect injection kinetics in the high temperature regime, *Appl. Phys. Lett.* **69**, 2725.
335. Herner, S.B., Krishnamoorthy, V., and Gossmann, H.J. (1997) Extrinsic dislocation loop behavior in silicon with a thermally grown silicon nitride film, *J. Appl. Phys.* **81**, 7175.
336. Dallaporta, H., Liehr, M., and Lewis, J.E. (1990) Silicon dioxide defects induced by metal impurities, *Phys. Rev. B* **41**, 5075.
337. Kasi, S.R. and Liehr, M. (1992) Preoxidation Si cleaning and its impact on metal oxide semiconductor characteristics, *J. Vac. Sci. Tech.* **10**, 795.
338. Higashi, G.S. and Chabal, Y.J. (1993) Surface chemical composition and morphology, in W. Kern (Ed.), *Handbook of Silicon Wafer Cleaning Technology*, Noyes Publications, NJ p. 433.
339. Depas, M., Heyns, M.M., Nigam, T., Kenis, K., Sprey, H., Wilhelm, R., Grossley, A., Sofield, C.J., Graf, D. (1996) Critical processes for ultrathin gate oxide integrity, in H. Z. Massoud, E. H. Poindexter and C. R. Helms (Eds.), *The Physics and Chemistry of SiO_2 and the Si-SiO_2 Interface - 3*, The Electrochem. Soc., NJ.
340. Weitzel, C.E., Palmour, J.W., Carter, C.H., Moore, K., Nordquist, K.J., Allen, S., Thero, C., and Bhatnagar, M. (1996) Silicon carbide high-power devices, *IEEE Trans. Electron Devices* **43**, 1732.
341. Collinge, J.P. (1997) Silicon-on-Insulator Technology: Materials to VLSI, *Kluwer Publishers*, Boston.
342. Hu, C. (1994) SOI for high speed ULSI, *Jpn. J. Appl. Phys.* **33**, 365.
343. Reichert, G., Raynaud, C., Faynot, O., Balestra, F., and Cristoloveanu, S. (1997) Submicron SOI-MOSFETs for high temperature operation (300 - 600 K), *Microelectronics Engineering* **36**, 359.
344. Dadap, J.I., Hu, X.F., Anderson, M.H., Downer, M.C., Lowell, J.K., and Aktsipetrov, O.A. (1996) Optical second-harmonic spectroscopy of a Si(001) metal-oxide-semiconductor structure, *Phys. Rev. B* **53**, R7607.

STUDY OF THE Si/SiO$_2$ INTERFACE USING POSITRONS:

Present Status and Prospects

J.M.M. DE NIJS and M. CLEMENT*
DIMES
Delft University of Technology
P.O. Box 5053
2600GB Delft
The Netherlands

In this paper we consider the potential of positrons for the study of defects in the MOS system. After a brief introduction on the fundamentals of the transport and annihilation of positrons in solids, we will deal with the techniques to drive the positrons towards the Si/SiO$_2$ interface. To illustrate the nature of the information obtained from positron experiments, we discuss experimental results that reveal a correlation between the positron annihilation data and the generation rate of interface state due to hydrogen release. We will also deal with the prospects of positron studies to provide more detailed information on the chemical nature and microscopic structure of defects.

1. Introduction

The reliability of the metal-oxide-silicon (MOS) systems has been extensively studied during the past two decades. These studies have led to a good operational understanding of MOS degradation processes from the electrical point of view; however, it has not resulted in a consistent and satisfactory understanding from the viewpoint of physics. A major cause for the poor progress on the physical understanding is the lack of sufficiently sensitive analytical techniques. Electrical techniques are very sensitive, one can even study individual defects [1,2], but they do not provide structural information. Electron spin resonance studies have contributed very valuable insights concerning trivalent silicon centers such as E' and P$_b$ [3], but these

---

*) IRI, Delft University of Technology, P.O. Box 5042, 2600AG Delft, The Netherlands.

centers do not determine the MOS reliability in a dominant fashion. On the other hand, the hydrogen-induced interface states and the bulk oxide neutral electron traps are not observed by this technique [4-6]. The structure of the interfacial region has also been investigated by high-resolution transmission electron microscopy [7] and X-ray photoemission spectroscopy [8,9]. Although the results from these studies have contributed to the understanding of the structure of the Si/SiO$_2$ interface, they have not provided specific insights in the relation between the structure and the reliability. In this paper we will consider the potential of positron annihilation (PA) to provide additional insights in the MOS system.

Positron annihilation spectroscopy is a technique that provides structural and chemical information on defects in layered systems. The method is very defect selective, which makes it an interesting technique to study the MOS system. Earlier work has shown that positrons are indeed a very sensitive technique to investigate defects in the MOS system; however, it has also revealed the limitations of the conventional approaches [10]. Positron beam measurements have a fairly good depth resolution, but not sufficient to selectively focus on the defects at the Si/SiO$_2$ interface. Furthermore, the interpretation of the experimental data (γ-spectra) is not straightforward.

In this paper we will show that the conventional positron technique can be adapted so that it only responds to the defects at the Si/SiO$_2$ interface. This adaptation works for MOS capacitors with a relatively thick (100nm) oxide layer and a thin aluminum gate and also for capacitors with a polysilicon gate. For the samples with an aluminum gate, we will discuss an example of the information one may obtain from PA studies. To conclude we will summarize the possibilities for expanding the experimental approaches used in positron studies.

2. Fundamentals of positron annihilation

In the following we will present a brief description of the PA technique. An extended treatment can be found in the review of Asoka-Kumar et al. [10] and in our own papers [11-13].

For the PA experiments, monoenergetic positrons are implanted into the system. Because of their small mass, the positrons are distributed over a rather broad depth region. A typical implantation profile is shown in Fig. 1. The average implantation depth is controlled by the implantation energy (E_{imp}), because of which the technique is used to study the depth distribution of defects. For systems consisting of two or more relatively thin layers with different defects, like the MOS system, the signal always contains information from various defects and special numerical programs are needed for their separation.

After implantation the positrons thermalize. Subsequently, they diffuse around until being captured by a defect. Finally, they annihilate with an electron, thereby producing two 511keV γ-quanta. There are various ways to extract information from γ-quanta, but here we will restrict ourselves to Doppler broadening measurements. The crux of this technique is that the energy of the γ-quanta is affected by the momentum of the annihilating particles and thus contains information on the structure of the annihilation site. The energy distribution affects the shape of the γ-peak which is characterized with the aid of the shape parameters S and W defined in Fig. 2.

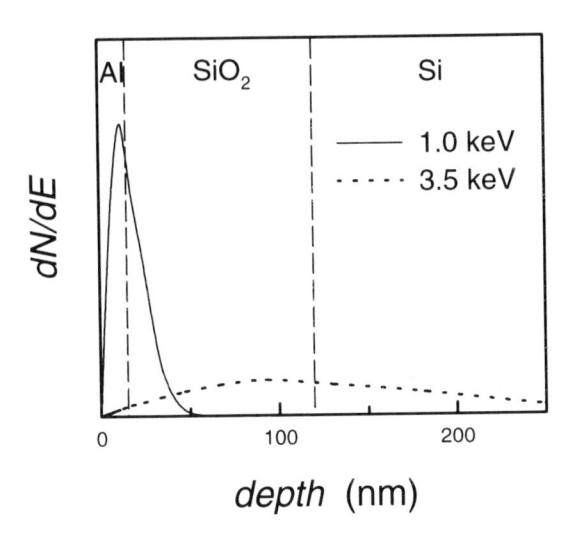

Fig. 1. Typical high en low energy positron implantation profiles. The vertical dashed lines indicate the Al/SiO$_2$ and SiO$_2$/Si interfaces.

The momentum of the interacting electron depends on the electronic properties of the defect. Defects contained in a particular layer thus can be characterized by specific (S,W) coordinates. The definitions of S and W offer the advantage of a linear formalism: if annihilation takes place at two defects, A and B, with characteristic coordinates (S_A,W_A) and (S_B,W_B), and with trapped fractions f_A and f_B, where $f_B = 1-f_A$, then the measured parameters (S_M,W_M) are given by:

$$S_M = f_A \times S_A + f_B \times S_B \tag{1a}$$

and

$$W_M = f_A \times W_A + f_B \times W_B. \tag{1b}$$

This linear relationship greatly facilitates the interpretation of experimental results when plotting the (S,W) data obtained in an implantation energy scan in the S-W plane, thereby using the positron implantation energy as a running parameter. If the defects A and B are situated in two different layers, located at different depth, the implanted positrons will be distributed over the layers containing the A and B defects. This distribution can be varied by changing the implantation energy. Such a variation corresponds to a straight trajectory in the S-W plane, running from coordinates (S_A, W_A) towards (S_B, W_B). The presence of an additional, third, trapping layer at another depth can now be easily detected: the trajectory will be curved or it exhibits a kink.

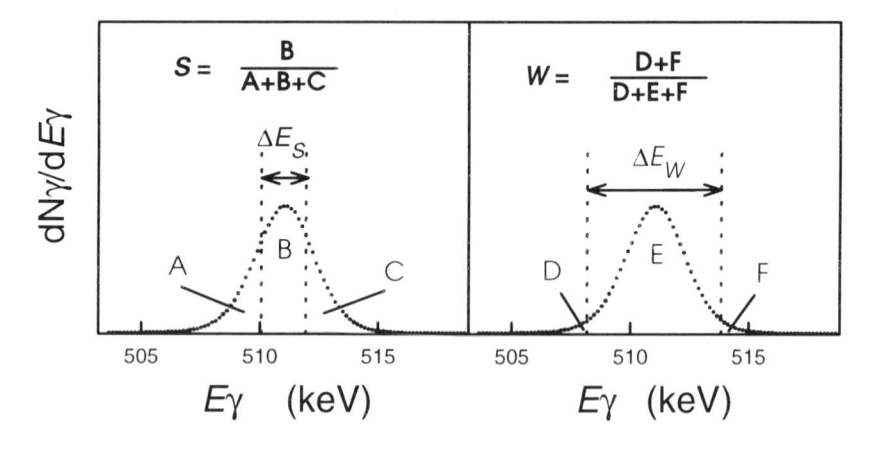

Fig. 2. Definition of parameters S and W of the γ-peak. The indicated area's (A to F) are determined with the aid of an integration window (ΔE_S and ΔE_W)

3. Experimental features of PA

3.1 POSITRON TRANSPORT THROUGH THE OXIDE

The ultimate goal of our investigations is the study of the defect properties of the Si/SiO$_2$ interface and of thin oxide layers beneath a polysilicon gate. Unfortunately, as pointed out above, implantation provides a moderate defect selectivity only; for this reason we guide the positrons towards the interface by controlling their transport after thermalization using an electric field. In a first experiment we used an MOS system with a 350nm thick thermal oxide and a 10nm aluminum gate. The positrons were implanted with an energy of 4keV, which corresponds to implantation in the middle of the oxide layer; only negligible fractions of the positrons will thermalize in the aluminum layer or the silicon substrate. Next, we monitored the evolution of the peak shape when applying different gate biases. The result is shown in Fig. 3 where also the characteristic (S, W) coordinates for silicon, SiO$_2$ and aluminum and for the Si/SiO$_2$ interface are indicated. The data clearly show that the positrons annihilate at different defect sites when changing the bias voltage, which demonstrates that the

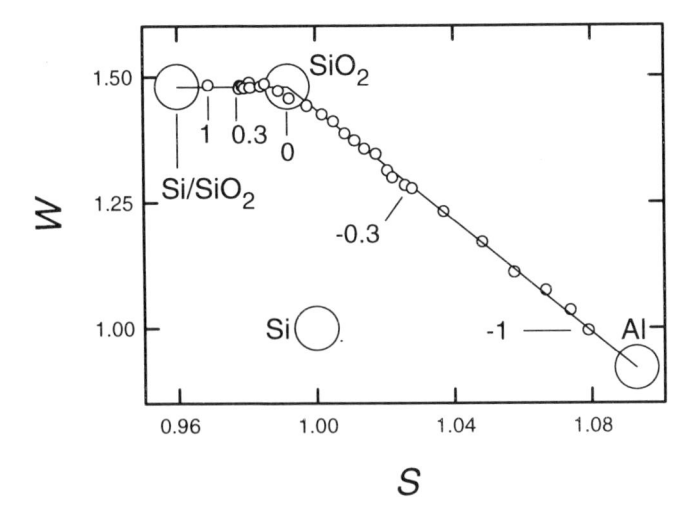

Fig. 3. Field dependence of the S and W parameter for 4keV positrons implanted into the thermal oxide of a MOS capacitor with a 15nm aluminum gate. The oxide layer thickness is 350nm. The numbers along the trajectory denote the applied electric field in MV/cm. The large circles indicate the S-W coordinates of aluminum, the oxide and the silicon substrate.

positrons drift through the oxide layer. The trajectory clearly exhibits two straight sections, one for positive and one for negative bias. Apparently only three annihilation sites are involved. For 0V bias, all positrons annihilate in the oxide layer, but when applying a negative bias, the positrons are transported towards the aluminum gate. The fraction that annihilates at the gate increases gradually with the electric field. For a field of -1MV/cm all

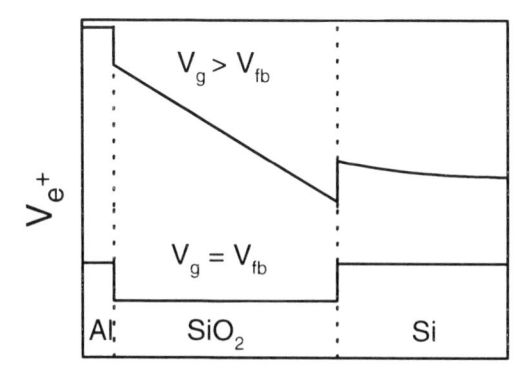

Fig. 4. Schematic potential diagram for positrons in the MOS system for the case of positive bias voltage and for flatband condition where the electric field approaches zero.

positrons are trapped at the gate. Interestingly, the positive bias trajectory is not directed towards the characteristic coordinates of the silicon substrate, which indicates that the positrons do not annihilate in the substrate. Apparently they are not injected from the oxide layer into the silicon but trapped at the Si/SiO$_2$ interface.

From the above observations it is possible to derive the potential diagram for positrons in the MOS system (Fig. 4). Since for positive bias the positrons are not injected into the substrate, it can be concluded that a potential step exists at the SiO$_2$/Si interface. More detailed analysis of the negative bias trajectory in combination with the use of different gate metals indicate a similar potential step a the SiO$_2$/gate interface [13].

3.2 ALUMINUM-GATED CAPACITORS

Factually, in the above method we have used the thick oxide layer to collect the implanted positrons. Subsequently, they are all driven from this layer towards the SiO$_2$/Si interface, thus permitting an effective characterization of the interface. Unfortunately, for thinner oxides, it is impossible to avoid implantation of substantial fractions of positrons in the gate and in the silicon substrate. Consequently, the interface positron signal cannot be isolated. This problem can be solved by using a positron energy scan at constant bias voltage.

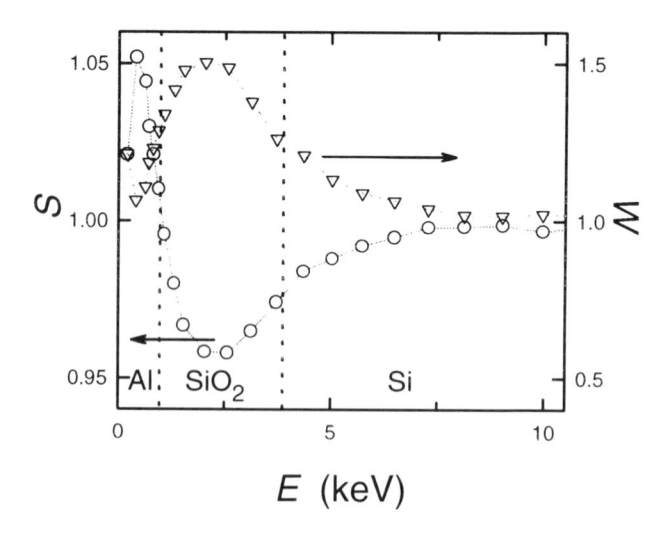

Fig. 5. S- and W- parameter versus E_{imp}. for MOS system with thin Al gate and 100nm oxide subjected to +15V bias. The vertical dotted lines indicate the positions of the interfaces.

In Fig. 5 we show the *S* and *W* data of a MOS capacitor with a 100nm oxide and a 10nm aluminum gate subjected to 15V positive bias. For low implantation energies the positrons are introduced into the aluminum layer where they are trapped immediately. For energies larger than ≈7.5keV all positrons are annihilated in the silicon substrate, in agreement with the asymptotical behavior of both curves. Implantation in the oxide layer produces the minimum in the $S(E_{imp})$ and the maximum in $W(E_{imp})$ curve.

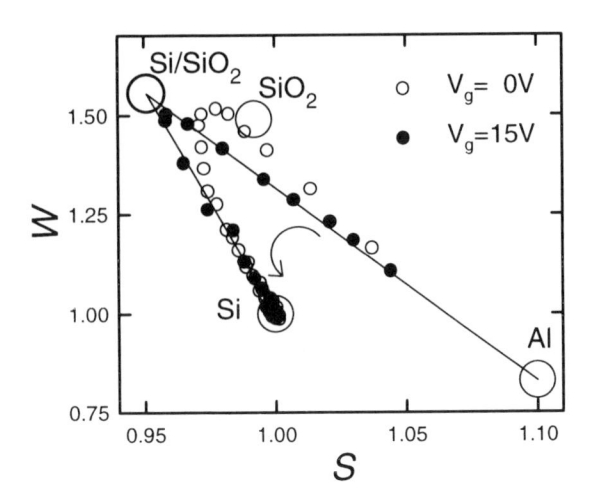

Fig. 6. *S-W* trajectories for MOS system with thin aluminum gate and 100nm oxide layer for 0V and +15V bias. The arrow indicates the direction of increasing E_{imp}. The large circles denote the characteristic (*S,W*) coordinates for the different layers and interfaces.

In Fig. 6 we have plotted the $S(E_{imp})$ and $W(E_{imp})$ data obtained at 0V and +15V bias from the same MOS capacitor in the *S-W* plane, using the implantation energy as a running parameter. The arrow indicates the direction of increasing energy. For large implantation energies, both trajectories approache the silicon substrate point, irrespective of the bias voltage. The trajectory for positive bias reveals two straight sections, which reflects that trapping in the oxide layer is completely suppressed and that the positrons only annihilate in the gate metal, at the SiO₂/Si interface, and in the substrate. Because of the favorable relative positions in the *S-W* plane of the aluminum, silicon and interface characteristics points, it is possible to accurately determine the interface coordinates from the point of intersection of the high and low implantation energy sections of the trajectory. The 0V trajectory makes an excursion towards the bulk SiO₂ coordinates, indicating that here a large fraction of the positrons is trapped in this layer.

For this 100nm oxide, it is seen that the positive bias trajectory closely approaches the point of intersection. Such a close approach is not needed for the precise determination of the interface coordinates as long as the trajectory exhibits clear straight low and high-energy sections. Therefore we expect that this method can be used for thinner oxide layers, for example down to 60nm, without a severe loss of accuracy.

3.3 POLYSILICON-GATED CAPACITORS

From a technological point of view, study of oxide layers with thickness of 100nm or more, like those discussed above, is not of particular interest. Instead, one would like to investigate much thinner oxides, which should be provided with a polysilicon gate [16]. As stated above, in our technique we used the oxide layer to collect the implanted positrons. Such a layer should be sufficiently thick in order to gather all implanted positrons. On the other hand, it should contain only very few defects so that the positrons are not immediately trapped after thermalization. Furthermore, the layer should permit the induction of an electric field to drive the positrons towards the Si/SiO_2 interface. Considering these requirements, we explored the possibility to use a 100nm polysilicon layer as implantation layer.

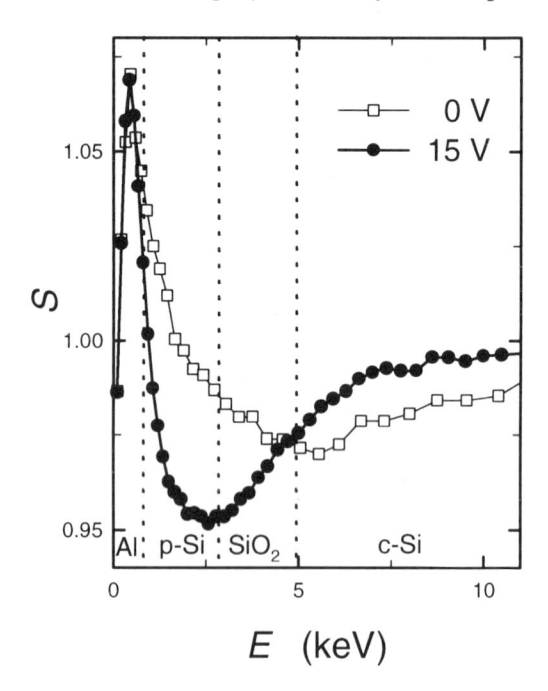

Fig. 7. $S(E_{imp})$ curves for MOS system with polysilicon gate subjected 0V and +15V bias.

In conventional polysilicon layers the grain boundaries will act as very efficient positron traps. To circumvent this trapping problem, silicon in an amorphous state was deposited by a low temperature chemical vapor deposition process (540°C). Subsequently, the silicon was annealed at a temperature of 900°C which results in the nucleation and growth of crystalline silicon islands. This way silicon islands with typical dimensions of 60nm or larger are obtained [17]. Next, the deposited silicon was implanted with 10^{13} of phosphorus atoms per cm^2 and subjected to a 900°C activation anneal and a 400°C passivation anneal in forming gas. Finally, the capacitors were completed with a 10nm aluminum top layer.

In Figs. 7 and 8 we give the results of the positron measurements on a capacitor with a 100nm oxide and a 60 nm polysilicon layer [16]. Fig. 7 shows that for positive bias conditions all positrons can be guided towards the Si/SiO_2 interface. Fig. 7 reveals a broad minimum for implantation energies of between 2 and 3keV which is associated with the intersection point of Fig. 8. The width of the minimum reflects the fact that although the implantation depth is substantially varied, all positrons annihilate at the interface. Moreover, an energy of 2keV corresponds with implantation in the middle of the polysilicon layer, which shows that at positive bias a very large fraction of the positrons implanted in the polysilicon layer are driven out of this layer into the oxide.

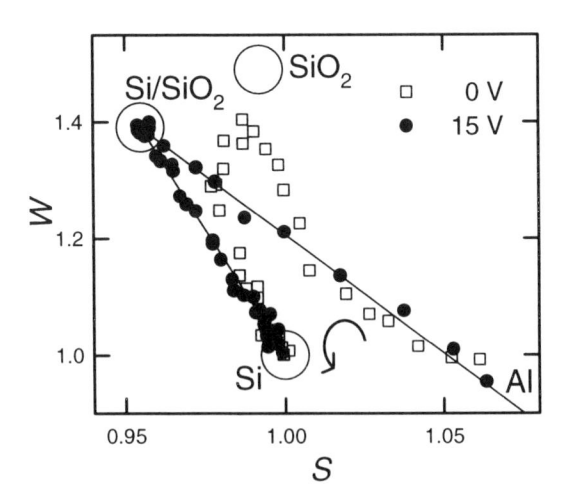

Fig. 8. *S-W* trajectories for MOS system with polysilicon gate subjected to 0 and +15V bias.

The present results show that a polysilicon layer can be used to collect the implanted positrons; from here they can be driven into the oxide, irrespective of the oxide thickness. This feature opens the way for studies on very thin oxides (<10nm), which are presently underway.

4. Example of positron studies: The effect of a post oxidation anneal

To demonstrate the capabilities of positron experiments we studied the effects of a high temperature (1000°C) post oxidation anneal (POA) [11] by means of PA. These experiments were performed on 100nm thick dry oxides thermally grown at 1000°C and provided with a thin aluminum gate. Two parameters of the POA were varied: time and ambient. Fig. 9 contains the corresponding positron data. Surprisingly, the data points of all the samples define two trajectories only; one (open symbols) for the samples subjected to a short anneal (2min.) in N_2 or to 20min. in argon and another one (solid symbols) for specimens subjected to a long anneal in N_2. We also subjected some samples to a 20min. anneal in N_2, followed by a 20min. anneal in argon. In this case the data points follow the trajectory of the samples that were not or only briefly annealed in N_2. Apparently the effect of the N_2 anneal was undone by the argon anneal. This implies that the effect of the N_2 anneal is reversible.

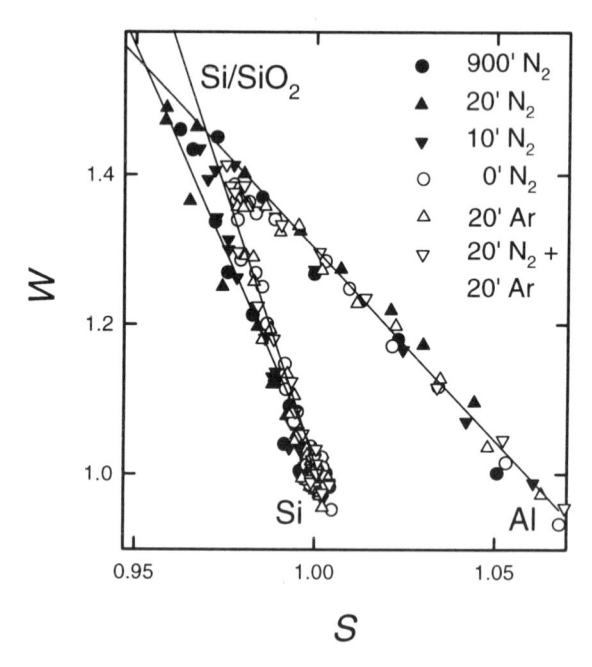

Fig. 9. S-W trajectories of specimens with a long (4, 6, 10, 20 or 900min.) POA in N_2 (solid symbols) and of samples with 0 or 2min. POA in N_2, 20min. POA in Ar, or 20min. POA in N_2 followed by 20min. in Ar (open symbols). The data were obtained at +15V bias.

In a parallel experiment, we determined the densities of interface-related defects for the different samples. The density of oxygen vacancy-associated hole traps was measured using hole injection experiments [11].

P_b-type interface states were determined from capacitance-voltage measurements. The furnace used in our experiments was equipped with a special unloading tube to prevent passivation of these centers by moisture in the ambient. In Fig. 10 we show the measured concentrations of these defects as a function of the POA time. The densities of both defects gradually changed with the POA anneal time. Furthermore, the annealing ambient (Ar or N_2) had no effect on the numbers of hole traps and P_b centers. Apparently, the change in the state densities only depends on the anneal time and not on the ambient. Clearly, the positrons are not trapped at P_b centers or oxygen vacancies since the annealing behavior of the two latter centers is different from that of the positron traps.

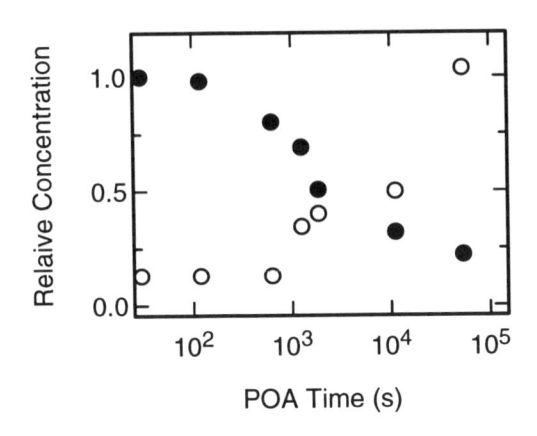

Fig. 10. Evolution with POA time of the numbers of O vacancy-related hole traps (open symbols) and of interface states in the as-grown oxide (P_b centers, solid symbols). The maximum values are 9×10^{12} cm^{-2} for the hole traps and 5×10^{11} cm^{-2} for the interface states.

We also measured the rate of generation of interface states due to release of hydrogen [4,6]. For these measurements, samples with a thin aluminum gate were exposed to vacuum ultraviolet (VUV) radiation at negative bias conditions. Such an exposure results in the release of hydrogen from the SiO_2/Al interface [6]. Fig. 11 shows for samples subjected to different anneals, the density of interface states versus the integrated photocurrent. The data points again define two clearly different types of behavior: a low interface state generation rate for samples with a short POA in N_2 or with one in argon, and a high rate for those subjected to a long anneal in N_2. Furthermore, also this effect is reversible: when applying a 20min. POA in argon after 20min. of N_2 anneal again a low generation rate is found. These data demonstrate that the positron data do correlate with the interface state generation rate, namely, a larger generation rate correlates with a smaller S and larger W value for the Si/SiO_2 interface.

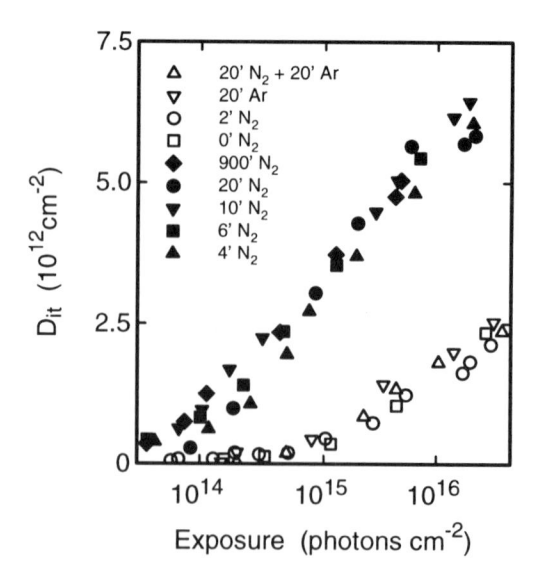

Fig. 11. Density of VUV-induced interface traps vs. VUV exposure. The oxides had received POA in N_2, in Ar, or an anneal in N_2 followed by one in Ar. POA times are indicated in minutes.

These findings lead to the following considerations regarding the nature of the positron traps: In metals and semiconductors it has been clearly established that positrons preferentially occupy so-called "open volume" defects like vacancies and nanocavities. In our case, the larger open spaces or interstices in the oxide network near the substrate could similarly trap the positrons. Conceivably, also mobile hydrogen atoms could be efficiently trapped at such interstices, after which they could dimerize with earlier arrived atoms. After a N_2 anneal nitrogen would be tied up in the interstices. In contrast, when applying POA in argon, they remain empty because argon is not bonded. In the case of a combined anneal, first in N_2, then in argon, the nitrogen is driven out of the interstices. Thus the filling and emptying behavior would satisfy the observed reversibility of the effect. The explanation implies that filling the interstices with N_2 inhibits the trapping of the positrons and hydrogen.

5. Prospects of positron studies.

Our results clearly indicate that the positrons reveal relevant material properties of the MOS system that are not detected by other techniques. This makes it attractive to extend the positron studies, for example, to investigate degradation phenomena for polysilicon-gated samples with low hydrogen content.

Using novel and more refined approaches like lifetime, correlation and two-dimensional angular correlation annihilation radiation (2D-ACAR) measurements, which are presently being developed, offer new perspectives. Doppler broadening data only provide an overall characterization of a defect layer; the contributions of different annihilation sites can not be distinguished. In contrast, using lifetime measurements one can discriminate between different defects in a single layer [18].

Conventional positron annihilation Doppler broadening experiments have a signal-to-noise ratio of approximately 100. When using two detectors to simultaneously record both γ-quanta, the background signal is reduced by 2-3 orders of magnitude. The improved sensitivity greatly extends the energy window towards high-momentum annihilation events with core electrons. Since core electrons have better defined electronic states as compared to valence electrons, element-specific features can be discerned in the high-energy tail of the γ-spectrum. This was recently demonstrated for vacancy-gold and vacancy-arsenic complexes in silicon implanted with gold and arsenic [19].

In 2D-ACAR experiments the momentum of an annihilating electron-positron pair is established by measuring the angle between the two γ-photons. After collection of sufficient data points, one thus obtains the momentum distribution in momentum space. As shown by the work of Peng et al., this momentum spectrum contains very detailed information on the electronic structure of the positron trapping site [20]. These more sophisticated measurements require improved calculations of the interaction of electrons and positrons at defect sites. Such studies are presently underway [21].

At the present time the positron annihilation technique uses rather large diameter, typically 10mm, positron beams which is a major drawback for experiments on thin oxides and device structures. Shortly, in our laboratories this diameter will be reduced to approximately 1mm without a severe loss of brightness. For the future, beam diameters as small as 1 μm are expected, so that also good resolution in the sample plane will be obtained.

Acknowledgement

The authors would like to thank prof. dr. P. Balk for a critical reading of the manuscript.

References

1. Cobden, D.H. and Uren, M.J. (1993) Random telegraph signals from liquid helium to room temperature, *Microelectron. Eng.* **20**, 163-170..
2. Saks, N.S., Groeseneken, G. and DeWolf, I. (1996) Characterization of individual interface traps with charge pumping, *Appl. Phys. Lett.* **68**, 1383-85.
3. Lenahan, P.M. (1993) Electron spin resonance and Instabilities in metal insulator semiconductor systems, *Microelectron. Eng.* **22**, 129-138.
4. Cartier, E., Stathis, J.H. and Buchanan, D.A. (1993) Passivation and depassivation of Si dangling bonds at the Si/SiO$_2$ interface by H^0, *Appl. Phys. Lett.* **63**, 1510-12.
5. DiMaria, D.J., Cartier, E. and Arnold, D. (1993), Impact ionization, trap creation, degradation and breakdown in SiO$_2$ films on silicon, *J. Appl. Phys.* **73**, 3367-84.
6. Nijs, J.M.M. de, Druijf, K.G., Afanas'ev, V.V., Drift, E. van der and Balk, P. (1994) Hydrogen-induced donor-type Si/SiO$_2$ interface states, *Appl. Phys. Lett.* **65**, 2428-30.
7. Hollinger, G., Saoudi, R., Ferret, P. and Pitaval, M. (1988) The microstructure of Si/SiO$_2$ interfaces investigated by XPS and HRTEM, in Helms, C.R. and Deal, B.E. (eds.), *The Physics and Chemistry of SiO$_2$ and the Si/SiO$_2$ interface*, Plenum Press, New York, pp. 211-218.
8. Grunthaner, F.J. and Grunthaner, P.J. (1986) Chemical and electronic structure of the Si/SiO$_2$ interface, *Mat. Sc. Rep.* **1**, 65-160.
9. Himpsel, F.J., McFeely, F.R., Taleb-Ibrahimi, A., Yarmoff, J.A. and Hollinger, G. (1988) Microscopic structure of the Si/SiO$_2$ interface, *Phys. Rev. B* **38**, 6084-96.
10. P. Asoka-Kumar, P., Lynn, K.G., and Welch, D.O., (1995) Characterization of defects in Si and Si-SiO$_2$ using positrons, *J. Appl. Phys.* 76, 4935.
11. Clement, M., de Nijs, J. M. M., van Veen, A., Schut H. and Balk, P., (1995) Effect of Post Oxidation Anneal on VUV Radiation-Hardness of the Si/SiO$_2$ System studied by Positron Annihilation Spectroscopy, *IEEE Trans. on Nucl. Sc.* **NS-42**, 1717-24.
12. Clement, M., de Nijs, J. M. M., Balk, P., Schut H. and van Veen, A., (1996) Analysis of positron beam data by the combined use of the shape and wing parameter, *J. Appl. Phys.* **79**, 9029-36.
13. Clement, M., de Nijs, J. M. M., Balk, P., Schut H. and van Veen, A., (1997) Transport of positrons in the electrically biased MOS system, *J. Appl. Phys.* **81**, 1943-55.
14. B. Nielsen, K.G. Lynn, Y.-C Chen and D.O. Welch, Appl. Phys. Lett. 51 (1987) 1022.
15. The first studies of the Si/SiO$_2$ system of Nielsen et al. [14] showed that the positrons are trapped at the Si/SiO$_2$ interface. For the numerical analysis of their data they assumed that the interface would be impermeable; every positron would be trapped. Subsequently, this assumption has been treated as a well-established fact, even in studies concerning a MOS system subjected to a bias [10].
16. Clement, M., de Nijs, J. M. M., van Veen, A., Schut H., Mallee, R., and Balk, P., Positron beam technique for the study of defect at the Si/SiO$_2$ interface of a polysilicon-gated MOS system, accepted for publication in MRS (1997).
17. French, P.J., Drieënhuizen, B.P., Poenar, D., Goosen, J.F.L., Mallee, R., Sarro, P.M. and Wolffenbuttel, R., (1996) The development of a low-stress polysilicon process compatible with standard device processing, *J. Microelectromechan. Sys.* **5** 187-96.
18. Uendono, A., Wei, L., Tanigawa, S., Suzuki, R., Ohgaki, H., Mikado, T. and Ohji, Y. (1993) Positron annihilation in a MOS system studied by using a pulsed monoenergetic positron beam, *J. Appl. Phys.* **74**, 7251-56.
19. Myler, U., Goldberg, R.D., Knights, A.P., Lawther, D.W. and Simpson, P.J., (1996) Chemical information in positron annihilation spectra, *Appl. Phys. Lett.* **69** 3333-35.
20. Peng, J.P., Lynn, K.G., Asoka-Kumar, P., Becker, D.P., and Harshman, D.R., (1996) Study of the Si/SiO$_2$ interface using variable energy positron two-dimensional angular correlation of annihilation radiation, *Phys. Rev. Lett.* **76** 2157-60.
21. Alatalo, M., Barbielini, B., Hakala, M., Kauppinen, H., Korhonen, T., Puska, M.J.,. Saarinen, K., Hautojärvi, P. and Nieminen, R.M. (1996) Theoretical and experimental study of positron annihilation with core electrons, *Phys. Rev. B* **54**, 2397-08.

MEDIUM ENERGY ION SCATTERING STUDIES OF SILICON OXIDATION AND OXYNITRIDATION

E. GARFUNKEL[1], E. P. GUSEV[1,2], H. C. LU[2], and T. GUSTAFSSON[2]
Departments of Chemistry[1] and of Physics and Astronomy[2], and Laboratory for Surface Modification[1,2],
Rutgers University, Piscataway, NJ 08855 - 0849, USA

M.L. GREEN
Bell Laboratories, Lucent Technologies, Murray Hill, NJ 07974, USA

The paper reviews some of our recent high resolution medium energy ion scattering (MEIS) experiments on mechanistic and structural aspects of ultrathin (<5 nm) dielectric films (oxides, SiO_2, and oxynitrides, SiO_xN_y) thermally grown on silicon surfaces. The growth mechanism of ultrathin films O using isotopic ($^{16}O_2/^{18}O_2$) labeling methods, the transition region near the oxide/substrate interface, and silicon oxynitridation in N_2O and NO are discussed.

1. Introduction

The purpose of this paper is to review some of our recent work on the structure and growth mechanism of ultrathin silicon oxide and oxynitride films on Si(100).[1-11] In the sub-5 nm regime, of interest in commercial devices in the very near future, a truly atomistic understanding of these phenomena is necessary, and novel probes to study them will have to be utilized. Our work has largely been performed using Medium Energy Ion Scattering (MEIS), a very powerful but not widely available tool for the study of surface and thin film structure and composition.[12-16] Using sequential adsorption of isotopes, we show that the growth of ultrathin oxide films is very different from that for thicker films and involves reactions in two distinct spatial regions. We present results that pertain to the mechanism for the reactions that occur both at the surface and near the interface. We also discuss the influence of nitrogen on film growth and properties, and present results for the location and amount of nitrogen obtained under some different processing conditions.

2. Silicon Oxidation and Oxynitridation: Background

The causal relationship between defects in the SiO_2/Si interface region and device reliability is by now well established.[17, 18] On the other hand, the exact atomic scale structure of all the electrical defects has yet to be confirmed. There are therefore many questions concerning the role and atomic configuration of defects, as well as the formation mechanisms of thin oxides, that need to be addressed. While the formation of thick oxide films (>20 nm) is well described by the Deal-Grove model[19] (molecular diffusion of oxygen to the SiO$_2$/Si interface and reaction with silicon at the interface), the oxidation kinetics for ultrathin films are different.[20-23] Various phenomenological models have been proposed to account for the deviation (in particular, the fast initial growth kinetics), although they often require a large number of fitting parameters with little direct experimental support.[18, 22, 24, 25]

The structure of the region between the crystalline silicon and the amorphous SiO$_2$ is also still under debate.[18, 26-30] As the thickness of gate dielectrics decreases, this *transition region* becomes a significant part of the dielectric. Reports on the thickness of this region vary from 0-7 nm, the variations resulting from either different oxidation procedures or different measurement techniques. Despite extensive work, neither the atomic-scale structure nor the composition (or gradient) in the transition region are well understood. There is currently no universally accepted model.

Oxynitrides are candidates to replace conventional silicon gate oxides for sub-0.25 µm devices. Nitrogen incorporated into the oxide has been shown to improve device reliability, in part by forming a barrier against boron diffusion into and through gate oxides.[31-38] The oxidation of silicon in N$_2$O is particularly attractive due to ease of processing. On the other hand, understanding oxynitridation in N$_2$O is complicated by the fast gas-phase decomposition of the molecule into N$_2$, O$_2$, NO, and O at typical oxidation temperatures (800-1100 °C), with the rate and branching ratio of the products strongly dependent on processing conditions such as temperature, residence time, gas flow, and oxidation reactor type.[32, 39-41] NO is believed to be responsible for nitrogen incorporation into the film. [10, 31, 42, 43] There is still no consensus about what is or should be the optimal nitrogen concentration and distribution in the dielectric. Accurate nitrogen depth profiles have been difficult to measure because of limitations of SIMS[44] and HF etch-back profiling methods. The growth mechanism of thin oxynitrides, in particular nitrogen incorporation into and concurrent removal from the film, is even less well understood.[38, 42, 45, 46]

3. Experimental

Our experiments were done using the technique of Medium Energy Ion Scattering (MEIS)[13], a high resolution version of conventional Rutherford Backscattering Spectroscopy (RBS).[47] The incident ions (in our case, protons) can suffer two kinds of energy losses in the target. A *large angle deflection* results from the (very rare) collisions with the ion cores of the target atoms. Such collisions are accompanied by a kinetic energy loss of the protons, the amount of which depends in a simple fashion on the mass of the target atom in question. This means that the technique discriminates between different elements in the sample (in our case silicon, different isotopes of oxygen, and nitrogen). Also, the incident ions may suffer energy losses through electronic excitations (electron-hole pair creation and plasmon excitations), resulting in *small angle deflections*. These collisions are very frequent, and can be taken into account in an average way through the concept of the *energy loss* (dE/dx). Thus, the energy distribution of the backscattered ions is a direct measure of the path lengths in the target and the mass of the scattering species. Compared to RBS, usually done at

MeV energies, MEIS uses a lower ion beam energy (97.2 keV for most of the data presented below), close to the maximum of dE/dx for protons in Si, which ensures optimal depth resolution. The use of a high resolution toroidal electrostatic energy analyzer[48, 49] ($\Delta E/E \sim 0.1\%$) results in almost monolayer resolution in the near surface region.[14, 49] Accurate depth distributions of target elements of interest (Si, ^{18}O, ^{16}O, and N) can be deduced from simulations of ion energy distributions.[3] The use of sequential exposure to different oxygen isotopes allow us to investigate the temporal dependence of the oxidation. More details about the MEIS set-up, data acquisition and analysis can be found elsewhere.[1-11, 16]

The samples discussed in this paper were grown in state-of-the art RTO systems or vertical furnaces. For the isotopic oxidation experiments (performed in a quartz furnace), both $^{18}O_2/^{16}O_2$ and $^{16}O_2/^{18}O_2$ sequential exposures were used. Silicon oxynitridation was performed in N_2O or NO ambients at 700 - 1000 °C. The final thickness of most films was below 5 nm.

4. Isotopic Labeling Studies

One advantage of isotopic labeling is that it allows us to determine the time evolution of the oxidation process; most kinetic studies only measure the oxide film thickness as a function of time, which is a much more integrated quantity. MEIS spectra for a thin $Si^{16}O_2$ film before and after reoxidation in $^{18}O_2$ are shown in Fig. 1. The ^{16}O spectra show a sharp leading edge, corresponding to the oxide-vacuum interface, while the trailing edge is broadened. The latter is in part due to broadening of the nearly monoenergetic incident beam from the statistical nature of the ion-solid interaction (straggling), but this is only a part of this broadening. A more detailed analysis shows contributions to the broadening from a gradient of the oxygen concentration near the film-substrate interface, as well as some roughness of the interface. [11] The spectrum changes drastically after ^{18}O exposure. For films greater than ~ 30 Å, there are now two additional peaks in the spectra in an energy region corresponding to ^{18}O. These correspond to ^{18}O atoms near the oxide surface (at higher

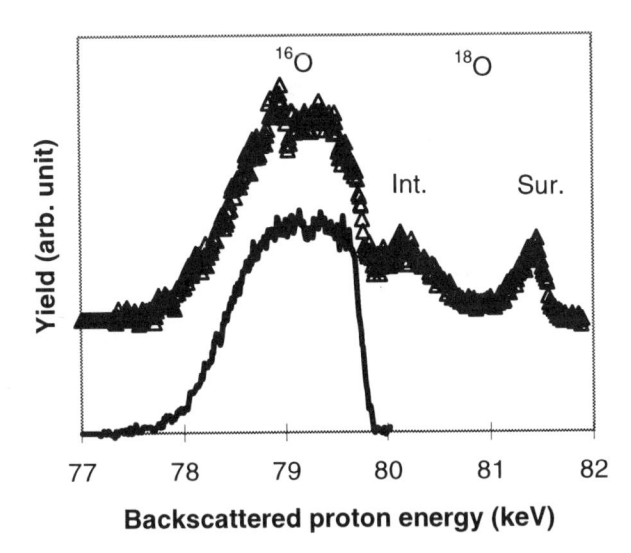

Fig. 1 MEIS spectra before (line) and after (Δ) reoxidation of a ~45 Å oxide film in 7 Torr $^{18}O_2$ at 920°C for 1h.

backscattering energies) and to ^{18}O atoms near the Si/SiO$_2$ interface. The ^{16}O peak changes also during reoxidation; the high energy part of the peak is depleted concurrent with the growth of the surface ^{18}O peak. This observation shows that the surface ^{18}O peak is due to (a) surface exchange reaction(s), i.e., as ^{18}O incorporates near the oxide surface, some ^{16}O leaves. Also, the low energy side of the ^{16}O peak broadens and shifts, implying that the thickness of the ^{16}O-containing oxide layer increases during the reoxidation. The ^{18}O peak at lower energies is broader than what can be expected based on straggling and roughness effects, indicating that ^{18}O incorporation takes place at least over a ~10 Å region.

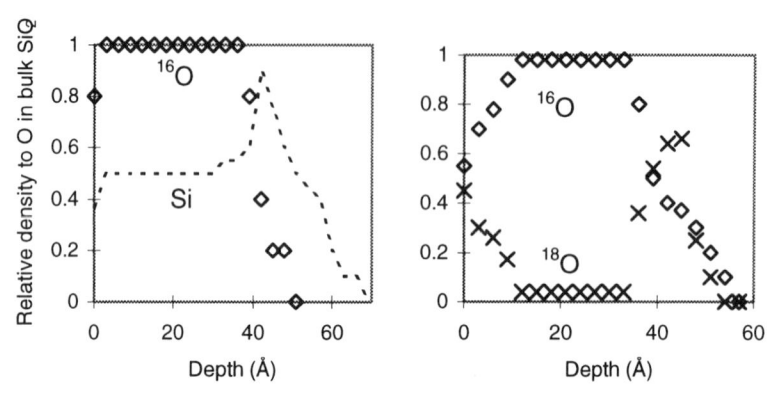

Fig. 2 a) Depth profiles of oxygen and silicon for the ~45Å oxide film in Fig. 1; notice the silicon density decreases below the interface due to the shadowing effect in the crystalline substrate. b) After reoxidation in 7 Torr ^{18}O$_2$ at 920°C for 1h. Notice that the ^{18}O near the surface almost equals the loss in total ^{16}O.

A quantitative evaluation of these data is shown in Fig. 2. As anticipated, the oxygen concentration at the oxide/substrate interface shows a transition region of width ~ 10 - 15 Å, which as noted above is due to roughness and a compositional gradient (suboxide states). The RMS value of the interface roughness for high-quality thermal oxides is usually below 0.5 nm, depending on the probing technique and processing conditions. X-ray diffraction experiments performed on oxide films grown in the same reactor and under conditions similar to those of our films show about 0.2 - 0.3 nm RMS roughness.[33] Surface roughness will also contribute to the MEIS results as our spectra average over a macroscopic beam spot.

For sub-5 nm gate dielectrics, the transition region between the well-ordered Si substrate and the amorphous oxide becomes a significant part of the film. As key electrical defects are located near the interface, this region is of great importance for device performance. In addition, our studies suggest that the transition region may also be important in understanding the growth reaction.

While the ^{16}O concentration at the surface decreases after reoxidation in ^{18}O (Fig. 2), and the ^{18}O concentration increases, their sum remains approximately constant. This is (part of) the basis for claiming that the surface reaction is not a growth reaction, but an exchange reaction. Near the oxide-substrate interface, the ^{18}O and ^{16}O profiles overlap. This is at variance with the traditional Deal-Grove model[19] in which the new Si^{18}O$_2$ oxide should grow right at the interface, below the initial Si^{16}O$_2$ layer. Our results show that the reaction occurs throughout the near-interfacial transition region, resulting in isotopic mixing near the interface. Both the surface reaction and the (near) interface reaction depend on the oxidation conditions.

Our results show that any model of silicon oxidation as a reaction occurring at a well-defined geometrical plane does not apply to ultrathin films. The reaction actually takes place throughout the near-interfacial (transition region). The presence of the surface exchange reaction should be considered in any complete model of oxidation. We believe that the near-interfacial reaction is a result of continued oxidation of incompletely oxidized silicon, i.e. suboxides, silicon interstitials, silicon monoxide and/or silicon clusters. In this model, the reaction of oxygen with the silicon substrate is not perfect; it occurs with some probability for the generation of the incompletely oxidized silicon atoms which are then consumed by the near-interfacial reaction.

5. The Surface Reaction

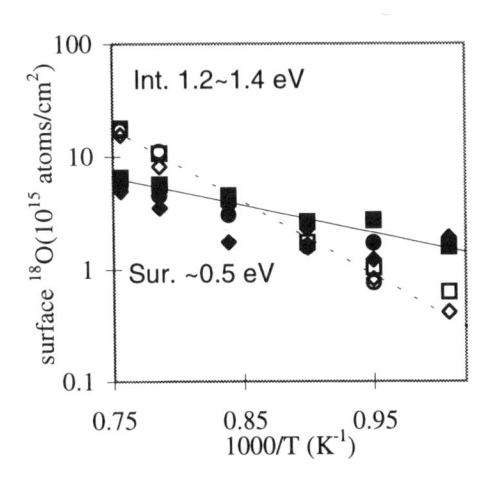

Fig. 3 The Arrhenius plot of the surface (solid symbols) and interfacial (open symbols) ^{18}O after samples with initial thickness of 30Å (square), 45Å (diamond), and 60Å (circle) are further oxidized in 7 Torr of ^{18}O$_2$ for 1h.

The surface exchange reaction, observed earlier by others as well[50-52], is not well understood. Although it appears not to be a growth reaction for films > ~3 nm, it may become very important in sub-3 nm oxide films; in this regime the surface and the near-interface reactions overlap in space and, as a result, should affect each other. Our results show that (for films > 3 nm) the surface reaction occurs well separated from the growth reaction near the interface. In Fig. 3 we show results of concentration measurements of ^{18}O as a function of temperature, plotted in Arrhenius coordinates. The (apparent) activation energy for the surface reaction is only some 0.5 eV, drastically different from what we measure for the interface reaction (1.2 - 1.4 eV) and even more different from the classical Deal-Grove value for thicker films (2.0 eV). The time dependence of the surface reaction is given in Fig. 4. Under our conditions, the ^{18}O concentration at the surface saturates in less than 1 hour. The interface reaction is slower, but shows continued (indefinite) growth. Finally, we show in Fig. 5 the pressure dependence of the surface reaction. It is quite weak, proportional to $p^{0.3}$. For thin films, we find an interface pressure dependence of ~ $p^{0.67}$, while a classic Deal-Grove mechanism yields a linear pressure dependence, as has been measured by some for much thicker films.

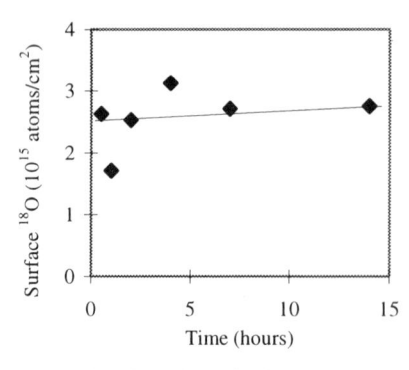

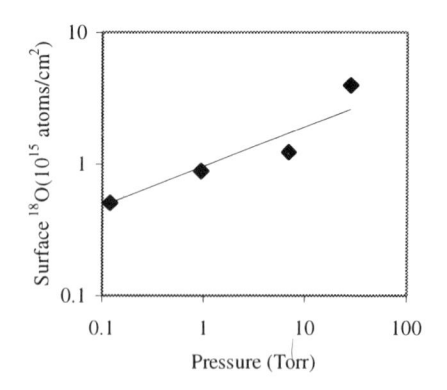

Fig. 4 Time dependence for the surface reaction: surface ^{18}O after 45Å samples are further oxidized in 7 Torr ^{18}O$_2$ at 840°C

Fig. 5 Pressure dependence: 30Å initial oxide, same oxidation conditions. The slope correspond to ~p$^{0.3}$

Taken together, the results in Fig. 3 - 5 show that the mechanism for the surface reaction is quite different from that of the interface reaction. In further studies, not presented here, the reaction kinetics were observed to be strongly influenced by the processing conditions and the presence of impurities. As water is commonly present in small amounts under processing conditions, we have studied the effects of deliberately increasing the partial pressure of water vapor on the surface reaction. We find that at relatively high partial pressures, water does advance the surface reaction. However, the activation energy (~0.8 eV) is different from the dry O$_2$ case, and it also displays a different pressure dependence (p^1). Exposure to elevated pressures of hydrogen also fails to reproduce our dry O$_2$ results. On the other hand, a deliberate exposure to CO gas gives more reasonable results (an apparent activation energy of 0.4 eV and a p$^{0.2}$ pressure dependence) and enhances the surface reaction considerably under processing conditions similar to those used by us. Still unexplored are the roles played by hydrocarbons and atomic oxygen. It is also still very much an open question to what extent the surface reaction can be explained by a single impurity and what the influence of defects in the substrate and at the surface is.

6. Ultrathin Oxynitride Films

Oxynitrides have improved hot-electron damage and boron penetration resistance than pure SiO$_2$.[31-37, 42, 46, 53] However, many fundamental questions remain, both concerning the mechanism of oxynitridation and the reason(s) for the improved properties. The distribution of the nitrogen atoms in the film is obviously an important factor but is difficult to determine accurately by conventional techniques. We have recently shown that MEIS can be used successfully to analyze such films.[5, 8-11] In particular, we have studied the growth mechanism and composition of ultrathin oxynitrides thermally grown on Si(100) in N$_2$O and NO, as well as sequentially grown in NO/O$_2$/NO and NO/O$_2$/N$_2$O.

A thin SiO$_2$ film, subsequently annealed in NO, shows a nitrogen peak located in the first ~ 1.5 nm on the oxide side of the SiO$_x$N$_y$/Si interface (Fig. 6). Interestingly, no significant amount of nitrogen is observed on the substrate side of the interface. If oxynitridation instead takes place in N$_2$O, the amount of N incorporated into the film is much lower (Fig. 6).

Using Si^{18}O$_2$ as the starting film, we find N^{16}O annealing results in ^{16}O incorporation at the interface, overlapping the region where N incorporation occurs (Fig. 7). This result lends strong support to the idea that NO enters SiO$_2$ and diffuses to the interface, analogous to the O$_2$ behavior in the classical Deal-Grove picture.

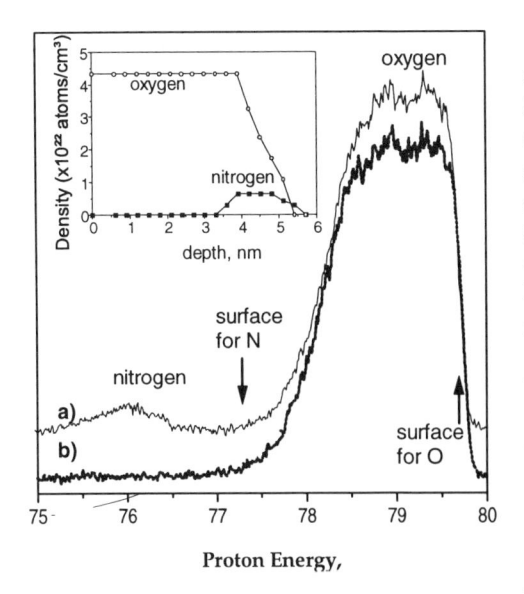

Fig. 6 MEIS spectrum for an O$_2$ (850°C, 5.5 min)/NO (950°C, 60 min.) furnace grown oxynitride (a), and a film grown in a furnace in N$_2$O at 850°C for 110 min.(b). The scattering angle is 125°. The oxygen and nitrogen profiles shown in the inset correspond to spectrum (a).

It has recently been shown that nitrogen removal from a nitrided film occurs during silicon oxynitridation in N$_2$O.[42, 53] The high temperature gas phase chemistry of N$_2$O is complex,[40] and both NO and (atomic) O are produced. At least two different models for the mechanism of N removal have been put forward: Carr et. Al [53] proposed that atomic oxygen causes nitrogen removal, whereas Saks et. al [42] argued that NO is responsible. To further elucidate the role of NO in nitrogen incorporation and removal, we have performed sequential exposures of Si (at 850 °C, furnace) to NO, O$_2$, and then NO (NO/O$_2$/NO) and analyzed the resultant depth profiles. This experiment clearly shows that NO does not efficiently remove N from the oxynitride. Atomic oxygen remains a likely candidate responsible for N removal,[53] although further studies are required to clarify its role. Exposure to ozone (an effective atomic oxygen source) results in nitrogen removal, supporting the atomic oxygen model.[53]

In a series of experiments including various sequences of NO, O$_2$, and N$_2$O reactions, we have also demonstrated that N-nanolayer engineering is possible using purely thermal methods (i.e., not deposited oxides).

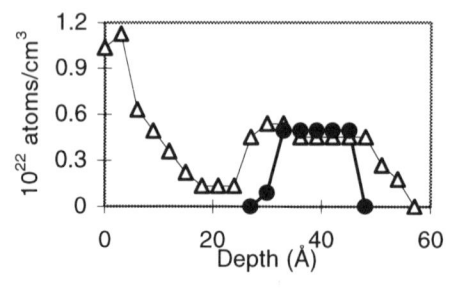

Fig. 7 ^{16}O (Δ) and N (●) profiles after a Si^{18}O$_2$ film is annealed in NO. Notice that most of ^{16}O near the surface is incorporated during the process of making the Si^{18}O$_2$ film.

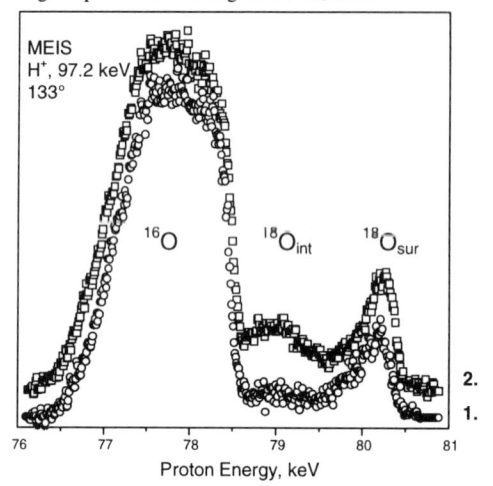

Fig. 8 MEIS spectra in the oxygen region of two ultrathin oxynitrides with different amount of nitrogen, after reoxidation in ^{18}O$_2$ (920°C, 7 Torr, 1 hour). Initial oxynitrided film grown in O$_2$/NO (spectrum 1) has a higher, by almost a factor of three, concentration of nitrogen near the interface than an N$_2$O grown oxynitride (spectrum 2). The scattering angle is 133° and the incident ion energy 97.2 keV.

The presence of nitrogen in oxynitrided films is known to retard the oxidation kinetics.[54] To understand the mechanism of this retardation, oxynitride films with different amounts of nitrogen near the SiO$_x$N$_y$/Si interface and pure ("control") SiO$_2$/Si films were reoxidized in dry ^{18}O$_2$ under equivalent conditions and the spatial distribution of the ^{18}O incorporated into the films were analyzed (Fig. 8). Analogous to the pure SiO$_2$ case, we observed two distinct regions where oxygen incorporation into the oxynitride films occurs during high-temperature oxidation: at/near the interface and near the outer oxide surface. The (near)interface growth reaction is found to be significantly retarded in the presence of the interfacial nitrogen (with a higher degree of the retardation for higher concentrations of nitrogen). The presence of nitrogen near the interface does not effect the surface exchange reaction (Fig. 8).

7. Summary

As gate dielectrics in ULSI devices shrink below 5 nm, an atomic scale understanding of the physical and chemical processes occurring in such ultrathin films becomes more critical. We have used high resolution medium energy ion scattering, a technique which provides sub-nm depth resolution, to study ultrathin silicon oxides and oxynitrides. Several new and interesting features of the growth and composition of these dielectrics have been observed. By using isotopic (^{18}O$_2$/^{16}O$_2$) labeling methods we directly observe oxidation reactions in two spatially distinct regions: an oxide growth reaction throughout the near-SiO$_2$/Si interface region, and an exchange reaction of the oxygen isotopes near the outer oxide surface. The former behavior is inconsistent with the traditional (Deal-Grove) model of silicon oxidation. Our results also support the existence of a thin (~ 1 nm) transition region between the silicon substrate and the amorphous stoichiometric oxide overlayer, although the detailed microstructure of this region is still unclear. Further, we have demonstrated the use of MEIS for silicon oxynitridation studies by accurately determining the nitrogen depth distribution. Finally, our ability to accurately image the depth at which N and O are incorporated have led us to propose that NO behaves very similarly to O_2 in oxynitridation, and to begin to nano-engineer N and O containing layered structures by thermal methods.

Acknowledgments

This work was supported by NSF (DMR-9701748, ECS-9530984) and SRC (97-BJ-451).

References

1. Gusev, E.P., Lu, H.C., Gustafsson, T., and Garfunkel, E. (1994) On the mechanism of ultrathin silicon oxide film growth during thermal oxidation, in S. P. Murarka, K. Rose, T. Ohmi and T. Seidel (eds.), *Interface Control of Electrical, Chemical, and Mechanical Properties*, MRS, vol. 318, p. 69.
2. Lu, H.C., Gustafsson, T., Gusev, E.P., and Garfunkel, E. (1995) An isotopic labeling study of the growth of thin oxide films on Si(100), *Appl. Phys. Lett.* **67**, 1742.
3. Gusev, E.P., Lu, H.C., Gustafsson, T., and Garfunkel, E. (1995) The growth mechanism of thin silicon oxide films on Si(100) studied by medium energy ion scattering, *Phys. Rev. B* **52**, 1759.
4. Gusev, E.P., Lu, H.C., Gustafsson, T., and Garfunkel, E. (1996) Initial oxidation of silicon: New ion scattering results in the ultrathin regime, *Appl. Surf. Sci.* **104/105**, 329.
5. Lu, H.C., Gusev, E.P., Gustafsson, T., Garfunkel, E., Green, M.L., Brasen, D., and Feldman, L.C. (1996) High resolution ion scattering study of silicon oxynitridation, *Appl. Phys. Lett.* **69**, 2713.
6. Lu, H.C., Gusev, E.P., Garfunkel, E., and Gustafsson, T. (1996) An ion scattering study of the interaction of oxygen with Si(111): Surface roughening and oxide growth, *Surf. Sci.* **341**, 111.
7. Gusev, E.P., Lu, H.C., Gustafsson, T., and Garfunkel, E. (1996) New features of silicon oxidation in the ultrathin regime: an ion scattering study, in H. Z. Massoud, E. H. Poindexter and C. R. Helms (eds.), *The Physics and Chemistry of SiO2 and the Si-SiO2 Interface - 3*, The Electrochemical Soc., Pennington, p. 49.
8. Lu, H.C., Gusev, E.P., Gustafsson, T., and Garfunkel, E. (1997) The effect of near-interfacial nitrogen on the oxidation behavior of ultrathin silicon oxynitrides, *J. Appl. Phys.* **81**, 6992.
9. Gusev, E.P., Lu, H.C., Gustafsson, T., Garfunkel, E., Green, M.L., and Brasen, D. (1997) The composition of ultrathin oxynitrides thermally grown in NO, *J. Appl. Phys.* **82**, 896.
10. Lu, H.C., Gusev, E.P., Gustafsson, T., Green, M.L., Brasen, D., and Garfunkel, E. (1997) Compositional and mechanistic aspects of ultratrhin oxynitride film growth on Si(100), *Microelectronic Eng.* **36**, 29.
11. Gusev, E.P., Lu, H.C., Gustafsson, T., and Garfunkel, E. (1997) Silicon oxidation and oxynitridation in the ultrathin regime: Ion scattering studies, *Brazil. J. Phys.* **27**, 302.
12. Tromp, R.M. and van Loenen, E.J. (1985) Ion beam crystallography of Si(111) (7x7), *Surf. Sci.* **155**, 441.
13. van der Veen, J.F. (1985) Ion beam crystallography of surfaces and interfaces, *Surf. Sci. Rept.* **5**, 199.
14. Vrijmoeth, J., Zagwijn, P.M., Frenken, J.W.M., and van der Veen, J.F. (1991) Monolayer resolution in medium-energy ion-scattering experiments on the NiSi$_2$(111) surface, *Phys. Rev. Lett.* **67**, 1134.
15. Copel, M., Tromp, R.M., Timme, H.J., Penner, K., and Nakao, T. (1996) Effects of surface oxide on room temperature nitridation of Si(100), *J. Vac. Sci. Technol.* **A 14**, 462.
16. Gustafsson, T., Garfunkel, E., Gusev, E.P., Haberle, P., Lu, H.C., and Zhou, J.B. (1996) Structural studies of oxide surfaces, *Surface Review and Letters* **3**, 1561.
17. Balk, P. (1988) *The Si-SiO$_2$ System*, Elsevier, Amsterdam.
18. Helms, C.R. and Poindexter, E.H. (1994) The Si-SiO2 system: its microstructure and imperfections, *Rep. Prog. Phys.* **57**, 791.
19. Deal, B.E. and Grove, A.S. (1965) General relationship for the thermal oxidation of Si, *J. Appl. Phys.* **36**, 3770.
20. Massoud, H.Z., Plummer, J.D., and Irene, E.A. (1985) Thermal oxidation of silicon in dry oxygen: Growth-rate enhancement in the thin regime, *J. Electrochem. Soc.* **132**, 2693.
21. Irene, E. (1988) Models for the oxidation of silicon, *Crit. Rev. Sol. St. Mat. Sci.* **14**, 175.
22. Deal, B.E. (1988) Historic perspectives of silicon oxidation, in C. R. Helms and B. E. Deal (eds.), *The Physics and Chemistry of SiO$_2$ and the Si-SiO$_2$ Interface*, Plenum Press, New York, p. 5.
23. Dimitrijev, S. and Harrison, H.B.(1996) Modeling of growth of thin silicon oxides, *J. Appl. Phys.* **80**, 2467.

24. Delarious, J.M., Helms, C.R., Kao, D.B., and Deal, B.E. (1989) Parallel oxidation model for Si including both molecular and atomic oxygen mechanisms, *Appl. Surf. Sci.* **39**, 89.
25. Sofield, C.J. and Stoneham, A.M. (1995) Oxidation of Si: the VLSI gate dielectric? *Semic. Sci. Techn.* **10**, 215.
26. Hattori, T. (1995) Chemical structure of the SiO_2/Si interface, *CRC Crit. Rev. Sol. State Mater. Sci.* **20**, 339.
27. Himpsel, F.G. (1994) Electronic structure of semiconductor interfaces, *Surf. Sci.* **299/300**, 525.
28. Batson, P.E. (1993) Simultaneous STEM imaging and electron energy loss spectroscopy with atomic-column sensitivity, *Nature* **366**, 727.
29. McFeely, F.R., Zhang, K.Z., Banaszak-Hall, M.M., Lee, S., and Bender-IV, J.E. (1996) An inquiry concerning the principles of Si2p core-level photoemission shift assignments at the Si/SiO_2 interface, *J. Vac. Sci. Technol. B* **14**, 2824.
30. Haight, R. and Feldman, L.C. (1982) Atomic structure of the $Si-SiO_2$ interface, *J. Appl. Phys.* **53**, 4884.
31. Carr, E.C. and Buhrman, R.A. (1993) Role of interfacail nitrogen in improving thin silicon oxides grown in N_2O, *Appl. Phys. Lett.* **63**, 54.
32. Tobin, P.J., Okada, Y., Ajuria, S.A., Lakhotia, V., Feil, W.A., and Hedge, R.I. (1994) Furnace formation of silicon oxynitride thin dielectrics in N_2O, *J. Appl. Phys.* **75**, 1811.
33. Green, M.L., et al. (1994) RTO of Si in N_2O between 800 and 1200°C: Incorporated nitrogen and roughness, *Appl. Phys. Lett.* **65**, 848.
34. Hattangady, S.V., Niimi, H., and Lucovsky, G. (1995) Controlled nitrogen incorporation at the gate oxide surface, *Appl. Phys. Lett.* **66**, 3495.
35. Bhat, M., Han, L.K., Wristers, D., Yan, J., Kwong, D.L., and Fulford, J. (1995) Effect of chemical composition on the electrical properties of NO-nitrided SiO_2, *Appl. Phys. Lett.* **66**, 1225.
36. Fleetwood, D.M. and Saks, N.M. (1996) Oxide, interface, and border traps in thermal, N_2O, and N_2O nitrided oxides, *J. Appl. Phys.* **79**, 1583.
37. Hill, W.L., Vogel, E.M., Misra, V., McLarty, P.K., and Wortman, J.J. (1996) Low Pressure Rapid Thermal CVD of Oxynitride Gate Dielectrics for N-channel and P-channel MOSFET's, *IEEE Trans. Electron Devices* **43**, 15.
38. Habraken, F.H.P.M. and Kuiper, A.E.T. (1994) Silicon nitride and oxynitride films, *Materials Sci. and Eng. Rept.* **R12**, 123.
39. Hartig, M. and Tobin, P.J. (1996) A model for N_2O gas phase chemistry, *J. Electrochem. Soc.* **143**, 1753.
40. Ellis, K.A. and Buhrman, R.A. (1996) Furnace gas-phase chemistry of silicon oxynitridation in N_2O, *Appl. Phys. Lett.* **68**, 1696.
41. Gupta, A., Toby, S., Gusev, E.P., Lu, H.C., Li, Y., Green, M.L., Gustafsson, T., and Garfunkel, E. (1997) Nitrous oxide gas phase chemistry during silicon oxynitride growth, *to be published*.
42. Saks, N.S., Ma, D.I., and Fowler, W.B. (1995) N depletion during oxidation in N_2O, *Appl. Phys. Lett.* **67**, 374.
43. Tang, H.T., Lennard, W.N., Zinke-Allmang, M., Mitchell, I.V., Feldmán, L.C., Green, M.L., and Brasen, D. (1994) Nitrogen content of oxynitride films on Si(100), *Appl. Phys. Lett.* **64**, 64.
44. Frost, M.R. and Magee, C.W. (1996) Characterization of nitrided SiO_2 thin films using SIMS, *Appl. Surf. Sci.* **104/105**, 379.
45. Baumvol, I.J.R., Stedile, F.C., Ganem, J.J., Trimaille, I., and Rigo, S. (1996) Nitrogen transport during rapid thermal growth of silicon oxynitride in N_2O, *Appl. Phys. Lett.* **69**, 2385.
46. Ganem, J.J., Rigo, S., Trimaille, I., Baumvol, I.J.R., and Stedile, F.C. (1996) Dry oxidation mechanisms of thin dielectric films formed under N_2O using isotopic tracing methods, *Appl. Phys. Lett.* **68**, 2366.
47. Feldman, L.C., Mayer, J.W., and Picraux, S.T. (1982) *Materials Analysis by Ion Channeling*, Academic Press, New York.
48. Smeenk, R.G., Tromp, R.M., Kersten, H.H., Boerboom, A.J.H., and Saris, F.W. (1982) *Nucl. Instr. and Methods* **195**, 581.
49. Tromp, R.M., Copel, M., Reuter, M.C., van Hoegen, M.H., Speidell, J., and Koudijs, R. (1991) A new 2-D particle detector for a toroidal electrostatic analyser, *Rev. Sci. Instrum.* **62**, 2679.
50. Han, C.J. and Helms, C.R. (1988) O^{18} tracer study of Si oxidation in dry O_2 using SIMS, *J. Electrochem. Soc.* **135**, 1824.
51. Rochet, F., Rigo, S., Froment, M., d'Anterroches, C., Maillot, C., Roulet, H., and Dufour, G. (1986) The thermal oxidation of silicon: The special case of the growth of very thin films, *Adv. Phys.* **35**, 339.
52. Trimaille, I. and Rigo, S. (1989) Use of ^{18}O isotopic labeling to study thermal dry oxidation of silicon as a function of temperature and pressure, *Appl. Surf. Sci.* **39**, 65.
53. Carr, E.C., Ellis, K.A., and Buhrman, R.A. (1995) Nitrogen profiles in thin SiO_2 in N_2O: the role of atomic oxygen, *Appl. Phys. Lett.* **66**, 1492.
54. Raider, S.I., Gdula, R.A., and Petrak, J.R. (1975) Nitrogen reaction at the $Si-SiO_2$ interface, *Appl. Phys. Lett.* **27**, 150.

SYNCHROTRON AND CONVENTIONAL PHOTOEMISSION STUDIES OF OXIDES AND N$_2$O OXYNITRIDES

Z.H. LU
Institute for Microstructural Sciences
National Research Council of Canada
Ottawa, Ontario K1A 0R6

1. Introduction

One of the most critical materials in developing deep submicron integrated circuits is high-quality ultrathin (~ a few nm) gate dielectric film. In such a thin film, the microscopic structure is expected to play a critical role in many aspects of the film's properties, device performance and reliability. The importance of the dielectric-silicon interface has been recognized for many years as one of the primary issues in the semiconductor industry. In fact, it is such a unique interface that it makes the silicon metal-oxide-semiconductor (MOS) transistors possible and the Si industry such a phenomenal success. Thus there has been a tremendous amount of research conducted into its electrical properties over the last few decades [1]. The microscopic structure of the interface, however, has not been studied in great detail partly because advanced probing techniques such as synchrotron radiation photoemission spectroscopy (PES) were not available until recent years, and partly because the importance of microstructural information became apparent only in recent years when the Si industry began pushing for a deep submicron ULSI device with the gate dielectric approaching several atomic layers. Between a crystalline Si substrate and an amorphous oxide overlayer, the interface transition region consists of a mixture of Si in various oxidation states denoted as Si^{+1}, Si^{+2}, and Si^{+3} where the Si atom has one, two, and three first-nearest-neighbor oxygen atoms, respectively. Si^{+x} (x=1,2,3) is also frequently referred to as suboxide. PES has so far been the most successful technique in identifying these suboxides [2-4]. Conventional photoemission using X-ray sources such as the Al K_α line, which is known as X-ray photoelectron spectroscopy (XPS), has been used but has less sensitivity in detecting the suboxides [4,5]. Other advantages such as a large probing depth and commercial availability, however, have made XPS a valuable tool for routine characterization of film thickness and composition. For most materials such as oxynitride, high-resolution XPS can resolve most chemical species in the bulk film. Thus XPS and PES make a useful combination to characterize both the film and the interface structures. Here recent progress and future opportunities for the application of XPS and PES in studying industrial gate oxides and N$_2$O oxynitride will be discussed.

E. Garfunkel et al. (eds.), Fundamental Aspects of Ultrathin Dielectrics on Si-based Devices, 49–63.
© 1998 *Kluwer Academic Publishers. Printed in the Netherlands.*

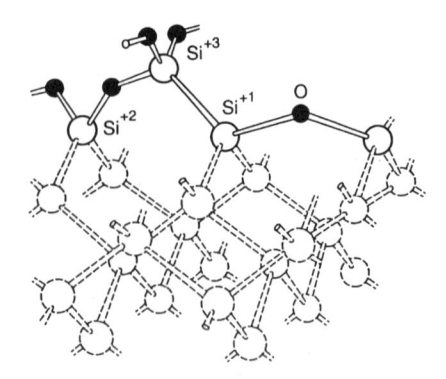

Figure 1. Schematic view of the interface structure at a SiO$_2$/Si(100) interface.

2. Photoemission Spectroscopy

2.1. INTERFACE CHEMICAL STATE ANALYSIS

In a photoemission process [6], the core shell electrons are excited by a photon into an empty band above the Fermi energy. Some of the excited photoelectrons would escape the surface into vacuum. An electron analyzer would then be used to analyze the photoelectron kinetic energy vs. its intensity. As the photon energy and the material work function are constant, the core-level binding energy can then be determined. Each element can therefore be identified by its distinct core-level energy. Due to strong electron-electron interactions within each element, any changes in the valence electron charge distribution cause the core shell electron to shift in binding energy. The shift in the binding energy is often referred to as a chemical shift in a case where the shift is caused by changes in the chemical bonding. The valence charge density of one element is primarily associated with the chemical nature and configuration of its first-nearest-neighbor atoms. The second-nearest-neighbor atoms may in principle affect the valence charge density but the effects would be secondary.

In the case of SiO$_2$/Si(100), the interface region consists of various Si suboxides as schematically shown in Figure 1. As compared with bulk Si, the first-nearest-neighbors around a Si^{+x} atom are (4-x) Si atoms and x oxygen atoms. The core levels of these suboxides are expected to have different binding energies.

Figure 2 shows a Si 2p$_{3/2}$ PES spectrum recorded from a SiO$_2$/Si(100) system with an oxide thickness of about 0.5 nm. A bulk Si peak, various suboxides, and dioxide are clearly resolved in this spectrum. The chemical shifts and the full-width at half-maximum (FWHM) of each of these peaks are listed in Table I. The interpretation of these suboxide peaks is rather straightforward according to well-established theory. As an O is more electronegative than a Si, the O would gain electrons from the Si in the formation of a Si-O bond. With a reduction in the valence charge, the repulsive electron Coulomb interaction within the Si atom would be reduced so that the Si 2p core-level binding energy would be shifted to a higher binding energy, as is observed. The magnitude of the shift is proportional to the number of Si-O bonds. Each bond is found to contribute to ~ 0.9 eV shift in the core level. Doubt about the uniqueness of assigning each of these shifted peaks to the Si^{+x} was raised by Holl and McFeely [7] who claimed, based on studies of a H$_8$Si$_8$O$_{12}$ cluster on a Si(100) surface, that the second-nearest-neighbor effect might be essentially coincident with those peaks arising from first-nearest-neighbor effects. Using first-principle calculations, Pasquarello, Hybertsen and Car [8,9] found that the contribution of the second-nearest-neighbors to the core-level shift, as compared with first-nearest-neighbors, is negligible. The calculated Si^{+x} core-level shifts agree well with the experimental observations.

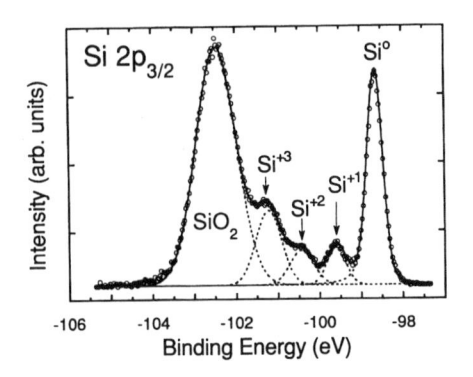

Figure 2. Si 2p$_{3/2}$ spectrum recorded at 130 eV photon energy from a SiO$_2$/Si(100) interface.

In Table I we observe that the FWHM increases progressively as the oxidation state increases. It is known that a variation in the Si-Si bond length and bond-angle, in the case of amorphous silicon, will lead to a fluctuation in the Si valence charge distribution. From a crystalline silicon lattice to an amorphous SiO$_2$ random network, the widths of Si-O bond-length and bond-angle distribution functions of the suboxides presumably increase with increasing oxidation state. This was thought [5] to be the physical basis for a gradual Si 2p peak width broadening with increasing oxidation state. This interpretation was supported by a tight-binding calculation [10]. Using the first-principle methods, Pasquarello et al. [8] have calculated the core-level shift as a function of the Si-O bond-length, and O-Si-O and Si-O-Si bond-angles. They found that the core level is rather insensitive to bond-angle distortions (in disagreement with ref. [10]), but depends weakly on Si-O bond-length. They suggested that structure disorder

TABLE 1. Chemical shifts, FWHMs, and relative cross-sections of Si^{+x} at 130 eV.

Chemical state	Chemical shift (eV)	FWHM (eV)	$\sigma_{SiOx}/\sigma_{Si}$
Si^0	0	0.41	1
Si^{+1}	0.97	0.55	1.0
Si^{+2}	1.80	0.68	1.1
Si^{+3}	2.60	0.70	1.7
Si^{+4}	3.82	1.13	2.2

may not be the only source of peak broadening; other effects such as phonon broadening may have a comparable influence on the suboxide peak width. The phonon effect has yet to be tested experimentally and theoretically.

With regard to the chemical shift of the Si^{+4} peak, it was found that it increases with increasing oxide thickness. This effect was used [11] to propose that structural changes such as strain-induced bond-angle distortions are a function of dioxide thickness. Recent data [12] from Si K-edge X-ray absorption, i.e., Si^{+4} 1s transition to the conduction band edge, showed no shift in the core level over a wide range of oxide thickness. More detailed XPS analysis [12] indicated that both the O 1s and Si^{+4} 2p core levels have the same amount of shift as a function of oxide thickness. This suggests that the shift is due to core-hole relaxation and surface "charging", typical behavior in an insulating film. Due to a large difference in dielectric constant among Si, SiO_2, and vacuum, a recent calculation [9] showed that, by adding a correction of core-hole relaxation energy to the original chemical shift, the Si^{+4} 2p core-level binding energy is found to increase with increasing oxide thickness.

2.2. QUANTITATIVE THICKNESS MEASUREMENT

When a photoelectron transports through a material, it will be scattered by other particles such as atoms and plasmons. There are two types of scattering: elastic scattering and inelastic scattering. For elastic scattering, the primary photoelectron wave and the scattered waves, in the case of a well-defined crystal, form a diffraction pattern. The diffraction pattern can be detected with a high-angular-resolution electron analyzer. This has been evolved into a new analytical technique known as X-ray photoelectron diffraction (XPD) [13]. The interaction between the primary wave and the scattered waves can also be modulated by photon energy, i.e., photoelectron kinetic energy. The latter has been used recently to deduce suboxide inter-layer spacing [14]. For most applications, the electron analyzer is set to sample photoelectrons from a large solid angle so that the elastic scattering effect is averaged out. Here we will consider only the inelastic scattering effect. As a specific core-level photoelectron losses its characteristic energy after inelastic scattering, its intensity is attenuated before escaping the sample surface into the vacuum. The intensity decay over a distance z is known [15] to take an exponential form, $e^{-z/\lambda}$, where λ is the attenuation length or the inelastic mean free path. Based on this exponential form, the equation for determining the SiO_2 thickness, d_{oxy}, can then be deduced to be the following,

$$d_{oxy} = \lambda_{oxy} \sin\alpha \ln[I^{+4}/(\beta I^0) + 1],$$

where λ_{oxy} is the inelastic scattering mean free path in the oxide, α is a photoelectron take-off angle, and I^{+4} and I^0 are the photoelectron area peak intensities of the oxide and bulk silicon, respectively. The constant β, the photoelectron intensity ratio of pure SiO_2 and pure Si is 0.75 for Al K_α XPS. λ_{oxy} is normally determined by measuring oxide films with thickness measured by other techniques. In recent SiO_2 thickness metrology studies involving XPS, transmission electron microscopy, spectroscopic ellipsometry, and capacitance-voltage analysis, we established that $\lambda_{oxy} = 2.96$ nm for Al K$_\alpha$ XPS [16].

The number of suboxides, N$_{SiOx}$, can also be determined by the following equation [3],

$$N_{SiOx} = n_{Si} \lambda_{Si} (I^{+x}/I^0) (\sigma_{Si}/\sigma_{SiOx}) \sin\alpha,$$

where $n_{Si} = 5.0 \times 10^{22}$ atoms cm^{-3} is the atomic density of bulk silicon. λ_{Si}, the inelastic scattering mean-free path in bulk silicon, is 2.11 nm for Al K$_\alpha$ and 0.33 nm for 130 eV photons, respectively. σ is the photoionization cross-section. The $\sigma_{SiOx}/\sigma_{Si}$ ratios for 130 eV photons estimated by Himpsel et al. [3] are listed in Table I.

3. Thermal Oxides

Thermal oxide has been the chief gate dielectric film. Understanding of this relatively simple material is important for current silicon technology and is also critical to the understanding of more complex new dielectric films such as N$_2$O oxynitride. One batch of 6" p-type 4-6 ohm-cm Si(100) wafers was used for oxide growth under identical conditions. About 7 nm SiO$_2$ was grown in a thermal furnace at 700 °C in dry oxygen.

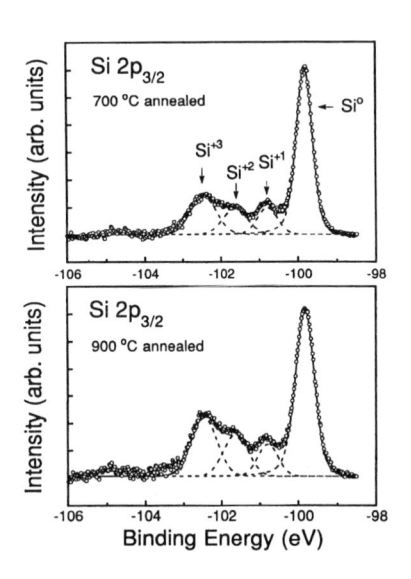

Figure 3. Si 2p$_{3/2}$ spectra recorded from the SiO$_2$/Si(100) samples annealed at 700 °C and 900°C. The open circles are as-recorded data after the subtraction of the SiO$_2$ and the p$_{1/2}$ components. The dashed lines show various suboxide peaks as labeled.

The oxides were then annealed for 30 minutes at 700, 800, 900, and 1000 °C in a nitrogen atmosphere. The growth conditions were described in detail elsewhere [17]. The high quality of the oxides annealed at temperatures > 900 °C has been confirmed by fabricating and measuring polysilicon-gated n-channel MOSFET devices [17]. This set of samples was grown under the same conditions, eliminating process variables, such as oxide thickness and oxidation time, which were reported to produce interfaces with different interface state densities [18]. PES measurements have a relatively short probing depth of about 1.5 nm. Therefore, thin SiO$_2$ films with a thickness of about 1 nm have to be used in order to probe the interface. The thin oxide films were obtained by etchback in diluted (1% HF) solution followed by a DI water rinse. The oxide thickness, homogeneity and surface contamination were monitored by XPS. The etchback oxide thickness was homogeneous across the sample (5x15 mm^2) used for the synchrotron PES measurements, which were carried out at the Synchrotron Radiation Center, University of Wisconsin-Madison [19]. A fraction of a monolayer of carbon, the same amount as that before etching, was detected on the surface. The C was present as a physisorbed hydrocarbon and did not affect the measurement of the buried SiO$_2$/Si interface.

Figure 3 shows Si $2p_{3/2}$ PES spectra measured on two different samples. For clarity, the Shirley background, the $p_{1/2}$ components and the SiO$_2$ contributions have been subtracted from the raw data. The open circles are the subtracted raw data. The dashed lines represent the $p_{3/2}$ peaks of the various chemical states as labeled. The data clearly show a difference in the suboxide structures. Based on these peak intensities, the amount of various suboxides can be calculated according to the above equation. The suboxide density as a function of anneal temperature T_a is shown in Figure 4. We found that the amount of both Si^{+2} and Si^{+3} increases while that of Si^{+1} is constant with increasing annealing temperature. The insensitivity of Si^{+1} to temperature may be explained by the fact that it is located right on the substrate surface. The surface structural parameters such as roughness and terraces that may affect the Si^{+1} intensity are not affected by the annealing treatment.

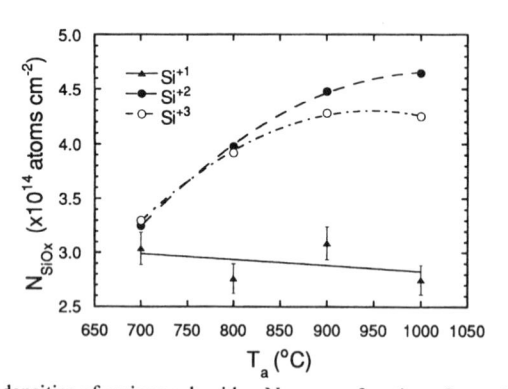

Figure 4. The densities of various suboxides, N_{SiOx}, as a function of anneal temperature, T_a.

The increase in Si^{+2} and Si^{+3} densities with increasing temperature can be explained by a strain-energy minimization principle. It is known that there is about a factor of two bond-density or "lattice" mismatch between bulk silicon and the SiO$_2$ overlayer. This places both materials under strain, tensile in the substrate and compressive in the oxide. It is also known that annealing will reduce the strain on both sides of the interface [17]. As the Si atomic density in the suboxides is between that of bulk Si (5×10^{22} atoms cm^{-3}) and that of SiO$_2$ (2.25×10^{22} atoms cm^{-3}), an increase in the amount of suboxides will enlarge the interface gradient region and therefore reduce strain on both sides of the interface. It is much like the growth of a buffer layer for epitaxy of a strain-free film on a lattice-mismatched substrate, a tactic commonly practiced in molecular-beam epitaxy technology. Here the suboxide functions as a buffer layer and its thickness is mediated by the temperature during the annealing, i.e., a thermodynamic process. Based on a strain-energy minimization principle, Ohdomari et al. [20] have found that a gradient interface transition is energetically favored over an abrupt interface.

Studies were also carried out on the effect of growth kinetics [21]. Experiments performed on furnace oxides showed that the densities of both Si^{+2} and Si^{+3} increase while that of Si^{+1} is constant with increasing growth temperature. As furnace oxidation proceeds at a very slow rate, it is concluded that the interfacial structure is controlled by, as in the post-oxidation annealing, thermodynamic considerations, i.e., structural relaxation during growth. It is possible that oxidation kinetics may dominate the interface structure as the time of oxide formation is significantly reduced to times shorter than that of structural relaxation in processes such as rapid-thermal oxidation or plasma oxidation. Further studies would be useful in clarifying this point.

4. N$_2$O Oxynitride

4.1. THE NATURE OF N CHEMICAL BONDING

Because of their superior properties and enhanced device reliability, nitrided oxides or oxynitrides are currently under intensive study for use in future generations of integrated circuits. There are many ways of fabricating oxynitrides such as direct oxidation in N$_2$O, NO, or NH$_3$ ambient on a bare or a pre-oxidized wafer. Here the discussion will be focused solely on the rapid-thermal N$_2$O oxynitrides formed on a bare Si surface (RTNS) and on a rapid-thermally oxidized (~10 nm) surface (RTNO). The RTO films were grown in dry O$_2$ at 1050 °C. Furnace oxides were also used as starting

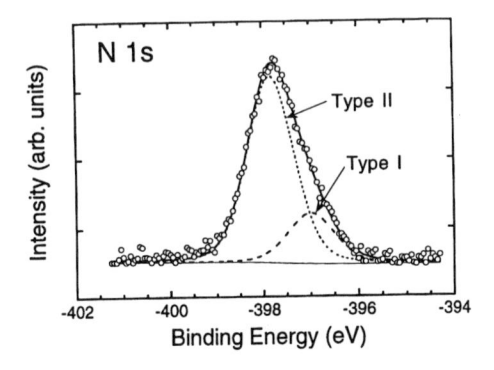

Figure 5. N 1s XPS spectrum recorded from a RTNS oxynitride/Si(100).

material but no apparent spectroscopic difference was noted.

The oxynitride films were grown on 6" Si(100) wafers by rapid thermal processing (RTP) using a system described in detail in reference [22]. 0.5 µm NMOS and PMOS circuits were fabricated using these oxynitride films. Electrical measurements confirmed the high electrical quality of these films. XPS was used to establish the nature and the distribution of N in the film. The measurements [23] were carried out on a PHI 5500 system which was equipped with a monochromated Al K$_\alpha$ source and a hemispherical electron analyzer. Figure 5 shows a selected N 1s core level recorded from a 1050 °C-formed RTNS oxynitride. The spectrum was taken with a 5.8 eV pass energy. As can be seen in the figure, there are two components in the spectrum indicating two types of N in the oxynitride. The same peak structure was also observed for the RTNO film. Curve-fitting analysis showed that one peak (type I) is located at 396.97 eV and another peak (type II) has a chemical shift of 0.85 eV at 397.82 eV. Measurements on Si_3N_4 films yield a N 1s peak at 397.0 eV. This suggests that the chemical bonding of the type I nitrogen in the oxynitride is similar to that of nitrogen in the nitride, i.e., one N bonded to three Si atoms. Type I nitrogen has been reported in all types of oxynitrides grown in various gas ambient by RTP or furnace method [24-27]. The Si_3N_4-like structure has been widely accepted.

The chemical shift of type II nitrogen suggests that it has a different bonding environment. The sign of the shift indicates that type II nitrogen has more electronegative atoms such as oxygen in its first or second shell. A recent first-principle calculation by Rignanese et al. [28] found that one oxygen atom in the first-shell would induce a 1.5 eV chemical shift, larger than the observed data. The calculation suggests that a core-hole relaxation and the second-nearest-neighboring oxygen atoms concur in yielding a chemical shift comparable to type II nitrogen.

Other types of nitrogen with larger chemical shifts have been reported in oxynitride films, especially in those films grown in a NO or a NH_3 ambient. Peaks with a shift of ~2.2 eV [24,25] and ~2.8 eV [26,27] have been reported. The interpretation of these peaks, however, is still speculative. Their large chemical shifts, however, suggest that the first-nearest-neighbor effects should be the main cause. In addition to N-O bonds, as suggested in these articles, N-H$_x$ species are also possible. Based on measurements on low-temperature Si(100) and Si(111) exposure to NH_3, Dufour et al. [29] found several N 1s peaks located at about 397 eV, 398 eV, 399 eV, and 400 eV. The chemical nature of these peaks is presumably associated with silicon nitride and various types of adsorbed NH$_x$ species. It is possible that the N 1s peaks with chemical shifts of ~2.2 eV and ~2.9 eV are due to NH$_x$ species incorporated in these types of oxynitride.

4.2 N DISTRIBUTION

The concentration of N in the film can be calculated from the XPS peak intensities of N 1s, Si^{+4} 2p, and O 1s. Their relative sensitivity factors were determined from thermal SiO_2 and Si_3N_4 samples. As there is an uncontrollable amount of residual surface contaminants such as hydrocarbons, here we calculated only the relative N concentration using N 1s and Si^{+4} 2p peaks to eliminate these surface artifacts. Thus the maximum possible N concentration in the oxynitride would be 57 at.%, which represents pure Si$_3$N$_4$. The depth distribution of N was obtained by measuring a set of samples with various film thicknesses thinned from the as-grown samples in a dilute (1%) HF solution. The oxynitride thickness was determined by the I^{+4}/I^0 ratio assuming the same photoelectron inelastic mean free path as for the oxide. As the N concentration is very small, such an assumption is reasonable. In order to increase the depth resolution, the XPS measurements were made at a 30° photoelectron take-off angle. The spectra were taken using a 5.85 eV electron pass energy. As the N concentration is very dilute, it took several hours to record each spectrum. Figure 6 show N concentration depth profiles obtained from a RTNS oxynitride film grown at 1050 °C for 3 minutes. The data show that type I nitrogen is distributed mainly within the first nm region of the interface while type II nitrogen is distributed mainly outside of the first 1 nm region. No apparent difference in the N distribution between the RTNS and the RTNO films was found. One possible explanation for the observed N distribution is strain-energy minimization. As we know and mentioned earlier, there is a large mismatch in the Si atomic density across the oxide/Si interface, from 2.2x10^{22} atoms cm^{-3} in the oxide to 5x10^{22} atoms cm^{-3} in the bulk silicon. The transition is realized through the formation of an extra sacrificial buffer layer of suboxides. It is also known that the Si density in Si$_3$N$_4$ is 4x10^{22} atoms cm^{-3}, a value between that of bulk Si and SiO$_2$. The inclusion of Si$_3$N$_4$ near the oxide/Si interface therefore creates a buffer layer which would minimize the mismatch-induced strain energy at the oxynitride/Si(100) interface.

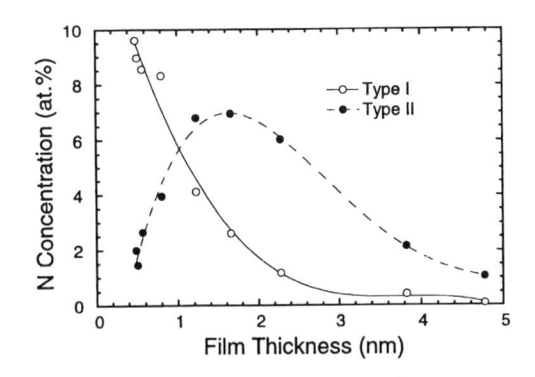

Figure 6. XPS etchback profile of N concentration from a RTNS oxynitride film.

N concentration as a function of growth temperature was also studied. Figure 7 shows type I and II N concentrations in both the RTNS and the RTNO films as a function of growth temperatures. The growth time for all films was 3 minutes. As can be seen from the figures, shapes of both types of nitrogen distribution remain unchanged over the temperature range studied. The concentrations are found to increase with increasing growth temperature. Both the RTNO and the RTNS films follow the trend as a function of temperature. The increase of N content at increased growth temperature was also observed by ion-beam analysis of the RTNS oxynitride [36].

The inclusion of N near the interface for the N_2O RTP oxynitride has been reported by various research groups [26,27,30,31]. It was also reported that N had a more uniform distribution for N_2O furnace oxynitride grown on bare Si surfaces [27,30,31]. The N was found, however, to build up again near the interface for N_2O furnace oxynitride formed on a pre-oxidized surface [26,30], and for NO furnace oxynitride formed on both bare and pre-oxidized surfaces [25,27]. Details regarding these types of oxynitride can be found in other review articles in this proceedings.

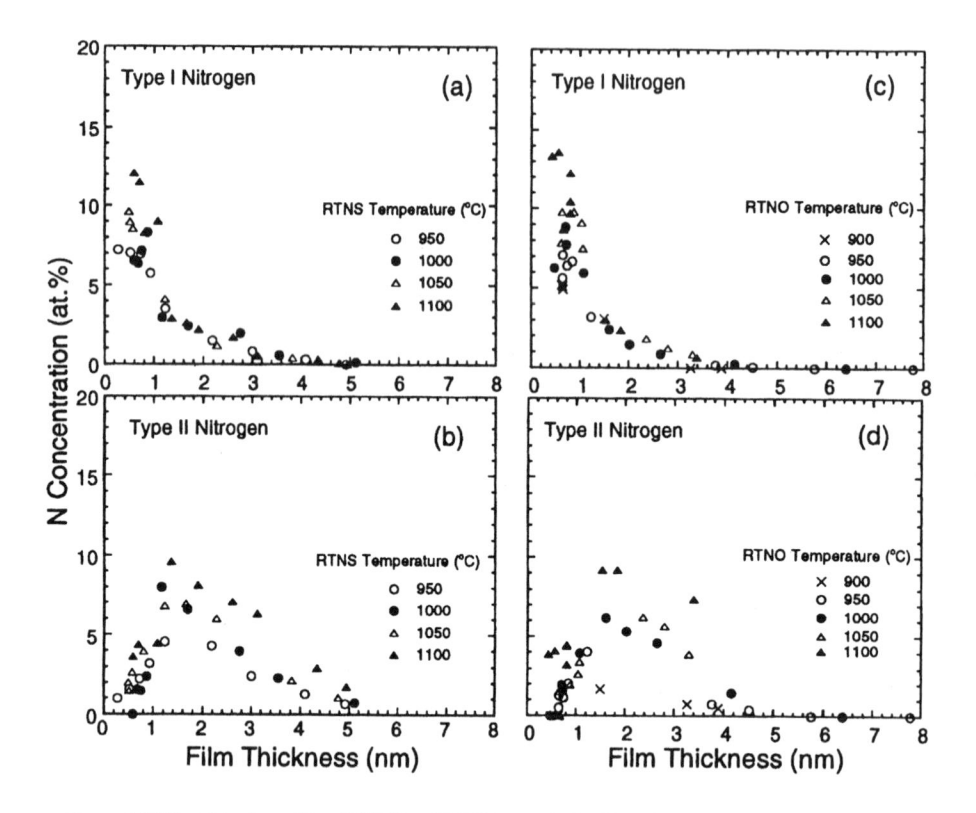

Figure 7. XPS etchback profile of RTNS and RTNO oxynitride films formed at various temperatures.

4.3 OXYNITRIDE/Si(100) INTERFACE MICROSTRUCTURE

The oxynitride/Si interface structure was studied using a synchrotron PES at a photon energy of 130 eV. The measurements were carried out at the Stanford Synchrotron Radiation Laboratory using a 6 m toroidal-grating monochromator (TGM) beamline. The photoelectrons were collected by an angle-integrated PHI double-pass cylindrical-mirror analyzer in an UHV chamber with a load-lock chamber. The photoelectron take-off angle was 45°. Details regarding the measurements are given in reference [23].

A reference rapid-thermal oxide (RTO) wafer was also grown using the same RTP apparatus in dry O_2. In order to focus on the effect of N inclusion on the interface microstructure, all PES samples were grown at 1050°C for three minutes. This is to avoid the complications from other parameters such as the temperature which also plays a critical role in shaping the interface structure as discussed in the thermal SiO_2 section.

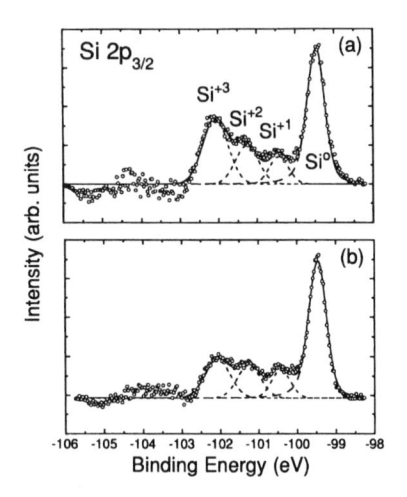

Figure 8. Si $2p_{3/2}$ PES spectra recorded from (a) the RTO SiO_2/Si(100) sample and (b) the RTNS oxynitride/Si(100) sample.

Figure 8(a) shows PES data taken from the RTO sample with an etchback oxide thickness of 1 nm. The suboxide structure is similar to other thermal oxides, as discussed in the previous section. The PES spectrum recorded from the RTNS sample is shown in Figure 8(b). Measurements on the RTNO sample yield a similar spectrum. Curve-fitting analysis shows that the chemical shift of the suboxide peaks at the oxynitride/Si interface is identical to that at a SiO_2/Si(100) interface. It was reported [32] that Si(-N), Si(-N_2), and Si(-N_3), have chemical shifts of 0.77 eV, 1.47 eV, and 2.49 eV, respectively. Thus it can be concluded that the interface transition at an oxynitride/Si interface comprises primarily Si-O bonds. The subnitrides, if they exist, are not detectable by the PES unless they are the dominant species at the interface.

TABLE II. The number of suboxide, N_{SiOx}, at various dielectric/Si(100) interfaces formed at 1050 °C for three minutes.

Sample	N_{SiOx} (10^{14} atoms cm^{-2})			Si^{+4} (nm)
	Si^{+1}	Si^{+2}	Si^{+3}	
RTO SiO_2	2.22	3.49	4.16	1.0
RTNS Oxynitride	2.25	2.98	2.43	0.8
RTNO Oxynitride	2.24	2.90	2.91	0.9

In Figure 8, we observe that the intensities of Si^{+2} and Si^{+3} states at the oxynitride/Si interface are significantly reduced as compared to those at the RTO SiO_2/Si interface. Based on their relative intensities, the densities of various suboxides were calculated and the results are listed in Table II. This table shows that the density of Si^{+1} is the same for all samples. This may not be surprising as Si^{+1} is located right on the substrate surface and therefore is not susceptible to structural changes in the dielectric film. The densities of both Si^{+2} and Si^{+3}, however, are found to be significantly reduced by about 1/5 for Si^{+2} and 1/3 for Si^{+3} at the oxynitride/Si interface. This can again be explained by a structural relaxation mechanism when N is incorporated near the oxynitride/Si interface. In the case of a thermal SiO_2/Si interface, we established that a large quantity of $Si^{+2,3}$ is produced at high temperature. This was attributed to structural relaxation through the introduction of extra sacrificial layers of $Si^{+2,3}$ as a strain buffer. Here in the case of the oxynitride/Si interface, the above XPS data showed that N is incorporated near the interface region. Due to the high Si atomic density in Si_3N_4, the nitrided region would have a high Si density and consequently would function as a buffer layer to help reduce the "lattice" mismatch stress. This means that the amount of sacrificial $Si^{+2,3}$ would be less at an oxynitride/Si interface than that at a SiO_2/Si interface at any given temperature. A schematic model comparing the interface microstructures at the SiO_2/Si and the oxynitride/Si interfaces is shown at Figure 9.

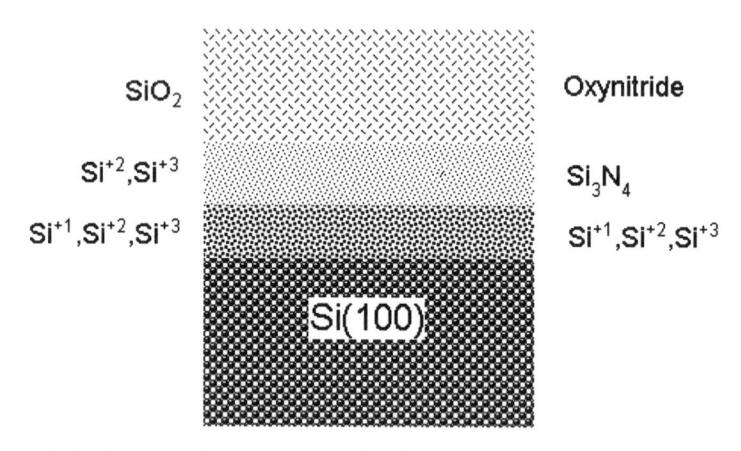

Figure 9. A schematic model comparing the interface microstructures at the SiO_2/Si(100) and the Oxynitride/Si(100) interface.

The advantage of a nitrided interfacial layer is that there is no need to strain the Si-N bonds to form a buffer layer. At a SiO_2/Si interface, Si-O would have to be strained so that the Si atomic density in suboxides could match on one end the bulk Si and on another end the SiO_2. The suboxides have been found to be highly strained as reported in a recent calculation [33]. The evidence of a less strained oxynitride/Si system may be found in electron spin resonance measurements which showed that local strain at a dangling bond site decreases upon nitridation [34]. One superior electrical property of the gate oxynitride is the enhanced reliability under current stress or radiation [35,36].

Should we consider the breaking of a strained Si-O suboxide bond as one of the leading microscopic causes of defect generation under current stress, it would be clear that the robust oxynitride performance could be attributed to a less strained oxynitride/Si interface.

5. Summary

Synchrotron radiation PES has been used to identify various species at the transition layer between a dielectric film and a Si surface. The transition layer was found to consist of a mixture of suboxides, which can be clearly detected by Si 2p core-level PES. The chemical shift of the Si 2p core-level is found to be ~ 0.9 eV per Si-O bond.

The suboxide distributions in the furnace thermal oxide and rapid-thermal oxynitride have been studied. For the furnace thermal $SiO_2/Si(100)$, it is found that the densities of both Si^{+2} and Si^{+3} increase while that of Si^{+1} is constant with increasing anneal temperature. The insensitivity of Si^{+1} to temperature is explained by the fact that it is located right on the substrate surface and therefore is not susceptible to structural changes in the dielectric film. The changes in $Si^{+2,3}$ densities can be understood by the fact that the suboxides play two critical roles at the interface. First, they function as a topological link between the crystalline substrate and the amorphous SiO_2 continuous-random network (CRN). Second, they act as a strain buffer to accommodate the very large "lattice" mismatch across the interface. Structural relaxation in the SiO_2 CRN at an increased temperature is achieved by generating an extra-buffer layer of $Si^{+2,3}$. In the case of oxynitride, the extra-buffer layer is created by the inclusion of Si_3N_4 near the interface. As the atomic density of Si in the nitride is between that in bulk Si and that in SiO_2, the strain on the Si-O or Si-N bond at an oxynitride/Si interface would be reduced and therefore these bonds are more robust under current stress or radiation.

It is found that the lowest amount of suboxide is ~ 1 ML (monolayer; defined as 6.78×10^{14} atoms cm^{-2}) for the oxynitride/Si(100) interface, and ~ 1.4 ML for the $SiO_2/Si(100)$ interface. This may be the minimum amount of suboxide required to form a topological link between a Si(100) surface and an oxynitride or a SiO_2 CRN. A recent theoretical $SiO_2/Si(100)$ model indicated that ~ 1.5 ML is required for a defect-free interface [33].

The use of XPS as a precise thickness measurement tool was discussed. A recent thickness metrology study involving XPS, TEM, spectroscopic ellipsometry, and capacitance-voltage measurements found that the inelastic photoelectron mean-free path in the SiO_2 film is 2.96 nm for the Al K_α photon [16].

High-resolution XPS core-level studies on the rapid-thermal N_2O oxynitride found two N 1s peaks with binding energy at 396.97 eV and 397.82 eV. The peak at 396.97 eV is attributed to N in the form of Si_3N_4. The peak at 397.82 eV may be due to N bonded to three Si atoms in the first-shell and nine O atoms in the second-shell, as suggested in reference [28]. For oxynitrides grown in NO and NH_3 ambients, peaks at higher binding energies were also reported. Possibilities including various nitrogen hydrides were discussed. Future experiments would help substantiate this interpretation.

XPS N concentration depth profiles were obtained by using a wet chemical (1% HF) etch. The nitrogen is found to distribute near the interface. The concentration was found to increase with increasing growth temperatures.

6. Acknowledgements

The author wishes to acknowledge contributions/technical support from many of his collaborators/colleagues, especially S.P. Tay, K.H. Tan, R. Cao, P. Pianetta, T. Miller, T.-C. Chiang, M.J. Graham, and J.R. Phillips. The synchrotron PES measurements were made at the Synchrotron Radiation Center, Univ. of Wisconsin-Madison, which is supported by the NSF under Award No. DMR-95-31009, and at the Stanford Synchrotron Radiation Laboratory, which is operated by DOE under contract DE-AC03-76SF00515.

7. References

1. Sze, S.M. (1985) *Semiconductor Devices: Physics and Technology*, John Wiley & Sons, Toronto.
2. Hollinger, G. and Himpsel, F.J. (1984) Probing the transition layer at the SiO_2-Si interface using core level photoemission, *Appl. Phys. Lett.* **44**, 93-95
3. Himpsel, F.J., McFeely, F.R., Taleb-Ibrahimi, A., Yarmoff, and Hollinger, G. (1988) Microscopic structure of the SiO_2/Si interface, *Phys. Rev.* **B 9**, 6084-6096.
4. Lu, Z.H., Graham, M.J., Jiang, D.T., and Tan K.H. (1993) SiO_2/Si(100) interface studied by Al K_α x-ray and synchrotron radiation photoelectron spectroscopy, *Appl. Phys. Lett.* **63**, 2941-2943.
5. Grunthaner, P.J., Hecht, M.H., Grunthaner, F.J., and Johnson, N.M. (1987) The localization and crystallographic dependence of Si suboxide species at the SiO_2/Si interface, *J. Appl. Phys.* **61**, 629-638.
6. Siegbahn, K., (1986) Photoelectron spectroscopy: retrospects and prospects, *Phil. Trans. R. Soc.Lond.* A **318**, 3-36.
7. Holl, M.M.B. and McFeely, F.R. (1993) Si/SiO_2 interface: New structures and well-defined model systems, *Phys. Rev. Lett.* **71**, 2441-2444.
8. Pasquarello, A., Hybertsen, M.S., and Car, R. (1995) Si 2p core level shifts at theSi(001)-SiO_2 interface: a first-principle study, *Phys. Rev. Lett.* **74**, 1024-1027.
9. Pasquarello, A., Hybertsen, M.S., and Car, R. (1996) Theory of Si 2p core-level shifts at the Si(001)-SiO_2 interface, *Phys. Rev.* **B 53**, 10942-10950.
10. Nucho, R.N. and Madhukar, A. (1980) Electronic structure of SiO_2: α-quartz and the influence of local disorder, *Phys. Rev.* **B 21**, 1576-1588.
11. Grunthaner, F.J., Grunthaner, P.J., Vasquez, R.P., Lewis, B.F., Maerjian, J., and Madhukar, A. (1979) High-resolution x-ray photoelectron spectroscopy as a probe of local atomic structure: application to amorphous SiO_2 and Si-SiO_2 interface, *Phys. Rev. Lett.* **43**, 1683-1685.
12. Tao, Y., Lu, Z.H., Graham, M.J., and Tay, S.P. (1994) X-ray photoelectron spectroscopy and x-ray absorption near-edge spectroscopy study of SiO_2/Si(100), *J. Vac. Sci. Technol.* **B 12**, 2500-2503.
13. Fadley, C.S. (1984) Angle-resolved x-ray photoelectron spectroscopy, *Progress in Surf. Sci.* **16**, 275-388.
14. Sieger, M.T., Luh, D.A., Miller, T., and Chiang, T.-C. (1996) Photoemission extended fine structure study of the SiO_2/Si(111) interface, *Phys. Rev. Lett.* **77**, 2758-2761.
15. Powell, C.J., Jablonski, A., Tanuma, and Penn, D.R. (1994) Effects of elastic and inelastic electron scattering on quantitative surface analyses by AES and XPS, *J. Electr. Spec. and Related Phenom.* **68**, 605-616.
16. Lu, Z.H., McCaffrey, J., Brar, B., Wilk, G.D., Wallace, R.M., Feldman, L.C., and Tay, S.P. SiO_2 film thickness metrology by X-ray photoelectron spectroscopy, submitted for publication in *Appl. Phys. Lett.*

17. Tay, S.P. and Ellul, J.P. (1992) *J. Electr. Mater.* **21**, 45-55.
18. Fukuda, H., Yasuda, M., Iwabuchi, T., Kaneko, S., Ueno, T., and Ohdomari, I. (1992) Process dependence of the SiO$_2$/Si(100) interface trap density of ultrathin SiO$_2$ films, *J. Appl. Phys.* **72**, 1906-1911.
19. Lu, Z.H., S.P. Tay, Miller, T., and Chiang, T.-C. (1995) Process dependence of the SiO$_2$/Si(100) interface structure, *J. Appl. Phys.* **77**, 4110-4112.
20. Ohdomari, I., Akatsu, H., Yamakoshi, Y., and Kishimoto, K. (1987) Study of the interfacial structure between Si(100) and thermally grown SiO$_2$ using a ball-and-spoke model, *J. Appl. Phys.* **62**, 3751-3754.
21. Lu, Z.H., Graham, M.J., Tay, S.P., Jiang, D.T., and Tan, K.H. (1995) Effects of growth temperature on the SiO$_2$/Si(100) interface structure, *J. Vac. Sci. Technol.* **B 13**, 1626-1629.
22. Tay, S.P., Ellul, J.P., Lu, Z.H., Hebert, K., and Irene, E.A. (1995) Characterization of ULSI gate dielectric films formed in a rapid thermal N$_2$O ambient, in R.B. Fair and B. Lojek (eds.), *Proceed. 3rd Inter. Rapid Therm. Proces. Conf.*, RTP'95, pp.135-140.
23. Lu, Z.H., Tay, S.P., Cao, R., and Pianetta, P. (1995) The effect of rapid thermal N$_2$O nitridation on the oxide/Si(100) interface structure, *Appl. Phys. Lett.* **67**, 2836-2838.
24. Bhat, M., Ahn, J., Kwong, D.L., Arendt, M., and White, J.M. (1994) Comparison of the chemical structure and composition between N$_2$O oxides and reoxidized NH$_3$-nitrided oxides, *Appl. Phys. Lett.* **64**, 1168-1170.
25. Hegde, R.I., Tobin, P.J., Reid, K.G., Maiti, B., and Ajuria, S.A. (1995) Growth and surface chemistry of oxynitride gate dielectric using nitric oxide, *Appl. Phys. Lett.* **66**, 2882-2884.
26. Bouvet, D., Clivaz, P.A., Dutoit, M., Coluzza, C., Almeida, J., Margaritondo, G., Pio, F. (1996) Influence of nitrogen profile on electrical characteristics of furnace or rapid thermally nitrided silicon dioxide films, *J. Appl. Phys.* **79**, 7114-7122.
27. Lu, H.C., Gusev, E.P., Gustafsson, T., Garfunkel, E., Green, M.L., Brasen, D., and Feldman, L.C. (1996) High resolution ion scattering study of silicon oxynitridation, *Appl. Phys. Lett.* **69**, 2713-2715.
28. Rignanese, G.-M., Pasqarello, A., Charlier, J.-C., Gonze, X., and Car, R. Nitrogen incorporation at Si(100)-SiO$_2$ interface: relation between N 1s core level shifts and microscopic structure, submitted for publication in *Phys. Rev. Lett.*.
29. Dufour, G., Rochet, F., Roulet, H., and Sirotti, F. (1994) Contrasted behavior of Si(100) and Si(111) surfaces with respect to NH$_3$ adsorption and thermal nitridation: a N 1s and Si 2p core level study with synchrotron radiation, *Surf. Sci.* **304**, 33-47.
30. Okada, Y., Tobin, P.J., Lakhotia, V., Feil, W.A., Ajuria, S.A., and Hegde, R.I. (1993) Relationship between growth conditions, nitrogen profile, and charge to breakdown of gate oxynitrides grown from pure N$_2$O, *Appl. Phys. Lett.* **63**, 194-196.
31. Carr, E.C., Ellis, K.A., and Buhrman, R.A. (1995) N depth profiles in thin SiO$_2$ grown or processed in N$_2$O: the role of atomic oxygen, *Appl. Phys. Lett.* **66**, 1492-1494.
32. Stober, J., Eisenhut, Rangelov, G., and Fauster, Th., (1994) Initial stage of the thermal nitridation of the Si(100) surface with NH$_3$ and NO: a surface sensitive study of Si 2p core-level shifts, *Surf. Sci.* **321**, 111-126.
33. Pasquarello, A., Hybertsen, M.S., and Car, R. (1996) Structurally relaxed models of the Si(001)-SiO$_2$ interface, *Appl. Phys. Lett.* **68**, 625-627.
34. Yount, J.T., Lenahan, P.M., and Wyatt, P.W. (1993) An electron spin resonance study of the effects of thermal nitridation and reoxidation on P$_b$ centers at (111) Si/SiO$_2$ interfaces, *J. Appl. Phys.* **74**, 5867-5870.
35. Hwang, H., Ting, W., Maiti, B., Kwong, D.L., and Lee, J. (1990) Electrical characteristics of ultrathin oxynitride gate dielectric prepared by rapid thermal oxidation of Si in N$_2$O, *Appl. Phys. Lett.* **57**, 1010-1012.
36. Okada, Y., Tobin, P.J., Hegde, R.I., Liao, J., and Rushbrook, P. (1992) Oxynitride gate dielectrics prepared by rapid thermal processing using mixtures of nitrous oxide and oxygen, *Appl. Phys. Lett.* **61**, 3163-3165.

STRESS IN THE SiO$_2$/Si STRUCTURES FORMED BY THERMAL OXIDATION

A. SZEKERES
Institute of Solid State Physics, Tzarigradsko Chaussee 72, 1784 Sofia, Bulgaria

1. Introduction

Contemporary microelectronics is based on silicon devices, which involve SiO$_2$/Si structure, and hence a thin SiO$_2$/Si interface existing between the oxide and the Si substrate. To obtain the optimum Si-based device performance the dimensions of the elements have been reduced significantly to almost the technological limit. This miniaturization of elements has required also a drastic reduction of the SiO$_2$ film thickness up to few tens of nanometers and, therefore, the interface becomes a significant part of the whole oxide. Further reduction of the oxide thickness, however, creates serious problems connected with device reliability. One of the problems is related to the high internal stresses induced in SiO$_2$ films during oxidation process. The effect of this stress on Si oxidation kinetics has received considerable attention in the 1980s. In 1986 a special Workshop on Oxidation Mechanisms was organized treating growth mechanism of thin SiO$_2$ and influence of stress on Si oxidation kinetics. Some of the papers presented at this Workshop were published in a special issue of Philosophical Magazine [1]. These topics still remain in the focus of extensive investigation due to anomalous phenomena observed at low oxidation temperatures and in the initial regime of Si oxidation [1-7]. The knowledge of structural strains and their eventual reduction and control gains growing technological importance, especially for ultrathin SiO$_2$ films, where the oxidation induced stress may deteriorate the device characteristics.

The present work reviews our recent results concerning the thermal oxidation of crystalline silicon. Some of these measurements are recently reported, others have appeared in the literature [8-13].

The stress behavior in thin SiO$_2$/Si structures and the influence of process conditions, such as oxidation temperature and oxidizing ambient, on the generation of structural stress in the oxide at the SiO$_2$/Si interface and in the Si substrate is presented. It is well known that post-oxidation annealing of SiO$_2$/Si structures in hydrogen ambient considerably reduces the electrically active defects connected with interface traps. This fact led to the idea to investigate the results of hydrogen introduction directly during the oxidation process. In this paper the role of hydrogen intentionally introduced in the oxides during the growth process on the magnitude of stress and its gradient towards SiO$_2$/Si interface is discussed. The electrical properties of SiO$_2$/Si structures formed in such an oxidation ambient are also investigated, but discussion of the results is beyond the scope of this paper. A possibility to introduce hydrogen species directly into the growing oxides using hydrogenated amorphous silicon (a-Si:H) as a hydrogen source during crystalline silicon oxidation is shown.

2. Experimental conditions

All experiments were performed using commercially available n-Si (111) and p-Si (100) wafers (~440 μm) with a resistivity of 4-10 Ωcm, oxidized at temperatures ranging from 800 to 1060°C at atmospheric pressure in dry O$_2$ (< 3 ppm H$_2$O content) and in H$_2$O steam. For the purpose of introducing hydrogen species in the oxide directly during the process of oxide growth, some oxidation runs were done in hydrogen-enriched dry O$_2$. In this case ~50-60 nm thick chemical vapor deposited a-Si:H films onto Si substrates with about 10 % hydrogen content served as hydrogen source. It has been observed that hydrogen effusion from a-Si layers starts at temperatures higher than 350°C and the hydrogen is released in molecular form [14]. Therefore, it is reasonable to assume that at the oxidation temperature hydrogen, released from the a-Si:H layer when entering the oxidation environment, takes part in the oxidation process. In these oxidation runs the bare c-Si substrates were loaded together with samples covered with a-Si:H films close to each other on the quartz boat and the oxidation proceeded in dry oxygen. A standard RCA procedure was used to clean all wafers prior to oxidation.

The oxides in different oxidation runs were grown up to a thickness ranging from 12 to 70 nm, the films' thickness being determined from ellipsometric measurements.

The oxidation-induced stress in the SiO$_2$ films was estimated by using ellipsometric and radius of curvature measurements described below. Since all measurements were carried out at room temperature, the total stress obtained directly by sample bending measurement or deduced from the oxide refractive index values consists of two components; intrinsic stress σ_i and thermal stress σ_{th}. The intrinsic oxide stress arises from lattice mismatch between the oxide and the silicon substrate and is a consequence of volume expansion occurring during oxidation. The thermal component appears during cooling the samples to room temperature due to the difference in the thermal expansion coefficients α of the oxide and the Si substrate. The thermal stress can be expressed as $\sigma_{th} = (T_m - T_{ox})(\alpha_{Si} - \alpha_{SiO2})E_{SiO2}/(1-\nu_{SiO2})$, where α_{Si} and α_{SiO2} are the thermal expansion coefficients of Si and SiO$_2$, T_{ox} is the oxidation temperature, and T_m is the temperature of the measurements. The slope of the total stress σ_t vs. T curve gives $d\sigma_t/dT = (\alpha_{Si} - \alpha_{SiO2})E_{SiO2}/(1-\nu_{SiO2})$ and is experimentally determined to be 2.7x10^5 Nm^{-2} °C^{-1} for wet SiO$_2$ [7] and 2.16x10^5 N m^{-2} °C^{-1} for dry SiO$_2$ [15]. All elastic constants for Si and SiO$_2$ used further in our calculations were taken from the literature data summarized in [7].

To obtain directly the oxidation induced stress in thin oxides, the sample bending was traced by measuring the radius of sample curvature before and after oxidation and by calculating its value by counting the Newton's rings produced by optical interference between the curved substrate and a reference with known curvature. With this method one can determine the magnitude as well as the direction of bending and, therefore, can distinguish between compressive and tensile stresses. By convention, the radius of curvature is negative for convex surfaces (compressive stress) and positive for concave surfaces (tensile stress). The sample bending measurement gives the magnitude of stress with an accuracy of about 1×10^8 Nm^{-2}. The stress calculations were performed by employing a linear elasticity theory equation, so-called Stoney's formula [7] given by

$$\sigma = \frac{E_{Si}}{6(1-\nu_{Si})} \cdot \frac{t_{Si}^2}{t_{ox}} \cdot \frac{1}{R}, \qquad (1)$$

where E_{Si} is Young's modulus of substrate, ν_{Si} is its Poisson's ratio, t_{Si} is the thickness of substrate, t_{ox} is the oxide film thickness, and R is the net radius of curvature. This formula can be applied to structures where the film thickness is much smaller than the substrate thickness and with the assumption that the stress in the plane of the wafer is isotropic and the wafer deformation is elastic. Indeed, the wafer bending measurements of Si substrates made by us, and many other researchers as well, before oxidation and after complete removal of the thermally grown oxide have shown no residual bending, indicating that during oxidation the Si substrates undergo plastic deformation.

Ellipsometric measurements, made at $\lambda = 632.8$ nm and at different angles of light incidence in the range of 50° -70° served for determining the oxide thickness and its refractive index used later for stress calculations. Assuming that the elastic compression is the cause for the changes in the density and the refractive index of the material, one can calculate the oxide stress from the first-order compressibility relationship [16] given by $n = n_0 + (\Delta n/\Delta\sigma)\sigma$, where $\Delta n/\Delta\sigma$ is the compressive coefficient and is 9×10^{-12} m^{2}N^{-1} for SiO_2 ; n_0 is the refractive index of a stress-free oxide (1.46 for silica).

It is worth mentioning that the large increase in the refractive index of oxides grown at low temperatures and its strong temperature dependence observed in a number of experiments [1,7,16] cannot be attributed to elastic deformation only, and therefore the first-order compressibility relationship should be taken with caution. We have considered the correlation between the results of the two types of stress determination and have found a satisfactory agreement of the stress values for oxides thinner than about 25 nm [10]. For thicker oxides, however, the stress level obtained from sample bending measurements was smaller than that calculated from the refractive index. This is discussed later in Section 3.2.

We also carried out spectroscopic ellipsometric measurements for the characterization of the SiO_2/Si interface properties with a Rudolph Research variable-angle-of-incidence ellipsometer in a spectral range of 280-430 nm. Different physical models for the SiO$_2$/Si interface were applied using the Bruggeman effective-medium approximation (EMA) [17].

We have applied an ellipsometric approach, developed by us and discussed in details elsewhere [13], for finding the interface parameters with considerably high accuracy. In brief, the real (tanΨcosΔ) and the imaginary (tanΨsinΔ) parts of the basic ellipsometric equation in a double-layer system (as the SiO$_2$/Si structure is) are represented by polynomials of the top oxide layer thickness. The absolute terms of the polynomials determine the ellipsometric angles of Ψ_i and Δ_i corresponding to a single-layer system, i.e. to the Si/interface [13]. One of the advantages of this approach is that the ellipsometric angles Ψ_i and Δ_i characterizing directly the Si/interface structure are obtained without having to remove completely the oxide top layer. In practice, such top layer removal is impossible without affecting the interface. This ellipsometric approach, however, requires a chemical thinning of the oxide step by step up to a few nanometers from the interface and after each step carrying out ellipsometric measurements. The chemical etching of the oxide was accomplished in an etchant solution consisting of 15 cm^3 HF (48%), 10 cm^3 HNO$_3$ (70%) and 300 cm^3 H$_2$O. From the obtained Ψ_i and Δ_i data the pseudodielectric function of the Si/interface structure is directly calculated. The spectral ellipsometric measurements were done at an incidence angle of 50°, where the error introduced by the polynomial approximation is negligibly small [13].

3. Results and discussion

Before studying the temperature and thickness dependence of stress in more detail a series of experiments were performed from the following consideration. During thermal oxidation, SiO$_2$ films grow on both sides of the Si substrate. Generally, Si wafers in IC technology are polished on the front side used for pattern formation, and are lapped or chemically etched on the back side. Therefore, the oxidation process might proceed quite differently on both Si sides and the back oxide may affect the stress in the front side oxide. In order to study this effect, we traced the change in the radius of curvature of Si substrates having oxides on both sides [8]. The oxidation was accomplished in pure dry O$_2$ at temperatures ranging from 850 to 1050°C. The results have shown that there is a tendency to a concave sample curvature for thin films, which is temperature dependent. For illustration, the thickness dependence of sample curvature K=1/R for initially flat (100)Si wafers, oxidized at 1050°C is given in Figure 1. As the sample bending is determined by the superposition of stress in both front and back-side oxides, the observed positive sample curvature is an evidence for a generation of higher compressive stress in the back-side oxides in the initial stage of oxidation. The experiments with (111)Si samples showed that the influence of the back-side oxide is weaker.

These results clearly demonstrate that the presence of back-side oxide obscures the real stress value in the front-side oxide, which must be taken under consideration when the Si back side is not protected against oxidation and during the study of oxidation kinetics by in situ measurements.

To avoid back-side oxide influence, at all further experiments the oxides on the Si back side were stripped completely away before the measurements.

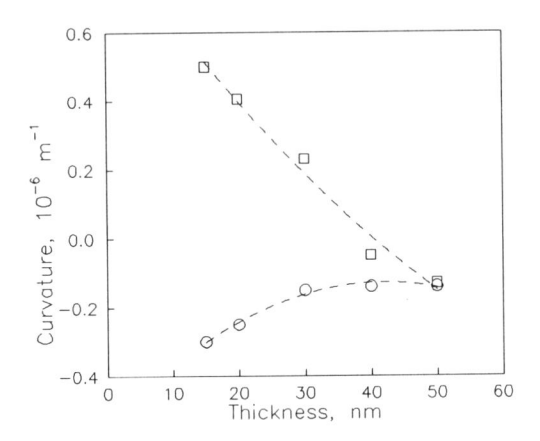

Figure 1 Curvature of the (100)Si substrates oxidized at temperature of 1050°C in dry O₂ ambient in dependence on front-side oxide thickness before (□) and after (o) removal of the back-side oxide.

3.1. TEMPERATURE DEPENDENCE OF OXIDE STRESS

The stress behavior at different oxidation temperatures was investigated in thick SiO_2 films (50-60 nm), for which the intrinsic oxide stress relaxation was expected to be accomplished. The oxides were grown on both types of Si substrates in a pure dry-oxygen ambient and a hydrogen-enriched one, as well as in H_2O steam.

For all kinds of samples, oxidized in dry ambient, high temperature oxidation resulted in nearly zero intrinsic stress. With the decrease of the oxidation temperature an increase in the compressive oxide stresses was observed. When the Si substrates were oxidized in dry O_2, but in a hydrogen-enriched environment, although the stress-temperature curve followed the shape of that obtained for dry SiO_2 films, the stress levels were higher. Only at 1060°C the total stress reached the value of the thermal component, and in this case the intrinsic stress was zero.

Contrary to that, in wet oxides at 800°C the total stress was equal to the thermal component, which is an evidence for a stress-free oxide growth. With increasing the temperature, however, the total stress became higher than the corresponding thermal stress and followed its rising.

The results are summarized in Figure 2, where the thermal stress component is presented by a solid line. Taking into account that the oxide viscosity changes with OH concentration [18] the thermal stress was calculated separately for dry (line 1) and wet oxides (line 2). The upper part of the Figure presents the results from the steam oxidation, while the lower one - for dry oxidation.

The lower stress in (111)Si samples observed contrary to that anticipated from the E/(1-v) orientation dependence ($\sigma_{111} > \sigma_{110} > \sigma_{100}$) could be explained by the surface step model proposed by Mott [2] and quantitatively explained by Leroy [7]. According to this model, for (111)Si, where the areal density is higher and, therefore, the surface steps are more pronounced, the resulting stress considerably reduces because part of the volume change is relieved by the receding Si step.

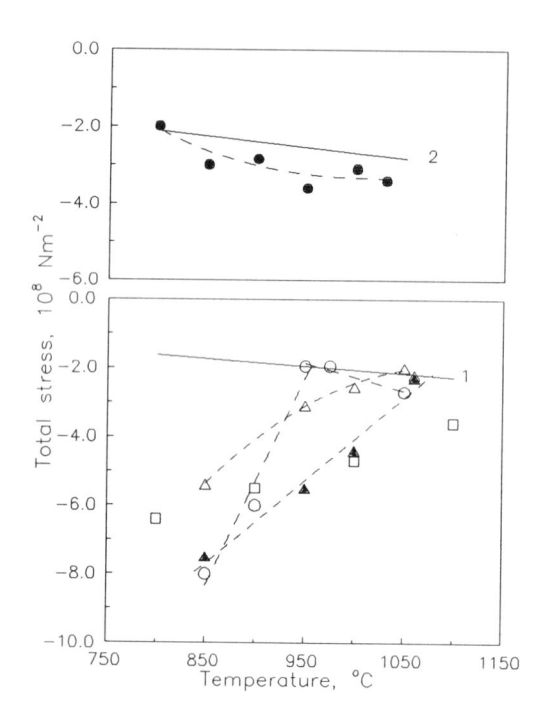

Figure 2. Total stress in SiO$_2$ films grown on (100)Si (o) and (111)Si (Δ, $\blacktriangle$ and $\bullet$) in dependence on oxidation temperature. The oxidation was performed in pure (o, Δ) and hydrogen-enriched ($\blacktriangle$) dry O$_2$ and in steam ($\bullet$). Some results of (100)Si oxidation in dry O$_2$ + 4.5 % HCl ambient ($\square$) from [22] are also given. The solid lines 1 and 2 represent the thermal stress in dry and wet oxides, respectively.

The results for oxidation in pure oxygen and in steam agree essentially with previous independent measurements [15,19,20]. Apparently, at 1000°C and above oxidation in pure O$_2$ results in conditions, relatively free of stress, and the corresponding intrinsic oxide stress is nearly zero. These results are consistent with viscous flow occurring above 965°C during SiO$_2$ growth, where viscoelastic relaxation of stress by viscous shear flow of SiO$_2$ takes place [20,21]. For wet oxides, an increase of total oxide stress with temperature has been also observed by EerNisse [20], who has found a significant residual Si wafer bending even after complete removal of the grown oxides. Experiments in [20] have proved that wet oxides also cannot sustain stress above 965°C and the observed higher compressive oxide stresses are due to the Si surface damage beneath the wet oxide.

As far as we know, experiments concerning the stress behavior during oxidations in dry oxygen in the presence of hydrogen have not been reported yet. It is interesting to note that the addition of a small amount of HCl to the oxygen has lead to similar effect on stress even at temperatures as high as 1100°C [22]. For illustration, some results from [22] are presented by square symbols in Figure 2. This phenomenon has not been explained yet. We believe that what these two experiments have in common is that in both cases free hydrogen in either molecular or atomic form exists in the oxidizing ambient.

Water molecules have been shown [23] to diffuse interstitially in SiO_2 network and to interact with already formed Si-O bonds, creating silanol groups [23]. Incorporation of SiOH groups in its turn leads to opening up of the SiO_2 network, which lowers the oxide viscosity and increases the diffusivity of the oxidizing species through the oxide[18]. Since the oxide network rearranges itself as fast as its viscosity allows, the intrinsic stress generated at the SiO_2/Si interface will relax more rapidly in wet oxides than in dry one, as it is evident from Figure 2. Introduction of hydrogen in oxides grown in a hydrogen-enriched O_2 ambient, however, does not lead to such effect. On the contrary, even higher stresses than for dry oxide are observed. Apparently, the formation of SiOH groups does not dominate in this case, but rather hydrogen diffuses through the growing oxide to the Si surface and creates hydrogen-related defects, increasing its structural strain. Diffusion of hydrogen atoms through the oxide and pile up at the SiO_2/Si interface has been experimentally observed [24,25]. The results of HCl oxidation [22] also support our suggestion. By analogy with wet oxidation, Cl in oxide also forms open bonds between Si and O atoms facilitating the oxidant diffusion. Indeed, the oxidation rate is enhanced, but the oxide stresses are higher than those characteristic for dry O_2 oxidation [22]. Moreover, our study of the radiation sensitivity of MOS structures, formed on the samples oxidized in such a way, has indicated that hydrogen is incorporated in the oxide, showing an initially more defective SiO_2/Si interface [26]. Another possible explanation for the observed higher stress in hydrogen-enriched oxides might be that hydrogen presence at the interface region affects the oxide viscosity.

3.2. THICKNESS DEPENDENCE OF OXIDE STRESS

Since SiO_2 film growth is accompanied by stress annealing, the stress in the oxide near the interface is greater than in the oxide further away. This thickness variation of the intrinsic oxide stress has been reported by many researchers [9,15,20,26]. Here we consider the cases when silicon is oxidized in pure dry and hydrogen-enriched O_2 ambients. As was shown in the previous section, water-containing ambients do not generate internal stresses in the growing oxides and therefore no thickness dependence is observed.

Typical results for (111)Si oxidation at 850° and 1060°C are presented in Figures 3(a) and 3(b) respectively. As is seen, at the initial stage of oxidation a considerably high intrinsic stress is generated which decreases as the oxide becomes thicker. The stress levels are somewhat higher than those previously reported [15,20,27]. These differences could arise from the different experimental conditions and measurement techniques.

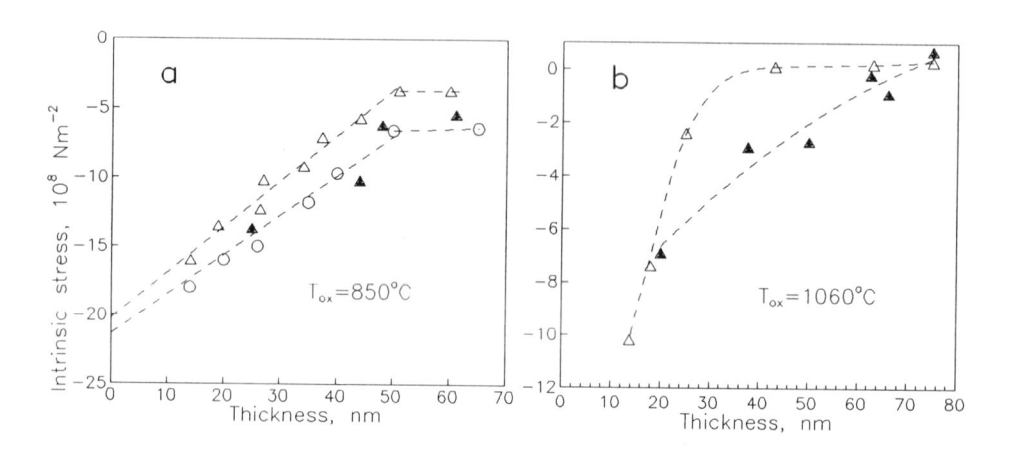

Figure 3. Intrinsic oxide stress in SiO_2, grown on (111)Si (Δ, $\blacktriangle$) and (100)Si (o) at temperature of 850°C (3a) and 1060°C (3b).

The stress relaxation process is different for the two kinds of oxidation ambient. While a rapid relief of structural strains is obtained in dry O_2 ambient, a gradual decrease of the stress in the hydrogen-enriched oxides is observed. This effect is more pronounced for high-temperature oxidation. The observed scatter in the experimental data is most probably due to the uncontrollable diffusion of hydrogen into the growing oxide.

Since SiO_2 is a viscoelastic material, the internal strains created at the SiO_2/Si interface release by viscous flow [7,28]. The relaxation time τ depends on the oxide viscosity η, $\tau = 2\eta[(1+\nu_{SiO2})/E_{SiO2}]$, which is an inverse function of temperature. For this reason the intrinsic oxide stress will relax much faster at high temperatures. As the difference in the oxidation conditions is only the hydrogen content in the oxidizing ambient, the lower relaxation rate observed in hydrogen-enriched oxides suggests that excess hydrogen at the as-forming SiO_2/Si interface might alter the oxide viscosity. Similar relaxation behavior was observed at the high-temperature oxidation of thin hydrogenated amorphous silicon films [9]. In order to prove this suggestion further investigations are needed.

While the experimental data do not extend to films thinner than 14 nm, a common extrapolated value of ~2×10^9 Nm^{-2} is obtained for the maximum intrinsic stress at zero oxide thickness. This stress value appears to be independent of oxidation temperature, which is in agreement with the common origin of stress arising from the volume expansion in silicon. Based on photoreflectance measurements of silicon dioxide films Fitch et al. [27] obtained a stress value of 4.6×10^8 Nm^{-2} at the interface. Higher values of about 1×10^9 Nm^{-2} and 9×10^8 Nm^{-2} have been estimated by nonlinear optical techniques [29] and by spectroscopic ellipsometry [30], respectively.

The refractive index profiles in the oxides grown at 850°C and 1060°C are shown in Figure 4. The calculation of the refractive index values from the corresponding oxide stress values, using the first-order compressibility relationship, are also shown in the Figure. The observed similarity in both dependences indicates that the stress has an essential influence on the refractive index in the oxide and, hence, on its density. The index value obtained from the two independent measurements, however, do not coincide for thick oxides. We believe that this discrepancy is due to structural changes, which occur during oxidation and cause additional oxide densification. The increase in density can be attributed to a mechanism other than the elastic deformation, i.e. denser domains as reported by Bruckner [18].

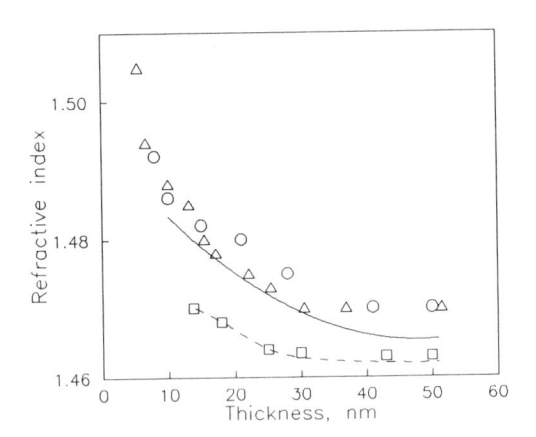

Figure 4. Refractive index in SiO$_2$ grown on (111)Si (Δ, □) and (100)Si (o) substrates at 850°C (Δ, o) and 1060°C (□) in dry O$_2$ ambient. The solid (850°C) and dashed (1060°C) curves correspond to refractive index values calculated from the oxide stress.

3.3. SiO$_2$/Si INTERFACE CHARACTERIZATION

Spectroscopic ellipsometry is one of the most powerful tools for investigation of interface properties [17,31,32]. We applied this technique to obtain information concerning the SiO$_2$/Si interface thickness and its physical composition. The measurement procedure was described earlier in Section 2. In these experiments 30 nm thick oxides grown at 850°C in dry ambient were thinned step by step down to about 4 nm and measured after each step in order to obtain the dependence of the real (tanΨcosΔ) and the imaginary (tanΨsinΔ) parts of the basic ellipsometric equation on the oxide thickness. From the approximation of these dependences the ellipsometric angles of Ψ_i and Δ_i corresponding to the Si/interface were determined [13]. Further, the pseudodielectric functions $<\varepsilon_1>$ and $<\varepsilon_2>$ of the Si/interface structure were directly calculated using these angles. In the computer modeling, by using Bruggeman effective-medium approximation, the interface was considered as a heterogeneous material with optically identifiable regions of amorphous silicon and silicon dioxide, $Si_x(SiO_2)_{1-x}$, where x is the atomic fraction. In order to find the interface parameters that fit the experimental data, we changed the interface thickness t_{int} and x in the theoretical model from 0.4 to 1.5 nm and from 0.4 to 0.8 nm, respectively.

Typical results of the spectral dependence of the pseudodielectric functions for Si/interface, formed on (111)Si substrate, is shown in Figure 5. In the Figure the theoretical models of the SiO$_2$/Si interface with thickness of 1 nm and with x=0.4, 0.6, and 0.8 are also given by dashed, solid, and dotted lines, respectively. As is seen, for $<\varepsilon_1>$ the physical model of t_{int} = 1 nm and x= 0.6 is in good agreement with the experimental data within the entire investigated spectral region. This model also fits well the experimental data for $<\varepsilon_2>$, although a discrepancy is observed around the peak at 365 nm.

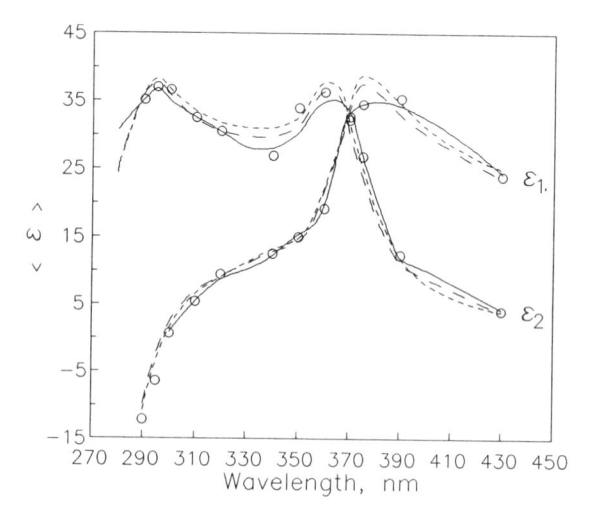

Figure 5. The pseudodielectric function of Si/interface structure vs. wavelength. The open circles represent the experimentally determined dielectric functions. The theoretical models for a Si surface with a 1 nm thick interface overlayer and with x=0.4, 0.6, and 0.8 are also given by dashed, solid, and dotted lines, respectively.

The imaginary part $\langle\varepsilon_2\rangle$ is related to the absorption in the material and can be used as an indication for the state of the material. Towards the near-uv region the absorption of the Si substrate drastically rises and the optical response carries information basically from the interfacial layer. It is well known that when an interfacial layer is present the magnitude of the peak around 295 nm decreases in comparison to that of Si substrate, but its position remains unchanged. Since around that peak the experimental points also lay on the theoretical curve, we can state that the model is correct and the interface in the 850°C dry SiO_2 can be characterized with a thickness of about 1 nm and physical composition $Si_{0.6}(SiO_2)_{0.4}$.

The discrepancy between the theoretical and experimental data around $\langle\varepsilon_2\rangle$ peak at 365 nm is most probably due to the fact that the optical response in this spectral range already contains information not only about the interface, but also about the state of the underlying Si substrate. This effect may come from the optical constants of the substrate, different from that used for the model. In our case, after the last etching step a thin (~4 nm) but still thermally grown oxide, which is under a rather high compressive stress, remains on the Si substrate. As it was shown in Section 3.2, the SiO_2/Si interface is under compressive stress of about 2×10^9 Nm^{-2}, and therefore, the Si substrate will be under considerable tension. In the theoretical model the refractive index and extinction of Si were taken from the experimentally found values for the stress-free bare Si substrate. In order to estimate the Si substrate optical constants in the presence of oxide, we searched for the corresponding dielectric functions of Si with the obtained interface parameters being fixed. The results are presented in Figure 6. For comparison, the experimentally obtained dielectric functions

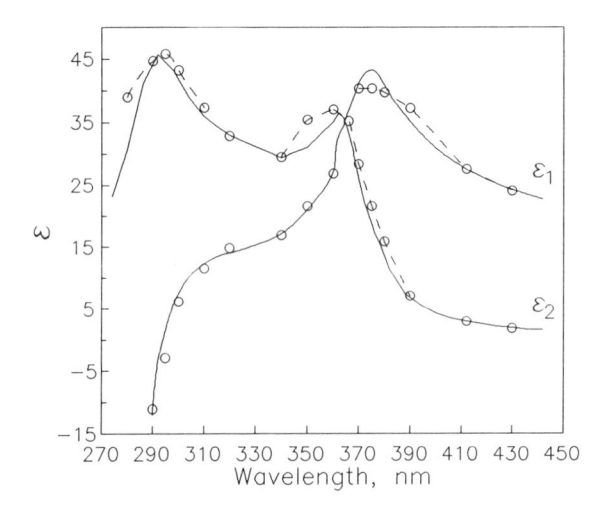

Figure 6. The dielectric functions of Si substrate before (solid line) and after (open circles) its oxidation.

of bare-Si substrate, measured after complete removal of the thermal oxide films, are also given by solid lines. As is seen, for strained Si substrate broadening of the characteristic peaks is observed. We suggest that this effect is due to the structural strains in the near Si surface region beneath SiO_2/Si interface, which results in a change of the Si optical constants in the surface region. As the penetration depth of light in Si in uv region is about 5 nm, this strained Si sublayer thickness should be at least of the same order. Direct evidence for the existence of such Si subsurface layer with expanded Si-Si bond lengths close to the interface has been found by means of nonlinear optical techniques [29,33].

The same kind of experiments carried out with oxides grown in hydrogen-enriched dry O_2 ambient at 850°C, showed an increase of the SiO_2/Si interface thickness up to 1.3 nm, whereas the atomic fraction of Si decreased to x=0.4. The influence of hydrogen on the interface thickness has been investigated by Dawson et al. [34]. In their experiments a hydrogen containing ambient led to an interface thickness of about 1.8 nm. Until now in the literature there is no explanation for this effect. It could be suggested that the Si surface microroughness is responsible for the interface thickening, but in order to prove this, additional investigations are necessary.

4. Conclusions

The results reported in this paper give some insight into the stress behavior in thermally grown SiO_2 at different temperatures and oxidizing ambient.

It has been shown that in SiO_2 films, grown in dry ambient, the stress level rapidly increases toward the SiO_2/Si interface, where it reaches the maximum value of $\sim 2 \times 10^9$ Nm^{-2}. This may cause problems in realization of up-to date devices with ultrathin oxide films.

The experiments with hydrogen-enriched oxygen ambient have shown that in contrast to wet oxidation, introduction of hydrogen species from the oxidizing ambient into growing oxide increases the intrinsic oxide stress and changes its relaxation behavior. It is suggested that higher relaxation time is caused by the change of the oxide viscosity.

The spectroscopic ellipsometric measurements have indicated the existence of a considerably strained region in Si substrate beneath the oxide with an estimated thickness of at least 5 nm. Dry oxidation of silicon at 850°C results in about 1 nm thick SiO_2/Si interface with a physical composition of $Si_{0.6}(SiO_2)_{0.4}$. Excess hydrogen in the oxide increases this thickness to 1.3 nm and changes the physical composition.

5. References

1. Workshop on oxidation processes (1987), *Phil. Mag.* B **55**(2) pp. 113-311. and **55**(6) pp. 631-763.
2. Mott, N.F. Rigo, S. Rochet, F. and Stoneham, A.M. (1989) Oxidation of silicon, *Phil. Mag.* B **60** 189-212.
3. Irene,E.A. (1987) New results on low-temperature thermal oxidation, *Phil. Mag.* B **55** 131-145.

4. Gusev, E.P. Lu, H.C. Gustafsson, T. Garfunkel, L. (1995) Growth mechanism of thin silicon oxide films on (100)Si studied by medium-energy ion scattering, *Phys. Rev.* B **52** 1759-1775. and references therein.
5. Landsberger, L.M. and Tiller, W.A. (1990) Two-step oxidation expeiments to determine structural and thermal history effects in thermally-grown SiO_2 films on Si, *J. Electrochem. Soc.* **137** 2825-2836.
6. Kouvatsos, D. Huang, J.G. and Jaccodine J.R. (1991) Fluorine-enhanced oxidation of silicon, *J. Electrochem. Soc.* **138** 1752-1755.
7. Leroy, B. (1987) Stresses and silicon interstitials during the oxidation of a silicon substrate, *Phil. Mag.* B **55** 159-199. and references therein.
8. Alexandrova, S., Szekeres, A. and Christova, K. (1988) Stress in silicon dioxide films, *Phil. Mag. Lett.* **58** 33-36.
9. Szekeres, A. and Danesh, P. (1996) Mechanical stress in SiO_2/Si structures formed by thermal oxidation of amorphous and crystalline silicon, *Semicond, Sci. Technol.* **11** 1225-1230.
10. Szekeres, A. Christova, K. and Paneva, A. (1992) Stress-induced refractive index variation in dry SiO_2, *Phil. Mag.* B **65** 961-966.
11. Danesh, P. and Szekeres, A.(1995) Electrical properties of hydrogen-rich Si/SiO_2 structures, *J. Non-Crystal. Solids*, **187** 270-272.
12. Alexandrova, S. Szekeres, A. and Koprinarova, J. (1989) The role of stress on silicon dry oxidation kinetics, *Semicond, Sci. Technol.* **4** 876-878.
13. Paneva, A. and Szekeres, A. (1993) Ellipsometric approach for evaluation of optical parameters in thin multileyer structures, *Surf. Interface Anal.* **20** 290-294.
14. Beyer, W. and Wagner, H. (1982) Determination of the hydrogen diffusion coefficient in a-Si:H from hydrogen effusion experiments, **53** 8745-8749.
15. Kobeda, E. and Irene, E.A. (1989) In situ stress measurements during thermal oxidation of silicon, *J. Vac. Sci. Technol.* B **7** 163-166.
16. Fargeix, A. and Ghibaudo, G. (1984) Densification of thermal SiO_2 due to intrinsic oxidation stressing, *J. Phys. D: Appl. Phys.* **17** 2331-2336.
17. Aspnes, E.S. and Theeten, J.B. (1980) Spectroscopic analysis of the interface between Si and its thermally grown oxide, *J. Electrochem. Soc.* **127** 1359-1365. and references therein.
18. Bruckner, R. (1970) Properties and structures of vitreous silica I. and II., *J. Non-Crystal. Solids.* **5** 123-175. and 177-216.
19. Kobeda, E. and Irene, E.A. (1987) Intrinsic SiO_2 film stress measurements on thermally oxidized Si, *J. Vac. Sci. Technol.* B **5** 15-19.
20. EerNisse, E.P. (1977) Viscous flow of thermal SiO_2, *Appl. Phys. Lett.* **30** 290-293.
21. EerNisse, E.P. (1979) Stress in thermal SiO_2 during growth, *Appl. Phys. Lett.* **35** 8-10.
22. Mack, L.M. Reisman,A and Bhattachacharya (1989) Stress measurements of thermally grown thin oxides on (100)Si substrates, *J. Electrochem. Soc.* **136** 3433-3437.
23. Hagon, J.P. Stoneham, A.M. and Jaros, M. (1987) Transport processes in silicon oxidation II Wet oxidation, *Phil. Mag.* B **55** 225-235.
24. Kuroda, T. and Iwakuro, H. (1993) Modification of silicon dioxide by hydrogen and deuterium plasmas at room temperature, *Jpn. J. Appl. Phys.* **32** L1273-L1276.
25. Gale, R. Feigl, F.J. Magee, C.W. and Young, D.R. (1983) Hydrogen migration under avalanche injection of electrons in Si metal-oxide-semiconductor capacitors, *J. Appl. Phys.* **54** 6938.
26. Danesh, P. Szekeres, A. and Kaschieva, S. (1995) Oxidation of a-Si:H (Si/SiO_2 interface properties), *Solid-State Electronics* **38** 1179-1182.
27. Fitch, J.T. Bjokman, C.H Lucovsky, G Pollak, F. H. and Yin, X (1989) Intrinsic stress and stress gradients at the SiO_2/Si interface in structures prepared by thermal oxidation of Si and subjected to rapid thermal annealing, *J. Vac. Sci. Technol.* B **7** 775-781.
28. Mrstik, B.J. Revesz, A.G., Ancona, M. Hughes, H.L. (1987) Structural and strain-related effects during growth of SiO_2 films on silicon, *J. Electrochem. Soc.* **134** 2020-2026.
29. Govorkov, S.V. Emel'yanov, V.I. Koroteev, N.I. Petrov, G.I. Shumay I.L. and Yakovlev, V.V. (1989) Inhomogeneous deformation of silicon surface layers probed by second-harmonic generation in reflection, *J. Opt. Soc. Am.* B **6** 1117-1124.
30. Nguyen, N.V. Chandler-Hotowitz, D. Amirtharaj, P.M. and Pellegrino, J.G. (1994) Spectroscopic ellipsometry determination of the properties of the thin underlying strained Si layer and the roughness at SiO_2/Si interface, *Appl. Phys. Lett.* **64** 2688-2690.
31. Jellison, Jr., G.E. (1991) Examination of thin SiO_2 films on Si using spectroscopic polarization modulation ellipsometry, *J. Appl. Phys.* **69** 7627-7634.

32. Irene, E.A. (1983) Applications of spectroscopic ellipsometry to microelectronics, *Thin Solid Films* **233** 96-111.
33. Daum, W. Krause, H.-J. Reichel, U. and Ibach, H. (1993) Identification of strained silicon layers at Si-SiO$_2$ interfaces and clean Si surfaces by nonlinear optical spectroscopy, *Phys. Rev. Lett.* **71** 1234-1237.
34. Dawson, J.L. Krisch, K. Evans-Lutterodt, K.W. Tang, M.-T. Manchanda, L. Green,M.L. Brasen, D. Higashi, G.S. and Boone, T. (1995) Kinetic smoothening: Growth thickness dependence of the interface width of the Si(001)/SiO$_2$ interface, *J. Appl. Phys.* **77** 4746-4749.

MODELLING THE OXIDE AND THE OXIDATION PROCESS
Can silicon oxidation be solved?

A M Stoneham
*Centre for Materials Research, Department of Physics and Astronomy
University College London
Gower Street, London WC1E 6 BT, UK*
a.stoneham@ucl.ac.uk

C J Sofield
*AEA Technology
Harwell Laboratory, Didcot, Oxon OX11 0RA, UK*
carl.sofield@aeat.co.uk

Abstract

The moves to miniaturisation and hence to thinner oxides (or successor materials), with the increased power of modelling, and with new information available from techniques like scanning probe methods, suggest it might be timely to face a major challenge. Is it possible to define what the best oxide would be like, to estimate its performance (especially its failure), and to predict the conditions which should lead to that optimum oxide? Clearly, for this to be achieved, it is essential to understand the key processes and to make accurate (but not necessarily first principles) quantitative predictions, as well as chosen experiments. This paper discusses some of the key questions and ideas: those of the growth processes, including the basic mechanisms (which cannot be Deal-Grove in character), how the observed layer by layer growth can be compatible with the observed growth at terraces, and the nature of the key degradation processes, like charge localisation and energy localisation.

1. Introduction

There are two broad questions which provide the context for silicon oxidation studies. First, *can silicon remain supreme?* The roadmaps of the semiconductor industry give a confident yes. Even in areas where new materials may become significant (self-organisation in III-Vs, organic and nitride photonic materials, II-VI nanodots, and so on) compatibility with silicon is crucial and, in some cases, silicon dioxide has new roles. The same roadmaps point to challenges. The 0.25 micron technology will need 4-10 nm gate oxide, accurate to 0.2 nm across a wafer, with negligible property variations. The technology beyond the 0.1 micron gate will need 2-6 nm oxide thickness. Operation at perhaps 3.5 V will be needed, with acceptable breakdown and wear-out behaviour. The breakdown and wearout of ultrathin oxides may be acceptable. However, the density of defects responsible for low field breakdown needs to be reduced significantly for 0.1 micron VLSI CMOS manufacture, probably by reducing the levels of particulate, metal and organic contamination. The control of charge and the noise associated with traps must be kept within defined limits. These trends, with thinner oxides and lower thermal budgets, can only be achieved with more sophisticated understanding and modelling. The basis of improved understanding comes partly from key ideas from other areas of interface science, partly from new experimental tools, such as the scanning probe microscopies (AFM, STM) and partly from the growth in the power of modelling.

The second broad question is *"Can the silicon oxidation problem be solved?"* Oxidation here includes processes involving nitrogen or other variants. What does "solved" mean? It means the ability to define what the best possible oxide would be like, the ability to estimate its performance, and the ability to define operations or processes so as to come acceptably close to that performance in practice. What would a solution be like? It would include realistic prediction of growth in the conditions intended (which will usually not be those conditions appropriate for the Deal-Grove mechanism). It would include a realistic understanding of failure mechanisms and their quantitative prediction. It would also imply a description of the "best" oxide (for example, its structure) and the routes to create that structure. Such a solution will only be possible with a major collaborative effort, replacing much of the present incremental work. One crucial development is the idea that solutions of this sort are possible.

2. General points about silicon oxidation

It is helpful to start by identifying some of the basic features of silicon oxidation. Full references are given in two recent reviews [1, 2]; here, the emphasis is on the issues.

Mismatch and stress One can create atomistic models which force the Si and its oxide to match, but it is not a comfortable match. This raises issues like interfacial stress, intermediate layers, viscous flow, and the density of defects like P_b centres.

Is it silicon <u>dioxide</u>? Is the oxide stoichiometric? Is there a reactive intermediate layer? Are there Si=Si bonds, or Si or O dangling bonds? Is the oxide amorphous, or is there some crystallinity; if so, how does the oxide relate to the many different silicon dioxide bulk phases?

What diffuses during oxidation? Isotope experiments show Si diffusion is negligible. For thick dry oxides, it is the interstitial oxygen molecule which moves. For thin oxides, it may be some oxygen ion. For wet oxidation, it is probably interstitial water. In any adequate description of silicon oxidation, it is crucial to include information from isotope experiments. Kinetic studies alone are more limited than many believe.

Regimes of silicon oxidation In the Deal-Grove regime, diffusion and interfacial processes can control the rate. These processes operate in series. To a first approximation, the diffusion constant and the interface reaction rate are both independent of oxide thickness.

We can determine when the Deal-Grove description fails by looking at the dimensionless parameter $g = - d[\log(dx/dt)] / d[\log x]$, where x is the oxide thickness at time t [3]. For diffusion control, $g = 1$; for interface control, $g = 0$; for the more general Deal-Grove situations, g lies between 0 and 1. *Experiment shows clearly that g is greater than 1 when the oxygen pressure is low or the temperature low.* Thin oxides are normally prepared under conditions for which the Deal-Grove picture fails. It is in this non-Deal-Grove regime that almost all scanning probe microscope studies and many transmission electron microscope studies have been made. It is in the non-Deal-Grove regime that low-field breakdown of capacitors occurs (see [2], fig 1). So, even if the Deal-Grove analysis remains an enormously useful reference case, our emphasis should be on alternatives.

One should not be surprised that the mechanisms are somewhat different for the thinner oxides. The problems of stoichiometry and mismatch are obvious factors. Growth itself is another factor, since oxidation at the Si/oxide interface occurs one event (involving one or perhaps two oxygen atoms) at a time. The striking features to emerge from experiment (for full references, see [2, 4]) are first that, initially, the oxide appears to grow *layer by layer*, and secondly that this initial growth occurs not at steps, but on *terraces*. Clearly, this description is simplistic. One cannot define layers unambiguously for amorphous oxides, but the picture is still broadly right for growth from a first layer to perhaps the fourth or fifth layer. The implications, we shall see, are that it is some species of oxygenic ion which moves, and that the rate of oxidation should depend strongly on the oxide thickness. We must be aware that rate-determining steps could occur at the oxide/gas interface, whether limited by the sticking probability, or by the incorporation step, in which adsorbed oxygenic species move into the network, usually interstitially. The rate should also depend on whether or not there is excitation, such as from low-energy electron beams (or even injection of electrons from the silicon) and indeed such effects are well-documented. In the presence of very low-energy electron beams, values of the g parameter (defined above) can be of order 10 (for references and analysis, see [1]).

3. Basic Analysis and Interpretation

An obvious issue concerns the nature of the oxide close to the silicon. A range of studies suggests differences between the first 1-2 nm and the next 10 nm or more of the oxide. The experiments include isotope experiments, pulsed laser atom probe analysis, the deviations from Deal-Grove behaviour, the effects of electron beams on oxidation, and various electronic structure and core-level spectroscopy measurements.

There is more than one way to describe what is seen. One picture is the *reactive layer model* [5]. The idea is that the oxide nearest to the silicon is not fully oxidised, so that species (perhaps O_2) which diffuse interstitially through the outer oxide react within this layer, prior to reaching the silicon itself. Diffusion continues by some other mechanism. The reactive layer is therefore opaque to interstitial O_2. The reaction does not have to be at the outside (stoichiometric oxide side) of the reactive layer, but this side will receive the largest flux of interstitial oxidising species. This model was first proposed to interpret isotope data, and is consistent with recent information (e.g.. [6, 7]).

A second picture emphasises the problems of matching oxide to Si. If one makes a cut through a silicon crystal and an amorphous oxide, the spacings of bonds over the two surfaces do not match. Suppose [1, 2] there is a deformable intermediate layer of thickness L and elastic constant c. For a given mismatch δx, the elastic energy per unit area is $\frac{1}{2} c (\delta x/L)^2 L$. There is an energy εL because this new phase would have a higher energy than other bulk phases. There are some tens of phases of SiO_2. A suitable value of ε can be estimated by looking at the ensemble of the data, which suggest the following. First, the lowest energy amorphous phases have an energy of about 0.25 eV per SiO_2 unit more than the lowest energy crystalline form (quartz). Secondly, the energies of these phases depend on molecular volume in much the same way as does vitreous silica under elastic pressure. Finally, there is an interfacial energy S from the two interfaces; S is lower than the interfacial energy when there is no intermediate layer. Minimising the sum of these terms yields an equilibrium value of L. Whether or not equilibrium can be achieved is another matter. The role of stress can be made very clear by looking at the oxidation of small Si particles, since the fact that the volume per Si in the oxide is perhaps twice that in the original Si means that the oxidation rate should be suppressed for smaller particles. This is indeed observed.

A third approach is to concentrate on the way the oxide deviates from a stoichiometric continuous random network. In effect, one concentrates on defects. Defects can mean differences in topology, such as different ring sizes. This can be a helpful description, but it is hard to quantify usefully for predictive purposes, even when there is some direct structural information, not least because many silicon-oxygen systems (such as silicate glasses) can equally successfully be described as close-packed oxygens with interstitial cations. However, we should recognise topological change near the Si/oxide interface as one mechanism for responding to mismatch.

Perhaps the best-known point defect is the P_b centre, the dangling bond pointing into the oxide from an interfacial Si. Experiment shows that the concentration of the P_b centres can be quite high (perhaps one percent of the interface silicons, once allowance has been made for those which have been hydrogenated [8]) and that the concentration correlates with the interfacial stress. Indeed, it may be that the P_b centres are the equivalent for this system of the misfit dislocations which compensate for mismatch in crystalline solids. A second point from hyperfine data is that there cannot be an oxygen in the oxide near to the axis of the dangling bond. By far the most successful fit of the hyperfine data suggests an Si-Si bond in the oxide above the dangling bond [9].

Even if this model is too simple, it raises the question for all interface defects, especially those with unpaired spins. As oxidation proceeds, what happens to these defects? Oxidation occurs at (or near) the Si/oxide interface. The defect may simply become embedded in the oxide, staying at a distance from the oxide/gas surface which remains constant (fig 1). Alternatively, it may react with the mobile oxidising species. Even in this case, a dangling-bond defect may continue as another dangling bond defect within the bulk oxide (this is clear if neutral oxygen molecules are the oxidising species, since they have an even number of electrons). The continuous random silica network will be defective. The places where it is defective are good candidates for charge localisation (as in charge trapping) or for energy localisation (as in electrical breakdown).

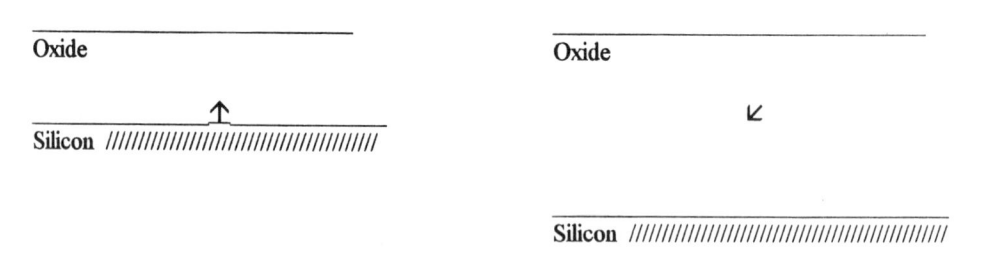

Fig 1. A defect at the silicon/oxide interface evolving as the oxide grows. The arrow shows the axis may change. If growth occurs by reactions at the silicon/oxide interface, the defect will stay at the same distance from the oxide/gas interface. The diagram has been drawn to emphasise this point.

4. Growth Processes and Alternatives to Deal-Grove

Just as the isotope experiments have forced a reconsideration of the mechanisms deduced from kinetics alone, so reconsideration is necessary following the observations of *layer by layer growth* and of *growth at terraces, not steps*. Good kinetic data remain important, and provide constraints on any model.

Some of the kinetic data are consistent with a rate which varies as exp (B/x) with thickness x [10]. This dependence can suggest the Mott-Cabrera mechanism [11], in which an interface process is enhanced by an electric field. The electric field itself is generated by tunnelling from the Si substrate to adsorbed oxygen, probably oxygen molecules; the process could be dissociative. We know from isotope experiments that Si does not move, so the interface process must be adsorbed oxygen moving into the oxide, and probably into interstitial sites in the network. For the electric field to have an effect, the injected species must be charged. This leaves problems. It is not obvious how the fields of charged adsorbed species can assist the injection of other ions on the same surface, nor is it clear how this helps with the achievement of layer-by-layer growth. The critical role of a *charged* oxidising species for the earliest stages of oxidation is a more general idea, and is more robust. It is observed in scanning tunnelling microscopy, for example, that oxidation occurs very differently if the STM tip is left in place while oxidation proceeds, rather than withdrawn and replaced only for measurement.

There is an alternative process which links more naturally to the experimental data [3, 4]. Like the Mott-Cabrera model, it needs the mobile oxidising species to be charged. Just how the species become charged is uncertain, but the known enhancement of rates when there is excitation, or when there is possible thermionic emission from the Si, suggest that charging is possible, at least for thin films. We recall that layer-by-layer growth is only observable for very thin oxides, up to perhaps four layers thick. What is needed is a mechanism which makes it more favourable than random for an incident oxygen to fill a gap in a terrace, rather than start a new plane. This can happen if there is a term in the energy which encourages the oxygens to get as close as possible to the Si substrate. Now Si has a much larger polarisability (dielectric constant) than the oxide. This leads to a significant image interaction, i.e. classical electrostatics is important (note that the interaction between two point electronic charges separated by a 4nm oxide film is about 3kT at room temperature).

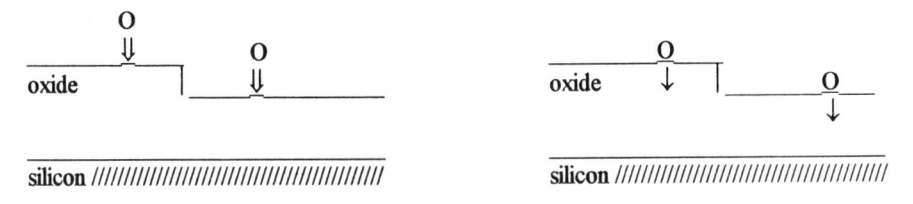

Fig 2. On the left, the sticking probabilities of the two incident ions will be different, since the one which lands closer to the silicon will have an extra attractive energy from the image interaction. On the right, the incorporation process, from an adsorbed to interstitial ion, is enhanced where the oxide is thinner. These enhancements encourage layer-by-layer growth.

We can identify at least three mechanisms (two illustrated in Fig 2) which have the right qualitative features for charged oxidising species, and both exploit the extra polarisability of the Si. In **Mechanism I**, the sticking probability is higher at the sites which allow the ion to be closer to the Si. In **Mechanism II**, it is the incorporation process, as the oxygen moved from adsorbed to interstitial, which is enhanced at sites closer to the Si. A further possibility, **Mechanism III**, which we will not discuss further here, is that the charge transfer process depends on the oxide thickness. These processes can be quantified for simplified model systems, both analytically and by Monte-Carlo calculations. These calculations need image interactions calculated for a three-layer system of substrate, oxide, and gas phase. The outcome is in good accord with what is observed. What the calculations show is that, if the dominant oxidising species is charged, layer-by-layer oxidation at terraces is to be expected for the first few layers of oxide. The results also show that the predicted kinetics (the value of the g parameter, defined above) also correspond to experiment. Further, the roughness should oscillate with thickness. Indeed, there is some evidence [12] of unexpected variations of roughness with thickness, and this might be related to the processes we are discussing. Thus our simple model to understand layer-by-layer growth on terraces (not steps) appears to explain in addition the observed kinetics and the roughness oscillations, without extra assumptions.

The roughness [2], whether that at the Si/oxide interface or the oxide/gas interface, is both a symptom of the oxidation process and a factor which influences oxide behaviour. It can be influenced in various ways, such as by contamination, or by heat treatment. Depending on the circumstances, it can increase or decrease on further oxidation. The description of roughness involves several possible length scales: the root mean square height, the correlation length across the wafer and, possibly more important when electrical properties are discussed, the curvature of the interface. Roughness has many consequences: on the local atomic arrangements in the oxide, on the growth mechanism itself, on the nature and dimensions of defects, on the possibility of easy transport channels, on electric field enhancement, and so on. Any "solution" of the oxide problem will have to include a description and model of roughening and its consequences.

Even the best current Monte-Carlo models [13], which do produce an amorphous oxide, consider only a flat surface, and exclude the known defects. Moving to rough interfaces, and to phenomena like graphoepitaxy, introduces new features [14]. Whilst these and other generalisations are demanding, they are by no means impossible. They do need a substantial effort in modelling, and will surely need to exploit the significant recent developments in electronic structure calculations.

5. Electrical Breakdown

Electrical strength is usually described by the field which can be sustained in certain standard conditions. In 1986, 8.5 MV/cm was the standard; by 1995, 13.5 MV/cm was needed. These are, of course, figures for an area of wafer which will have good and bad regions, for breakdown is basically a local phenomenon. The description in terms of a field, or of a charge to breakdown, is practical but simplistic. What is still needed is a route from a known oxide structure (if need be, with every atomic position defined, and the roughness fully specified) to a propensity to failure. This sounds impossibly ambitious, even as an exercise in modelling. Yet this route could be mapped out in the next few years. One has only to look at the parallel area of fracture studies, where the there has been a revolution from the developments in understanding dislocation behaviour and from the major changes in software and hardware for modelling. Moreover, developments in the femtosecond spectroscopy of excited states of bulk oxides are offering some of the key insights into electrical breakdown phenomena.

In the earlier sections, we have suggested the following points. First, the mechanisms of oxidation for ultra-thin oxides are not simply limited by thickness-independent diffusion and interface reaction steps. It is likely that charged oxygen species are involved; it is likely that reactions occur within the oxide network close to the Si/oxide interface; it is probable that there are different local topologies, and that there are altered local stoichiometries, at least in the vicinity of point defects. Secondly, many of the key steps can be modelled, at least in isolation. Thus, in principle, one might hope to calculate a number of representations of the oxide structure (i.e., precise atomic positions at different stages; clearly, a statistically significant number of realisations would be needed for credible predictions).

Achieving these realisations in practice would be challenging. The next stage is still harder but, again, I believe possible. How does one predict the degradation processes occurring in electrical breakdown? The inputs here are of three sorts. First, there is the recognition that the basic process of electrical breakdown is *energy localisation*, which is often (but not necessarily) preceded by *charge localisation*. It is not enough to stroe charge: for breakdown, energy must be released in an effective way (bending a paperclip stores energy, but releasing that energy is non-trivial). Energy localisation means, in particular, that it is possible to provoke damage, some process in which an atom or ion is moved for an extended period. For thin films, we must always be aware that it is the available energy which matters, and that a threshold field is deceptive. Energies to displace atoms are typically a few electron volts, even for the most efficient of processes. A field as large as 10 MV/cm provides a potential drop of only 2 V in 2 nm; even if it were to accelerate an electron without energy loss, the 2 eV electron would only cause displacements by some indirect mechanism, probably one of low efficiency.

The second input recognises that much is known about energy localisation and the ways in which atoms are displaced. It is known from the extensive femtosecond spectroscopic studies of excited electronic states of bulk silica, notably quartz, but including vitreous silicas [15]. In quartz, the exciton self-traps; essentially, the hole component localises on an oxygen, which is displaced, and the electron is localised (albeit less strongly) and a neighbouring silicon. Neither the electron nor the hole separately self-traps in quartz, although it is a delicate balance; both carriers are localised by substitutional Ge, and the extra polarisability of the Si may be sufficient to make self-trapping stable near the Si/oxide interface [3]. The hole self-traps in vitreous silica, which shows that fluctuations from site to site in oxide properties matters. A fair amount is known both experimentally and theoretically about ion displacement processes in silicas and analogous materials. For our purposes, the points are that what happens in the bulk is probably very similar to what happens in thin films, apart from certain effect of the high field in the film. that electron-hole recombination provides one means of efficient energy localisation, such that energy up to a maximum of the band gap (8-10 eV for silicas) might be available for degradation processes. Thus we should look at models in which one carrier is localised (charge localisation, perhaps hole self-trapping at a favoured site) and where another carrier recombines with it, releasing energy and displacing an ion. The ways by which ions are displaced are several: the excited state my be antibonding; ions may change their charge states; there may simply be an enhanced local vibrational amplitude, speeding normal thermal processes. It is not clear whether the initial carrier capture is intrinsic (e.g., self-trapping), associated with a statistically-rare feature of a continuous random network, or linked to some well-defined defect, such as one which developed from a P_b centre earlier in oxidation. If it were associated with a statistically-rare feature, then this would put severe demands on any predictions of the atomic structure of the oxide.

The third input comes from the successful experiments on breakdown and wearout mechanisms, including statistical analysis. At this point, we must recognise two levels of energy localisation. There is the localisation on an atomic scale, in which individual point defects are created. But there is also localisation on a larger scale, such that the defects are generated in proximity to each other, perhaps because of a region of oxide with different properties or because of an enhancement of the electric stress because of interface roughness or the defects which have already been created.

At the point defect level, the observed degradation mechanisms do appear to have parallels with those in bulk silicas. The electric field ensures carrier injection, and some process of electron-hole recombination drives a defect generation step. It is clear that injected carriers can also lose energy in generating heat, and the fraction of electrons contributing to breakdown (defect production) steps is quite small. On the larger scale, there seem to be useful analogies with the electrical behaviour of those other oxide films which show switching from one conduction state to another [16]. Thus something akin to a conducting path (filament) evolves, since defects are formed more readily in the high-field region ahead of the filament tip. Formation of electron traps and conduction via these traps leads to soft breakdown (the capacitor is not destroyed thermally). Such a filament may be destroyed in a discharge event. Clearly, the noise spectrum contains information on the numbers of filaments and each spike is a measure of charge transferred in a discharge event. Analysis of the noise spectrum [17] suggests that each such discharge event itself could suffice to discharge the available stored charge [18].

Just how the filamentary breakdown paths are established, and which defects are involved, is complex. One natural idea would be that the necessary energy localisation occurs through an excitonic process, as known from may studies in quartz. This would involve electron (or hole) trapping followed by hole (electron) capture, with the recombination energy sufficing to lead to displacement of an oxygen ion. An analysis of several studies [19] suggests that it is slow interface traps which build up and lead to filamentary breakdown. Certainly there have been suggestions [20] that hole trapping which gave electron traps and which determined the breakdown behaviour. Recent work [21] argues that the traps are E' centres, consisting of a hole trapped at an oxygen vacancy. This is an area in which detailed atomistic theory could be deployed with value.

For thicker films (more than 10nm), DiMaria et al [22] conclude that critical fields are of order 5MV/cm. These films are thicker than the 6nm above which Fowler-Nordheim injection overwhelms tunnelling contributions [23, 24]. For thin films, a critical *voltage* is needed (as remarked above), which is about 8 V, corresponding to an electron with energy of order 5 eV near the anode. Tunnelling is important for thicknesses below about 3 nm [23, 24], and sensible values of parameters (3.15 eV barrier height, effective mass 0.32 electron masses [23]) are found. These are, of course, averages over the oxide in some sense, whereas the defect events which lead to breakdown are local. The Weibull statistics of breakdown fit well with realisations from a model in which point defects build up into filaments across the oxide [23].

A series of experiments on c-Si/oxide/p-Si systems [25] showed a number of defect generation events. These experiments showed high reliability and yield for thermal oxidation in steam at lowish temperatures (650-700°C). Some defects were generated at the c-Si/oxide interface, irrespective of polarity: positive charges were created, and slow states were produced. The creation of the slow interface states is a precursor in the breakdown of thin oxides. Other events occurred at the anode (and so for different interfaces when the polarity is reversed), where electron traps were created, and electrons were indeed trapped.

DiMaria [26] concludes that hot electrons produce neutral traps at the anode, and that the critical factor is not the electrode material, but the position of the Fermi level at the anode. The generation of the neutral traps [27] is the dominant cause of the leakage currents introduced in the low-field, direct-tunnelling regime for thin oxides. The other obvious possibilities (anode hole injection, or oxide nonuniformities) are deemed unlikely. It is likely that these leakage currents are related to H-induced defects. However, Kimura and Ohmi [28] identify neutral traps (1.17 eV from the cathode conduction band, about 2 eV from the oxide conduction band) uniformly generated in the oxide which they associate with distortion of the oxide bond structure during stressing. Hydrogen appears to be important in many ways [28, 29]; for example, electrons with more than 2 eV at the Si/oxide interface appear to release H. This leaves behind a centre which can be an carrier trap. Clearly, much needs to be done to link the thin film defects to established defect spectroscopies.

Conclusions

This paper analyses the implications of some of the technological challenges [30] arising from moves to miniaturisation and to thinner oxides, and the associated scientific (and ultimately technological) opportunities. The challenges are clear. One is control, both of oxide nature and of its quality. Another challenge, underlying control, is that the key rate-determining processes are not yet identified unambiguously, and are clearly not just those of the Deal-Grove model. The opportunities are that new experiments and new approaches in modelling atomic processes may enable progress beyond incremental steps. With the increased power of modelling, and with new information available from femtosecond spectroscopies and from scanning probe methods, it may be possible to define what the best oxide would be like, estimate its performance (especially its failure), and predict the conditions which should lead to that optimum oxide. These are not easy goals, but they are no longer wholly fantasy. Certainly we shall need to understand the key growth processes. We have noted that observations like layer-by-layer growth (from core level spectroscopies) not at steps (from novel electron microscopy) together can be understood in terms of atomic processes of a particular class. This class of processes, without further assumptions, also offers an explanation of the observed (non-Deal-Grove) kinetics and the periodicities in roughness seen in atomic force microscopy. Even more, we need to understand the nature of the critical degradation processes, like charge localisation and energy localisation, and here the information from femtosecond spectroscopies may be crucial. We believe there is a growing need to understand the processes by which oxide is formed and degraded, and not merely the possible structures and properties of specific oxide films recorded in standard experiments.

References

1. Mott, N.F., Rigo, S., Rochet, F., and Stoneham, A. M., (1989), Oxidation of Silicon, *Phil Mag* **B60** 189-212.
2. Sofield, C.J., and Stoneham, A.M., (1995), Oxidation of Silicon: the VLSI gate dielectric? *Semicond Sci Tech* **10** 215-244.
3. Stoneham, A.M. and Tasker, P.W., (1987) Image charges and their influence on the growth and the nature of thin oxide films, *Phil Mag* **B55** 237-252.
4. Torres, V.J.B., Stoneham, A.M., Sofield, C.J., Harker, A.H., and Clement, C.F., (1995), Early stages of silicon oxidation, *Interface Science* **3** 131-144.
5. Stoneham, A.M., Grovenor, C.R.M., and Cerezo, A., (1987), Oxidation and the structure of the silicon/silicon oxide interface, *Phil Mag* **B55** 201-210.
6. Gusev, E.P., Lu, H.C., Gustafson, T., and Grafunkel, E., (1996) The initial oxidation of silicon: new ion-scattering results in the ultra-thin regime, *Appl Surf Sci* 329-334.
7. Devine, R.A.B., (1996) Structural nature of the Si/SiO2 interface through infrared spectroscopy, *Appl Phys Lett* **68** 477-491.
8. Stesmans, A., (1993) Relationship between Stress and Dangling Bond Generation at the (111)Si/SiO$_2$ Interface, *Phys Rev Lett* **70** 1723-1726.
9. Ong, C.K., Stoneham, A.M., and Harker, A.H., (1993), Environment of the P_b centre at the Si(111)/SiO$_2$ interface, *Interface Science* **1** 139-146.
10. Kamohara, S. and Kamigaki, Y., (1991) Activation energy enhancement during initial silicon oxide growth in dry oxygen, *Journal of Applied Physics* **69**, 7871-7875. See also Massoud, H.Z., Plummer, J.D. and Irene, E.A., (1985), *J Electrochem Soc* **132** 2685, 2693.

11. Cabrera, N., and Mott, N.F., (1948), Theory of the Oxidation of Metals, *Rep Prog Phys* **12** 163-184.
12. Niwa, M., Kouzaki, T., Okada, K., Udagawa, M., and Sinclair, R., (1994), Atomic-order planarization of Ultrathin SiO$_2$/Si(100) Interfaces, *Jap J Appl Phys* **33** 388-394.
13. Carniato, S., Boureau, G., and Harding, J.H., (1997), Modelling oxygen vacancies at the Si(100)-SiO$_2$ interface, *Phil Mag* **A75** 1435-1445.
14. Torres, V.J.B., and Stoneham, A.M., 1997, in preparation.
15. Itoh, N. and Stoneham, A.M., (1997) Transient Defects and Electronic Excitation, to appear in *Structure and Imperfections in Amorphous and Crystalline SiO$_2$* (R Devine, ed) J Wiley, New York.
16. Dearnaley, G., Stoneham, A.M., and Morgan, D.V., (1970) Electrical phenomena in amorphous oxide films, *Rep Prog Phys* **33** 1129-1192.
17. Depas, M., Nigam, T., and Heyns, M.M., (1996) Definition of dielectric breakdown for ultra-thin (< 2 nm) Gate Oxides, *Sol State Electronics* in press. See also Depas, M., Nigam, T, and Heyns, M.M. (1996) Soft breakdown of Ultra-Thin Gate Oxide layers, *IEEE Trans Electron Devices* **43** 1499-1502.
18. Harker, A.H., (1996), private communication.
19. Depas, M., Vermeire, B., and Heyns, M.M., (1996) Breakdown and defect generation in ultrathin gate oxide, *J Appl Phys* **80** 382.
20. Chen, I.C., Holland, S.E., and Hu, C (1986) *IEEE Trans Electron Device Lett* **EDL-37** 146.
21. Conley, J.F., Lenahan, P. M., Lelis, A.J., and Oldham, T.R. (1995) *Appl Phys Lett* **67** 2179.
22. DiMaria D.J., Cartier, E., and Buchanan, D.A., (1996) Anode hole injection and trapping in silicon dioxide, *J Appl Phys* **80** 304-317.
23. Degraeve, R., Groseneken, G., Bellens, R., Depas, M., and Maes, H., 1996, A consistent model for the thickness dependence of intrinsic breakdown in ultra-thin oxides, to be published.
24. Wolters, D.R., and Zegers-van Duijnhoven, (1996), Tunnelling in thin SiO$_2$, Phil Trans Roy Soc **A354** 2327-2350.
25. Heyns, M. M. and von Schwerin, A. (1991), Charge Trapping and degradation of thin dielectric layers, in *Insulating Films on Semiconductors*, W Eccleston and M J Uren (eds), Institute of Physics Publishing, Bristol, UK.
26. DiMaria D.J., (1996) Explanation for the polarity dependence of breakdown in ultrathin oxide films, *Applied Physics Letters* **68** 3004-3006.
27. DiMaria D.J., and Cartier, E., (1995) Mechanism for stress-induced leakage currents in thin silicon dioxide films, *J Appl Phys* **78** 3883-3894.
28. Stesmans, A., (1996), Revised interpretation of hydrogen passivation of P$_b$ centers, *Appl Phys Lett* **68** 2723-2726.
29. Stathis, J.H., and Cartier, E., (1994) Atomic hydrogen reactions with P$_b$ centers at the (100) Si/SiO$_2$ interface, Phys Rev Letters **72** 2745-2748.
30. Krautschneider, W.H., Kohlhase, A., and Terletzki, H (1997) Scaling down and Reliability Problems of Gigabit CMOS Circuits *Microelectron Reliab* **37** 19-37.

CORE-LEVEL SHIFTS IN Si(001)-SiO$_2$ SYSTEMS: THE VALUE OF FIRST-PRINCIPLE INVESTIGATIONS

ALFREDO PASQUARELLO[1,2,3], MARK S. HYBERTSEN[3],
G.-M. RIGNANESE[1,4] AND ROBERTO CAR[1,2]
[1] *Institut Romand de Recherche Numérique en Physique des Matériaux (IRRMA),*
Ecublens, CH-1015 Lausanne, Switzerland
[2] *Department of Condensed Matter Physics,*
University of Geneva, CH-1211 Geneva, Switzerland,
[3] *Bell Laboratories, Lucent Technologies,*
700 Mountain Avenue, Murray Hill, New Jersey 07974, USA,
[4] *Unité de Physico-Chimie et de Physique des Matériaux,*
Université Catholique de Louvain,
1 Place Croix du Sud, B-1348 Louvain-la-Neuve, Belgium.

Abstract. A first-principle approach allows the study of relaxed structural models for surfaces and interfaces. This is a powerful tool for the study of the local bonding. The utility of the first-principle theory is substantially extended through the calculation of core-level shifts. Such results can be used in conjunction with measured photoemission spectra to make progress in understanding the local atomic structure at interfaces. We review the quantitative comparison of the calculated core level shifts with experiment for a series of molecules. We then describe results for the Si(001)-SiO2 interface.

1. Introduction

Core-level photoemission spectroscopy stands out as one of the most successful tools for providing direct information on the local bonding environment of specific atoms at surfaces and interfaces. The interpretation of photoemission spectra is often straightforward, and comparison with molecular counterparts or simple charge-transfer models appear to be sufficient to extract the relevant structural information. However, in some cases core-hole relaxation effects are sizeable and cannot be neglected. Furthermore, at surfaces and interfaces, the bonding environments may differ significantly from those in molecular counterparts, making quantitative predictions of core-level shifts as deduced from molecular analogs very difficult. Because of these reasons, the availability of a reliable theoretical framework could represent an invaluable tool for the assignment of unidentified spectral features.

First-principle approaches in which the electronic structure is described within density functional theory (DFT) could serve this purpose [1, 2]. Such approaches have several advantages. Because the forces acting on the ions can also be calculated within these approaches [3, 4], a structural optimization can be performed, providing in this way reasonable configurational models even when the actual structures are unknown. Then, core-level shifts, including core-hole relaxation effects, can be evaluated consistently within the same theoretical framework. An additional advantage of this approach derives from the fact that core-shifts are sensitively affected only by modifications of the electronic structure in the valence band. Such changes can accurately be described within a pseudopotential approach in which only valence electrons are treated explicitly [5]. This in turn allows the application of this approach to relatively large systems containing up to a hundred atoms given the current computational means.

The purpose of this work is to assess the current status of this theoretical method by describing its merits and its limitations in a series of applications. These range from small molecules to more complex systems such as surfaces and interfaces. More specifically, we focus here on core-level shifts induced by adsorbates forming Si-O bonds on Si(001) surfaces and on the relation between core-level shifts and bonding environments at Si(001)-SiO$_2$ interfaces.

This paper is organized as follows. A brief description of our approach for calculating core-level shifts is given in Section 2. In Section 3, a comparison between calculated and measured core-level shifts in a series of molecules provides a quantitative measure of the degree of accuracy that can be achieved with this approach. In Section 4, we calculate the shifts induced by dissociated water and spherosiloxane molecules on Si(001) surfaces. In Section 5, calculated Si $2p$ core-level shifts for all the different oxidation states of silicon at Si(001)-SiO$_2$ interfaces are presented. Finally, in Section 6, we apply our approach to nitrided Si(001)-SiO$_2$ interfaces and establish a relation between the N $1s$ core-level shifts and the bonding configurations of the incorporated N atoms.

2. Theoretical approach

The atomic relaxations and the core-shift calculations were performed within the local density approximation (LDA) to DFT [6]. Our approach accounts for the electronic structure and consistently provides atomic forces [3, 4]. Only valence electrons are explicitly considered using pseudopotentials (PPs) to account for core-valence interactions. The core-level shifts can be calculated both within the initial state approximation and including core-hole relaxation [7, 2]. The difficulty of using a PP approach is overcome as follows. Initial-state shifts are obtained in first order perturbation theory by evaluating the expectation value of the local potential on the atomic core-orbital, while final state effects are included by considering differences in total energies [5]. Following these procedures *relative* core-level shifts can be determined. Throughout this paper, we only considered vertical excitations, i.e. the molecular structure in the final state is the same as in the initial state. A detailed description of our approach is given in Refs. [7, 2].

3. Small molecules

The accuracy of LDA-DFT for the calculation of core-level shifts [1, 2, 8] can be examined by considering a series of small molecules for which experimental data are available [9]. We considered small molecules containing Si and N atoms. After full relaxation of the molecular structures, we calculated Si $2p$ and N $1s$ core-level shifts both in the initial-state approximation and including core-hole relaxation. The calculated shifts are reported in Table 1 and Table 2, together with the experimental results. The technical aspects of these calculations are given in Refs. [2, 8].

Overall agreement between theory and experiment is very good. In Table 1, the linear additivity of the shifts observed in the experimental data is also found in the calculated values. Another significant case is that of $Si(CH_3)_4$. On the grounds of simple electronegativity arguments a negative Si $2p$ shift is expected. However, the experimental shift is positive and theory correctly reproduces it. For molecules in which the neighboring atoms to the excited silicon atoms are hydrogen, silicon, chlorine or carbon atoms the error is smaller than a few tenths of an eV.

The necessity of considering core-hole relaxation effects appears more evidently in the case of the N $1s$ shifts given in Table 2. The agreement with experiment of these shifts is noticeably improved when final states are accounted for. In the N series, theoretical and experimental shifts differ by less than 0.3 eV over a broad range of shifts.

This level of accuracy is not found for the Si $2p$ shifts of molecules containing oxygen atoms: $O(SiH_3)_2$, $O(SiCl_3)_2$ and $Si(OCH_3)_4$. Although the

TABLE 1. Relative Si 2p initial-state and full shifts for a series of molecules. Experiment from [9]. Shifts are in eV. Negative shifts indicate higher binding energies.

	Initial State	Full	Expt.
SiH_4	0.00	0.00	0.00
Si_2H_6	−0.03	0.47	0.42
$SiClH_3$	−0.88	−0.95	−0.83
$SiCl_3H$	−2.28	−2.45	−2.16
$SiCl_4$	−2.89	−3.13	−3.11
$O(SiH_3)_2$	−0.42	−0.79	−0.53
$O(SiCl_3)_2$	−2.75	−3.24	−2.86
$Si(CH_3)_4$	1.00	1.21	1.32
SiH_3CH_3	0.31	0.40	0.46
$SiCl_3CH_3$	−1.97	−2.10	−1.97
SiH_2ClCH_3	−0.55	−0.54	−0.48
$Si(OCH_3)_4$	−0.15	−1.67	−0.42

TABLE 2. Same as in Table 1, but for N 1s core-level shifts.

	Initial State	Full	Expt.
NH_3	0.00	0.00	0.0
NH_2CH_3	0.06	0.57	0.5
$NH(CH_3)_2$	−0.08	0.58	0.7
$N(CH_3)_3$	−0.33	0.61	0.8
NH_2COH	−1.73	−1.13	−0.8
NO_2	−6.91	−7.24	−7.3
N_2O (N*NO)	−3.46	−3.30	−3.1
(NN*O)	−6.77	−7.08	−7.0
ClNO	−6.53	−5.68	−5.8

calculated shifts for these three molecules still show the correct qualitative trend, a quantitative discrepancy is observed. Calculated Si 2p core-shifts of silicon atoms which have oxygen neighbors, systematically overestimate the shifts towards larger binding energies relative to experiment. Using linear additivity arguments, one can estimate the excess negative shift to be about −0.3 eV per oxygen neighbor.

These results together with similar results obtained for C 1s shifts [1] confirm that LDA-DFT is overall quite successful in reproducing the observed core-level shifts. In the particular case of Si 2p core-level shifts in the presence of oxygen neighbors, the difference with experiment has been demonstrated to be systematic. This therefore allows useful conclusions even in these cases as we illustrate in the following sections.

4. Adsorbates on Si(001) interfaces

In this section, we consider dissociated water and spherosiloxane molecules on Si(001) surfaces, which are two different adsorbates forming Si-O bonds with the surface. The former gives rise to a well understood microscopic structure and therefore allows a stringent comparison between theory and experiment [10]. The microscopic structure resulting from the adsorption of spherosiloxane molecules can only be hypothesized by comparison with similar chemical reactions [11, 12]. In this case, the application of our approach allows one to critically examine the proposed structural model by comparison with experimental spectra [13].

4.1. DISSOCIATED WATER MOLECULES ON Si(001)

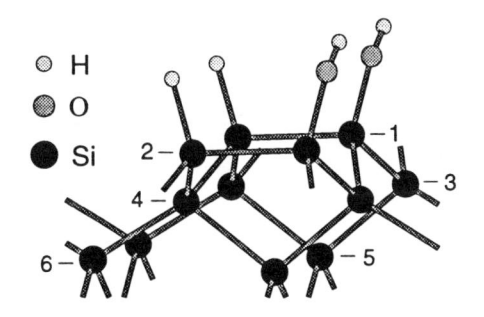

Figure 1. Ball and stick model of the relaxed positions resulting from dissociative water adsorption on Si(001). Distinct Si positions are labelled.

It is by now well established that H_2O is adsorbed dissociatively on Si(001)2×1 by saturating the free dangling bonds of the clean surface with −OH and −H groups [14]. In Fig. 1, we show the structure that results from our structural optimization process [10]. The Si 2p shift of the Si atom (1) directly bonded to the −OH group is found to be −1.1 eV with respect to the bulk value [10]. This compares rather well with the measured shifts of −1.0 eV (Ref. [15]) and −0.9 eV (Ref. [16]). The slight overestimation should be related to the same systematic effect as observed in the molecules. Core-hole relaxation is crucial to obtain this agreement, since the initial state shift calculated for the Si atom (1) is only −0.5 eV. The full shifts of the other Si atoms [(2) to (6)] are all within 0.2 eV from the bulk value. This agreement further strengthens the reliability of our theoretical approach.

4.2. DISSOCIATED SPHEROSILOXANE MOLECULES ON Si(001)

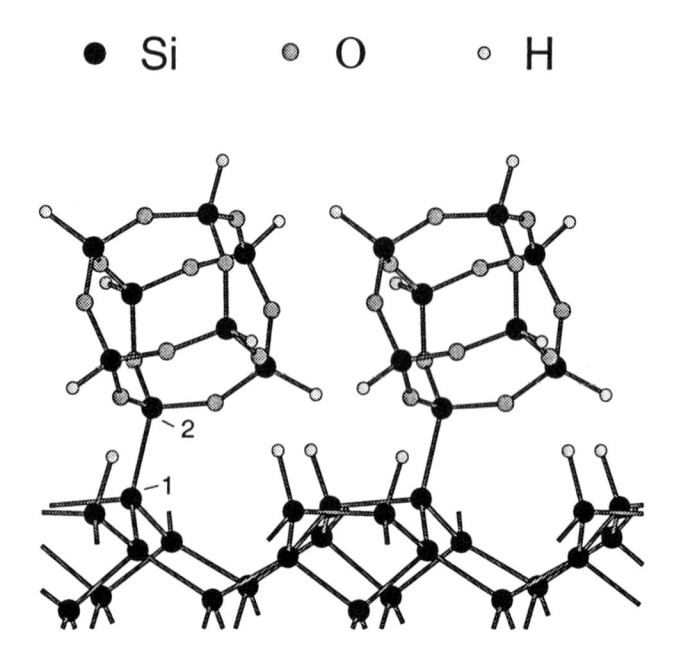

Figure 2. Ball and stick model of the relaxed surface structure of chemisorbed $H_8Si_8O_{12}$ clusters on Si(001).

In Fig. 2, we show the relaxed configuration corresponding to a spherosiloxane ($H_8Si_8O_{12}$) molecule attached at the surface through dissociation of one of its Si-H bonds [13]. This structural model was presumed for the experimental assignment of peaks in the measured photoemission spectrum corresponding to the new interface structure resulting from spherosiloxane clusters and Si(001) [11, 12]. In particular, in the experiment Si $2p$ peaks were observed at −1.04 eV and at −2.19 eV [11]. These peaks were attributed to the substrate silicon atom (1) and the cluster atom (2), respectively. Such assignments radically contrast with the more traditional interpretation which attributes a shift of approximately −1 eV for every nearest-neighbor oxygen atom [16–18]. Si atoms (1) and (2) would be in a Si^0 and Si^{+3} oxidation state with expected shifts of ≈ 0 and ≈ -3 eV, respectively.

[Figure: Calculated Si 2p core-level spectrum, intensity (arb.) vs core-level shift (eV) from -5 to 0]

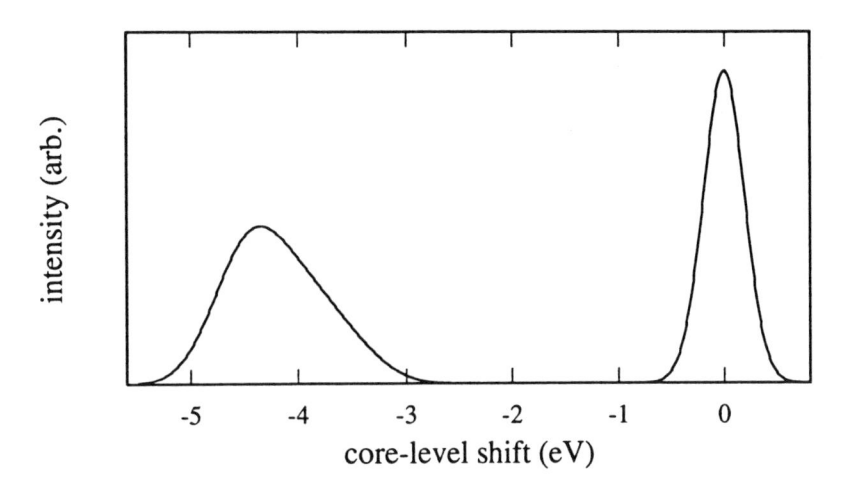

Figure 3. Calculated Si 2p core-level spectrum for chemisorbed $H_8Si_8O_{12}$ clusters on Si(001), corresponding to the model structure shown in Fig. 2. The relative intensities of the two peaks are arbitrary.

The calculated Si 2p shifts [13] for the model in Fig. 2 are represented schematically in Fig. 3, in which a Gaussian broadening was introduced to account for dynamical effects: a FWHM of 0.4 eV for Si^0 and 0.7 eV for Si^{+3}. All the Si^0 atoms merge in a single peak which coincides with the bulk reference peak. The Si^{+3} atoms belonging to the cluster give rise to an asymmetric peak with a maximum at −4.3 eV and with a FWHM of 1.1 eV. This peak agrees well with the experimental spectrum, where at high binding energies a well defined peak is found at −3.64 eV with a FWHM of 1.2 eV. The overestimation is again attributed to the systematic effect oberved in Si-O systems. The asymmetric shape of the high-energy peak results from core-hole relaxation which depends on distance to the screening silicon substrate. This calculation shows that the presumed model cannot account for the experimentally observed peaks at −1.04 eV and at −2.19 eV, and suggest that the actual interface formed by spherosiloxane molecules and Si(001) is more complex.

5. The Si(001)-SiO$_2$ interface

Photoemission spectra indicate that the three intermediate oxidation states of silicon occur in roughly comparable amounts at the Si(001)-SiO$_2$ interface [16–18]. We generated several structural models of this interface with the minimal transition region required to be consistent with the photoemission data [7, 19, 20]. Further constraints come from electrical measurements which indicate the presence of an extremely low density of defects

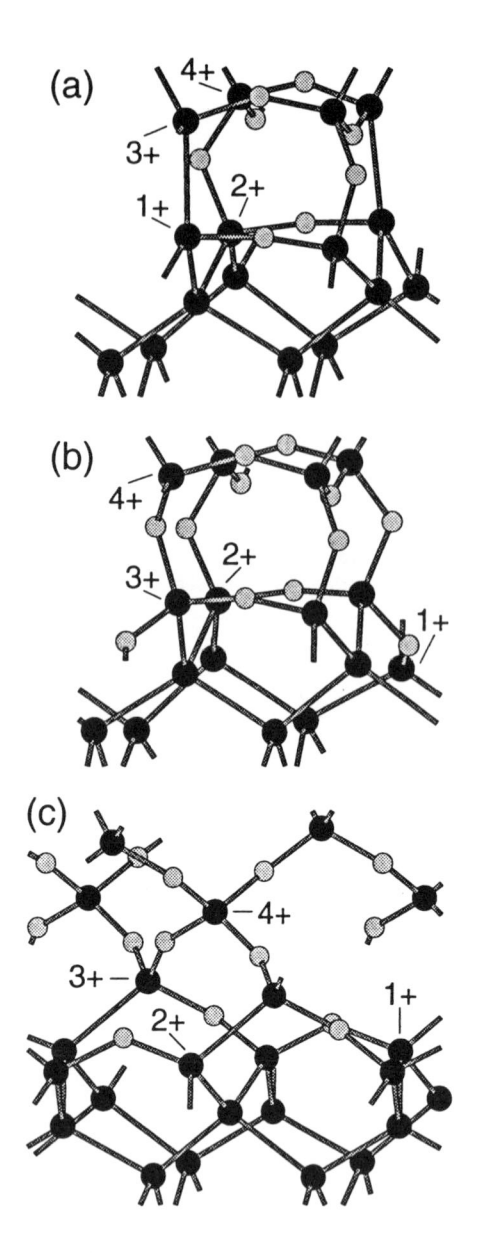

Figure 4. Ball and stick models showing the relaxed positions of several interface structures: (a) Si-Si bond pointing into the oxide, (b) O in the backbond connecting an interface Si atom to the substrate, and (c) β-cristobalite with the construction proposed in Ref. [21] at the interface. The formal Si partial oxidation states are indicated.

states. We therefore did not allow for any unsaturated dangling bond in our interface models. The models are obtained by attaching a crystalline form of SiO$_2$, such as tridymite or β-cristobalite, to Si(001). The initial structures obtained in this way are then allowed to relax fully within our theory (see Fig. 4).

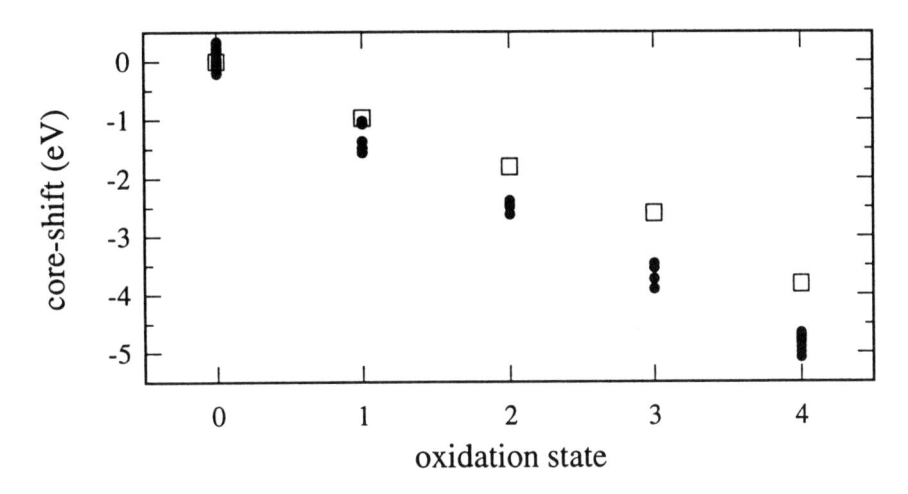

Figure 5. Si 2p core-level shifts at the Si(001)-SiO$_2$ interface as a function of oxidation state: models (solid circles) and experiment (open squares) from Ref. [18].

In Fig. 5, calculated Si 2p core-level shifts are plotted versus the oxidation state of the silicon atoms to which they belong, and compared to experimental values [18]. Apart from the usual systematic shift to higher binding energies the agreement with experiment is rather good. The calculated Si 2p shifts show a linear dependence on the number of first nearest neighbor oxygen atoms. This result confirms the interpretation based on charge-transfer models. In fact, core-hole relaxation effects account for about 50% of the shifts shown in Fig. 5, but their correction also scales linearly with the number of O neighbors. This increase of the final state effect with O neighbors can be understood as caused by a reduced valence screening in oxidized Si atoms. The large role of core-hole relaxation effects has also been supported experimentally through a careful comparison of photoemission and Auger spectra [22].

6. Nitrided Si(001)-SiO$_2$ interfaces

The incorporation of a low concentration of nitrogen atoms at the Si(001)-SiO$_2$ interface has promising applications in the electronic device industry [23, 24, 25]. Further improvement of the dielectric properties of such systems also relies on the availability of detailed microscopic information on the situation of the incorporated N atoms. In this section, we charaterize these N atoms by establishing a correspondence between their bonding environment and the N 1s shifts measured in photoemission experiments.

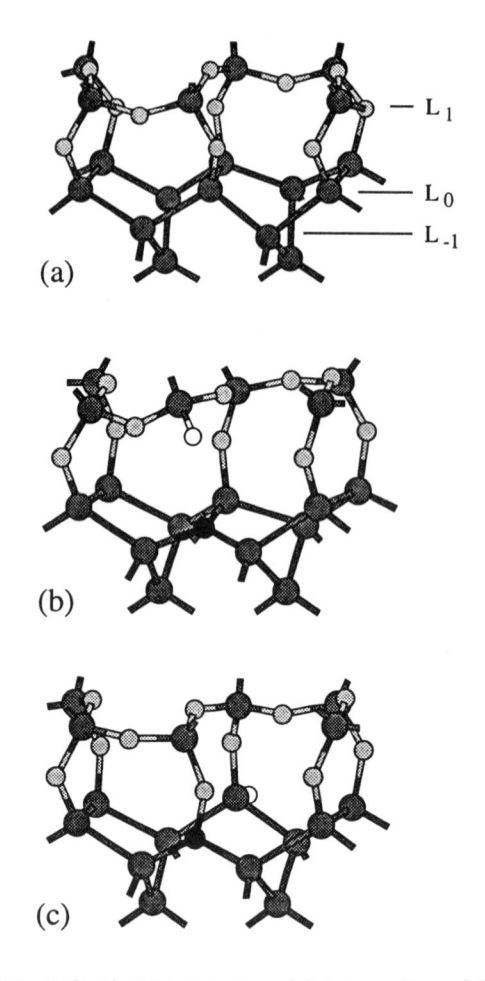

Figure 6. Models of the Si(001)-SiO$_2$ interface (a) before (from [7]) and after N (black atoms) incorporation, in a (b) N–Si$_3$ and (c) in a N-Si$_2$O configuration. L$_{-1}$, L$_0$ and L$_1$ label the Si layers. H atoms (white) saturate residual dangling bonds.

The experimental N 1s spectra show a broad principal peak (FWHM=$\sim$ 1.5 eV), approximately at the same energy as in bulk Si$_3$N$_4$, which appears to shift to larger binding energies for samples of increasing oxide thickness [24, 26, 27]. A deeper analysis of the shape of this peak, has led to the recognition of two components [28, 29]. The one closest to the bulk Si$_3$N$_4$ energy arises from the interfacial region and is generally attributed to N atoms bonded to three Si atoms (N-Si$_3$) [28, 29]. The second component, which is shifted by Δ=0.85 eV to larger binding energies [28], mainly results from N atoms distributed throughout the oxide [28, 29]. The contribution of the latter component increases for thick oxides, accounting for the observed shift of the principal XPS feature. Note, however, that a general consensus has not yet been reached neither on such a decomposition nor on its interpretation. In the presence of this uncertain experimental situation, we use our first-principle approach to explore some plausible bonding configurations.

Using as a starting point of our nitrided interface study one of the Si(001)-SiO$_2$ interface models obtained previously [see Fig. 6(a)] [7], we generated several nitrided interface models [two of them are shown in Figs. 6(b) and (c)] containing N atoms in different bonding configurations and at varying distances from the interface plane [8].

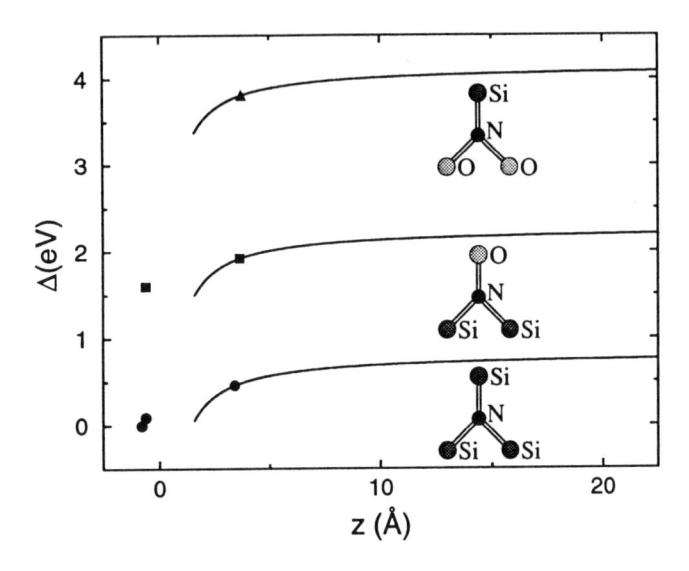

Figure 7. N 1s core-level shifts Δ at Si(001)-SiO$_2$ interfaces calculated for N-Si$_3$ (circle), N-Si$_2$O (square), and N-SiO$_2$ (triangle) configurations at different distances z from the interface. Continuous lines result from classical electrostatics.

N 1s shifts for these models were calculated including core-hole relaxation effects and are given in Fig. 7 as a function of the relaxed position z of the N atom with respect to the interface plane [8]. We took as a reference the shift of a N-Si$_3$ configuration in which the N atom is located most deeply in the Si substrate. From Fig. 7, it is evident that the shifts are strongly affected by first nearest neighbors. The presence of an O nearest neighbor yields shifts to higher binding energies of $\Delta=1.5$ eV with respect to corresponding N-Si$_3$ configurations. A second oxygen nearest neighbor brings this shift to $\Delta=3.5$ eV. The large separation between the calculated shifts of N-Si$_3$ and N-Si$_2$O configurations virtually rules out the possibility that both configurations contribute to the principal XPS peak.

With increasing distance from the interface the silicon substrate is less effective in screening the core-hole and an increase of the binding energy is observed (Fig. 7) [7]. An extrapolation obtained within classical electrostatics [7] illustrates this effect at large distances in Fig. 7. However, this dependence on distance to the interface only partially explains the observation of two components of different spatial origin in the principal XPS peak. Carr and Buhrman [24] found that as a function of oxide thickness the N 1s peak shifted to larger binding energies by 0.4 eV more than other oxide peaks. We attribute this shift to a chemical change in the second nearest neighbors. In the neighborhood of the substrate the enviroment around a N-Si$_3$ center is assumed to be predominantly rich in Si and N atoms [25, 30, 31], whereas an O-rich environment is expected in the oxide. For simulating such environments we studied auxiliary test-molecules in which the central N atom was always kept in a N-Si$_3$ configuration, while changes in second nearest neighbors were investigated [8]. Calculated shifts in these molecules show that a residual shift of 0.4 eV can indeed be explained in terms of such second nearest neighbor changes. We also explored the possibility of N-Si$_2$H configurations. The calculated shifts were found to be very close to those of corresponding N-Si$_3$ configurations. We can therefore not exclude that also N-Si$_2$H configurations contribute to the principal XPS feature. However, a H concentration of the same order as the N concentration, i.e. of about one monolayer [25, 30, 31], appears unlikely at such interfaces.

In summary, we propose an interpretation of the N 1s photoemission spectra, in which the N atom is found in a *single* first-neighbor configuration (N-Si$_3$), both in the interfacial region close to the substrate and in the oxide. The dependence of the shape of the principal photoemission feature on oxide thickness is explained in terms of core-hole relaxation and second-nearest neighbor effects. The experimental spectra do not show a peak in correspondence of the calculated shift for an oxygen first-nearest neighbor. This suggests that such configurations are absent at nitrided Si(001)-SiO$_2$ interfaces.

7. Conclusions

We presented a series of applications in which calculated core-level shifts provided useful support to the interpretation of photoemission spectra. We used a first-principle approach based on density functional theory, which provides a consistent framework both for performing structural relaxations and calculating core-level shifts. The combination of these two features together with the overall accuracy of the theory, makes of such a scheme a reliable tool to be used in conjonction with experiment. We expect that in the near future theoretical investigations as the ones exposed in this work will routinely be called for to assign unidentified spectral features.

Acknowledgements

We acknowledge fruitful discussions with M.L. Green, E. Garfunkel, E.P. Gusev, and Z.H. Lu.

References

1. L. Pedocchi, N. Russo, and D.R. Salahub, Phys. Rev. B **47**, 12992 (1993).
2. A. Pasquarello, M.S. Hybertsen, and R. Car, Physica Scripta **T66**, 118 (1996).
3. R. Car and M. Parrinello, Phys. Rev. Lett. **55**, 2471 (1985).
4. A. Pasquarello et al., Phys. Rev. Lett. **69**, 1982 (1992); K. Laasonen et al., Phys. Rev. B **47**, 10142 (1993).
5. E. Pehlke and M. Scheffler, Phys. Rev. Lett. **71**, 2338 (1993).
6. We use formulae for the exchange and correlation energy as given in J.P. Perdew and A. Zunger, Phys. Rev. B **23**, 5048 (1981).
7. A. Pasquarello, M.S. Hybertsen, and R. Car, Phys. Rev. Lett. **74**, 1024 (1995); Phys. Rev. B **53**, 10942 (1996).
8. G.-M. Rignanese, A. Pasquarello, J.-C. Charlier, X. Gonze, and R. Car, to be published.
9. W.L. Jolly, K.D. Bomben, and C.J. Eyermann, Atomic Data and Nuclear Data Tables **31**, 433 (1984).
10. A. Pasquarello, M.S. Hybertsen, and R. Car, J. Vac. Sci. Technol. B **14**, 2809 (1996).
11. M.M. Banaszak Holl and F.R. McFeely, Phys. Rev. Lett. **71**, 2441 (1993).
12. S. Lee, S. Makan, M.M. Banaszak Holl, and F.R. McFeely, J. Am. Chem. Soc. **116**, 11819 (1994).
13. A. Pasquarello, M.S. Hybertsen, and R. Car, Phys. Rev. B **54**, R2339 (1996).
14. Y.J. Chabal and S.B. Christman, Phys. Rev. B **29**, 6974 (1984).
15. C.U.S. Larsson, A.S. Flodström, R. Nyholm, L. Incoccia and F. Senf, J. Vac. Sci. Technol. A **5**, 3321 (1987).
16. F.J. Himpsel, F.R. McFeely, A. Taleb-Ibrahimi, J.A. Yarmoff and G. Hollinger, Phys. Rev. B **38**, 6084 (1988).
17. P.J. Grunthaner, M.H. Hecht, F.J. Grunthaner, and N.M. Johnson, J. Appl. Phys. **61**, 629 (1987).
18. Z.H. Lu, M.J. Graham, D.T. Jiang, and K.H. Tan, Appl. Phys. Lett. **63**, 2941 (1993).
19. A. Pasquarello, M.S. Hybertsen, and R. Car, Appl. Phys. Lett. **68**, 625 (1996).
20. A. Pasquarello, M.S. Hybertsen, and R. Car, Appl. Surf. Sci. **104/105**, 317 (1996).
21. I. Ohdomari, H. Akatsu, Y. Yamakoshi, and K. Kishimoto, J. of Non-Crystal. Solids **89** 239 (1987); J. Appl. Phys. **62** 3751 (1987).
22. A. Iqbal, C.W. Bates Jr., and J.W. Allen, Appl. Phys. Lett. **47**, 1064 (1985).
23. M.Y. Hao, W.M. Chen, K. Lai, and C. Lee, Appl. Phys. Lett. **66**, 1126 (1995); S.B. Kang, S.O. Kim, J.-S. Byun, and H.J. Kim, Appl. Phys. Lett. **65**, 2448 (1994); S.T. Chang, N.M. Johnson, and S.A. Lyon, Appl. Phys. Lett. **44**, 316 (1984).
24. E.C. Carr and R.A. Buhrman, Appl. Phys. Lett. **63**, 54 (1993).
25. M.L. Green et al., Appl. Phys. Lett. **65**, 848 (1994).
26. D.G.J. Sutherland et al., J. Appl. Phys. **78**, 6761 (1995).
27. A. Kamath et al., Appl. Phys. Lett. **70**, 63 (1997).

28. Z.H. Lu, S.P. Tay, R. Cao, and P. Pianetta, Appl. Phys. Lett. **67**, 2836 (1995).
29. S.R. Kaluri and D.W. Hess, Appl. Phys. Lett. **69**, 1053 (1996).
30. H.C. Lu *et al.*, Appl. Phys. Lett. **69**, 2713 (1996).
31. H.T. Tang *et al.*, J. Appl. Phys. **80**, 1816 (1996).

A SIMPLE MODEL OF THE CHEMICAL NATURE OF BONDS AT THE Si–SiO$_2$ INTERFACE AND ITS INFLUENCE ON THE ELECTRONIC PROPERTIES OF MOS DEVICES

Hisham Z. Massoud

Semiconductor Research Laboratory
Department of Electrical and Computer Engineering
Duke University, Durham, NC 27708-0291, U.S.A.

1. Introduction

Thermal and chemical treatments of metal-oxide-semiconductor (MOS) devices have resulted in a substantial collection of experimental observations which are still unexplained by existing physical models of MOS devices. For example, the exact physical and chemical nature of oxide-fixed charges is still largely undetermined. It has long been observed that electronic properties of MOS devices such as oxide-fixed charge density, interface traps, and barrier heights are changed by thermal treatments in vacuum, thermal treatments in inert gases, thermal treatments in reactive gases, electrical stressing, and exposure to radiation or photons [1]. The simple and direct influence of annealing in a nitriding ambient on the electrical characteristics of MOS devices is still not fully understood on an atomic level.

To achieve an *Atomic-Scale Understanding* of MOS devices, it is essential to model the different contributions of heat, chemistry, electron transport, radiation and light to the properties of such devices. The goal of this paper is to attract attention to the sources and contribution of dipoles in MOS devices in general, and to the role played by partial-charge transfer in interface bonds at the Si–SiO$_2$ interface in particular. A simple model was developed from first principles to take the contribution of dipoles due to partial-charge transfer in interface bonds on the electrical properties of MOS devices [2]. This model is used to explain some experimentally characterized trends in MOS devices and to take their contribution into account when developing a more detailed description of MOS device physics.

2. Experimental Observations

It has long been observed that the oxide-fixed charge density N_f in MOS devices and their threshold voltage depend on the temperature and duration of annealing in an inert atmosphere such as Ar, or in a nitriding atmosphere such as N_2, NH_3, N_2O, or NO [3]. The workfunction difference in MOS devices was observed to depend on the orientation of the subsrate and on its thermal/chemical processing history [4]. Fowler-Nordheim tunneling currents in MOS devices at the same electric field in oxides with identical thicknesses depend on the orientation of the substrate [5]. The tunneling barrier height in MOS devices depends on the orientation of the silicon substrate [6]. All such observations have not been modeled and accounted for in the current physical chemistry of MOS device theory.

To illustrate thermal and thermo-chemical effects in MOS device annealing, we discuss the dependence of the oxide-fixed charge density on the annealing time in nitrogen. For this purpose, we consider the results obtained by Revitz et al. [3] on oxides grown at 925 °C in dry oxygen on (100) silicon and annealed in nitrogen at 1050 °C. As can be seen in Fig. 1, oxides with thicknesses ranging from 100 to 500 Å were characterized.

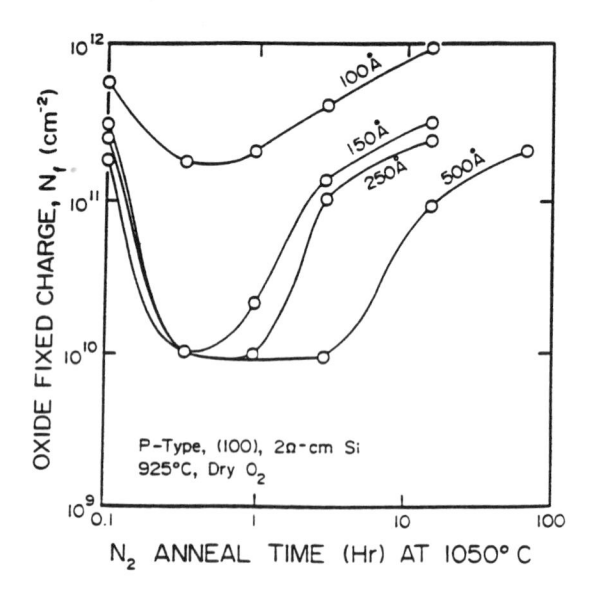

Fig. 1. The dependence of the oxide-fixed charge density N_f at the Si–SiO$_2$ interface on annealing time in nitrogen at 1050 °C [3].

The results indicate that the oxide-fixed charge density starts at a high value in the 10^{11}–10^{12} cm^{-2} range and decrease with the annealing time in nitrogen. N_f was calculated from the flat-band voltage obtained experimentally from high-frequency $C(V)$ measurements and assuming a fixed value of the workfunction difference. After a period that depends proportionally on the oxide thickness, N_f increases again. These results suggest that nitrogen species are transported through the oxide and upon reaching the interface, they react and are incorporated into this interface in such a fashion that the calculated density of oxide-fixed charge is thought to decrease then increase. These results are consistent with numerous reports in the literature that deal with nitrided and reoxidized nitrided oxides.

The results shown in Fig. 1 also show the important annealing kinetics occuring at short times where N_f decreases. In this early domain, there is a different mechanism for the reduction of N_f, one that is independent of the oxide thickness. This early-time annealing kinetics are thought to be purely structural and thermally driven in nature. It is identical to the dependence of N_f on annealing time and temperature in a chemically inert ambient such as argon. A physically based model that translates changes in the physical and chemical structures of the oxide and its interfaces into changes in its electronic charge densities is essential to advance the understanding of MOS devices, especially with thin and ultrathin oxides.

The workfunction difference ϕ_{MS} of MOS devices has been carefully measured and was found to depend on the substrate orientation [4]. These result are shown in Fig. 2 for (100) and (111) silicon. The workfunction difference was obtained from oxides of different thicknesses oxidized in identical conditions, except for the oxidation time by plotting the flat-band voltage V_{FB} as a function of the oxide thickess X_{ox}, fitting the results to a straight line and finding its intercept at zero oxide thickness.

These results are not explained by current MOS device theory where ϕ_{MS} is supposed to be independent of the silicon orientation because it is defined as the difference between two Fermi levels. The difference in ϕ_{MS} on (100) and (111) silicon shown in Fig. 2 is on the order of 0.2 eV and approximately the same on n-type and p-type silicon substrates, with the ϕ_{MS} value on the (100) orientation being larger than that on the (111) orientation. The spread in ϕ_{MS} values reported in the literature between different orientations of silicon ranges from 0 to 0.5 eV [4].

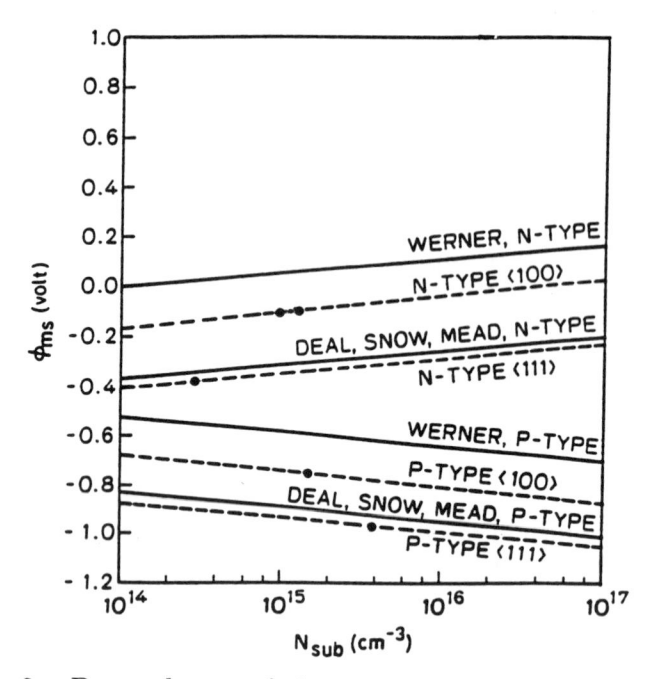

Fig. 2. Dependence of the metal-semiconductor workfunction difference on the substrate doping concentration for (100) and (111) silicon [4].

The characterization of Fowler-Nordheim (FN) tunneling in MOS structures has yielded two interesting observations. First, the tunneling current density in the FN tunneling regime measured at the same field in the oxide in MOS capacitor structures built on (100) and (111) silicon depends on the substrate orientation, with the current density on (100) being larger by approximately one order of magnitude than that measured on (111) silicon [5]. Second, the characterization of tunneling in the FN regime on (100), (111), and (110) silicon indicate an orientation-dependent tunneling barrier [6]. These results are shown in Fig. 3. Again, for the same field in the oxide and according with present descriptions of tunneling in silicon dioxide, we expect that there should be no orientation dependence of the current, as seen in Fig. 3(a), or of the tunneling barrier, as seen in Fig. 3(b). It should be mentioned at this point, that the analysis of tunneling in the FN regime relies itself on a derivation based on a number of assumptions which are not satisfied in the case of tunneling in silicon MOS devices.

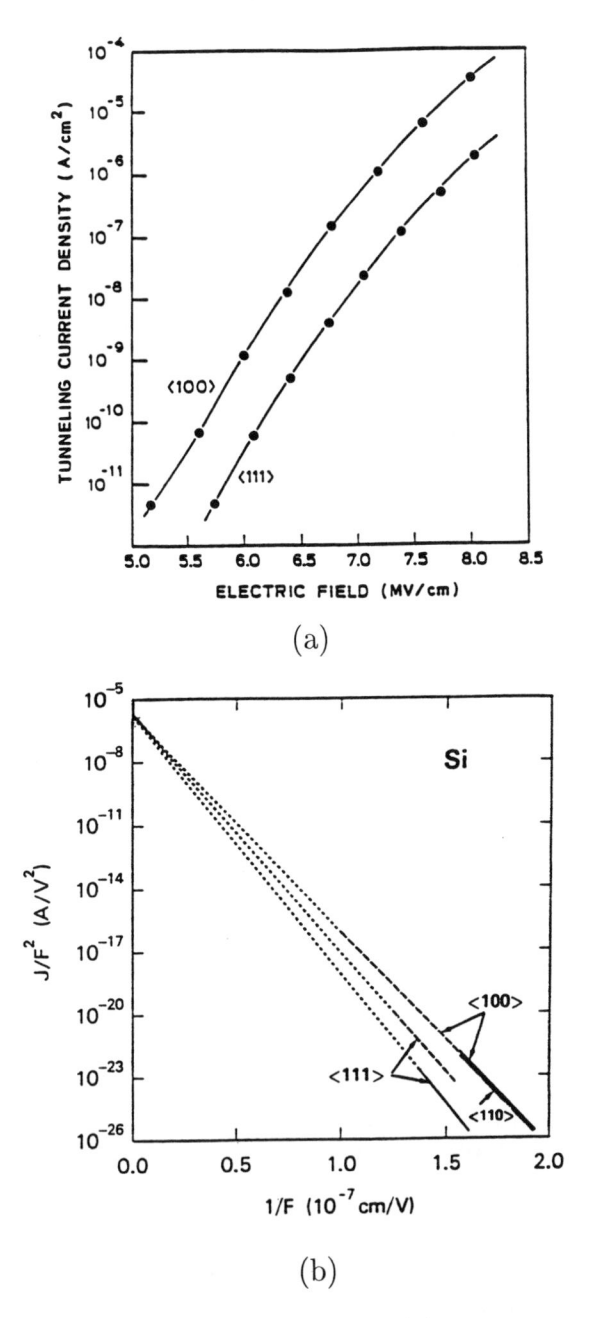

Fig. 3. Orientation dependence of (a) the current density [5], and (b) the tunneling barrier in the Fowler-Nordheim regime in MOS devices [6].

3. Sources of Dipoles in MOS Devices

There are several sources of electric dipoles in MOS structures whose influence has not been previously integrated in device models. These dipoles are either resulting from the behavior of free carriers or from the behavior of electrons in chemical bonds. The dipoles that result from the behavior of free carriers are due to the carrier wave penetration into dielectric barriers. These dipoles are called barrier-penetration (BP) dipoles. The dipoles that result from the behavior of electrons in chemical bonds are called partial-charge-transfer (PCT) dipoles.

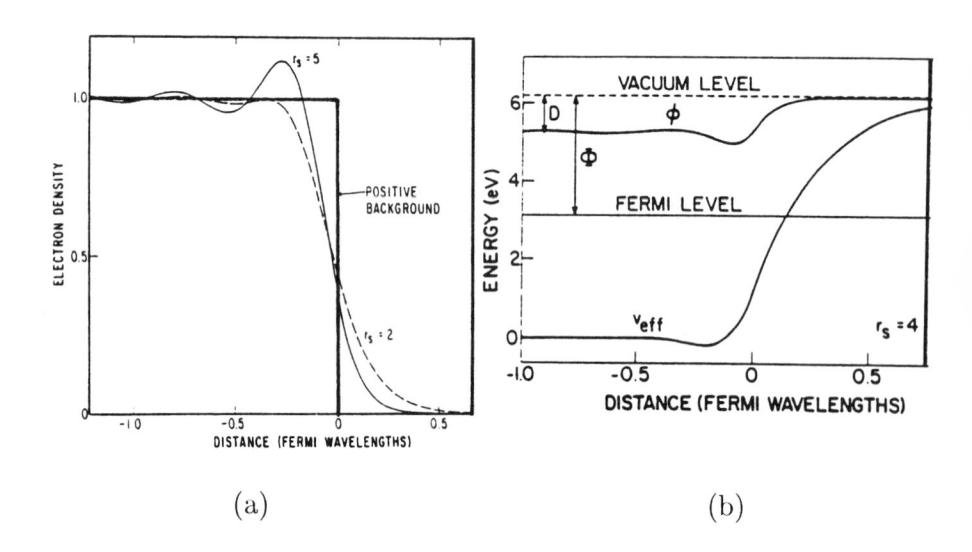

(a) (b)

Fig. 4. Barrier-penetration dipoles at the surface of metals. (a) The charge distribution, and (b) the dipole moment at the surface [8].

Barrier-penetration (BP) dipoles are electronic dipoles that exist as a result of electron or hole wave penetration into the dielectric barriers. The resulting dipole moment depends on the barrier height, the electron or hole concentration, the applied field, and the electron or hole effective mass in the silicon and the silicon dioxide. These dipoles are the same as those treated at the surface of metals using the jellium model [7] or the density functional theory [8], and illustrated in Fig. 4. For example, in the jellium analysis, the positive charge density of the core ions is simplified to a constant density whose concentration drops abruplty to zero at the surface of the metal. The free electrons in the structure are allowed to penetrate the surface barrier because of the extension of their wave beyond the edge of the surface as a direct result of the finite-energy barrier at the surface. The electron concentration and the positive charge distribution are shown in Fig. 4(a) and the dipole that results from the extension of the negative charge into the barrier leaving behind a net positive charge is shown in Fig. 4(b). The analysis of barrier-penetration dipoles in silicon MOS devices can be done by combining a quantum-mechanical solution in the dielectric with a simplified classical or complete quantum-mechanical solution in the silicon substrate [9]. The approach to the treatment of holes is not clear at this point. The changes in BP dipoles upon the application of an electric field in the silicon dioxide are also treated within the jellium-model approach [7].

Partial-charge-transfer (PCT) dipoles are chemically induced dipoles that result from the bonding of dissimilar atoms in interface bonds. Its magnitude depends on the bonding elements, the bond density, the bond orientation, and the charge separation. A model that calculates the dipole moment at the interfaces of silicon in MOS devices was developed based on the principle of electronegativity equalization [2]. This model estimates the magnitude of partial-charge transfer, the influence of PCT dipoles on MOS measurements, and its influence on the metal-semiconductor workfunction difference. The model is briefly presented next, followed by a discussion of the orientation and processing dependence of ϕ_{MS}, PCT dipoles on (100) and (111) silicon, partial-charge transfer in bonds near the interface, and at the gate/oxide interface.

4. Partial-Charge-Transfer Dipole Model

This model has as its starting point the principle of electronegativity equalization [10]. This principle states that when two or more different atoms combine to form a chemically stable compound, they adjust to an intermediate electronegativity. This intermediate electronegativity is estimated using the geometric mean of the individual electronegativities. Thus the electronegativity S_m of a compound is given by

$$S_m \left(\prod_{i=1}^{N} S_i \right)^{1/N}, \tag{1}$$

where S_i is the electronegativity of the i-th atom among the N atoms that form the compound. The electronegativity of common atoms and compounds of interest in MOS devices are shown in Table 1.

Table 1. Electronegativities of elements S_i (taken from the relative compactness scale [10] and of compounds calculated using Eq. (1) [2].

Element	S_i	Compound	S_m
H	3.55	SiO_2	4.256
N	4.49	SiH	3.175
O	5.21	SiN	3.571
F	5.75	SiOH	3.740
Al	2.22	SiO	3.847
Si	2.84	Si_2O	3.477
Cl	4.93	Si_2O_3	4.087

Electronegativity is the quality of an atom that determines, in its covalent bonding with another atom, how the electrons are unevenly shared. A more electronegative atom acquires more than half the charge of the bonding electrons. This definition is extended to the sharing of the bonding-electrons charge between atoms and compounds at the Si–SiO$_2$ interface. As seen from Table 1, all atoms and compounds of interest have electronegativities larger than that of silicon. Consequently, partial-charge transfer takes place in which electrons will be partially displaced towards the oxide, thus generating a dipole where the positive charge is on the silicon side of the interface and the negative charge on its silicon-dioxide side.

The magnitude of the partial-charge transfer ρ_i which affects the i-th atom in a compound whose electronegativity is S_m is given by

$$\rho_i = \frac{S_m - S_i}{2.08 \sqrt{S_i}}. \qquad (2)$$

To calculate the partial-charge transfer in the bond between a substrate silicon atom and a silicon dioxide molecule in the oxide, we calculate first the electronegativity of a hypothetical molecule of SiO$_2$.

Using Eq. (1), we find that [2]

$$S_{SiO_2} = (S_{Si} \, S_O \, S_O)^{1/3} = 4.256$$

$$S_{Si-SiO_2} = (S_{Si} \, S_{SiO_2})^{1/2} = 3.477$$

and the partial-charge transfer in the Si–SiO$_2$ bonds is

$$\rho_{Si-SiO_2} = \frac{S_{Si-SiO_2} - S_{Si}}{2.08 \sqrt{S_{Si}}} = 0.182 \text{ electron/bond}. \qquad (3)$$

This result is schematically illustrated in Fig. 5. The charge separation and the bond orientation determine the dipole moment in the partial-charge transfer at the interface. The charge separation was assumed to be the sum of the atomic and molecular radii of Si and SiO$_2$, respectively. The charge separation d was found at the Si–SiO$_2$ interface to be 2.88 Å [2]. Table 2 shows the magnitude of the partial-charge transfer ρ_i in bonds of interest at the Si–SiO$_2$ interface.

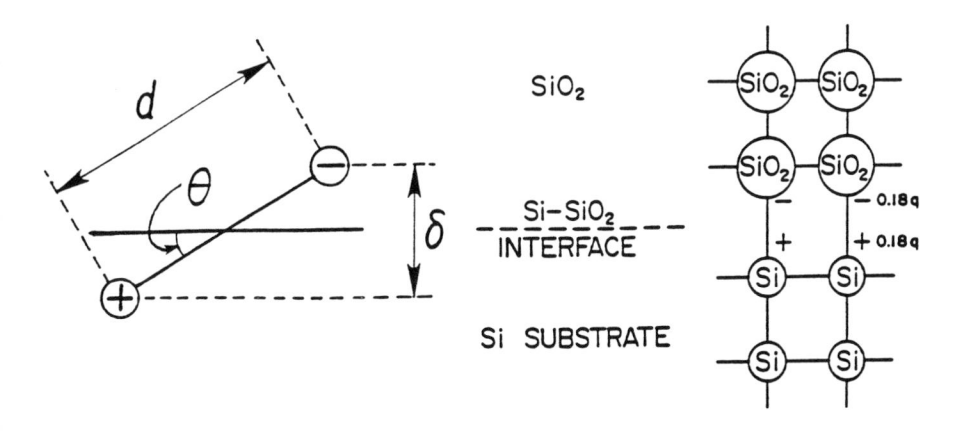

Fig. 5. Polarity of the partial-charge-transfer dipole at the Si–SiO$_2$ interface [2].

The dipole moment is calculated simply as follows. If the magnitude of partial-charge transfer in a bond of type j is ρ_j, then the charge transfer in these bonds is $q\rho_j$. If there $N_{B,j}$ bonds of type j per unit area, then the partial-charge density is $q\rho_j N_{B,j}$. If the separation between the positive and negative charges in the partial-charge transfer in a direction perpendicular to the interface is δ_j, then the dipole moment μ_j per unit area is $q\rho_j N_{B,j} \delta_j$. In terms of the charge separation d_j and the bond angle θ_j with the interface, i.e. for $\delta_j = d_j \sin\theta_j$, the dipole moment μ_j per unit area of interface bonds of type j is given by

$$\mu_j = q\,\rho_j\,N_{B,j}\,d_j\,\sin\theta_j\,. \tag{4}$$

Table 2. The magnitude of the charge transfer in interface bonds at the Si–SiO$_2$ interface [2].

Bond	Charge transfer ρ_j (electron/bond)
Si—O	0.288
Si—H	0.096
Si—N	0.209
Si—OH	0.258
Si—SiO$_2$	0.182
Si—SiO	0.133
Si—Si$_2$O	0.086
Si—Si$_2$O$_3$	0.162

Partial-charge transfer also occurs in bonds near the interface, both in the silicon and the silicon dioxide. By using the same approach, we can calculate the electronegativity and magnitude of charge transfer in bonds removed from the interface [2]. The contribution of partial-charge transfer in bonds in the silicon substrate to the net dipole moment can be calculated because of the crystalline ordering and orientation of the bonds. However, the statistical distribution in the orientation of bonds in the amorphous silicon dioxide layer results in a statistical distribution in the contribution of partial-charge transfer in bonds in the silicon dioxide layer to the net dipole moment. The approach to estimate the contribution from the oxide bonds should be based on structural simulations.

A derivation of the dipole moment in bonds between SiO$_2$ molecules indicates that the electronegativity of the n-th layer away from the interface in the oxide is [2]

$$S_{n,\mathrm{SiO}_2} = S_{\mathrm{Si}}^{(1/2^n)}\,S_{\mathrm{SiO}_2}^{(2^n-1)/2^n}\,. \tag{5}$$

A similar derivation of the electronegativity of the n-th layer in the silicon substrate yields [2]

$$S_{n,\text{Si}} = S_{\text{Si}}^{(2^n-1)/2^n} \, S_{\text{SiO}_2}^{(1/2^n)}. \tag{6}$$

The magnitude of partial-charge transfer in the n-th bond removed from the Si–SiO$_2$ interface in the silicon substrate and the silicon dioxide layer are given, respectively, by [2]

$$\rho_{n,\text{Si}} = \frac{S_{\text{Si}-[(n-1)\text{Si}-\text{SiO}_2]} - S_{\text{Si}}}{2.08\sqrt{S_{\text{Si}}}} \tag{7}$$

and

$$\rho_{n,\text{SiO}_2} = \frac{S_{[\text{Si}-(n-1)\text{SiO}_2]-\text{SiO}_2} - S_{\text{SiO}_2}}{2.08\sqrt{S_{\text{SiO}_2}}}. \tag{8}$$

It was found that the charge transfer in the n-th bond away from the interface is given by a simple relationship of the form [2]

$$\rho_n = \frac{1}{\alpha + \beta \, n} \tag{9}$$

where $\alpha_{\text{Si}} = 2.467$, $\beta_{\text{Si}} = 3.053$, $\alpha_{\text{SiO}_2} = 3.064$ and $\beta_{\text{SiO}_2} = 2.486$.

5. Influence of PCT Dipoles on MOS Measurements

It is seen from the schematic illustration in Fig. 6 that the partial-charge transfer in interface bonds at the Si–SiO$_2$ interface will generate two charge sheets of charge density $q \rho N_B$ separated by a distance δ.

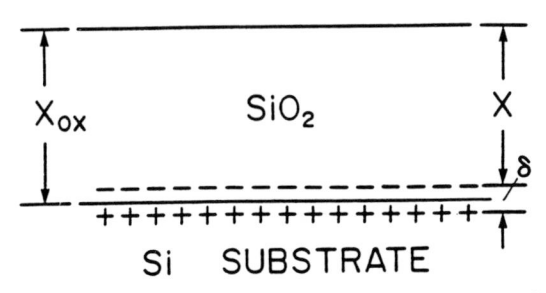

Fig. 6. Partial-charge transfer dipole at the Si–SiO$_2$ interface [2].

It can be easily shown that the flat-band voltage V_{FB} of a MOS capacitor changes from $V_{FB} = \phi_{MS} - qN_f/C_{ox}$ when no dipoles are taken into account to [2]

$$V_{FB} = \phi_{MS} + \frac{q\rho N_B X}{\epsilon} - \frac{q\rho N_B(X+\delta)}{\epsilon} - \frac{qN_f}{C_{ox}},$$

$$= \phi_{MS} - \frac{q\rho N_B \delta}{\epsilon} - \frac{qN_f}{C_{ox}}, \qquad (10)$$

where ϵ is approximated as $(\epsilon_{Si} + \epsilon_{ox})/2$. Equation (10) suggests that we can write that

$$V_{FB} = \phi_{MS}^{eff} - \frac{qN_f}{C_{ox}} \qquad (11)$$

where

$$\phi_{MS}^{eff} = \phi_{MS} - \frac{q\rho N_B \delta}{\epsilon}. \qquad (12)$$

It is clear from Eqs. (10)–(12) that the flat-band voltage expression is different when taking PCT dipoles into account, that the calculated value of N_f should reflect the presence of PCT dipoles, and that, if we determine the metal-semiconductor workfunction difference by measuring V_{FB} on similarly fabricated MOS capacitors with different oxide thicknesses, that what we measure is actually ϕ_{MS}^{eff} and not ϕ_{MS}.

The workfunction difference commonly measured on MOS devices includes the influence of PCT dipoles and their dependence on surface chemistry, bond angles, and bond concentrations. The substrate orientation dependence of ϕ_{MS}^{eff} results from its dependence on N_B and θ. For example, if we assume that all surface silicon atoms on the (100) and (111) orientations of silicon are bonded to SiO_2 molecules at the Si–SiO_2 interface, calculate the PCT dipole moments, and compare the differences in their flat-band voltages, we find that the electron barrier formed at the surface due to PCT dipoles is higher on (111) than (100) silicon by nearly 0.5 eV. This result is in qualitative agreement with the FN tunneling observations in Fig. 3. It also suggests the possibility of devices built on (111) being more resistant to hot-carrier injection [2]. With continuous scaling of MOSFET channel length and the reduction in the degradation of its characteristics due to interface traps, it is possible that

MOSFET fabricated on (111) wafers might benefit from the higher barrier height.

The thermal and chemical processing conditions of the Si–SiO$_2$ interface determine the types and concentrations of bonds N_{Bj}, the charge transfer ρ_j and separation d_j in such bonds. In addition, the effective dielectric constant reflects the local chemistry of the interface. The processing dependence of the oxide-fixed charge shown in Fig. 1 is a direct result of the influence of PCT dipoles on the value of ϕ_{MS}^{eff} which, in turn, enters into the calculation of N_f.

A more detailed description of charges in MOS devices should include the charges resulting from partial-charge transfer in bonds at the Si–SiO$_2$ interface, at the gate/oxide interface, and in bonds away from both interfaces. The net contribution to the flat-band voltage of MOS devices is more likely to result from PCT dipoles at both interfaces and those in bonds away from the Si–SiO$_2$ interface into the silicon wafer because the random nature of bond orientation in the amorphous oxide would have a statistical nature that would likely average to a negligible contribution.

The contribution of BP dipoles at both interfaces must be taken into account when evaluating the flat-band voltage but most importantly when modeling tunneling in MOS devices. The barrier penetration phenomenon is accompanied by two important changes. First, the concentration of carriers at the interface is reduced. Second, the barrier height is affected by the BP dipole. These changes occur in the two most important parameters that determine the tunneling current in MOS devices. At the present time, several approaches are being compared for the estimation of BP dipole moments [9].

6. Conclusions

We have described the role played by dipoles in MOS devices. The sources of dipoles of importance here are barrier-penetration (BP) dipoles that result when free carriers quantum-mechanically penetrate dielectric barriers, and partial-charge-transfer (PCT) dipoles related to electron clouds in bonds. The contribution of PCT dipoles to the electrical characteristics of MOS devices explains the dependence of the work-function difference on substrate orientation and processing conditions, the dependence of oxide-fixed charge on annealing in an inert *vs* nitriding atmosphere, and the orientation dependence of tunneling current density and tunneling barrier on substrate orientation. This model offers a framework to explore the role of dipoles in MOS device physics, process modeling, and a starting point for barrier engineering.

References

1. E. H. Nicollian and J. R. Brews, *MOS (Metal Oxide Semiconductor) Physics and Technology*, John Wiley and Sons, New York, 1982.
2. H. Z. Massoud, J. Appl. Phys., **63**, 2000 (1988).
3. M. Revitz, et al., J. Vac. Sci. Technol., **16**, 345 (1979).
4. H. Z. Massoud and J. D. Plummer, p. 251, *The Physics and Chemistry of SiO₂ and the Si–SiO₂ Interface*, C. R. Helms and B. E. Deal, Ed., Plenum Publishing, New York, 1988.
5. G. Krieger and R. M. Swanson, Appl. Phys. Lett., **39**, 818 (1981).
6. Z. A. Weinberg, J. Appl. Phys., **53**, 5052 (1982).
7. See for example, M.-C. Desjonquères and D. Spanjaard, *Concepts in Surface Physics*, Springer Series in Surface Sciences, Vol. 30, Springer Verlag, Berlin, 1993.
8. N. D. Lang, "Density Functional Approach to the Electronic Structure of Metal Surfaces and Metal-Adsorbate Systems," in *Theory of of the Inhomogeneous Electron Gas*, S. Lundqvist and N. H. March, Editors, Plenum Press, New York, 1983.
9. H. Z. Massoud, to be published.
10. R. T. Sanderson, Science, **114**, 670 (1951).
11. R. T. Sanderson, *Inorganic Chemistry*, Reinhold, New York, 1967.

CHEMICAL PERSPECTIVES ON GROWTH AND PROPERTIES OF ULTRATHIN SiO$_2$ LAYERS

G.F. CEROFOLINI[1], N. RE[2]
[1]*SGS-THOMSON Microelectronics, 20041 Agrate MI, Italy*
[2]*Dipartimento di Chimica dell'Università, 06100 Perugia PG, Italy*

1. Introduction

Silicon provides the material hardware for the electronic and optical industries. The high-volume silicon production is actually possible because of the following reasons:

- relatively pure SiO$_2$ ores are found in large fields;
- a cheap and reliable technology is known for producing 'metallurgical' silicon with impurity content around 5 − 10%; and
- metallurgical silicon is transformed into volatile, easily purifiable, compounds via a single process.

This is done via gas-phase attack of HCl or CH$_3$Cl to powdered metallurgical silicon.

The HCl attack to silicon produces all kinds of chlorosilanes SiH$_{4-n}$Cl$_n$ with $n = 0, \ldots, 4$; among them the most important ones are SiHCl$_3$ (the precursor for single crystalline silicon) and SiCl$_4$ (precursor for optical-grade SiO$_2$). The relative abundance of each SiH$_{4-n}$Cl$_n$ compound is controlled by the reaction conditions (temperature, pressure, silicon dispersion, etc.) and the chlorine amount in each compound can be varied by adding or subtracting hydrogen.

The CH$_3$Cl attack to silicon in the presence of a copper catalyst in elemental or chloride forms (Rochow process [1]) produces almost uniquely Si(CH$_3$)$_2$Cl$_2$ (the monomer for silicone), with minor quantities of SiCH$_3$Cl$_3$ (monomer for branched silicone), Si(CH$_3$)$_3$Cl (chain-terminator monomer for silicone) and polysilanes [2].

Figure 1 sketches the major high-tech silicon or silicon-based materials and how they are produced.

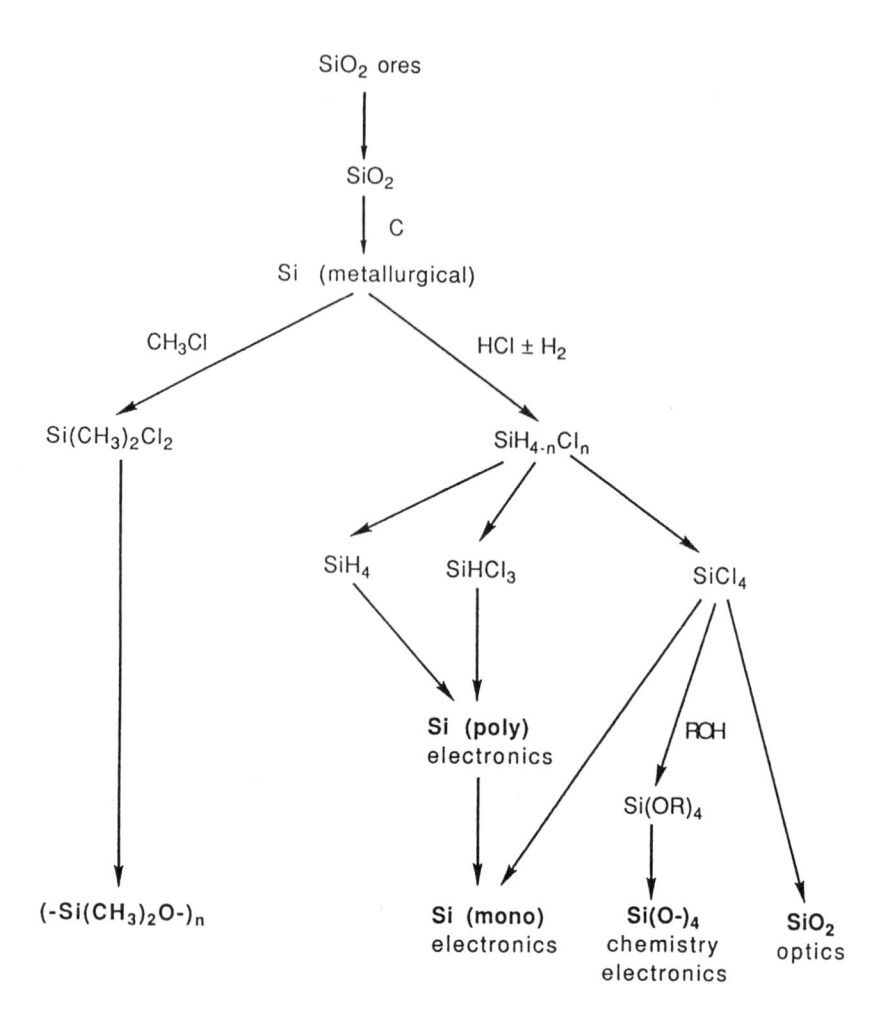

```
                    SiO₂ ores
                        │
                        ▼
                      SiO₂
                        │ C
                        ▼
                 Si (metallurgical)
             CH₃Cl  ╱         ╲  HCl ± H₂
                   ╱           ╲
                  ▼             ▼
          Si(CH₃)₂Cl₂        SiH₄₋ₙClₙ
                           ╱     │     ╲
                          ▼      ▼      ▼
                        SiH₄   SiHCl₃   SiCl₄
                           ╲    │    ╱    ╲
                            ▼   ▼   ▼     ▼ ROH
                           Si (poly)       │
                           electronics     ▼
                              │         Si(OR)₄
                              ▼
        (-Si(CH₃)₂O-)ₙ     Si (mono)   Si(O-)₄    SiO₂
                           electronics  chemistry  optics
                                        electronics
```

Figure 1: The major silicon-based materials and the pathways through which they are produced. Materials indicated in bold are end products.

The Si − Si σ bond has energy and reactivity which resemble quite strictly the π bond in olefins [3]. The major difference is that while the RCl addition to olefins is limited to the cleavage of the π bond,

$$H_2C = CH_2 + RCl \longrightarrow RH_2C - CH_2Cl \quad (R = H, Me, Et, etc.),$$

the addition to silicon exposes new cleavable Si − Si bonds to the atmosphere, so that this process proceeds with the formation of volatile $SiR_{4-n}Cl_n$ and new exposed sites.

A somewhat similar reactivity toward silicon is displayed by HX, with X strongly electronegative group (*e.g.*, X = F or OR). Of course, since all $SiR_{4-n}Cl_n$ compounds are volatile while a few $SiH_{4-n}X_n$ are not (*e.g.*, $Si(OH)_4$), the previous arguments may be applied only if the reaction is carried out in a medium in which the reaction product is dissolved (*e.g.*, H_2O in the considered example).

Both the Si − F and Si − O bonds have a high ionic character. While the polar character of the Si − F bond is simply understood in terms of electronegativity difference (actually, the Si − F bond has the highest known bond energy, 6.9 eV, usually explained as resulting from a strong electrostatic reinforcement to the covalent contribution), the electronegativity difference between silicon and oxygen does not justify by itself a large polar character of the Si − O bond. This fact is however confirmed by the large chemical shifts toward higher binding energy of the Si 2p core levels after oxidation (around 1.0 eV per Si − O bond) and by the $\widehat{ROR'}$ angle in methyl-silyl ethers, increasing from 111.5° in $H_3C - O - CH_3$ to 144.1° in $H_3Si - O - SiH_3$ through 120.6° in $H_3C - O - SiH_3$ [3].

The large electron transfer from silicon to the electronegative group X makes silicon sensitive to nucleophilic attack, whose effects are often paradoxical. For instance, as first observed by Ubara *et al.*, the nucleophilic attack is eventually responsible for the hydrogen termination of oxidized surfaces after etching in buffered HF aqueous solution [4, 5]. Ascribing the increased sensitivity to nucleophilic attack to a reduction of the bond energy, Cerofolini *et al.* have also explained the large enhancement of the silicon oxidation rate in $O_2:F_2$ (with F_2 concentration in the 10−100 ppm range) at moderate temperature (400−550 °C), as well as the deviations from the Deal–Grove linear-parabolic kinetics in the early stages of silicon oxidation in dry O_2 at high temperature (800 − 1100 °C) [6].

2. Choice of the Theoretical Framework

The concept of backbond weakening because of electron transfer from silicon to electronegative groups X (X = F, OH or O−) explains a lot of different phenomena, like hydrogen termination after attack to SiO_2 with aqueous HF or even alkaline solutions, the enhancement of the oxidation rate at moderate temperature by F_2, and the enhanced rate in the early stages during oxidation at high temperature.

The term 'weakening' has however been used in a rather vague manner, and an analysis of what this term does actually mean has not been given yet. Though the *effects* of this putative weakening are large (increase of the oxidation rate by one or more orders of magnitude) and often counterintuitive (*e.g,* hydrogen termination after alkaline etching), the energetic causes for them are expectedly modest, well below 1 eV [6].

Any theoretical setting for the prediction of such causes must therefore be extremely accurate, the target being a better accuracy than 0.1 eV. Quantum mechanical calculations to within such an accuracy are practically feasible only for sufficiently small molecules.

In this light, we mimic:

- the Si − Si bond at the silicon surface by the molecule $H_3Si - SiH_3$;
- the $Si - Si(O-)_n$ and $Si - SiF_n$ ($n = 1, 2, 3$) bonds at the interface with the molecules $H_3Si - SiH_{3-n}(OH)_n$ and $H_3Si - SiH_{3-n}F_n$, respectively; and
- the $(-O)_3Si - Si(O-)_3$ bond of the oxygen-vacancy defect in the oxide with the molecule $(HO)_3Si - Si(OH)_3$.

The theoretical framework used for the description of these molecules is based on density functional theory. The calculations reported in this paper have been performed by the Gaussian 94 program package [9] and were done on IBM RISC/6000 workstations. The B3LYP hybrid exchange-correlation functional was used for all the calculations. This functional is based on the Becke's three-parameter functional [10] including a Hartree-Fock exchange contribution with the non-local correction for the exchange potential proposed by Becke in 1988 [11], together with the non-local correction for the correlation energy provided by the functional of Lee, Young and Parr [12]. Molecular structures were optimized using the B3LYP expression above. It has been already demonstrated that this hybrid functional gives accurate optimized geometries and, for large basis sets, accurate atomization energies for a wide range of molecules [13]. A 6-311G** basis set [14] was used for all the atoms. Full geometry optimizations were performed on all the molecules with a Si − Si bond, while in fragments with only one silicon atom the Si-O or Si-F distance only was optimized in the moieties $Si^+\leftarrow:OH_2$ or $Si^+\leftarrow:FH$, unless differently stated. In this scheme it is possible to calculate the quantum mechanical state of each of the considered model molecule with an energy accuracy better than 0.1 eV [7], the time required being of a few hours.

3. Homolytic Cleavage

First of all, we have calculated the dissociation enthalpy $\Delta H_{vac}^{\bullet\bullet}$ for the homolytic cleavage of the above molecules (here and in the following the lower index denotes the medium in which the dissociation occurs, while the upper index is related to the dissociation mode).

The data (calculated ignoring fragment relaxations) are summarized in table 1, which shows a quite surprising result: *the substitution of an electronegative group for hydrogen does not weaken, but rather strengthens, the Si − Si bond with respect to the homolytic cleavage.* This conclusion is also sustained by the decrease of the bond length resulting from the substitution of X for H.

Table 1: Si − Si bond length d and dissociation enthalpy $\Delta H_{vac}^{\bullet\bullet}$ for the homolytic cleavage.

molecule	d (Å)	$\Delta H_{vac}^{\bullet\bullet}$ (eV)
$(HO)_3Si - Si(OH)_3$	2.325	3.84
$H_3Si - SiH_3$	2.366	3.14
$H_3Si - SiF_3$	2.344	3.49
$H_3Si - SiHF_2$	2.358	3.29
$H_3Si - Si(OH)_3$	2.335	3.52
$H_3Si - SiH(OH)_2$	2.346	3.35
$H_3Si - SiH_2OH$	2.357	3.21

4. Heterolytic Cleavage

The concept of backbond weakening explains so many facts that we have to exclude all plausible cleavage mechanisms before discarding it. Rather, if we recognize which kind of dissociation is made easier by the electronegative substituent, we have a hope to identify the attack mechanism.

For, we have calculated the influence of this substitution on the heterolytic cleavage. When one considers asymmetric molecules like the last five of table 1, one has to consider two possible heterolytic dissociations:

$$H_3Si - SiH_{3-n}X_n \xrightarrow{\Delta H_{vac}^{\mp}} H_3Si^+ + {}^-SiH_{3-n}X_n \quad (\text{`}\mp\text{ dissociation'}) \quad (1)$$

and

$$H_3Si - SiH_{3-n}X_n \xrightarrow{\Delta H_{vac}^{\pm}} H_3Si^- + {}^+SiH_{3-n}X_n \quad (\text{`}\pm\text{ dissociation'}). \quad (2)$$

Elementary considerations suggest that the $\mp$ dissociation (1) is energetically favored with respect to the $\pm$ dissociation (2). Calculations performed for the molecule $H_3Si - SiF_3$ (summarized in table 2) show that this prediction is correct, the difference being higher than 2.5 eV.

Table 2: Comparison of the dissociation enthalpies for the two heterolytic dissociations of $H_3Si - SiF_3$.

molecule	products	enthalpy
$H_3Si - SiF_3$	$H_3Si^+ + {}^-SiF_3$	$\Delta H^{\mp}_{vac} = 11.09$ eV
$H_3Si - SiF_3$	$H_3Si^- + {}^+SiF_3$	$\Delta H^{\pm}_{vac} = 13.60$ eV

Limiting therefore our attention to the heterolytic dissociation (1), we have calculated the dissociation enthalpies for all the molecules considered in table 1. Table 3 shows that even in this case it is energetically more difficult to cleave all the considered molecules, except $H_3Si - SiF_3$, than $H_3Si - SiH_3$. This conclusion forces us to look another cleavage process if we want to continue to use the concept of backbond weakening.

Table 3: Dissociation enthalpies $\Delta H^{\pm}_{vac}$ for the heterolytic cleavage of the Si $-$ Si bond.

molecule	products	$\Delta H^{\mp}_{vac}$ (eV)
$(HO)_3Si - Si(OH)_3$	$(HO)_3Si^+ + {}^-Si(OH)_3$	12.54
$H_3Si - SiH_3$	$H_3Si^+ + {}^-SiH_3$	11.32
$H_3Si - Si(OH)_3$	$H_3Si^+ + {}^-Si(OH)_3$	11.94
$H_3Si - SiH(OH)_2$	$H_3Si^+ + {}^-SiH(OH)_2$	11.99
$H_3Si - SiH_2OH$	$H_3Si^+ + {}^-SiH_2OH$	11.69
$H_3Si - SiF_3$	$H_3Si^+ + {}^-SiF_3$	11.09
$H_3Si - SiHF_2$	$H_3Si^+ + {}^-SiHF_2$	11.48
$H_3Si - SiH_2F$	$H_3Si^+ + {}^-SiH_2F$	11.41

5. Heterolytic Dissociation Stabilized by the Formation of an Adduct with a Lewis Base

Having discarded both homolytic and heterolytic dissociations, we must search for a more complex dissociation mode. Clearly enough, heterolytic dissociation alone cannot occur via simple thermal activation because the required energy is too high. This energy might be drastically reduced (even ignoring the electrostatic pairing) in the presence of a Lewis base able to form an adduct with the cation.

We shall start our analysis considering $H_3Si - SiH_{3-n}(OH)_n$ molecules and a water molecule as Lewis base. In the presence of water, the possible dissociation reactions are

$$H_3Si - SiH_{3-n}(OH)_n + H_2O \xrightarrow{\Delta H^{\mp,H_2O}_{vac}} H_3Si^+ \leftarrow :OH_2 + {}^-SiH_{3-n}(OH)_n \quad (3)$$

and

$$H_3Si - SiH_{3-n}(OH)_n + H_2O \xrightarrow{\Delta H^{\pm,H_2O}_{vac}} H_3Si^- + H_2O:\rightarrow {}^+SiH_{3-n}(OH)_n \quad (4)$$

While it was possible to choose on naive considerations which reaction between (1) and (2) is energetically more convenient, choosing the most stable configuration between $H_3Si^+ \leftarrow :OH_2 + {}^-SiH_{3-n}(OH)_n$ (the product of reaction (3)) and $H_3Si^- + H_2O:\rightarrow {}^+SiH_{3-n}(OH)_n$ (the product of reaction (4)) is not trivial and can be decided only after detailed calculations. Calculations performed for the molecule $H_3Si - Si(OH)_3$ (shown in table 4) show that dissociation (4) is energetically favored over dissociation (3), the difference being almost 1.5 eV.

Table 4: Dissociation enthalpies for the two possible water-assisted Si − Si heterolytic cleavages of $H_3Si - Si(OH)_3$.

reactants	products	enthalpy
$H_3Si - Si(OH)_3 + H_2O$	$H_3Si^+ \leftarrow :OH_2 + {}^-Si(OH)_3$	$\Delta H^{\mp,H_2O}_{vac} = 8.42$ eV
$H_3Si - Si(OH)_3 + H_2O$	$H_3Si^- + H_2O:\rightarrow {}^+Si(OH)_3$	$\Delta H^{\pm,H_2O}_{vac} = 6.96$ eV

Limiting therefore the attention only to the $\pm$ dissociation, we obtain the data of table 5.

Table 5: Water-assisted dissociation enthalpy of various $H_3Si - SiH_{3-n}X_n$ molecules.

reactants	products	$\Delta H^{\pm,H_2O}_{vac}$ (eV)
$H_3Si - SiH_3 + H_2O$	$H_3Si^- + H_2O:\rightarrow {}^+SiH_3$	7.80
$H_3Si - SiH_2OH + H_2O$	$H_3Si^- + H_2O:\rightarrow {}^+SiH_2OH$	7.36
$H_3Si - SiH(OH)_2 + H_2O$	$H_3Si^- + H_2O:\rightarrow {}^+SiH(OH)_2$	7.04
$H_3Si - Si(OH)_3 + H_2O$	$H_3Si^- + H_2O:\rightarrow {}^+Si(OH)_3$	6.96
$(HO)_3Si - Si(OH)_3 + H_2O$	$(HO)_3Si^- + H_2O:\rightarrow {}^+Si(OH)_3$	7.53

At last this table seems to indicate which attack to Si−Si bond is sensitive to the inductive effects — the attack of a Lewis base to the silicon bonded to electronegative groups (this is just the meaning of nucleophilic attack!).

Table 6 shows however that this feature does not hold true when X is a fluorine rather than hydroxyl termination. This is due to the fact that the increase with n of the binding energy of H_2O (or HF) to $^+SiH_{3-n}F_n$ (from 3.52 eV (2.23 eV) for $n = 0$ to 5.01 eV (3.38 eV) for $n = 3$, as deduced from tables 2 and 6) is more than counterbalanced by the corresponding increase of the $\pm$ dissociation enthalpy (from 11.32 eV for $n = 0$ to 13.60 eV for $n = 3$, as deduced from tables 2 and 3).

Table 6: Fluorine terminations inhibit the nucleophilic attack of very weak Lewis bases like H_2O or HF.

reactants	products	$\Delta H_{vac}^{\pm,H_2O}$ (eV)
$H_3Si - SiH_3 + H_2O$	$H_3Si^- + H_2O:\to{}^+SiH_3$	7.80
$H_3Si - SiH_2F + H_2O$	$H_3Si^- + H_2O:\to{}^+SiH_2F$	7.78
$H_3Si - SiHF_2 + H_2O$	$H_3Si^- + H_2O:\to{}^+SiHF_2$	8.05
$H_3Si - SiF_3 + H_2O$	$H_3Si^- + H_2O:\to{}^+SiF_3$	8.62
$H_3Si - SiH_3 + HF$	$H_3Si^- + HF:\to{}^+SiH_3$	9.09
$H_3Si - SiHF_2 + HF$	$H_3Si^- + HF:\to{}^+SiHF_2$	9.34
$H_3Si - SiF_3 + HF$	$H_3Si^- + HF:\to{}^+SiF_3$	10.22

Understanding the origin of the observed terminations after attack of HF in gas phase or in aqueous solution requires a combination of all the above results. While fluorine surface termination resulting from gas-phase attack of HF to silicon [15] are understood in terms strength of the Si−F bond, the hydrogen termination observed after etching in buffered aqueous solution of HF [4] is understood in terms of F^- attack. Since F^- is a Lewis base much stronger than HF or H_2O, it attacks nucleophilically $-SiF_3$. The attack is presumed to proceed with the following pathway:

$$\begin{aligned} \equiv Si - SiF_3 + F^- &\longrightarrow \equiv Si^- + SiF_4 \\ \equiv Si^- + HF &\longrightarrow \equiv SiH + F^- \\ \hline \equiv Si - SiF_3 + HF &\xrightarrow{F^-} \equiv SiH + SiF_4 \end{aligned} \qquad (5)$$

The catalytic role of F^- explains why hydrogen termination is favored in a NH_4F buffered solution, though other strong bases, like OH^- or Cl^-, have similar effects. An additional factor explaining the role of fluorine in determining hydrogen termination is the high anion affinity of SiF_n groups. (see table 7, taken from ref. [8]).

Manifestly, the addition in the first step of reaction (5) is kinetically favored by the vicinity of the fluoride ion. At last we note that the first step of reaction (5) is possible only if the electron affinity of $\equiv$ Si$^\bullet$ is sufficiently high (that happens at the silicon surface) while the second step is possible only because of the solvation energy of F^-.

Table 7: Fluoride anion affinity of $Si(CH_3)_{4-n}F_n$ molecules ($n = 0, \cdots, 4$).

molecule	anion affinity (eV)
$(CH_3)_4Si$	1.3
$(CH_3)_3SiF$	1.7
CH_3SiF_3	2.2
SiF_4	2.8

6. From Gas-Phase Chemistry to Chemistry in SiO$_2$

The previously mentioned calculations show that the considered ionized states have a formation energy so high as to make extremely improbable their formation via thermal excitation. This state of affairs, however, changes significantly in condensed matter.

In bulk SiO_2 the $(HO)_3Si - Si(OH)_3$ molecule can be used to mimic the $Si - Si$ bond associated with an oxygen vacancy. In the vicinity of siloxanic oxygen the $(-O)_3Si - Si(O-)_3$ center is known to undergo heterolytic dissociation:

$$(-O)_3Si - Si(O-)_3 + O\begin{matrix}\diagup Si \equiv \\ \diagdown Si \equiv\end{matrix} \longrightarrow (-O)_3Si^+ \leftarrow :O\begin{matrix}\diagup Si \equiv \\ \diagdown Si \equiv\end{matrix} + {}^-Si(O-)_3. \qquad (6)$$

This mechanism has been clarified in refs. [16, 17]. Extended *ab initio* molecular dynamics simulations for α quartz have shown that the enthalpy difference $\Delta H_{SiO_2}^{\pm, \equiv Si_2O}$ ($= \Delta H_{SiO_2}^{\mp, \equiv Si_2O}$) between the rhs and lhs configurations in reaction (6) is 2.3 eV, with an activation energy of 0.3 eV to decay from the excited ionized state to the fundamental one [18].

Interpreting the enthalpy difference as the energy required for the base-assisted cleavage of the $(-O)_3Si - Si(O-)_3$ bond, we observe that this value is appreciably reduced with respect to that calculated for $(HO)_3Si - Si(OH)_3 + H_2O$. The following factors have been identified as responsible for this difference:

- water is a weaker base than $\equiv Si_2O$;
- electrostatic anion-cation pairing stabilizes the ionized configuration; and
- the silica matrix effects (band structure, energy lanscape, and topological constraints) affect the calculated energy levels.

Assuming that these contributions are mutually independent, one has

$$\Delta H_{SiO_2}^{\pm,\equiv Si_2O} = \Delta H_{vac}^{\pm,H_2O} - (\Delta W + \Delta U + \Delta E), \tag{7}$$

where ΔW is the energy difference due to the different strengths of the bases H_2O and $\equiv Si_2O$, ΔU is the electrostatic energy stabilizing the ionic pair, and ΔE is the energy correction related to the matrix effects. These energies are estimated with the following considerations.

Base strength. The quantity ΔW is given by the difference of the dissociation enthalpies in a vacuum of the adducts $(\equiv Si)_2O{:}{\rightarrow}^+Si\equiv$ and $H_2O{:}{\rightarrow}^+Si\equiv$,

$$\Delta W := \Delta H_{vac}^{diss}((\equiv Si)_2O{:}{\rightarrow}^+Si\equiv) - \Delta H_{vac}^{diss}(H_2O{:}{\rightarrow}^+Si\equiv).$$

For the calculation of this difference we assume the following scaling rule:

$$\frac{\Delta W}{\Delta H_{vac}^{diss}(H_2O{:}{\rightarrow}^+Si\equiv)} = \frac{\Delta H_{vac}^{diss}((\equiv Si)_2O{:}{\rightarrow}^+H) - \Delta H_{vac}^{diss}(H_2O{:}{\rightarrow}^+H)}{\Delta H_{vac}^{diss}(H_2O{:}{\rightarrow}^+H)}.$$

Taking: (i) $\Delta H_{vac}^{diss}(H_2O{:}{\rightarrow}^+Si\equiv) = 4.3$ eV (mean value of $\Delta H_{vac}^{diss}(H_2O{:}{\rightarrow}^+SiH_3) = 3.52$ eV and $\Delta H_{vac}^{diss}(H_2O{:}{\rightarrow}^+Si(OH)_3) = 5.01$ eV, values obtainable from the previously presented data via Born–Haber cycles), (ii) $\Delta H_{vac}^{diss}(H_2O{:}{\rightarrow}^+H) = 7.50$ eV (the gas-phase proton affinity of H_2O), and (iii) $\Delta H_{vac}^{diss}((\equiv Si)_2O{:}{\rightarrow}^+H) = 8.0$ eV (estimated from the gas-phase proton affinity of disilyl ether, 7.98 eV [19]), one has $\Delta W = 0.3$ eV. The smallness of this value guarantees that H_2O mimics well $(\equiv Si)_2O$.

Electrostatic pairing. The electrostatic contribution ΔU is estimated by assuming that the charges are pointlike and concentrated on the two silicon atoms. For a charge separation of 4.4 Å (as expected in α quartz from the *ab initio* molecular dynamics simulations of ref. [18]) the Coulomb law gives $\Delta U = 3.3$ eV.

Matrix effects. For $\Delta H_{SiO_2}^{\pm,\equiv Si_2O} = 2.3$ eV, $\Delta H_{vac}^{\pm,H_2O} = 7.53$ eV, $\Delta U = 3.3$ eV and $\Delta W = 0.3$ eV, eq. (7) gives the influence of the matrix effects on the dissociation enthalpy: $\Delta E \simeq 1.6$ eV.

Assuming that only $\Delta H_{vac}^{\pm,H_2O}$ varies with n in the moiety $Si - Si(O-)_n$ while ΔW, ΔU and ΔE remain unchanged, the enthalpy for the base-assisted heterolytic cleavage of $Si - Si$ bonds of small silicon clusters embedded in SiO_2 is (1.9 ± 0.2) eV in relation to the oxidation number of silicon (from 1.7 eV for $-\text{Si}(O-)_3$ to 2.1 eV for $\equiv \text{Si}(O-)$).

The same arguments which discarded the $\mp$ dissociation in a vacuum are extended to bulk SiO_2 if one assumes the constancy of ΔW, ΔU and ΔE with respect to the dissociation mode.

7. From Gas-Phase Chemistry to Chemistry at the Silicon Surface

While in bulk SiO_2 the base-assisted $\pm$ dissociation only is energetically possible, at the oxidized silicon surface the $\mp$ dissociation becomes possible too, though the $\pm$ dissociation remains the favorite one.

To demonstrate this statement, consider first the $\pm$ dissociation. While ΔW and ΔU may be taken from bulk SiO_2, for estimating ΔE one must take into account the matrix effects at the silicon-SiO_2 interface. If one assumes that ΔE is essentially determined by band-structure effects, they may be taken into account by substituting the electron affinity of crystalline silicon (4.0 eV) for the electron affinity of $H_3Si^\bullet$ (1.3 eV) [20]. This procedure gives $\Delta H_{Si-SiO_2}^{\pm, \equiv Si_2O} = (0.9 \pm 0.2)$ eV, where the uncertainty is associated with the different oxidation numbers of silicon.

For the $\mp$ dissociation we have: (i) to keep ΔW and ΔU unchanged; (ii) to consider $\Delta H_{vac}^{\mp, H_2O}$ instead of $\Delta H_{vac}^{\pm, H_2O}$; and (iii) to substitute the ionization energy of crystalline silicon (5.0 eV) for the ionization energy of $H_3Si^\bullet$ (8.1 eV). This procedure gives $\Delta H_{Si-SiO_2}^{\mp, \equiv Si_2O} = 1.7$ eV, close to the activation energy in the initial stages of silicon oxidation.

8. Discussion

Ab initio quantum mechanical calculations, being based on more or less approximate solutions of the Schrödinger equation, do not need any interpretation. Their understanding, however, requires they are interpreted in terms of atomic or group properties (like electronegativity) or mutual relationships among atoms or groups (like nature of the bond or charge transfer, however loose these concepts are). Though the interpretation adopted in this work (mainly based on eletronegativity) does not 'explain' the data of table 1, it seems however rather satisfactory. Other interpretations are however possible.

For instance, the very high Si – F bond energy might be explained in terms of lone-electron-pair donation from fluorine to silicon. A similar phenomenon is known to be responsible for conjugation in organic compounds like vinyl halides [21] whose structure, $H_2C^{\frac{1}{2}-} \cdots C(H) \cdots X^{\frac{1}{2}+}$, results from the resonating formulas

$H_2C = C(H) - X$ and $H_2C^- - C(H) = X^+$, where the standard notation for dative bonds has been used, with formal charges resulting from Lewis's octet theory. The equivalent structure for $H_3Si-SiH_2X$ would be

$$H_3Si^{\frac{1}{2}-}\cdots Si(H_2) \stackrel{...}{-} X^{\frac{1}{2}+},$$

resulting from the resonating formulas $H_3Si - Si(H_2) - X$ and $H_3Si^-\ Si(H_2) = X^+$. The existence of these resonating states is made plausible by the strong similarity of the $C = C\ \pi$ bond in olefins to the $Si - Si\ \sigma$ bond [3].

Even this picture does not explain the increase of the $Si - Si$ bond energy with n in $H_3Si-SiH_{3-n}X_n$ — the existence of n equivalent configurations would indeed be responsible for a larger charge delocalization,

$$H_3Si^{\frac{1}{2}-}\cdots Si(H_{3-n}) \stackrel{...}{-} X_n^{\frac{1}{2n}+},$$

and hence for a weaker electrostatic binding energy. However, it explains in a natural way the fluoride affinity of SiF_n moieties (table 7) and why the base-assisted dissociation enthalpy increases with n for $X = F$, while has an opposite behavior for hydroxyl terminations. In the latter case, indeed, conjugation effects are negligible (as suggested by the 'normal' $Si - O$ bond energy) and the base-assisted dissociation enthalpy decreases with n because of the already discussed reasons; in the former case, instead, the base-assisted dissociation enthalpy increases with n because the electrostatic pairing between the base and silicon is reduced because of charge delocalization over n bonds.

An explanation of the whole set of data seems to require electronegativity effects for compounds involving $Si(OH)_n$ moieties and conjugation effects for compounds involving SiF_n moieties.

9. Conclusions

We have performed extended quantum mechanical calculations to study the energy required to cleave the $Si - Si$ bond when one silicon is bonded to hydrogen (and thus mimics elemental silicon) and the other is bonded to electronegative groups. We have observed that when the silicon is bonded to OH groups the water-assisted heterolytic cleavage becomes progressively easier the higher is the number of OH terminations. These considerations are extended to bulk silica (to calculate the cleavage energy of the oxygen vacancy) and to the surface of oxidized silicon (to study the possible dissociation modes).

10. References

1. Rochow, E.G. (1945) J. Am. Chem. Soc. **67**, 963
2. Kuivila, C.S., Zapp, R.H., Wilding, O.K., and Hall, C.A. (1996) in: H. A. Øye, H. M. Rong, B. Cecccaroli, L. Nygaard and J. K. Tuset (eds.), *Silicon for the Chemical Industry III*, Tapir, Trondheim, p. 227
3. Barton, T.J., and Boudjouk, P. (1990) in J.M. Zeigler and F.W. Gordon Fearon (eds.), *Silicon-Based Polymer Science. A Comprehensive Resource*, Adv. Chem. Ser. No. 224, Am. Chem. Soc., Washington, DC, p. 1
4. Ubara, II., Imura, T., and Hiraki, A. (1984) Solid St. Commun. **50**, 673
5. Cerofolini, G.F., and Meda, L. (1995) Appl. Surf. Sci. **89**, 351
6. Cerofolini, G.F., La Bruna, G., and Meda, L. (1995) Appl. Surf. Sci. **89**, 361
7. Damrauer, R., and Hankin, J. (1995) Chem. Rev. **95**, 1137
8. Frisch, M.J., Trucks, G. W, Schlegel, H. M., Gill, P.M.W., Johnson, B.G., Wong, M.W., Foresman, J.B., Robb, M.A., Head-Gordon, M., Replogle, E.S., Gomperts, R., Andres, J.L., Raghavachari, K., Binkley, J.S., Gonzales, C., Martin, R.L., Fox, D.J., Defrees, D.J., Baker, J., Stewart, J.J.P., and Pople, J.A. (1994) *Gaussian 94*, Revision A.1, GAUSSIAN Inc., Pittsburgh, PA
9. Becke, A.D. (1993) J. Chem. Phys. **98**, 5648
10. Becke, A.D. (1988) Phys. Rev. **A 38**, 2398
11. Lee, C., Young, W., and Parr, R.G. (1988) Phys. Rev. **B 37**, 785
12. Bauschlicher,C. W. Jr. (1995) Chem. Phys. Lett. **246**, 40
13. Frisch, M.J., Pople, J.A., and Binkley, J.S. (1984) J. Chem. Phys. **80**, 3265; and references therein
14. Re, N., Sgamellotti, A., and Cerofolini, G.F. (1997) J. Phys. Chem., to be published
15. Miki, N., Kikuyama, H., Kawanabe, I., Miyashita, M., and Ohmi, T. (1990) IEEE Trans. Electron Devices **37**, 107
16. Rudra, J.K, and Fowler, W.B. (1987) Phys. Rev. **B 35**, 8223
17. Allan, D.C., and Teter, M.P. (1990) J. Am. Ceram. Soc. **73**, 3247
18. Boero, M., Pasquarello, A., Sarnthein, J., and Car, R. (1997) Phys. Rev. Lett. **78**, 887
19. Blake, J.F., and Jorgensen, W.L. (1991) J. Org. Chem. **56**, 6052
20. Electron affinities A_{el} and ionization energies E_{ion} are by-products of the calculations reported above. In particular, we calculated the following electron affinities: 1.28 eV for $H_3Si^\bullet$, 1.47 eV for $(HO)_3Si^\bullet$ and 2.27 eV for $F_3Si^\bullet$, and ionization energies: 8.14 eV for $H_3Si^\bullet$ and 6.82 eV for $(HO)_3Si^\bullet$. These quantities are significantly affected by relaxation effects and the reported values have been calculated for completely relaxed species. To give an idea of the importance of relaxation, we mention that the electron affinity of $(HO)_3Si^\bullet$ increases from 0.61 eV for the unrelaxed tetrahedral configuration to 1.47 eV for the relaxed configuration. Even the $H_2O:\rightarrow{}^+SiX_3$ bond energy (X = H, OH) is significantly affected by relaxation effects.
21. J. March, *Advanced Organic Chemistry*, 4th ed. (Wiley, New York, NY, 1992) p. 32

A theoretical model of the Si/SiO₂ interface

A. Markovits and C. Minot
Laboratoire de Chimie Théorique, UPR 9070 CNRS
Université P. et M. Curie. Boîte 137, Tour 23-22 p 114
4 Place Jussieu 75252 Paris Cédex 05, FRANCE.

Abstract

We present preliminary quantum chemical results for the Si(100)/SiO$_2$ interface. The interface is modelled by the superposition of three slabs : 1) four layers of silicon crystal to represent the silicon part; the bottom layer is saturated by hydrogen atoms whereas the top layer is at the interface. 2) a layer of oxygen atoms "adsorbed" on this top silicon layer; the interface may thus be primarily viewed as an oxidative adsorption of oxygen. Reconstruction at saturation is weak. The oxidation of the surface silicon atoms under oxygen adsorption leads to Si atoms that have different oxidation numbers, in agreement with XPS results. 3) a few layers of silica added epitaxially; the silica distorts to adapt to the geometry of the silicon crystal beneath. Only half of the oxygen layer is covered by the Si(+IV) ions of the silica. The interface model that results is close to that proposed by Ohdomari *et al.*[1,2] but differs by fine geometric details and does not proceed from the same construction.

1. Introduction

For many years, the adsorption of oxygen on Si(100) has been the object of many studies[3]. Indeed, it is a fundamental matter in semiconductor technology[4,5]. Silicon is the most used semiconductor; one of the reason is the good quality of its oxide[6]. A better knowledge of the microscopic structure could help to improve the performances of electron devices by reducing the thickness of the Metal-Oxide-Semiconductor gate of the CMOS transistors. The Si(100)/SiO$_2$ interface structure remains still unclear. The atomic arrangement at the interface is controversial. Calculations from quantum chemistry requires a simplification that may be excessive. Our models have to be periodic, this does not allow to consider amorphous arrangements. The size of unit cell must remain reasonable since it determines the size of the calculations; this forbids to consider defects. However we expect from these calculations a better understanding of the physics that governs the interface.

2. The model

```
_____                          _____
       silica layers                              silica layers
_____        Interface         ---------------
       O_ads                                      silicon layers
_____                          ---------------
       silicon layers
_____

         A)                                             B)
```

Figure 1. A) Our model for the interface SiO$_2$/Si(100); in this model, the face down (at the interface) of the silica layers is made of Si^{+4} ions whereas the face up (external) is made of O^{2-} ions. B) The abrupt model where the silica phase is directly attached to the silicon phase.

We present preliminary results of Hartree-Fock Periodic calculations (CRYSTAL program) for the Si(100)/SiO$_2$ interface. We consider a slab model, defined by a unit cell infinitely repeated in two dimensions by two cell vectors. The interface is in the middle of the slab and is sketched in figure 1; it results from the coupling of the O atoms that oxidise the (100) surface of a silicon slab with the Si(+IV) ions of the silica. The top face belongs to silica and is made of a layer of oxygen ions. The face at the bottom of the slab belongs to the silicon phase and the dangling bonds (DBs) are saturated by hydrogen atoms.

Our strategy to build a model implies the superposition of three thinner slabs :

1) the first one at the bottom represents the silicon part; it consists of four layers of Si atoms; the side of the slab represents the Si(100) surface at the interface whereas the dangling bonds of the bottom side are saturated by hydrogen atoms (see fig.2). The Si-H distance is 1.48 Å. We will show in section 4.2 that the clean surface reconstruction is important. Surface atoms form dimers; this induces a reorganisation of the sublayers. However (section 4) under adsorption at saturation, reconstruction becomes weak. In this preliminary work, the reorganisation of the sublayers will be neglected for the construction of the interface.

2) a layer of oxygen atoms "adsorbed" on the top silicon layer; the interface may thus be primarily viewed as an oxidative adsorption of dioxygen molecules. We have not allowed the migration of the oxygen atoms within the interspace of the first sublayer; a full relaxion of the sublayers would be very computer-time consuming. In the present paper, we have only considered the p(2x1) unit cell. The clean Si(100) surface is reconstructed. At partial coverage the reconstruction remains important : a dimerisation of the surface Si atoms associated to a reorganisation of the sublayer atoms. This one is much less significant at saturation. In section 5 we will also justify the adsorption modes by their analogues in carbon chemistry. We will also discuss the cleavage of the dioxygen molecule on the surface.

3) a few layers of silica in epitaxy on top of the silicon surface saturated by oxygen atoms; the silica distorts to adapt to the geometry of the silicon crystal beneath.

Our model (figure 1A) differs from an abrupt model (figure 1B) where the two phases are placed side to side without oxidation of the silicon layers. A difference concerns the order of the silica layers : in our model, the bottom layer of the silica

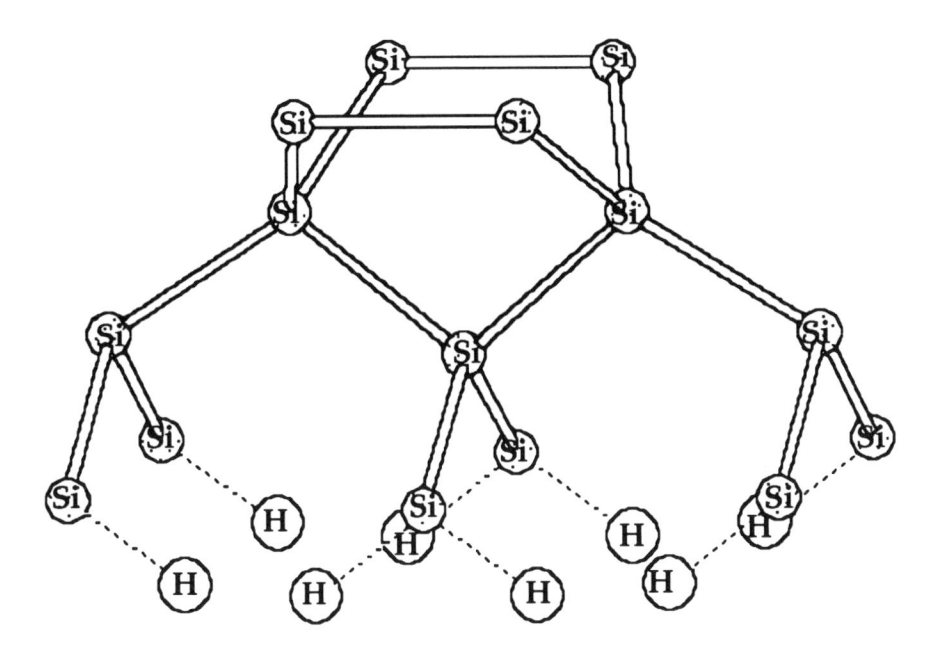

Figure 2. Perspective view of Si(100) slab model. The model contains four layers of Si atoms (small circles); the dangling bonds of one of the side of the slab are saturated with hydrogen atoms (5th layer, $d_{Si-H}=1.48$ Å). One side represents the 100 surface, the other side is saturated with H atoms.

phase is made of silicon ions while the upper one (external) is made of oxygen ions. It is the opposite for model B. This can be viewed as a problem of slab termination. The main difference however concerns the total electron count. As the O_{ads} take electrons to the silicon atoms, our model contains less electrons than model B to be distributed among the silicon atoms. This might be important to build the bonds of the interface unless the electrons required to form them could be adjusted by taking or removing electrons from DBs on the ions at the external face of the silica.

3. Method of calculation.

Calculations have been made with two versions of the crystal program. The last issue allows more possibilities and we are starting to repeat some of the calculations that we are presenting here.

The CRYSTAL 92[7] *ab initio* periodic-Hartree-Fock program uses a restricted HF Hamiltonian whereas the last release CRYSTAL 95[8] allows using an UHF Hamiltonian and estimates correlation effect by using a density functional correction after the SCF calculation[9]. The method is described by Pisani *et al.*[10]. Molecular calculations have been performed to generate optimised structures; these structures have been used as initial guess for the periodic calculations. They are also considered to justify the analogy between the structures of adsorption modes with molecules. They were performed using the MONSTERGAUSS program[11]. For the oxygen atom, we have used the PS-31G or the PS-31G* basis set derived from Bouteiller et al.[12,13]. For the polarisation function, the d orbital exponent is 0.8. For the silicon atom, we have modified the two most diffuse sp shells of the PS-211G or PS-211G* basis sets used for the bulk stishovite, SiO_2 [13] to improve the description of the silicon bulk. Two different exponents were used for the silicon and the silica. We have used the Durand and Barthelat effective core pseudopotentials[14].

For the largest calculated systems, the symmetry is lowered and we cannot afford polarisation functions. In this case, the cohesive energy of the diamond Si bulk is reduced by 15 kcal/mol per Si atom.

The adsorption energies have been calculated according to the expression:
$$E_{ads} = E_O + E_{slab} - E_{(O+slab)}$$
where $E_{(O+slab)}$ is the total energy of the optimised oxidised slab, E_O is the Unrestricted Hartree-Fock energy of an isolated oxygen atom in the triplet state (-15.6319 a.u.) according the calculations of Bouteiller et al.[12]) and E_{slab} is the energy of the optimised reconstructed slab. A positive energy corresponds to a stable adsorbate/substrate system.

The interface energies are defined by
$$E_{inter} = E_{O/Si(100)} + E_{SiO_2} - E_{SiO_2/O/Si(100)}$$
and refers to the binding between the silica phase and the oxidised silicon phase. This corresponds to the building of a SiO bond per unit cell. To make a comparison, the energy per bond of the β-cristobalite structure is 128.6 kcal/mol with the same basis set (145.5 kcal/mol with a correction for the correlation); this represents a cohesive energy of 89.25 eV per unit cell.

4. The Si(100) surface.

In the diamond structure each Si atom is in a sp^3 hybridisation and is bonded to four neighbours. The surface atoms of the (100) face are only bound to two Si atoms of the first sublayer and have two so-called "dangling bonds".

4.1 THE UNRECONSTRUCTED Si(100) SURFACE.

We find with the same basis set that we have used for the solid that the SiH_2 singlet ($^1A_1$ state) is more stable than the triplet ($^3B_1$) by 14.4 kcal/mol at the SCF level at variance to the carbon analogue; this is in agreement with experimental, and theoretical results (a difference of 21. kcal/mol) [15,16]. The σ_z level with more s character is lower than the p level. The unreconstructed surface dangling bonds also have two different components. One doubly occupied $\sigma(s+p_z)$ and one vacant p_x orbital [17,18]. The system is an insulator. We find that the singlet state is more stable than the triplet state by 8.7 kcal/mol per dimer at the Hartree-Fock level. After Perdew

91[9]correction, this energy difference increases (to 12.2 kcal/mol) since the correlation is more important when electrons are paired.

4.2 THE RECONSTRUCTED Si(100) SURFACE.

The general assumption is that the reconstruction occurs to eliminate half of the dangling bonds[19] by dimerisation. The complete coupling that would generate a double bond ($\sigma^2\pi^2$ state) is not favourable since the π bonds are weak for the atoms from the third row of the periodic table and when the dimer remains symmetric (no buckling) the most stable state is a triplet state ($\sigma^2\pi\pi^*$). The optimised distance of the dimer is 2.43 Å with the UHF hamiltonian and the energy for the dimerisation represents 55.8 kcal/mol (63.7 kcal/mol after Perdew 91 correction). These values have to be compared to those obtained for the $\sigma^2\pi^2$ state with the RHF hamiltonian, 2.17 Å and 33.1 kcal/mol only (51.9 kcal/mol after Perdew 91 correction). As the dimerisation requires an important motion of the Si atoms (the Si-Si distance in the unreconstructed surface is 3.84 Å), it is necessary that the sublayer atoms also move to follow these surface atoms. To reduce the numbers of parameters to vary, we have used the results obtained by Roberts[20] and we have coupled all the atomic motions, assuming them proportional to a single parameter, x, the dimerisation degree. x equal zero corresponds to the unreconstructed surface, x equal 1 to the reconstruction assumed by Roberts (see figure 3).

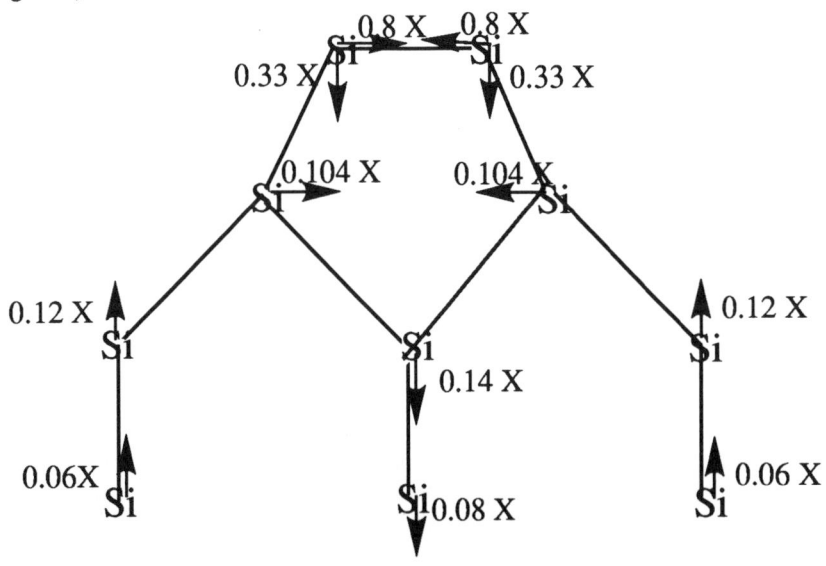

Figure 3. Roberts parameter used for our derived model. X is the dimerisation degree. With such a constraint, the most stable system for the clean surface is obtained by our UHF calculations when there are two parallel spins on the surface atoms. The Si-Si distance for the UHF calculation corresponds to a single Si-Si bond 2.43 Å; the distance obtained within the local density approximation by Roberts[20] is shorter, 2.23 Å. These dimer distances are very short in comparison with the distance in the unreconstructed surface, 3.84 Å.

5 The Oxygen adsorption on the Si(100) surface.

We are now going to oxidise the Si(100) surface with an atomic oxygen layer at two different coverages.

5.1 THE ADSORPTION OF ATOMIC OXYGEN ON THE Si(100) SURFACE AT A COVERAGE θ=1/2.

At a coverage θ=1/2, there is one oxygen atom per Si surface dimer. The most favourable oxygen adsorption is at the bridge site[21, 22] above a dimer and not between two dimers(see figure 4).

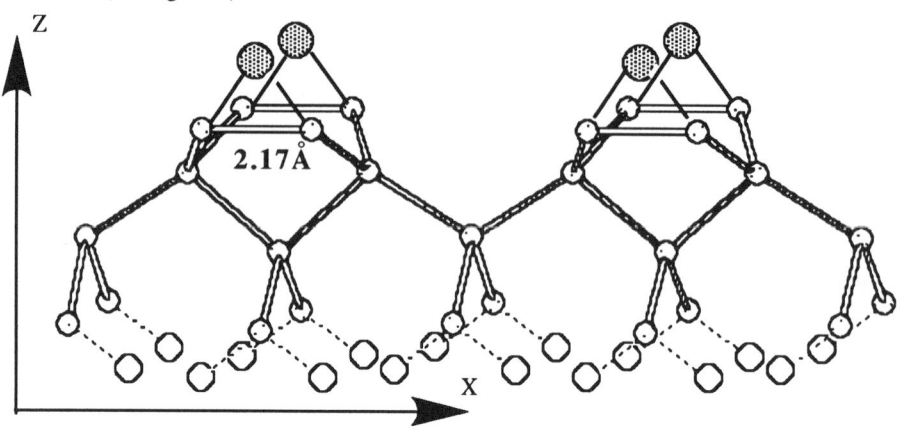

Figure 4. The adsorption of the atomic oxygen at θ=1/2. the oxygen atoms "on the dimers" are represented by ●; this system is reminiscent of the epoxide structure.

This can be surprising since at the bridge site, it would saturate the dangling bonds; the reason is that the distance between the dimers (or even the distance between the surface atoms in the unreconstructed surface) is too large to accommodate a bridging oxygen atom in the middle. The oxygen atom would move toward one of its neighbour and move over a dimer. The resulting species has all his electrons in pairs and is analogous to an epoxide; this analogy explains the stability. The optimal Si-Si distance is 2.17 Å. The adsorption energy is 47 kcal/mol. A secondary minimum for an elongated Si-Si distance (3.23 Å) corresponds to silano-ethers ..Si=Si-O-Si=Si.. with a linear arrangement Si-O-Si; it is unstable relative to the "silano-epoxide" structure.

5.2 THE ADSORPTION OF ATOMIC OXYGEN ON THE Si(100) SURFACE AT A COVERAGE θ=1.

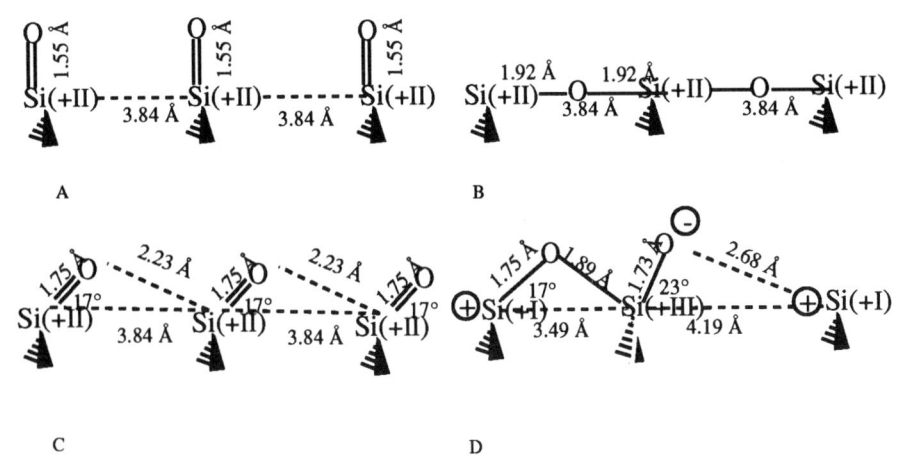

Figure 5. The adsorption modes for the atomic oxygen. The adsorption on top (A) isolobal to formaldehyde. The bridging adsorption modes for the unreconstructed surface, symmetric (B) or asymmetric (C), and for the reconstructed surface (D). Oxidation numbers are indicated in parentheses

In both the A and B modes (figure 5), silicon atoms are tetravalent. The SiSi dimer bonds are broken and there is *a priori* no reason for a surface reconstruction. These two modes are not the best, mode B because of the too large space between the silicon atoms as explained in the previous section and mode A, in spite of the analogy to the formaldehyde, since the atoms from the third period of the Mendeleev table do not easily form double bonds

The most favourable adsorption mode for the atomic oxygen is a **bridging mode D**. As compared to structure B, this structure is stabilised by two modes :

i) the bridging becomes asymmetric (see structure C) with a short SiO bond and a long one (a dative bond)

ii) the system is weakly reconstructed. The dimerisation of the surface (3.49Å) allows to approach the distance of 3.29 Å of silano-ethers.

It results that the silicon atoms are not equivalent. In the zwitterionic state, they have different formal charges; the triply coordinated silicon atom has a positive charge whereas the singly coordinated oxygen atom bears a negative charge. The SiO distance of 1.73 Å is short revealing some double bond character. In figure 5, we have indicated the oxidation states as I for the silicon atom bound to one oxygen atom and III for the other silicon atom. They have indeed very different local environments which have consequences on the Si 2p core levels seen in photoelectron spectroscopies. However, formal oxidation states of II would be still justified since each Si has only two electrons arising from the partition of the electrons of the two SiSi bonds (no electrons are counted for the ionic SiO bonds); an oxidation number of I for the Si bound to one oxygen atom would only appear in the diradicalar species. The adsorption energy (per O atom at θ=1) is 55.5 kcal/mol. This represents the saturation since the adsorption from θ=1/2 to θ=1 is exothermic.

5.3 THE ADSORPTION OF DIOXYGEN MOLECULES ON THE Si(100) SURFACE AT SATURATION.

The identification of O_2 adsorbed species raises the question of the initial stage of the oxidation[23]. Experimentally, both O and O_2 species can be identified on Si surfaces[3, 24]. Many experimental studies concluded that there is a metastable molecular precursor[25, 26] in the initial stage of oxidation of Si surfaces. Its structure is however still unclear [27].

We have found two adsorption modes for O_2 allowing the dimerisation of the surface atom at $\theta=1$ (one O_2 per dimer).

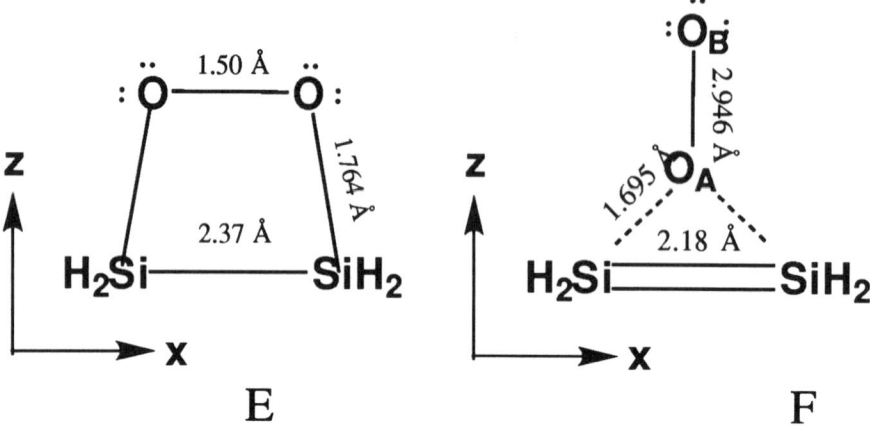

Figure 6. Cluster models for the molecular adsorption. The peroxo mode (E) is more stable than the perpendicular mode (F). Similar values are obtained for the periodic calculations.

In structure E, O_2 is parallel to the surface; it corresponds to peroxo compound and cluster models in the singlet state[23, 28] can reproduce a similar pattern; however for the periodic calculation at saturation, the heat of adsorption, 49.5 kcal/mol, is weaker than that for the $Si_2H_4O_2$ cluster model, 67.9 kcal/mol.. In the periodic model, the dimerisation remains and the adsorption mode "on the dimer" is the best. The SiSi distance, 2.34 Å is typical of a single SiSi bond.

Structure F with O_2 normal to the surface is also stable; the triplet state for the $Si_2H_4O_2$ cluster is 1 kcal/mol above the parallel mode (singlet state). This state corresponds to the final products obtained by Hoshino [27]. It has a very long O-O distance and the SiSi dimerisation is important. The periodic calculations with the UHF Hamiltonian leads to similar results. The conversion from the parallel mode to the perpendicular mode goes through a very high barrier 75.1-75.7 kcal/mol (85.6-94.75 kcal/mol with Perdew91 correction) for the two paths that we have investigated[18, 22]; this geometric motion implies a conversion singlet to triplet. Even if dissociation from the triplet state is easy, it is thus not likely to start by a conversion to the perpendicular mode.

5.4 THE O$_2$ DISSOCIATION ON THE Si(100) SURFACE.

We have investigated several modes for the O$_2$ dissociation on the Si(100) surface. One O atom has to pass over a silicon atom to move from the space "on the dimer" (as imposed in structure E) to the adjacent one "between the dimers ". If this O atom migrates on the surface, we can thus optimise the intermediate configurations where it lies right above one silicon atom. There are several possibilities i) a *symmetric* one, with the two O atoms over the adjacent Si atoms; this leads to structure D where the surface is unreconstructed, ii) an *asymmetric* one, where the O atom that remains above the dimer space has rotated and has moved between the Si atoms becoming nearly aligned with them (structure G, figure 7).

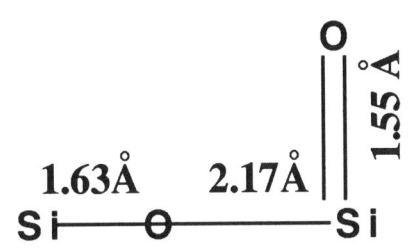

Figure 7. Asymmetric structure (G) for an oxygen atom above one silicon atom. This intermediate is more stable than the peroxo structure (E) by 1.4 kcal/mol.

The intermediates have energies within same range than as of the peroxo compound. We have found very high barriers to reach these intermediates, whereas the migration from there is exothermic and without a significant activation barrier. The lowest path that we have found preserve the symmetry and goes through structure D. At first, Si-Si is elongated; assuming constant SiO bonds, OO approaches the surface and the OO overlap population decreases and becomes zero at the distance of 3.13 Å; then, O$_2$ dissociates with no activation barrier; finally the SiSi distance is elongated to 3.84 Å; the top mode of O adsorption (structure A) represents an intermediate before the evolution to the final structure (structure D).

The activation barrier, 97.3 kcal/mol, is still very high. Estimation for the correlation using the Perdew91 correction does not decrease this value. The correlation is large for the peroxo compound that has adjacent lone pairs on the oxygen atom and thus, the correlation decreases when the O-O bond increases. The activation barrier is then slightly larger, 109 kcal/mol. This value exceeds the initial adsorption energies, meaning that this transition structure should no longer be adsorbed. The atomic oxygen migrates being in a triplet state. We have verified that the UHF calculation for the postulated transition state was lowered when a triplet state was assumed. This motion leads to atomic desorption (an atomic adsorption at $\theta=1/2$ for the O that remains); the O atom expelled in the gas phase is then available for an adsorption on a different site. Such adsorption-desorption process is fundamentally different from an atomic migration of the adsorbed O species generated from an O$_2$ precursor. Adsorbed O$_2$ generates an atomic oxygen desorbed and readsorbed later on.

Our calculations show that, in spite of the weak energy differences between the different adsorption modes, the dioxygen is not mobile on the surface and does not cleave well. Difficulties to find a dissociation on perfect surfaces are consistent with the coexistence of molecular and atomic oxygen on the surfaces; it also suggests that dissociation would not occur on perfect surfaces but would take place on defect sites.

6 The epitaxial growth of silica.

6.1 THE CHOICE OF THE SILICA, THE β–CRISTOBALITE.

We now have to find a silica suitable for epitaxial growth; the unit cell has to be a multiple of the unit cell of O/Si(100) system and of the silicon structure beneath. Many different SiO_2 phases exist. We choose the β–cristobalite for two main reasons. The first one is that it has a cubic structure close to that of the diamond. β–cristobalite structure may be thought as derived from the silicon diamond structure by expanding the cell parameter and inserting O atoms within Si-Si bonds. The second reason is the vicinity of the cell parameter, 5.20 Å, with that required for a multiple cell of the Si(100) layer : the $\sqrt{2} \times \sqrt{2}$ R 45° cell; $\sqrt{2}$ times 3.84 Å gives 5.43 Å. Such distortion seems affordable. The in-plane expansion goes along with a compression normal to the plane (c decreases from 7.35 Å to 6.88 Å). This only costs 3.1 kcal/mol per SiO_2. An analogous distortion for the α form of the cristobalite costs 14.7 kcal/mol.

6.2 THE SUPERPOSITION.

At the interface, we have to build a bond between the O_{ads} atoms, those "adsorbed" on the silicon phase, and the silicon ions of the silica. Note that the

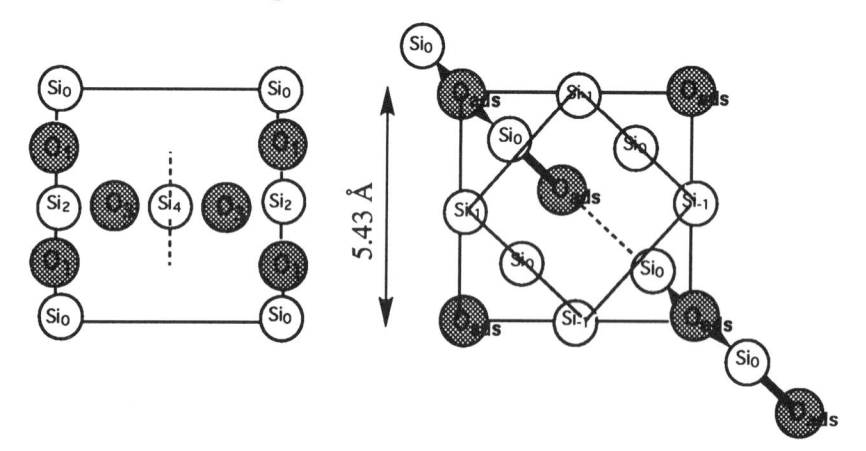

Figure 8. Left hand side: the unit cell of the silica; only the four first layers are represented; doted lines indicated the missing bonds to the silicon atoms of the fifth layer. Right hand side : the $\sqrt{2} \times \sqrt{2}$ R 45° cell of the silicon surface with the Oads. One diagonal is extended to represent the top view of pattern D (figure 5). The primitive cell is shown in doted lines. The two cells (top and bottom) are centered to generate O-Si bond by superposition with no shift. The atom labels refer to the position of the layer with respect to the interface.

oxidation states are already determined on the fragment layers and do not vary. The Si ions of the silica already have reached the maximum oxidation number (IV) in the silica.

Due to the choice of the unit cell, the $\sqrt{2} \times \sqrt{2}$ R 45° cell, the O atoms of the silica layer are rotated by 45° relative to the O_{ads} atoms "adsorbed" on the silicon phase. It is not possible to place the Si ions at the bridging sites between two O_{ads} atoms; the rotation does not allow a tetrahedral arrangement and the distances would not fit without excessive repulsion. The Si ions are therefore "above one O_{ads} atom". Only half of the oxygen layer is covered by the Si(IV) ions of the silica; indeed, the unit cell for the silica has one silicon atom per layer whereas the unit cell for O/Si(100), the $\sqrt{2} \times \sqrt{2}$ R 45° cell, contains two O_{ads} per layer.

6.3 THE INTERFACE THAT CORRESPONDS TO θ=1.

The study of the atomic adsorption, see section 5.2 and figure 5, concluded to the existence of two different O_{ads} atoms, one bridging atom and one singly coordinated atom. The reactivity of the singly coordinated atom is the strongest and therefore we expect the Si ions over these O_{ads} atoms as sketched in figure 5, structure D. This is indeed the most favourable case. Starting from there, we have optimised the new vertical distance of the O_1 to the dimer (1.76 Å), the L_1 distance (0.31 Å), the θ_1 and θ_2 angles (5° and 14° respectively). This values leads to a Si_0-O_1 bond length of 1.85 Å and to a Si_0-O_1-Si_2 angle of 109°.

Because of the symmetry loss, we had to remove the d polarisation functions. With these new basis sets, the interface energy stabilisation is 56.2 kcal/mol. This value is inferior to the energy per SiO bond in the various silica but remains important.

6.4 THE INTERFACE THAT CORRESPONDS TO θ=1/2.

At θ=1/2, the O_{ads} atom is bridging the dimer, the adsorption pattern being analogous to an epoxide structure[18, 21, 22, 29]. Similarly, the best adsorption mode is for a silicon atom tricoordinated, above these bridging O_{ads} atom. The structure at θ = 1 is more stable than the structure at θ = 1/2 plus one atomic oxygen by 67.7 kcal/mol. Thus, the most stable interface does not have vacancies of oxygen atom in the wide Si-Si spacing.

6.5 THE MODEL FOR THE INTERFACE.

The model contains Si atoms with different oxidation numbers as for the surface adsorption. This in agreement with XPS[30, 31]. The interface model that we have obtained can appear as rather close to that proposed by Ohdomari et al.[1, 2]. The similitude in the final geometry hides the differences in the approaches. We justify the geometry from a construction that follows an oxidative process instead of merely relaxing the atomic positions at the interface. Ohdomari's model was based on strain-energy minimisation; there was no relaxation of the sublayers for the Si(100) surface. Our calculations show that an important energy gain is associated to the surface reconstruction even if the model is closer to the unreconstructed system than to the dimerised structure. We obtain a Si$_0$-O$_1$-Si$_2$ angle of 109°, whereas the value in the simplified scheme by Engels[3] is 180° as in the β—cristobalite. Ohdomari considers that there is a bond between Si$_2$ and Si$_0$; this is important since a donation from Si$_0$ to Si$_2$ modifies the oxidation states of these two silicon atoms. In our model, we have not linked the Si$_2$ to the Si$_0$ atoms since both are positively charged. Results of the calculations are not unequivocal; the calculated distance is 2.70 Å is short but longer than an usual Si-Si single bond and the overlap population is weak, 0.058, but not negligible. The atomic positions of Ohdomari's model have been recently fully relaxed within the local density approximation to the density functional theory[32] using model B, the silicon ions at the external face being saturated by H atoms. These positions are very close to those derived by our construction with a shorter Si$_2$-Si$_0$ bond (2.38 Å).

The Si-O-Si angles are very relaxed (135° and 144°) close to the value of the α—quartz, 144°[2, 33]. This difference are probably due to the constraints that we have imposed to our relaxations at this stage of our work. They also can result from the model. As shown in section 2, in our model the silicon at the interface have been oxidized and the construction of the SiSi bond requires that two electrons are removed from the silica outmost layer to be transferred to the σ$_{SiSi}$ bonding orbital when these two atoms overlap. Model B, that has the electron pair for the SiSi bond, requires an opposite transfer to fix the oxygen ions. In both cases, this electron transfer partially neutralises the charge of the ions at the surface. The optimal electron count may be imposed by the saturation of the external face of the silica. Pasquarello, thus, has saturated the silicon ions of model B. This could be also imposed to our model by the hydrogenation of the outmost oxygen ions. Ohdomari's model seems from these models to require the use the silica layers as electron reservoir to adjust the electron count at the interface. If this transfer can be possible when the electron count is not appropriate (no saturation of the external face of the silica), the choice of the model is not very important and the different models would give the same interface pattern. This implies the exchange of electrons between the interface and the top layer of the silica.

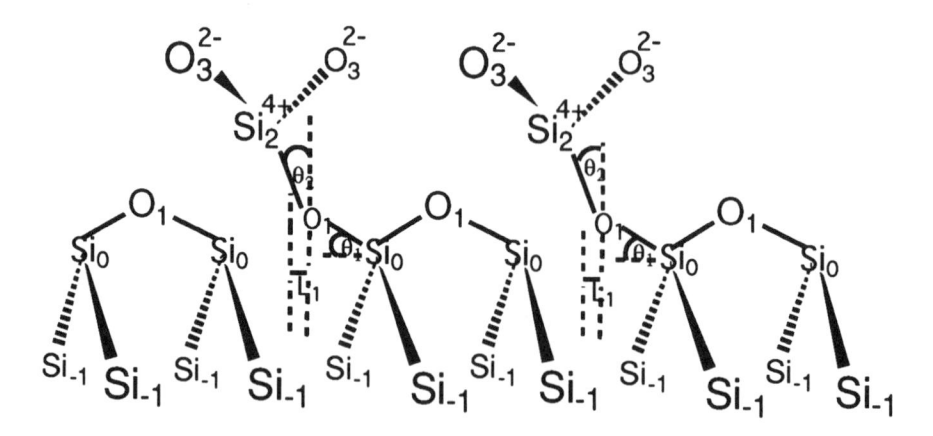

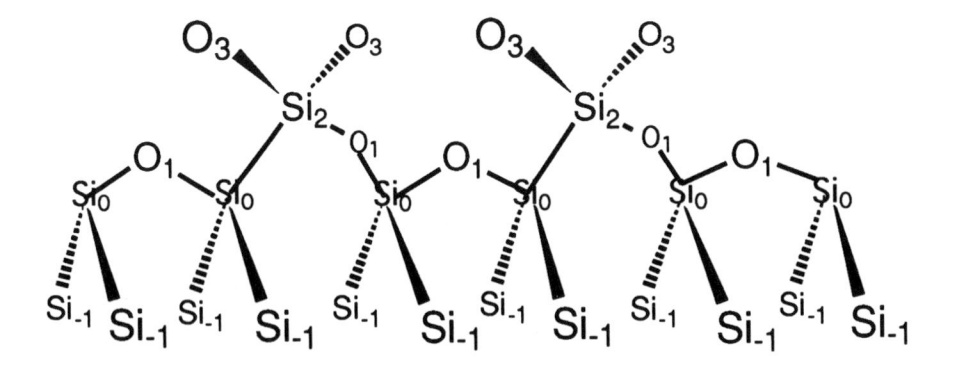

Figure 9. Our model for the interface after optimisation and the model by Ohdomari.

7 Conclusion

Our calculations show that the reconstruction of the clean surface is important but becomes weak when the surface is covered by oxygen atoms. Both atomic and molecular oxygen coexist on the Si(100) surface. Large activation barriers are required for the dissociation on the perfect surface and the presence of defects is probably required. The most likely path is a desorption-readsorption mechanism for an atomic species. Adsorption at saturation involves two different atomic oxygen atoms; one is bridging and the other one is singly coordinated; there are also two different surface silicon atoms with different oxidation states (I and III) in agreement with photoelectron spectroscopy [30, 31, 34]. This pattern remains for the interface model. The interface energy is large.

References

[1] Ohdomari, I., Akatsu, H., Yamakoshi, Y. and Kishimoto, K. (1987) Study of the interfacial structure between Si(100) and thermally grown SiO_2 using a ball-and-spoke model *J. Appl. Phys.*, **62**, 3751-3754.
[2] Ohdomari, I., Akatsu, H., Yamakoshi, Y. and Kishimoto, K. (1987) The structural models of the Si/SiO_2 interface *J. of Non-Crystalline Solids*, **89**, 239-248.
[3] Engel, T. (1993) The interaction of molecular and atomic oxygen with Si(100) and Si(111) *Surface Science Reports*, **18**, 91-144.
[4] Balk, P. (1988) The Si/SiO2 System *Material Science Monographs*, **32**,
[5] Fair, R. B. *Microelectronics Processing: Chemical Engineering Aspects* American Chemical Society, Whasington, DC,(1989).
[6] Sofield, C. J. and Stoneham, A. M. (1995) Oxidation of silicon: the VLSI dielectric? *Semiconductor Science and Technology*, **10**, 215-244.
[7] Dovesi, R., Pisani, C., Roetti, C., Causa, M. and Saunders, V. R. (1988)*Crystal*, QCPE Program n°577, Indiana University, Indiana,
[8] Dovesi, R., Saunders, V. R., Roetti, C., Causa, M., Harrison, N. M., Orlando, R. and Apra, E. (1996)*Crystal95, users manual*, University of Torino, Torino,

[9] Perdew, J. P. *Unified Theory of Exchange and Correlation Beyond The Local Density Approximation* Nova Science,New York (1991).
[10] Pisani, C., Dovesi, R. and Roetti, C. *Hartree-Fock Ab Initio Treatment of Crystalline Systems 48*; Lecture Notes in Chemistry, Springer Verlag,Berlin Heildeberg New York London Paris Tokyo (1988).
[11] Poirier, R. A. and Peterson, M. (1989) *Monstergauss*, Dept. of Chemistry Memorial University of New foundland, St. John's, Dept. of Chemistry Memorial University of New foundland, St. John's, Newfoundland, Canada.,
[12] Bouteiller, Y., Mijoule, C., Nizam, M., Barthelat, J. C., Daudey, J. P., Pelissier, M. and Silvi, B. (1988) Extended gaussian-type valence basis sets for calculations involving non-empirical core pseudopotentials *Mol. Phys.*, **65**, 295.
[13] Silvi, B., Jolly, L.-J. and D'Arco, P. (1992) Pseudopotential periodic Hartree-Fock study of the cristobalite to stishovite phase transition *J. Mol. Struct.*, **260**, 1-9.
[14] Durand, P. and Barthelat, J. C. (1975) A theoretical Method to Determine Atomic Pseudopotentials for Electronic Structure Calculations of Molecules and Solids *Theor. Chim. Acta*, **38**, 283-302.
[15] Becke, A. D. (1993) A new mixing of HF and LDF theories *J. Chem. Phys.*, **98**, 1372-1377.
[16] Curtiss, L. A., Raghavachari, K. and Pople, J. A. (1993) Gaussian-2 theory using reduced Moller-Plesset orders *J. Chem. Phys.*, **98**, 1293-1298.
[17] Liu, Q. and Hoffmann, R. (1995) The Bare and Acetylene Chemisorbed Si(001) Surface, and the Mechanism of Acetylene Chemisorption *J. Am. Chem. Soc.*, **117**, 4082-4092.
[18] Markovits, A. and Minot, C.; (1996)*The oxidation of the perfect Si(100) surface. Ab initio periodic pseudopotential Hartree-Fock calculations.*, Varna, World Scientific Press, Ed.J. Marshall.
[19] Schlier, R. E. and Farnsworth, H. E. (1959) Structure and Adsorption Characteristics of Clean Surfaces of Germanium and Silicon *J. Chem. Phys.*, **30**, 917-926.
[20] Roberts, N. and Needs, R. J. (1990) Total energy calculations of dimer reconstructions on the silicon (001) surface *Surface Science*, **236**, 112-121.
[21] Miyamoto, Y. and Oshiyama, A. (1991) Energetics in the initial stage of oxidation of silicon *Phys. Rev. B*, **43**, 9287-9290.
[22] Markovits, A. and Minot, C. (1997) Ab initio periodic pseudopotential Hartree-Fock calculations of O_2 dissociation on perfect Si(100) surface *J. Mol. Catal.A*, **119**, 185-193.
[23] Hoshino, T., Tsuda, M., Oikawa, S. and Ohdomari, I. (1993) Theoretical consideration on dimer vacancy images in the STM observations of Si(001) surfaces in terms of the adsorption of O_2 molecules *Surf. Sci. Letters*, **291**, L763-L767.
[24] Silvestre, C. and Shayegan, M. (1991) Initial Stages of the Reaction of Oxygen with Si(100) *Solid State Commun.*, **77**, 735-738.
[25] Ibach, H., Bruchmann, H. D. and Wagner, H. (1982) Vibrational Study of the Initial Stages of the Oxidation of Si(111) and Si(100) Surfaces *Appl. Phys. A*, **29**, 113-124.
[26] Höfer, U., Morgen, P., Wurth, W. and Umbach, E. (1985) Metastable Molecular Precursor for the Dissociative Adsorption of Oxygen on Si(111) *Phys. Rev. Lett.*, **55**, 2979-2982.

[27] Hoshino, T., Tsuda, M., Oikawa, S. and Ohdomari, I. (1994) Mechanisms of the adsorption of oxygen molecules and the subsequent oxidation of the reconstructed dimers on Si(001) surfaces *Phys. Rev. B*, **50**, 14999-15008.

[28] Zheng, X. M. and Smith, P. V. (1990) The Chemisorption Behavior of Oxygen on the Si(100) Surface *Surf. Sci.*, **232**, 6-16.

[29] Miyamoto, Y. and Oshiyama, A. (1990) Atomic and electronic structures of oxygen on Si(100) surfaces: Metastable adsorption sites *Phys. Rev. B*, **41**, 12680-12686.

[30] Hollinger, G. (1983) Multiple-Bonding Configurations for Oxygen on Silicon Surfaces *Phys. Rev. B*, **28**, 3651-3653.

[31] Himpsel, F. J., McFeely, F. R., Taleb-Ibrahimi, A., Yarmoff, J. A. and Hollinger, G. (1988) Microscopic structure of the SiO_2/Si interface *Physical review B*, **38**, 6084-6096.

[32] Pasquarello, A., Hybertsen, M. S. and Car, R. (1996) Comparison of structurally relaxed models of the Si(001)-SiO_2 interface based on different crystalline oxide forms. *Applied Surface Science*, **104/105**, 317-322.

[33] Hane, M., Miyamoto, Y. and Oshiyama, A. (1990) Atomic and electronic structures of an interface between silicon and β-cristobalite *Phys. Rev. B*, **41**, 12637-12640.

[34] Lu, Z. H., Graham, M. J., Jiang, D. T. and Tan, K. H. (1993) Si/SiO_2 interface studied by Al Kα x-ray and synchroton radiation photoelectron spectroscopy *Appl. Phys. Lett.*, **63**, 2941-2943.

SPATIALLY-SELECTIVE INCORPORATION OF BONDED-NITROGEN INTO ULTRA-THIN GATE DIELECTRICS BY LOW-TEMPERATURE PLASMA-ASSISTED PROCESSING

GERALD LUCOVSKY

Departments of Physics, Electrical and Computer Engineering, and Materials Science and Engineering, North Carolina State University, Raleigh, NC 27695-8202, USA; e-mail: gerry_lucovsky@ncsu.edu

Incorporation of nitrogen atoms into gate dielectrics: i) reduces defect generation at the Si-SiO$_2$ interface when incorporated at monolayer levels; ii) permits use of physically-thicker oxide-equivalent gate thicknesses when incorporated in the body of the dielectric; and iii) reduces boron penetration from p$^+$ poly-silicon gate electrodes through the dielectric films when incorporated at the poly-Si-dielectric interface, or in the body of the dielectric. This paper demonstrates that nitrogen atoms can be selectively incorporated into these different parts of device-quality gate dielectrics by low-thermal budget remote plasma assisted processing followed by rapid thermal annealing.

1. Introduction and Background

As device dimensions are scaled into the deep submicron to achieve higher levels of integration, there must be corresponding decreases in the oxide-equivalent thickness of dielectrics, $t_{ox,eq}$, to maintain current flow levels needed for circuit operation. Selective incorporation of nitrogen atoms (N-atoms) into ultra-thin silicon dioxide (SiO$_2$) dielectrics provides improvements in device reliability which are especially important when direct tunneling is the transport mechanism through the dielectric: $t_{ox,eq} < 3$ nm. Spatially selective incorporation of N: i) reduces defect generation at the Si-SiO$_2$ interface when incorporated at monolayer concentrations [1-4]; ii) allows the use of physically-thicker gate dielectrics when incorporated into the body of the dielectric as in oxide-nitride-oxide, ONO, or oxide-nitride, ON composites [5,6]; and iii) reduces boron atom (B-atom) penetration from p$^+$ polycrystalline silicon (poly-Si) gate electrodes through the dielectric films when incorporated at the interface the poly-Si gate electrode-dielectric interface or in the body of the dielectric [7-9]. This paper presents results from a research program at NC State University that has successfully demonstrated separate and independent control of N-atom incorporation in different parts of oxide gate dielectrics through combined use of low-temperature plasma-assisted processing at 300°C, and low-thermal-budget rapid thermal annealing (RTA), e.g., 30 s at 900°C [1,4-7,9,10].

1.1. Background: Remote Plasma-Assisted Processing

Figure 1 is a schematic diagram of a remote plasma processing chamber used for interface formation, film growth, and nitridation. The chamber provides gas introduction through a plasma excitation tube, as well as downstream injection through showerhead dispersal rings. Remote plasma processing is differentiated from direct plasma processing in three ways: i) it allows selective excitation of source and carrier gases; ii) the substrate is outside of the plasma glow; and iii) source gases injected downstream from the plasma generation region are prevented from back-streaming into the plasma

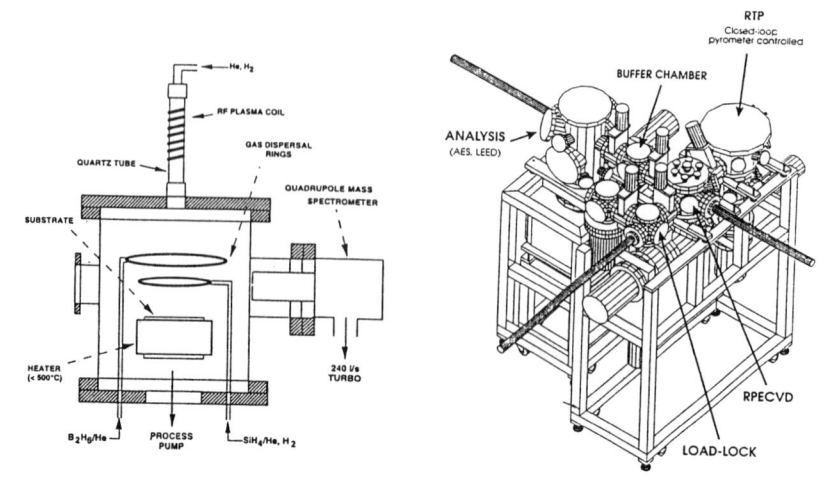

Fig. 1. Remote plasma chamber.   Fig. 2. Multichamber processing system.

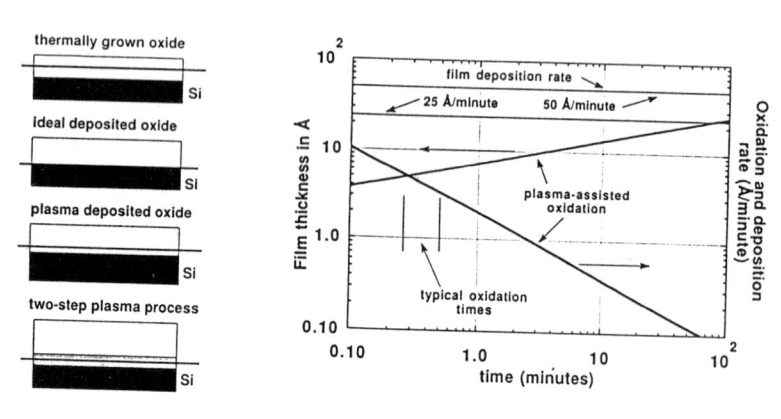

Fig. 3. Interface formation by oxidation and deposition.

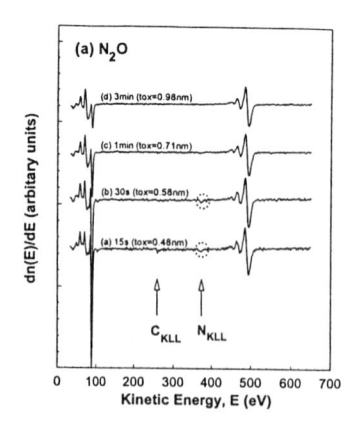

Fig. 4. Kinetics of oxidation and deposition.

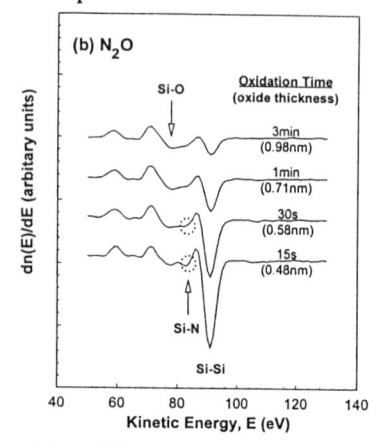

Figure. 5. Differential AES of N_2O oxidation (a) to 600 eV and (b) Si_{LVV} region.

generation region by gas flow and process pressure. Plasma processing chambers have been integrated into UHV-compatible multichamber systems as in Fig. 2, where the system also includes chambers for i) rapid thermal processing (RTP), and ii) on-line analysis, AES and low energy electron diffraction (LEED). The plasma chamber is fixtured for in-situ process diagnostics, including mass spectrometry (MS) and optical emission spectroscopy (OES). This multichamber system makes it possible to: i) interrupt plasma-assisted, oxidation, nitridation and/or deposition processes and perform on-line chemical analysis by AES; and ii) integrate on-line sequences of plasma processing with rapid thermal annealing (RTA). Table I illustrates the application of this processing sequence to the formation of device quality gate oxides. Figure 3 indicates several aspects of Si-SiO_2 interface formation: i) during thermal oxidation, substrate consumption occurs so that the Si-SiO_2 interface is buried below the top surface of the Si wafer; ii) in a ideal deposition process the metallurgical boundary between the Si and SiO_2 is at the original surface of the Si, iii) in a 'real' deposition

Table I Process Sequence for Forming Device-Quality Stacked Gate Oxide Dielectrics

Process Step	Processing Conditions	Processing Results
a) RPAO	Substrate temperature 300°C Process pressure: 300 mTorr Plasma Excited Mixture: He/O_2 (200 sccm He, 20 sccm O_2) Time: ~ 15 to 30 seconds	In-situ substrate cleaning (reduces C and F level) Forms Si-SiO_2 interface Grows passiviting oxide: ~0.5 nm Introduces sub-oxide bonding at Si-SiO_2 interface
b) RPECVD	Substrate temperature 300°C Process pressure: 300 mTorr Plasma Excited Mixture: He/O_2 (200 sccm He/ 20 sccm O_2) Down-stream Mixture: He/SiH_4 (20 sccm He/0.4 sccm SiH_4)	Forms body of dielectric film Deposition rate: 2..5-5.0 nm/min Stoichiometric SiO_2 No-IR detectable Si-H or Si-OH Low Si-OH (<< 5 at.% H)
c) RTA	Temperate: 900°C Time: ~ 30 seconds Low pressure or atmospheric inert gas ambient (e.g. Ar)	Reduces oxidation induced sub-oxide bonding at Si-SiO_2 interface Promotes densification of oxide films Reduces bonded-H (mostly in nitrides)

process, the activated oxygen species interact with the Si substrate so that oxidation takes place at least during the initial stages of the deposition (this has been designated as 'subcutaneous oxidation' and is discussed in Refs. 11 and 12); and iv) in the two step plasma oxidation/deposition sequence: the passivation layer of SiO_2 formed by the RPAO step is sufficient to prevent further oxidation during the RPECVD deposition [13,14]. Figure 4 displays rates for RPAO oxidation and RPECVD deposition. Kinetics for the RPAO process have been deduced from experimental studies of film thickness as a function of oxidation time using on-line AES (see Ref. 13), where it was shown that $t_{ox} = \alpha\, t^\beta$: α and β are experimentally determined parameters, t is the time in minutes, and t_{ox} is the oxide thickness in nm. For typical oxidation times of 15 to 30 s, the oxidation rate after 0.5 - 0.6 nm of oxide is grown is ~0.3 to 0.5 nm/minute, whereas deposition rates are higher, ~ 2.5 - 5.0 nm/minute, so that plasma-activated O-species

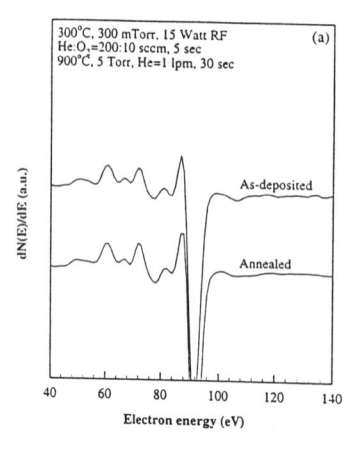

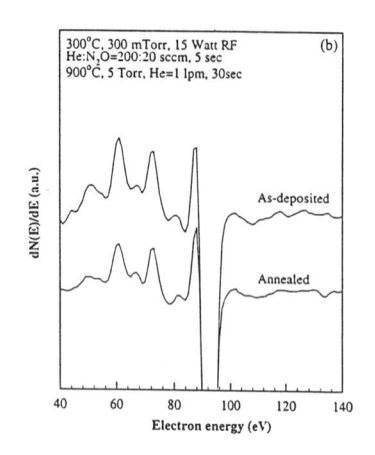

Fig. 6. AES spectra in Si_{LVV} region for RPAO in (a) O_2 and (b) N_2O as-deposited and after 30 s 900°C RTA.

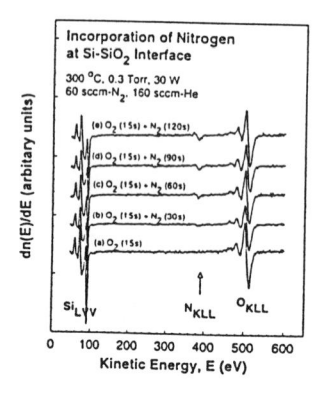

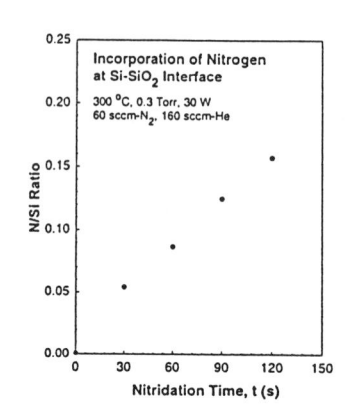

Fig. 7. AES for post oxidation nitridation.

Fig. 8. Linearity of and AES versus nitridation time for N_2 process.

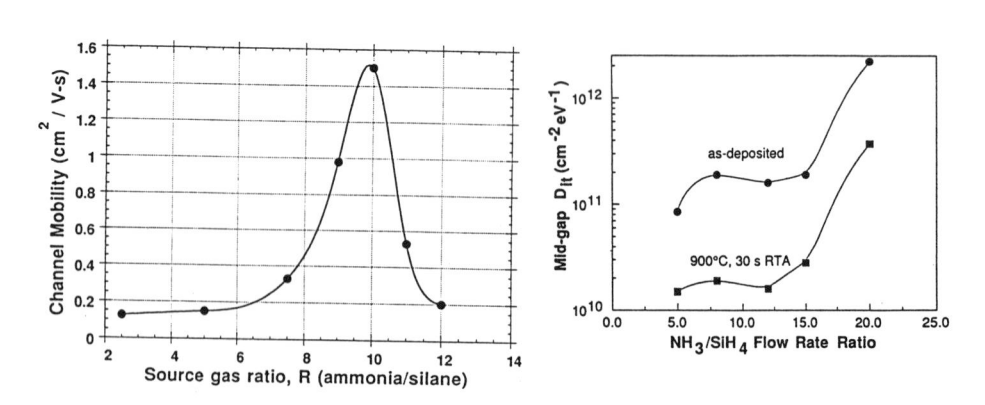

Fig. 9. TFT mobility versus R.

Fig. 10. D_{it} versus R for ONOs.

are consumed faster by the deposition reaction with SiH$_4$, than by oxidation at the buried Si-SiO$_2$ interface. Figure 4 explains results presented in Ref. 14, where it was shown that a pre-deposition RPAO at 250 to 300°C and an RPECVD oxide deposition at 400°C resulted in lower interface defect levels than a 400°C oxide deposition directly on a H-atom terminated Si-surface and associated with subcutaneous oxidation.

2. Si-SiO$_2$ Interfacial Nitridation

2.1. THE RPAO PROCESS

Interface nitridation has been achieved by an RPAO step using excited species from a remote He/N$_2$O plasma [1]. Figure 5(a) indicates the time evolution of differential AES spectra for this process, displaying Si$_{LVV}$, N$_{KLL}$ and O$_{KLL}$ features. Changes in Si$_{LVV}$ spectra shown in Fig. 5(b) coupled with changes in N$_{KLL}$ and O$_{KLL}$ features in Fig. 5(a) establish that N-atoms are incorporated in the *immediate vicinity* of the Si-SiO$_2$ interface. Three features in the Si$_{LVV}$ region establish interfacial nitridation: the Si-O feature at ~76 eV, the Si-N feature at ~83 eV, and the Si-Si substrate feature at ~91 eV. The 83 eV feature also includes a contribution from Si-suboxide bonding (see below); the 76 eV feature is assigned to Si-atoms with 4 O-neighbors, and the 83 eV feature has contributions from Si-atoms bonded to < 4 O-atoms and > 1 Si-atom, as well as Si bonded to at least 1 N-atom, and the 91 eV feature is assigned to Si-atoms with 4 Si neighbors. The oxide layer thickness is obtained from the relative signal strengths of the 76 eV and 91 eV features using an electron escape depth of approximated 0.6 nm as described in Ref. 14. This procedure is internally consistent in the sense that a plot of the Si-Si substrate signal as a function of the calculated thickness is exponential with a characteristic decay length equal to the electron escape depth. The Si-N and N$_{KLL}$ features are present at the initial stages of the oxidation process but are not observed once t$_{ox}$ > the electron escape depth. Combining AES data, with SIMS data of Ref. 1 and analysis of optical second harmonic generation (SHG) data (see Table III) [13], establishes that the N$_2$O RPAO produces essentially monolayer interfacial nitridation.

Figures 6(a) and (b) indicate derivative Si$_{LVV}$ AES spectra for oxide layers approximately 0.5 nm thick grown in (a) He/O$_2$ and (b) in He/N$_2$O, respectively, that are as grown at 300°C and after a 30 s 900°C RTA [15]. Based on AES amplitudes (I) in Fig. 6, the relative signal amplitude ratio, R, given by:

$$R = \{I(83 \text{ eV})/I(76 \text{ eV})\}_{\text{after RTA}} / \{I(83 \text{ eV})/I(76 \text{ eV})\}_{\text{before RTA}}, \quad (1)$$

decreases after the anneal. Comparison between these spectra with those in Figs. 5(a) and (b), combined with an observation that no N$_{KLL}$ signal is present in the samples prepared by RPAO in O$_2$, establishes that the 83 eV feature has contributions from both Si-N and Si-suboxide bonding. Similar spectra were obtained for films grown by rapid thermal oxidation (RTO) in O$_2$ and N$_2$O at 800°C, and after a 30 s 900°C RTA [15]. The sample prepared by RTO using N$_2$O does not display a detectable N$_{KLL}$ feature. To confirm that changes in the amplitude ratio were not due to oxidation during the RTA, relative amplitudes of the Si-Si Si$_{LVV}$ feature at ~ 92 eV, and the O$_{KLL}$ feature at ~ 510 eV were also monitored and found to be the same before and after the RTA. Table II contains normalized amplitude ratios, R, for the data presented in Figs. 6, as well as for the samples produced by RTO. Ratios less than one indicate that there is a relative decrease in the amplitude of the 83 eV sub-oxide bonding feature after the 900°C anneal.

TABLE II Changes in the relative intensities of the SiO$_x$ and SiO$_2$ AES at approximately 83 eV and 76 eV, respectively [15]

Interface Formation	R
Plasma O$_2$	0.88 ± 0.02
Plasma N$_2$O*	0.79 ± 0.02
RTO O$_2$	0.80 ± 0.02
RTO N$_2$O#	0.86 ± 0.02

*N-atom at the monolayer concentration range at Si-SiO$_2$ interface
#no N-atoms detected at Si-SiO$_2$ interface by N$_{KLL}$ AES

The effect of the 900°C anneal in reducing suboxide bonding also explains results from optical SHG measurements on oxidized vicinal Si(111) wafers off-cut ~ 5 degrees in the 11$\underline{2}$ direction [see Table III]. The SHG signal comes predominantly from the Si-SiO$_2$ interface [16]. The exciting source is a Nd:YLF laser at 1.17 eV. SHG performed on thermally grown oxides subjected to rapid thermal annealing at 900°C indicated a significant change in the relative phase of the SHG signals associated with the steps and terrace regions [16]. The SHG signal for the vicinal wafers is given by:

$$E(2\omega) = (A_1\cos\Phi + A_3\cos3\Phi(\exp(-i\Delta_{13}))), \qquad (2)$$

where A$_1$ and A$_3$ are the amplitudes of the harmonic signal components at Φ and 3Φ, Φ is the angle between the 11$\underline{2}$ direction and the incident electric field, and Δ_{13} is the relative phase. The relative phase is determined by the difference in resonance energies between the two harmonic components of the SHG signal. From Table III, the relative

TABLE III - Summary of results from studies of vicinal Si(111) surfaces off-cut ~ 5 degrees in the 11$\underline{2}$ direction [16]

(a) plasma processed interfaces	Phase, $\Delta\Phi_{13}$#	A$_1$/A$_3$#
O$_2$ - 15 s - 300°C	68	0.20 ±0.02
after 30 s 900°C RTA (0.5%O$_2$/Ar)*	23	0.35 ±0.03
N$_2$O - 30 s - 300°C	65	0.17 ±0.02
after 30 s 900°C RTA (Ar)*	11	0.35 ±0.03
(b) thermally-grown interface		
Furnace Oxidation at 850°C	72	0.19 ±0.02
after 30 s at 950° RTA (0.5%O$_2$/Ar)	23	0.33 ±0.03

see discussion below
• Processing conditions for optimum electrical properties [1].

phase of the as-grown thermal oxide is 72 degrees, and decreases significantly to ~ 23 degrees after the 900°C anneal. Similarly, relative phases for interfaces formed by 300°C RPAO also change markedly after a 30 s 900°C RTA. Prior to annealing, there are no significant differences between Δ_{13} for RPAO interfaces formed in N$_2$O and O$_2$, even though SIMS and on-line AES show a N-terminated interface for the N$_2$O process. After the RTA, Δ_{13} is equal to 23 degrees for the O-terminated interfaces, the same as obtained after annealing a thermally grown interface, but is reduced to approximately 11 degrees for the N-terminated interface. Combining the SHG results with the AES results we conclude that the Δ_{13} values of 67-72 degrees are characteristic of suboxide bonding arrangements, whereas the smaller values of Δ_{13} of 23 and 11 degrees are indicative, respectively, of more nearly idealized and planar interface bonding arrangements.

2.2. THE POST OXIDATION N_2 NITRIDATION PROCESS

Figure 7 presents an AES study of the second nitridation process in which an ultra-thin (~0.5-0.6 nm) oxide is first grown by the RPAO O_2 process, and then N-atoms are introduced at the interface by post deposition nitridation using a remote N_2 discharge [4]. The N_{KLL} signal is observed after the first post-oxidation exposure to the remote N_2 discharge, and grows with increasing time. Figure 7 indicates the AES N-signal strength as a function of nitridation time. The SIMS signals obtained from analysis of CsN+ also increase linearly with increasing time; similar SIMS data have been obtained by analyzing SiN- ions. Calibration of the SIMS signals is still a research issue, however, combining SIMS data of Ref. 1 with AES data, and the SIMS data in this study indicate that monolayer interface coverage by N occurs for an exposure time of approximately 90 s. The AES data presented in Fig. 8 is for the ratio of the N_{KLL} amplitude divided by the Si-Si_{LVV} amplitude, and also indicate linearity of N-signal strength with exposure time.

2.3. ELECTRICAL RESULTS

The results in Table IV are for field effect transistors (FETs) with interfaces nitrided by the plasma oxidation in N_2O and have 5.5 nm gate dielectrics [1]. The incorporation of interfacial N does not change the peak channel transconductance, g_m, but does reduce hot-electron induced changes in peak transconductance and threshold voltage, V_t. Differences in V_t are due to dopant atom redistribution for different interface processing.

Table IV Electrical Properties of FETs: Interfaces by N_2O RPAO [1,2]

(a) Interface Defects in MOS Capacitors

Interface Preparation	Interface traps, D_{it} (10^{10} cm^{-2}eV^{-1})
(a) plasma processed - 300°C	
oxygen terminated	1.0 ± 0.5
nitrogen terminated	1.0 ± 0.5
N-H terminated	20 ± 4
N-terminated (after RTA)	1.0 ± 0.5
(b) thermal oxidation	1.0 ± 0.5

(b) Initial Performance of FETs - Threshold Voltage (V_t) and Mobility

Interface Preparation	Threshold Voltage (V)	Peak Mobility (cm^{2}V^{-1}s^{-1})
Thermal Oxidation 900°C	0.33 ± 0.03	450 ± 5
Plasma Oxidation 300°C + RTA	0.38 ± 0.03	460 ± 5
Plasma nitridation 300°C + RTA	0.41 ± 0.03	460 ± 5

(c) Reliability of Short Channel FETs
5.5 nm gate oxide - 0.5 μm channel length

Interface Preparation	$\Delta g_m/g_m$	$\Delta V_t/V_t$
Thermal Oxidation 900°C	-16.2%	+9.9%
Plasma Oxidation 300°C + RTA	-20.5%	+14%
Plasma Nitridation 300°C + RTA	-12.5%	+6%

3. Nitride and Oxynitride Dielectric Films

Two different plasma-assisted deposition processes have been used to form nitride and oxynitride dielectric thin films [5,6]. The as-deposited nitrides produced by each of these processes are heavily hydrogenated, ~ 20 to 30 at.% H, whereas the concentration of the bonded-H in the as-deposited oxynitride alloys decreases with decreasing nitrogen content. The distinction between the two plasma-assisted processes is in the source gases used for the nitride deposition, and or nitride component of the oxynitride alloys. In the first nitride process the N-atom source gas is ammonia, NH_3 [5], and in the second process it is N_2 [6].

3.1. NITRIDE FILMS

3.1.1 *The NH_3 RPECVD Process*

The as-deposited nitride films produced by a 300°C RPECVD process using SiH_4 and NH_3 as the respective Si- and N-atom source gases are heavily hydrogenated. Activation of processes gases is by species from a remote He plasma. Details of the deposition process are discussed in Ref. [5]. The chemical composition and local bonding of the as-deposited nitrides vary with the source gas flow ratio, R, of NH_3 to SiH_4. Optimized electrical properties have been obtained in films deposited from a source gas ratio of R ~ 10; however, in order to obtain this level of performance, a post deposition RTA for 30 s at 900°C is required. Studies of the as-deposited nitride films by AES have established that films prepared with R < 10 are sub-nitrides, in the sense that they display Si-Si as well as Si-N AES features, whereas films prepared with R ≥ 10 are *quasi-stoichiometric* nitrides with no detectable Si-Si bonding, but with Si-N, as well as Si-H and/or SiN-H bonding [17]. Combining IR data which gives the distribution between Si-H and SiN-H, with i) the model calculation of Ref. 17 and ii) the on-line AES measurements, the atomic concentrations in as-deposited nitride films with R ~ 10 are [N] ~0.42, [Si] ~0.28 and [H] ~ 0.3. This alloy composition corresponds to an average number of bonds/atom < N > ~ 2.68, essentially the same as SiO_2 (< N > = 2.67), and following the discussion in Ref. 18, this explains why the as-deposited films provide optimized performance when used as gate dielectrics in thin film transistors (TFTs) (see Fig. 9) [17]. The electrical properties, e.g., the density of interface states, D_{it}, of the annealed nitrides incorporated in ONO composite dielectrics show a weaker compositional dependence near the optimum value of R = 10 (see Fig. 10) than for the TFTs so that process latitude for ONO and ON depositions is not as tightly constrained as for the as-deposited gate dielectrics used in the TFT application.

Returning to the bonding properties of the annealed films, additional spectroscopic studies have shown: i) for *quasi-stoichiometric* nitrides, H-atoms are incorporated predominantly in mono-hydride Si-H and SiN-H groups, where of H atom *pairs* occur on nearest-neighbor, i) SiN-H groups, which interact through H-bonding interactions between the H-atom of one of the SiN-H groups and the N-atom of the other, and ii) Si-H and SiN-H groups, which interact via a repulsive coulomb interaction between the near-neighbor H-atoms [19]. Another study has demonstrated that H-atoms are readily eliminated from these paired-bonding environments by one minute rapid thermal anneals at temperatures above the deposition temperature, and as low as 400°C. Additionally, Si- and N-atoms that have lost H-atoms in the temperature range below ~600°C, form new Si-N bonds as annealing proceeds to higher temperatures, e.g., to 900°C [20]. As shown in Fig. 11, films annealed at 600°C show high defect densities, whereas films annealed at 900°C are essentially indistinguishable from SiO_2. The plot in Fig. 12 indicates that as absorption in the Si-N bond-stretching mode increases the flat band

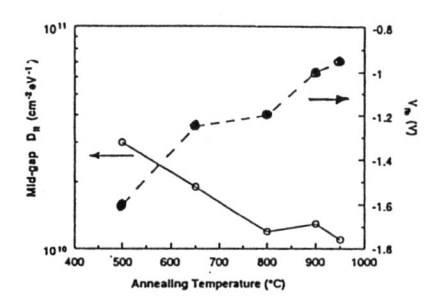

Fig. 11. Q$_f$ versus annealing temperature for ONO capacitors.

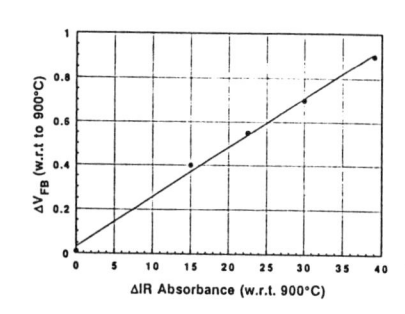

Fig. 12. V$_{FB}$ versus change in absorbance for Si-N stretching vibration IR model.

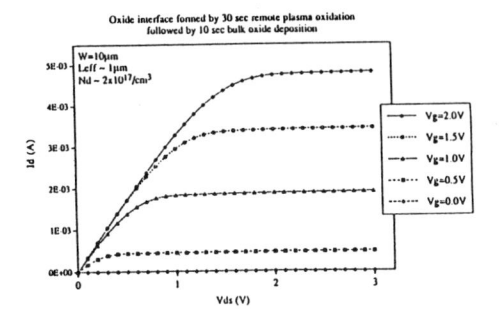

Fig. 13(a). I-V of FET with plasma-oxide, t$_{ox,eq}$ = 1.7 nm

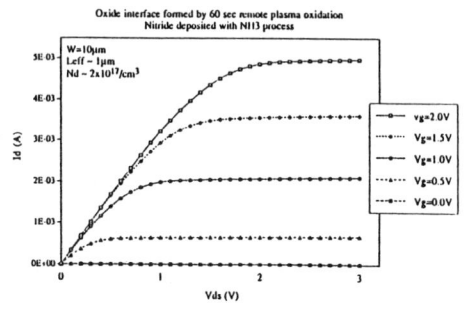

Fig. 13(b). I-V of FET with plasma-ON, t$_{ox,eq}$ = 1.8 nm

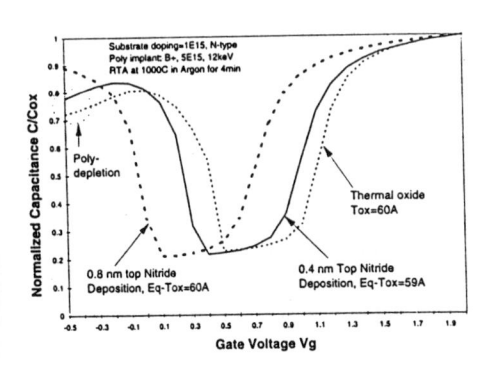

Fig. 14 (a) C-V for top surface nitride layer: tox ~ 6 nm.

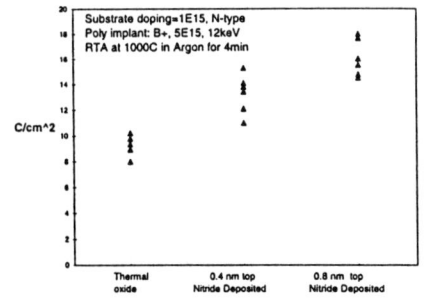

Fig. 14 (a) QBD for top surface nitride layer: tox ~ 6 nm.

voltage, V_{FB}, of an ONO capacitor structure decreases in absolute value indicating a decrease in fixed positive charge near the composite dielectric. These changes in bonding, the loss of H-atoms at temperatures in excess of the deposition temperature of 300°C, as well as the forming of additional Si-N at higher temperatures account for the excellent performance of these nitride films in MOS devices that have been subjected to post-deposition 900°C annealing or processing [10,21].

Data in Table V are for an ONO with 2.0 nm films for each of constituent layer [21]. Properties are equivalent to those with ONO structures processed at higher temperatures.

Table V Reliability Data for ONO Capacitor ($t_{ox,eq}$ = 5.0 nm) [21]

Dielectric Layer	QBD* C-cm^{-2}	Injected Q C-cm^{-2}	ΔV_{th} (V)	ΔD_{it} x10^{11} cm^{-2}-eV^{-1}
Plasma ONO	17	0.17	0.15	6.5
Thermally Processed ONO		0.1	0.1	1.9
Thermal Oxide	7-10	0.1	0.35	8.6

*Charge to Breakdown

Figures 13(a) and (b) display drive current characteristics for FETs fabricated with RPECVD dielectrics: (a) SiO$_2$ dielectric with accumulation capacitance equivalent to 1.7 nm of oxide, and (b) and ONO dielectric with accumulation capacitance equivalent to 1.8 nm [10]. The physical thicknesses of O and N layers are each ~1.2 nm.

3.1.2. The N$_2$ Process

Preliminary studies of ON devices with nitride layers fabricated by a high power RPECVD process that results in a relatively high concentration (~20 at.%) of bonded-H in SiN-H groups also indicate excellent FET current drive characteristics [22]. Finally, ON structures with O layers ~ 6 nm thick, and nitride layers of varying thickness between ~ 0.4 and 0.8 nm prepared by a low power N$_2$ process [9] have shown excellent suppression of B-atom transport from p$^+$ poly-Si through the oxide layer and into the underlying lightly doped n-type Si substrate (10 Ω-cm). The oxide was grown in dry O$_2$ at 800°C, and the ON structure was annealed at 900°C following the nitride deposition, but before the application of the poly-Si layer. The poly-Si layer for the implant was 200 nm thick and prepared by conventional LPCVD at ~520°C; the implant dose was 5x10^{15} B-atoms/cm^{-2} at 12 KeV. The B-atom driven in and activation was done at 1000°C for 4 minutes, resulting in a B-concentration of ~ 2x10^{20} cm^{-3} in the poly-Si. Figure 14(a) shows C-V traces that indicate the effectiveness of the thin Si$_3$N$_4$ films in stopping Boron transport out the of poly-Si film. The oxide reference sample showed a significant shift of the flat band voltage (V_{FB}) to ~1.1 V from an expected value of ~0.7 V. The 0.4 nm film provides some improvement (V_{FB} ~ 0.8), whereas the 0.8 nm essentially stops B-transport as indicated by the position of the flat band voltage as well as the lack of a depletion effect in the poly-Si gate electrode. Fig. 14(b) presents QBD results which correlate with the C-V data in Fig. 14(a); i.e., stopping B-penetration out the poly-Si gate electrode promotes larger QBD values.

3.2. COMBINED RPECVD/RTA PROCESSING OF NITRIDES

As noted above, nitride films deposited at 300°C and integrated in ONO structures showed relatively high defect densities after a 400-450 °C PMA. This was attributed to loss of hydrogen and the formation of Si- and N-dangling bond defects. Annealing to higher temperatures, beginning at about 600°C reduced the defect densities, and annealing to 900°C produced device-quality films with properties at least as good as thermally grown oxides. An interesting issue concerns nitride films deposited at higher temperatures, with significantly lower bonded-H concentrations, << 10 at. %. For example < N > and < C > for such a film with 5 at. % H are respectively, < N > = 3.3 and < C > = 5. These films fall outside the range of CNRs, and as such the dangling bond densities are expected to be high [23], > 10^{18} cm^{-3}, and more importantly, the defects that are generated will be randomly distributed throughout the thin film structure. Annealing to higher temperatures will not result in Si- and N-atom dangling bonds combining, simply because, Si_3N_4 does not display a viscoelastic transition which would be required for significant bond rearrangement. In contrast, in the high H-content films deposited at lower temperatures, the H-atoms are spatially correlated as nearest-neighbors as identified by IR studies, so that Si- and N-atom dangling bonds created by H-atom evolution at temperatures greater than the deposition temperature of 300°C are nearest neighbors in the network structure. As noted above, at temperatures > 600°C these nearest neighbor dangling bonds recombine to form Si-N bonds, thereby reducing the density of defects and accounting for the unique properties of nitride films produced by the combined low temperature RPECVD deposition at 300°C, and the low thermal budget RTA for 30 s at 900°C.

3.3. OXYNITRIDE FILMS

Oxynitride films as defined here are homogeneous alloys that lie on the join line from SiO_2 to Si_3N_4 in a ternary compositional diagram [5,6]. The compositions are characterized by the formula $(SiO_2)_x(Si_3N_4)_{1-x}$. Even though this representation neglects bonded hydrogen, it provides a useful way of defining oxynitride alloys, prepared by two different RPECVD processes, one using NH_3 [5] as the N-atom source gas, and a second using N_2 [6]. Both processes yield device quality films over a substantial portion of the alloy composition range after a 900°C RTA; however, the N_2 process has a greater process latitude with respect to the variation of the alloy composition as a function of the process gas flow rate ratios. Films produced in this way have been evaluated in MOS capacitor structures and show comparable properties; e.g., similar concentrations of interface defects, D_{it}, and flatband voltage shifts that correspond to similar densities of fixed positive charge. The discussions will emphasize two aspects of this research: i) the process latitude, and ii) (a) high defect levels in devices with films deposited at 300°C, but subjected to a PMA at 400°C, but (b) reduced defect levels in devices subjected to 30 s 900°C RTA..

For the NH_3 process, the source gases were 10% SiH_4 diluted in He for Si, N_2O for O and NH_3 for N. The SiH_4 was introduced through a downstream dispersal ring while N_2O and NH_3 were introduced upstream through a fused silica tube that passed through the plasma-excitation electrode. The plasma power was He 50 W at 13.56 MHz. The chamber pressure was maintained at 300 mTorr during the deposition and the substrate temperature was held at 300°C. The flow rate of the 10% SiH_4 mixture was fixed at 2 sccm so that the actual SiH_4 flow was only 0.2 sccm. The NH_3 flow rate was varied from 5 to 200 sccm, while the flow rate of N_2O was held fixed at 20 sccm. Using these gas flow rates, the composition of the deposited Si oxynitride, e.g., the relative N-fraction, could be controlled by changing the flow rate of the NH_3 gas (see Fig. 15 (a)). The oxynitride deposition rate varied between 10 and 20 Å per minute increasing with increasing NH_3 flow.

For the N_2 Process, the Si-source gas, SiH_4, is introduced downstream through a gas-dispersal ring, whilst the O- and N-atom source gases, 1%-N_2O-in-N_2 and N_2, respectively, along with diluent He are introduced upstream through a quartz plasma tube. The process conditions are: i) a process pressure of 0.2 Torr, a plasma power of 30 W and a substrate temperature of 300°C; ii) a He flow rate of 200 sccm; iii) a flow rate of 10 sccm SiH4 (diluted to 2 % in He); and iv) and variable flow rates of N_2 and N_2O: (a) 60 - z sccm of N_2, and (b) z sccm of N_2) (diluted to 1 % in N_2). Following in vacuo transfer, an RTA is performed at 900°C for 30 s in 0.1 Torr O_2 which is sufficient to prevent decomposition of the Si-SiO2 interface at 900°C. On-line AES analysis has been used to determine film composition from quantitative analysis of the relative spectral intensities in the differential-AES spectra. Elemental [Si], [O], and [N] concentrations (in terms of atomic percent) were derived from the Si_{LVV}, O_{KLL}, and N_{KLL} intensities.

Figures 15(a) and (b) include the film compositions in terms of an oxynitride pseudo-binary alloy notation: $(SiO_2)_x(Si_3N_4)_{1-x}$. For x = 0, the films are nitrides and for x=1, they are oxides. The trace in Fig. 15(a) demonstrates the limited process latitude of the NH_3 process. In the low NH_3 flow rate ratio range, the alloy composition changes from SiO_2 at R = 0, to an oxynitride alloy with x = 0.2 at R = 2. In contrast, for a flow ratio between R = 2 and R =12, there only a very small change in the allow composition, ~0.05 in the alloy notation. The trace in Fig. 15(b) indicates the increased process latitude in the N_2, process, where the alloy composition changes more gradually with gas flow rates. This is mostly due to the high dilution of N_2O in the N_2 process.

Oxynitride films prepared by both processes showed bonded hydrogen only in the form of N-H groups. The bending and stretching vibrations of the SiN-H group (~1200 and 3350 cm^{-1}, respectively), both increase in intensity as the N-concentration increases. There is also a small shift in the stretching mode frequency, from ~3390 cm^{-1} for high x to 3350 cm^{-1} for small x, i.e., in the Si_3N_4 rich alloys. This is indicative of transition from isolated to paired SiN-H groups [19]. The concentration of bonded-H, as estimated by the area under the N-H bond-stretching peak, is a factor of about 2 higher for the films grown from the NH_3 source gas.

Figures 16(a) and (b) present data for MOS capacitor structures. The capacitors formed by the NH_3 process were formed by a two step sequence: i) an in-situ RPAO process as described above, creating the Si-SiO2 interface and 0.5 to 0.6 nm of a passivating layer of SiO_2; followed by 15 nm of the oxynitride alloy [5]. The capacitors formed the N_2 process were O-oxynitride-O structures in which: i) the interface was formed by the RPAO process, and ii) the O, oxynitride and O layers were respectively, 3, 6 and 3 nm thick [6]. All plasma processing was done at 300°C, and all devices had Al electrodes, and were subjected to a 400°C PMA. For capacitors subjected to the 900°C RTA, Al electrodes were deposited after the RTA and the PMA then performed.

The fixed charge and interface trap densities have been determined by analysis of C-V data. Fig. 16(a) shows that the fixed charge density (Q_f) varies significantly with x in the as-deposited films [5]; Q_f decreases from ~1×10^{13}/cm^2 for x = 0.23, to ~2×10^{11}/cm^2 for x = 0.93. These high fixed charge densities are reduced significantly by the RTA process to a level of ~$(3-5) \times 10^{11}$/cm^2, which is essentially independent of x, and for the lowest values of x this reduction in Q_f is almost two orders of magnitude. Fig. 16(b) shows qualitatively similar data for the N_2 process [6]. Here the data is plotted as the flat band voltage rather than as Q_f; however, the magnitudes of Q_f computed from the flat band voltage shifts relative the calculated values for an idealized Al-O-Si structure yield essentially the same values of Q_f shown in Fig. 16(a).

Similar results are obtained D_{it} versus x, before and after the RTA. D_{it} has been determined by the conventional high frequency/quasi-static C-V technique. For as-deposited films prepared by the NH_3 process, the mid-gap D_{it} increases from $\sim 7 \times 10^{10}$ cm^{-2}eV^{-1} at x = 0.93 to > 2×10^{11}cm^{-2}eV^{-1} at x = 0.25. For comparison, the corresponding D_{it} levels for Si-SiO$_2$ interfaces formed by the same pre-deposition plasma-assisted oxidation process, and then overcoated with SiO$_2$ layers deposited by remote PECVD at 300°C are typically $(1-3) \times 10^{10}$ cm^{-2}eV^{-1}. The general trend for the D_{it} values to decrease as x increases; i.e., as the SiO$_2$ component of the alloy increases. The mid-gap D_{it} values decreased to $(2-6) \times 10^{10}$ cm^{-2}eV^{-1} with the higher values of D_{it} for lower values of x. Qualitatively similar data have been obtained for the N_2 process.

As in the case of the nitride films used in ONO structures, the RTA step reduces bonded-H in the oxynitride alloy films, and leads to bonding rearrangements that eliminate dangling bond defects. The RTA step leads to structural and chemical changes at the Si-SiO$_2$ interface; reducing suboxide bonding, and decreases interfacial transition regions.

4. Top Surface Nitridation

Substrates for these studies were prepared by forming a 10 nm thick thermally-grown oxide on the Si substrate [7]. The oxidation temperature was 900°C, and the oxidizing gas was dry O$_2$. Plasma nitridations were performed at ambient temperature ($\sim$23 °C) and at 300°C. A remote He-N$_2$ discharge was initiated using 200 sccm He, 60 sccm N$_2$, a process pressure of 0.1 Torr, and an RF power input of 30W at 13.56 MHz. Rapid thermal annealing (RTA) in N$_2$ or N$_2$O was performed in the RTP chamber at a pressure of 0.5 Torr for a duration of 40 s. On-line AES was performed at various intermediate stages of processing. Ex situ XPS was performed separately using an Al-K X-ray source (1.486 keV) in both the normal (detector at $\sim$80 degrees with respect to sample surface) and glancing-angle modes (detector at $\sim$20 degrees with respect to sample surface). SIMS analyses were performed at EVANS EAST, NJ, using Cs$^+$ ions, and detecting positive ions of CsSi, CsN, and CsSi.

Increasing duration of plasma exposure and/or wafer temperature results in a greater amount of nitridation of the oxide surface (see Fig. 17(b)). Fig. 17(a) which shows AES spectra from the oxide surface following exposure to the He-N$_2$ plasma treatment at 23 °C/50 min and 300°C/50 min. Nitridation is confirmed by the observation of a N$_{KLL}$ feature and concomitant decreases in the O$_{KLL}$ intensity. The Si$_{LVV}$ AES feature from the oxide surface before and after a 300°C remote He-N$_2$ plasma treatment shows a shift in the peak position in the direction of higher energy, from $\sim$76 eV (in SiO$_2$) towards $\sim$83 V (in Si$_3$N$_4$). This attests to a change in the bonding environment implying that the nitrogen exists in a bonded state and is bonded to silicon [7].

Atomic N-concentrations have been obtained from an analysis of the AES intensities and Fig. 20(b) shows the variation of [N]-content with the duration of the He-N$_2$ plasma exposure at the two temperatures, 23° C and 300°C. The following observations are made: (i) nitridation is greater at 300°C, and (ii) the [N] content tends to saturate with increasing exposure, and results in an ultra-thin surface oxynitride layer. The fact that the nitridation is confined to the near surface region is confirmed by *ex situ* XPS analysis. Fig. 18 shows high resolution spectra in normal- and glancing-angle modes, featuring the Si-2p and N-1s signatures, respectively. The observation of a higher N/Si ratio for glancing angle analysis than for normal incidence analysis, establishes that the N-atoms reside in the immediate vicinity ($\sim$1nm) of the surface.

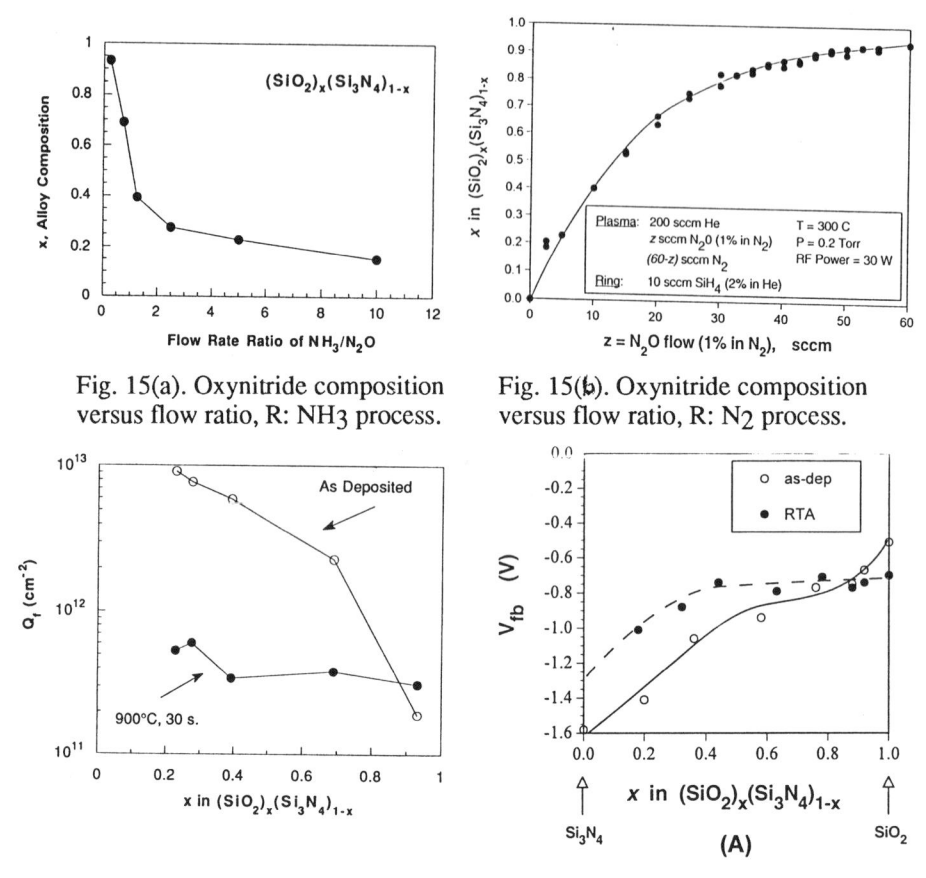

Fig. 15(a). Oxynitride composition versus flow ratio, R: NH3 process.

Fig. 15(b). Oxynitride composition versus flow ratio, R: N2 process.

Fig. 16. Compositional dependence of (a) Q_f for NH3, and (b) V_{FB} for N2 processes for oxynitrides.

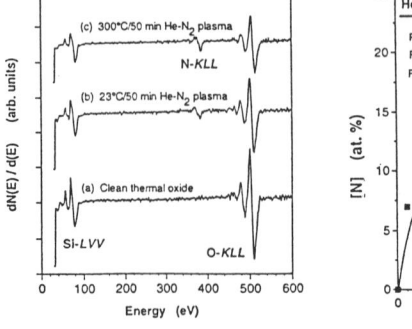

Fig. 17(a). Differential AES for top surface nitridation process.

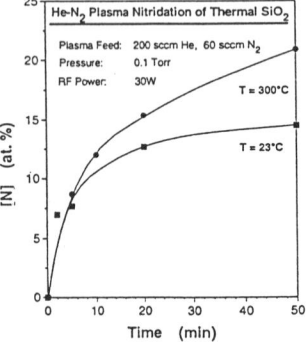

Fig. 17(b). Nitrogen concentration versus time for top surface nitridation process.

The AES analysis, which is surface-sensitive, shows a nitrogen concentration of the order of 10-20 at.%. SIMS analysis shows that the nitrogen is near the surface with areal densities of 3×10^{14} cm^{-2} and 5×10^{14} cm^{-2} respectively for 10 min and 50 min exposures at 300°C. SIMS analysis was performed and reported in Ref. 6 for the 300°C/50 min nitrided surface, capped with a remote-PECVD oxide. The results of SIMS and AES analyses give essentially the same atomic fractions, and establish that the nitrogen is confined to a ~1nm surface region. Thus the AES, XPS, and SIMS analyses together present a coherent picture of the concentration and spatial distribution of nitrogen atoms.

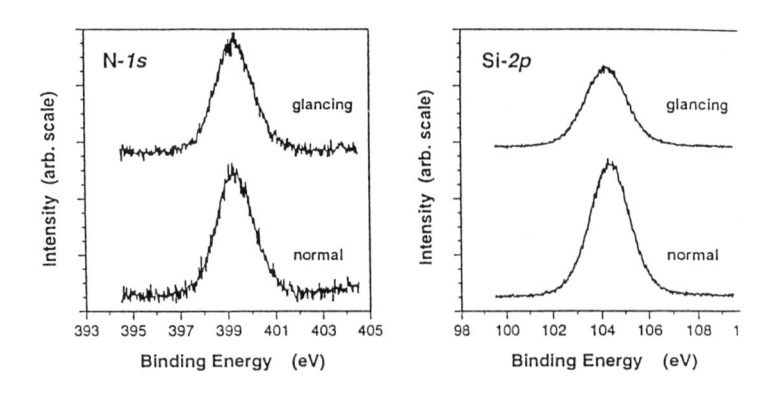

Fig. 18. XPS spectra, N_{1s} and Si_{2p}, for top surface nitridation process.

This process can be used to form an ultra-thin oxynitride barrier layer to suppress B-diffusion from the B-doped poly-Si gate electrode. In a recent research report, Hattangady and coworkers at TEXAS INSTRUMENTS have implemented a top surface nitridation process similar to that described above and have demonstrated that it can be used to stop B-atom transport out of p+ poly-Si gate electrodes [8].

5. Summary

The research results presented in this paper show that a combination of low-temperature (300°C) remote plasma-assisted processing, combined with low-thermal budget (30 s at 900°C) rapid thermal annealing yields nitrided stacked gate dielectrics that display device-quality performance and reliability as determined by comparisons with thermally-grown oxide dielectrics. The processing is divided into three in-situ steps that provide independent and separate control of the bonding chemistry at the interface, and the bonding chemistry of the deposited dielectrics films: i) RPAO to (a) eliminate residual hydrocarbon contaminants, (b) form a nitrided Si-SiO$_2$ interface, and (c) grow a passivating oxide layer ~ 0.5 to 0.6 nm thick; ii) RPECVD depositions to form (a) homogeneous oxide, or (b) stacked ON, ONO or O-oxynitride-O dielectrics; and iii) RTA at 900°C to provide chemical and structural relaxations. e.g., (a) reduction of suboxide bonding in interfacial Si-SiO$_2$ transition regions; and (b) elimination of H from nitride and oxynitride alloys, as well as bonding changes after dehydrogenation that reduce Si- and N-atom dangling bond densities. In addition nitridation of interfaces can also be obtained by performing the RPAO in O$_2$ followed by a remote plasma assisted nitridation. Finally, two approaches to stopping B-atom transport out of p+ poly-Si gate electrodes have been identified: i) the use of ultra-thin (~ 0.8 nm) RPECVD nitrides, and ii) top surface nitridation of oxide films; each of these processes is compatible with the three processing steps described immediately above. Nitrogen incorporation can also be achieved in the different parts of the gate dielectric by thermal processing; however, this invariably requires higher temperatures, and larger thermal budgets.

6. Research Issues

The are several issues remaining to addressed with respect to nitridation of gate dielectric structures. It is obvious that nitrogen is beneficial in several different ways, e.g.: i) reducing defect generation at Si-SiO$_2$ interfaces; ii) allowing the use of physically thicker dielectrics with oxide-equivalent thicknesses needed for scaling; and iii) blocking B-atom transport from p+ poly-Si gate electrodes.

The oxide equivalent thickness of an ON or ONO structure is very nearly equal to t_{ox} + 0.5t_n, where t_{ox} and t_n are the respective thicknesses of the oxide and nitride layers in the multilayer structure. The SIA Roadmap calls for $t_{ox,eq}$ to be at most 1.0 nm in the year 2012. Recent results, have shown that single layer oxide dielectrics can be used down to thicknesses of ~2 nm' projections suggest this could be reduced to 1.7 nm. One important limitation in NMOS devices will be in the off-state leakage current from the unbiased gate to the positively biased drain. Calculations have been performed at NC State University on composite structures [24], which indicate significant improvements in current reduction are possible in ON structures in spite of the reduced barrier heights at the poly-Si-nitride interface, ~ 2.1 eV as opposed to 3.1 eV at the Si-SiO$_2$ interface. The experimental results discussed above in Section 3.1.1. indicate a modest reduction in gate injection in an ON as compared to an oxide dielectric, whereas the calculations, based on a WKB approximation, and recently modified to include an exact treating of the tunneling probability, would predict significantly larger reductions, by factors of at least about 5 [25]. Based on the calculations, and results of Section 3.1.1., the following scenario is presented in Table VI for gate dielectric technology.

Table VI Gate Dielectric Technology

Gate Dielectric Material	Thickness Range	Limitation
oxide	to 1.7 - 2.0 nm	direct tunneling in off-state
oxide-nitride	2.5 to 1.2 nm	barrier height lowering at poly-Si interface
alternate dielectrics (e.g., Ta$_2$O$_5$)	1.5 to 0.8 nm	barrier height lower at poly-Si interface

It is therefore likely that composite ON structures will play a role in the device technology. In thicker gate dielectrics (~ 2 to 4 nm), the primary benefit will be in eliminating B-atom transport out of p+ poly-Si gate electrodes, and in the thinner gate dielectrics it will be in reducing off-state leakage. Some research has been done on alternative dielectrics such as TiO$_2$ and Ta$_2$O$_5$, however, it remains a research issue as whether the increased dielectric constant which allows for physically thicker films, is not be compromised by increased tunneling currents due to *intrinsically* smaller barrier bandgaps and barrier heights with respect to SiO$_2$.

There are two other significant issues in this stacked dielectrics: i) accurate determinations of N-concentrations, particularly at Si-SiO$_2$ and poly-Si-SiO$_2$/Si$_3$N$_4$ interfaces, and ii) accurate measurements of oxide and nitride layer thicknesses. Particle techniques such as MEIS and NRA offer great promise in this area, and calibration of these techniques may benefit from standards produced by plasma processing and evaluated independently by on-line AES. SIMS is useful, but very difficult to use in a quantitative manner, mostly because of i) significantly different relative sensitivity factors (RSF) for N in an SiO$_2$ matrix and N in a Si matrix. For example, very different SIMS profiles have been obtained for determining N at SiO$_2$ interfaces using CsN$^+$ ions and assuming the N-atoms are in the SiO$_2$ and SiN$^-$ ions are driven into the Si substrate during the sputtering process [4]. Top surface nitridation is equally difficult using SIMS because of difficulties in removing poly-Si gate materials by selective etching processes. Spectroscopic techniques such as IR and XPS hold some promise for secondary characterization, but again primary standards are not easily evaluated.

The second issue concerns thicknesses of dielectric layers. There are limitations most approaches including: i) transmission electron microscopy imaging; and ii) ellipsometry. The most useful approaches are in electrical measurements, wherein $t_{ox,eq}$ values obtained from accumulation C-V can be directly compared. A figure of merit, that takes into account direct tunneling can then determined from the ratio of the tunneling current under a specified bias condition and the effective capacitance. This pragmatic approach is also closely linked to the increases in performance that must accrue with scaling down to thinner gate dielectrics.

There are at least four areas where more work is necessary on basic microscopic mechanisms. These include: i) why do nitrided interfaces show reliability?: (a) is it intrinsic to the N-atom bonding arrangements as suggested by our group [25], or (b) is it simply related to a reduction in the number of Si atoms dangling bond defects that must be terminated by H (the reduced dangling bond density could be related the fact that N-atoms are *smaller* than O-atoms; ii) what is the mechanism of stopping B-atom migration out of poly-Si?; e.g., is it a chemical attachment of acceptor like B-atoms to donor like N-atoms?, or a limitation that derives from a more compact nitride structure; iii) what are the effects of N-atom incorporation on channel mobilities in NMOS and PMOS structures?; and iv) what are the trade-offs between increased dielectric constants in nitride and other alternative gate dielectric materials and reductions in interfacial barrier heights?; e.g., image charge effects decrease with increasing dielectric constant, so that there may be important effects which compensate for changes in barrier heights.

Acknowledgments

Supported by the NSF Engineering Research Center (ERC) for Advanced Electronic Materials Processing (AEMP), the Office of Naval Research (ONR), and the Semiconductor Research Corporation (SRC). The author acknowledges contributions of his former and present graduate students and post doctoral fellows: B. Chaflin, S. Habermehl, S. Hattangady, B. Hinds, K. Koh, Z. Lu, Y. Ma, H. Niimi, P. Santos-Filho, Y. Wu, H.-Y. Yang, T. Yasuda, and Z. Jing.

References

1. D.R. Lee, G. Lucovsky, M.R. Denker and C. Magee, J. Vac. Sci. Technol. A **13**, 607 (1995); D.R. Lee, C.R. Parker, J.R. Hauser and G. Lucovsky, J. Vac. Sci. Technol. B **13**, 1778 (1995).
2. M.L. Green, D. Brasen, K.W. Evens-Lutterodt, L.C. Feldman, K. Krisch, W. Lennard, H.T. Tang, L. Manchanda, and M.T. Tang, Appl. Phys. Lett. **65**, 848 (1994).
3. J. Ahn, J. Kim, G.Q. Lo and D.-L. Kwong, Appl. Phys. Lett. **60**, 2089 (1992).
4. H. Niimi and G. Lucovsky, unpublished.
5. Y. Ma, T. Yasuda, and G. Lucovsky, J. Vac. Sci. Technol. A **11**, 952 (1993); Y. Ma, T. Yasuda, S. Habermehl and G. Lucovsky, J. Vac. Sci. Technol. B **11**, 1533 (1993). Y. Ma, and G. Lucovsky, J. Vac. Sci. Technol. B **12**, 2504 (1994).
6. S.V. Hattangady, H. Niimi, and G. Lucovsky, J. Vac. Sci. Technol. A **14**, 3017 (1996).
7. S.V. Hattangady, H. Niimi, and G. Lucovsky, Appl. Phys. Lett. **66**, 3495 (1995).
8. S.V. Hattangady, R. Kraft, D.T. Grider, M.A. Douglas, G.A. Brown, P.A. Tiner, J.W. Kuehne, P.E. Nicollian and M.F. Pas, IEDM Tech. Dig., 495 (1996).
9. Y. Wu and G. Lucovsky, submitted to IEEE Device Letters.
10. C.R. Parker, G. Lucovsky and J.R. Hauser, submitted to IEEE Device Letters.
11. G.G. Fountain, S.V. Hattangady, R.A. Rudder, M.J. Markunas, G. Lucovsky, S.S. Kim and D.V. Tsu, J. Vac. Sci. Technol. A **7**, 576 (1989).
12. G. Lucovsky, S.S. Kim, D.V. Tsu, G.G. Fountain and R.J. Markunas, J. Vac. Sci. Technol. B **7**, 861 (1989).
13. G. Lucovsky, H, Niimi, K. Koh, D.R. Lee and Z. Jing, The Physics and Chemistry of SiO_2 and the Si-SiO_2 Interface, Ed. by H.Z. Massoud, E.H. Poindexter and C.R. Helms (Electrochemical Soc., Pennington, 1996), p. 441.
14. T. Yasuda, Y. Ma, S. Habermehl and G. Lucovsky, Appl. Phys. Lett. **60**, 434 (1992).
15. G. Lucovsky, A. Banerjee, B. Hinds, B. Claflin, K. Koh and H. Yang, J. Vac. Sci. Technol. B **15** (1997), in press.
16. C.H. Bjorkman, C.E. Shearon, Jr., Y. Ma, T. Yasuda, G. Lucovsky, U. Emmerichs, C. Meyer, K. Leo and H. Kurz, J. Vac. Sci. Technol. A **11**, 964 (1993); C.H. Bjorkman, T. Yasuda, C.E. Shearon, Jr., U. Emmerichs, C. Meyer, K. Leo and H. Kurz, J. Vac. Sci. Technol. B **11**, 1521 (1993).
17. G. Lucovsky, Y. Ma, S. S. He, T. Yasuda, D. J. Stephens and S. Habermehl, Mater. Res. Soc. Symp. Proc. **284**, 34 (1993); G. Lucovsky, S.S. He, M.J. Williams and D. Stephens, Microelectronic Eng. **25**, 329 (1994) and refs. therein.
18. G. Lucovsky and J.C. Phillips, J. Non-Cryst. Solids (1998), in press.
19. G. Lucovsky, Z. Jing, P. Santos-Filho, G. Stevens and A. Banerjee, J. Non-Cryst. Solids **198-200**, 19 (1996).
20. Z. Lu, M.J. Williams, P.F. Santos-Filho and G. Lucovsky, J. Vac. Sci. Technol. A **13**, 607 (1995).
21. Yi Ma, T. Yasuda, and G. Lucovsky, Appl. Phys. Lett. **64**, 2226, (1994).
22. C.R. Parker and J.R. Hauser, unpublished
23. J.C. Phillips, J. Non-Cryst. Solids **34**, 153 (1979); J.C. Phillips, J. Non-Cryst. Solids **47**, 203 (1983).
24. H.-Y. Yang, H. Niimi and G. Lucovsky, J. Appl. Physics (1997), in press.
25. G. Lucovsky, Z. Jing, and D. R. Lee, J. Vac. Sci. Technol. B **14**, 2832 (1996).

ISOTOPIC LABELING STUDIES OF OXYNITRIDATION IN NITRIC OXIDE (NO) OF Si AND SiO$_2$

I. TRIMAILLE, J.-J. GANEM, L.G. GOSSET, S. RIGO, I.J.R. BAUMVOL[1], F. C. STEDILE[2], F. ROCHET[3], G. DUFOUR[3] and F. JOLLY[3]
Groupe de Physique des Solides, Université Paris 7, Université Paris 6, UMR 75-88 CNRS
Tour 23, 2 place Jussieu, 75251 PARIS CEDEX 05, FRANCE
[1]Instituto de Fisica-UFRGS
91540-000 Porto Alegre, RS, Brazil
[2]Instituto de Quimica-UFRGS
91540-000 Porto Alegre, RS, Brazil
[3]Laboratoire de Chimie-Physique, Université Paris 6,
11 rue Pierre et Marie Curie, 75231 PARIS CEDEX 05, FRANCE

Abstract

Rapid thermal oxynitridation in nitric oxide (NO) of a thick (14 nm) SiO$_2$ film grown on Si(001) is studied as a first stage towards understanding of atomic transport mechanisms occuring during NO annealing of thin SiO$_2$ films. The SiO$_2$ films were grown in an ultra high vacuum rapid thermal processing (RTP) furnace in static pressure of natural O$_2$ (^{16}O$_2$). These films were then annealed in ^{15}N and ^{18}O-enriched NO (^{15}N^{18}O) for 20 and 80 s. Total amounts of nitrogen and oxygen (areal densities in at.cm^{-2}) and heavy isotopes depth distribution were measured using non resonant and resonant nuclear reactions analysis. The results are discussed in terms of atomic depth profiles and growth mechanisms. These first results are more likely explained by two mechanisms occuring in parallel. In the first one, NO diffuses through the silica network without reacting with it and both N and O are fixed in the near interface region. In the second one, ^{18}O is fixed near the oxide surface due to a mechanism related with a step-by-step motion of network oxygen atoms, by a simple diffusion process, induced by the presence of network defects, involving O only. This latter mechanism leads mostly to an exchange of oxygen atoms between the oxide network and the gas phase.
  Direct oxynitridation of Si(001) in nitric oxide (NO) is studied as a function of gas pressure. The dielectric films were grown in the RTP furnace in static pressures of ^{15}N and ^{18}O-enriched NO (^{15}N^{18}O). The nuclear reactions techniques mentioned above were employed to analyse the dielectric films. The thicknesses of the oxynitrides formed in NO never exceeded 3 nanometers, in our thermal treatments conditions. Moreover, the growth rates of these films are lower compared to N$_2$O growth rates, due to the higher amount of nitrogen atoms fixed in the films. At 1050 °C, for isochronal thermal treatments, the amounts of nitrogen fixed in the films decreases as the pressure P of NO increases (in the range 1 to 100 hPa) suggesting that nitrogen atoms may be fixed via a vacancy mechanism. The amount of nitrogen atoms was found to support a P$^{-1/4}$ law, whereas in N$_2$O the nitrogen amount varies as P$^{1/2}$. The areal densities of oxygen atoms are consistent with a P$^{1/4}$ law, as in the case of N$_2$O oxynitridation. Angle Resolved

X-Ray Photoelectron Spectroscopy (AR-XPS) for different incident angles was used as a complementary technique to provide informations on bonding structures and their distributions. AR-XPS results on the analysed samples show no evidence of N-O bonds. Furthermore, AR-XPS indicates a lower concentration of nitrogen near the external surface of the dielectric film, in agreement with resonant nuclear reactions analysis.

1. Introduction

The scaling down of ultralarge-scale integration circuits requires, among other things [1,2,3] like improved lithography, highly reliable ultrathin gate dielectrics. In fact, according to the last edition of the National Technology Roadmap for Semiconductors [4], the thicknesses of these dielectric films will be reduced to below 4 nm for the 0.1 µm technology, scheduled for year 2007. More precisely, 2 nm-thick gate dielectrics are required for metal oxide semiconductor field-effect transistors (MOSFETs) with 40 nm gates lengths [3] which have already been built [5].

The end of gate oxide scaling as a means of enhancing performance is projected at 3.4 nm [4], and therefore a gate dielectric more reliable than SiO_2 is required. Thermal oxynitrides grown in nitrous oxide (N_2O) and nitric oxide (NO) or N_2O and NO-annealed oxides have drawn considerable attention as gate dielectrics and tunnel dielectrics for devices of the next technology generations (see the large number of references in [6]). Compared to SiO_2 gate dielectrics, these oxynitrides increase charge-to breakdown and hot carrier immunity, improves N channel transconductance, and reduce boron diffusion from p^+-gates (see references in [6]). These improvements result from accumulation of nitrogen atoms at the silicon/dielectric interface. Furthermore, these oxynitrides exhibit lower growth rates than silicon dioxide [7,8], allowing better thickness uniformity as demonstrated in [9] for N_2O-based oxynitrides, and thus lower threshold voltage variation.

However fabrication of these oxynitrides has still to be optimized to achieve best device performance [10]. The nitrogen atoms depths profiles, the chemical and physical properties of these films, the point defect injection kinetics [11], are still not well known or understood. These issues, which strongly depend upon the growth parameters, should be addressed in order to understand why nitrogen atoms have a beneficial influence and to make a step towards the understanding of the growth mechanisms. Then a model to be implemented in Technology Computer Aided Design (TCAD) could be proposed and these oxynitrides would be succesfully used in technology.

In this paper, NO-based oxynitrides are studied using isotopic labelling, Nuclear Reactions Analyses (NRA), and AR-XPS. Owing to these techniques, we obtain informations on atomic depth profiles, atomic transport occuring during growth and bonding structures.

2. Experimental

2.1 SAMPLES PREPARATION

We used monocrystalline (001) silicon wafers, phosphorus-doped with nominal resistivity 3-6 Ω.cm. We cleaned the wafers into HF diluted (4 %) in ethanol for 30 s and rinsed them in ethanol for 30 s just before the thermal treatment performed in an ultra-high vacuum technology rapid thermal processing (RTP) setup described in [12] and specially designed for isotopic tracing experiments. Two kind of thermal treatments were performed : oxidation in O_2 followed by an anneal in NO and direct oxynitridation in NO. We used natural O_2 ($^{16}O_2$) and NO highly enriched (99.9 %) in ^{15}N and in ^{18}O ($^{15}N^{18}O$). The growth parameters are given in sections 3.1 and 3.2.

2.2 NUCLEAR REACTION ANALYSIS

2.2.1 *Measure of atomic areal densities*
Non resonant nuclear reaction analysis was used to measure the atomic areal densities of ^{14}N, ^{15}N, ^{16}O and ^{18}O in the oxynitride films. General principles of dosing by NRA have been detailed by Amsel et al. [13]. The following reactions were used in the present work at non resonant energies:
$^{15}N(p,\alpha\gamma)^{12}C$, induced by a deuteron beam of 1.00 Mev, with a detection at 150°
$^{18}O(p,\alpha\gamma)^{15}N$, induced by a proton beam of 0.73 MeV, with a detection at 150°
$^{16}O(d,p_0)^{17}O$, induced by a deuteron beam of 0.81 MeV, with a detection at 90° [14].

The number of detected particles (α or p) is proportional to the atomic areal densities of the atomic species in the film. The absolute values are determined by comparison with standards of atomic areal densities known within 5 %. The detection limit is 10^{13} at.cm^{-2} for ^{18}O and ^{16}O [15] and better than 10^{12} at.cm^{-2} for ^{15}N [16]. The areal densities of oxygen and nitrogen can be converted into estimated thicknesses X of the films by making the following approximation: we consider the films formed solely by the Si_3N_4 ($\rho = 3.1$ g.cm^{-3}) and SiO_2 phases ($\rho = 2.26$ g.cm^{-3}), then, for areal densities expressed in 10^{15} at.cm^{-2},
X (nm) = (0.188*^{15}N-areal density) + (0.226*($^{16}O + ^{18}O$) areal densities).

2.2.2 *^{15}N and ^{18}O depth profiling*
To achieve high resolution depth profiling, we used the very narrow resonances of the $^{18}O(p,\alpha\gamma)^{15}N$ (width $\approx$ 100 eV) and $^{15}N(p,\alpha\gamma)^{12}C$ (width $\approx$ 120 eV) nuclear reactions at 152 keV and 429 keV respectively [17-19]. The number of detected particles (respectively α and γ) as a function of the energy of the incident proton beam (excitation curve) is recorded for energies around the energy of the resonance. This excitation curve is an image of the concentration depth profile of ^{18}O ([^{18}O]/[$^{18}O+^{16}O+^{15}N$]), or of ^{15}N ([^{15}N]/[$^{18}O+^{16}O+^{15}N$]). These depth profiles can be extracted using SPACES [20,21], a personal computer implementation of the stochastic theory of energy loss, allowing also to perform the convolutions of the various components contributing to an excitation curve. The entries to this program, the stopping power for protons in the dielectric film and the straggling constant (that describes the dispersion of energy loss for the protons in the film) used in this study are those calculated for SiO_2 at 152 keV and 429 keV. This approximation is justified as follows :i) for SiO_2 films annealed in NO the nitrogen contents compared to the oxygen content represents less than 2 % ; ii) both stopping power and straggling constant are very close for SiO_2 and Si_3N_4 and in the case of thin (<6 nm) NO grown films the choice of the medium (SiO_2 or Si_3N_4) has very little effect on the calculated excitation curves, in a depth scale in µg.cm^{-2}. A depth resolution better than 1 nm near the surface can be achieved in a grazing incidence geometry [18].

2.3 XPS MEASUREMENTS ON NO GROWN SAMPLES

We used a Mg Kα (hν = 1253.6 eV) radiation to scan the O 1s, N 1s and Si 2p regions. The total instrumental broadening was set to give a Si 2p3/2 full width at half maximum of about 0.6 eV. The data treatment consisted in: i) substracting the background (Shirley Type) ; ii) stripping numerically the Si 2p1/2 component from all the Si 2p spectra [22] ; iii) then approximating the experimental spectra by a sum of Gaussian curves, each of them corresponding to a different chemical state.

We performed all the measurements using different angles (0°, 65° and 75°) from the sample normal to change the sampling depth.

3. Results and discussion

3.1 OXYNITRIDATION IN NITRIC OXIDE OF SiO$_2$ FILMS GROWN ON Si(001)

First, approximately 14 nm of thermal oxide was grown on silicon wafers in a static pressure of 140 hPa of natural oxygen, at 1050 °C, in the RTP furnace. This relatively high thickness was chosen as a first step before studying NO-annealing of thinner oxides. Then, a control oxide was kept (sample NO80) and two NO oxynitrides were grown by annealing this oxide at 1050 °C and under a static pressure of 20 hPa of $^{15}N^{18}O$ for 20 s (sample NO81) and 80 s (sample NO82).

In order to obtain a better in-depth resolution throughout the nitrided oxide, we used step-by-step dissolution of the films combined with nuclear depth profiling. Two pieces of each nitrided samples underwent chemical etching for 20 s or 35 s, in a B.E. 40-1 solution (FNH4 40 %:HF 50 %;40:1). On all samples (either etched or not), excitation curves of both resonances (on ^{15}N and ^{18}O) were recorded and areal densities of ^{15}N, ^{16}O, and ^{18}O were measured by non resonant NRA. The atomic areal densities and the estimated thickness obtained from non resonant NRA are summarized in table 1.

TABLE 1. Atomic areal densities in ^{15}N, ^{16}O, and ^{18}O and estimated thicknesses for 14 nm-thick Si^{16}O$_2$ films annealed in 20 hPa of NO at 1050 °C then etched for various durations.

samples	annealing in $^{15}N^{18}O$	etch time	^{15}N areal densities*	^{16}O areal densities*	^{18}O areal densities*	estimated thickness
			(x10^{15} at.cm^{-2})			(nm)
NO80	none	none	-	63±1.5	-	14
NO81	20 s	none	0.25±0.02	56±1.3	1.35±0.04	13
NO8120	20 s	20 s	0.20±0.01	23±2	0.37±0.03	5
NO8135	20 s	35 s	0.25±0.01	13±1	0.33±0.03	3
NO82	80 s	none	1.05±0.05	51±1.3	4.95±0.07	13
NO8220	80 s	20 s	1.00±0.02	44±2	1.29±0.04	10
NO8235	80 s	35 s	0.95±0.02	15±1	0.73±0.02	4

To enhance clarity, boldface is used for the unetched samples.
*The ± sign refers to *statistical error only*.

Figure 1 shows the excitation curves of the $^{15}N(p,\alpha\gamma)^{12}C$ nuclear resonance at 429 keV on the 80 s nitrided sample, and for etch times of 0, 20, and 35 s. The areas under the curves, proportional to the areal densities of ^{15}N in the samples, are the same for the unetched and 20-s etched sample. The area under the curve of the 35 s-etched sample is a few percent less. This result, confirmed by NRA measurement, evidences that nitrogen is located essentially near the oxynitride/Si interface. For longer etching time, the ^{15}N peak moves towards lower energies, because the ^{15}N depth profile comes closer to the surface sample, and becomes sharper, because of less sensitivity to straggling effects.

The simulation of the excitation curves of both resonances were performed on 35 s-etched samples first. Then, the deduced profiles were used for the simulation of near interface peaks of excitation curves recorded on the 20 s-etched samples. The latter profiles were then used for the simulation of the unetched sample. Finally we obtained the ^{15}N and ^{18}O depth profiles on the unetched samples with a depth resolution better than 2 nm in the whole films.

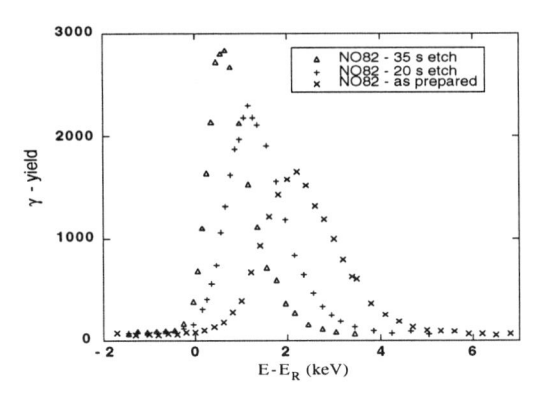

Figure 1. Excitations curves of the ^{15}N(p,αγ)^{12}C nuclear reaction at 429 keV, for an oxide film grown in natural oxygen up to 14 nm and annealed in a static pressure of 20 hPa of ^{15}N^{18}O for 80 s at 1050 °C (sample NO82), for various etching times. A tilt angle of 65° between the normal to the sample and the direction of the beam was used to improve depth resolution. These curves were recorded with charge of 30 μC per point.

Figures 2 and 3 show the excitation curves of both resonances recorded on samples NO81 and NO82, as well as the simulated excitation curves calculated by SPACES corresponding to the depth profiles also shown in these figures. For the sake of simplicity, let's separate the oxynitride into two regions, one " near interface " region where nitrogen is located and one " near surface " region where no nitrogen is detected. These profiles show also that i) the areal densities of ^{18}O and ^{15}N atoms incorporated in the " near interface " region are roughly the same; ii) most of ^{18}O atoms are located in the " near surface " region. These profiles suggest two hypotheses for transport mechanisms:

First hypothesis: NO (or product of decompostion of NO) moves interstitially through the silica network and reacts with the network, exchanging his oxygen atom. When arriving in the near interface region, NO (or product of decompostion of NO) poorer in ^{18}O fixes O (^{16}O and ^{18}O) and to a lesser extend N (see [23] for a precise description of the mechanism in the case of silicon oxidation) by reacting with silicon atoms. In that case, the fact that, for both anneal durations, ^{15}N and ^{18}O concentrations are similar in the " near interface region " is purely coincidental. The areal densities of atoms (^{15}N and O) corresponding to growth are in that case higher than the sum of the ^{18}O and ^{15}N areal densities found near the interface, when in fact we didn't measure significant increase in thickness. Furthermore, this means that this mechanism leads to little isotopic exchange [23], and cannot explain the observed ^{16}O loss (see table 1). For these reasons, this hypothesis is not satisfactory.

Second hypothesis: two mechanisms occuring in parallel, as in the case of dry oxidation of silica [24].

Mechanism i): NO (or products of decomposition of NO) diffuses through the silica network without reacting with it and reacts at the oxynitride/Si interface to fix both N and O. The fact that in the " near interface " region ^{15}N and ^{18}O depth profiles are very close for both 20 s and 80 s annealing, is in favour of this mechanism with NO as the diffusing species. Our results also show that the ^{15}N and ^{18}O areal densities are approximatively proportional to the annealing duration, revealing a linear regime for ^{15}N and ^{18}O incorporation at the interface, in our experimental conditions. This hypothesis in favour of NO reacting in the "near interface region" is consistent with the interpretation of Gusev *et al.* [25] based on a comparison, in terms of nitrogen depth profiles, between pre-oxides films annealed in NO, studied by various authors, and their NO grown films.

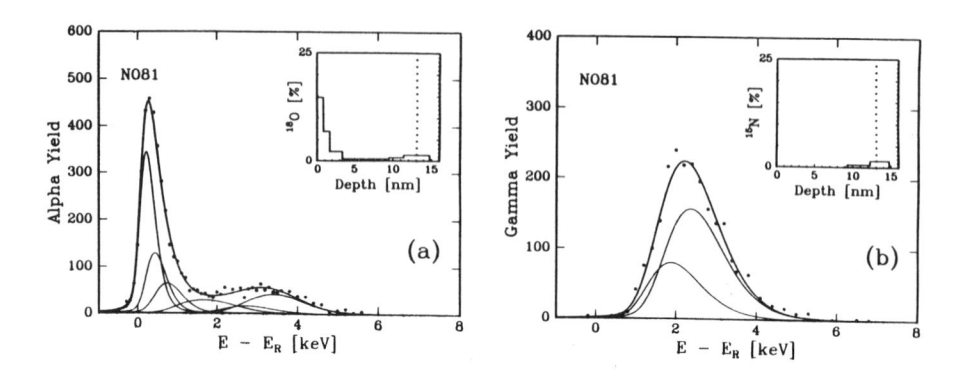

Figure 2. Excitation curves of the ^{18}O(p,αγ)^{15}N (a) and ^{15}N(p,αγ)^{12}C (b) nuclear reactions at 152 keV and 429 keV respectively, for an oxide film grown in natural oxygen up to 14 nm and annealed in a static pressure of 20 hPa of ^{15}N^{18}O for 20 s at 1050 °C (sample NO81). The ^{18}O and ^{15}N depth profiles used for the simulation are shown in inset. The dottted line correspond to the oxinitride/silicon interface, as estimated by NRA. A tilt angle of 65° between the normal to the sample and the direction of the beam was used to improve depth resolution. 30 μC (a) and 20 μC (b) per point.

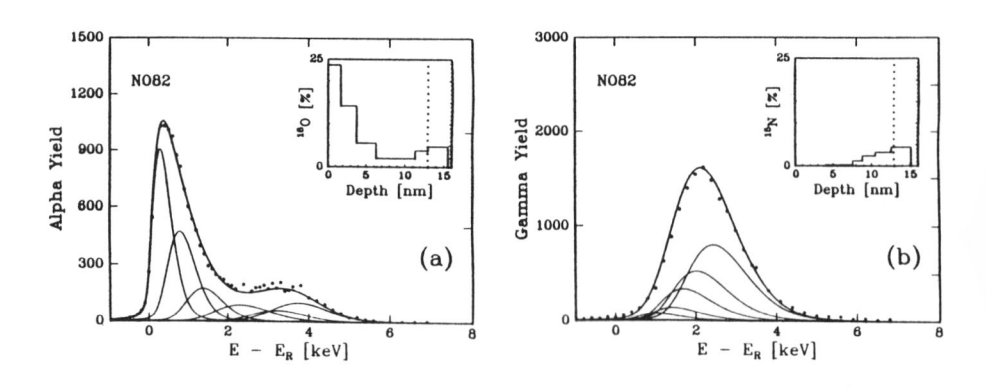

Figure 3. Same as figure 2 but measured on the sample NO82. 30 μC per point.

Mechanism ii) : the fixation of ^{18}O near the oxide surface occurs due to a mechanism related with a step-by-step motion of network oxygen atoms, by a simple diffusion process, induced by the presence of network defects, involving O only, like a peroxyl bridge for example, since no nitrogen is found in this region, N being released through a non reacting gas (in the sense of nitrogen fixation) like N_2 or NO_2. There is no evidence that N is fixed in the film and then removed as it is the case in N_2O anneal of oxides or nitrided oxides [26,27,28]. However, let us quote the conclusive remark by Gusev et al. [25] : "NO (when directly introduced as a key reactant) has a very low reactivity compared to N_2O " in removing nitrogen on nitrided oxides. Mechanism ii) leads to an exchange of oxygen atoms between the oxide network and the gas phase and explains the loss in ^{16}O atoms observed after NO anneal (see table 1). In our experimental conditions, most of ^{18}O (~ 70 %) is fixed " near surface region ", suggesting then, that within the framework of this study, mechanism ii) is the dominant one for ^{18}O fixation.

Mechanism ii) has been clearly demonstrated in the case of dry oxidation of silicon using sequential $^{16}O_2/^{18}O_2$ oxidations, for relatively thick oxides, where the ^{18}O depth profiles exhibits an erfc profile [29]. The erfc depth profile corresponds to diffusion in a semi-infinite solid, with constant ^{18}O network surface concentration C_s. However this boundary condition can not always be used, since the introduction of the diffusion species may take place by means of chemical reactions which are rate limiting. In that case, the incoming flux of ^{18}O atoms into the solid, J, has to be written as follows :

$$J(x=0,t) = K[C_s^0 - C_s(t)], \qquad (1)$$

where K is the transfer coefficient or reaction rate constant (the reaction is assumed to be first order), $C_s(t)$ is the surface concentration in ^{18}O of the oxide network and C_s^0 is the concentration of network oxygen atoms multiplied by M^{18} the labelling in ^{18}O of the gas phase ($M^{18} = 0.999$ for the samples considered here) and t the duration of NO anneal. If $C(x,0)=0$, then the solution of Fick's equation becomes [30]:

$$\frac{C(x,t)}{C_s^0} = M^{18}\left[erfc(\frac{x}{2\sqrt{Dt}}) - \exp(hx + h^2 Dt)erfc\left(\frac{x}{2\sqrt{Dt}} + h\sqrt{Dt}\right)\right] \qquad (2)$$

where $h = K/D$, D being the apparent diffusion coefficient [31] and where $\frac{C(x,t)}{C_s^0}$ is the ^{18}O concentration depth profile in the sample. For the limit case $h \rightarrow \infty$, the erfc solution is obtained.

Since SPACES allows only the use of boxes or erfc as the different layers in the depth profiles for the simulation of the excitation curves, we fitted the boxes-shaped profiles of ^{18}O in the " near surface region ", with Equation 2 (see figure 4), by varying D and h. The extracted values for D (1.2×10^{-15} cm^2.s^{-1}), h (1.2×10^6 cm^{-1}), and therefore K (1.4×10^{-9} cm.s^{-1}), are not intented to be precise measurements but orders of magnitude.

Figure 4 clearly shows that we are able to fit both ^{18}O near surface boxes-shaped profiles, with h and D as parameters, and this favours hypothesis 2. However, the conclusive experiment is to anneal in $^{15}N^{18}O$ a significantly thicker oxide. In case of hypothesis 1, the $^{18}O/^{15}N$ ratio at the interface would be less than 1, as it is the case here, due to a stronger impoverishement in ^{18}O of the mobile species for a greater diffusion length. In case of hypothesis 2, the $^{18}O/^{15}N$ ratio would still be 1.

For comparison with silicon oxidation in dry oxygen, we extracted h and D in the case of a wafer oxidized at 1050 °C under oxygen pressures of 84 hPa first in $^{16}O_2$ for 60 s and then in $^{18}O_2$ for 150 s (see figure 3a in [32]). In that case we obtain the following approximate values: $D \approx 3 \times 10^{-16}$ cm^2.s^{-1}, $h \approx 2 \times 10^6$ cm^{-1}, $K \approx 6 \times 10^{-10}$ cm.s^{-1}. For a 20 hPa pressure, we would expect a value of D twice lower [24]. Since all these values are only orders of magnitude, it is hazardous to compare them, even if it seems that D corresponding to oxygen fixation in NO annealing is higher than D in further oxidation in O_2. For a 20 nm-thick $Si^{16}O_2$ film grown in conventional furnace and reoxidized at 930 °C for 5 hours [29], it not surprising that the limit condition $h \to \infty$ is attained, when considering the lower D (2.7×10^{-18} cm^2.s^{-1}).

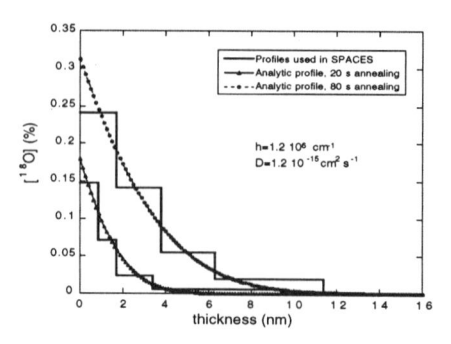

Figure 4. Near surface ^{18}O concentrations depth profiles for 14 nm-thick $Si^{16}O_2$ film annealed in 20 hPa of $^{15}N^{18}O$ for 20 s and 80 s at 1050 °C, used as an entry to the SPACES program to fit the measured excitation curves in figures 3a and 4a (solid line, no marker), and fits of theses boxes-shaped profiles using Equation 2 (solid lines, dots and triangles).

3.2 DIRECT OXYNITRIDATION IN NITRIC OXIDE OF Si(001)

Direct oxynitridation in NO of Si(001) is studied as a fonction of the gas pressure. The NO grown films were realized in the RTP setup, at 1050 °C for 80 s, for various pressures of $^{15}N^{18}O$: 0.1 hPa (sample TNO19), 1 hPa (sample TNO20), 5 hPa (sample TNO21), 20 hPa (sample TNO22), 60 hPa (sample TNO23) and 100 hPa (sample TNO24).

The ^{15}N and ^{18}O areal densities for each film is presented as a function of the $^{15}N^{18}O$ gas pressure in figure 5. The areal density of nitrogen decreases when the reacting gas pressure increases, and nearly follows a $P^{-0.25}$ law when pressure is above 1 hPa. Such a dependence with the gas pressure suggests that the nitrogen incorporation may depend on a mechanism of vacancies creation favoured at low pressure treatment. This result is in contrast to N_2O oxynitridation where the nitrogen amount varies as $P^{1/2}$ [33]. On the contrary, and as in the case of oxynitridation in N_2O we find a $P^{0.25}$ dependence for the oxygen fixation in the film when the pressure is above 1 hPa. This suggests that interstitial sites are more likely involved in the case of oxygen. The mechanisms of incorporation of nitrogen and oxygen seem to be symmetrical. However, for an NO pressure below 1 hPa (TNO19), we find less nitrogen and less oxygen compared to that found in TNO20. This could be due to an etching process occuring below a gas pressure limit. At 1050 °C, this phenomenon may start for NO pressures below 1 hPa.

The thicknesses of the films were estimated through non resonant NRA and were found to vary from 1.4 nm (TNO19) to 2.2 nm (TNO24). Spectroscopic Ellipsometry (SE) operating between about 2 and 4 eV, was also carried out at the Institut Universitaire de Technologie of Orsay. The measured dielectric spectra, performed at three incident angles (66°, 68° and 70°) to minimise the noise effect, were modelled using the Bruggeman Effective Medium Approximation (BEMA) [34]. The thickness and the chemical composition of the layer (volumic distribution of SiO_2 and Si_3N_4) were fitted using a linear regression analysis to provide the optimum agreement with experiment. The SE measured thicknesses were quite in agreement with NRA equivalent thicknesses as, at most, we only get a 15 % difference between both techniques.

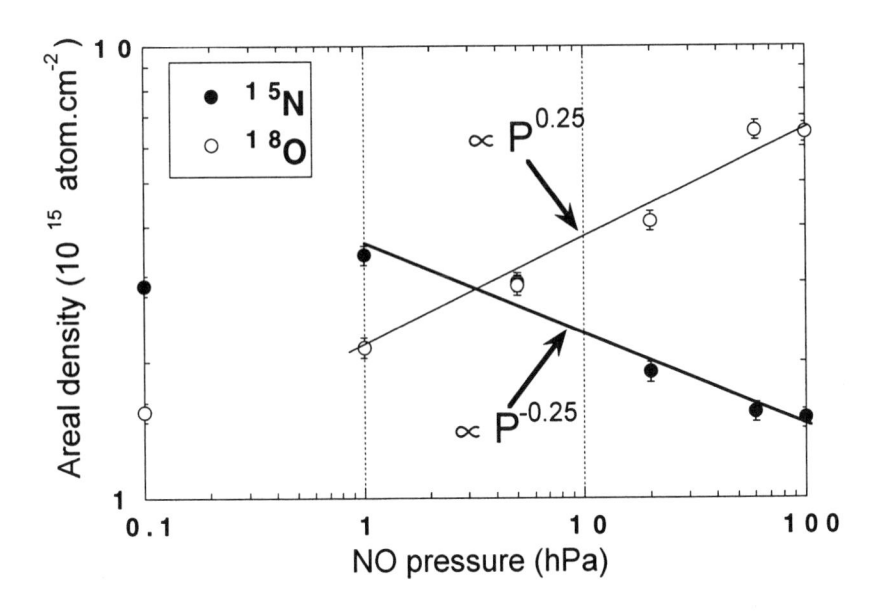

Figure 5. NRA areal densities of ^{15}N and ^{18}O measured on oxynitrides thermally grown on silicon at 1050 °C for 80 s under various ^{15}N^{18}O pressures.

For the XPS measurements in the oxynitrides, we used a weighted average photoelectron escape depth between those in a pure oxide [35] and in a pure nitride [36]. Actually, this value λ is estimated to be 2.6 nm at 0° and 0.7 nm at 75°.

Remembering that the area of each peak depends on the number of emitting atoms, we determined by this way, using the O 1s and N 1s peak, the nitrogen concentration present in the different films. The nitrogen concentration N/(N+O) both measured by XPS (for 0° and 75°) and by NRA are represented in figure 6. One can appreciate that generally speaking, both techniques are in agreement. For all the samples, prepared at different NO pressure, the grazing XPS geometry (65° or 75°) measures display a lower nitrogen concentration compared to the normal geometry. These results show that the surface region of the films is richer in oxygen, which confirms previous AR-XPS studies [37] and MEIS studies [38]. This will be discussed later. It is the first time that such a high nitrogen concentration was found in silicon oxynitrides grown in NO, from 15 to 45 % (NRA results: ^{15}N/^{15}N+O), in comparison with the silicon oxynitrides in N_2O [39], for which not more than 5 % of nitrogen was found. The maximum nitrogen concentration at lower pressure (45 %) was measured in the oxynitride film grown at 1 hPa. However, if we extrapolate the concentration we found in this work to the higher processing pressure used in the study of Gusev *et al.* [25], our results and those of [25] would agree.

We show in figure 7 the Si 2p spectra measured on five different samples (TNO19 to TNO23). All the spectra exhibit two structures, the low binding energy one corresponding to the silicon substrate and the high energy one to the chemically bond silicon in the film. The latter structure moves towards lower binding energies when the concentration of silicon nitride is increased in the film, in part because the energy shift for SiO_2 is larger than the corresponding shift for Si_3N_4. Figure 7 shows that the lower the NO pressure treatment is, the richer the film in nitrogen is, confirming the results presented in figure 5.

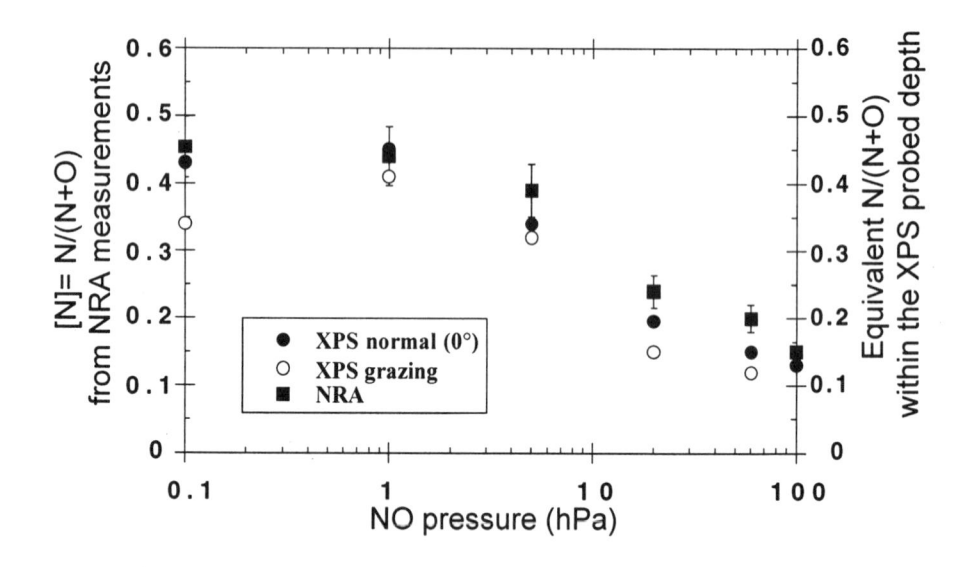

Figure 6. Nitrogen concentration ([N]=N/(N+O)) measured, in the oxynitride samples for various nitric oxide pressures, by NRA, XPS in normal emergence (0°), and XPS in grazing geometry (65 ° or 75 °).

First we considered that the films were constituted by a mixture of SiO_2 and Si_3N_4. However, to be in agreement with the N concentration obtained using the O 1s and N 1s peaks, we then admitted the presence of an intermediate compound SiO_xN_y (oxynitride), which has an intermediate composition between SiO_2 and Si_3N_4. Moreover, considering the N 1s and O 1s profiles, we found that neither the peak location nor the FWHM change noticeably for all samples and for both measurement geometry. The values found are also in agreement with standards (SiO_2 and Si_3N_4). In contrast with a previous study [38], where the samples were prepared at lower temperature and at higher gas pressure (T = 850° to 1000 °C, P = 1 atm, t = 60 min) in a classic furnace (Joule effect), we have no evidence for N-O bonds in our samples. Nevertheless, we are in agreement with another study [40] where the silicon oxynitrides were grown in comparable conditions as in the present work (T = 1000 to 1150 °C, t = 1,2,5 min, RTP system) but at 1 atm pressure. Let us remark that N-O bonds were also found for N_2O grown oxynitrides either in a classical furnace [26] or in a RTP system [41].

The Si 2p3/2 spectra of the high binding energy were then considered as the sum of the contribution of SiO_2, SiO_xN_y, and Si_3N_4. The SiO_2 peak (I[Si 2pox]) position was fixed to be at 429.45 eV (FWHM = 1.41 eV) from the O 1s peak according to a silicon oxide standard. The Si_3N_4 peak (I[Si 2pnitr]) position was fixed to be at 2.51 eV (FWHM = 1.35 eV) from the Si 2p peak of the silicon substrate according to a silicon nitride standard. The oxynitride peak (I[Si 2poxynit]) was fixed to be between the oxide and nitride ones. The relative intensity of each contribution brings information about the chemical bonding composition of the film. The results are summarised in table 2.

The ultra-thin oxynitrides films can therefore be divided in two regions: a near surface region (A Region) and a near oxynitride/silicon interface region (B Region). Figure 8 displays the relative concentration of each compound in both regions for two samples TNO19 (low pressure) and TNO23 (higher pressure). For all the samples, the A Region is richer in SiO$_2$ than the B Region, which could be due to the air exposure between the thermal treatments and the sample analyses. At low pressure, Si$_3$N$_4$ is the dominant compound in the film and it is located principally in the B Region. At intermediate pressure, all the chemical species are more likely to be uniformly distributed in the films, with no particular dominant specie. At higher pressure, the film is mostly constituted by SiO$_2$, with a very weak concentration of Si$_3$N$_4$ essentially located in B Region.

In summary, XPS results reveal the presence of three chemical structures, besides the Si-Si bonds of the bulk ; i) four nitrogen atoms bounded to one Si (Si-4N in Si$_3$N$_4$), ii) four oxygen atoms bounded to one Si (Si-4O in SiO$_2$), and iii) four atoms (O and N) bounded to one Si (in SiO$_x$N$_y$).

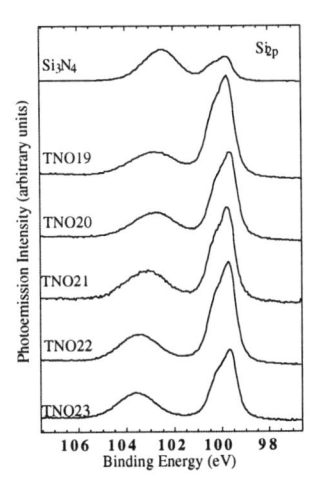

Figure 7. Si 2p spectra measured on a Si$_3$N$_4$ standard, and on five oxynitrides grown on silicon under various gas pressures (0.1 hPa (TNO19), 1 hPa (TNO20), 5 hPa (TNO21), 20 hPa (TNO22), and 60 hPa (TNO23)).

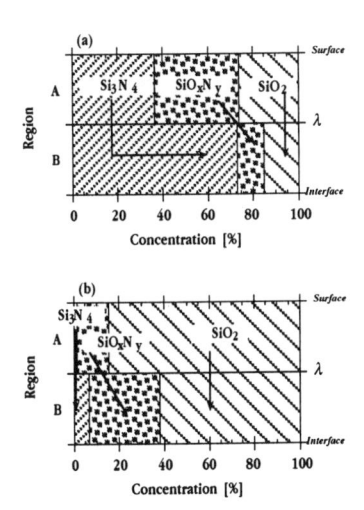

Figure 8. Relative concentrations measured on TNO19 (a) and TNO23 (b) by AR-XPS of the 3 different compounds (SiO$_2$, Si$_3$N$_4$, SiO$_x$N$_y$) in the superficial region (A region) and in the near oxynitride/silicon region (B region).

TABLE 2. Relative concentrations of each compound in the 5 different samples for normal and grazing emergence angle, in XPS measurements.

Sample	Emergence angle	SiO$_2$ %	SiO$_x$N$_y$ %	Si$_3$N$_4$ %
TNO19	0°	20.5	24.6	54.9
	75°	26.0	37.5	36.5
TNO20	0°	22.0	51.0	27.0
	65°	24.4	45.3	30.3
TNO21	0°	34.5	49.0	16.4
	65°	40.0	48.2	11.8
TNO22	0°	64.4	26.9	8.7
	75°	75.4	21.5	3.7
TNO23	0°	69.0	26.0	5.0
	75°	84.5	14.0	1.5

Figure 9 (a and b) represents the excitation curves measured on TNO19 sample (0.1 hPa of $^{15}N^{18}O$) around the energy resonances at 152 keV and 429 keV respectively.

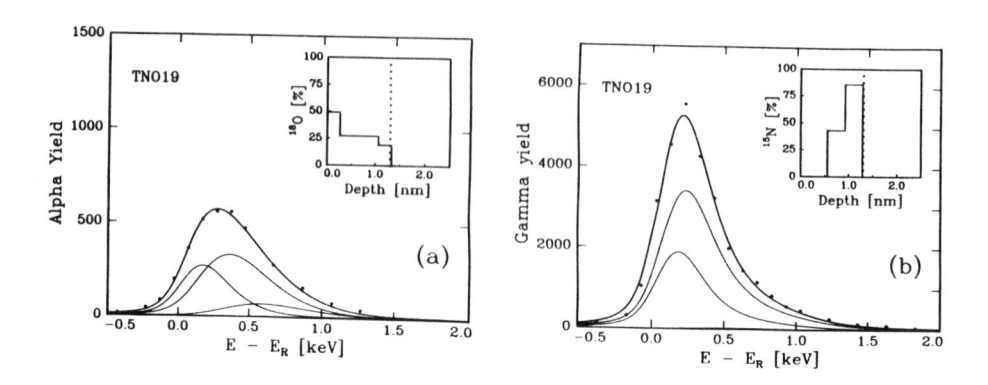

Figure 9. Excitation curves of the $^{18}O(p,\alpha\gamma)^{15}N$ (a) and $^{15}N(p,\alpha\gamma)^{12}C$ (b) nuclear reactions at 152 keV and 429 keV respectively measured on the sample TNO19 around the energies of the resonances. The experimental data points are fitted using two uniform box distributions. In insets are represented the related concentration depth profiles. The dashed line indicates the oxynitride/silicon interface. A tilt angle of 65° between the normal to the sample and the direction of the beam was used to improve depth resolution. 30 µC (a) and 10 µC (b) per point.

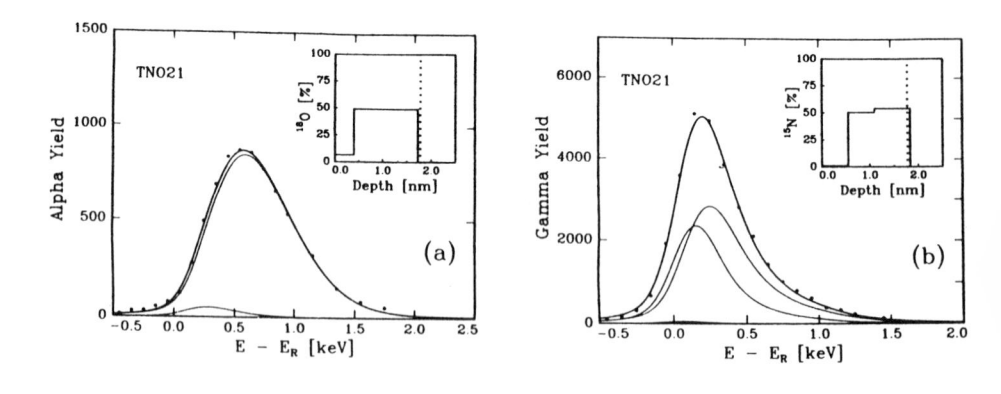

Figure 10. Same as figure 9 but measured on the sample TNO21.

These excitation curves have been decomposed in various boxes distributions of different concentrations. In inset are presented the related depth concentration profiles. At this pressure, the nitrogen concentration in the film is in majority present in the near oxynitride/silicon interface region. The ^{18}O is mostly incorporated near the surface region during the thermal treatment. Although no nitrogen is measured in the surface region, the results evidence that ^{18}O is not only atoms present near the surface. We suspect the presence of natural oxygen which should be the result of air contamination, leading to an ^{18}O/^{16}O mixture at the surface. It is clear that nitrogen preferentially fixes the interface oxynitride/silicon where fresh silicon is available and where oxygen vacancies should be created.

Figure 10 (a and b) represents the same as figure 9 (a and b) but for a sample grown at higher NO pressure (TNO21, 5 hPa of ^{15}N^{18}O). At this intermediate pressure, the insets of figures 10a and 10b show that the ^{15}N and ^{18}O concentrations are quite the same in the sample. In the superficial region, again we find again a mixture of ^{18}O/^{16}O probably due to the air exposure.

All the samples were submitted to the same analyses and they all display the same phenomenon of air contamination in the near air/oxynitride surface. The NRP analyses are in agreement with the XPS results.

Conclusion

In the case of NO anneal of SiO$_2$ film, we suggest that two mechanisms occur in parallel. In the first one, NO diffuses through the silica network without reacting with it and the same amount of N and O are fixed in the near interface region. In the second one, ^{18}O is fixed near the oxide surface due to a mechanism related with a step-by-step motion of network oxygen atoms, by a simple diffusion process, induced by the presence of network defects, involving only O. This latter mechanism, dominant within the framework of our study, leads mostly to exchange of oxygen atoms between the oxide network and the gas phase. However further studies have to be carried out to confirm our hypothesis, and to precise the nature of defects involved during NO anneal.

In the case of NO grown films, we suggest that, at low pressures, nitrogen incorporation could be driven by a more effective vacancy creation where the nitrogen is preferentially fixed. The oxygen incorporation is more likely due to the presence of interstitial sites favoured at higher pressure. The schemes of nitrogen and oxygen incorporations seem to be symmetrical. The presence of nitrogen in the film seems to limit the films thicknesses which never exceed 2.5 nm. The overall concentrations of nitrogen in the films measured by NRA and XPS are very close, reinforcing the validity of the results. Nuclear reactions analyses using low energy resonances are in agreement with the AR-XPS results. The nitrogen preferentially fixes in the near interfacial region, this mechanism becoming dominant at low treatment pressure. At higher pressure, the oxidation mechanism becomes dominant and the oxygen seems to incorporate in the film from the air/oxynitride interface. To shed light on the mechanisms involved in this case, sequential ^{14}N^{16}O/^{15}N^{18}O on Si(001) should be investigated.

Acknowledgments

The authors want to thank B. Agius and M.-C. Hugon, from the Laboratoire d'Étude de Matériaux en Films Minces (Orsay, France), for Spectroscopic Ellipsometry measurements and for their help in the spectra interpretations. This work is supported by the french Centre National de la Recherche Scientifique, GDR 86.

References

1. Terman, L.M. (1997) The impact of device scaling limits, *Applied Surface Science* **117/118**, 1-10.
2. Hara H. (1997) Some issues yet to be solved for the age of 0.1 µm technologies, *Applied Surface Science* **117/118** 11-19.
3. Bohr M.T. (1996) Technology development strategies for the 21st century, *Applied Surface Science* **100/101** 534-540.
4. The National Technology Roadmap to Semiconductors (1994), Semiconductor Industries Association.
5. Ono M. et al. (1995) A 40 nm gate length N-MOSFET, IEEE *Trans. Electron Devices* **ED-42** 1822-1830.
6. Hegde R.I., Maiti B., and Tobin P.J. (1997) Growth and Film Characteristics of N_2O and NO Oxynitride Gate and Tunnel Dielectrics, *J. Electrochem. Soc.* **144** 1081-1086.
7. Kim K., Hee Lee Y., Suh M.-S., Youn C.-J., Lee K.-B, and Jae Lee H. (1996), Thermal Oxynitridation of Silicon in N_2O Ambients, *J. Electrochem. Soc.* **143** 3372-3376.
8. Fukuda H., Koyama N., Endoh T., Nomura S. (1997) Growth kinetics of nanoscale SiO_2 layer in a nitric oxide (NO) ambient, *Applied Surface Science* **113/114** 595-599.
9. Wrixon R.,Twomey A., O'Sullivan P., and Mathewson A. (1995), Enhanced Thickness Uniformity and Electrical Performance of Ultrathin Dielectrics Grown by RTP Using Various N_2O-Oxynitridation Processes, *J. Electrochem. Soc.* **142** 2738-2742.
10. Arakawa T., Matsumoto R., (1997), Impact of nitrogen concentration profile in silicon oxynitride films on stress-induced leakage current, *Applied Surface Science*, **113/114**, 605-609.
11. Tsamis C., Kouvastos D.N., and Tsoukalas D. (1996) Influence of N_2O oxidation temperature on point defect injection kinetics in the high temperature regime, *App. Phys. Lett.* **69** 2725-2727.
12. Trimaille I., Ganem J-J., (1997), Isotopic tracing of oxygen during thermal growth of thin films of SiO_2 on Si in dry O_2, *Brazilian Journal of Physics*, **27**, 293-302.
13. Amsel G., Nadai J.P., d'Artemare E., David D., Girard E., and Moulin J. (1971) Microanalysis by the direct observation of nuclear reactions using a 2 MeV Van de Graaff, *Nucl. Instrum. Meth.* **92** 481-498.
14. Ganem J-J., Rigo S., Trimaille I., Lu G-N., Molle P., (1992), Deuteron beam analysis of rapid thermal nitridation of silicon and thin SiO_2 films, *Nucl. Instrum. Meth.* **B64**, 778-783.
15. Rigo S. (1992) Nuclear microanalysis study of the growth of thin dielectric films on silicon by classical and rapid thermal treatments, *Nucl. Instrum. Meth.* **B64**, 1-11.
16. Maillot C., Roulet H., Dufour G., Rochet F., and Rigo S. (1986) Study of atomic transport mechanisms during thermal nitridation in ammonia using ^{15}N and D labelled gas, *Applied Surface Science* **26** 326-334.
17. Battistig G., Amsel G., d'Artemare E., and Vickridge I. (1991) A very narrow resonance in $^{18}O(p,\alpha)^{15}N$ near 150 keV. Application to isotopic tracing. I. Resonance width measurement, *Nucl. Instrum. Meth.* **B61**, 369-376.
18. Battistig G., Amsel G., d'Artemare E., and Vickridge I. (1992) A very narrow resonance in $^{18}O(p,\alpha)^{15}N$ near 150 keV. Application to isotopic tracing. II. High resolution depth profiling of ^{18}O, *Nucl. Instrum. Meth.* **B66**, 1-10.
19. Amsel G., Maurel B., (1983), High resolution techniques for nuclear reaction narrow resonance width measurements and for shallow depth profiling,*Nucl. Instrum. Meth.* **218**, 183-196.
20. Vickridge I. and Amsel G. (1990) SPACES: a PC implementation of the stochastic theory of energy loss for narrow-resonance depth profiling, *Nucl. Instrum. Meth.* **B45**, 6-11
21. Amsel G., and Vickridge I. (1990) Analytic calculation for some useful depth profiles of the linear expansion coefficients used in SPACES, *Nucl. Instrum. Meth.* **B45**, 12-15.
22. Himpsel, J., McFeely F. R., Taleb-Ibrahimi, A., Yarmoff J.A., and Hollinger, G. (1988) Microscopic structure of the SiO_2/Si interface, *Phys. Rev.* **B38** 6084-6096.
23. Rigo, S. (1988) Si oxidation mechanisms as studied by oxygen tracer methods, in C.R. Helms and B.E. Deal (eds.), *The Physics and Chemistry of SiO_2 and the $Si-SiO_2$ Interface*, Plenum Press, New York, pp. 75-84.
24. Trimaille I., Rigo S., (1989), Use of ^{18}O isotopic labelling to study thermal dry oxidation of silicon as a function of temperature and pressure, *Applied Surface Science* **39**, 65-80.
25. Gusev E.P., Lu H.C., Gustafsson T., Garfunkel E., Green M.L. Brasen D., (1997), The composition of ultrathin silicon oxynitrides thermally grown in nitric oxide, *J. Appl. Phys.* **82**, 896-898.
26. Bouvet, D., Clivaz, P.A., Dutoit, M., Coluzza, C., Almeida, J., Margaritondo, G, Pio, F. (1996) Influence of nitrogen profile on electrical characteristics of furnace or rapid thermally nitrided silicon dioxide films, *J. Appl. Phys.* **79**, 7114-7122.

27. Baumvol, I.J;R., Stedile, F.C., Ganem, J-J., Trimaille, I., Rigo, S., (1997), Isotopic tracing during rapid thermal growth of silicon oxynitride films on Si in O_2, NH_3, and N_2O, *Appl. Phys. Lett.* **70**, 2007-2009.
28. Carr, E.C., Ellis, K.A., and Buhrman, R.A. (1995) N depth profiles in thin SiO_2 grown or processed in N_2O: The role of atomic oxygen, *App. Phys. Lett.* **66**, 1492-1494.
29. Trimaille, I., Stedile, F.C., Ganem, J.-J., Baumvol, I.J.R., and Rigo, S. (1996) Mechanisms of thermal growth of very thin films of SiO_2 on Si(001) in dry O_2, in H.Z. Massoud, E.H. Pointdexter, and C.R. Helms (eds.), *The Physics and Chemistry of SiO_2 and the Si-SiO_2 Interface-3*, The Electrochemical Society, Pennington, pp. 59-71.
30. Philibert J. (1991) *atoms movements diffusion and mass transport in solids*, Les Editions de Physique, Les Ulis. The reader may refer also to Crank J. (1975) *The Mathematics of Diffusion*, Clarendon Pess, Oxford.
31. Ganem, J-J., Trimaille, I., André, P., Rigo, S., Stedile, F.C., Baumvol, I.J.R., (1997) Diffusion of near surface defects during the thermal oxidation of silicon, *J. Appl. Phys.* **81**, 8109-8111.
32. Stedile, F.C., Baumvol, I.J.R., Ganem, J.-J., Rigo, S., Trimaille, I., Battistig, G., Schulte, W.H., Becker H.W. (1994), IBA study of the growth mechanisms of very thin silicon oxide films:the effect of wafer cleaning, *Nucl. Instrum. Meth.* **B85**, 248-254.
33. Gosset, L.G., Ganem, J.-J., Trimaille, I., Rigo, S., Baumvol, I.J.R. (unpublished).
34. Bruggeman, D.A.G. (1935), Berechnung verschiedener physikalischer Konstanten von heterogenen Substanzen. I. Dielektrizitätskonstanten und Leitfähigkeiten der Mischkörper aus isotropen Substanzen, *Ann. Phys.* **24** (Leipzig) 636-679.
35. Tao, Y., Lu, Z.H., Graham, M.J., Tay, S.P., (1988) X-ray photoelectron spectroscopy and x-ray absorption near-edge spectroscopy study of SiO_2/Si(100), *J. Vac. Sci. Technol.* **B12**, 2500-2503.
36. Maillot, C., Roulet, H., and Dufour, G. (1984) Thermal nitridation of silicon:An XPS and LEED investigation, *J. Vac. Sci. Technol.* **B2** 316-319.
37. Kamath A., Kwong D.L., Sun Y.M., Blass P.M., Whaley S., White J.M., (1997) Oxidation of Si(100) in nitric oxide at low pressures: an x-ray photoelectron spectroscopy study, *Appl. Phys. Lett.* **70**, 63-65.
38. Lu H.C., Gusev E.P., Gustafsson T., Garfunkel E., Green M.L., Brasen D., Feldman L.C. (1996), High resolution ion scattering study of silicon oxynitridation, *Appl. Phys. Lett.* **69**, 2713-2715.
39. Ganem, J.-J., Rigo, S., Trimaille, I., Baumvol, I.J.R., and Stedile, F.C. (1996) Dry oxidation mechanisms of thin dielectric films under N_2O using isotopic tracing methods, *Appl. Phys. Lett.* **68**, 2366-2368.
40. Yao, Z.-Q. (1995) The nature and distribution of nitrogen in silicon oxynitride grown on silicon in a nitric oxide ambient, *J. Appl. Phys.* **78** 2906-2912.
41. Bhat, M., Yoon, G.W., Kim, J., Kwong, D.L., Arendt, M. and White J.M. (1994) Effects of NH_3 nitridation on oxides grown in pure N_2O ambient, *Appl. Phys. Lett.* **64**, 2116-2118.

THERMAL ROUTES TO ULTRATHIN OXYNITRIDES

M. L. GREEN, D. BRASEN, L. C. FELDMAN[*], E. GARFUNKEL[+],
E. P. GUSEV[+], T. GUSTAFSSON[+], W. N. LENNARD[#], H. C. LU[+] and
T. SORSCH

Bell Laboratories/Lucent Technologies
Murray Hill, New Jersey, USA

[*]*Department of Physics and Astronomy*
Vanderbilt University
Nashville, Tennessee, USA

[+]*Department of Physics and Laboratory for Surface Modification*
Rutgers University
Piscataway, New Jersey, USA

[#]*Interface Science Western*
The University of Western Ontario
London, Ontario, Canada

1. Introduction

Ultrathin (< 6 nm) silicon oxynitrides are desirable as gate dielectrics for present and future ultra-large-scale-integrated (ULSI) circuits due to their improved reliability [1-3] and boron penetration resistance [4,5] compared to SiO_2. Over the history of integrated circuit processing, many nitridation and oxynitridation chemistries have been used to form such dielectrics. Direct nitridation via reaction of Si with N_2 [6,7] required very high temperatures (T>1200°C) and therefore too high a thermal budget. The use of NH_3 as a nitridation agent, especially to nitridize the Si/SiO_2 interface [8-11], resulted in excessive H incorporation and the potential for degradation of dielectric reliability. More recently, oxynitrides have been grown in either N_2O [12,13] or NO [14,15]. These dielectrics contain a small but significant amount of nitrogen [16], almost invariably concentrated near the Si/SiO_2 interface [17].

Oxynitrides can and have been grown under a wide range of oxidation potentials, i.e., oxygen partial pressures. For example, oxynitrides grown from N_2O or NO form under high oxygen partial pressures, since one of the by-products of the decomposition of these gases is oxygen. On the other hand, oxynitrides have also been grown in ostensibly pure N_2, where trace amounts of oxidants are inevitably present, resulting in oxynitride formation under very low oxygen partial pressures. In this paper, we hope to make the distinction between these cases. In particular, incorporated nitrogen is not in chemical equilibrium when growth occurs at high oxygen partial pressures. Therefore, one must rationalize the presence of nitrogen in NO and N_2O-oxynitrides. We will point out that oxynitrides grown under extremely low oxygen partial pressures can be in chemical equilibrium, and therefore may contain more nitrogen than NO or N_2O-oxynitrides, and further, it may be distributed more uniformly. It is necessary to understand nitrogen incorporation phenomena because they will influence the way nitrogen is distributed throughout the film

2. Oxynitridation

2.1. IDEAL NITROGEN DISTRIBUTION

A ULSI gate oxynitride ideally should have nitrogen peaks at the Si/SiO_2 interface, to improve hot carrier resistance, and at the SiO_2/polysilicon interface, to prevent the penetration of boron from the heavily doped gate electrode. Thus, the nitrogen distribution depicted in Fig. 1 might be considered ideal. More nitrogen is required at the polysilicon interface, since the boron flux can be very large, depending upon thermal budget. At the Si interface, only enough nitrogen to improve the hot carrier resistance is needed, as more might also lower the channel mobility. Very little nitrogen is desired in the interior of the dielectric, since a large total amount of incorporated nitrogen might raise the fixed charge. The actual ideal amounts of nitrogen required at each interface are not known, but typical interfacial concentrations range from 1 to 5% at each interface.

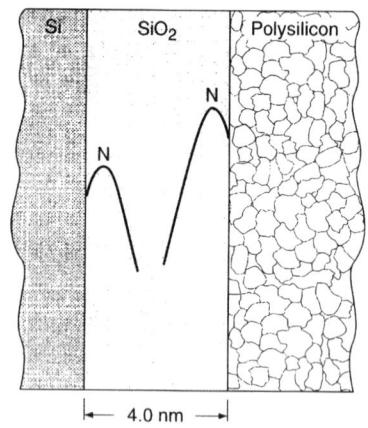

Figure 1. Schematic representation of an ideal nitrogen distribution in a ULSI gate dielectric.

2.2. SILICON-OXYGEN-NITROGEN PHASE EQUILIBRIUM

After N$_2$O or NO oxynitridation, one typically observes the incorporated nitrogen to be segregated very close to the Si/SiO$_2$ interface [12, 17], as is depicted in Fig. 2. As will be seen later, the N peak can be shifted to other areas of the film, but the processing required to accomplish that must take into account the metastability of the incorporated nitrogen in the SiO$_2$. In fact, according to chemical equilibrium, nitrogen should not incorporate into an SiO$_2$ film that is grown in almost any partial pressure of oxygen, i.e., > 10^{-20} atm., depending upon temperature [18]. This can be seen from Fig. 3, which shows that the Si$_2$N$_2$O/SiO$_2$ phase boundary exists at about 10^{-18} atm. for T=1400K. At any oxygen partial pressure greater than that, which surely exists during N$_2$O or NO decomposition, only SiO$_2$ phases (crystalline or amorphous), should form. Therefore, it is hard to explain why nitrogen is incorporated in SiO$_2$ during reaction with N$_2$O or NO, as it cannot be stable from a chemical thermodynamic standpoint. However, at least two reasons for its presence in the SiO$_2$ can be suggested. First, nitrogen atoms, by-products of the decomposition reaction of N$_2$O or NO, may simply be kinetically trapped at the reaction zone near the interface. Alternatively, the nitrogen at the interface may indeed be thermodynamically stable, due to the presence of free energy terms that are not yet understood. For example, if the nitrogen plays a role in lowering the interfacial strain, there

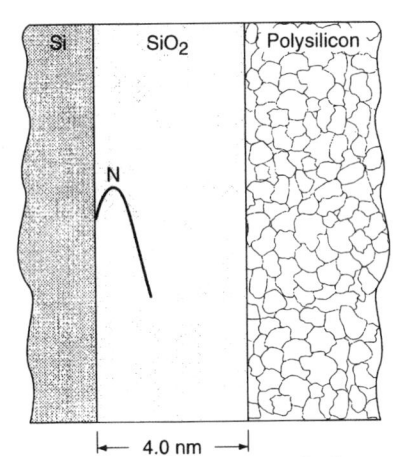

Figure 2. Schematic representation of nitrogen profile typically observed after oxynitridation in N$_2$O or NO. For the case of NO, the film exhibits self limiting behavior and only grows to about 2 nm thickness.

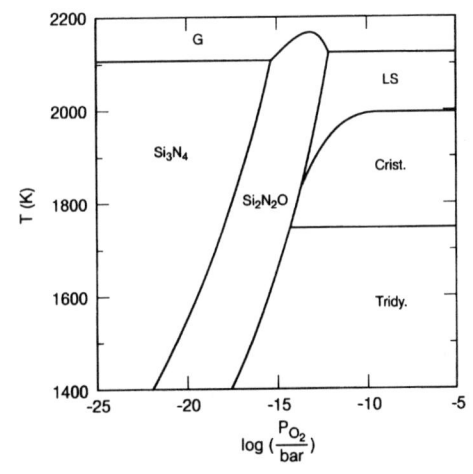

Figure 3. Pressure-temperature phase diagram for the Si-O-N system, from [18].

may be a strain free energy contribution when nitrogen is incorporated at the interface. This term would be negative and may be larger in magnitude than the chemical free energy term, especially for very thin films. Incorporated nitrogen in N_2O or NO-oxides is almost invariably associated with the Si/SiO_2 interface, consistent with a special, stabilizing role of the nitrogen at the interface. Even when nitrogen is implanted into Si [19], it tends to migrate to the Si/SiO_2 interface and be incorporated in the SiO_2. Therefore, there is some evidence to suspect that the nitrogen plays a specific role at the interface, but also that it is not stable away from it. It has been shown that nitrogen exhibits a different bonding chemistry as it is moved away from the interface [20, 21], and further, that when in that position, nitrogen is not stable with respect to N_2O [22] or O [23].

2.3. OXYNITRIDE GROWTH

Therefore, in this paper we look at two ways of growing oxynitrides. The first, illustrated in Fig. 4a, occurs under extremely low partial pressures of oxygen. The second, illustrated in Fig. 4b, occurs under much higher oxygen partial pressure. We will discuss the nitrogen incorporation and distribution in each case.

2.3.1. *Low partial pressures of oxygen*

In this case we grow oxynitrides in nominally pure N_2. We use the term "nominally" to mean that although the input N_2 gas stream is purified at the point of use and therefore extremely free of contaminants such as H_2O, H_2 and O_2, the growth chamber, a cold wall rapid thermal processing (RTP) module, contributes impurities to the ambient through outgassing from the walls. There is much evidence from the ceramic literature that reactions between Si and N_2 are mediated by gas phase impurities such as H_2O, H_2 and O_2 [24-26]. Therefore, although the Si/N_2 system may be relatively inert for T<1200°C, the $Si/N_2/H_2O/H_2/O_2$ system is not. Thus, we find that N_2 reacts with Si at moderate temperatures (850-1050°C) in an RTP module, due to the presence of gas phase impurities.

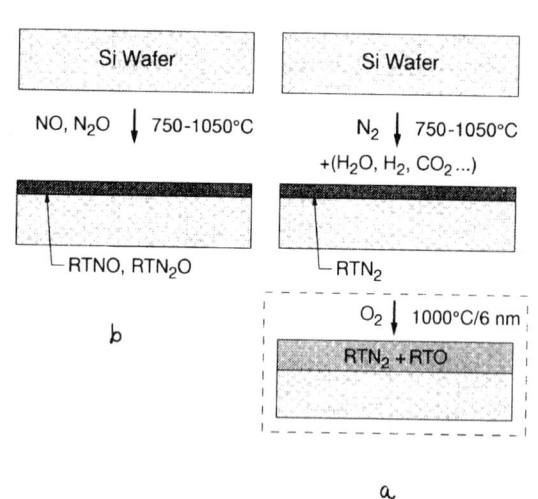

Figure 4. Process flows for growing oxynitrides in a) low oxygen partial pressure (the last reoxidation step is optional), and b) high oxygen partial pressure (e.g., NO or N_2O oxynitridation).

The low oxygen partial pressure oxynitridation experiments were carried out in an integrated rapid thermal oxidation (RTO) tool. 150mm CZ Si (100) wafers, P-type (boron doped, 4-6 Ω-cm), were given an RCA clean and a 15:1 HF dip. The hydrophobic wafers were immediately loaded into the loadlock, and pumped down to a pressure of 5×10^{-7} Torr. They were then transferred to the RTO chamber, which was held at 10^{-8} Torr. The chamber was then backfilled with point-of-use purified N_2 (after such purification through a resin column, the N_2 stream contains less than 1 ppb each of H_2O, O_2, CO_2 and CO) to atmospheric pressure. The wafers were heated in this ambient for times between 20 and 1200 seconds at temperatures between 850 and 1050°C. Such treatment, called RTN_2, resulted in the formation of ultrathin SiO_xN_y films. All films were less than 1.2 nm, as measured by ellipsometry (using η =1.47) and cross-sectional TEM. The fact that SiO_xN_y, and not Si_3N_4, forms is due to the affinity of silicon for oxygen, and the ubiquity of small amounts of oxidizing impurities such as H_2O and O_2, outgassing from the quartz and stainless steel surfaces of the RTO chamber. The nitrogen and oxygen contents of the films were determined by nuclear reaction analysis (NRA), described fully elsewhere [27], using the $^{14}N(d,\alpha_{0,1})^{12}C$ and $^{16}O(d,p_1)^{17}O$ nuclear reactions. We also used high-resolution (ΔE/E ~0.1%) medium energy ion scattering (MEIS) [17, 28], to accurately obtain the nitrogen depth profile.

Figure 5 illustrates nitrogen content as a function of temperature for RTN_2 and RTN_2O films. The RTN_2O films are all 10 nm thick, and the RTN_2 films were grown for

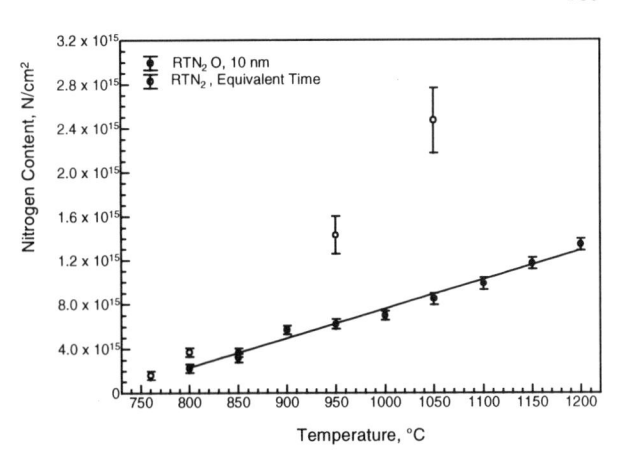

Figure 5. Nitrogen content as a function of N_2 nitridation time for silicon wafers given RTN_2 and RTN_2O oxynitridations for equivalent time periods.

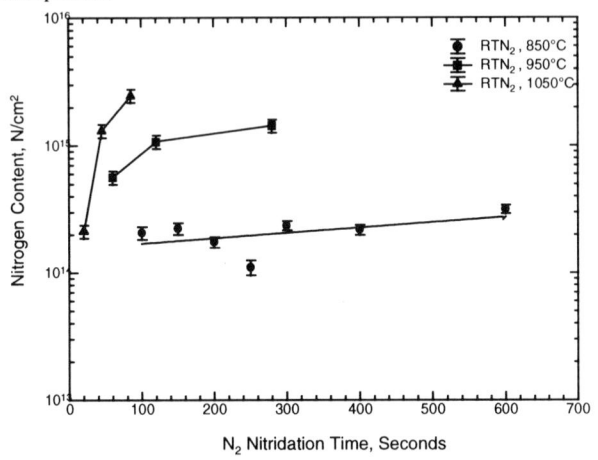

Figure 6. Nitrogen content as a function of N_2 nitridation time for silicon wafers given RTN_2 treatment at temperatures between 850 and 1050°C.

the same times required to grow the RTN$_2$O films. Notice that the nitrogen content of the layers is the same for lower temperatures, but above 950°C, the RTN$_2$ films actually contain more nitrogen than the RTN$_2$O films. This is surprising, but perhaps explainable because the RTN$_2$ films contain nitrogen in equilibrium, i.e., grown under conditions described by the left hand corner of the phase diagram, Fig. 3. On the other hand, in the RTN$_2$O films, grown under conditions all the way to the right of that point, nitrogen may only be stable at the interface, for reasons suggested earlier. The rest of the incorporated nitrogen was pushed into the bulk of the growing N$_2$O-oxynitride, at which point it was removed by a reverse reaction. Therefore, it is not surprising that more nitrogen is present in the equilibrium case. Fig. 6 illustrates nitrogen content as a function of N$_2$ nitridation time for RTN$_2$ treatment at various temperatures. At 850°C, nitrogen content is almost constant with time, whereas at the higher temperatures it increases and then may saturate with time. It is convenient to express nitrogen content in terms of monolayer equivalents on the Si (100) surface, even though the nitrogen does not exist as such. If one monolayer (ML) is 6.8×10^{14} N/cm^2, it can be seen from Fig. 6 that anywhere from 0.2 to 3 ML of nitrogen can be incorporated, depending upon temperature and time.

Figure 7 is a plot of the ratio N/(N+O) as a function of RTN$_2$ time at 850, 950 and 1050°C. The ratio increases with increasing temperature and time, and appears to approach that of the high temperature stoichiometric compound Si$_2$N$_2$O [18, 29, 30]. The RTN$_2$ films may consist of amorphous mixtures of SiO$_2$ and Si$_2$N$_2$O phases. Mixtures of Si$_2$N$_2$O, Si$_3$N$_4$ and SiO$_2$ phases have similarly been observed in plasma deposited oxynitrides [31].

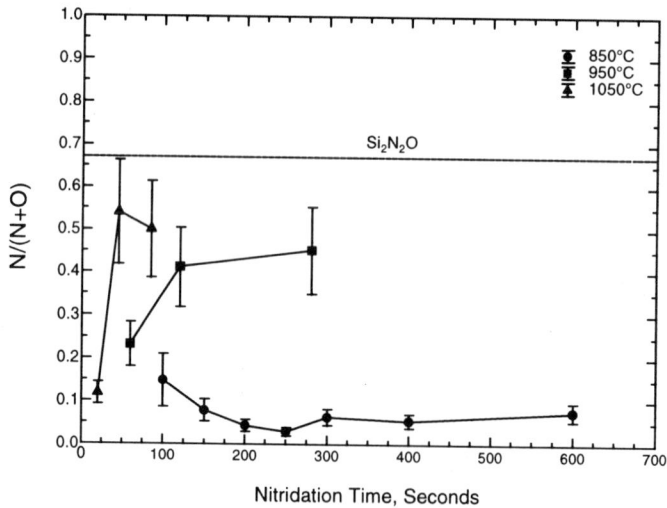

Figure 7. N/(N+O) ratio after RTN$_2$, as a function of time at 850, 950 and 1050°C.

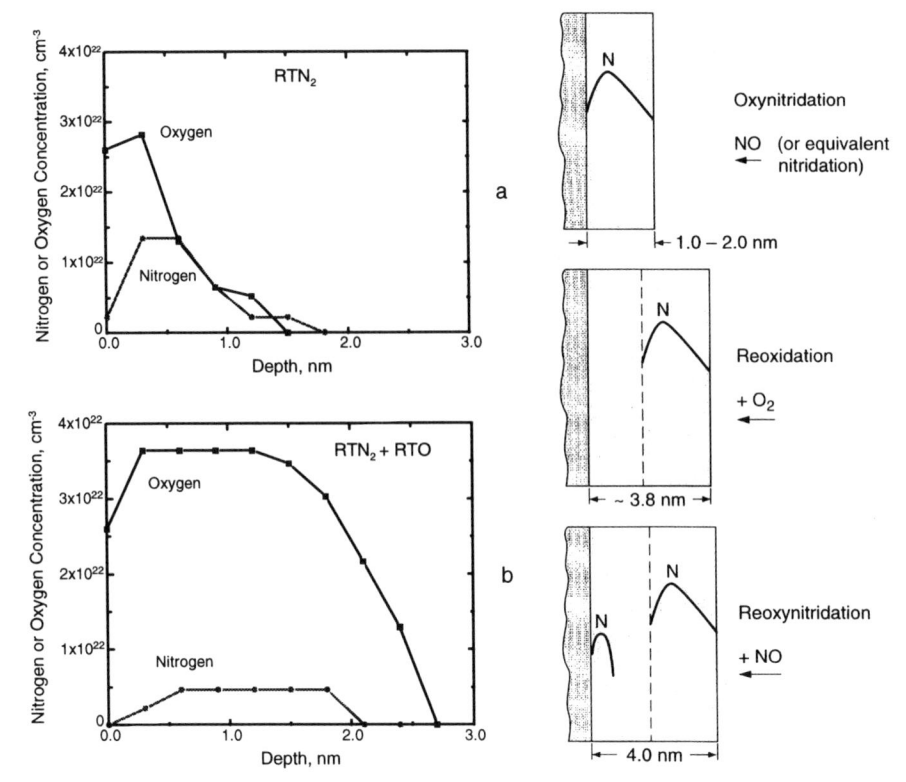

Figure 8. Nitrogen depth profiles, determined by MEIS, for a) an RTN$_2$ film grown at 950°C for 280 seconds, and b) a similar RTN$_2$ film reoxidized in O$_2$ at 1000°C for 25 seconds.

Figure 9. Depiction of the process flow for creating the ideal nitrogen profile using NO gas.

The nitrogen distribution in the dielectric is an important consideration. Figs.8a and b show nitrogen depth profiles in RTN$_2$ as well as RTN$_2$ + RTO dielectrics, determined from MEIS experiments. Oxygen and nitrogen are seen to be concentrated at the top of the RTN$_2$ film. However, they are fairly uniformly distributed in the RTN$_2$ + RTO film.

2.3.2. *High partial pressures of oxygen*

Nitrogen incorporated during NO or N$_2$O oxynitridation is not in chemical equilibrium, as dictated by the phase diagram of Fig. 3. The nitrogen closest to the Si/SiO$_2$ interface may be in thermomechanical equilibrium due to its ability to lower the interfacial stress, or it may just be kinetically trapped there. Therefore, it is difficult to grow an NO or N$_2$O oxynitride in which the nitrogen is incorporated anywhere else but the interface.

Figure 9 depicts a scheme for achieving the ideal nitrogen profile suggested in Fig. 1. First, a thin oxide layer containing a high concentration of nitrogen is grown. This can be conveniently done using NO, as is depicted in Fig. 10, but other surface nitridation techniques such as ion-implantation would also work [19]. Notice from Fig. 10 that it is quite easy to incorporate a monolayer equivalent of nitrogen at temperatures as low as 750°C. Such NO oxynitrides tend to saturate growth at about 1.5 -2.0 nm. This thin layer is then reoxidized in O_2, which moves it intact to what will become the top of the film, since new oxide has grown beneath it. The reoxidation rate is very strongly dependent

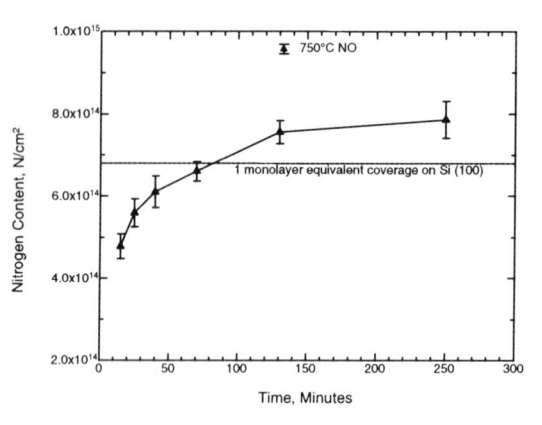

Figure 10. Nitrogen concentration, as measured by nuclear reaction analysis, as a function of NO exposure time at 750°C.

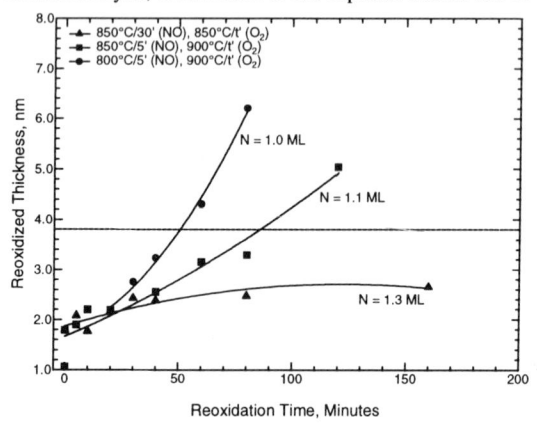

Figure 11. Reoxidation behavior (NO/O_2 thickness as a function of O_2 reoxidation time) for NO films containing various initial concentrations of nitrogen.

upon the amount of nitrogen initially present, as is shown in Fig. 11. Nitrogen contents differing by only 10% (1.0 vs 1.1 monolayer equivalents) have a profound effect on reoxidation kinetics. The time required to reach a target reoxidized thickness of 3.8 nm is 50 minutes for the former case, but 85 minutes for the latter. Finally, the third step, reoxynitridation in NO, is a key step. It has been found [32] that NO does not remove nitrogen from the interior of the film, as does atomic oxygen, generated either from the decomposition of N_2O [22], or from ozone [23]. Therefore, the top nitrogen peak is undisturbed. Further, the last step reincorporates nitrogen at the Si/SiO_2 interface, but with little oxide regrowth. This provides for excellent thickness control. Fig. 12 depicts medium energy ion-scattering results that confirm that an NO/O$_2$/NO sample has nitrogen concentrations at both the top and the bottom of the oxynitride film.

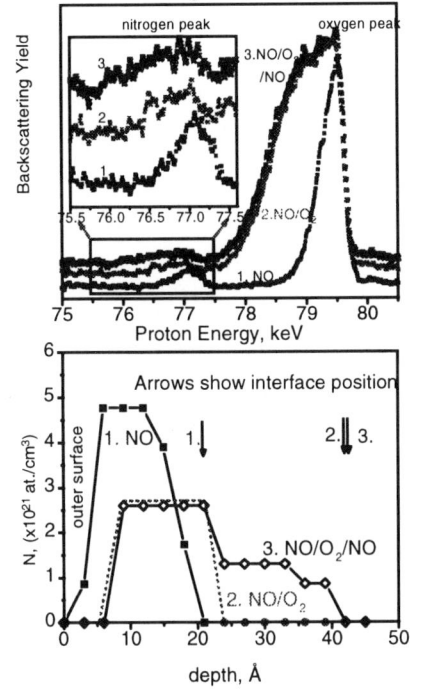

Figure 12. Medium energy ion scattering spectra, and deconvolved nitrogen distributions, for 1)NO(850°C, 60'), 2)NO+O$_2$(850°C, 240') and 3)NO/O$_2$+NO(850°C, 45') process steps.

3. Summary

We have shown that silicon oxynitrides can be grown under both high and low oxygen partial pressure conditions. Those grown at high partial pressures are not in chemical equilibrium, and the amount of nitrogen incorporated in them is typically less than is incorporated during a similar thermal cycle under low oxygen partial pressure conditions. Understanding the effects of O$_2$, N$_2$O and NO on the nitrogen incorporation and removal reactions allows one to tailor useful nitrogen profiles in the oxynitrides.

Acknowledgments

The authors would like to acknowledge valuable discussions with H. Du and W. Reents. Experiments performed at Rutgers University were supported by NSF (DMR-9701748, ECS-9530984) and SRC/Lucent (97-BJ-451).

4. References

1. Matsuoka, T., Taguchi, S., Ohtsuka, H., Taniguchi, K., Hamaguchi, C., Kakimoto S. and Uda, K., (1996), IEEE Trans. Electron Dev., **43**, 1364.
2. Hao, M.-Y., Chen, W.-M., Lai, K., Lee, J. C., Gardner, M. and Fulford, J. , (1995), Appl. Phys. Lett., **66**, 1126.
3. Okada, Y., Tobin, P. J., Rushbrook, P. and DeHart, W. L., (1994), I EEE Trans. Electron Dev., **41**, 191.
4. Krisch, K. S., Green, M. L., Baumann, F. H., Brasen, D., Feldman, L. C. and Manchanda, L., (1996), IEEE Trans. Electron Dev., **43**, 982.
5. Mathiot, D., Straboni, A., Andre E. and Debenest, P., (1993), J. Appl. Phys., **73**, 8215.
6. Ito, T., Hijiya, S., Nozaki,T., Arakawa, H., Shinoda, M. and Fukukawa, Y., (1987), J. Electrochem. Soc., **125**, 448.
7. Atkinson, A., Moulson A. J. and Roberts, E. W., (1976), J. Amer. Ceramic Soc., **59**, 285.
8. Hori, T., Iwasaki H. and Tsuji, K., (1989), IEEE Trans. Electron Dev., **36**, 340.
9. Moslehi, M. M. and Saraswat, K. C., (1985), IEEE Trans. Electron Dev., **32**, 106.
10. Hayafuji, Y. and Kajiwara, K., (1982), J. Electrochem. Soc., **129**, 2102.
11. Ito, T., Nozaki, T. and Ishikawa, H., (1980), J. Electrochem. Soc., **127**, 2053.
12. Hwang, H., Ting, W., Kwong, D.-L. and Lee, J., (1990), IEDM Tech. Dig., 421.
13. Fukuda, H., Arakawa, T. and Ohno, S., (1990), Jap. J. Appl. Phys., **129**, L2333.
14. Yao, Z.-Q., Harrison, H. B., Dimitrijev, S., Sweatman, D. and Yeow,,Y. T., (1994), Appl. Phys. Lett., **64**, 3584.
15. Okada, Y., Tobin, P. J., Reid, K. G., Hegde, R. I., Maiti, B. and Ajuria, S. A., (1994), IEEE Trans. Electron Dev., **41**, 1608.
16. Green, M. L., Brasen, D., Evans-Lutterodt, K. W., Feldman, L. C., Krisch, K., Lennard, W. N., Tang, H. T., Manchanda, L. and Tang, M.-T., (1994), Appl. Phys. Lett., **65**, 848.
17. Lu, H. C., Gusev, E. P., Gustafsson, T., Garfunkel, E., Green, M. L., Brasen, D. and Feldman, L. C., (1996), Appl. Phys. Lett., **69**, 2713.
18. Hillert, M., Jonsson, S. and Sundman, B., (1992), Z. Metallkunde, **83**, 648.
19. Liu, C. T., Lloyd, E. J., Ma, Y., Du, M., Opila, R. L. and Hillenius, ,S. J., (1996), IEDM Tech. Dig., 499.
20. Lu, Z. H., Hussey, R. J., Graham, M. J., Cao, R. and Tay, S. P., (1996), J. Vac. Sci. Technol. B, **14**, 2882.
21. Gusev, E., Green, M. L., Morino, K. and Hirose, M., to be published.
22. Saks, N. S., Ma, D. I. and Fowler, W. B., (1995), Appl. Phys. Lett., **67**, 374.
23. Carr, E. C., Ellis, K. A. and Buhrman, R. A., (1995), Appl. Phys. Lett., **66**, 1492.
24. Jennings, H. M., (1983), J. Mat. Sci., **18**, 951.
25. Mitomo, M., (1976), J. Amer. Ceramic Soc., **58**, 527.
26. Messier, D. R., Wong, P. and Ingram, A. E., (1973), J. Amer. Ceramic Soc., **56**, 171.
27. Tang, H. T., Lennard, W. N., Zinke-Allmang, M., Mitchell, I. V., Feldman, L. C., Green, M. L., Brasen, D., (1994), Appl. Phys. Lett., **64**, 3473.
28. Gusev, E. P., Lu, H. C., Gustafsson, T., Garfunkel, E., Green, M. L. and Brasen, D., (1997), J. Appl. Phys., **82**, 896.
29. Wada, H., (1991), J. Mater. Sci., **26**, 2590.
30. Du, H., Tressler, R. E. and Spear, K. E., (1989), J. Electrochem. Soc., **136**, 3210.
31. Ito, T., Kitayama, D. and Ikoma, H., (1997), Jap. J. Appl. Phys., **36**, 612.
32. Lu, H. C., Gusev, E. P., Gustafsson, T., Brasen, D., Green, M. L. And Garfunkel, E., (1997), Microelec. Eng, **36**, 29.

NITROGEN IN ULTRA THIN DIELECTRICS

H.B. HARRISON, H.-F. LI, S. DIMITRIJEV and P. TANNER
Griffith University, Nathan, Queensland, Australia 4111.

Abstract

Gas phase growth or annealing of pre-grown oxides in nitrous oxide (N_2O) on silicon produces desirable dielectric layer properties. However since this gas is really a dilute form of nitric oxide (NO) it could be expected that it would give such properties as well. In this paper we consider the use of NO as both a growth environment and as an annealing gas firstly on silicon and then briefly on silicon carbide. We note that unlike N_2O, NO has some unique properties for silicon applications that may be useful for future generation integration. Its as grown self limiting growth properties for example may be helpful in layer thickness control. However we also see that the NO can have deleterious effects through the introduction of too much nitrogen at the interface transition layer. The paper explores both the physical and electrical properties of as grown and annealed layers exposed to NO for silicon. We also proposed a growth model that has particular relevance, and then presents some results for fine geometry MOSFET devices. Finally a brief consideration of silicon carbide shows some similarities in properties to that of silicon when oxides grown on it are annealed in NO and N_2O. In the case of SiC only annealing is considered, but in this case NO does show superior properties to those gained through the use of N_2O.

1. Introduction

This paper is effectively broken into two parts, the first considers mainly dielectrics on silicon and then a brief overview of some work done on SiC is presented. As a consequence unless otherwise specified the comments and discussion is directed at silicon. Particularly in this introductory section.

It is now well established that nitrogen appropriately incorporated at the Si/SiO_2 interface can provide very desirable properties in ultra thin (<5 nm) silicon dioxide dielectrics. Such dielectric layers are often referred to as oxynitride even though the percentage of nitrogen incorporated is usually only very small with peak atomic concentrations less than around ten atomic percent [1, 2].

Nitrogen is the most compatible atom to oxygen with regard to electronegative, bond dissociation energy and bond length [3]. It is felt that nitrogen bonds with silicon and thus reduces strain at the transition layer of the Si interface. It is also believed that this bonding slows the oxidation rate down and improves the oxide properties by reducing the number of sub oxide states and relaxing the strained chemical bonds in the interface region, thus reducing the trap assisted leakage current and device degradation due to reliability effects on the dielectric [4].

A common way of introducing the nitrogen species is by exposure to a nitrogen rich gas either during growth or subsequent to it as part of a total heat treatment [5, 6]. Other techniques such as ion implantation have been used with varying degrees of success [7].

The two most popular nitrogen rich gases to be used to date are nitrous (N_2O) and nitric (NO) oxides, even though there are claims lately that nitrogen itself may under certain circumstances give desirable properties without resorting to the two former more toxic gases [8].

NO has been demonstrated to form high quality gate dielectrics using NO-nitridation of SiO_2 [9]. It has been shown that NO has a significantly higher N incorporation efficiency compared to all other gases. MOS capacitors with NO-nitrided gate dielectrics also show highly suppressed charge-trapping, high immunity to interface state generation and excellent breakdown characteristics again compared to other gases particularly wet or dry oxygen [10].

In this paper we concentrate our attention on some physical and electrical characteristics of dielectrics affected by NO gas in a thermal environment on both Si and SiC. For Si we consider some properties of these dielectrics for MOSFET type device structures at micron dimensions. However we show initially that N_2O is really a subset of NO as a consequence we introduce results obtained from this gas as a comparison from time to time.

2. Gas-Phase Chemistry of Heated N_2O

On heating N_2O gas it decomposes into NO as well as other gases. Thus in comparing these two gases we need to consider the gas-phase chemistry of N_2O. Hartig and Tobin have reviewed this area in detail [11]. They show that the N_2O gas goes through a series of thermal decomposition steps. We consider here only the simple model first proposed [12] as this is seen to be sufficient to explain the differences experienced when comparing N_2O and NO results.

A simple model was formulated around the reactions:

$$N_2O \rightarrow N_2 + O \qquad (1)$$

$$N_2O + O \rightarrow 2NO \qquad (2)$$

$$O + O \rightarrow O_2 \qquad (3)$$

It was assumed that atomic oxygen resulting from the reaction of equations 1 and 2 and recorded in 3, recombined immediately to O_2 and thus it was concluded that after reaching a temperature of decomposition N_2O reduced to 64.3% N_2, 31.0% O_2 and only 4.7% NO.

It is unclear as to what happens with the reaction of NO with O_2. In [13] it is claimed that the reaction of equation 4 transpires whilst [14] disputes this and claims that this reaction is immediately reversible above 600 C.

$$2NO + O_2 \longrightarrow 2NO_2. \qquad (4)$$

Whatever occurs it is clear that the N_2O environment is really a dilute form of NO and that the carrier gases N_2 and particularly O_2 will play a major role in any surface reaction along with the NO gas.

Thus when comparing the results of NO heat treatments to those of N_2O it is not unreasonable to expect different resulting properties.

3. Physical Properties of NO Films

3.1. AS GROWN FILMS

It is well known that the growth rate of dielectric layers in an N_2O environment are much lower than that which occurs in pure oxygen under similar conditions. Furthermore the growth in NO is even slower and displays distinctly limiting thickness characteristics. Figure 1(a) shows the curves of rapid thermally processed oxynitrided silicon samples. The growth conditions as a function of temperature are outlined, and the solid line curves were obtained using N_2O gas [14]. In figure 1(b) the peak nitrogen concentration (atomic %) as a function of time for the same conditions as figure 1(a) are shown. The percentages of nitrogen were gained from AES measurements for the N_2O samples and by XPS for the NO and are therefore only relative. Also shown in figure 1(a) are other thicknesses of NO grown layers gained by our group [15] and other groups [20].

These results show that for the N_2O at least there is a saturation of nitrogen at the interface with temperature. That the peak concentration of this nitrogen increases with temperature. For this gas the nitrogen at the interface grows with time and slows the growth of the dielectric layer down as the nitrogen accumulates at the interface. It would appear that the nitrogen build up is caused by the NO species diffusing through the growing oxide [16] that is growing mainly due to the O_2 in the decomposed gas (it is not clear what happens to the oxygen in the NO). Since the growth rate does not drop to zero even for long nitridation times, the reaction of the oxidant is never completely blocked. We have referred to these sites as growth sites [17] and consider that since there is a modification to the number of these sites that the growth rate will be modified accordingly. This has lead to a very accurate growth model for dielectrics grown in N_2O developed later in this paper.

In an experiment performed by our group [18] to gain information about the type of bonds at the interface, oxide layers were grown in N_2O and NO to the same layer thickness. XPS was used and it was shown that the nitrogen in the NO grown layer showed evidence of much stronger bonding than the N_2O. The binding energy peak was clearly evident for all samples grown in NO but not evident at all with N_2O samples, suggesting a stronger bonding to the silicon. Hence with different nitrogen concentrations and different bonding it is to be expected that growth rate and physical properties of the two growth species will be different. It is also seen that the thermal budget for NO processing is less than that for N_2O since it incorporates larger amounts

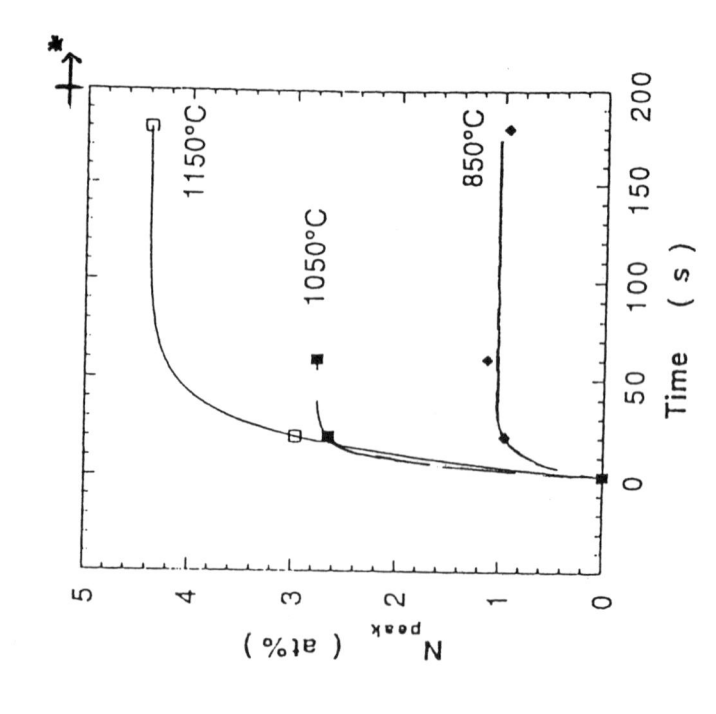

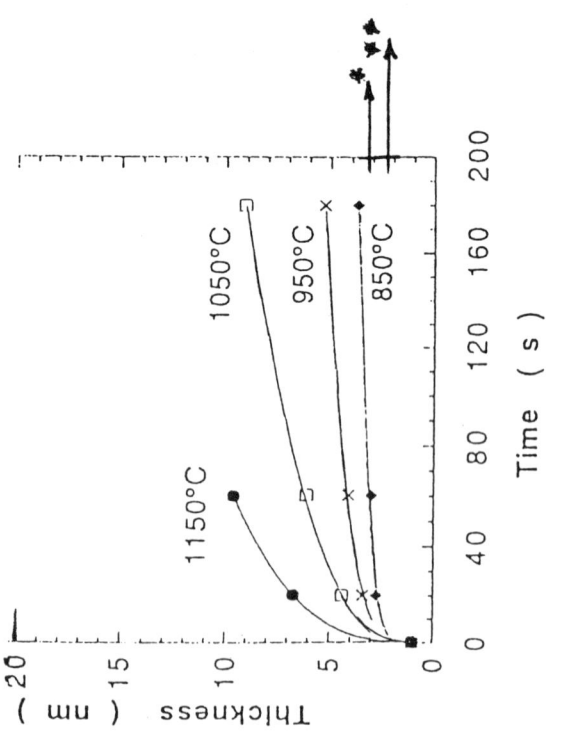

Figure 1 (a) Thickness (nm) as a function of time (seconds). Solid lines in N2O gas, from [14]. Other data, * our work, [15] ** [20]. (b) Peak atomic percent of nitrogen at the interface. Source the same as for 1(a).

of nitrogen at the interface with presumably stronger bonding at lower temperatures and for shorter processing times.

There is also sufficient evidence to suggest there is a difference in properties of films grown in N_2O in RTP systems compared to conventional hot walled (chw) furnaces. Certainly different percentages of nitrogen and nitrogen concentrations have been recorded for similar thermal conditions. One could expect that decomposition (including exothermic) effects would be contained closer to the silicon surface using RTP, and therefore more tightly controlled than in a (chw) furnace [19]. Similar effects would not be expected for NO as there would be no further decomposition expected and hence less dependence on processing conditions.

3.2. FILMS ANNEALED IN NO

The limiting growth thickness of NO grown layers with time and temperature provides little scope for variable thickness layers particularly when the maximum thickness is only a few nano meters (for realistic processing times and temperatures). It could be argued though that this property is an excellent one for future generation integration where strict thickness control will be required over for example the entire area of a 400 mm diameter wafer. It is also interesting to note that these layers appear to have an offset, that is a non zero thickness at time zero, this property is not new as it portrays a similar problem to Deal and Grove but with much thicker layers [23].

It may be for this reason that NO has been more commonly used to anneal pre grown oxide layers. Usually an oxide is firstly grown in oxygen and then subjected to a heat treatment (anneal) in NO. There are some similarities between this case and the as grown in NO, in that during the anneal there is little further growth of the film layer. The grown layer quickly saturates with nitrogen at the interface that then slows down any further growth as in the as grown case.

Analysis shows that the percentage of nitrogen at the interface is very high and concentrated at the interface rather than spread throughout the layer as it appears to do with the as grow as shown below in figure 2 [20]. The apparent spread throughout the layer may just be that as the layers as grown are so thin that we are really only observing the interface layer where we know there is a significant nitrogen concentration. However the physical characteristics are sufficiently similar to expect that the annealed layers would have similar electrical characteristics to their as grown counterparts but with some small difference.

J. Kuehne et. al [21] used an Applied Materials reactor with pre grown dielectrics of 2.5 and 5.5 nm and annealed in NO in the same chamber. They annealed for different times and temperatures. SIMS data showed that the N concentration varied from 1% through to 9% from low (900 C) to high temperature (1000 C) and small (15 s) to long (120 s) times. They showed that the oxide regrowth was a strong function of N incorporation and ranged from 0.2 to 1.1nm. They claimed the NO had no effect on the chamber, which is significant from a machine manufacturer point of view.

In an attempt to introduce even more nitrogen into the interface Sun et. al [22] first grew an oxynitride layer in N_2O and then annealed by in-situ rapid thermal NO nitridation. This had the advantage that it could provide a wider range of thicknesses because the N_2O growth is not as limiting. They claim the resultant properties were

196

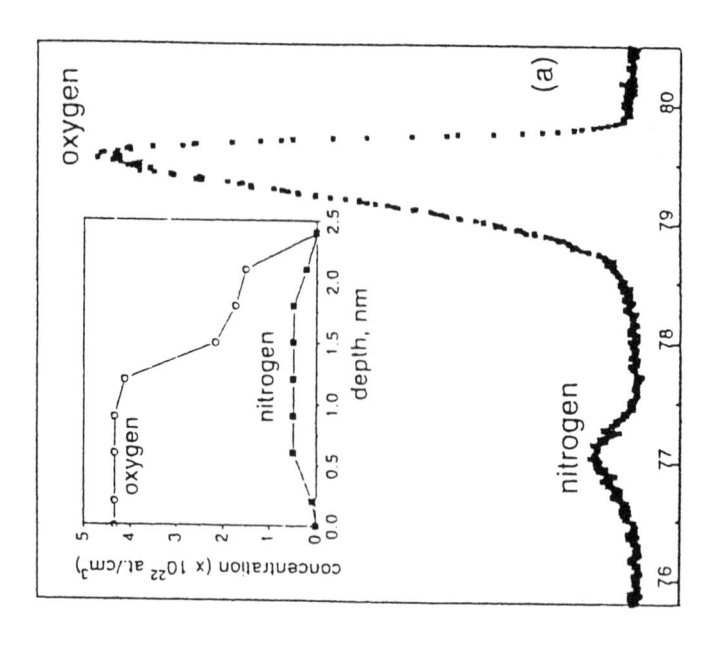

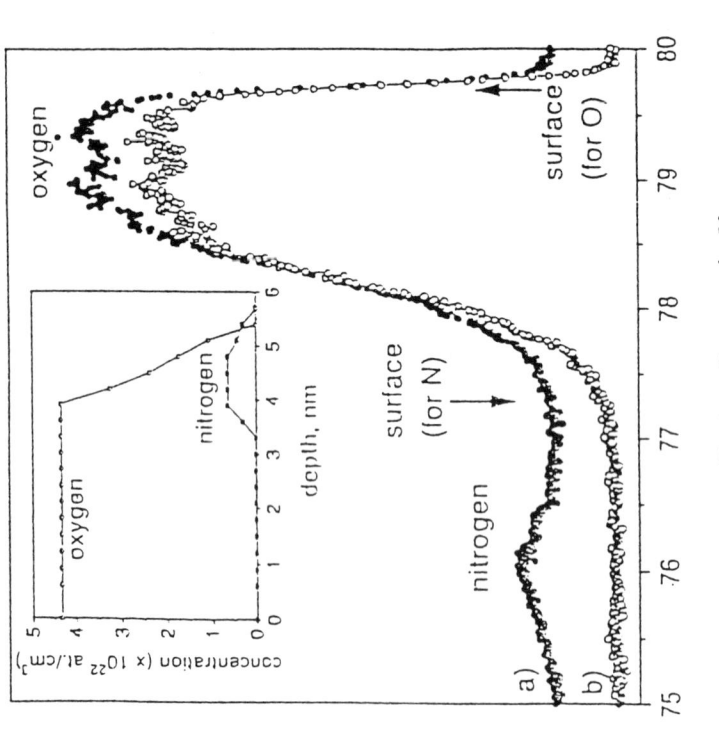

Figure 2. Medium Energy Ion Scattering (MEIS) Studies of nitrogen in Silicon Dioxide. The insert shows the nitrogen and oxygen profiles.
(a) Growth in NO in a furnace for 60 minutes
(b) Annealing an 850C, O2 grown layer (5 ins) in NO at 950C for 50 mins.

better than those obtained with NO annealing of oxygen grown layers. Whilst they gave mainly the results for electrical characterisation they did show that the NO on annealing locked up any further growth in a similar manner to that of NO onto oxide grown layers which would suggest similar resultant electrical characteristics.

4. Modelling the Growth of Dielectrics on Si in Nitrogen Rich Environments

We develop here an extension of the widely accepted Deal-Grove approach to oxide growth modelling [23] In the extension we include the effect of the nitrogen at or near the silicon/silicon dioxide interface by modifying some of the basic parameters based on experimental observation. We have reported on this approach in the past [24, 25, 26] and consequently we provide only a brief review on this before presenting some newer results that include a comparison of growth parameters for N_2O, NO and $NO+O_2$ gas environments.

In equilibrium it is assumed that there are three consecutive fluxes of oxidising species viz.: that due to diffusion of the oxidising species from the surface of the already grown oxide at the interface, J_s, that due to the chemical reaction at the interface, J_r, and that due to the incorporation of the oxidising species in the oxide, J_o.

$$\text{where} \quad J_s = D \cdot (C_o - C_s)/x \quad (1)$$
$$J_r = K_r \cdot C_s, \quad (2)$$
$$\text{and} \quad J_o = C_{ox} \cdot dx/dt. \quad (3)$$

(C_o, C_s and C_{ox} are concentrations of the oxidising species, at the surface of the oxide, at the interface and in the grown oxide respectively and D is the diffusion coefficient. K_r is the interface reaction rate constant and x the oxide thickness).

Under equilibrium conditions $J_s = J_r = J_o$ and the following expression results

$$dx/dt = C_o/C_{ox} \cdot (D/(x+D/K_r)) \quad (4)$$

This is often expressed as

$$dx/dt = B/(2x+A) \quad (5)$$

Where $A = 2D/K_r$ and $B = 2D \cdot C_o/C_{ox}$.

The condition $A/2 >> x$ results in the so called linear growth regime, whereas other conditions are related to the parabolic growth. We consider here only the linear case and referred to B/A as the linear growth rate constant. Thus equation (5) reduces to

$$dx/dt = B/A = K_r \cdot C_o/C_{ox} \quad (6)$$

Using this equation and comparing to experimental results produces significant differences. However, we have shown [27] that this equation can still be used successfully if modified slightly to account for physical occurrences mainly at or very near to the interface. If in equation 2, K_r is replaced by a growth site concentration of C_{gr} and rate constant r such that

$$J_r = rC_{gr}C_s \qquad (7)$$

Furthermore if the growth site concentration C_{gr} is modelled by:

$$C_{gr} = l_{gr} \cdot \exp(-E_A/kT) + a_{gr} \cdot \exp(-t/C) \cdot \exp(-E_A/kT) \qquad (8)$$

where E_A is a growth site activation energy, l_{gr} and a_{gr} are time and temperature independent constants and C is a constant linking the recombination of growth sites to the growth rate constant C_{gr}.

This then leads to a more accurate linear rate model which is given by

$$dx/dt = l_o \cdot \exp(-E_A/kT) + a_o \cdot \exp(-E_A/kT) \cdot \exp(-t/C) \qquad (9)$$

Where $a_o = r(C_o/C_{ox}) \cdot a_{gr}$ and $l_o = r(C_o/C_{ox}) \cdot l_{gr}$.

Which when integrated and initial conditions applied leads to

$$x = \exp(-E_A/kT) \cdot (t \cdot l_o + a_o \cdot C \cdot (1 - \exp(-t/C))) \qquad (10)$$

From [28] growth thickness figures are given for various nitrogen ambient as well as pure oxygen. These figures repeated below are for four minutes (240 seconds) of growth time in a conventional furnace at temperatures ranging from 850 C through 950 C. They are summarised below in table 1.

TABLE 1.

Temperature (C)	O_2	$NO+O_2$	N_2O	NO
		(thickness in angstroms)		
850	25.5	21	24.5	18.5
900	31	23.5	25.5	19
950	37.5	26.5	27.5	19.5
E_A (eV)	0.45	0.26	0.21	0.08

Additional to these figures are the growth values for NO and $NO+O_2$ above those already obtained for N_2O and O_2. From observation the growth rates seem to reduce as the percentage of nitrogen in the gas increases. That is as the nitrogen percentage increases it would appear that the number of growth sites reduces and thus slows down the growth rate in comparison. The resultant activation energies as calculated from these figures are also shown in the table, and on a relative basis support the locking up of growth model. However the absolute values of activation energies particularly for N_2O are much lower than our previously reported values (2x). We are currently investigating this in light of the different temperature regime and the possibility of multiple activation energies, or the different thermal environment (furnace not RTP) or a combination of these effects.

5. Electrical Properties of NO Films

5.1 AS GROWN

There is little published on the as grown electrical characteristics of NO. This is presumably because the self limiting nature of the grown layer and the uncertainty of the resultant extremely thin film thickness. However we have carried out some preliminary measurements on as grown layers and present here the results. The films have generally excellent electrical breakdown and other properties that suggest an ultra smooth interface which appears as another positive characteristic of NO films.

Measurements were made on circular aluminium gate capacitors with an area of 0.01cm^2 on layers grown on <100> silicon in NO at 1150°C for 5 minutes. The resultant layer is 3.0 nm in thickness as calculated from the C-V measurements. The C-V and I-V characteristics as measured are shown in figure 3. Also shown in this figure are results for N_2O and O_2 as well. The conditions for growth are outlined in table 2 for all gases.

TABLE 2. Film thickness, flatband voltage and interface state density before and after stressing for various growth and annealing conditions.

growth conditions	film thickness (nm)	interface state density (before stress) ($\times 10^{+11}$/cm^{2}eV)	interface state density (after stress) ($\times 10^{+11}$/cm^{2}eV)	flatband voltage shift (mV)
at 1150°C in NO for 5 min.	3.0	2.14	2.40	46
at 1150°C in NO for 5 min & annealed in N$_2$ for 1 min.	3.0	1.78	1.81	61
at 1150°C in NO for 5 min.. & reoxidized at 1150°C in O$_2$ for 2 min.	3.8	3.25	3.42	126
at 950°C in N$_2$O for 5 min.	4.8	4.12	-	-
at 950°C in O$_2$ 30 seconds	4.0	6.26	-	-

It can be seen from figure 3 that the leakage current of the 3 nm thick NO grown film is at least four-orders of magnitude lower than that of the 4.8 nm N_2O grown film. These leakage figures would suggest the layers are capable of withstanding fields of greater than 20 MV/cm. The C-V curves indicate that both N_2O and O_2 layers appear to have a

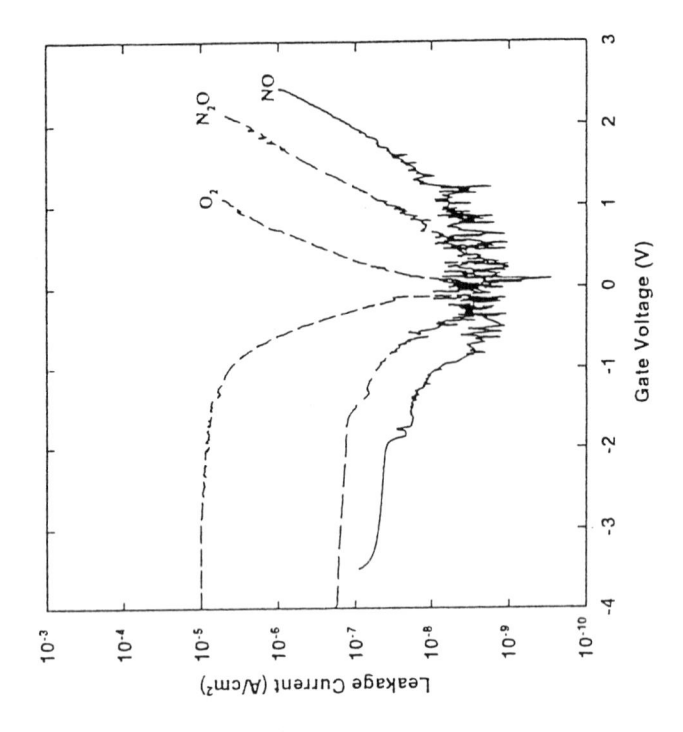

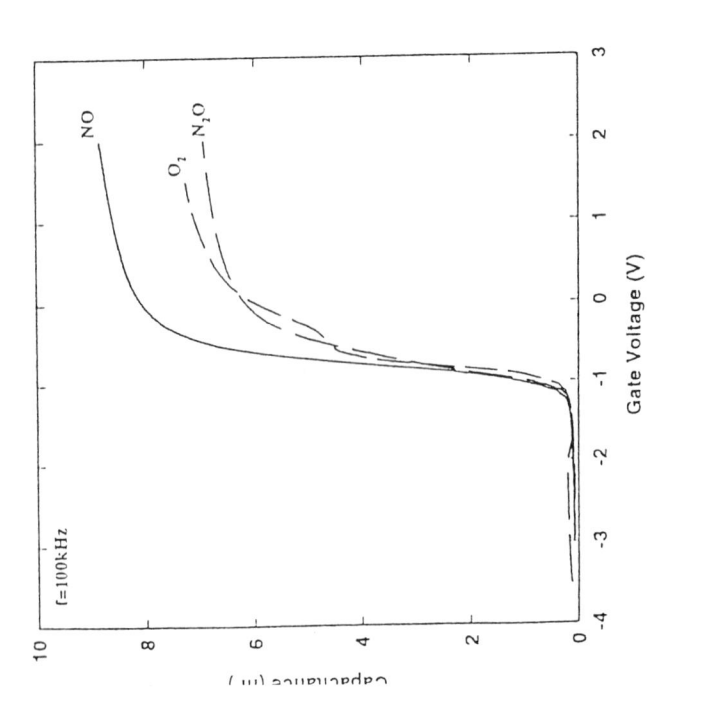

Figure 3. A comparison of CV and IV characteristics for layers grown in NO, O2 and N2O. (a) High frequency CV and (b) Current Voltage for the layers grown as per the text.

proportion of slow interface traps whilst that of NO seems relatively free. Also table 2 shows that NO layers have lower density of interface states. Thus the resultant electrical properties of the as grown NO are extremely impressive when compared to those of N_2O and O_2.

5.2 LAYERS ANNEALED IN NO

In a continuation of the above experiment pre-grown oxide layers were annealed in NO. The layers were grown in O_2 at 950°C for 30 secs in an RTP system. They were then subjected to an NO anneal for 5 mins at 1000°C and 1150°C. The oxide thickness of the pre grown layer was measured by ellipsometry to be 4 nm and after anneal this thickness had changed by only 1 nm for both cases which was within the measurement resolution. Again aluminium dots were etched and C-V and I-V measurements made. The results are in figure 4 and show that the leakage is considerably improved for both annealing conditions but with a higher breakdown for the higher temperature which is consistent with higher nitrogen concentrations at the interface. The C-V curves are interesting showing the higher temperature curve to have a larger flat band shift. However neither of the annealed C-V curves show evidence of traps as is the case with the dry oxide. These threshold shifts have also been observed on small dimension devices [15]

The flatband voltage shift in the C-V curve of figure 4 is also interesting since the surface state density for both annealing conditions remained about the same. To examine this effect further we took some pre grown oxides, grown in dry oxygen at 1150 C in a cwh furnace for 5 mins. The resultant layer thickness was around 20.0 nm as gained from C-V measurements. The samples were then annealed in N_2 for 5 mins at 1150 C in a furnace, and various percentages of NO/N_2 under the same conditions. In each case the interface density was measured to be roughly $2.0 \times 10^{11}/cm^2 eV$ using conductance measurement techniques. This value did not change even with the change of flat band voltage.

However the CV characteristics as per figure 5 show a continuous movement of flat band voltage with increased NO concentration. This would suggest that the more nitrogen incorporated in the oxide layer the larger the flat band shift. We expect that as far as the electrical characteristics are concerned that there is an optimum amount of nitrogen for surface state effects above which we then experience an increase in fixed charge density. This effect has also been reported for NH_3 and N_2O [29].

Slow trap profiling [30] of these samples in which the voltage is scanned from accumulation to inversion, and back, in small steps and the resultant decaying current is sampled enables the slow near surface traps to be profiled. In this case a slow but steady build up in trap density measured from depletion to accumulation and the introduction of a level at -5 volts when scanning the other way, (see figure 6) is seen to develop as the N_2 gas concentration is decreased. These introduced traps whilst outside the CV profiling limit are important in that they support the thesis that too much nitrogen (or oxygen from the NO or both) as a result of the decrease in dilution of N_2 have a deleterious effect. The shift in threshold and build up in slow trap density would appear

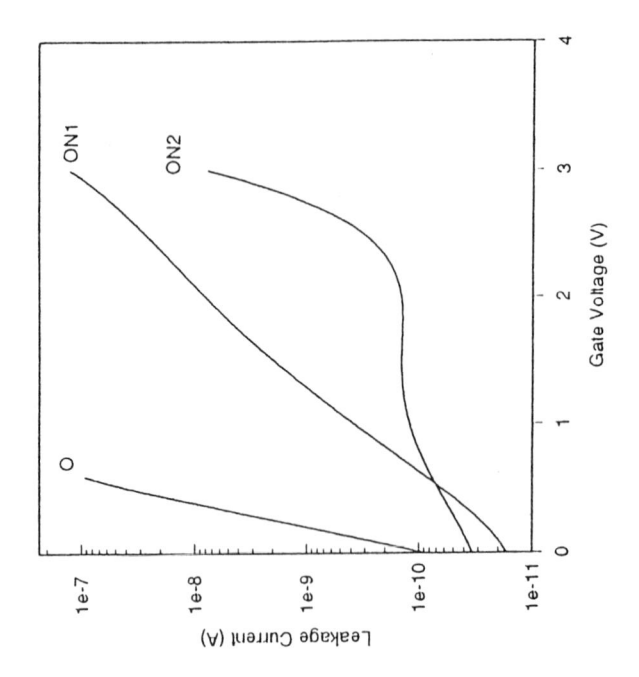

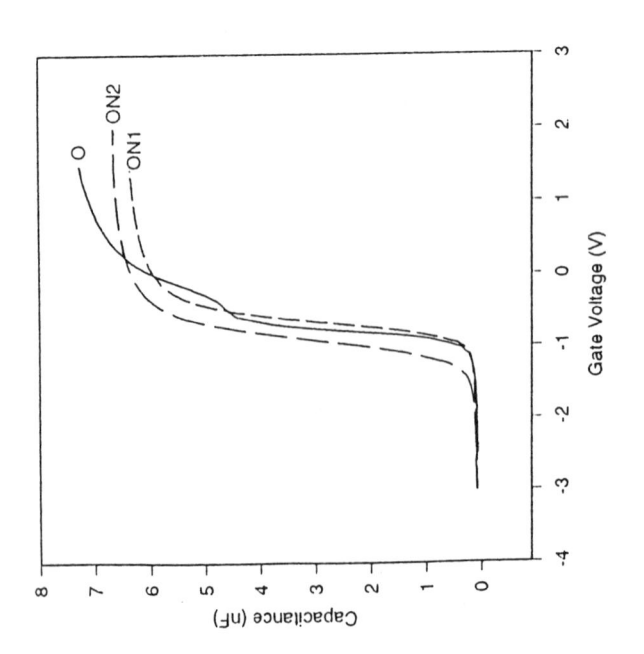

Figure 4. A comparison of annealed to as grown oxide layers. ON1 and ON2 annealing in NO for same temperature 1150°C for different times 1 and 5 mins. (a) high frequency CV and (b) current voltage.

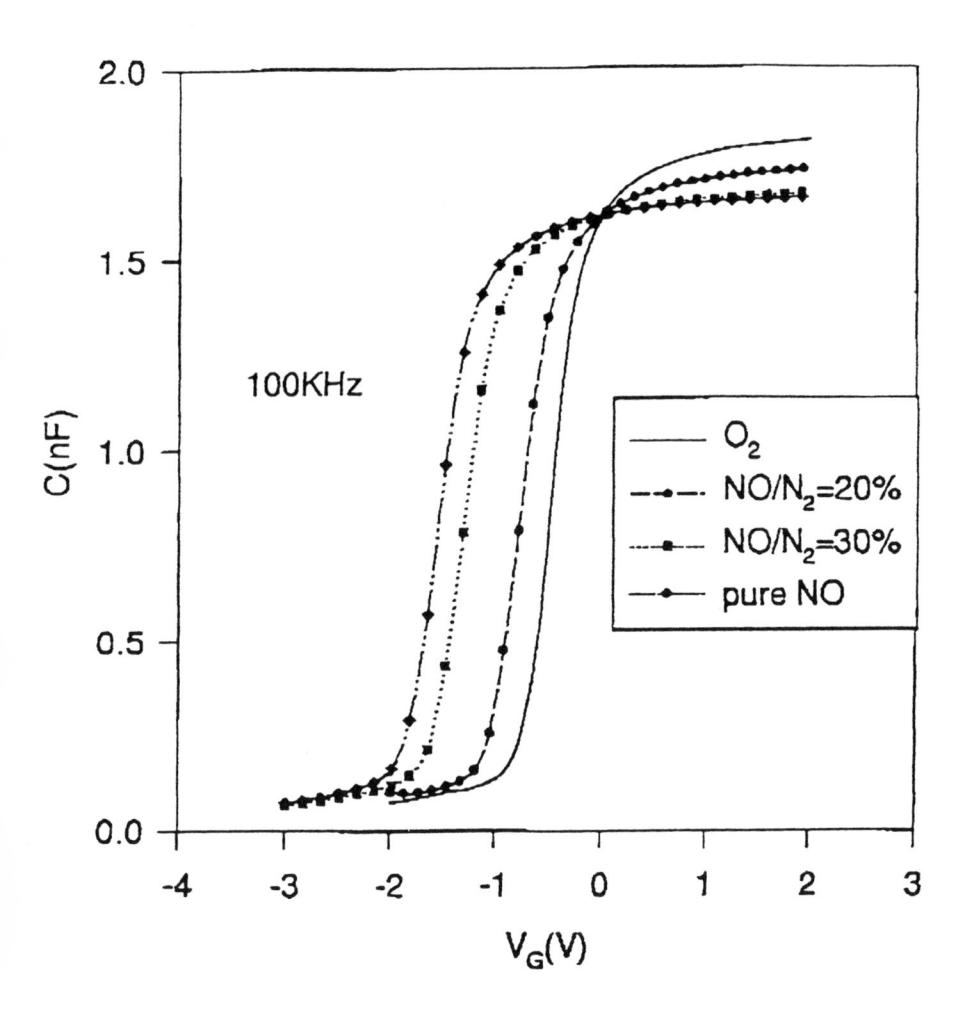

Figure 5. High frequency CV characteristics of annealed oxide layers. Annealing was with different ratios of NO to nitrogen as per the insert in the figure. Note the shift in threshold voltage as the percentage of NO increases.

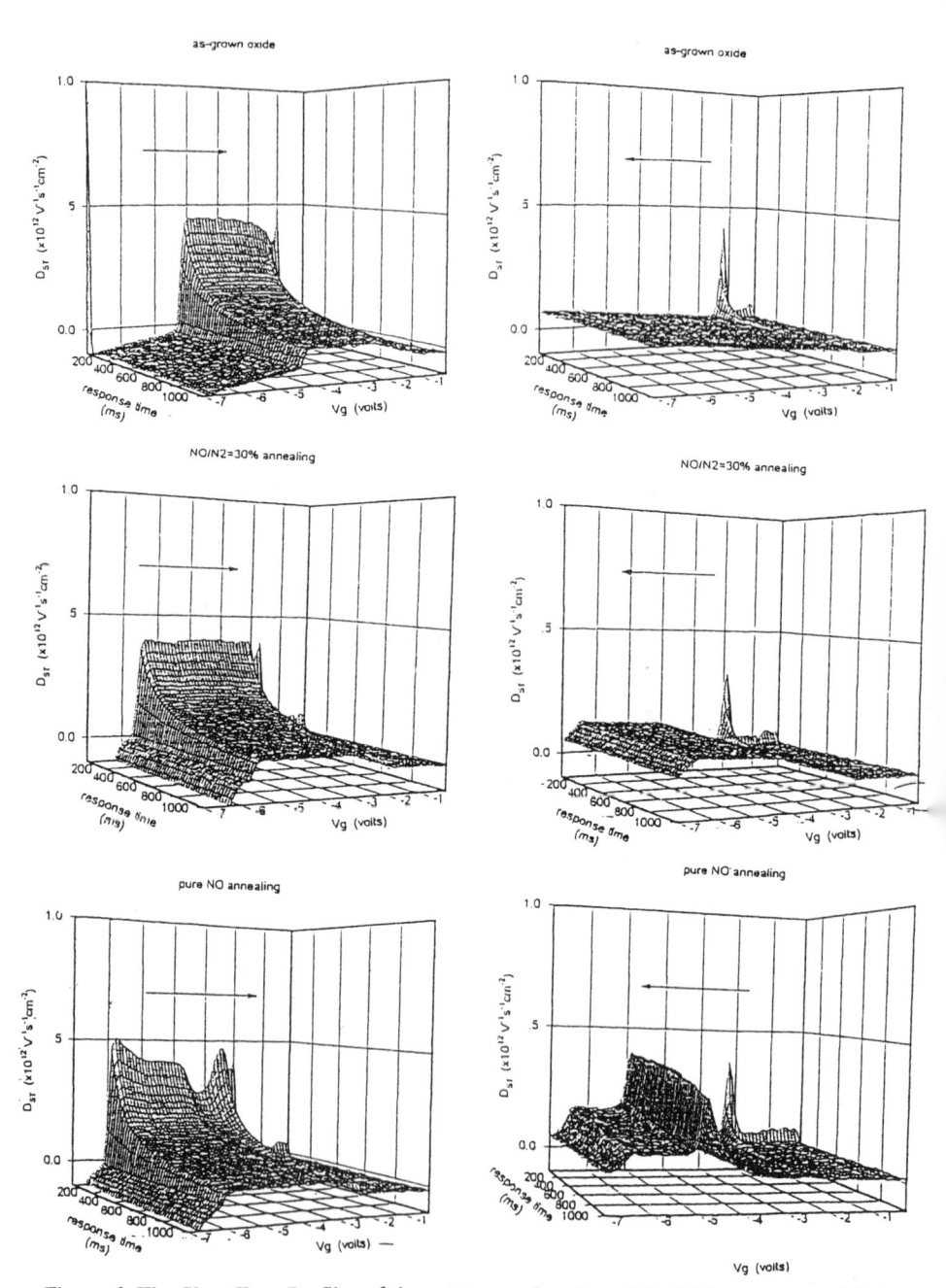

Figure 6. The Slow Trap Profiles of the oxide as a function of the NO percentage. (a) As grown oxide (b) 30% NO/N2 and (c) pure NO. Note the build up of slow traps in both cases ie from inversion to accumulation and from the opposite direction.

to be as a result of this. Further these slow traps appear in the inversion regime which is the main region of operation of MOSFETS.

6. Device Properties of NO Modified Oxides

To examine the effects of annealing pre grown oxides in nitric oxide gas we had some MOSFETs structures made for us. The conditions of fabrication were as follows. LDD n-channel type MOSFETs were fabricated on p-/p+ epitaxial silicon substrates using a twin well CMOS technology. After a local oxidation for the field oxide and active area definition, control or initial oxides were grown in dry oxygen at 800°C followed by annealing in N_2 for 30 minutes at 900°C. This lead to an oxide layer thickness of 12 nm. Following this step a number of wafers were annealed in 99% chemically pure nitric oxide ambient at 1150°C for 5 minutes in an RTP unit. After the gate oxide formation a layer of polysilicon was deposited in an LPCVD system and doped in POCL3 gas o form n+ polysilicon gates. Next n+ implantation, spacer, S/D implantation and BPSG deposition was performed, after which contact areas were opened and a layer of aluminium was deposited followed by sintering and annealing to form ohmic contacts and interconnects.

An HP 4145B semiconductor parametric analyser, HP 4284 LCR meter and HP 3314 function generator were used to measure the electrical properties of devices before and after hot-carrier stress. This stressing was performed with V_g=1.5 Volts and V_d=9.5 Volts for nominally 2000 seconds.

Figure 7 shows the Id versus V_d (insert in figure) and the gate to drain capacitance, C_{gd}, versus gate voltage curves for the n-channel MOSFETs (W/L$_{eff}$=25 µm/0.85 µm) with gate oxide thickness of 12 nm for the conventional devices and 12.9 nm for the NO annealed devices. It can be seen from figure 7 that the NO annealed I_d-V_d characteristics do not change noticeably upon stressing. However this is in contrast to the non annealed devices which show significant degradation.

It has been reported (31, 32) that the gate-to-drain capacitance measurements can be used to characterise hot carrier degradation of MOSFETs. It is believed that the decrease in C_{gd} in the strong inversion region is due to the generation of acceptor type interface states in the top half of the silicon bandgap, whilst the increase in C_{gd} in the accumulation and depletion regions is due to the trapping of holes during stress. In the figure the conventional gate oxide devices show considerably greater change in capacitance in both inversion and accumulation. In fact for the NO annealing in accumulation there is little of stress degradation but more evident in the inversion area. One should however consider the slow trapping properties of the NO annealing as discussed earlier in this presentation for a more complete understanding of the annealing on this property. It does appear however that NO annealing of such structures significantly improves the "hot carrier" endurance.

On the 10 nm gate thickness devices, transconductance g_m measurements resulted in the curves of figure 8. It is obvious that the NO annealed devices have a lower peak gm than the conventional oxide devices but an improved high field value. This is an indication of the effect of nitrogen at the interface on the channel electron mobility. This degradation has been proposed as a major obstacle to the use of nitrogen. However most

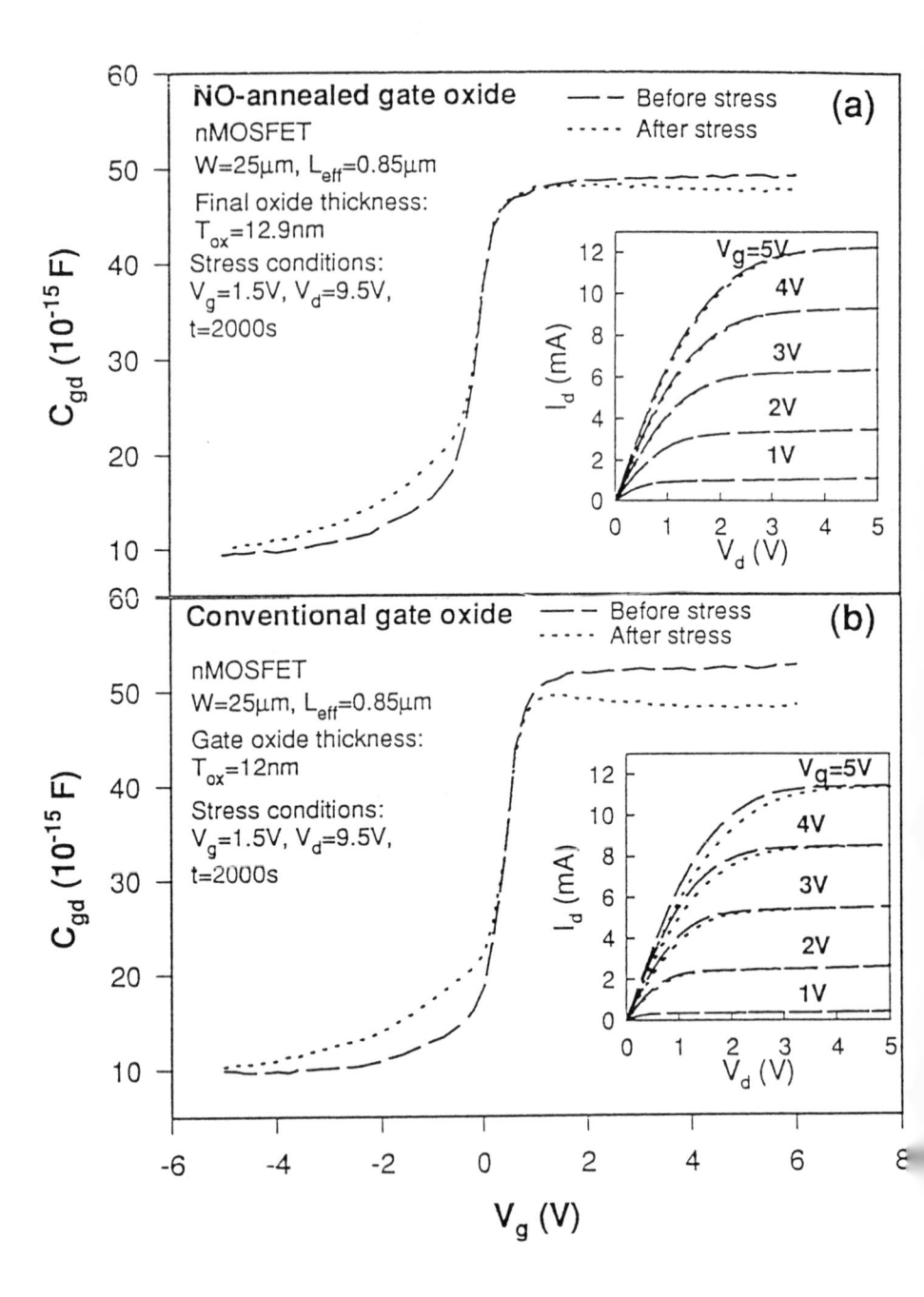

Figure 7. Id vs Vd characteristics (insert) and CV characteristics of MOS structure
(a) NO-annealed
(b) Non-annealed

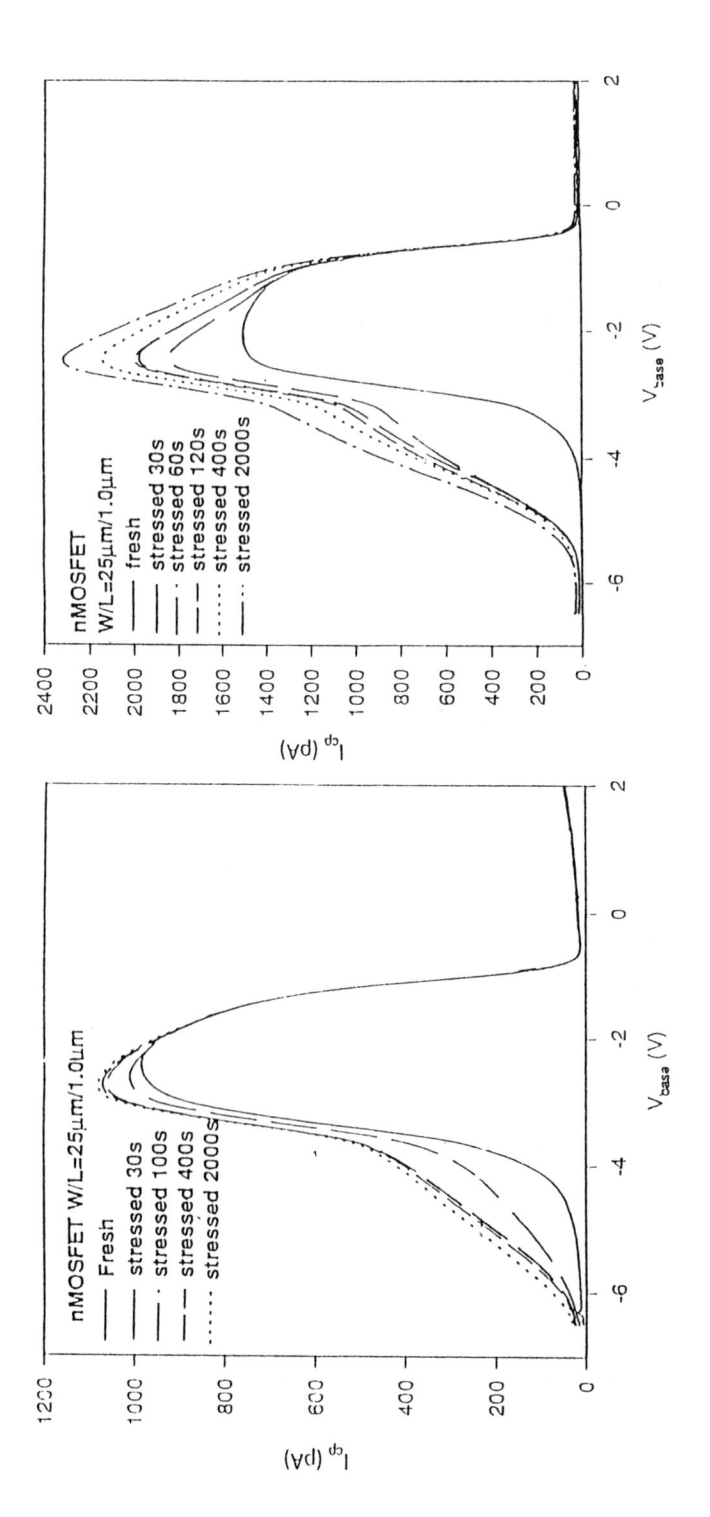

Figure 8(a). 100A NO-annealed gate oxide stressed at different conditions. Stress conditions: Vg=1.5V, Vd=9.5V.

Figure 8(b). 100A thermal gate oxide stressed at different conditions. Stress conditions: Vg=1.5V, Vd=9.5V.

devices will be working well into the high field regime and therefore we would not consider this to be a problem. It is also evident from that curve that stressing has less effect on the NO annealed devices than on conventional.

Similar devices (10nm oxides) were subject to charge pumping (CP) measurements. The results are shown in figure 9. These measurements were carried out using a square wave signal with an amplitude of 3 volts and a frequency of 200 kHZ. Both the conventional and NO annealed devices showed an increase of peak current (CP) and a shift of characteristic to the negative potential region. It is believed that the peak CP current is proportional to the total number of fast interface states located in the degraded part of the channel while the shift of the left edge (negative potential) is due to the positive charge in the gate oxide after hot carrier injection. From the figure it can be concluded that the initial thermal oxide (conventional device) has higher initial interface trap density, higher interface trap generation rate and more hot carrier induced positive trapped charge after hot-carrier stressing.

7. Dielectrics with Nitrogen on Silicon Carbide (SiC)

SiC has some intrinsic advantages over Si for some applications. It is these applications that may see SiC replace or supplement silicon. Si technology is so mature and well established that SiC will have to offer not only technical advantages but commercial as well if it is to be widely accepted. Table 3 provides a comparison of properties of Si and SiC. The wide bandgap and higher thermal conductivity of SiC would suggest its replacement of silicon in higher temperature operations for example.

TABLE 3 - Comparison of properties of SiC and Si

Semiconductor	6H-SiC	Si
Bandgap (eV at $27\,^{\circ}$C)	2.9	1.12
Thermal conductivity (W cm^{-1} C^{-1} at $27\,^{\circ}$C)	5.0	1.5
Saturated elec drift vel (cm S-1, at E$>$ 2x 10^5 V cm^{-1})	2×10^7	1×10^7
Electron mobility (cm^2 V^{-1} S^{-1} at $27\,^{\circ}$C)	250	1400
Hole mobility (cm^2 V^{-1} S^{-1} at $27\,^{\circ}$C)	50	600
Breakdown electric field (V cm^{-1})	40×10^5	3×10^5
Dielectric constant (dimensionless)	10	11.8

Unlike other semiconductors such as Ge and GaAs, SiC does have an oxide growing capability. To date thermal oxides have been grown on SiC to enable MOSFETs to be produced. Significantly p-channel devices have been produced with acceptable characteristics however the complementary device has usually suffered from a resultant poorer oxide. Initial speculation surrounded the incorporation of the p-type dopant which was mainly Al into the oxide leading to significant traps and an inferior layer [33]. Later studies infer that by replacing the Al with B the problem does not diminish which would suggest the problem may be more fundamental in nature [34]. In an effort to provide more insight into this and after speculating that nitrogen may well be the

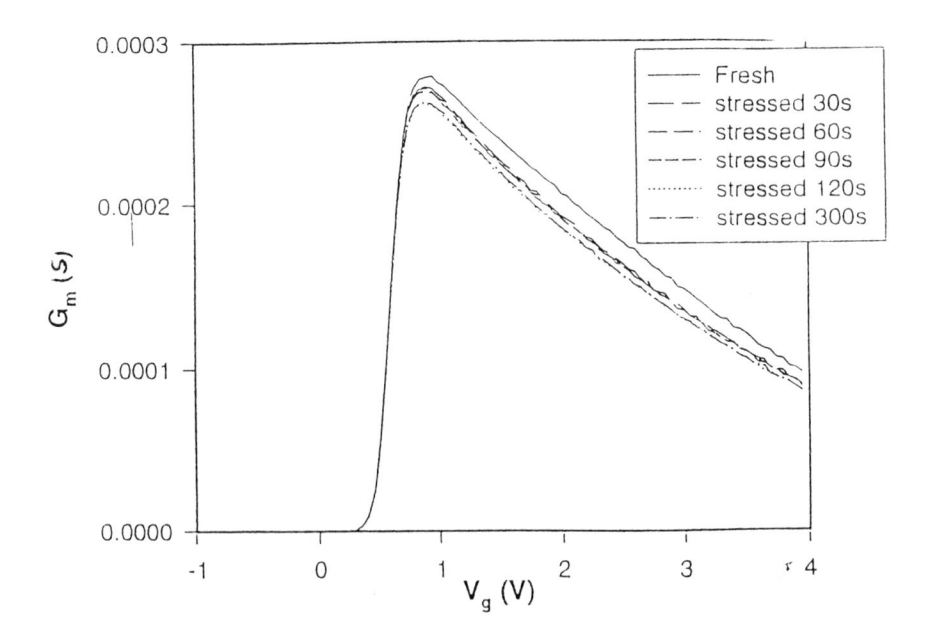

Figure 9(a). 10nm thermal oxide nMOSFET W/L=25/1.0
Stress conditions: Vg=1.5V, Vd=9.5V.

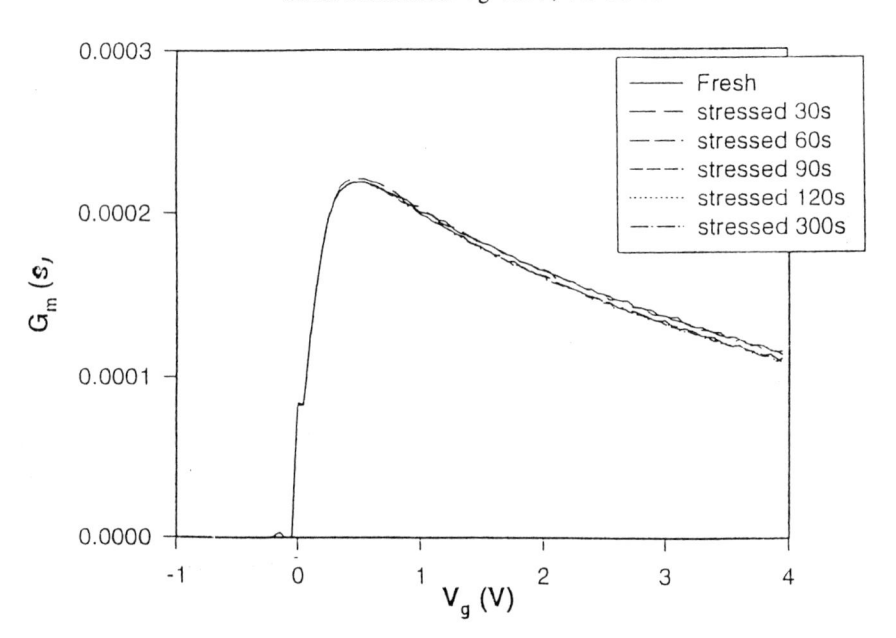

Figure 9(b). NO-annealed 10nm thermal oxide nMOSFET W/L=25/1.0
Stress conditions: Vg=1.5V, Vd=9.5V.

species that is providing the improvement in the p-type device we have performed a series of studies using nitrogen rich gases. These being a natural extension to the work performed on Si.

Our initial result show that by using NO as compared to N_2O that an improvement in dielectric characteristics for n-type device fabrication can be obtained. The following is essentially a condensing of the reference, " Nitridation of Silicon-Dioxide Films grown on 6H Silicon Carbide" Dimitrijev, Li, Harrison and Sweatman. IEEE Electron Device Letters, Vol. 18, No 5, May 1997.

Si-faced 6H-SiC, wafers, manufactured by CREE Research, were used in this experiment. The concentration of the nitrogen doped N-type wafer was 4.8×10^{17} cm^{-3}, while the concentration of the aluminium doped P-type wafer was 2.5×10^{18} cm^{-3}. The wafers were cut into approximately 1.5 x 1.5 cm pieces, and cleaned by both an $H_2SO_4:H_2O_2$ solution and RCA cleaning process. Immediately before oxidation, the samples were dipped in 1% HF for 60 s. The SiC carbide pieces were placed onto a 6-inch silicon wafer, to perform the oxidation in an AG610 rapid-thermal processing (RTP) unit. The oxidation was performed in six 5-minute steps (to allow cooling of the RTP unit) in high-purity O_2 at around 1100 C. After the oxidation, two sets of samples received an additional 5-minute treatment at the same temperature: one set of samples was exposed to a 99% (chemical pure grade) NO, while the other was exposed to a high-purity N_2O environment. Following the oxidation and the nitridation of the samples, aluminium was evaporated at the top and the back of the samples, and circular 0.3 cm dots defined at the top by a photolithography process to create MOS capacitors. No post-metal annealing is performed, to avoid masking the defects created during the oxidation and/or nitridation. MOS capacitor characterisation was performed by high-frequency (100 kHz) capacitance-voltage and conductance-voltage measurements, using a computer-controlled HP 4284A LCR meter.

While both techniques lead to incorporation of nitrogen at the interface, the annealing in N_2O leads to new oxide growth, as opposed to the annealing in NO which nitrides the interface with virtually no new oxide growth [35]. The nitrogen accumulation at the interface is related to observed improvements of the electrical characteristics of oxide/silicon interface[35, 36, 37, 38, 39, 40, 41]. Similar effects have been observed when the oxide is grown on nitrogen implanted polysilicon [42].

It is interesting to compare these silicon-related results to the results of Palmour [43], showing a similar accumulation of nitrogen at the oxide/silicon-carbide interface during thermal oxidation of nitrogen-doped N-type SiC substrate. This leads us to a hypothesis that the nitrogen incorporation improves the oxide/N-type silicon-carbide interface, which would otherwise remain inferior compared to the oxide/silicon interface. Furthermore, this suggests that the quality of oxides grown on P-type SiC could be improved by an appropriate nitridation process.

De Meo [44] showed that oxides on SiC substrate can directly be grown in N_2O rather than O_2 or H_2O. However, they could not demonstrate an improvement in the interface characteristics of N_2O grown oxides on N-type SiC. Having observed a number of differences in the effects of NO and N_2O on oxides grown on Si [6, 45], we decided to anneal oxides grown on SiC in both NO and N_2O atmosphere. We also included both N-type and P-type SiC substrates. The main results are presented in figure 10 and figure 11, and table 4.

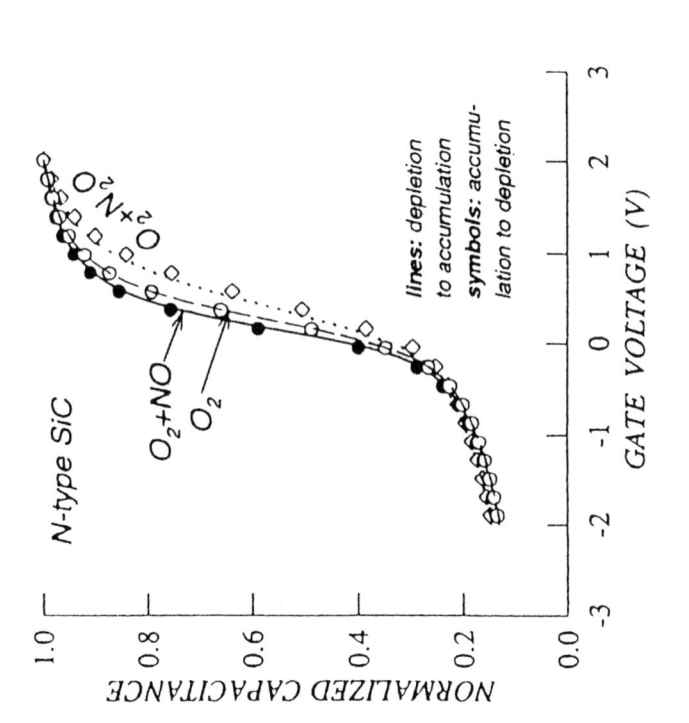

Figure 10. High-frequency (100 kHz) C-V measurements of MOS capacitors created on N-type 6H SiC.

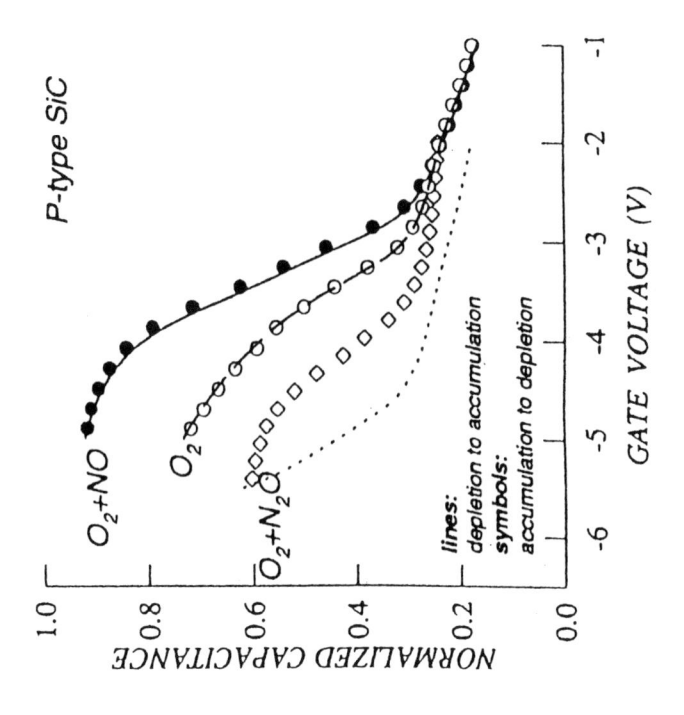

Figure 11. High-frequency (100 kHz) C-V measurements of MOS capacitors created on P-type 6H SiC.

TABLE 4

EFFECTS OF NO AND N_2O ANNEALING ON THE OXIDE CHARGE AND
INTERFACE TRAP DENSITY OF MOS CAPACITORS CREATED ON 6H SiC

	O_2	$O_2 + NO$	$O_2 + N_2O$
	\multicolumn{3}{c}{N-type ($\phi_{ms} = 0.15V$)}
t_{ox} [nm]	3.5	3.5	3.9
C_{FB}/C_{ox}	0.62	0.62	0.65
V_{FB} [V]	0.3	0.2	0.5
Q_{ox} (@V_{FB}) $\times 10^{12} cm^{-2}$	−0.9	−0.3	−1.9
D_{it-max} $\times 10^{11}\ cm^{-2}eV^{-1}$	1.1	0.3	1.3
	\multicolumn{3}{c}{P-type ($\phi_{ms} = -2.70V$)}
t_{ox} [nm]	3.5	3.5	3.9
C_{FB}/C_{ox}	0.79	0.79	0.81
V_{FB} [V]	< −5.0	−3.9	< −5.5
Q_{ox} (@V_{FB}) $\times 10^{12} cm^{-2}$	> 14	7.3	> 15

The oxide thicknesses, shown in table 4, were calculated from the accumulation capacitances of the N-type samples, and assumed to be the same for the corresponding P-type samples. Direct calculations of the oxide thicknesses on P-type samples would not be meaningful as the C-V curves of P-type samples do not saturate at negative voltages. The HF C-V curves are normalised accordingly, and shown for the both sweeping directions. To calculate the work-function differences (øms) shown in Table 4, the following parameters are used: the work-function of aluminium øms=4.1 eV [46], the silicon-carbide electron affinity =3.85 eV [47], and the intrinsic carrier concentration of 6H SiC=1.6 x10e-6 cm -3 [48]. The flat-band voltages were determined from the flat-band capacitances, calculated by the standard expression [49]. The oxide-charge densities were calculated as $Q_{ox}=-C_{ox}(V_{FB}-\text{øms})/q$. The interface-trap densities D_{it} were determined using the conductance technique [50].

Analysing the presented results, the following two observations are made: (i) the effects of annealing in either NO or N_2O are much more pronounced in aluminium-doped (P-type) SiC samples, compared to nitrogen-doped (N-type) SiC samples; (ii) NO annealing improves, while N_2O annealing deteriorates the electrical characteristics of thermally grown oxides on either P- or N-type SiC substrate.

The first point directly supports the hypothesis on the beneficial role of nitrogen, as the stability of the oxides grown on the nitrogen-doped SiC substrates can be explained by the nitrogen present at the interface before any nitridation treatment was applied. As opposed to the relative stability of N-type samples, the electrical characteristics of the oxides grown on P-type substrates were significantly improved by the NO annealing. The C-V curve of the NO-annealed sample almost reaches the accumulation value observed on the N-type samples. It is much closer to the ideal C-V curve than the C-V curve of O_2 (not-nitrided) sample, which shows the problems observed by other researchers . It appears the presence of a significant amount of interface traps causes such a stretch that the C-V curve cannot reach the accumulation capacitance up to -5 V, which is the breakdown voltage.

The results clearly demonstrate a deterioration of the interface characteristics during the N_2O annealing. Again, this is especially pronounced for the P-type samples. Different C-V curves appear for the different sweep directions during the measurements, indicating that the equilibrium cannot be achieved due to the action of slow traps. In addition, a negative voltage shift is obvious, indicating the presence of a significant amount of positive charge in the oxide or at the interface with SiC. We believe it would be very interesting to study these effects of N_2O annealing, as it may help to understand better the nitridation of thermally grown oxides not only on SiC, but also on Si substrates.

8. Conclusion

We have considered the use of NO both as a growth gas and as an annealing environment for silicon. It is now clear that the use of NO results in a semi limited thickness layer but with some excellent electrical properties. These properties are a result presumably of the trapped nitrogen at the silicon/silicon dioxide interface. In this case however the nitrogen is not just confined to the interface but also fills the dielectric layer.

For the annealed case there is an equivalent improvement in dielectric quality however it is questionable as to the amount of nitrogen needed to achieve these properties. After saturation of the interface or maybe even before the electrical properties seem to be optimum and further increases cause a degradation. This degradation is observed at least in part by an increase in slow traps that could be important in device operation.

The use of NO or NO and other carrier gases would appear to offer advantages over the use of N_2O in terms of what could be referred to as dielectric engineering.

Our results on NO and N_2O annealing of oxides grown on N-type and P-type 6H-SiC substrates were also presented. The results show that the annealing significantly affected the P-type oxides, while the effects in N-type oxides were marginal. This indicates that a stable nitrided interface is created during the oxidation in the case of nitrogen-doped N-type SiC substrates. It is also demonstrated that the oxides grown on P-type can be improved by NO annealing, but not by N_2O annealing.

9. References

1. T. Ito, T. Nakamura, and H. Ishikawa, (1982) IEEE Trans. Electron Devices **ED-29**, 498.
2. S.K. Lai, J. Lee, and V.K. Dham, (1983) IEDM Tech. Dig. 190.
3. Z.-Q. Yao, (1994) PhD Thesis, Griffith University.
4. Z.-Q. Yao, (1995) J. Appl. Phys. **78**, 2906.
5. W. Ting, H. Hwang, J.Lee, and D.L. Kwong, (1991) J. Appl. Phys. **70**, 1072.
6. Z.-Q.Yao, H.B. Harrison, S. Dimitrijev, D. Sweatman, and Y.T. Yeow, (1994) Appl. Phys. Lett. **64**, 3584.
7. J.A. Diniz, P.J. Tatsch, M.A.A. Pudenzi, (1996) Appl. Phys. Lett. **69**, 2214.
8. M.L. Green, L.C. Feldman, T.W. Sorsch, W. Lennard, E.P. Gusev, E. Garfunkel, H.C. Lu, and T. Gustafsson, (1997, November 17) Appl. Phys. Lett., in press.
9. Z.-Q. Yao, H.B. Harrison, S. Dimitrijev, Y.T. Yeow, (1995) IEEE Electron Device Lett. **16**, 345.
10. M. Bhat, D.J. Wristers, L-K. Han, J. Yan, H.J. Fulford and D-L. Kwong, (1995) IEEE Trans. on Elec. Dev., **42**, 907.
11. M.J. Hartig, and P.J. Tobin, (1996) J. Electrochem. Soc., **143**, 1753.
12. P.J. Tobin, Y. Okada, S.A. Ajuria, V. Lakhotia, W.A. Feil and R.I. Hedge, (1994) J. Appl. Phys. **75**, 1811.
13. K.A. Ellis and R.A. Buhrman, (1996) Appl. Phys. Lett. **68**, 1698.
14. G. Weidner, D. Kruger, M. Weidner and T. Grabolla, (1996) Microelectronics Journal **27**, 647.
15. Z.-Q. Yao, R. Ghodsi, H.B. Harrison, S. Dimitrijev, T.Y. Yeow (1996) in The Physics and Chemistry of SiO_2 and Si-SiO_2 Interfaces, ECS Proc. 96-1.
16. I.J.R. Baumvol, F.C. Stedile, J.-J. Ganem, I. Trimaille and S. Rigo. (1996) Appl. Phys. Lett. **69**, 2385.
17. S. Dimitrijev and H.B. Harrison, (1996) J. Appl. Phys. **80**, 2467.
18. Z.-Q. Yao, H.B.Harrison, S. Dimitrijev and Y.T. Yeow, (1994) SPIE Conference Proceed. Vol. 2335.
19. D. Bouvet, P.A. Clivaz, M. Dutoit, C. Coluzza, J. Almeida, G. Margaritondo and F. Pio, (1996) J.Appl.Phys.**79**, 7114.

20. H.C. Lu, E.P. Gusev, T. Gustafsson, E. Garfunkel, M.L. Green, D. Brasen and L.C. Feldman, (1996) Appl. Phys. Lett. **59**, 2713.
21. J. Kuehne, S. Hattangady, J. Piccirillo, G. Xing, G. Milner. (1997) MRS Spring Conf., San Francisco.
22. S.C. Sun, C.H. Chen, and J.C. Lou, (Oct 1995) IEEE-EDS Proceedings, China.
23. B.E. Deal and A.S. Grove, (1965) J. Appl. Phys. **36**, 3770.
24. S. Dimitrijev, D. Sweatman, and H.B. Harrison, (1993) Appl. Phys. Lett. **62**, 1539.
25. S. Dimitrijev, H.B. Harrison, and D. Sweatman (1994) in Mictoelectronics Technology and Process Integration, edited by F.E. Chen and S.P. Murarka, Proc. SPIE 2335 (SPIE Bellingham).
26. S. Dimitrijev, H.B. Harrison, and D. Sweatman, (1996) IEEE Trans. Electron Dev., **43**, 267.
27. S. Dimitrijev and H.B. Harrison, (1996) J. Appl. Phys., **80**, 2467.
28. Y. Ma and Y. Ono, ECS Meeting Abstracts, number 372, Montreal 1997.
29. H. Fukuda, Myasuda, T. Iwabuchi, and S. Ohno, (1991) IEEE Elec. Dev. Lett., **12**, 587.
30. P. Tanner, S. Dimitrijev, H.B. Harrison, (1995) Electronic Letters, **31**, 1880.
31. R. Ghodsi, Y.T. Yeow and M.K. Alam, (1994) Appl. Phys. Lett., **65**, 1139.
32. R. Ghodsi, Y.T. Yeow, C.H. Ling and M.K. Alam, (1994) IEEE Trans. Elec. Devices, **ED-41**, 2423.
33. J.N. Shenoy, G.L. Chindalore, M.R. Melloch, J.A. Cooper, Jr., J.W. Palmour, and K.G. Irvine, (1995) J. Electronic Materials, **24**, 303-309.
34. S. Sridevan, P.K. McLarty, and B.J. Baliga, (1996) IEEE Electron Device Lett., **17**, 136-138.
35. Y. Okada, P.J. Tobin, K.G. Reid, R.I. Hedge, B. Maiti, and S.A. Ajuria, (1994) IEEE Trans. Electron Dev., **41**, 1608-1613.
36. Z.-Q. Yao, H.B. Harrison, S. Dimitrijev and Y.-T. Yeow, IEEE Electron Device Lett., (1995) **16**, 345-347.
37. M. Bhat, D.J. Wristers, L.-K. Han, J. Yan, H.J. Fulford, and D.-L. Kwong, (1995) IEEE Trans. Electron Dev., **42**, 907-914.
38. M. Bhat, J. Kim, J. Yan, G.W. Yoon, L.K. Han, and D.L. Kwong, (1994) IEEE Electron Device Lett., **15**, 421-423.
39. A. Uchiyama, H. Fukuda, T. Hayashi, T. Iwabuchi, and S. Ohno, (1990) Electron. Lett., **26**, 932.
40. H. Fukuda, T. Arakawa, and S. Ohno, (1990) Electron. Lett., **26**, 1505.
41. Z. Liu, H.-J. Wann, P.K. Ko, C. Hu, and Y.C. Cheng, (1992) IEEE Electron Device Lett., **13**, 402.
42. C.K. Yang, T.F. Lei, and C.L. Lee, (1995) IEEE Trans. Electron Dev., **42**, 2163-2169.
43. J.W. Palmour, R.F. Davis, H.S. Kong, S.F. Corcoran, and D.P. Griffis (1989) J. Electrochem. Soc., **136**, 502-507.
44. R.C. De Mao, T.K. Wang, T.P. Chow, D.M. Brown, and L.G. Matus, (1994) J. Electrochem. Soc., **141**, L150.
45. Z.-Q. Yao, H.B. Harrison, S. Dimitrijev, and Y.-T. Yeow, (1994) IEEE Electron Dev. Lett., **15**, 516-518.
46. R.S. Muller, T.I. Kamins, (1986) Device Electronics for Integrated Circuits, 2nd. ed., Wiley, New York, p.380.
47. V.V. Afanas'ev, M. Bassler, G. Pensi, and M.J. Shulz, (1996) J. Appl. Phys., **79**, 3108-3114.
48. M. Ruff, H. Mitlehner, and R. Helbig, (1994) IEEE Trans. Electron Dev., **41**, 1040-1054.
49. S.M. Sze, (1981) Physics of Semiconductor Devices, 2nd ed., Wiley, New York, p.372.
50. D.K. Schroder, (1990) Semiconductor Materials and Device Characterization, Wiley, New York.

ENDURANCE OF EEPROM-CELLS USING ULTRATHIN NO AND NH$_3$ NITRIDED TUNNEL OXIDES

A. MATTHEUS, A. GSCHWANDTNER, G. INNERTSBERGER,
A. GRASSL, A. TALG

SIEMENS AG
Otto-Hahn-Ring 6
81739 München
Germany

1. **Introduction**

EEPROMs (electrical erasable programmable read only memory) actually are used for various applications in microcontrollers or chip card products. For cell reliability improvement, reduction of internal programming voltages (<12V) is favorable which would also simplify the peripheral circuit design related to internal voltage generation. New card technologies require more memory capacity and therefore a higher integration density with thinner dielectric layers. In floating gate tunnel oxide (FLOTOX) type EEPROMs this reduction of the tunnel oxide (TOX) also leads to a desired lower programming voltage. But leakage current through the tunnel oxide limits the thickness reduction.

An established approach for improving the TOX quality is thermal nitridation with NH$_3$, N$_2$O or NO.

By nitridation with ammonia for example the cycle endurance of the storage cells can be improved by one order of magnitude for today used TOX thicknesses [1,2,3]. Although the nitridation improves the reliability of oxides, a further scaling down leads to a significant increased stress induced leakage current (SILC) – a major concern of data retention.

This paper describes the properties of NO and NH$_3$ nitrided tunneloxides produced in a rapid thermal processing system (RTP). In a first step time of flight (TOF)-SIMS was applied for high resolution nitrogen depth profiling. The influence of the nitrogen content on the oxide growing is described in detail.

For better comprehension of the break down mechanisms in the oxide during programming and erasing of the EEPROM cell our investigations are focused on the influence of an annealing step after 7 million cycles applied on the EEPROM cell.

A very favorable method to obtain better TOX quality is to perform the whole gate stack process under stringent controlled conditions. The best approach to realize these requirements is achieved by using a cluster tool.

2. Experimental

In this chapter the formation and TOF-SIMS analysis of NO and NH_3 nitrided oxides are described.
Figure 1 shows a standard NH_3 nitridation [4] scheme for thin oxides fabricated in a RTP- system.

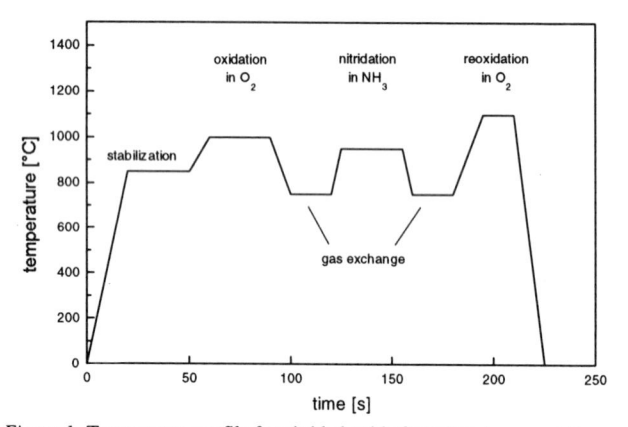

Figure 1: Temperature profile for nitrided oxide formation in ammonia.

The procedure for NO nitrided oxides is similar (figure 2), but there is no reoxidation step necessary since no hydrogen is incorporated as in the previous case. The elimination of the reoxidation step significantly reduces the thermal budget of the whole process.

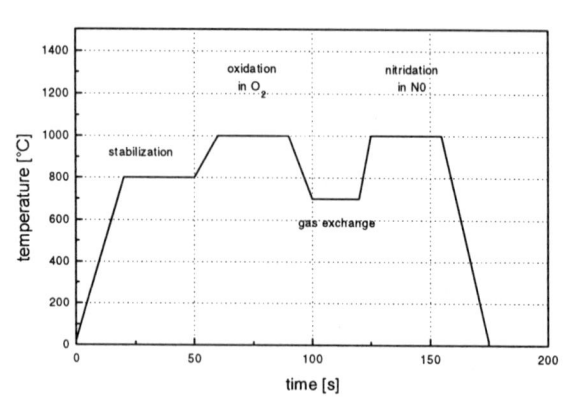

Figure 2: Temperature profile for NO nitrided oxide formation.

For analysis of nitrogen incorporation, TOF-SIMS was applied for depth profiling. Due to the sputtering being carried out at low energy (Cs$^+$, 1keV) at a tilt angle of 45°, a depth resolution on the nanometer scale is obtained. The SiN$^-$ secondary ion species was chosen as a measure for the nitrogen in the layers. The result in Figure 3a,b show an enhancement of nitrogen-concentration at the Si/SiO$_2$-interface. It is noted that the SiN-"tail" (about 4 nm per order of magnitude signal drop) is due to the residual sputter-mixing effect and does not indicate significant nitrogen diffusing into silicon in this regime.

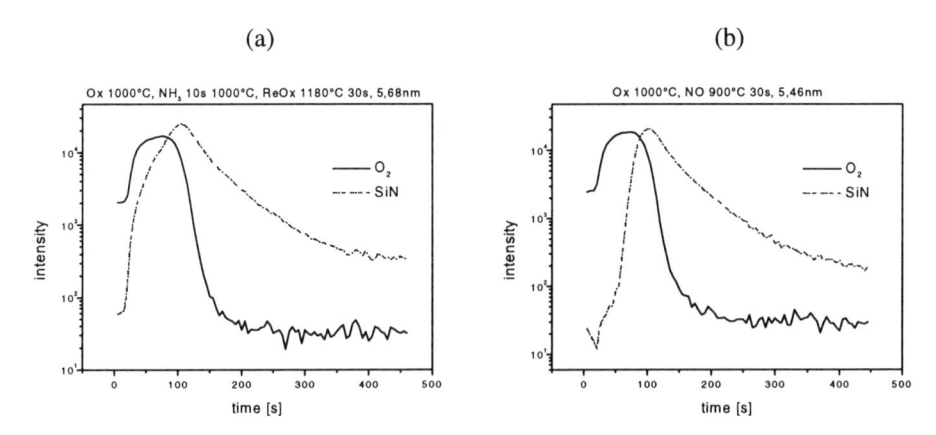

Figure 3. TOF-SIMS of NH$_3$ and NO nitrided oxides

Figure 3 shows also that the amount of incorporated nitrogen is very similar for both NH$_3$ and NO nitridation species.

A sensitive check for the nitrogen concentration in the nitrided oxide is an additional reoxidation, as shown in figure 4. The lower the additional thickness the higher the amount of nitrogen in the layer.

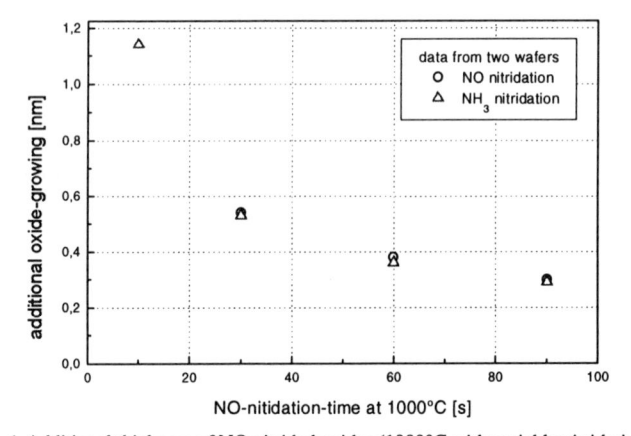

Figure 4. Additional thickness of NO nitrided oxides (1000°C with variable nitridation-time) caused by 90s-reoxidation at 1050°C.

3. Electrical Measurements

3.1 EEPROM CELL

The electrical measurements are performed using individual FLOTOX-EEPROM cells as shown in the schematics of figure 5.

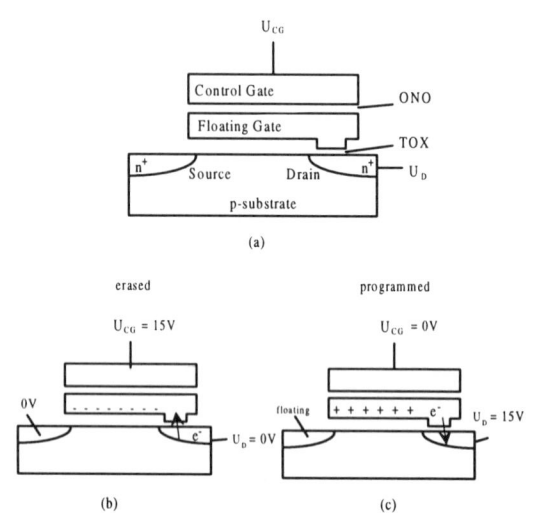

Figure 5. FLOTOX-EEPROM-cell storage transistor [5] (a) schematic, (b) erased state, (c) programmed state

In the erase mode (figure 5(b)) the control gate (CG) is at high voltage (U_{CG} =15V), while the drain is grounded (U_D=0V). Electrons tunnel into the floating gate, (FG) and the threshold voltage (V_T) shifts in *positive* direction.

In the program mode (figure 5(c)) the CG is grounded (U_{CG} =0) and the drain is connected to high voltage (U_D =15V). Electrons tunnel out of the FG into the drain leaving the FG positively charged. This shifts V_T in *negative* direction.

Fowler-Nordheim (FN) tunneling occurs only for a sufficiently high potential difference between CG and drain, i.e. a sufficiently high electrical field across the TOX (the tunnel current has an exponentially increasing dependence on the electrical field across the TOX [6]).

The following measurements deal with the reliability of these thin TOX.

3.2 CYCLE ENDURANCE

A standard method for characterization of the cell quality is the cycle endurance test. This means setting the EEPROM in the programmed and afterwards in the erased state continuously. The programming window (see figure 6) is defined by the difference of the threshold voltages of these two states. Between 10^6 and 10^7 cycles a reduction of the programming window occurs because charges are trapped within the TOX and shift the threshold voltage V_T.

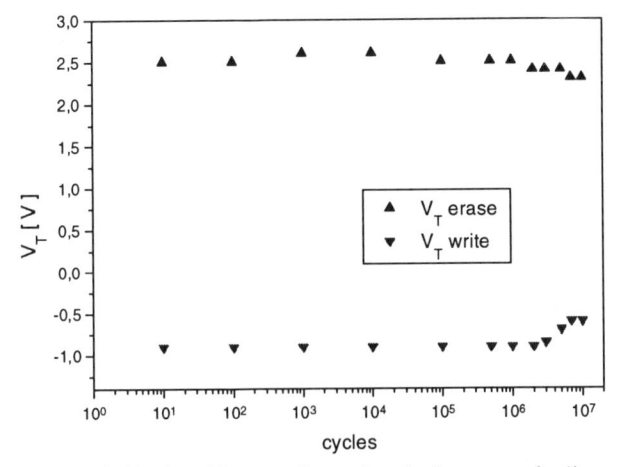

Figure 6. Threshold voltage V_T versus the number of write-erase cycles (1ms pulses) of 5.5 nm NH$_3$ nitrided oxide.

3.3 DATA RETENTION

An additional test for the TOX quality is the standard data retention test (24h, 250°C bake) performed for the nitrided oxides with different thicknesses. The cells with the thinner TOX being cycled 10^5 times before the bake show a great reduction of the programming window. For the 7.5nm TOX there is no influence of the bake for cycled and fresh cells (see figure 7).

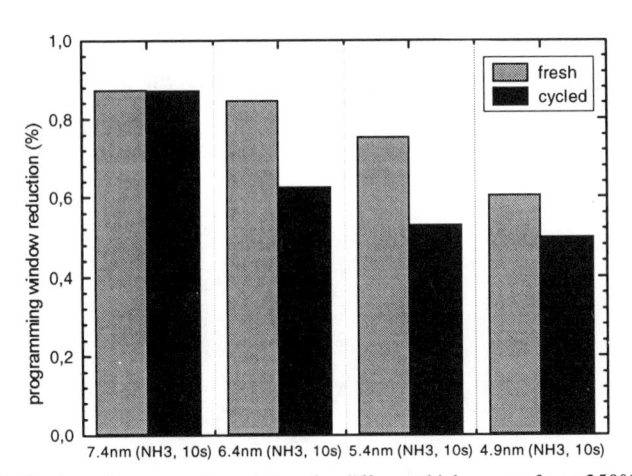

Figure 7. Reduction of programming window for different thicknesses after a 250°C 24h bake.

3.4 STRESS INDUCED LEAKAGE CURRENT

The stress induced leakage current (SILC) remains a major concern for data retention. Therefore Fowler-Nordheim tunneling characteristics of planar capacitors have been investigated. Figure 8.a shows FN-characteristics of ultrathin fresh oxides for various oxide thickness values. For characterization a voltage |V| is considered which causes a current level of $J>10^{-9} A/cm^2$, e.g. |V|>4V for the 5.5nm NH_3 nitrided oxide. This value decreases for thinner oxides. A leakage current below 3V occurring for fresh oxides, is caused by direct tunneling.

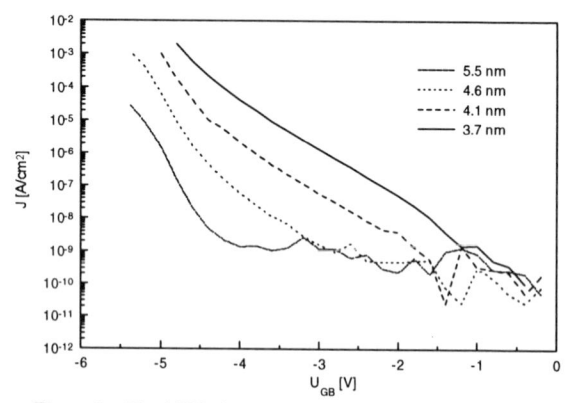

Figure 8.a. Fresh FN-characteristics for ultrathin NH_3 nitrided oxides.

The susceptibility of NO nitrided oxides to stress induced leakage current is tested by dc charge injection of $-1C/cm^2$ (J=0.01 A/cm2 for the 5.5nm oxide, 0.1 A/cm^2 for the 6.5nm and 7.5nm oxide). In that case (open symbols in figure 8.b) only the 7.5nm oxides show a sufficiently low leakage current. The same behavior is found for NH_3 nitrided and reoxidized films [7]. These measurements are an indication for reliability problems of thinner oxides.

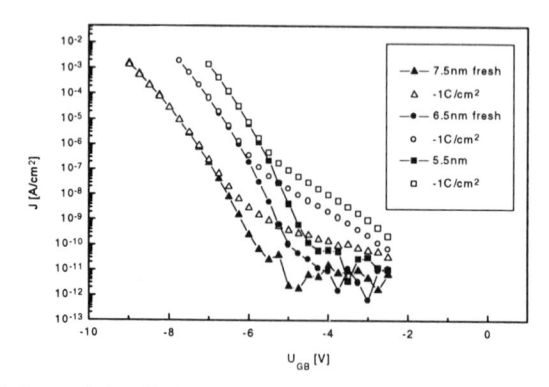

Figure. 8.b. FN-characteristics of NO nitrided oxides before and after current injection of $-1C/cm^2$.

3.5 ANNEALING EXPERIMENTS

For TOX characterization annealing experiments have been performed using 7.5nm NO nitrided films. The cells were cycled $7 \cdot 10^6$ times, then annealed and cycled again to check if the stress induced damage can be reduced. The applied temperatures were 250°C, 350°C and 450°C and the ambients were N_2, H_2 or D_2. The annealing temperature is limited to about 450°C by the AlSiCu metalization. The temperature effect on the program and erase voltages is shown in figure 9.

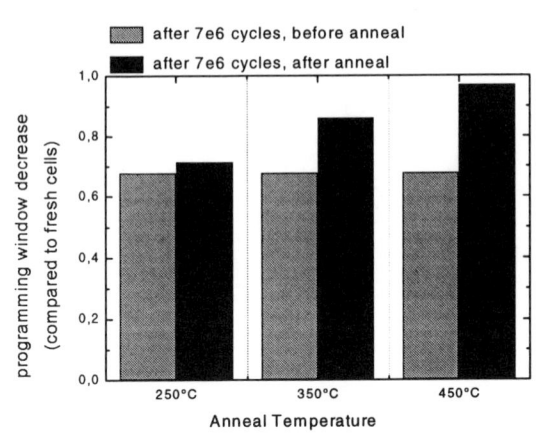

Figure 9. Programming window decrease after cycling and annealing for a 7.5nm NO nitrided TOX.

After cycling the cells $7 \cdot 10^6$ times the program window is reduced to 68%±3% compared to fresh cells (gray bars in figure 9). As an industrial standard cell failure is usually assumed for a reduction of the programming window by 50%. After annealing the programming window increases to 72%, 86% and 95% compared to fresh cells for 250°C, 350°C and 450°C respectively. This means that trapped charges within the 7.5nm TOX – induced during the program/erase stress (ac-stress) – are freed by the annealing. Annealing at elevated temperatures reveal more efficient reduction of stress induced oxide damage. However after further cycling (10^6 times) the charges return into the already existing oxide traps very fast and the programming window of annealed and not annealed cells show identical behavior.

This means that the stress induced oxide damage is an irreversible effect and cannot be effectively removed by annealing.

It is reported that for MOS Transistors the threshold voltage shift is improved one order of magnitude by using deuterium as an annealing species [8]. The deuterium accumulates at the interface and replaces the hydrogen. Due to the higher mass of the deuterium the hot carrier reliability is improved.

In order to investigate the influence of H_2 and D_2 ambient on EEPROM cells the annealing at 400°C to 450°C has been tested.

First results show that the deuterium anneal has no influence on the threshold voltage shift compared to H_2 or N_2 anneal. The results are preliminary because the Si_3N_4 passivation has not been removed as reported in [8]. D_2 annealing experiments without the passivation layer are in preparation.

It is on concern to answer the question to the deuterium effect and to see whether an effective improvement of the reliability can be achieved by using D_2 already during production and before capping the wafer with SiN_4. In this way the nature of the damage could be clarified.

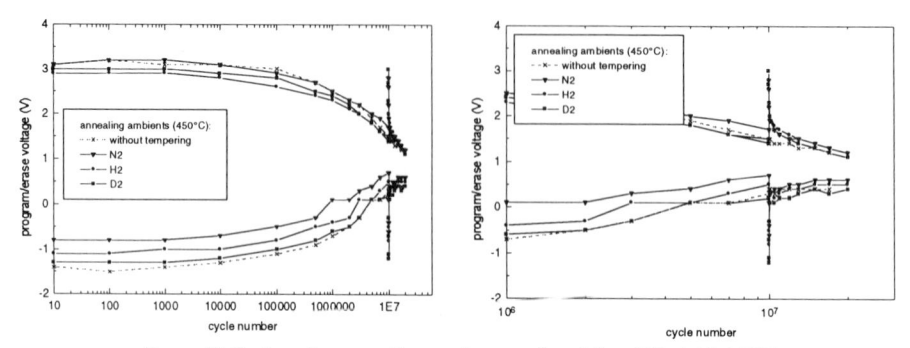

Figure 10. Cycle endurance with annealing step for a 7.5nm NO nitrided TOX.

4. Conclusion and outlook

Endurance, data retention and SILC emphasize that an improvement of the thin TOX is mandatory. To get high quality Si/TOX-interfaces the surface and ambient conditions have to be controlled carefully. This leads to the approach of integrated processing in a cluster tool.

The HOT CLUSTER (see figure 11) is the first commercial use of MESC (Module Equipment Sub Committee of the Information and Controls committee of SEMI) and CMTC (Cluster Tool Module Communication) standards for cluster tool hardware and software interfaces [9, 10].

The tool comprises a vapor phase cleaning module (VPC), a RTP-module for thermal TOX growth and nitridation, a RTCVD poly module and a RTCVD nitride module.

This cluster tool enables the gate stack process for EEPROMs – TOX, floating gate and ONO interpoly dielectrics – to be performed in an integrated sequence.

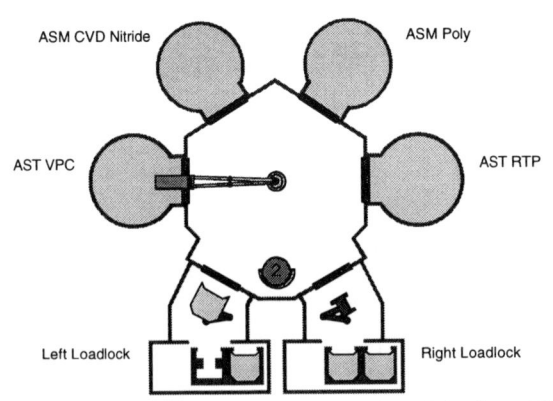

Figure 11. HOT CLUSTER (Advance 800 Polygon) –with modules from ASM and AST.

For this gate stack process the wafer has to be cleaned in the VPC module with HF and methanol vapor to remove the native oxide.

Then the tunnel oxide is thermally grown and nitrided in the RTP module. The standard sequence covers an initial thermal oxidation, a nitridation with ammonia and a reoxidation step for removing incorporated hydrogen. This last step can be skipped by using NO as nitridation species. Additionally both thermal budget and process time are reduced significantly. After that the in-situ doped amorphous silicon is deposited in the poly module.

In a final step the interpoly dielectric is fabricated by thermal growth of the bottom oxide followed by a CVD silicon nitride deposition in the nitride module.

Table 1 shows a comparison of a standard production process flow on stand alone equipment with the advanced integrated process sequence on the HOT CLUSTER tool.

TABLE 1. Comparison of an EEPROM gate stack process.

Standard	HOT CLUSTER
cleaning	vapor phase cleaning
time linkage	--------
RTO/N TOX(tunnel oxide)	RTO/N TOX(tunnel oxide)
time linkage	--------
CVD amorphous silicon undoped	RTCVD in-situ doped amorphous silicon
POCl$_3$ doping and diffusion	--------
etching NF3	--------
cleaning	--------
time linkage	--------
bottom oxide	RTO bottom oxide & RTCVD silicon nitride in one chamber
CVD silicon nitride	--------

The cluster tool working in stringent ambient conditions is a good approach for growing high quality TOX layers and in situ doped poly silicon.

5. References

1. H. Fukuda, M. Yasuda, T. Iwabuki (1993) Novel Single Step Oxynitridation Technology for Forming Highly Reliable Eeprom Tunnel Oxide Films, *Jpn. J. Appl. Phys.* **32,** 447-451.

2. M. Dutoit et al. (1994) Thin SiO2 films nitrided by RTP in NH3 or N2O for applications in EEPOM's, *Microelec. J.,* **25,** 539-551.

3. H.G. Pomp et al. (1993) Lightly N2O nitrided dielectrics grown in a conventional furnace for EEPROM and 0.25µm CMOS, *IEEE-Meeting Washington 5-8 Dec 1993,* 463-465.

4. R. Kakoschke (1995) Nitrided Tunnel Oxides for Improved Endurance of EEPROM CELLS, RTP-Conference 1995.

5. Betty Prince (1991) Semiconductor Memories, Wiley, (New York).

6. R.H. Fowler, L. Nordheim (1928) *Proc. Roy. Soc.* **Ser A, 119**, (London), 173.

7. A. Mattheus, A. Gschwandtner, R. Kakoschke, M. Kerber, A. Talg (1996) New Results of NO grown Dielectrics for FLOTOX/FLASH-EEPROM applications, RTP-Conference, 1996.

8. I. C. Kizilyalli, J. W. Lyding, K. Hess (1997) *IEEE Electron Device Letters,* **Vol. 18, No. 3**, 81.

9. L.Deutschmann, F. Glowacki, T. Knarr, N. Verhaar, C. Werkoven (1996) Advanced cluster tooltechnology: current status, *Proc. 4th Int. Conf. RTP'96.*

10. C. Werkoven, N. Verhaar, T. Bergman, L. Deutschmann (1996) A MESC/CTMC-based »best of breed« cluster tool for ONO applications, *Proc. 4th Int. Conf. RTP'96.*

EFFECTS OF THE SURFACE DEPOSITION OF NITROGEN ON THE OXIDATION OF SILICON

T.D.M. SALGADO[1], I.J.R. BAUMVOL[2], C. RADTKE[1], C. KRUG[1], AND F.C. STEDILE[1]
[1]*Instituto de Química and* [2]*Instituto de Física - UFRGS*
Av. Bento Gonçalves, 9500 - Porto Alegre - RS - Brasil - 91509-900

We have studied the influence of the surface deposition of N on the oxidation of Si(100) in dry O_2. This problem has been addressed before by implanting 30 keV N ions through a 20 nm-thick SiO_2 film, followed by thermal oxidation. In the present work the deposition of N was performed by ion implantation on the Si(100) surface at a very low energy (approximately 20 eV), at fluences between 1 and 10×10^{14} cm^{-2}. Oxidations were performed in dry O_2 at 1000 °C, under 50 mbar, during time intervals between 15 and 120 min. In order to allow for high resolution depth profiling, the implanted isotope was ^{15}N, and the oxidations were mostly performed in 97% ^{18}O-enriched O_2. The areal concentrations of ^{15}N (before and after oxidation), and of ^{18}O were determined by nuclear reaction analysis, while the ^{15}N and ^{18}O profiles were determined by means of low energy nuclear resonance profiling. The results show that: i) the retained amounts of ^{15}N just after ion beam deposition stayed in the range 3×10^{13} - 7×10^{14} cm^{-2}; ii) the oxide growth rate is strongly influenced by the presence of nitrogen, decreasing with the increase of the areal concentration of nitrogen; iii) before oxidation, the implanted ^{15}N atoms occupy the very near surface layers of Si; and iv) after oxidation, ^{15}N is distributed within the oxide film, with a maximum concentration near the oxide surface, which is a desired feature. Further studies on isotopic tracing of oxygen during thermal growth of silicon oxide on N-deposited Si surfaces are also reported.

1. Introduction

There has been a great interest in the thermal oxynitridation of Si in N_2O and in NO in the last few years, mainly because of the superior electrical properties of ultrathin (less than 10 nm) films of silicon oxynitride on Si as compared to SiO_2 films of similar thicknesses [1,2] for gate dielectrics applications. N can be introduced in the oxynitride films on Si by a variety of thermal treatments, leading to different N contents and depth distributions in the films: i) nitridation of ultrathin SiO_2 films in NH_3, and reoxidation of the oxynitride films in O_2 [3]; ii) direct growth in N_2O and NO, nitridation of ultrathin

SiO$_2$ films in N$_2$O and NO, and reoxidation of the resulting oxynitride films in O$_2$ [1,4,5]; iii) other combinations of the previous gas sequences, like direct growth in N$_2$O followed by nitridation of the oxynitride films in NH$_3$ [6]; iv) ion implantation of N through SiO$_2$ films, and subsequent annealing in different atmospheres [7]. In particular, the reoxidation of ultrathin oxynitride films in O$_2$ has been largely explored as a route to adequately trim the film thickness, as well as the N concentration in different regions of the film. Previous studies have established that the presence of rather small concentrations of N (1% and less) either near the oxynitride/Si interface, or in the bulk, or even near the film surface, slows down the rate of film growth under thermal reoxidation in O$_2$ [8]. A slower growth rate due to the introduction of N is also observed during direct thermal growth in N$_2$O and NO [9].

In spite of the wealth of well established empirical facts concerning the growth kinetics, the depth distribution of the different species, and the electrical characteristics of the oxynitride films, there remains a considerable lack of knowledge about the growth mechanisms of silicon oxide films in the presence of N. Many authors derived mathematical expressions [9,10] capable of fitting the different growth kinetics, remaining mainly in the framework of the Deal and Grove model [11]. Dimitrijev et al. [9] determined the kinetics of growth in N$_2$O, attributing the limiting step to the neutralization of the oxide growth sites at the interface by N atoms. On the other hand, Ting et al. [10] attributed the slow down of the growth rate (with respect to oxide growth in O$_2$) to the effect of N in hampering the diffusion of the species responsible for growth, namely O$_2$, to the oxide(oxynitride)/Si interface. However, these and many other authors were not able to answer fundamental questions like: i) how does the presence of N alter the atomic transport and growth mechanisms of silicon oxide films ? Is it by acting as a diffusion barrier, by reducing the concentration of reaction sites at the oxide(oxynitride)/Si interface, or even by a different mechanism ? and ii) what are the relationships between growth mechanisms and the distribution of N in the resulting oxynitride films ? Some efforts have been undertaken recently in the direction of answering these fundamental questions. Trimaille et al. [7] have implanted N in Si through SiO$_2$ films, and performed isotopic tracing studies during the further thermal growth in O$_2$ of the implanted sample, concluding that N affects only the near oxide/silicon interface reaction where growth takes place, leaving the incorporation of oxygen near the film surface unchanged. This result was corroborated by Lu et al. [12] in the case of ultrathin (5 nm) films, when N is introduced in the oxynitride films by means of thermal growth in N$_2$O and by remote plasma deposition. Ganem et al. [13] have studied the reoxidation in O$_2$ of thicker oxynitride films (~ 10 nm) grown in N$_2$O, reporting modifications on the near interface and near surface oxygen incorporation, and on the N distribution in the oxynitride film. Tang et al. [4] showed that at the initial stages of film growth in N$_2$O, the nitridation of less than one monolayer, forming silicon nitride, inhibits the oxide growth; that silicon oxynitrides grow in the intermediate thickness region between 0.5 and 3 nm; and that oxidation is the dominant process of growth above 3 nm. Sutherland et al. [14] complemented this picture, demonstrating that one monolayer of N initially existent at the oxynitride/Si interface remains essentially intact after the growth mode has switched to the production of SiO$_2$ (thicker films). Lu et al. [15] remarked the dramatic effect of N concentrations far below the monolayer range in reducing the growth rate of silicon oxynitride films.

A complete review of the literature is far beyond the scope of the present work. Nevertheless, the few comments made above indicate that understanding how nitrogen affects the growth of silicon oxide films is a complex task, and no direct microscopic evidence is yet available. As an attempt to shed some light on this subject, we report here on the thermal oxidation in O$_2$ of Si(100) substrates whose surfaces were intentionally contaminated with controlled amounts of N, aiming to investigate: i) the influence of the surface concentration of N (in the range 1/30 to 1 monolayer) on the rate of thermal growth of silicon oxide films on Si; ii) the transport of the chemical species O and N during growth; iii) the distribution of N in the resulting silicon oxynitride films; and iv) the modifications on the growth mechanisms with respect to the mechanism of thermal growth of SiO$_2$ films in O$_2$ in the absence of N. The deposition of N on the Si(100) wafers surface was performed by ion implantation at the extremely low energy of 20 eV. In order to allow for highly sensitive and selective determination of the amounts of N and O in the films, as well as high depth resolution profiling of these species, the less abundant isotopes ^{15}N, introduced by very low energy ion implantation, and ^{18}O, by performing thermal oxidations in ^{18}O$_2$, were used. Furthermore, in order to perform isotopic tracing of O during film growth, sequential oxidations in ^{16}O$_2$ and ^{18}O$_2$ were accomplished. The growth kinetics and the depth profiles were determined, respectively by nuclear reaction analysis and narrow nuclear resonance depth profiling.

2. Experimental

Si(100) wafers were cleaned in a 4 % HF solution in ethanol (30 s) and rinsed in ethanol (30 s) just before being introduced in the vacuum of the ion implantation chamber, in order to minimize the presence of native oxide. The base pressure in the ion implantation chamber was 10^{-8} mbar. N ions were extracted from the ion source at a voltage of 30 kV, and mass analyzed by a 90° magnet. The plasma in the ion source was produced using a ^{15}N-enriched N$_2$ gas. Thus, ^{15}N$^+$ ions at an energy of 30 keV were accelerated towards the silicon wafer in the implantation chamber, which is electrically isolated from the rest of the ion implanter. The same voltage supply used to extract the beam from the ion source was used to polarize the samples at + 30 keV, so reducing the effective ion energy arriving in the target to zero. A battery was then used to polarize the target at - 20 V, in order to slightly focus the ion beam on the Si wafers. Figure 1 gives a sketch of the experimental setup used for these very low energy implantations. Typical current densities of ^{15}N$^+$ were of the order of 1 µA.cm^{-2}. Under these conditions we expect to have a very low density of defects generated in the surface region, which should be annealed during the subsequent thermal treatment. The retained doses of ^{15}N on the Si(100) wafers were measured as a function of the implanted doses by nuclear reaction analysis (NRA) as explained below, and the results are shown in Fig. 2. Hereafter, we refer to the ^{15}N implantations in terms of the retained doses.

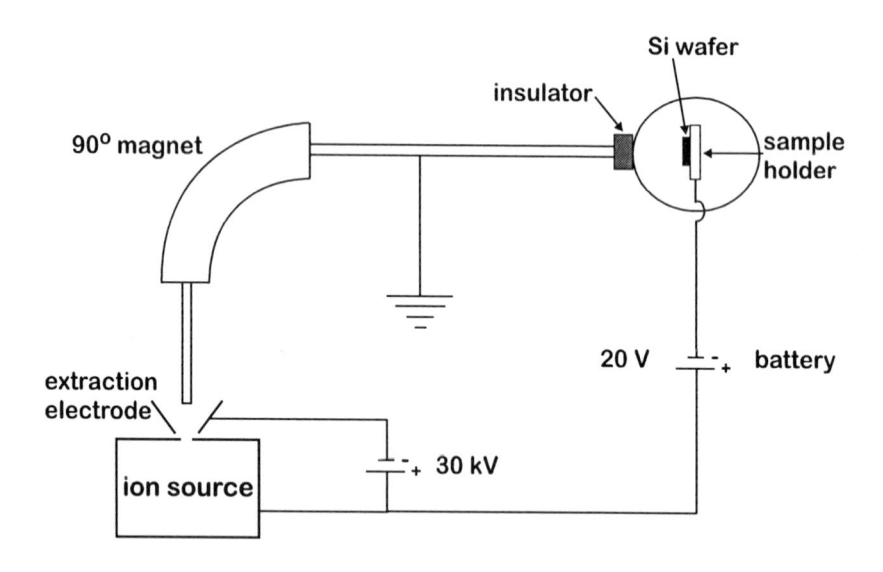

Figure 1. Schematic of the experimental setup used for the very low energy implantations.

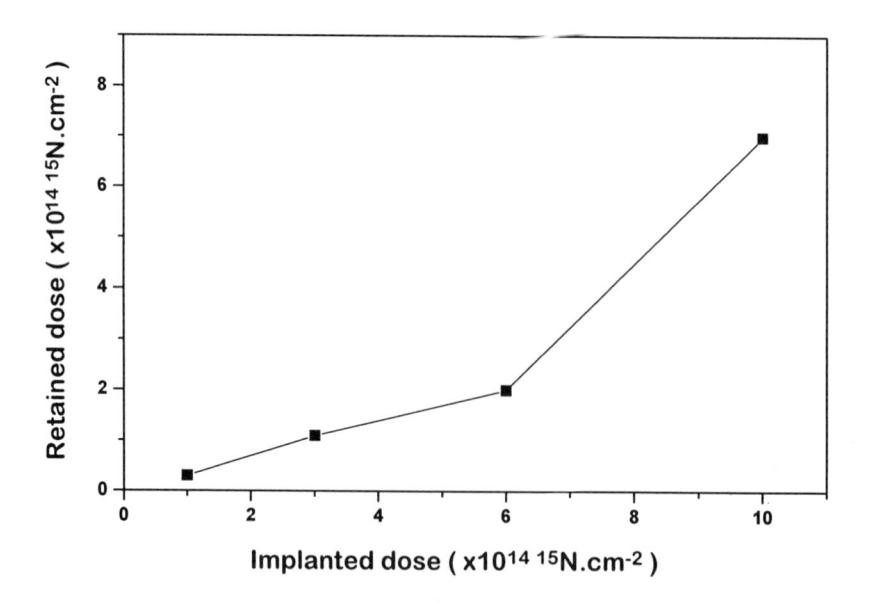

Figure 2. Retained doses of ^{15}N deposited on the Si(100) wafers as a function of the implanted doses.

The total amounts of ^{15}N, ^{16}O, and ^{18}O in the samples were determined using, respectively, the cross section plateaus of the nuclear reactions: ^{15}N(p,αγ)^{12}C at 1000 keV, ^{16}O(d,p$_0$)^{17}O at 810 keV, and ^{18}O(p,α$_0$)^{15}N at 730 keV, using convenient standards [3]. The advantage of analyzing rare isotopes by NRA is the superior sensitivity to detect them as compared to the most abundant ones: 10^{12} cm^{-2} for ^{15}N as compared to 10^{14} cm^{-2} for ^{14}N, and 10^{12} cm^{-2} for ^{18}O as compared to 10^{14} cm^{-2} for ^{16}O. The film thicknesses can be obtained using the equivalent relationship 10^{15} O atoms.cm^{-2} ↔ 0.226 nm of SiO$_2$. This relationship is strictly valid only for SiO$_2$ films, with an assumed density of 2.21 g/cm^3, but since the N concentration in the oxynitride films of the present work is only of a few percent, it can be used as a good approximation. On the other hand, since the N depth distribution in the films is not homogeneous, the film density varies and so the above conversion factor may be locally incorrect. Therefore, we prefer to express the film thicknesses in terms of atomic areal densities.

The ^{15}N and ^{18}O depth profiles were obtained by nuclear resonance profiling (NRP), using the relatively strong, narrow, and isolated resonances in the cross sections of the nuclear reactions ^{18}O(p,α)^{15}N at 151 keV and ^{15}N(p,αγ)^{12}C at 429 keV [3]. A tilted sample geometry ($\Psi = 65°$) was used in order to increase the depth resolution. The measured excitation curves (i.e. α or γ yields versus incident proton energy) around the resonance energy E$_R$ can be converted into concentration depth distributions by means of the SPACES simulation program [16]. Under these conditions, the profiling method assures a depth resolution of approximately 1 nm near the film surface. The experimental excitation curve, the simulated excitation curve, and the corresponding ^{15}N profile in a sample as-implanted with 7×10^{14} ^{15}N.cm^{-2} are shown in Fig. 3.

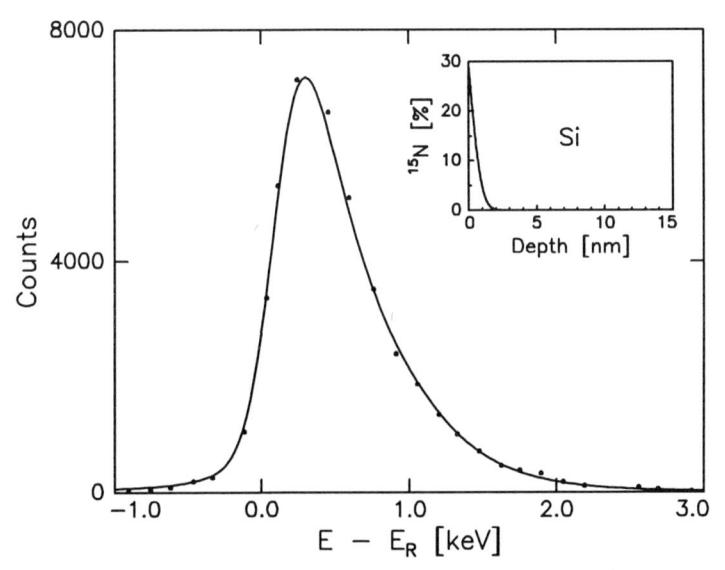

Figure 3. Experimental excitation curve (points) of the ^{15}N(p,αγ)^{12}C nuclear reaction around the resonance at 429 keV, its simulation (line), and the correspondent ^{15}N depth profile (inset) in the case of 7×10^{14} ^{15}N.cm^{-2} in a Si(100) wafer.

The oxidations were performed in a conventional furnace under static pressures (50 mbar) of either pure natural O_2 (called $^{16}O_2$, which in fact contains 0.2 % of ^{18}O), or 97% ^{18}O-enriched oxygen ($^{18}O_2$), at a temperature of 1000 °C. The base pressure in the quartz tube before pressurization was 10^{-7} mbar.

3. Results and Discussion

Isotherms of ^{18}O incorporation by Si(100) wafers containing different doses of ^{15}N at their surface, oxidized in $^{18}O_2$ are shown in Fig. 4. The film thicknesses of all samples were seen to increase with the oxidation time, as expected, but the amount of incorporated ^{18}O was seen to decrease with the increase of the ^{15}N dose (see also the isochrones shown in Fig.5).

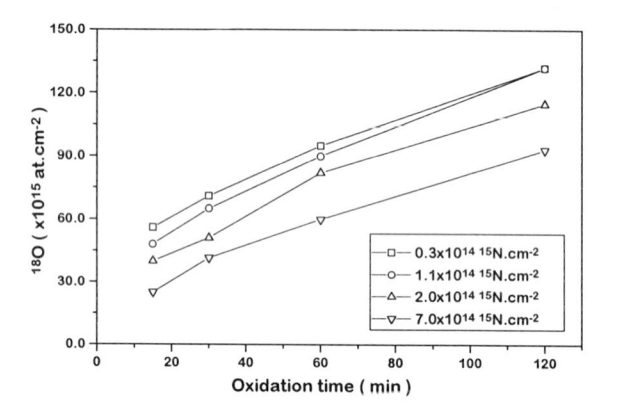

Figure 4. Isotherms of ^{18}O incorporation during thermal oxidation in $^{18}O_2$ of Si(100) wafers containing different doses of ^{15}N at their surface.

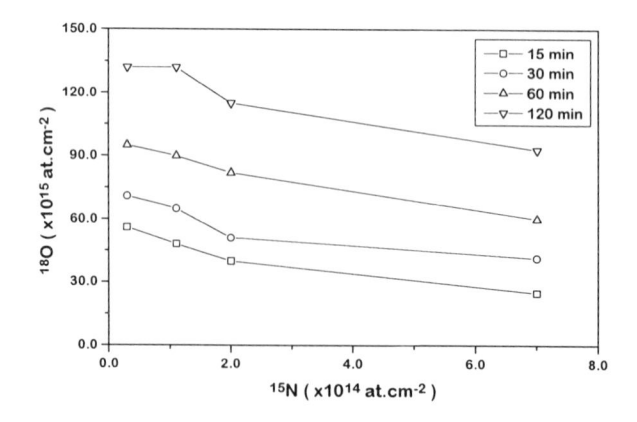

Figure 5. Dependence of the ^{18}O incorporated in the oxides on the dose of ^{15}N deposited on the Si(100) surface before oxidation.

Excitation curves of the $^{18}O(p,\alpha)^{15}N$ nuclear reaction around the resonance energy of 151 keV, for samples implanted with different doses of ^{15}N and oxidized in $^{18}O_2$ for 15 min, are shown in Fig. 6a. The decrease of the film thickness as the ^{15}N dose increases is once more clear (the solid arrows indicate the energy position of the SiO$_2$/Si interface in each sample). Fig. 6b shows the experimental excitation curve, the simulated excitation curve, and the corresponding ^{18}O depth profile for the sample having (before oxidation) an areal density of 1.1×10^{14} $^{15}N.cm^{-2}$, and then oxidized during 15 min.

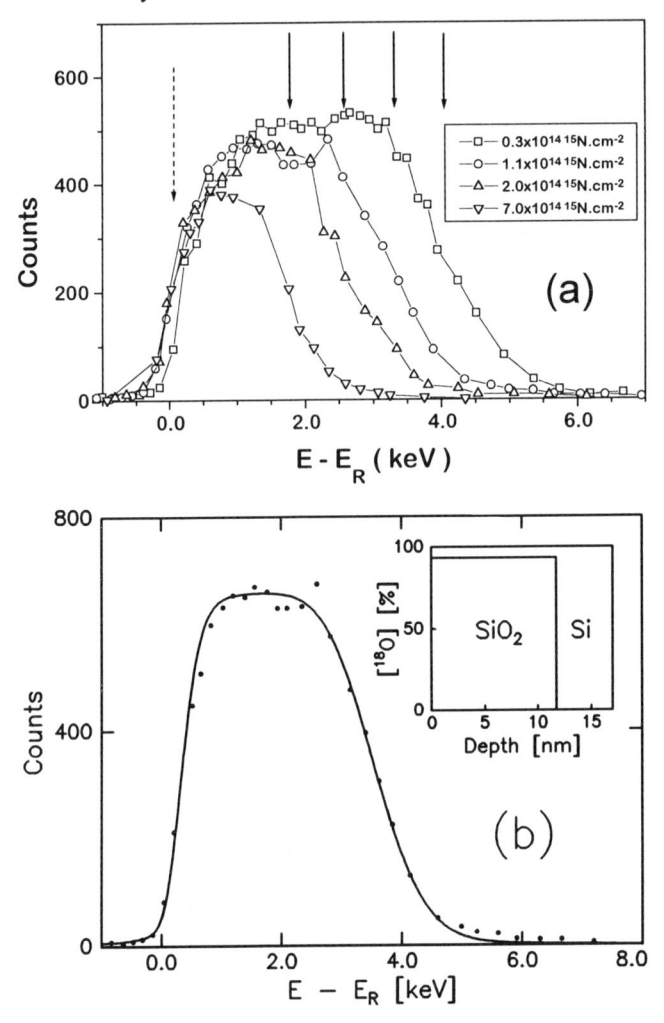

Figure 6. (a) Excitation curves of the $^{18}O(p,\alpha)^{15}N$ nuclear reaction around the resonance at 151 keV for Si wafers surface implanted with different doses of ^{15}N and oxidized during 15 min. The arrows indicate the energies corresponding to the surface (dashed line), and to the SiO$_2$/Si interface (solid line) in each sample; (b) The excitation curve (points) for the oxidized sample which contained 1.1×10^{14} $^{15}N.cm^{-2}$ before oxidation, its simulation (line), and the correspondent ^{18}O profile.

Excitation curves of the ^{15}N(p,αγ)^{12}C nuclear reaction around the resonance energy of 429 keV, for samples implanted with 1.1×10^{14} ^{15}N.cm^{-2} and oxidized for 30 and 60 min, as well as for samples implanted with 7.0×10^{14} ^{15}N.cm^{-2} and oxidized for 60 and 120 min are given in Fig. 7. The experimental excitation curve from the sample having (before oxidation) an areal density of 1.1×10^{14} ^{15}N.cm^{-2} in the surface, and then oxidized during 30 min, its correspondent simulated excitation curve, and ^{15}N profile are shown in Fig. 8a, while the complete set of ^{15}N profiles obtained from the simulation of the excitation curves of Fig. 7a are shown in Fig. 8b. We notice from Figs, 7 and 8 that the ^{15}N atoms are redistributed during thermal oxidation, occupying the regions of the oxide film between approximately 2 nm below the surface and the SiO$_2$/Si interface. Furthermore, a net loss of ^{15}N during oxidation is observed.

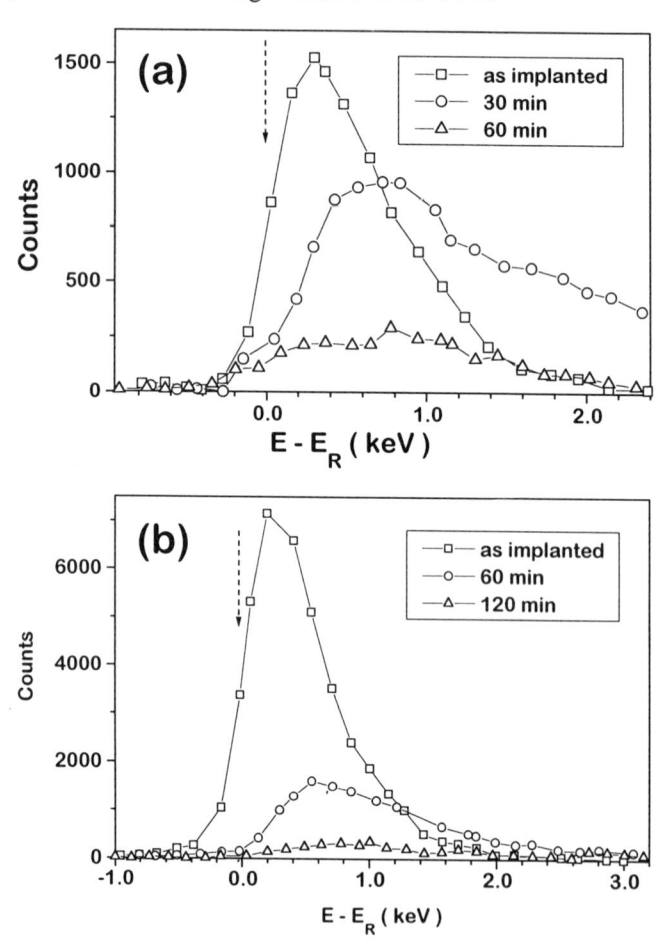

Figure 7. Excitation curves of the ^{15}N(p,αγ)^{12}C nuclear reaction around the resonance at 429 keV, for samples: (a) implanted with 1.1×10^{14} ^{15}N.cm^{-2} and oxidized in ^{18}O$_2$ for 30 and 60 min; and (b) implanted with 7.0×10^{14} ^{15}N.cm^{-2} and oxidized for 60 and 120 min. The arrows indicate the energy corresponding to the oxide surface.

235

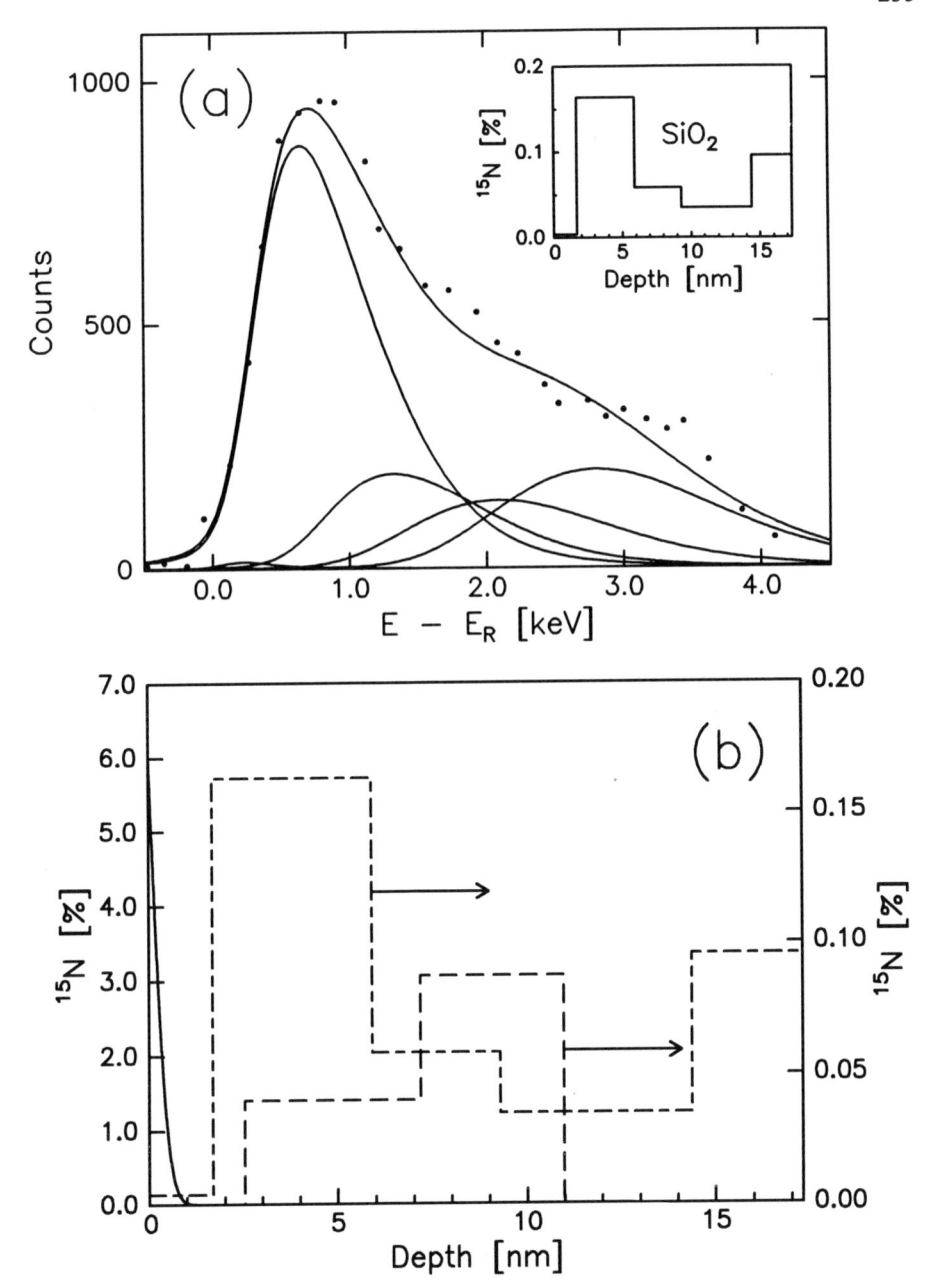

Figure 8. (a) The experimental excitation curve (points) of the ^{15}N(p,αγ)^{12}C nuclear reaction around the resonance energy at 429 keV for the sample having (before oxidation) an areal density of 1.1×10^{14} ^{15}N.cm^{-2}, oxidized during 30 min, the simulated excitation curve (line), and the corresponding ^{15}N profile (inset); (b) ^{15}N profiles obtained from the simulation of the excitation curves of Fig. 7a: solid line: as-deposited (left-hand vertical scale); long dash-short dashed line: 30 min oxidation (right-hand vertical scale); dashed line: 60 min oxidation (right-hand vertical scale).

Sequential oxidations in $^{16}O_2$ (60 min) followed by $^{18}O_2$ (90 min) were performed on samples deposited with ^{15}N, in order to investigate the incorporation of ^{18}O in the different regions of the oxide during the second oxidation step [17], as well as its dependence on the ^{15}N implanted dose. The experimental excitation curve of one of these samples, together with its simulation, and the ^{18}O profile are shown in Fig. 9. The complete set of experimental excitation curves for the $^{18}O(p,\alpha)^{15}N$ nuclear reaction are shown in Fig. 10a, while the ^{18}O profiles in the oxide films are shown in Fig. 10b. The incorporation of ^{18}O at the near-surface and near-interface regions is noticeable for all samples, corroborating previous results [7,11]. However, differently from what was observed by previous authors, we also found ^{18}O incorporated in the bulk of the oxides, in concentrations that increase with the ^{15}N dose.

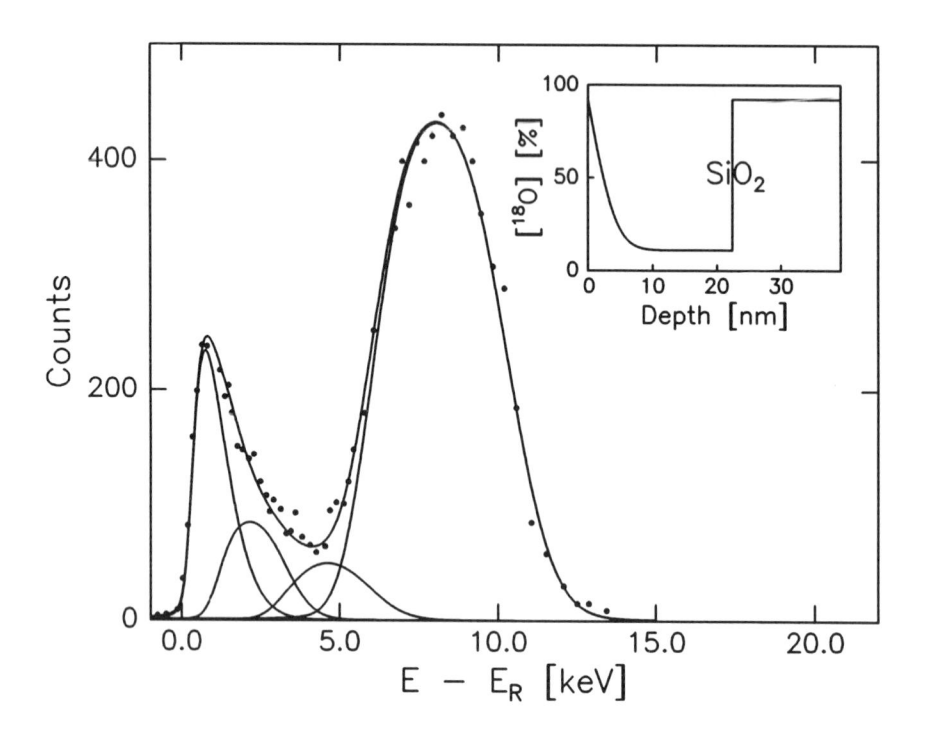

Figure 9. Excitation curve (points) of the $^{18}O(p,\alpha)^{15}N$ nuclear reaction around the resonance energy at 151 keV for a sample containing, before oxidation, 2×10^{14} $^{15}N.cm^{-2}$ and then submitted to sequential oxidations in $^{16}O_2$ (60 min) followed by $^{18}O_2$ (90 min), its simulation (line), and the ^{18}O profile (inset).

237

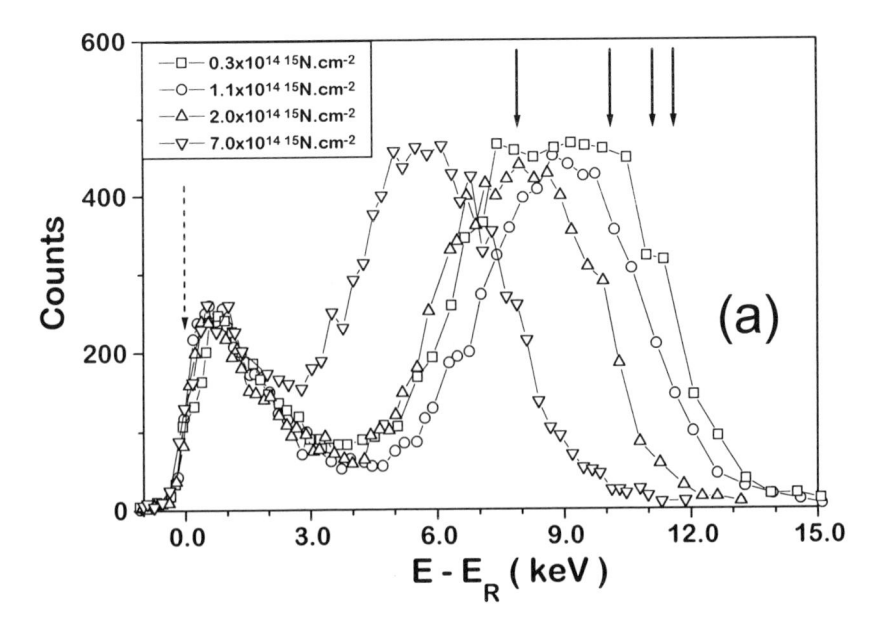

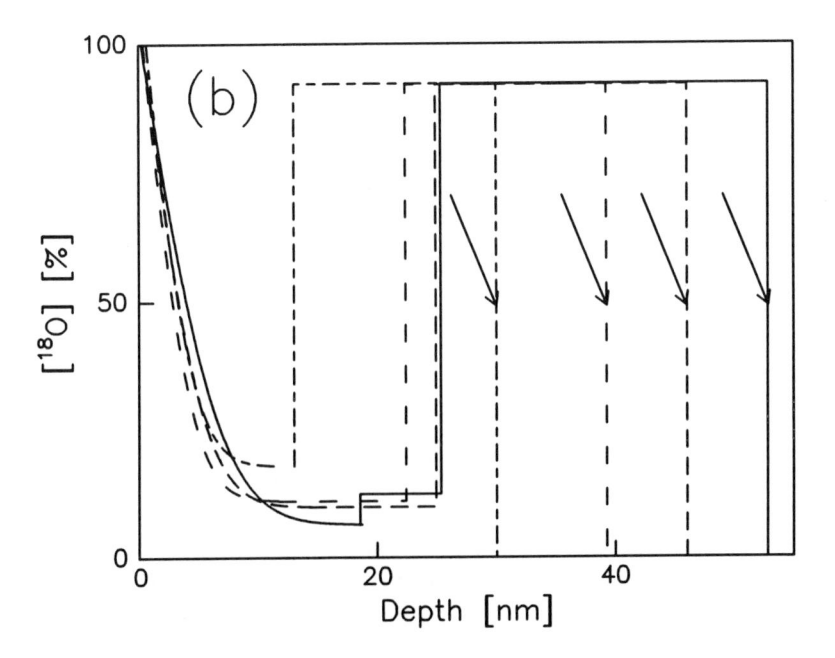

Figure 10. (a) Experimental excitation curves of the $^{18}O(p,\alpha)^{15}N$ nuclear reaction for samples having different ^{15}N areal densities, sequentially oxidized in $^{16}O_2$ (60 min) and in $^{18}O_2$ (90 min). The arrows indicate the energy position of the surface (dashed) and of the SiO_2/Si interface (solid) in each sample; (b) the correspondent ^{18}O profiles. The arrows indicate the position of the SiO_2/Si interface in each sample.

4. Conclusions

It was seen that the rate of thermal growth of SiO_2 films on Si(100) in dry O_2 depends on the areal density of N deposited on the surface of the Si wafers prior to thermal oxidation: the growth rate decreases with the increase of the areal density of deposited N. This effect is valid for all oxidation times, and consequently for all SiO_2 film thicknesses used in the present work, which go from 4 to more than 30 nm. Since N is found in the near surface, bulk and interface regions after the oxidation step, its presence in any of these regions could hamper the oxidizing species diffusion, supporting Ting's model [10], while the presence of N at the interface could neutralize the oxide growth sites, supporting Dimitrijev's model [9]. However, more specific experiments, capable to rule out one (or none, or both) theories are still missing.

The N atoms initially deposited in the surface layers of Si are seen to be distributed within the oxide film during oxidation, presenting a marked depletion at the oxide surface, and a maximum concentration near the surface region, for medium oxidation times. This behavior is analogous to that observed in the thermal reoxidation in O_2 of N_2O-grown silicon oxynitride films [13]. Besides, a significant loss of N is observed as thermal oxidation proceeds: the amount of N lost increases with the increase of the oxidation time.

The $^{16}O_2$-$^{18}O_2$ sequential oxidation studies here performed in Si wafers which had N deposited on their surfaces, revealed some departures from the previously known picture: the ^{18}O introduced in the second oxidation step is seen to occupy not only the near interface region (as predicted by the Deal and Grove model [11]), and the near surface region (as observed by Trimaille et al. [7,17], who attributed its origin to isotopic exchange due to network defects in this region), but it is also fixed in the bulk of the oxide films. The concentration of ^{18}O fixed in the bulk of the oxide films increases with the increase of the areal density of N deposited on the Si surface. This new feature can be attributed to two different causes, both related to the presence of N: i) the reduction of the concentration of reaction sites at the SiO_2 / Si interface due to the presence of N will reject part of the ^{18}O atoms arriving to the interface (in the form of $^{18}O_2$ molecules), forcing them to migrate back towards the surface. These rejected ^{18}O can be fixed in the bulk of the oxynitride by exchange with N atoms, which will themselves migrate towards the surface and desorb, a mechanism which can explain the loss of N during oxidation. It is noteworthy that the amount of ^{18}O incorporated during the second oxidation step near the SiO_2/Si interface decreases with the increase of the areal density of deposited N, as it happens with the total film thickness; ii) the presence of N in the bulk of the oxide film alters the nature and concentration of defects, leading to several possible kinds of sites for ^{18}O fixation which would not exist in the absence of N. This is also supported by the fact that the ^{18}O profiles in the samples sequentially oxidized in $^{16}O_2$-$^{18}O_2$ are erfc-like near the oxide surface (indicating a diffusion against the concentration gradient of defects characteristic of the SiO_2 surface, as discussed by Trimaille et al. [17]), whereas they depart from this diffusion pattern going into the oxide bulk.

In summary, we have investigated the effect of the surface deposition of N on the thermal oxidation of Si, and we found experimental evidences that N reduces the growth rate and is redistributed within the oxide film with a higher concentration near the oxide surface. These facts lead us to conclude that N might influence the diffusion and the reaction of the oxidant species. The kind of N distribution found here for medium oxidation times was recently reported as ideal for the gate dielectric material in sub-micrometric MOS devices [18,19].

In order to have a more complete picture of the atomic transports taking place during these thermal treatments, we are presently investigating the transport of Si atoms, during thermal growth of SiO_2 films on Si wafers surface deposited with Si and N, which will be reported in a forthcoming publication.

Acknowledgments

The authors would like to thank Odile Kaytasov and Jacques Chaumont from the CSNSM, Université Paris XI, Orsay, France, for the kind support with the surface depositions of ^{15}N.

References

1. Fukuda, H., Endoh, T., and Nomura, S. (1996) Characterization of the SiO_2/Si Interface Structure and the Dielectric Properties of N_2O-Oxynitrided Ultrathin SiO_2 Films, in H.Z. Massoud, E.H. Poindexter, and C.R. Helms (eds.), *The Physics and Chemistry of SiO_2 and the Si-SiO_2 Interface-3*, The Electrochemical Society, Pennington, USA, pp. 15-27.
2. Kumar, K., Chou, A.I., Lin, C., Choudhoury, P., and Lee, J.C. (1997) Optimization of sub 3 nm gate dielectrics grown by rapid thermal oxidation in a nitric oxide ambient, *Appl. Phys. Lett.* **70**, 384-386.
3. Baumvol, I.J.R., Stedile, F.C., Ganem, J.-J., Trimaille, I., and Rigo, S. (1996) Thermal Nitridation of SiO_2 Films in Ammonia, *J. Electrochem. Soc.* **143**, 2939-2952.
4. Tang, H.T., Lennard, W.N., Zhang, C.S., Griffiths, K., Li, B., Feldman, L.C., and Green, M.L. (1996) Initial growth studies of silicon oxynitrides in a N_2O environment, *J. Appl. Phys.* **80**, 1816-1822.
5. Baumvol, I.J.R., Stedile, F.C., Ganem, J.-J., Trimaille, I., and Rigo, S. (1996) Nitrogen transport during rapid thermal growth of silicon oxynitride films in N_2O, *Appl. Phys. Lett.* **69**, 2385-2387.
6. Baumvol, I.J.R., Stedile, F.C., Ganem, J.-J., Trimaille, I., and Rigo, S. (1996) Isotopic tracing during rapid thermal growth of silicon oxynitride films on Si in O_2, NH_3 and N_2O, *Appl. Phys. Lett.* **70**, 2007-2009.
7. Trimaille, I., Raider, S.I., Ganem, J.-J., Rigo, S., and Penebre, N.A. (1993) Use of ^{18}O Labelling to Study Growth Mechanisms in Dry Oxidation of Silicon, in C.R. Helms and B.E. Deal (eds.), *The Physics and Chemistry of SiO_2 and the Si-SiO_2 Interface-2*, Plenum Press, New York, USA, pp. 7-13.
8. Raider, S.I., Gdula, R.A., and Petrak, J.R. (1975) Nitrogen reaction at a silicon-silicon dioxide interface, *Appl. Phys. Lett.* **27**, 150-152.
9. Dimitrijev, S., Sweatman, D., and Harrison, H.B. (1993) Model for dielectric growth on silicon in a nitrous oxide environment, *Appl. Phys. Lett.* **62**, 1539-1541.
10. Ting, W., Hwang, H., Lee, J., and Kwong, D.L. (1991) Growth kinetics of ultrathin SiO_2 films fabricated by rapid thermal oxidation of Si substrates in N_2O, *J. Appl. Phys.* **70**, 1072-1074.

11. Deal, B.E. and Grove, A.S. (1965) General Relationship for the Thermal Oxidation of Silicon, *J. Appl. Phys.* **36**, 3770-3778.
12. Lu, H.C., Gusev, E.P., Gustafsson, T., and Garfunkel, E. (1997) Effect of near-interfacial nitrogen on the oxidation behavior of ultrathin silicon oxynitrides, *J. Appl. Phys.* **81**, 6992-6995.
13. Ganem, J.-J., Trimaille, I., Rigo, S., Baumvol, I.J.R., and Stedile, F.C. (1996) Reoxidation in O_2 of silicon oxynitride films grown in N_2O, *Appl. Phys. Lett.* **68**, 2366-2368.
14. Sutherland, D.G.J., Akatsu, H., Copel, M., and Himpsel, F.J. (1995) Stoichiometry reversal in the growth of thin oxynitride films on Si(100) surfaces, *J. Appl. Phys.* **78**, 6761-6769.
15. Lu, Z.H., Hussey, R.J., Graham, M.J., Cao, R., and Tay, S.P. (1996) Rapid thermal N_2O oxynitride on Si(100), *J. Vac. Sci. Technol.* **B14**, 2882-2887.
16. Vickridge, I. and Amsel, G. (1990) SPACES: A PC Implementation of the Stochastic Theory of Energy Loss for Narrow-Resonance Depth Profiling, *Nucl. Instrum. Meth.* **B45**, 6-12.
17. Trimaille, I. and Rigo, S. (1989) Use of ^{18}O Isotopic Labelling to Study Thermal Dry Oxidation of Silicon as a Function of Temperature and Pressure, *Appl. Surf. Sci.* **39**, 65-80.
18. Ellis, K.A. and Buhrman, R.A. (1996) The removal of nitrogen during boron indiffusion in silicon gate oxynitrides, *Appl. Phys. Lett.* **69**, 535-537.
19. Arakawa, T., Hayashi, T., Ohno, M., Matsumoto, R., Uchiyama, A., and Fukuda, H. (1995) Relationship between nitrogen profile and reliability of heavily oxynitrided tunnel oxide films for flash electrically erasable and programmable ROMs, *Jpn. J. Appl. Phys.* **34**, 1007-1015.

SURFACE, INTERFACE AND VALENCE BAND OF ULTRA-THIN SILICON OXIDES

Takeo Hattori
Department of Electrical and Electronic Engineering,
Musashi Institute of Technology
1-28-1 Tamazutsumi, Setagaya-ku, Tokyo 158, Japan

1. Introduction

In 1987, silicon-based metal-oxide-semiconductor field-effect transistors (MOSFETs) with gate oxide film thickness of 3.3 nm were shown to operate at liquid nitrogen temperature.[1] As a result of continuing progress in microfabrication technology since that time, MOSFETs with gate oxide film thickness of 1.5 nm were shown to operate at room temperature[2] and the fabrication of 1 Gbit dynamic random access memory (DRAM) was reported.[3,4] The mass production of 1 Gbit DRAM using MOSFETs with gate oxide film thickness of about 5 nm must be realized at the beginning of the next century. In this case the thickness of one-molecular-layer of SiO_2 corresponds to 6 % of gate oxide film thickness. Therefore, it is necessary to control the formation of SiO_2 and SiO_2/Si interface on an atomic scale by improving the cleanliness and flatness of Si surfaces before the oxidation The control of oxide formation on an atomic scale is important for the formation of high quality SiO_2/Si interfaces for future metal-oxide-semiconductor (MOS) technology .[5]

Because the formation of native oxide can be suppressed by terminating Si surface with hydrogen atoms,[6,7] hydrogen-terminated Si surface (abbreviated as H-Si surface herafter) must be used instead of a clean Si surface for this purpose. Here, an atomically flat H-Si(111)-1 × 1 surface can be obtained by the treatment[8] in 40% NH_4F solution or that[9] in boiling water, while an atomically flat H-Si(100)-2 × 1 surface can be obtained by the annealing[10] in a hydrogen atmosphere under pressures higher than 0.2 Torr at 700℃ or by the hydrogen termination and surface reconstruction[11] of (100) silicon annealed at high temperature (>1100℃) in a H_2 atmosphere at 1 bar. If the H-Si can be oxidized without breaking Si-H bonds, the flatness of the Si surface must be roughly preserved. It was found from the study on the stability of nearly one monolayer thick oxide (abbreviated as preoxide hereafter) formed on Si(100) surface at 300°C without breaking Si-H bonds that preoxide is stable up to 900°C in dry argon gas.[12]

The oxidation studies described in the following chapters were prepared as follows. The nearly 0.5 nm thick preoxides were formed at 300°C in 1-4 Torr dry oxygen by oxidizing atomically flat hydrogen-terminated Si(111)-1 × 1 and Si(100)-2 × 1 surfaces. Through these preoxides the oxidation at 600-900°C in 1 Torr dry oxygen were performed. A rather high pressure of oxygen was used in order to minimize the effect of impurities in the oxidizing atmosphere on the oxidation process. The amount of water vapor in oxygen gas used was below 37 ppb. In order to heat Si wafers in oxygen under high pressure Si wafers were only heated optically.

Interface and valence band structures of ultra-thin silicon oxides were studied from the measurement of photoelectron spectra excited by monochromatic AlK α radiation with an acceptance angle of 3.3 degrees, using ESCA-300 manufactured by Scienta Instrument AB,[13] while surface structures of ultra-thin silicon oxides were studied from the observation of noncontact-mode atomic force microscope (NC-AFM) images with a force constant of 39 N/m and resonant frequency of about 300 kHz, and a single-crystalline silicon probe, using instrument manufactured by OMICRON Vakuum Physik GmbH.[14] Other experimental details and analytical procedure of Si 2p photoelctron spectra were described elsewhere.[15]

Recently, it was found that the $SiO_2/Si(111)$ interface structures are abrupt on an atomic scale and changes periodically with progress of oxidation.[16] Using this finding as a guide, it will be shown in the following that the surface and valence band structures of ultra-thin oxides are affected by the interface structures.

2. SiO_2/Si Interface Structures

2.1. STRUCTURAL ORIGIN

Figure 1 shows changes in Si $2p_{3/2}$ spectra with oxide film thickness (abbreviated as thickness hereafter)as a parameter. In this figure the spectral intensities of the silicon substrate are adjusted to be equal to each other in order to show the oxidation-induced changes in interface structure consisting of intermediate oxidation states (abbreviated as intermediate states hereafter)[17], Si^{1+}, Si^{2+} and Si^{3+}. Here, Si^{1+}, Si^{2+} and Si^{3+} denote a Si

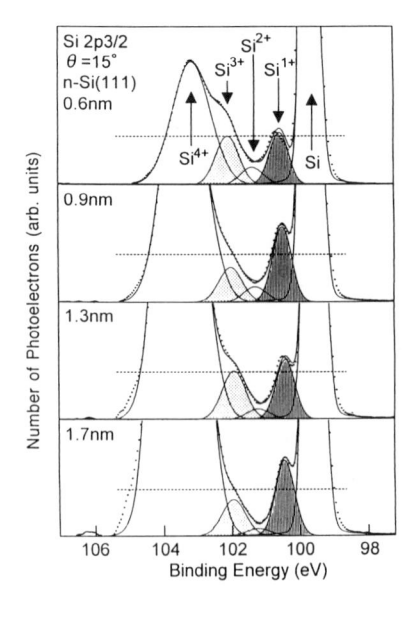

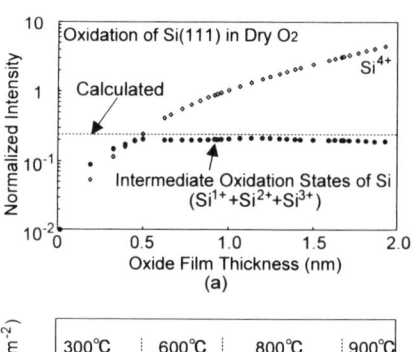

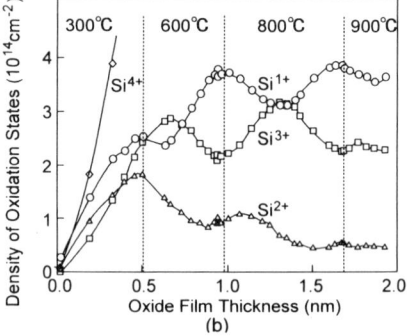

Fig. 1. Oxidation induced changes in Si 2p spectra obtained for photoelectron take-off angle of 15 degrees with oxide film thickness as a parameter. The dashed line on each spectrum shows the average of the amounts of Si^{1+} and Si^{3+}.

Fig. 2. (a) Normalized spectral intensity of Si^{4+} and summation of intensities for all intermediate oxidation states are shown as a function of oxide film thickness. (b) Dependence of areal densities of Si^{1+}, Si^{2+}, Si^{3+} and Si^{4+} on thickness.

atom bonded to one oxygen atom and three Si atoms, a Si atom bonded to two oxygen atoms and two Si atoms and a Si atom bonded to three oxygen atoms and one Si atoms, respectively. According to Fig. 1, the amounts of Si^{1+} and Si^{3+} for two thicknesses of 0.6 and 1.3 nm, and those for two thicknesses of 0.9 and 1.7 nm are almost the same. This implies that the interface structure changes periodically with the progress of oxidation. Figure 2(a) shows the spectral intensity of Si^{4+} and that of intermediate oxidation states normalized by the Si 2p spectral intensity of silicon substrate as a function of thickness. Here, Si^{4+} denotes a silicon atom bonded with four oxygen atoms. According to Fig. 2(a), the normalized spectral intensity of intermediate states saturates at a thickness of nearly 0.5 nm, while the normalized spectral intensity of Si^{4+} does not saturate at this thickness. Furthermore, the saturated normalized spectral intensity of intermediate states is in good agreement with that calculated for an abrupt compositional transition at the interface[15], which is shown by the dashed line in Fig. 2(a). Here, an abrupt compositional transition is defined by the existence of intermediate states only at the interface. Therefore, once the interface is formed, the amount of intermediate oxidation states is not affected by the further oxidation. The abrupt compositional transition was also realized on Si(100) surface by the oxidation at 600-800°C through nearly 0.5 nm thick preoxided formed by the oxidation of H-Si(100)-2 × 1 at 300°C.[18] Figure 2(b) shows the areal densities of Si^{1+}, Si^{2+}, Si^{3+} and Si^{4+} as a function of thickness. According to this figure, although the total amount of intermediate oxidation states does not changes with progress of oxidation, the areal density of Si^{1+} and that of Si^{3+} repeatedly increase and decrease with a period in thickness of nearly 0.7 nm for thickness less than 1.7 nm. Furthermore, with progress of oxidation, the areal density of Si^{1+} changes in opposite phase with the areal density of Si^{3+}. These findings clearly demonstrate that the interface structure changes periodically with progress of oxidation for thickness less than 1.7 nm.

If the oxidation reaction occurs at atomically flat interface, the SiO_2/Si(111) interface structures consisting of Si^{1+} and Si^{3+} should appear alternately as a result of periodic change in bonding nature of Si crystal at the interface. Here, the period in thickness of 0.7 nm in Fig. 2(b) is in excellent agreement with the thickness of two molecular layers of silicon dioxide calculated assuming oxidation-induced volume expansion factor of bulk Si.[16] Therefore, at 800°C, oxidation reactions occur monolayer by monolayer at the interface until 1.7-nm-thick oxide film is formed. For atomically flat interfaces, the minimum amounts of Si^{1+} and Si^{3+} must be zero and the amount of Si^{2+} must be close to zero at every stage of oxidation. However, this is not the case. In order to explain the coexistence of Si^{1+} and Si^{3+} in Figs. 1 and 2(b), it is necessary to consider the existence of monoatomic steps at the interface at every stage of oxidation. The decrease in areal density of Si^{2+} with the increase in thickness in Fig. 2(b) implies that the monoatomic step density decreases with the progress of oxidation. In order to explain Fig. 2(b), it is necessary to consider non-uniform oxide film as shown in Fig. 3. If we describe this distribution of thickness by Gaussian function[19], Fig. 4 is obtained for various full width half maximum(FWHM). Best fit of thickness dependence of calculated density of oxidation states to experimental data in Fig. 2(b) is obtained for FWHM of 0.5 nm. The almost same amount of changes in Si^{1+} and Si^{3+} are obtained at thickness of 1.0 and 1.7 nm. This implies that the non-uniformity is not enhanced by the subsequent oxidation.

2.2. LATERAL SIZE OF ATOMICALLY FLAT INTERFACE

The lateral size of an atomically flat oxidized region on Si(111) surface was determined as follows from the effect of terrace width on the layer-by-layer oxidation.[20] For this

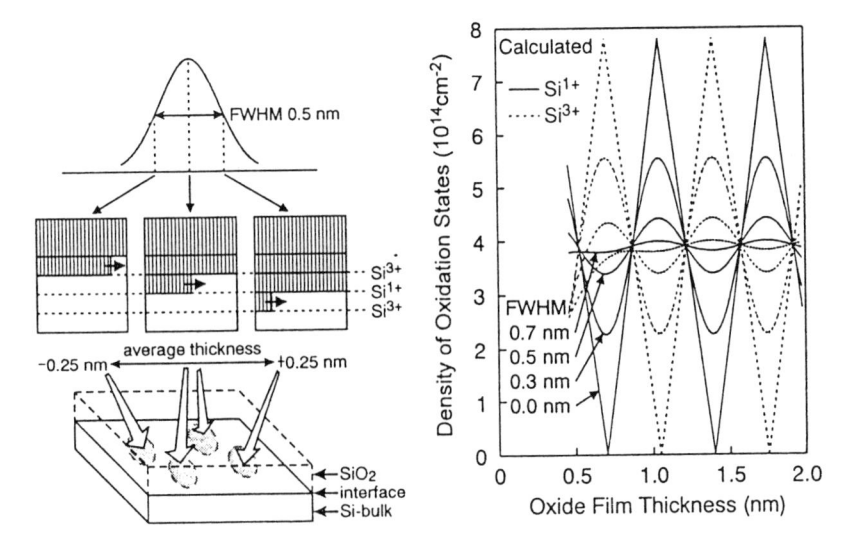

Fig. 3. Distribution of oxide film thickness expressed by Gaussian function.

Fig. 4. Density of oxidation states calculated for Gaussian distribution of oxide film thickness with FWHM as a parameter.

study the vicinal(111) 0.02-0.03 degrees, 3degrees and 6 degrees surfaces with terrace width of 360-600 nm, 6 nm and 3 nm, which are illustrated in Fig. 5, respectively, were used. The spectral intensity of Si^{4+} and that of intermediate oxidation states normalized by the Si $2p_{3/2}$ spectral intensity of silicon substrate as a function of thickness are shown for three kinds of vicinal surfaces in the upper part of Fig. 6. According to this figure, the saturated normalized spectral intensities of intermediate states are in good agreement with those calculated for an abrupt compositional transition at the interface,

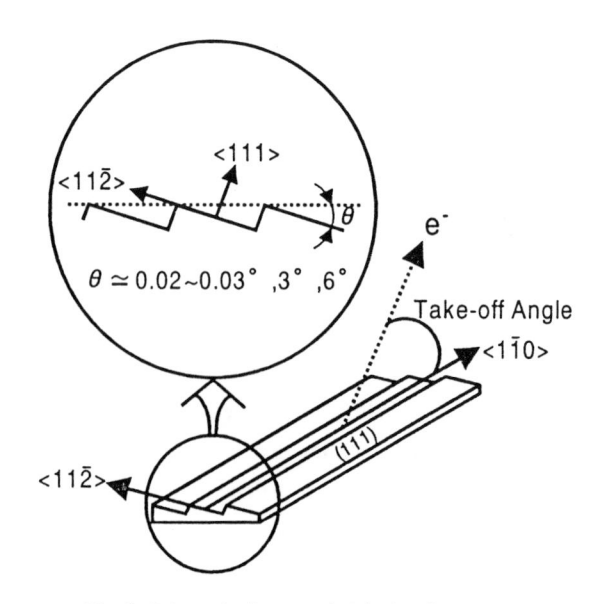

Fig. 5. Schematic diagram of vicinal surfaces.

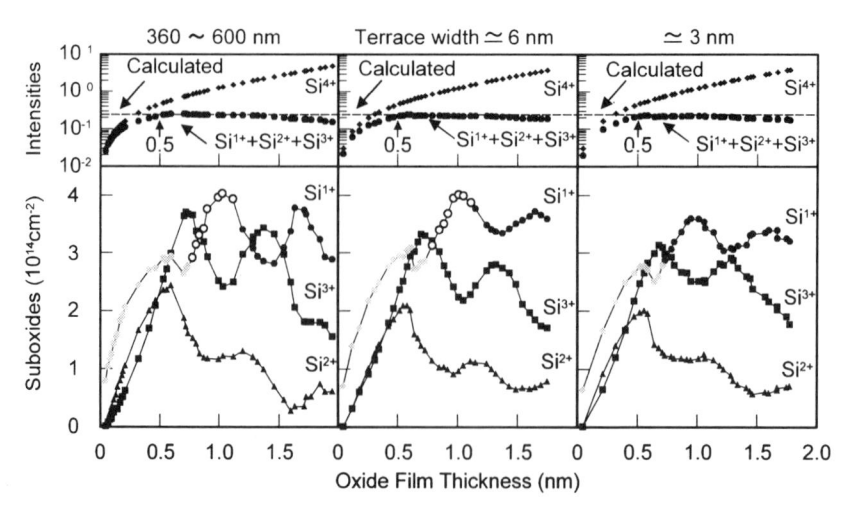

Fig. 6. Upper part shows normalized spectral intensity of Si^{4+} and sum of intensities of all intermediate oxidation states are shown as functions of oxide film thickness, while lower part shows the dependence of densities of Si^{1+}, Si^{2+} and Si^{3+} on thickness for three kinds of vicinal surfaces.

which are shown by the dashed lines in this figure.[16] Because almost the same results are obtained for three kinds of vicial surfaces, the atomic steps on the Si surface before the oxidation do not affect the abrupt compositional transition . The areal densities of Si^{1+}, Si^{2+}, Si^{3+} and Si^{4+} as a function of thickness are shown in the bottom of Fig. 6. On vicinal (111) 0.02-0.03 degrees, the oxidation reaction occurs layer-by-layer at the interface, for thickness less than 1.7 nm.

The effect of terrace width on the interface structures appears on the thickness dependence of amount of Si^{1+} in the thickness range from 0.7 to 1.0 nm because of following two reasons: 1) In the thickness range from 0.7 nm to 1.0 nm with progress of oxidation the amount of Si^{1+} increase, while that of Si^{3+} decrease. Therefore, in this thickness range the formation of Si^{1+} interface proceeds with progress of oxidation; 2) On the vicinal (111) surfaces there are double atomic steps of silicon, because silicon monohydrides can be only observed on these vicinal H-Si surfaces.[21] The oxidation of two Si layers from the top produces nearly 0.7 nm thick silicon oxide, which is equal to the the period in thickness observed for the periodice changes in interface structures.

Then, until the thickness reaches 0.7 nm, the oxidation of third silicon layer from the top can not interfere with the oxidation on adjacent terrace on vicinal (111) 3 and 6 degrees. This results in the almost same results below the thickness of 0.7 nm for three kinds of vicinal surfaces. However, on vicinal (111) 6 degrees the size of oxidized region , which is produced by layer-by-layer oxidation reaction at the interface, is limited

Terrace width ≃ 6 nm

Si^{1+} → ≃0.7nm <111>

Thickness　　0.7 nm　　0.7 nm ~ 1.1 nm　　< 1.1 nm

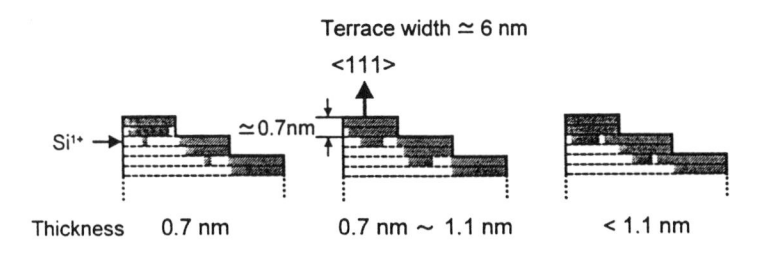

Fig. 7 Oxidation process on vicinal(111) 3 degrees surface.

by the terrace width of 3 nm and the oxidation of the third silicon layer from the top starts to interfere with the oxidation on adjacent terrace above the thickness of 0.7 nm, because the thickness dependence of amount of Si^{1+} for vicinal (111) 6 degrees starts to deviate from that for vicinal (111) 0.02 - 0.03 degrees above the thickness of 0.7 nm. On the other hand, on vicinal (111) 3 degrees the size of oxidized region at the thickness of 0.7 nm is smaller than the terrace width of 6 nm and the oxidation of the third silicon layer from the top starts to interfere with the oxidation on adjacent terrace above the average thickness of 1.1 nm as shown in Fig. 7, because the thickness dependence of amount of Si^{1+} for vicinal (111) 6 degrees starts to deviate from that for vicinal (111) 0.02 - 0.03 degrees above the thickness of 1.1 nm. Therefore, the lateral size of an atomically flat oxidized region on Si(111) surface must be in the rage from 3 to 6 nm at the thickness of 0.7 nm.

3. Oxidation-induced Surface Microroughness

Recently, it was found that the surface microroughness changes in accordance with the changes in interface structure.[22] In order to investigate the correlation between surface and interface structures of ultra-thin oxides formed on Si(111) surface, we measured non-contact mode atomic force microscope (NC-AFM) images and Si 2p photoelectron spectra with progress of oxidation. Here, the effect of atomic steps on the surface microroughness was minimized by using vicinal (111) 0.017 degrees from <110> and 0.008 degrees from <112> with an average terrace area of about 1 μ m × 2 μ m.

Figure 8 shows averaged values of two kinds of surface microroughness, Ra and Rms, of oxide film, measured at three positions on the oxide surface, as a function of thickness. It can be seen from this figure that the surface microroughness changes periodically with the progress of oxidation and can be correlated with the periodic changes in the amount of Si^{1+} as explained in the following. As illustrated in Fig. 9, in the case of forming Si^{1+} the insertion of an oxygen atom between two Si atoms consisting of a Si-Si bond oriented along the <111> direction at the interface expands the oxide network only along the <111> direction, while in the case of forming Si^{3+} the insertion of an oxygen atom between two Si atoms consisting of a Si-Si bond at the interface expands the oxide network mostly along the direction perpendicular to <111>. Therefore, the formation of Si^{1+} at the interface results in the increase in surface microroughness caused by the formation of protrusions on the oxide surface.

Figure 10 shows the surface morphology of oxide films with thicknesses of 1.0 and 1.7 nm measured over an area of 200 nm × 200 nm. According to this figure, small protrusions with a lateral size of about 5 nm diameter are present on the 1-nm-thick oxide film, while ones with a lateral size of about 20 nm diameter are observed on the 1.7-nm-thick oxide film. These observations support the appearance of oxidation-induced protrusions discussed above. The

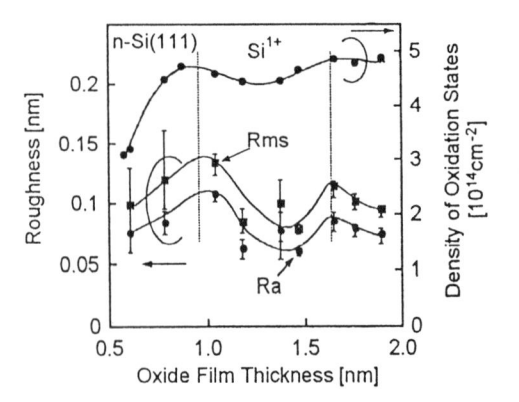

Fig. 8. Two kinds of surface microroughness of oxide formed on Si(111) and the density of Si^{1+} shown as functions of oxide film thickness.

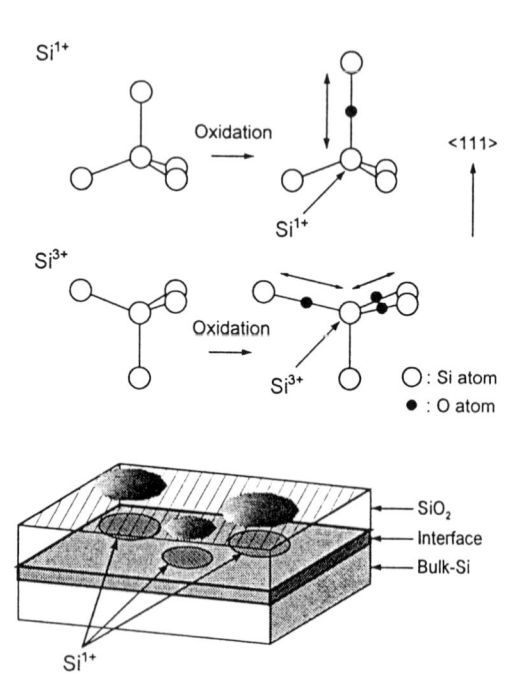

Fig. 9. Upper part shows direction of volume expansion produced by the formation of Si^{1+} and Si^{3+} at the interface. Lower part shows a diagram of relation between the oxidation-induced protrusions and the interface structure consisting of Si^{1+}.

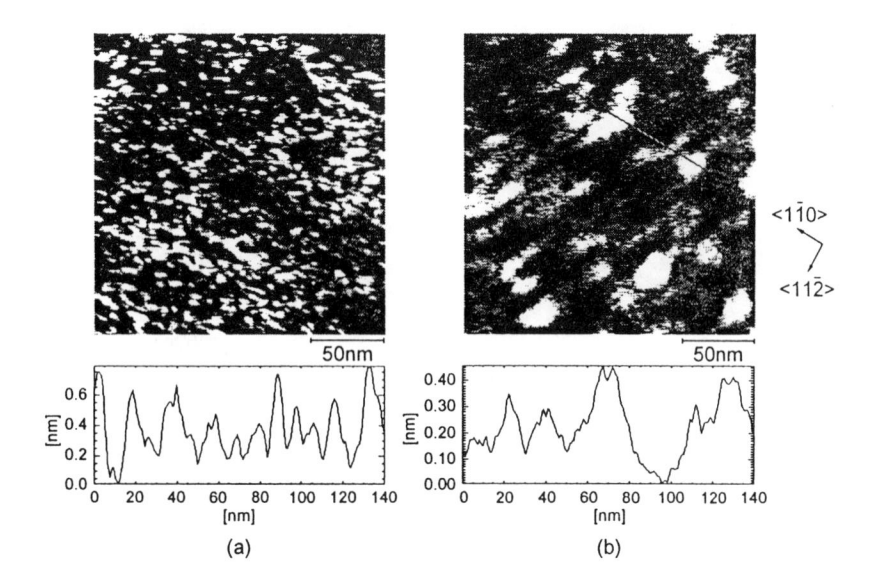

Fig. 10. AFM images and cross section of oxide films with thicknesses of (a) 1.0 nm an (b) 1.7 nm. Cross-sectional profiles were obtained along the lines shown in the AFM images.

lateral size of protrusions at a thickness of 1.0 nm is close to the lateral size of the atomically flat oxidized region determined by the effect of terrace width on the layer-by-layer oxidation described in chapter 2.2.

The same kind of study was also performed on Si(100) surface.[23] Figure 11 shows the surface morphology of oxide films with thicknesses of 0.8 and 1.2 nm measured over an area of 200 nm × 200 nm. The upper part of Fig. 8 shows no significant change in surface morphology produced by the change in thickness from 0.8 to 1.2 nm, while the lower part of this figure shows an extremely smooth surface. Two kinds of surface microroughness, Rms and Ra, are shown in Fig. 12 as a function of thickness. According to this figure, surface microroughness repeatedly increase and decrease with a period in thickness of 0.18 nm. This period in thickness is in good agreement with the periodic increase in thickness observed for the oxidation of Si(100) surface in pure water at room temperature.[24] This periodic change in surface microroughness implies the layer-by-layer oxidation. Figure 12 also shows the gradual increase in the surface microroughness with the increase in thickness, which must be produced by the gradual increase in interface roughness with the increase in thickness of thermal oxide film.[25]

In order to correlate the surface microroughness with the interface structure, the change in interface structure was measured with progress of oxidation. The upper part of Fig. 13

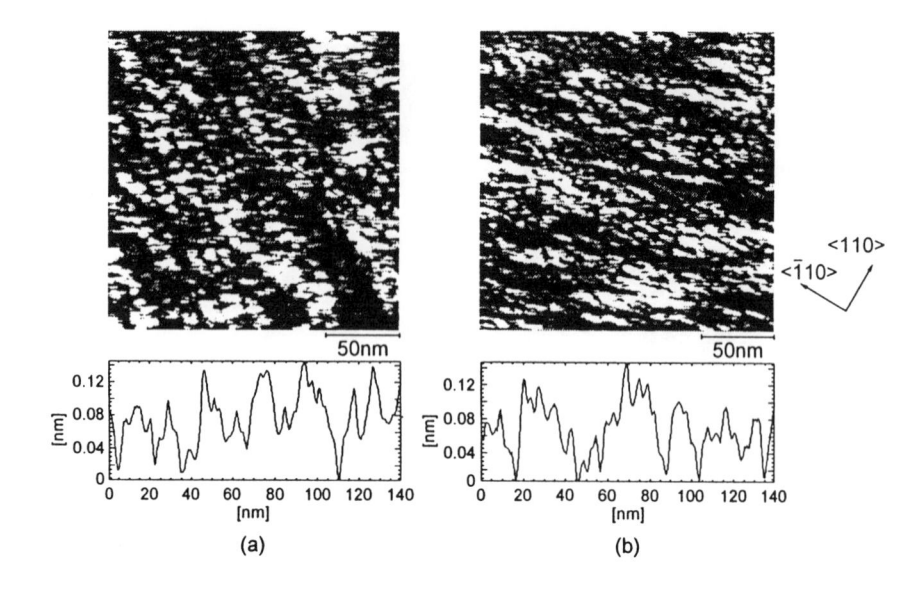

Fig. 11. AFM images and cross section of oxide films with thicknesses of (a) 0.8 nm and (b) 1.2 nm. Cross-sectional profiles were obtained along the lines in the AFM images.

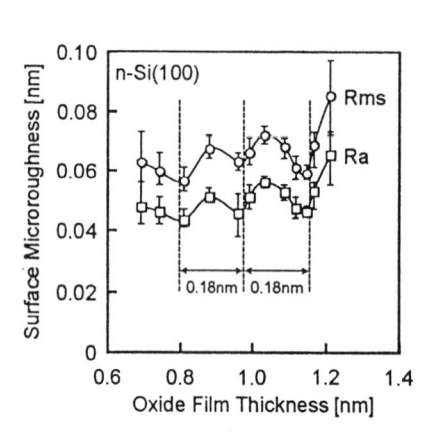

Fig. 12. Two kinds of surface microroughness of oxide formed on Si(100) and the density of Si^{1+} shown as functions of oxide film thickness.

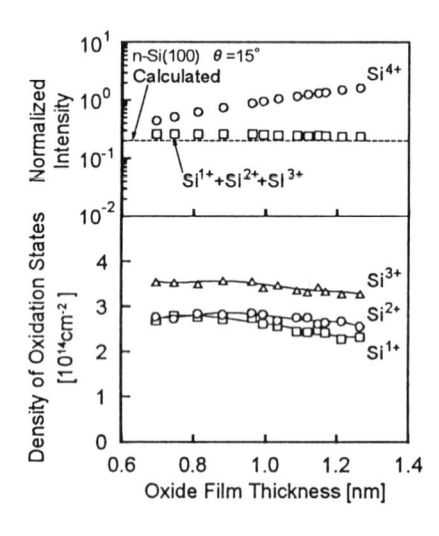

Fig. 13. Upper part shows dependence of normalized spectral intensity of Si^{4+} and summation of intermediate states on oxide film thickness, while lower part shows dependence of areal densities of Si^{1+}, Si^{2+}, Si^{3+} and Si^{4+} on oxide film thickness.

shows the Si 2p spectral intensity of Si^{4+} and that of intermediate states consisting of Si^{1+}, Si^{2+} and Si^{3+} normalized by the spectral intensity of Si substrate as a function of thickness. According to this upper part of Fig. 13, the saturated normalized spectral intensity of suboxides is slightly larger than that calculated, which is shown by the dashed line in Fig. 13, for an abrupt interface[26] consisting only of Si^{2+}. The lower part of Fig. 13 shows the areal densities of Si^{1+}, Si^{2+} and Si^{3+} as a function of thickness. According to the lower part of Fig. 13, the interface consists not only of Si^{2+}, but also of Si^{1+} and Si^{3+}. Therefore, the interface structure must be the combination of the interface consisting of Si^{2+}, the interface[27] consisting of Si^{1+} and Si^{2+}, the interface[28] containing Si^{3+} and the interface[29] consisting of Si^{1+} or Si^{2+}. Therefore, the layer-by-layer oxidation must occurs locally.

4. Valence Band Discontinuities at near the SiO$_2$/Si Interface

More than thirty years ago Williams[30] determined the energy band diagram of Si-SiO$_2$ system using internal photoemission. Later, it was found from the study [31] of the impurity-effect on the energy band discontinuity at SiO$_2$/Si interface formed on Si(100) surface that the impurity-induced decrease in conduction band(abbreviated as the C. B. hereafter) discontinuity and the impurity-induced increase in valence band(abbreviated as the V. B. hereafter) discontinuity are almost the same. Namely, they found that the C. B. discontinuity of 3.5 eV at the interface for the ultra-cleanly prepared oxide is larger than that of 3.2 eV for the conventionally prepared oxide, while the V. B. discontinuity of 4.5 eV at the interface for the ultra-cleanly prepared oxide is smaller than that of 4.7eV for the conventionally prepared oxide. The impurity effect on the conventionally prepared interface is equivalent to the effect of terminating dangling bonds with hydrogen atoms because almost the same C. B. and V. B. offset was obtained if the ultra-cleanly prepared oxide is annealed in the mixture of hydrogen and argon gases. Therefore, the increase in C. B. discontinuity and the decrease in V. B. discontinuity at the interface observed for the ultra-cleanly prepared oxide must be correlated with dangling bond-induced dipole layer. The results are summarized in Table 1. The impurity-induced stabilization of SiO$_2$ on Si(100) [33] must be also caused by the termination of dangling bonds by hydrogen atoms. It will be shown in the following that V. B. discontinuities exist not only at the interface, but also in the oxide near the interface.[34]

The changes in V. B. spectra of oxide formed on Si(111) surface with the progress of oxidation are shown in Fig. 14 and analyzed by the following two kinds of analytical

Table 1. Conduction and valence band discontinuity at SiO$_2$/Si(100) interface

Oxidation	Discontinuity[eV] C.B.	V.B.	Sum[eV]	m*/m$_0$	Reference
Conventional	3.2	4.7	7.9	0.42	31)
Superclean	3.5	4.5	8.0	0.42	31)
a)		4.7			31)
b)		5.1			31)
	3.25	4.49	7.74	0.34	32,33)

a) Superclean oxidation followed by annealing in H$_2$/Ar, 450℃, 30min
b) Superclean oxidation followed by annealing in N$_2$, 800℃, 60min

procedures. First analytical procedure is to take difference in V. B. spectrum measured for two thicknesses, which is close enough to each other, so as to eliminate the V. B. spectrum of Si substrate. From this analysis the V. B. spectrum of oxide surface can be obtained. Namely, if we assume that the V. B. discontinuity does not depend on the interface structure, but depend on the distance from the interface, the change in V. B. spectrum produced as a result of small increase in oxide film thickness corresponds to the V. B. spectrum of the oxide near the surface. Second analytical procedure is to take difference in V. B. spectra measured for two photoelectron take off angles, that is, 15 and 90 degrees, so as to eliminate the V. B. spectrum of Si substrate. Because the electron escape depth in silicon dioxde for these two photoelectron take off angles is 0.9 and 3.4 nm, respectively, the relative contribution of silicon oxide and silicon to the V. B. spectrum is different for two photoelectron take offangles. Therefore, from the difference in two spectra for two take off angles, the V. B. spectrum of silicon oxide existed mostly within 0.9 nm from the oxide surface can be obtained.

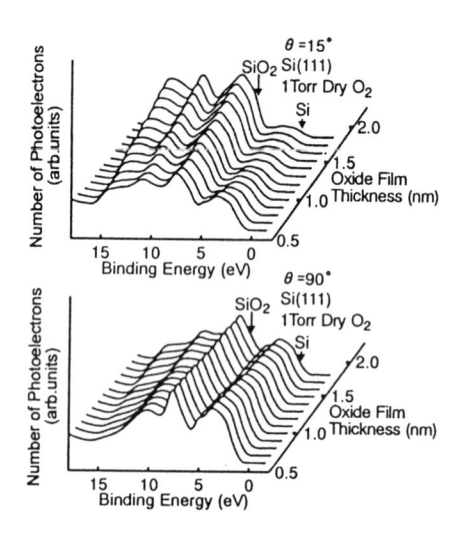

Fig. 14. Changes in valence band spectra of silicon oxide formed on Si(111) surface with the progress of oxidation for two photoelectron take-off angles.

Figure 15(a) shows the difference in energy between the top of V. B. of oxide surface and that of silicon as a function of thickness determined from the first procedure with the assumption that density of state near the V. B. edge follows parabolic dependence on the binding energy. Here, the thickness is an average of two thicknesses, where difference between two spectra is taken. According to this figure, the top of valence band of oxide surface increases by about 0.2 eV near the thickness of 0.9 nm. In order to explain the spectral difference it is assumed that the top of V. B. of oxide surface changes abruptly at the thickness of 0.9 nm as indicated by the dashed line in Fig. 15(a). This assumption implies that the oxide layer consists of transition layer and bulk layer (referred as two layer model hereafter). The energy of top of V. B. of silicon oxide within 0.9 nm from the interface can be determined also from the second procedure and is also shown in Fig. 15(a). The critical thickness of 0.9 nm above which the energy level of top of V. B. changes abruptly is very close to the thickness of structural transition layer determined from the analysis of infrared absorbance and X-ray reflectance.[36, 37] Therefore, the present results suggest that the oxidation-induced stress in the interfacial transition layer produces the change in the energy level of top of V. B. of silicon oxide near the interface. The difference in V. B. spectra obtained from the second procedure can be explained if the two layer model described above and the non-uniform oxide film expressed by Gaussian function with FWHM of 0.5 nm are considered. If we see Fig. 15(a) in details, the difference between the experimental data and dashed line exhibits the periodic changes in the energy level of top of V. B. with progress of oxidation, which exactly corresponds to the periodic changes in interface structures.[16] The difference in energy between the top of V. B. of oxide surface and that of silicon determined for Si(100) is also shown in Fig. 15(b). The critical distance from the interface, above which the energy level of V. B. edge of oxide agrees with that

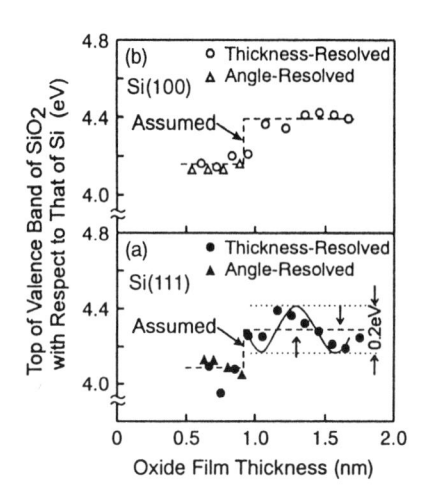

Fig. 15. Changes in the top of valence band edge of oxides formed on Si(111) and Si(100) surfaces, which were determined using two kinds of analytical procedures described in the text, are shown in (a) and (b), respectively, as a function of oxide film thickness.

of bulk SiO$_2$, for Si(111) and Si(100) are almost equal to each other. The saturated values of V. B. discontinuities in Fig.15 almost agrees with that obtained by Alay et al. [33,38]

5. Summary

The studies on the surface, interface and valence band structures of ultra-thin silicon oxides at the initial stage of oxidation are reviewed. On Si(111) surface the periodic changes in interface structures appear as a result of bonding nature of Si crystal at the interface. An abrupt compositional transitions can be realized on Si(111) and (100) surfaces and are weakly affected by the atomic steps on the initial surface. The lateral size of atomically flat oxidized region on Si(111) surface is 3-6 nm at the thickness of 0.7 nm and is roughly agrees with that observed using non-contact mode atomic force microscope. The surface microroughness of oxide surface on Si(111) surface was found to change periodically with progress of oxidation. This changes can be correlated with the changes in interface structures. Namely, the formation of Si^{1+} results in the formation of protrusions on oxide surface. The periodic changes in surface microroughness was also found on Si(100) surface and must arise from the monolayer by monolayer growth of thermal oxide. The top of valence band of silicon oxide within 0.9 nm from the interface was found to be different from that of bulk SiO$_2$. Furthermore, it was found that on Si(111) surface the valence band discontinuity at the interface is affected by the interface structure.

Acknowledgements

The present work was partially supported by Ministry of Education, Science, Sports and Culture through a Grant -in-Aid for Scientific Research on Priority Areas, "Ultimate Integration of Intelligence on Silicon Electronic Systems" (Head Investigator: Tadahiro Ohmi, Tohoku University).

References

1. Sai-Halasz, G. A., Wordeman, M. R., Kern, D. P., Ganin, E., Rishton, S., Zicherman, D. S., Schmid,

H., Polcari, M. R., Ng, H. Y., Restle, P. J., Chang, T. H. and Dennard, R. H. (1987) Design and Experimental Technology for 0.1- μ m Gate-Length Low-Temperature Operation FET's, *IEEE Electron. Device Lett.* **8**, 463-466.

2. Sasaki Momose, H., Ono, M., Yoshitomi, T., Ohguro, T., Nakamura, S., Saito, M. and Iwai, H. (1994) Tunneling gate oxide approach to ultra-high current drive in small-geometry MOSFETs, *IEDM* 94, pp. 593-596.
3. Horiguchi, M., Sakata, T., Sekiguchi, T., Ueda, S., Tanaka, H., Yamasaki, E., Nakagome, Y., Aoki, M., Kaga, T., Ohkura, M., Nagai, R., Murai, F., Tanaka, T., Iijima, S., Yokoyama, N., Gotoh, Y., Shoji, K., Kisu, T., Yamashita, H., Nishida, T. and Takeda, E. (1995) An experimental 220 MHz 1Gb DRAM, *Int..Solid-State Circuits Conf., Digest of Technical Papers*, pp. 252-253.
4. Sugibayashi, T., Naritake, I., Utsugi, S., Shibahara, K., Oikawa, R., Mori, H., Iwao, S., Murotani, T., Koyama, K., Fukazawa, S., Itani, T., Kasama, K., Okuda, T., Ohya, S. and Ogawa, M. (1995) A 1Gb DRAM for file applications, *IEEE Int. Solid-State Circuits Conf., Digest of Technical Papers*, pp. 254-255.
5. Hattori, T. (1995) Chemical structures of the SiO_2/Si interface, *Critical Rev. Solid State Mat. Sci.* **20**, 339-382.
6. Takahagi, T., Nagai, I., Ishitani, A. and Kuroda H. (1988) The formation of hydrogen passivated silicon single-crystal surfaces using ultraviolet cleaning and HF etching, *J. Appl. Phys.* **64**, 3516-3521.
7. Sakuraba, M., Murota, J. and Ono, S. (1994) Stability of the dimer structure formed on Si(100) by ultraclean low-pressure chemical-vapor deposition, *J. Appl. Phys.* **75**, 3701-3703.
8. Higashi, G. S., Becker, R. S., Chabal, Y. J. and Becker, A. J. (1991) Comparison of Si(111) surfaces prepared using aqueous solutions of NH_4F versus HF, *Appl. Phys. Lett.* **58**, 1656-1658.
9. Watanabe, S., Nakayama, N. and Ito, T. (1991) Homogeneous hydrogen-terminated Si(111) surface formed using aqueous HF solution and water, *Appl. Phys. Lett.* **59**, 1458-1460.
10. Aoyama, T., Goto, K., Yamazaki, T. and Ito, T. (1996) Silicon (001) surface after annealing in hydrogen ambient, *J. Vac. Sci. Technol.* **A14**, 2909-2915.
11. Bender, H., Verhaverbeke, S., Caymax, M., Vatel, O. and Hynes, M. M. (1994) Surface reconstruction of hydrogen annealed (100) silicon, *J. Appl. Phys.* **75**, 1207-1209.
12. Ohmi, T., Morita, M., Teramoto, A., Makihara, K. and Tseng, K. S. (1992) Very thin oxide film on a silicon surface by ultraclean oxidation, *Appl. Phys. Lett.* **60**, 2126-2128.
13. Gelius, U., Wannberg, B., Baltzer, P., Fellner-Feldegg, H., Carlsson, G., Johansson, C. -G., Larsson, J., Munger, P. and Vergerfos, G. (1990) A new ESCA instrument with improved surface sensitivity, fast imaging properties and excellent energy resolution, *J. Electron Spectrosc. Relat. Phenom.* **52**, 747-785.
14. Guthner, P. (1996) Simultaneous imaging of Si(111)7 × 7 with atomic resolution in scanning tunneling microscopy, atomic force microscopy, and atomic force microscopy noncontact mode, *J. Vac. Sci. & Technol.* **B14**, 2428-2431.
15. Nohira, H., Tamura, Y., Ogawa, H. and Hattori, T. (1992) Initial stage of SiO_2/Si interface formation on Si(111) surface, *IEICE Trans. Electron.* **E75-C**, 757-763.
16. Ohishi, K. and Hattori, T. (1994) Periodic changes in SiO_2/Si(111) interface structures with progress of thermal oxidation, *Jpn. J. Appl. Phys.* **33**, L675-L678.
17. Hollinger, G. and Himpsel, F. J. (1984) Probing the transition layer at the SiO_2-Si interface using core level photoemission, *Appl. Phys. Lett.* **44**, 93-95.
18. Aiba, T., Yamauchi, K., Shimizu, Y., Tate, N., Katayama, M. and Hattori, T. (1995) Initial stage of oxidation of hydrogen-terminated Si(100)-2 × 1 surface, *Jpn. J. Appl. Phys.* **34**, 707-711.
19. Ohishi, K. and Hattori, T. (unpublished).

20. Omura, A., Sekikawa, H. and Hattori, T. (1997) Lateral size of atomically flat oxidized region on Si(111) surface, *Appl. Surf. Sci.* **117/118**, 127-130.
21. Lyo, I.-W., Avouris, Ph., Schubert, B. and Hoffmann, R. (1990) Elucidation of the initial stage of the oxidation of Si(111) using scanning tunneling microscopy and spectroscopy, *J. Phys. Chem.* **94**, 4400-4403.
22. Ohashi, M. and Hattori, T. (1997) Correlation between surface microroughness of silicon oxide film and SiO$_2$/Si interface structure, *Jpn. J. Appl. Phys.* **36**, L397-L399.
23. Hattori, T., Fujimura, M., Yagi, T. and Ohashi, M. (1997) Periodic changes in surface microroughness with progress of thermal oxidation of silicon, reported at *6th Int. Conf. on Formation of Semiconductor Interfaces*, Cardiff.
24. Yasaka, Y., Uenaga, S., Yasutake, H., Takakura, M., Miyazaki, S. and Hirose, M. (1992) Cleaning and oxidation of heavily doped Si surfaces, *Mater. Res. Soc. Symp. Proc.* **259**, 385-390.
25. Niwa, M., Kouzaki, T., Okada, K., Udagawa, M. and Sinclair, R. (1993) Atomic-order planarization of ultrathin SiO$_2$/Si(001) interfaces, *Jpn. J. Appl. Phys.* **33**, 388-394.
26. Pantelides, S. T. and Long, M. (1978) Continuous-random-network models for the Si-SiO$_2$ interface, *The Physics of SiO$_2$ and its Interface*, S. T. Pantelides, Ed., Pergamon, New York, pp. 339-343.
27. Herman, F., Batra, I.P. and Kasowski, R.V. (1978) Electronic structure of a model Si-SiO$_2$ interface, *The Physics of SiO$_2$ and its Interface*, S. T. Pantelides, Ed., Pergamon, New York, pp. 333-338.
28. Banaszak Holl, M. M., Lee, S. and McFeely, F. R. (1994) Core-level photoemission and the structure of the Si/SiO$_2$ interface : A reappraisal, *Appl. Phys. Lett.* **65**, 1097-1099.
29. Pasquarello, A., Hybertsen, M. S. and Car, R. (1995) Si 2p core-level shifts at the Si(001)-SiO$_2$ interface: a first-principles study, *Phys. Rev. Lett.* **74**, 1024-1027.
30. Williams, R. (1965) Photoemission of electrons from silicon into silicon dioxide, *Phys. Rev.* A**140**, 569-575.
31. Ohmi, T., Morita, M. and Hattori, T. (1988) Defects and impurities in SiO$_2$ interface for oxides prepared using superclean methods, *The Physics and Chemistry of SiO$_2$ and the Si-SiO$_2$ Interface*, Plenum Press, New York, pp. 413-419.
32. Yoshida, T., Imafuku, D., Alay, J. L., Miyazaki, S. and Hirose, M. (1995) Quantitative analysis of tunneling current through ultrathin gate oxides, *Jpn. J. Appl. Phys.* **34**, L903-L906.
33. Heimlich, C., Kubota, M., Murata, Y., Hattori, T., Morita, M. and Ohmi, T. (1990) ARUPS study of an impurity-induced stabilization of SiO$_2$ on Si(100), *Vacuum* **41**, 793-795.
34. Nohira, H. and Hattori, T. (1997) SiO$_2$ valence band near the SiO$_2$/Si(111) interface, *Appl. Surf. Sci.* **117/118**, 119-122.
35. Ishikawa, K., Ogawa, H., Oshida, S., Suzuki, K. and Fujimura, S. (1995) Thickness-deconvolved structural properties of thermally grown silicon dioxide film, *Ext. Abstr. of Int. Conf. on Solid State Devices and Materials*, Osaka, pp. 500-502.
36. Sugita, Y., Awaji, N. and Watanabe, S. (1996) Transient oxide layer at a thermally grown SiO$_2$/Si interface, interpreted based on local vibration and X-ray reflectivity, *Ext. Abstr. of Intern. Conf. on Solid State Devices and Materials*, Yokohama, pp. 380-382.
37. Alay, J. L., Fukuda, M., Bjorkman, C. H., Nakagawa, K., Sasaki, S., Yokoyama, S. and Hirose, M. (1995) Determination of valence band alignment at ultrathin SiO$_2$/Si interface by high-resolution X-ray photoelectron spectroscopy, *Jpn. J. Appl. Phys.* **34**, L653-L656.
38. Alay, J. L., Fukuda, M., Nakagawa, K., Yokoyama, S. and Hirose, M. (1995) The valence band alignment at ultra-thin SiO$_2$/Si(100) interfaces determined by high-resolution X-ray photoelectron spectroscopy, *Ext. Abstr. of Intern. Conf. on Solid State Devices and Materials*, Osaka, pp. 28-30.

LOW TEMPERATURE ULTRATHIN DIELECTRICS ON SILICON AND SILICON CARBIDE SURFACES: FROM THE ATOMIC SCALE TO INTERFACE FORMATION

PATRICK G. SOUKIASSIAN [*]

*Commissariat à l'Énergie Atomique, DSM - DRECAM - SRSIM, Bât. 462, Centre d'Etudes de Saclay, 91191 Gif sur Yvette Cedex, France
and
Département de Physique, Université de Paris-Sud, 91405 Orsay Cedex, France*

Abstract

Oxides, nitrides and oxynitrides are among the most important passivation layers of semiconductor surfaces and are generally grown through molecular (O_2, H_2O, N_2, NH_3, N_2O or NO) interaction with the surface. The reaction could be promoted by surface electronic modification using e.g. a catalyst, by elevated temperatures, by photoreaction using unmonochromatized synchrotron radiation, or by surface structure modification. Some of the latest developments in low temperature ultrathin dielectric growth on representative silicon and silicon carbide surfaces are presented. Such important issues as interface formation, atomic scale initial and self-propagating oxidation, influence of surface structure and composition, role of defects, oxide/oxynitride stoichiometry, and reaction micromechanisms are addressed in this review article. The presented investigations are based on photoelectron spectroscopies using second and third generation synchrotron radiation light sources and atom-resolved scanning tunneling microscopy experiments.

[*] Also: *Department of Physics, Northern Illinois University, DeKalb, IL 60115, U.S.A.*

1. Introduction and historical background

The formation of insulator/semiconductor interfaces is very important from both fundamental point of view and technological aspects. For silicon surfaces, the most common insulators are silicon oxides and nitrides, and more recently oxynitrides [1-3]. The actual trend in microelectronics is towards much higher integration densities which subsequently also requires the ability to grow very thin insulating layers which would be very useful in ULSI technology [1,2]. Oxides, nitrides and oxynitrides are among the most important passivation layers of semiconductor surfaces. They are grown through molecular (O_2, H_2O, N_2, NH_3, N_2O or NO) interaction with the surface generally maintained at elevated temperatures. However, achieving oxides (and other insulating thin films) formation at low temperatures generally results in products that are more resistant to radiation damages. Such an interesting characteristic could be very useful in providing electronic devices having lifetimes significantly increased and able to operate in hostile environments with many applications in advanced microelectronics.

Primarily due to the rather exceptional properties of silicon dioxide, silicon has been by far, the most commonly used semiconductor in the electronics industry until now [1-3]. Furthermore, SiO_2 could be grown on silicon surfaces through a simple process as thermal oxidation at atmospheric pressure with molecular oxygen or oxygen containing molecules (H_2O, NO, N_2O). The oxidation reaction is enhanced by temperature. The situation is generally very different for other semiconductors such as III-V compounds, which do not have high quality native oxides and mandatory requires other passivation layers. In this context, the situation of silicon carbide (SiC) is of special interest.

SiC is a IV-IV compound semiconductor having a wide band gap (from 2.3 eV to 3.3 eV depending on the crystallographic phase) and existing in hexagonal (α) and cubic (β) phases. In addition, SiC has a high thermal stability ($\approx 600°C$ instead of 150°C for Si), a breakdown electric field 10 times larger than Si and a high thermal conductivity (comparable to Cu). Its average merit factor is about ≈ 400 compared to 1 and 1.5 for silicon and gallium arsenide respectively [4-6]. This makes SiC a very promising semiconductor for high temperature, high power, high frequency and high voltage electronic devices and sensors [4-8]. Furthermore, this ceramic is also rather inert chemically and resistant to radiation damages which makes SiC an electronic material especially suitable to work in hostile environments [4-8]. Within the next decade, SiC is expected to challenge silicon in advanced technology with numerous applications in space, aeronautics or nuclear industries [8].

Some of the key issues in successful SiC devices include obtaining high quality and well characterized surfaces, surface metallization and surface passivation [4-8]. The latter could be achieved through the growth of SiO_2 on the SiC surfaces. However, this generally results in the formation of mixed Si-C oxides having poor electrical properties [4]. Recently, due to investigations based on advanced experimental tools as synchrotron radiation experiments and scanning tunneling microscopy measurements, very significant progresses have been made in the control of SiC surfaces, especially for its cubic (100) surface, including the recent discovery of highly stable atomic lines which are of strong interest in nanotechnologies, especially in the field of micro/nano-electronics of the future [9-14].

The recent progress and understanding of oxide/nitride growth on surfaces result from the availability, these last two decades, of novel and advanced experimental probes such as synchrotron radiation related techniques including core level and valence band photoemission and surface/photoemission extended x-ray absorption fine structures (SEXAFS, PEXAFS), scanning tunneling (STM) and atomic force (AFM) microscopies, and medium energy ion scattering [15-27]. Theory has also brought some significant insights into the knowledge of surface oxidation [28-31].

In this article, I will review some of the recent progress in the field ultrathin dielectrics (oxides, nitrides and oxynitrides) grown directly on silicon and silicon carbide surfaces at low temperatures and also through photoinduced promoted reactions. The presented results are based on core level and valence band photoemission spectroscopies using synchrotron radiation and atom-resolved scanning tunneling microscopy experiments. I will focus on some important issues such as interface formation and composition, initial oxidation, atomic scale molecular interaction, development and propagation of the oxidation reaction on a surface, role of surface composition, structure and temperature in oxide product and stoichiometry, and finally describe a representative example of photoinduced surface reaction.

2. Experimental details

The synchrotron radiation experiments have been performed at the Synchrotron Radiation Center (SRC), University of Wisconsin-Madison using Mark II and V Grasshopper beam lines and at the Advanced Light Source (ALS), Lawrence Berkeley National Laboratory using the spectromicroscopy beam line 7.0. A double-pass angle-integrating cylindrical mirror analyzer (CMA) was used at SRC while a high resolution hemispherical analyzer having an acceptance angle of ± 6° was used at ALS. Both instruments were designed and manufactured by PHI Instruments. The STM experiments have been performed at Laboratoire de Photophysique Moléculaire (CNRS), Université de Paris-Sud/Orsay using an Omicron scanning tunneling microscope at working pressures better than 4×10^{-11} Torr. The STM topographs (filled and empty electronic states) have been recorded in the constant current mode, are reproducible and not tip dependent. Single domain β-SiC thin films have been prepared at LETI (CEA-Technologies Avancées, Grenoble) by chemical vapor deposition (C_3H_8, SiH_4) growth on vicinal (4°) Si(100) wafer surfaces. The surface quality is checked to exhibit sharp diffraction patterns by low energy diffraction LEED attached to each experimental system. Other experimental details concerning data acquisition, sample surface preparation, data analysis and curve fitting procedures are available elsewhere [4, 9-14].

3. Atomic scale initial oxidation
a - O_2/Si(111)7x7

Oxide growth is generally achieved through the interaction of molecular oxygen (or oxygen containing molecules) with the surface. The very initial step of such a reaction is of special interest since the knowledge of oxygen adsorption site and the role of defects may give useful information on how the oxidation starts and develops on the surface. The oxidation of silicon surfaces has been studied using many experimental probes and theoretical approaches [19-31]. Among them, scanning tunneling microscopy

Si(111)7x7

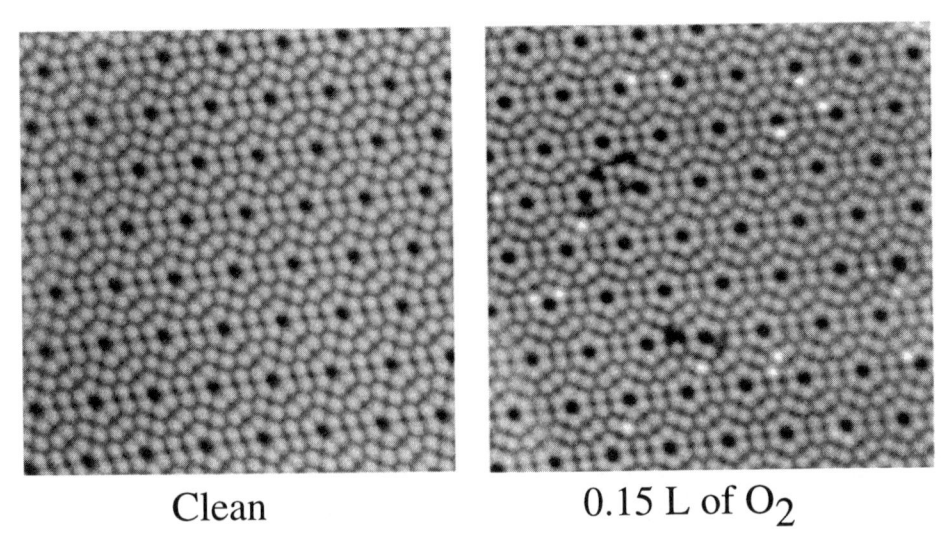

Clean   0.15 L of O_2

Figure 1: 190 Å x 190 Å STM topographs of the same area of clean (left) and 0.1 L of O_2 exposed (right) Si(111)7x7 surface (from Ref. 26). The tip bias is V_t = - 2.0 V with a tunneling current of 1.0 nA *(Courtesy of Gérald Dujardin)*.

(STM) is a very powerful and appropriate tool to explore the initial molecular interaction with the surface. In this context, the Si(111)7x7 surface is of special interest since it is a model surface having a very low defect density. As a matter of fact, the initial oxygen interaction on the Si(111)7x7 surface is probably one of the most investigated chemisorption case [23-27]. Figs. 1a and 1b display STM topographs of a clean Si(111)7x7 surface and covered with 0.15 Langmuir of oxygen (1 Langmuir = 1 L = 10^{-6} torr.s) respectively [25]. As can be seen from Fig. 1, the oxygen molecules (bright spots) do not adsorb on defect sites. Valence band photoemission spectroscopy indicates that this low coverage oxygen adsorption on the Si(111)7x7 surface is not dissociative which allows to identify bright spots (Fig. 1b) as representing oxygen molecules [25,26]. However, with time (few hours), the bright sites of this oxygen covered Si(111)7x7 surface become dark which indicates reaction between oxygen and the surface and implying molecular dissociation [25]. Interestingly, low H_2O deposition leads to the promotion of the reaction with the silicon surface which is relevant in wet oxidation process [27]. In contrast to oxygen behavior on the Si(111)7x7 surface, STM measurements for Si(100)2x1 have shown that oxygen adsorption preferentially takes place at specific defect sites resulting from half dimers [24].

b - O_2/β-SiC(100)3x2

Let us now turn to the very initial oxidation of a silicon carbide surface using the same STM probe. The interaction of oxygen with the β-SiC(100)3x2 surface is of special fundamental interest since its structure has been recently determined by STM [9].

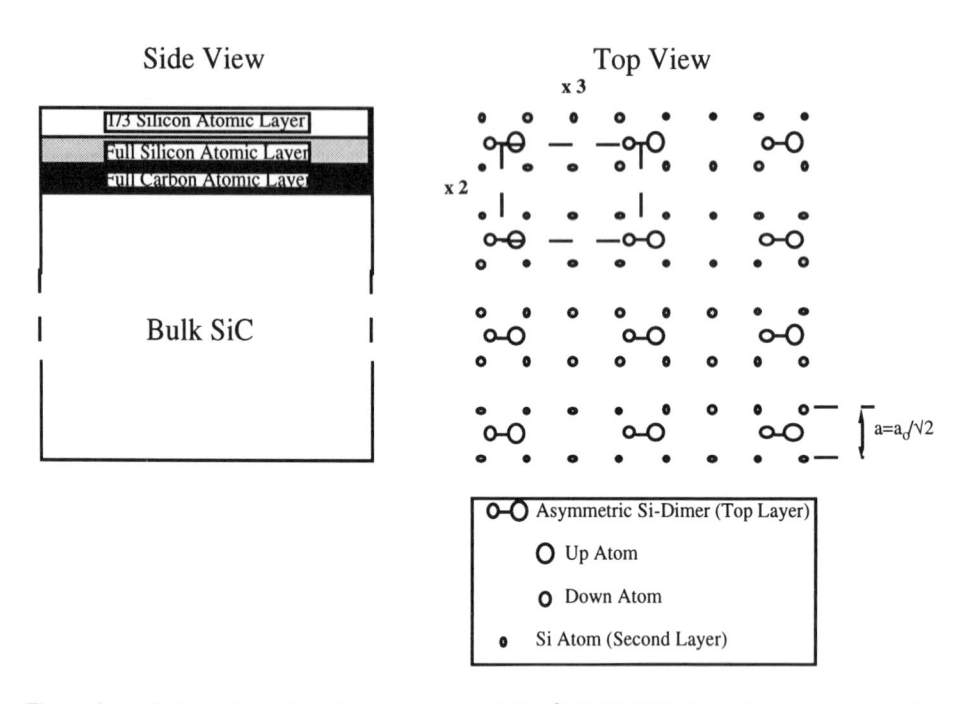

<u>Figure 2</u>: Schematic and surface structure of the β-SiC(100)3x2 surface reconstructions showing the parallel asymmetric Si-dimer ordering (PAD model). Only the two top layers are shown in the top view, the second layer being presented with a bulk array (after Ref. 9). Dashed contours correspond to surface unit cells and a_0 (4.36 Å) is the β-SiC lattice parameter.

In fact, this surface reconstruction is a Si-rich surface terminated by 1.3 Si monolayer (ML) including a full 1 Si ML covered by rows of parallel asymmetric Si dimers (PAD model) that are all tilted in the same direction perpendicularly to the row. This structural arrangement forms a 3x2 surface reconstruction resulting from PAD rows at a 1/3 Si

ML coverage (Fig. 2) [9]. The latter array makes this 3x2 surface a very "open" one, especially when compared to the Si(111)7x7 and Si(100)2x1 surfaces.

One can follow the initial oxygen interaction in Fig. 3 which displays STM topographs of clean and 0.5 L O_2 covered β-SiC(100)3x2 surface. The clean surface

clean β-SiC(100)3x2   0.5 L O_2 / β-SiC(100)3x2

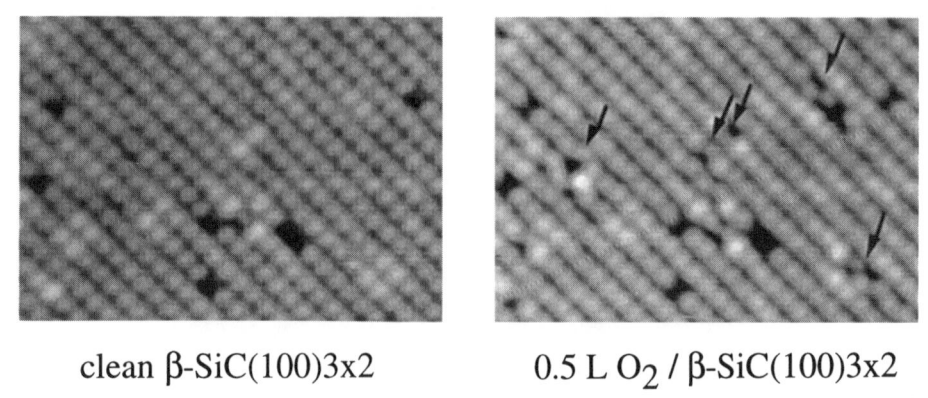

Figure 3: Clean β-SiC(100)3x2 (left) and covered by 0.5 L of O_2 (right) surfaces 100 Å x 150 Å STM topographs obtained by tunneling into the filled states (from Ref. 9). The tip bias is V_t = + 3.5 V with a 0.2 nA tunneling current. Arrows indicate some characteristic changes after O_2 deposition.

shows oval spots that represent Si-Si dimers with some defects including missing dimers, dimers pairs [9] and dimers that appears as brighter oval spots (Fig. 3). The latter defects are of special interest since they appear to be oxygen adsorption sites upon 0.5 L of O_2 exposure. In fact, the bright sites turn to dark or half dark [9]. Fig. 4 (top) gives additional details about this interesting process: first, one can better see that the bright active site become dark after O_2 exposure and second, two new bright sites have been induced by oxygen in the neighboring (Fig. 4) [32]. Similarly, in Fig. 4 (bottom), the two bright active defect sites have turned into two half-dark dimers with for new bright sites nearby. These new bright sites become active sites upon further oxygen deposition [32]. Scanning tunneling spectroscopy (STS) probing filled and empty electronic states shows that the new active bright sites have the same spectroscopic signature than the bright sites already present on the clean surface [32]. This very interesting feature further demonstrates that the initial and the oxygen-induced active sites have the same nature. In this way, further oxygen deposition results in the

β-SiC(100)3x2

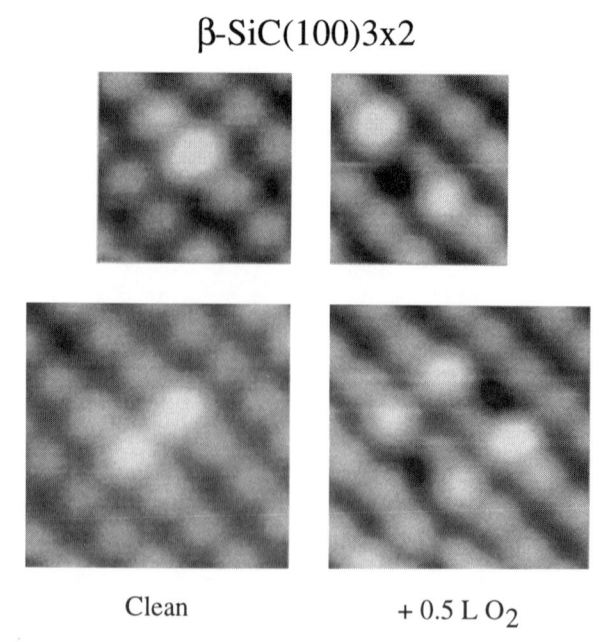

Clean   + 0.5 L O$_2$

Figure 4: STM topographs of two selected area of clean β-SiC(100)3x2 surfaces (left) and the same after 0.5 L of O$_2$ exposure (right) (after Ref. 32). The tip bias is $V_t = +3.5$ V with a 0.2 nA tunneling current.

development of surface oxidation in a self propagated manner [32]. Also, the creation of oxygen-induced new bright active sites is about 40% higher along Si-dimer rows than perpendicularly to these rows [32].

Fig. 5 shows a more global view of this initial oxidation process: STM topographs (filled and empty electronic states) for the clean β-SiC(100)3x2 and exposed to 1 and 5 L of O$_2$ indeed shows patchworks of oxidized Si islands that develop with increasing oxygen exposures [33]. Such a local process spreads over the surface as expanding cluster patchworks of oxidized sites, through a self-propagating nano-mechanism [32]. The fact that surface oxidation seems to develop along step edges result from higher defect densities in these areas. This self-propagated reaction is very well reproduced by numerical simulations taking into account the slightly anisotropic propagation of new active sites [32]. In fact, from these STM results, one can deduce that the initial adsorption of oxygen molecules on the β-SiC(100)3x2 surface is taking place at particular active defect sites which plays a central role in subsequent surface

Clean β-SiC(100)3x2

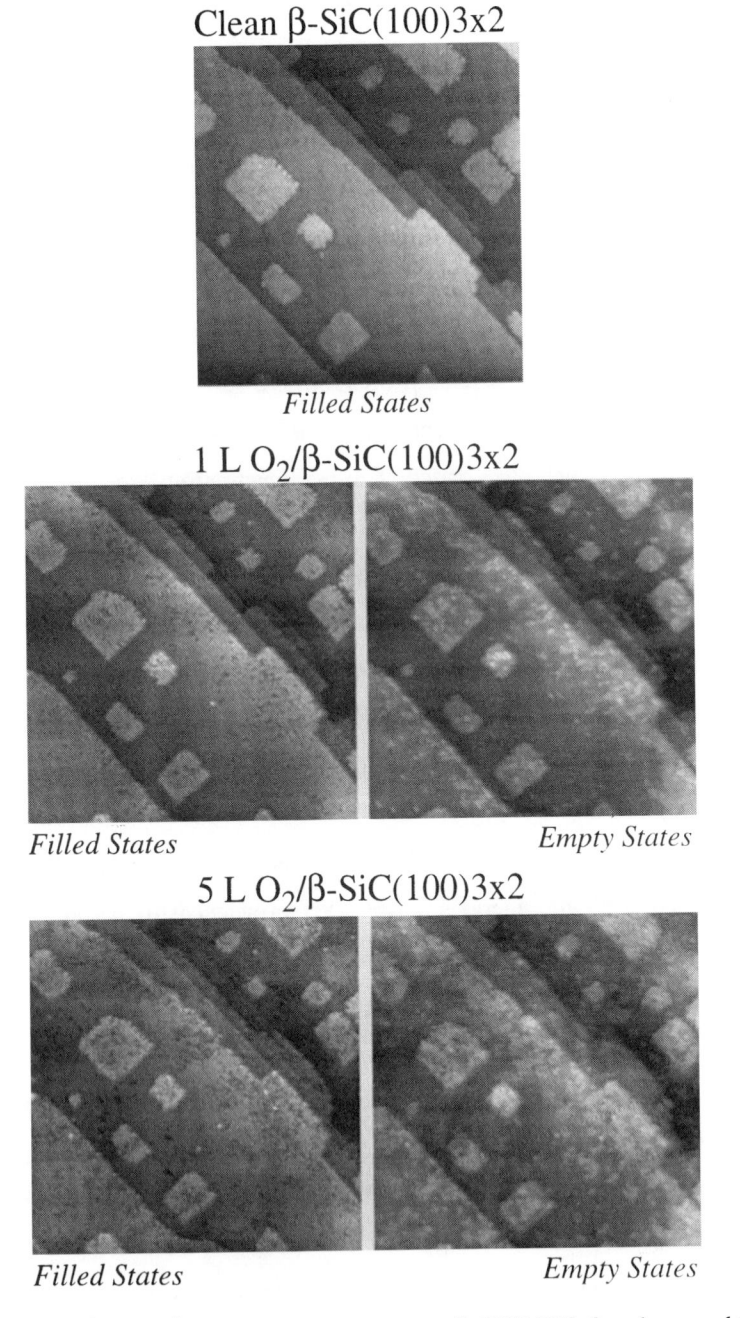

Filled States

1 L O$_2$/β-SiC(100)3x2

Filled States   *Empty States*

5 L O$_2$/β-SiC(100)3x2

Filled States   *Empty States*

<u>Figure 5</u>: 800Å x 800Å STM topographs of clean β-SiC(100)3x2 and covered by 1 L O$_2$ and 5 L O$_2$ (filled and empty electronic states) surfaces (after Ref. 33). The tip bias is $V_t = \pm 2.5$ V at a 0.2 nA tunneling current.

oxidation. The behavior of this silicon carbide surface is very different from that of Si(111)7x7 where oxygen adsorption does not take place at defect sites, unlike the case of Si(100)2x1 surface for which defects at Si-dimers are active sites playing a significant role for oxygen adsorption [24].

4. Direct insulator/semiconductor (SiC) interface formation
a - SiO_2/β-SiC(100)3x2 interface

We now look the oxidation process at a larger scale to focus on oxide growth and interface formation for the same β-SiC(100)3x2 surface at much higher oxygen exposures. The main experimental technique used here is core level and valence band photoemission using synchrotron radiation at the Synchrotron Radiation Center (SRC, University of Wisconsin-Madison) and at the Advanced Light Source (ALS, Lawrence Berkeley National Laboratory). These probes bring very useful information about chemical environment and bonding configurations of Si, C and O atoms. Important issues include the effect of surface temperature on the oxidation process and products, and interface composition and structure.

Fig. 6 displays representative Si 2p (a), valence band (b) and C 1s (c) spectra for a β-SiC(100)3x2 surface exposed to 10,000 L of molecular oxygen at various surface temperatures ranging from 25°C to 500°C [34]. For the clean surface, the Si 2p core level is recorded at photon energies of 150 eV (surface sensitive) and 115 eV (bulk sensitive) and exhibits 3 components A, B and C related to surface Si-Si, bulk Si-C and surface Si-C bonding respectively [34]. One can follow the oxidation steps through the Si 2p core level spectra which provide information about the Si atom environment. When this surface is exposed at room temperature to 10,000 L of oxygen, a core level shift M at 3.8 eV binding energy (B.E.) indicates the formation of silicon oxides mainly as SiO_2 [34]. At surface temperatures of 300°C and 500°C, one can see in Fig. 6a that the same O_2 exposure results in larger amount of SiO_2 at a thickness of about 10 Å. Furthermore, this oxide shows a higher binding energy (over 4 eV) suggesting SiO_2 formation already having bulk-like properties despite the fact that, at this oxygen exposure, a very thin oxide layer has grown ($\approx$ 10 Å) [34,35].

$O_2/\beta\text{-SiC}(100)3\times 2$

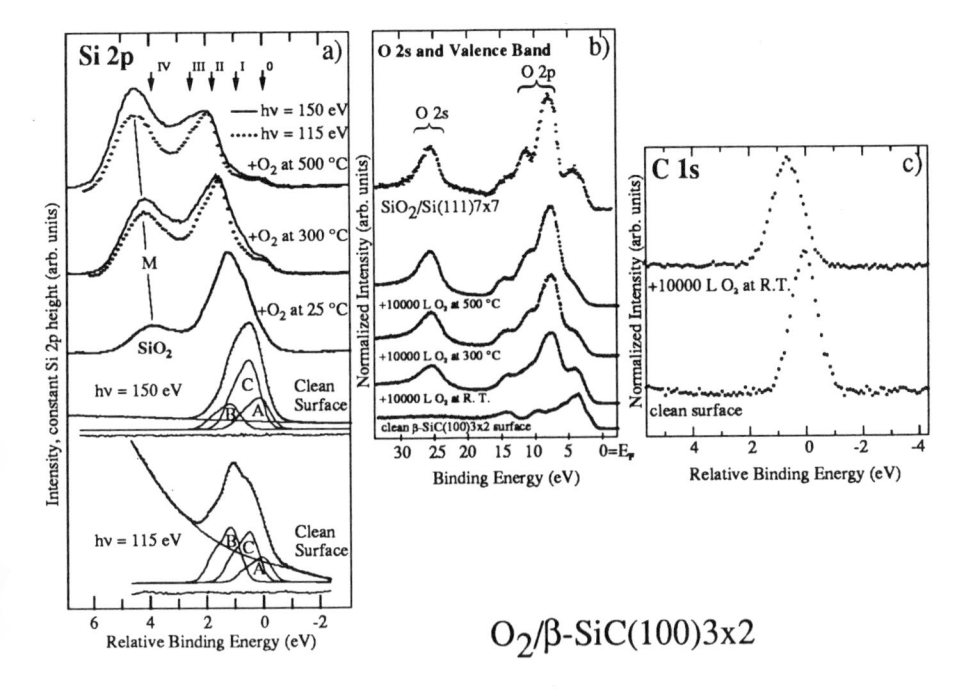

Figure 6: a) Si 2p core level for clean β-SiC(100)3x2 surface and exposed to 10^{+4} L of O_2 at various temperatures ranging from 25°C to 500°C (after Ref. 34). The photon energies are 150 eV and 115 eV for surface and for bulk sensitive modes respectively. For comparison, bars (0 to IV) indicates the various Si oxidation states (Si^0, Si^+, Si^{2+}, Si^{3+} and Si^{4+}) binding energies in the case of very thin Si oxides. The two bottom spectra show the Si 2p core level decomposition, fits, background subtraction and residual background for the clean β-SiC(100)3x2 surface in bulk- (bottom) and surface-sensitive (second from bottom) modes.
b) Valence band and O 2s core level for β-SiC(100)3x2 surfaces exposed to 10^{+4} L of O_2 at various temperatures from 25 °C to 500°C (after Ref. 34). For comparison, the VB of a SiO2/Si(111)7x7 interface is also presented. The photon energy is 150 eV.
c) C 1s core level for clean and exposed to 10^{+4} L of O_2/β-SiC(100)3x2 surfaces at room temperature (after Ref. 34). The photon energy is 340 eV.

The valence band data for the same sequence exhibit (Fig. 6b) a strong similarity when compared to that of SiO2/Si(111)7x7 interface, with same shape and binding energy for O 2p electronic levels and O 2s core levels indicating that the oxide products are similar [34]. The C 1s core level (Fig. 6c) provides some insights about the C environment. One can see that, upon oxygen exposure, the C 1s core level is slightly affected with a 0.5 eV chemical shift and a ≈ 10% broadening. This suggests that, during the oxidation process, the carbon atoms resulting from Si-O bond establishments are also oxidized as CO or CO_2 which are desorbed from the surface [34]. The presence of carbon species as e.g. graphite would result in a large chemical component shifted by ≈ 2 eV to higher binding energy, as previously observed elsewhere [36]. Here, such a structure is not observed [34]. However, the slight C 1s broadening and energy shift upon oxidation seems to indicate that some bonding between oxygen and carbon atoms.

High resolution core level photoemission spectroscopy experiments (overall energy resolution ≈ 70 meV) performed recently at ALS (Berkeley) provide deeper insights about the SiO_2/β-SiC(100)3x2 interface composition and structure. Fig. 7 displays Si 2p core level for a β-SiC(100)3x2 surface exposed to O_2 at 500°C. Surface (150 eV) and bulk (340 eV) sensitive photon energies were used at grazing incidence (very surface sensitive, ≈ several Å at hv = 150 eV) and normal incidence (very bulk sensitive, few ten Å at hv = 340 eV) [37]. Conventional peak decomposition allows identification of the various oxidation states as previously performed for silicon oxidation [37]. One can see from Fig. 7 that the core level shifts are located at higher binding energy by 1 eV, 1.8 eV, 2.6 eV and 4 eV relative to the energy of Si 2p core level at 99.2 eV for pure silicon. Apparently, this configuration seems very similar to what is known for SiO_2/Si(100)2x1 and SiO_2/Si(111)7x7 interfaces where it corresponds to Si^+, Si^{2+}, Si^{3+} and Si^{4+} oxidation states, the latter being related to silicon dioxide [19-22]. This means that the corresponding interface is apparently not abrupt with decreasing oxidation states when approaching the substrate surface [19-22]. However, one can notice that, in the bulk sensitive modes (Figs. 7a and 7c), the component located at 1.8 eV has a larger intensity that what is observed for corresponding SiO_2/Si(100)2x1 and SiO_2/Si(111)7x7 interfaces [19-22]. This is very likely to result from the establishment of C-O bonds in a configuration close to Si-O-C, where the oxygen atoms would be bonded to both Si and C atoms [34,37]. Therefore, the 1.8 eV core level shifted component observed at Si 2p (Fig. 7) would result from both Si^{2+} and Si-O-C contributions. Actually, such a bonding configuration would give a core level shift close to what is observed for an oxygen atom bonded to two Si atoms as Si-O-Si (Si^{2+}). This 1.8 eV component is not dominant in the surface sensitive spectrum (Fig. 7b) which suggests that the Si-O-C bonding configuration is taking place deeper, at the interface between the oxide and the SiC surface. This indicates that

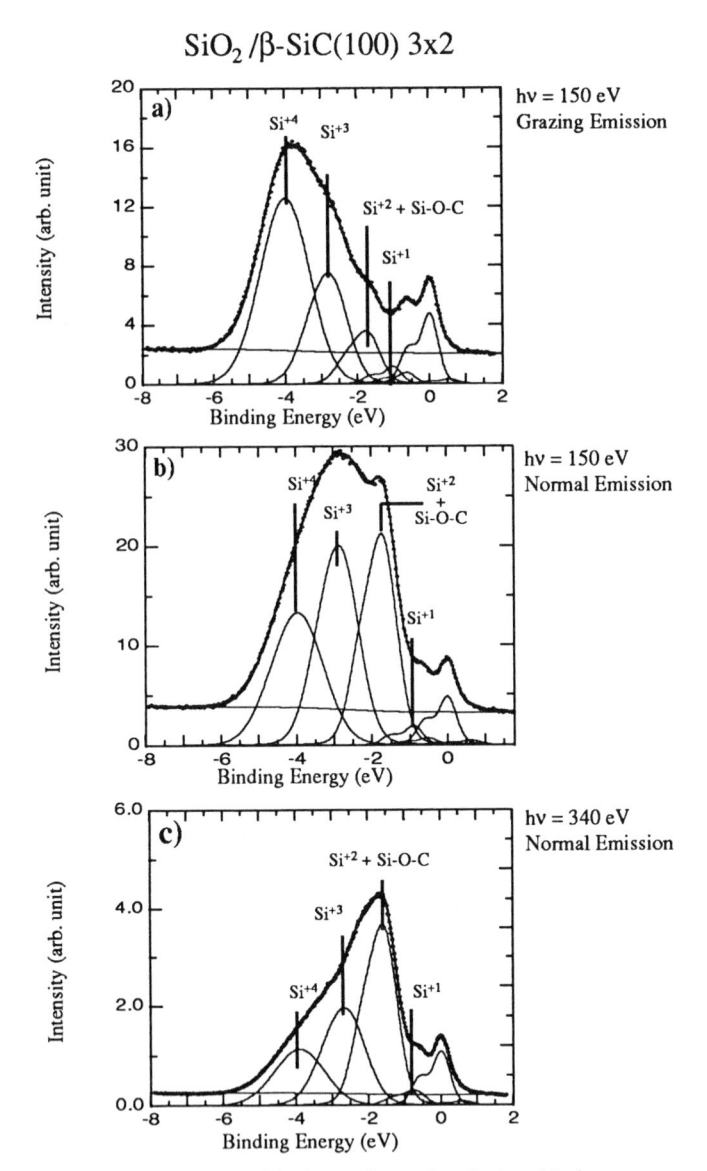

Figure 7: Si 2p core level spectra with the various chemical oxidation states components for a 10^{+4} L of O_2 exposure at 500°C of the β-SiC(100)3x2 surface at normal and grazing incidence emission angles. The photon energies are 150 eV (surface sensitive) and 340 eV (bulk sensitive).

the very thin oxide layer is probably lying on the substrate through a C-terminated plane, with oxygen atoms bonded to Si atoms above and to C atoms underneath. This bonding configuration suggests a rather abrupt SiO_2/β-SiC(100)3x2 interface, contrary to what could be expected for the oxidation of compound semiconductors. In addition to the fact that this is a silicon rich surface, the relatively high reactivity of this β-SiC(100)3x2 surface to oxidation is also likely resulting from its very open surface structure, especially when compared to Si(100)2x1 and Si(111)7x7 surfaces.

b - Oxide/β-SiC(100)2x1-c(4x2) and β-SiC(100)1x1 interfaces

In order to explore the influence of surface composition and surface structure, let us now turn to other SiC surface reconstructions including the stoichiometric Si-terminated β-SiC(100)2x1-c(4x2) and C-rich β-SiC(100)1x1 surfaces. Fig. 8 displays Si 2p core level for such surfaces together, for the shake of comparison, with the β-SiC(100)3x2 exposed to 10,000 L of O_2 at room temperature and at 500°C, following the same sequence as in Fig. 6. As can be seen from Fig. 8, the exposure to molecular oxygen does not result in a large ≈ 4 eV core level shift indicating SiO_2 formation on β-SiC(100)3x2 [33,38]. Instead, the ≈ 2 eV core level shift observed for the β-SiC(100)2x1 surface rather indicates mixed Si and C oxide formation having low oxidation states. For the C-rich β-SiC(100)1x1 surface, the oxygen interaction appears even more limited with a Si 2p broadening indicating very small oxide amounts having very low oxidation states. The C 1s core level spectra for the same surfaces and the same room temperature oxygen exposure sequence are also displayed in Fig. 8. Oxygen deposition results in a 1 eV core level for the stoichiometric 2x1/c(4x2) surfaces which further stresses the formation of mixed SiOC oxides as already mentioned above. In contrast to the other β-SiC(100) surface reconstructions, the same C 1s spectral line for the 1x1 surface remains almost unaffected by oxygen, which correlates with the lack of large oxide product formation. It is likely that the absence of large amounts of Si as for the β-SiC(100)3x2 [4,9] naturally limits silicon oxide formation. In addition, the presence of a C layer at the surface seems to decrease significantly the oxygen sticking probability and/or limit oxygen diffusion below the surface, which explains the subsequent rather important lack of reactivity of the C-rich β-SiC(100)1x1 to the oxidation sequence. These results stresses the crucial importance of surface composition and surface ordering in successful oxidation.

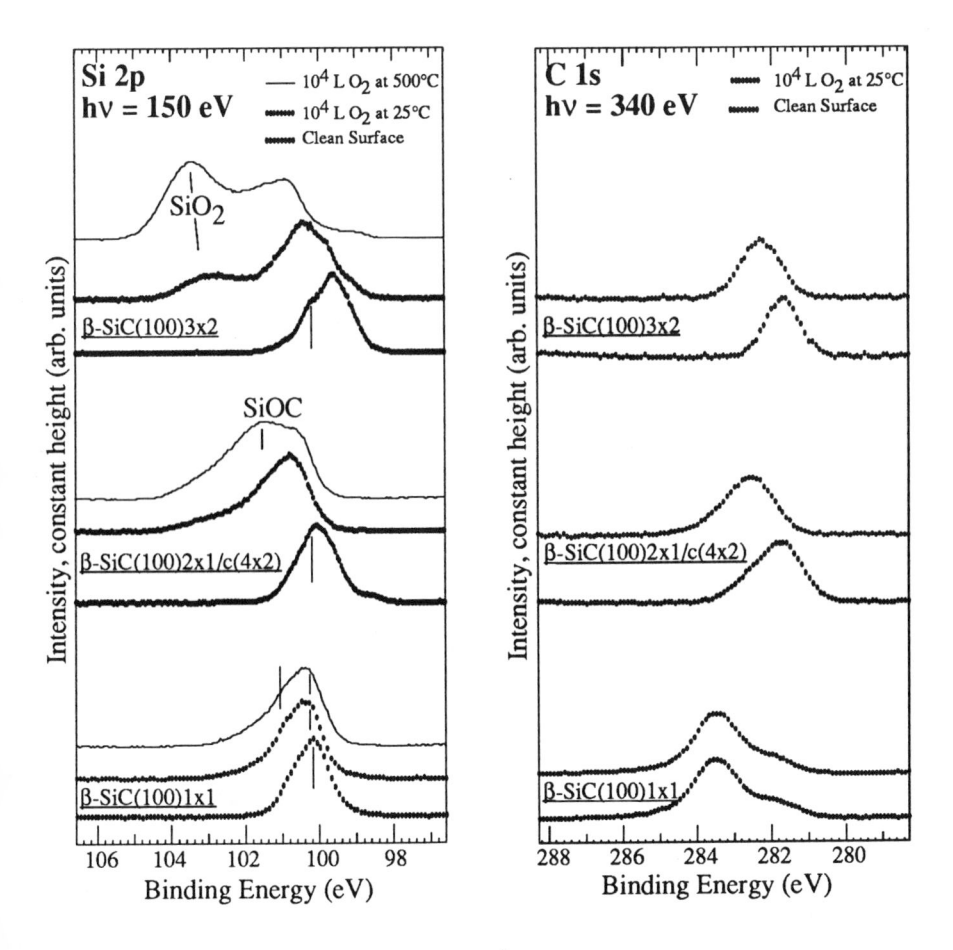

Figure 8: Si 2p core level spectra (left) for 10^{+4} L of O_2 exposures at room temperature and at 500°C of the β-SiC(100)3x2, β-SiC(100)2x1-c(4x2) and β-SiC(100)1x1 surfaces. The photon energies are 150 eV (surface sensitive mode).
C 1s core level spectra (right) for 10^{+4} L of O_2 exposures at room temperature of the β-SiC(100)3x2, β-SiC(100)2x1-c(4x2) and β-SiC(100)1x1 surfaces. The photon energy is 340 eV (surface sensitive mode).

5. Photoinduced silicon nitridation and oxynitridation

Another approach to low temperature surface passivation is the use of a radiation-promoted reaction. For instance, the role of an electron beam in the room temperature oxidation of silicon surfaces is known since a decade [39,40]. In fact, when a silicon surface is placed in an oxygen atmosphere at a ≈ 10^{-7} torr pressure and exposed to a 3 KV electron beam, one can observed SiO_2 formation on the surface [39,40]. This promoted oxidation effect results from enhanced dissociation of the oxygen molecules, probably by charge transfer into the antibonding molecular orbitals, resulting in atomic oxygen and probably also negative oxygen ion species that are more reactive with the surface. Room temperature exposure of a Si(111)2x1 surface to ammonia (NH3) does not result in significant silicon nitride products. However, the NH3 molecule has a high sticking probability on surfaces especially on those of silicon. When the NH3/Si(111)2x1 surface is exposed to unmonochromatized synchrotron radiation (white light), a silicon nitride layer having high nitridation states grows on the surface as Si3N4 [41]. This photoinduced process results from molecular dissociation leading to the formation of active nitrogen species that bonds more easily to substrate atoms [41].

In addition to oxides and nitrides, oxynitrides form another important class of thin film dielectrics. Oxynitridation could be achieved by thermal nitridation of silicon dioxide films or by the interaction of molecules containing nitrogen and oxygen as N2O and NO directly with the surface [42-47]. We now look at the effect of an unmonochromatized synchrotron radiation beam on the surface of Si(111)2x1 (cleaved) exposed to 1000 L of NO using core level photoemission spectroscopy at the Si 2p core level (Fig. 9) in the surface sensitive mode (hv = 130 eV). The room temperature deposition of NO on the surface results in a large core level shift located at about 2.8 eV higher binding energy (lower kinetic energy) which indicates oxynitride formation as SiO_xN_y [48]. This means that the NO adsorption on the surface is dissociative, at least in part. When this surface is irradiated with unmonochromatized synchrotron radiation (USR) during 5 and 30 minutes, one can see in Fig. 9 that the oxynitride related core level has increased in intensity and is shifted by 3.5 eV from the Si 2p core level line [48]. The latter feature also indicates that, upon irradiation, the oxynitride has become oxygen-rich with amount increasing with irradiation time. This aspect is further stresses by looking in Fig. 10 at the N 2s and O 2s core levels for the same sequence where the O 2s/N 2s intensity ratio is increasing with irradiation time [48]. At such a high NO exposure at room temperature, only part of the molecules are dissociated. This means the USR dissociates remaining NO molecules favoring their interaction with the surface which leads to the formation of an oxygen rich SiN_xO_y/Si(111)2x1 interface at room

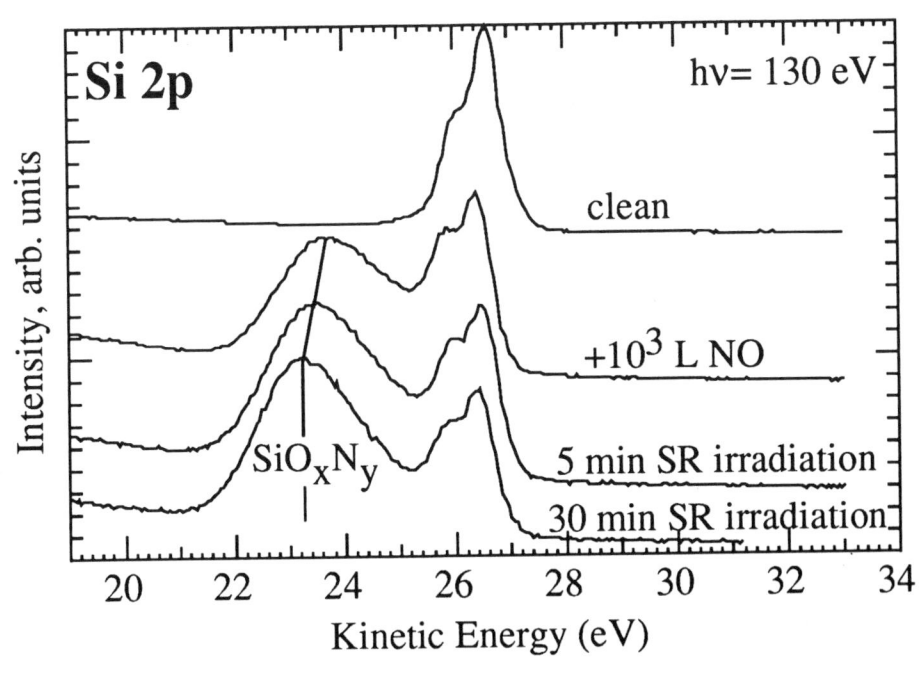

Figure 9: Si 2p core levels spectra for the clean Si(111)2x1 surface, after 1000 Langmuirs of NO exposure, followed by 5 and 30 minutes irradiation to the white light of unmonochromatized synchrotron radiation (from Ref. 48). The photon energy is 130 eV.

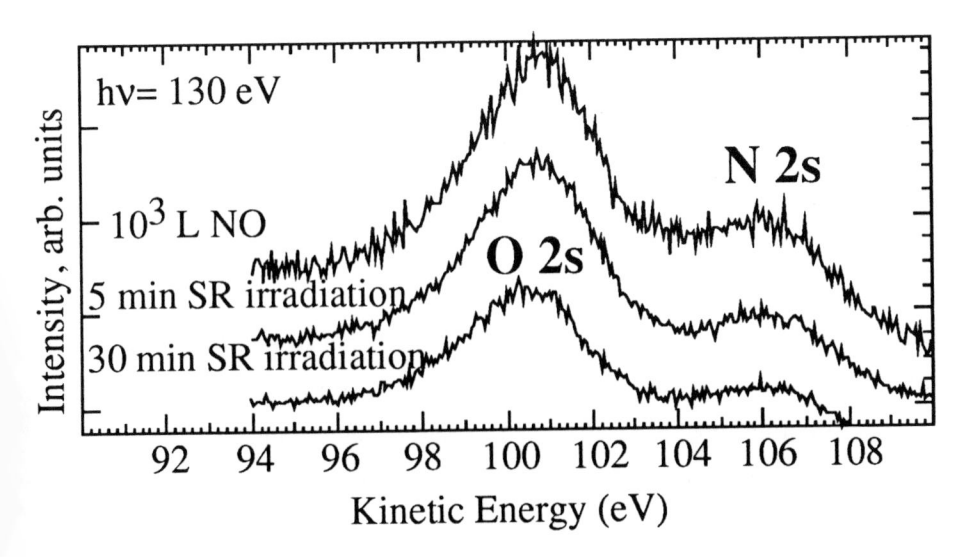

Figure 10: O 2s and N 2s spectra for 1000 L NO/Si(111)2x1 followed by 5 and 30 minutes irradiation to the white light of unmonochromatized synchrotron radiation (after Ref. 48).

temperature [48]. However, this process also likely results in selective N desorption through a photo-stimulated electronic process which would explain oxygen-rich oxynitride formation [48]. While the effect of USR seems to be similar to that of alkali metal catalysts which also promote silicon oxynitridation by favoring oxygen-rich oxynitride formation [49], it interestingly appears to have the opposite effect of thermal annealing which favors the formation of nitrogen-rich oxynitrides [50]. Therefore the combination of both photo-induced and thermal oxynitridation could allow the monitoring of oxynitride having a desired O/N stoichiometry.

6. Conclusions

To conclude, in contrast to the behavior of the Si(111)7x7 surface, scanning tunneling microscopy reveals the role of active defect sites in the oxidation of β-SiC(100)3x2 surface, a refractory wide band gap semiconductor. Oxygen adsorption results in additional similar defect sites that are active upon further oxygen deposition. The initial oxidation is therefore developing in a self-propagated manner. Core level and valence band photoemission spectroscopies using synchrotron radiation show that the SiO_2/β-SiC(100)3x2 develops already at very low temperatures (below 500°C) which results from the particular open surface structure of this Si-rich reconstruction of silicon carbide. In strong contrast, the other surface reconstructions are less reactive to oxygen exposures (between 25°C and 500°C) with mixed Si and C oxide formation having low oxidation states for the stoichiometric β-SiC(100)2x1-c(4x2) surfaces, and very small amount of oxide products for the C-rich β-SiC(100)1x1 surface reconstruction. The SiO_2/β-SiC(100)3x2 interface structure is composed of an oxide layer, primarily SiO_2, associated with lower oxidation states lying on a C-terminated surface with oxygen atoms bonded, at the interface, to both Si and C atoms. Room temperature nitridation and oxynitridation could be achieved using a photoinduced reaction with unmonochromatized synchrotron radiation. Such a reaction results from photo-induced enhanced molecular dissociation resulting in active atomic and/or ionic oxygen species that react with the substrate. Photoinduced oxynitridation leads to promoted reaction with formation of oxygen-rich oxynitride products.

Acknowledgments

This work was supported by the U.S. National Science Foundation (NSF) under contracts N° DMR 88-07754 and 92-23710, the U.S. Department of Energy (DOE), the Northern Illinois University Graduate School Funds and Ministère de l'Éducation Nationale, de la Recherche et de la Technologie (MENRT, France). The author is grateful to his students, postdoctoral research associates and collaborators that were involved at various degrees in this work and especially acknowledges Fabrice Amy, Jonathan Denlinger, Ludovic Douillard, Gérald Dujardin, Douglas Dunham, Alain Glachant, Sung-Tae Kim, Pawitterjit S. Mangat, Andrew Mayne, Manfred Riehl-Chudoba, Sandrine Rivillon, Elie Rothenberg, Fabrice Semond and Brian Tonner, and thanks Lea di Cioccio, Claude Jaussaud and Catherine Pudda at LETI (CEA - Technologies Avancées, Grenoble, France) for providing β-SiC(100) thin film samples.

References

1. S. Wolf et R.N. Tauber, Silicon Processing for VLSI Era, Lattice, Sunset Beach, California (1986); S.M. Sze, Physics of Semiconductor Devices, Wiley-Interscience, Wiley, New York (1981).
2. M. Jaros, Physics and Applications of Semiconductor Microstructures, Oxford Science Publications, Series on Semiconductor Science and Technology - **1**, Clarendon Press, Oxford (1989).
3. J.L. Leray, P. Paillet and J.L. Autran, *J. Physique III*, **6** (1996) 1625.
4. Properties of Silicon Carbide, G. Harris editor, EMIS Datareviewseries, INSPEC (London), Vol. **13**, (1995).
5. P.A. Ivanov and V.E. Chelnokov, *Semicond.* **29** (1995) 1003.
6. H. Morkoç, S. Strite, G. B. Gao, M. E. Lin, B. Sverdlov and M. Burns, *J. Appl. Phys.* **76** (1994) 1363.
7. R.F. Davis, *J. Vac. Sci. Technol.* A **11** (1993) 829.
8. Silicon Carbide Electronic Materials and Devices, Materials Research Society Bulletin, Vol. **22**, March (1997).
9. F. Semond, P. Soukiassian, A. Mayne, G. Dujardin, L. Douillard and C. Jaussaud, *Phys. Rev. Lett.* **77** (1996) 2013.
10. P. Soukiassian, F. Semond, L. Douillard, A. Mayne, G. Dujardin, L. Pizzagalli and C. Joachim, *Phys. Rev. Lett.* **78** (1997) 907.
11. V.M. Bermudez and J.P. Long, *Appl. Phys. Lett.* **66** (1995) 475.
12. M.L. Shek, *Surf. Sci.* **349** (1996) 317.
13. J.P. Long, V.M. Bermudez and D.E. Ramaker, *Phys. Rev. Lett.* **76** (1996) 991.
14. P. Soukiassian, F. Semond, A. Mayne and G. Dujardin, *Phys. Rev. Lett.* **79** (1997) 2498.
15. F.J. Himpsel, B.S. Meyerson, F.R. McFeely, J.F. Morar, A. Taleb-Ibrahimi and J.A. Yarmoff in Photoemission and Adsorption Spectroscopy of Solids and Interfaces with Synchrotron Radiation, M. Campagna and R. Rosei editors, *North Holland*, Amsterdam (1990); and references therein.

16 G. Margaritondo, Introduction to Synchrotron Radiation, Oxford University Press, New York (1988); and references therein.
17 A. Flodström, R. Nyholm and B. Johansson in Advances in Surface and Interface Science, Synchrotron Radiation Research, R.Z. Bachrach editor, Plenum, New York, Vol. 1 (1992) 199; and references therein.
18 G.M. Rothberg, K.M. Choudhary, M.L. den Boer, G.P. Williams, M.H. Hecht and I. Lindau, Phys. Rev. Lett. 53 (1984) 1183.
19 F.J. Himpsel, F.R. McFeely, A. Taleb-Ibrahimi, J.A. Yarmoff and G. Hollinger, Phys. Rev. B 38 (1988) 6084; and references therein.
20 M. Tabe, T.T. Chiang, I. Lindau and W.E. Spicer, Phys. Rev. B 34 (1986) 2706.
21 W. Braun and H. Kuhlenbeck, Surf. Sci. 180, 279 (1986).
22 Z.H. Lu, S.P. Tay, R. Cao and P. Pianetta,Appl. Phys. Lett. 67 (1995) 2836
23 Ph. Avouris and D. Cahil, Ultramicroscopy 42 (1992) 838.
24 M. Udagawa, Y. Umetani, H. Tanaka, M. Itoh, T. Uchiyama, Y. Watanabe, T. Yokotsuka and I. Sumita, Ultramicroscopy 42 (1992) 946.
25 G. Dujardin, A. Mayne, G. Comtet, L. Hellner, M. Jamet, E. Le Goff and P. Millet, Phys. Rev. Lett. 76 (1996) 3782.
26 I.S. Hwang, R.L. Lo and T.T. Tsong, Phys. Rev. Lett. 78 (1997) 4797.
27 R. Kliese, B. Röttger, D. Badt and H. Neddermeyer, Ultramicroscopy 42 (1992) 824.
28 C.J. Sofield and A.M. Stoneham, Semicond. Sci. Technol. 10 (1995) 215.
29 N.F. Mott, S. Rigo, F. Rochet and A.M. Stoneham, Phil. Mag. B 80 (1989) 189.
30 A. Pasquarello, M.S. Hybersten and R. Car, Phys. Rev. Lett. 74 (1995) 1024.
31 V.J.B. Torres, A.M. Stoneham, C.J. Sofield, A.H. Harker and C.F. Clement, Interf. Sci. 3 (1995) 133.
32 A. Mayne, F. Semond, G. Dujardin and P. Soukiassian, recent STM results.
33 F. Semond, Ph.D. Thesis, Université de Paris-Sud/Orsay, 19 December (1996).
34 F. Semond, L. Douillard, P. Soukiassian, D. Dunham, F. Amy and S. Rivillon, Appl. Phys. Lett. 68 (1996) 2144.
35 F.J. Grunthaner, P.J. Grunthaner, R.P. Vasquez, B.F. Lewis, J. Maserjian and A. Madhukar, J. Vac. Sci. Technol. B 16 (1979) 1443.
36 M. Riehl-Chudoba, P. Soukiassian, C. Jaussaud and S. Dupont, Phys. Rev. B 51 (1995) 14300.
37 D. Dunham, P. Soukiassian, J.D. Dennlinger, B.P. Tonner and E. Rothenberg, recent ALS high energy resolution photoemission experiments. Silicon Carbide and III-Nitrides, edited by H. Morkoç and G. Pensl, in press
38 F. Semond, F. Amy, L. Douillard, P. Soukiassian, D. Dunham and Z. Hurych, recent results
39 B. Carrière, J.P. Delville, A. El Maachi, Phil. Mag. B 55 (1987) 721.
40 B. Carrière, A. Chouiyakh and B. Lang, Surf. Sci. 126 (1983) 495.
41 F. Cerrina, B. Lai, G.M. Wells, J.R. Willey, D.G. Kilday and G. Margaritondo, Appl. Phys. Lett. 50 (1987) 533.
42 S. Matsuo and K. Kikuchi, Jap. J. Appl. Phys. 22 (1980) L 210.
43 M.M. Moslehi and K.C. Saraswat, IEEE Trans. Electron Devices 32 (1985) 1.
44 T. Ito, T. Nakamura and H. Ishikawa, IEEE Trans. Electron Devices 29 (1982) 498.
45 A. Ronda, A. Glachant, C. Plossu and B. Balland, Appl. Phys. Lett. 50 (1987) 171.
46 B. Balland, A. Glachant, A. Ronda and J.C. Dupuy, Phys. Stat. Sol. 100 (1987) 187.
47 A. Glachant, B. Balland, A. Ronda, J.C. Bureau and C. Plossu, Surf. Sci. 205 (1988) 287.
48 A. Glachant, P. Soukiassian, S.T. Kim, S. Kapoor, A. Papageorgopoulos and Y. Baros, J. Appl. Phys. 70 (1991) 2387.
49 M. Riehl-Chudoba, L. Surnev and P. Soukiassian, Surf. Sci. 306 (1994) 313; and references therein.
50 M.D. Wiggins, R.J. Baird and P. Wynblatt, J. Vac. Sci. Technol. 18 (1981) 965.

INTERACTION OF O_2 AND N_2O WITH Si DURING THE EARLY STAGES OF OXIDE FORMATION

A. A. SHKLYAEV[*]
Joint Research Center for Atom Technology (JRCAT),
Angstrom Technology Partnership (ATP), Tsukuba, Ibaraki 305, Japan

Abstract

The interaction of O_2 and N_2O with silicon surfaces is studied using optical methods such as ellipsometry and second-harmonic generation. These methods give the integral characteristics of the process in the wide temperature and pressure range. The extended precursor model is considered to analyze the temperature dependence of initial oxidation. Oxide formation is characterized by the progressively decreasing growth rate with gas pressure approaching the transition to Si etching. This new result directly reflects the oxide island nucleation process. We consider the model for oxide island nucleation and growth, in which the nucleation proceeds through the interaction in the layer of intermediately adsorbed species. In contrast to that in epitaxy, the critical oxide island size is found to be dependent on O_2 pressure. The surface morphology is discussed with respect to a size of oxide islands.

1. Introduction

The recent advance in understanding of the silicon oxide formation has been performed due to an application of different experimental methods [1-6]. At elevated temperatures between about 400 and 800°C the interaction of O_2 and N_2O with silicon is characterized by the following reactions: adsorption and desorption from the intermediate adsorbed state, dissociation of adsorbed molecules leading to Si-etching with SiO desorption, stable oxide formation proceeding with the interaction in the layer of intermediate adsorbed species (adspecies) [7]. In these reactions, N_2O and O_2 show the similar behavior [8]. The competition between the reactions results in existence of the so-called critical conditions [9-10] describing the temperature dependence of the critical gas pressure P_{tr}: if the gas pressure P_{ox} is kept below P_{tr}, then gas phase etching of the silicon surface takes place and the surface remains clean, whereas if $P_{ox} > P_{tr}$, stable oxide islands form on the surface with simultaneous Si-etching between those islands until the whole surface is covered with oxide.

 The present work bases on the kinetic data obtained with optical methods such as ellipsometry and second-harmonic generation. These methods are very sensitive to the early stage of oxide formation and provide us with the integral characteristics of the process in the wide temperature range. We consider the extended precursor model which is able to describe the temperature dependence of the initial oxidation. Kinetic parameters

---
[*] On leave from the Institute of Semiconductor Physics, Novosibirsk, 630090, Russia.

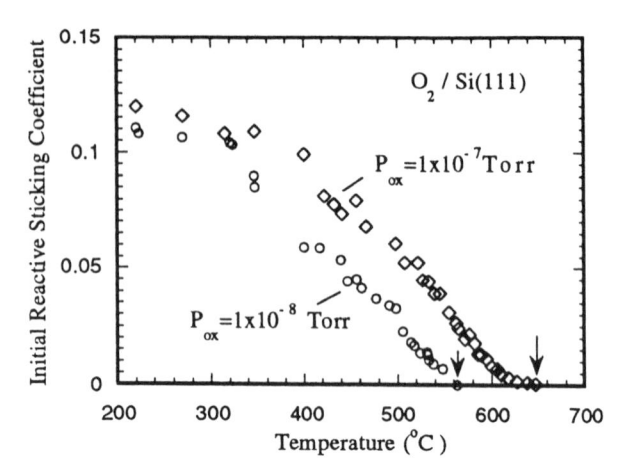

Figure 1. Initial reactive sticking coefficient of O$_2$ on Si(111)-7×7 versus surface temperature at two O$_2$ pressures [7]. The arrows show critical temperatures for the corresponding O$_2$ pressures.

of the reactions in this model were obtained by the fits to the experimental data. In order to describe the pressure dependence of the initial oxide growth rate the model of oxide island nucleation and growth is considered. This model allows us to describe the oxide formation at pressures around and above P_{tr} from the viewpoint of a size of oxide islands.

2. Precursor model for initial oxidation

The transition from oxide growth to Si etching is characterized by the gradual decrease of the oxide growth rate with increasing temperature. Figure 1 shows that the initial reactive sticking coefficient (S_o) describing the probability for incident molecules to form stable oxide species on the surface, decreases with increasing temperature and reaches zero at the transition temperature to Si etching without oxide growth. Such temperature behavior of S_o is consistent with the precursor model in which the first stage of adsorption involves trapping of incident molecules on the precursor state. In the simple model the adsorption species on the precursor state may either react to form stable oxide species or desorb into the gas phase [2]. At elevated temperatures, formation of stable SiO$_2$-like species is accompanied by Si etching with SiO desorption. The last reaction might be described into two stages: dissociation of adsorbed molecules to form SiO-like species on the surface and then their desorption into the gas phase. Figure 2 shows the most probable configurations of the intermediately adsorbed species [11].

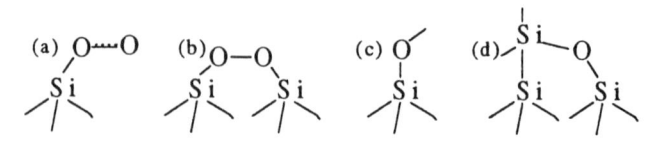

Figure 2. Probable configurations of intermediately adsorbed molecular and atomic oxygen on Si(111): (a) molecular peroxy radical, (b) molecular peroxy bridge, (c) atom in on-top, (d) atom in short bridge.

Thus, the interaction of O_2 with the Si surface might be expressed by the following reactions:

$$O_2 (g) \xrightarrow{\alpha\phi} O_2 (prec) \quad (1)$$

$$O_2 (g) \xleftarrow{k_d} O_2 (prec) \quad (2)$$

$$\tfrac{1}{2} O_2 (prec) + Si (s) \longrightarrow SiO (ad) \quad (3)$$

$$SiO(g) \longleftarrow SiO (ad) \quad (4)$$

$$SiO (ad) + \tfrac{1}{2} O_2 (prec) \longrightarrow SiO_2 (s), \quad (5)$$

where O_2 (prec) is the molecular precursor state, α the trapping coefficient of the precursor state, ϕ the flux of O_2 molecules incident on the surface, k_d the rate constant for O_2 desorption. Here, we assume that all reaction sites are vacant, that is a good approximation for the early stage of oxide formation. Modification of the reactions (1)-(5) to the case of N_2O concerns the reactions (3) and (5) in which the dissociation of N_2O is accompanied by the desorption of N_2 [12,13], because of molecular nitrogen does not react with the Si surface at temperatures below 800°C [14].

At $T < 500°C$ where the desorption probability of SiO is negligibly small, the reaction (3)-(5) may be expressed as

$$O_2(prec) + Si(s) \xrightarrow{k_r} SiO_2 (s), \quad (6)$$

where k_r is the rate constant for SiO_2-like species formation. The detailed studies of the initial oxidation at temperatures between -100 and 300°C has been performed by Gupta et al. [2]. With the precursor model involving the reactions (1), (2) and (6), they have obtained that in the steady state approximation the slow decrease of S_o with temperature is described by

$$S_o = \frac{\alpha}{1 + (k_d^o / k_r^o) exp[-(E_d - E_r)/kT]} \quad (7)$$

with $E_d - E_r \approx 0.04$ eV and $\alpha = 0.45$, where the rate constants are presented in the form $k_i = k_i^o \, exp(-E_i / kT)$.

At $T > 500°C$, SiO desorption is significant [7,15]. To express S_o in this case, the reaction (3) and (4) for SiO(g) formation have been written into one stage [7]

$$SiO(g) \xleftarrow{k_{dr}} \tfrac{1}{2} O_2 (prec) + Si (s), \quad (8)$$

where k_{dr} is the rate constant for volatile SiO formation. Then, for reactions (1), (2), (6) and (8) S_o can be described as

$$S_o = \frac{\alpha}{1 + k_d / k_r + k_{dr} / k_r}. \quad (9)$$

The parameters of the rate constants were determined by fitting Eq. (9) to the experimental data for S_o in the temperature range between about 320 and 560°C. Figure 3 shows the result of the fit. It was obtained that $E_d - E_r = 0.24 \pm 0.03$ eV, $E_{dr} - E_r = 3.0 \pm 0.2$ eV and $\alpha \approx 0.5$ [7]. The value of $E_{dr} - E_r$ is close the activation energy for SiO formation in O_2 beam scattering experiments [16]. Figure 4 shows the schematic adiabatic potential energy curves for O_2 on Si(111) for the two temperature ranges.

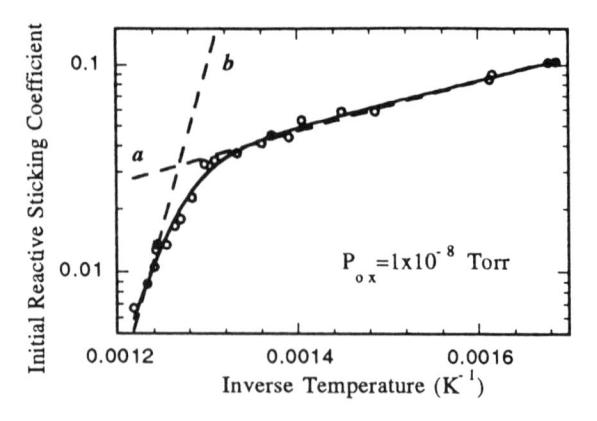

Figure 3. Initial reactive sticking coefficient of O_2 on Si(111)-7×7 versus inverse temperature. The dashed lines (*a*) and (*b*) are the approximation of the experimental data by exponent $S_o = S_o^{(o)} \exp(E/kT)$ with $E=0.25$ and 3.09 eV, respectively. The solid line is the fit of Eq. (9) to the experimental data for the O_2 pressure of 1×10^{-8} Torr [7].

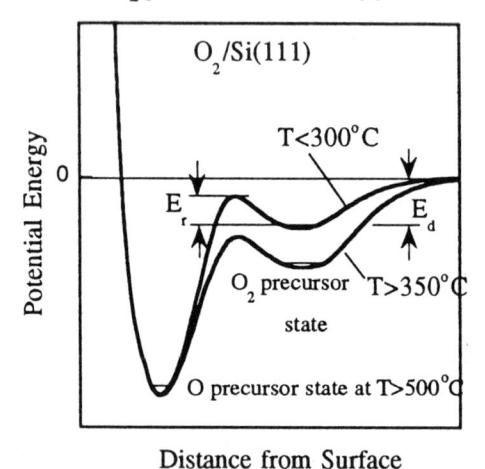

Figure 4. Schematic adiabatic potential energy curves as a function of the distance of an O_2 molecule from a Si(111) surface for the two temperature ranges. The deep minimum corresponds to the dissociated adsorption state which at $T>500°C$ becomes intermediate preceding either volatile SiO or stable SiO_2-like species formation.

Spectroscopic studies [3] show an existence of the intermediately adsorbed states of O_2 on the Si(100) surfaces even if the kinetic measurements do not indicate a significant temperature dependence of the initial oxidation [17]. This feature can be easy explained by the precursor model with $E_d - E_r \approx 0$. Although a decrease of S_o with increasing temperature is usually associated with the precursor state, recently it was shown that the adsorption data could be successfully described with a reaction proceeding through the precursor state for O_2 adsorption on Si(111)-2×1 where the increase of S_o with temperature have been observed [18]. Therefore, it is reasonably to assume that the weakly activated adsorption of N_2O on Si(100) [8] can also be consistent with the precursor model and might be characterized by the adiabatic potential energy curve like shown in Fig. 2, being modified for the case $E_r > E_d$.

3. Reversible adsorption of N_2O and O_2 on an oxidized silicon surface

The number density of adspecies on the precursor state is determined by the balance between the incident adsorption flow and the expense of adsorbed molecules for their desorption, Si etching and stable oxide formation. As a result, the adspecies coverage is too small to be undoubtedly measured at small gas pressures. As will be shown below, despite an increase of the adspecies coverage with increasing gas pressure, this dependence remains significantly weaker than the pressure dependence of the oxide growth rate. This makes measurements of the adspecies coverage on the bare silicon surface problematic also at big gas pressures. However, the coverage of intermediately adsorbed molecules can be measured on an oxidized silicon surface. Figure 5 shows the time dependence of the coverage for the reversible adsorption of N_2O and O_2 on SiO_2 obtained with ellipsometry. With gas admittance the adspecies coverage increases in time reaching the balanced value which depends on temperature. Figure 6 shows that the balanced adspecies coverage θ_{ad} increases with gas pressure. It has been determined that the adsorption isotherms could be well described by the Langmuir formula: $\theta_{ad} = aP_{ox} / (1 + aP_{ox})$, where a is a coefficient depending on temperature [8]. This result directly shows an existence of adsorption-desorption processes on the oxidized silicon surface.

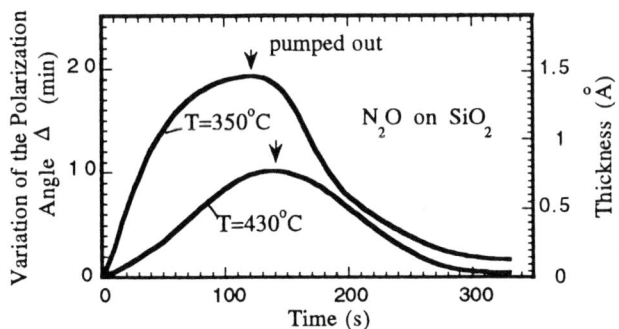

Figure 5. Time variation of the polarization angle Δ (the adsorption layer thickness is given on the right-hand axis) during reversible N_2O adsorption on an oxidized silicon surface at a pressure of 2.3×10^{-2} Torr for two temperatures [8]. The arrows mark when exposure was finished.

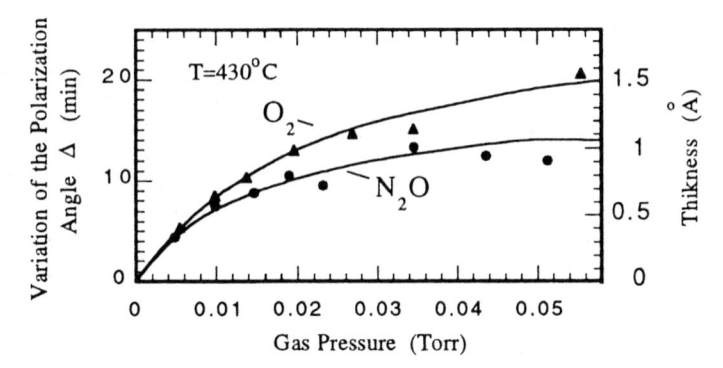

Figure 6. Pressure dependence of the polarization angle Δ measured in saturation of reversible adsorption of N_2O and O_2 on an oxidized silicon surface [8]. Solid lines show the approximation of the experimental data by the Langmuir formula.

4. Model for oxide-cluster nucleation and growth

At elevated temperatures, the oxide growth proceed via oxide cluster nucleation. This was firstly suggested by Smith and Ghidini [9] and have been confirmed by other researches with several experimental methods [4,6,10]. Using scanning tunneling microscopy (STM), Feltz, Memmert and Behm have directly observed homogeneous nucleation of oxide clusters at P_{ox} above P_{tr} [6]. Figure 7 shows that the initial oxide growth rate and subsequent oxide decomposition rate obtained at the same temperature after oxygen evacuation, decrease progressively with pressure approaching P_{tr} [19]. Figure 8 shows that the initial reactive sticking coefficient S_o also decreases with pressure decreasing to P_{tr}. The pressure dependence of S_o is not usual behavior for a gas-solid interaction and reflects an interaction in the layer of intermediately adsorbed species. As is seen in Fig. 7 this interaction arises near P_{tr} showing the increased influence of nucleation stage on the kinetics of oxide formation. However, the precursor model describing the temperature dependence of initial oxidation does not include the interaction between adspecies in the reaction (5), and therefore another model should be considered for the pressure dependence observed near P_{tr}.

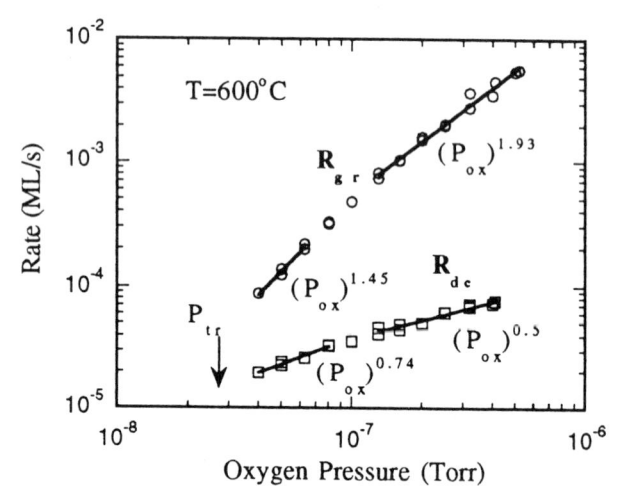

Figure 7. Pressure dependence of the initial oxide growth rate R_{gr} and the oxide decomposition rate R_{dc}. The arrow marks the transition pressure $P_{tr}(T)$. The pressure dependence of the rates are fitted by a power function P_{ox}^{ε} (solid lines), where ε is a fitting parameter [19].

Figure 8 shows that at 600°C S_o increases with O_2 pressure, but even at $P_{ox} \approx 20 P_{tr}$, $S_o \approx 2.2 \times 10^{-2}$ that is much smaller than $S_o = 0.14$ at room temperature (RT). At elevated temperatures, two reactions of oxygen removal from the surface are responsible to the small value of S_o. One of them is the reaction (2), oxygen desorption from the precursor state. An increase of the desorption rate with temperature is characterized by the activation energy $E = E_d - E_r$. In the wide temperature range around RT, $E \approx 0.04$ eV [2,5] and 0.24 eV at $T \geq 350°C$ [7] were determined. These values of the activation energy give an increase of the ratio $R_d / R_{gr} = \alpha / S_o$ by a factor of 10 with temperature increased from RT to 600°C, thereby providing the relationship

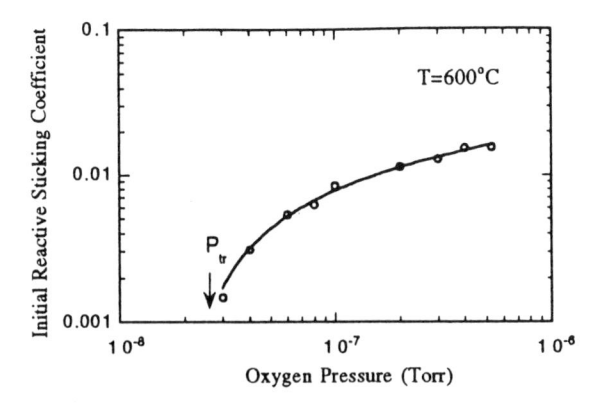

Figure 8. Pressure dependencies for the initial reactive sticking coefficient at 600°C. The arrow marks the transition pressure P_{tr} [22]. The solid line is an approximation of the experimental data by a smooth curve.

$R_d \gg R_{gr}$ at 600°C, where R_d is the oxygen removal rate from the precursor state due to oxygen desorption, and α is the trapping coefficient of O_2 on the precursor state, which value is about 0.5 [2,7]. The other is the reaction (4), Si etching with SiO formation. The probability of this reaction for O_2 molecule impinged the surface is about 0.1 obtained with different experimental techniques [6,7,15]. This value is much bigger than $S_o \approx 0.01$ at 600°C. Thus, the each channel for oxygen removal provides $R_d \gg R_{gr}$ thereby showing that an expense of oxygen from the precursor state for oxide formation is much smaller than for O_2 and SiO desorption. As a result, the adspecies coverage θ_{prec} is determined by a dynamic balance between adsorption and desorption.

For the condition where the adsorption rate $R_{ad} \sim P_{ox}$ and $R_d \sim \theta_{prec}$, the balance gives

$$\theta_{prec} \sim P_{ox} \qquad (10)$$

The relationship (10) is valid not only for small θ_{prec} corresponding to P_{ox} near P_{tr}, but also for P_{ox} well above P_{tr}, if the interaction between adspecies does not affect the adsorption and desorption. In (10), θ_{prec} is formed by either O_2 adsorbed on the precursor state or SiO-like species appeared in the reaction (3) for O_2 dissociation.

Due to intensive adsorption and desorption, adspecies coverage θ_{prec} is maintained in a period of time much shorter than that for oxide growth. Therefore, during oxide growth, θ_{prec} can be treated as a constant with a homogeneous distribution on the surface, whose value is proportional to O_2 pressure according to (10). This allows the application of a steady state approximation to determine relations between the fast reactions which accompany the slow oxide growth.

Nucleation and growth in epitaxy are usually described by rate equations [20] in which an approximate form for the number density of oxide islands may be expressed as

$$dN/dt = k\,\theta_{prec} N_c, \qquad (11)$$

where N and N_c (cm^{-2}) are the number densities of stable oxide islands and critical ones, respectively, and k is the rate constant for capturing adspecies by the critical islands to form a stable one. The relation

$$N_c \sim \theta_{prec}^i \quad (12)$$

can be obtained for our case in the steady state approximation for N_c with the exponent i showing the number of adspecies nucleating an island. Since (10) is valid for each kind of adspecies, we assume that the formation of the critical island may include different adspecies. For example, SiO-like species capturing an O_2 molecule forms the oxide complex, i.e., the critical island of a size $i=2$ which could be then transformed into the stable SiO_2-like island by additional oxidation.

At elevated temperatures, oxide growth and decomposition are known to mainly occur on a perimeter between an oxide island and the bare silicon surface [6,21]. Therefore the oxide growth rate is proportional to the number density L (cm^{-2}) of reaction sites for capturing adspecies on the perimeter and is given by

$$R_{gr} = k_r \theta_{prec} L , \quad (13)$$

where k_r is the rate constant for oxide growth. Note that the description of the oxide growth rate by (13) does not take into account oxide decomposition during oxide growth. Such approximation is valid where R_{gr} is much higher than R_{dc}, i.e., for the kinetic data shown in Fig. 7.

For growth due to a reaction on an island perimeter, it has been derived [22] that R_{gr} is expressed as a function of stable oxide coverage θ_{ox} as

$$R_{gr} \sim \theta_{prec} (\theta_{ox}^2 N_c)^{1/3} . \quad (14)$$

For the oxide growth rates determined at an identical value of oxide coverages for different O_2 pressures, i. e., when θ_{ox} becomes only a numerical coefficient, the insertions of (10) and (12) into Eq. (14) gives the scaling relationship

$$R_{gr} \sim P_{ox}^{i/3+1} \quad (15)$$

among the critical island size i, P_{ox} and R_{gr}. The relationship (15) allows us to draw out i from our kinetic data even if we do not know exactly the kinetic parameters of the reactions.

To obtain the pressure dependencies of i, the experimental data for R_{gr} have been approximated by the power function $P_{ox}^{\varepsilon_{gr}}$ in several narrow pressure ranges [19,22]. Then, according to (15), the critical cluster size was determined from

$$i = 3(\varepsilon_{gr} - 1). \quad (16)$$

Figure 9 shows that the contribution from critical islands of larger sizes gradually increases with O_2 pressure approaching P_{tr} giving the increase of the average critical oxide island size from 1.3 to 2.8. To our knowledge, the possibility of the pressure dependence of i has not been considered for epitaxial growth. In our case it can be related with the existence of the critical conditions for silicon oxidation. With O_2 pressure decreasing to P_{tr}, the oxide growth rate decreases progressively (Fig. 7). At a decreased rate of oxide growth, only those islands with a longer lifetime can grow. Since islands of a larger size have a longer lifetime, the pressure dependence of i may thus reflects the dependence of the lifetime of the critical island on its size [22].

Future determination of the critical oxide island size for the interaction of N_2O with silicon is of a definite interest. Although, there is no difference between O_2 and N_2O in the reactions (1)-(4), an essential difference might exist in the mechanism of critical island formation. The reaction probability for stable oxide formation is smaller for N_2O

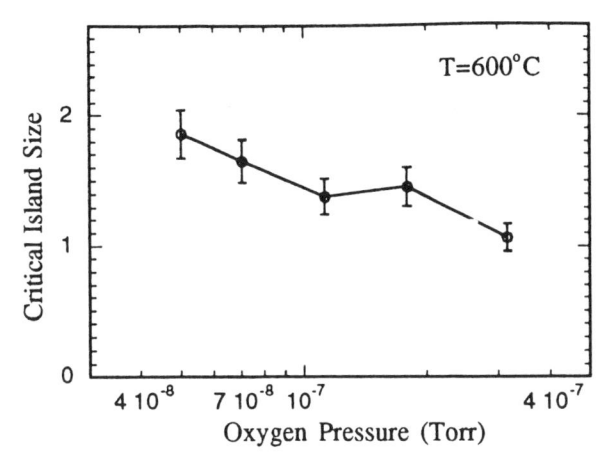

Figure 9. Pressure dependence of the critical oxide cluster size at 600°C [22].

than that for O_2 [8,12]. This might be due to introduction of two oxygen atoms into island nucleation by O_2 dissociation, instead of one in the case of N_2O. If the both oxygen atoms take part in island nucleation, then the critical island size i should be smaller for O_2 than that for N_2O. The expected difference between values of i would show a number of oxygen atoms which involve in island nucleation due to dissociation of molecules on the precursor state.

5. Oxide island size in the critical conditions for silicon oxidation

As shown in Figure 10, two branches of the critical conditions for silicon oxidation have been observed. The branch P_{tr} shows the minimal P_{ox} for the nucleation of oxide islands on the bare silicon surface, whereas the branch $P_c (< P_{tr})$ gives the minimal P_{ox} for the existence of large oxide islands under a dynamic equilibrium between the growth

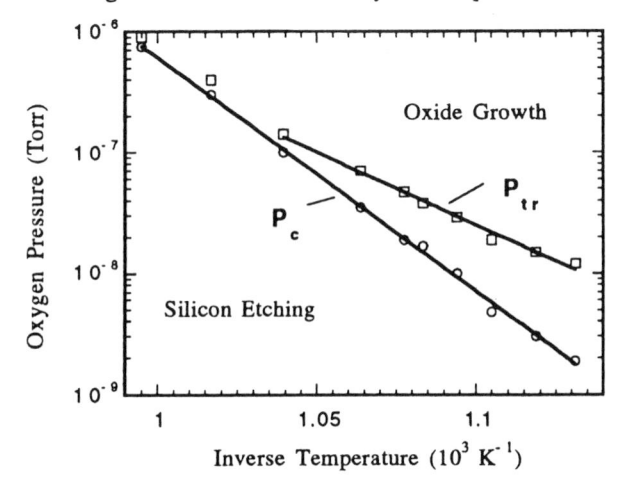

Figure 10. Plot of the critical conditions for the interaction of O_2 with Si(111)-7×7. The solid line P_{tr} is the boundary for oxide nucleation on the surface. The solid line P_c presents the boundary where growth and decomposition of oxide clusters are balanced [10].

and decomposition processes [10]. For P_{ox} between P_{tr} and P_c, the sense of the critical cluster may be borne to an *equilibrium* island whose size stays constant at a fixed P_{ox}. Since the critical oxide island size increases with P_{ox} decreasing to P_{tr}, a further decrease of P_{ox} to P_c is expected to cause an increase of size, in this case, of *equilibrium* islands. The clustering is a process which itself produces structures of a nanoscale size. Oxidation near the critical conditions forms surfaces covered with oxide islands. Such surfaces are convenient substrates for development of the selective growth on nanoareas of either oxidized or bare silicon. Another aspect of oxidation near the critical conditions concerns the existence of the two boundaries P_{tr} and P_c (Fig. 8) which shows that the initial stage of oxide formation requires O_2 pressure well above that requires for oxide growth. This feature may be employed for Si etching in nanowindows in ultrathin SiO_2 used as a mask.

6. Conclusion

Intermediate adsorption states are the key feature of the interaction of O_2 and N_2O with silicon at the early stage of oxide formation. Competing reactions such as desorption, Si etching with SiO production and stable oxide formation give rise to the critical condition for silicon oxidation at elevated temperatures. Near the critical conditions, stable oxide forms through island nucleation proceeding with an interaction of intermediately adsorbed species. The silicon oxide clustering would be a process applicable for nanostructure fabrications.

Acknowledgments

The optical second-harmonic generation studies were performed in the Institute of Physical and Chemical Research (RIKEN), Japan. This work was partly supported by the New Energy and Industrial Technology Development Organization (NEDO).

References

1. Engel, T. (1993) The interaction of molecular and atomic oxygen with Si(100) and Si(111), *Surf. Sci. Rep.* **18**, 91-144.
2. Gupta, P., Mak, C. H., Coon, P. A., and George, S. M. (1989) Oxidation kinetics of Si(111) 7×7 in the submonolayer regime, *Phys. Rev.* B **40**, 7739-7749.
3. Silvestre, C. and Shayegan, M. (1991) Initial stages of the reaction of oxygen with Si(100), *Solid State Communications* **77**, 735-738.
4. Borman, V. D., Gusev, E. P., Lebedinski, Yu. Yu., and Troyan, V. I. (1994) Mechanism of submonolayer oxide formation on silicon surfaces upon thermal oxidation, *Phys. Rev.* B **49**, 5415-5423.
5. Bratu, P., Kompa, K. L., and Höfer, U. (1994) Kinetics of oxygen dissociation on Si(111)7×7 investigated with optical second-harmonic generation, *Phys. Rev.* B **49**, 14 070-14073.
6. Feltz, A., Memmert, U., and Behm, R. J. (1994) High temperature scanning tunneling microscopy studies on the interaction of O_2 with Si(111)-(7×7) surface, *Surf. Sci.* **314**, 34-56.
7. Shklyaev, A. A. and Suzuki, T. (1996) Initial reactive sticking coefficient of O_2 on Si(111)-7×7 at elevated temperatures, *Surf. Sci.* **351**, 64-74.

8. Baklanov, M. R., Kruchinin, V. N., Repinsky S. M., and Shklyaev, A. A., (1989) Initial stages of the interaction of nitrous oxide and oxygen with the (100) silicon surface under low pressures, *React. Solids* **7**, 1-18.
9. Smith, F. W. and Ghidini, G. J. (1982) Reaction of oxygen with Si(111) and (100): Critical Conditions for the Growth of SiO_2, *Electrochem. Soc.* **129**, 1300-1306.
10. Shklyaev, A. A. and Suzuki, T. (1995) Branching of critical conditions for Si(111)-(7×7) oxidation, *Phys. Rev. Lett.* **75**, 272-275.
11. Schubert, B., Avouris, Ph., and Hoffman, R. (1993) A theoretical study of the initial stages of Si(111)-7×7 oxidation, *J. Chem. Phys.* **98**, 7593-7605.
12. Keim, E. G., Wolterbeek, L., and van Silfhout, A. (1987) Adsorption of atomic oxygen (N_2O) on a clean Si(100) surface and its influence on the surface state density; a comparison with O_2, *Surf. Sci.* **180**, 565-598.
13. Uno, K., Namiki, A., Zaima, S., Nakamura T., and Ohtake, N., (1988) XPS study of the oxidation process of Si(111) via photochemical decomposition of N_2O by an UV excimer laser, *Surf. Sci.* **193**, 321-335.
14. Schrott, A. G. and Fain, S. C., Jr. (1981) Nitridation of Si(111) by nitrogen atoms, *Surf. Sci.* **111**, 39-52.
15. Shimizu, N., Tanishiro, Y., Takayanagi, K., and Yagi, K. (1987) On the vacancy formation and diffusion on the Si(111)7×7 surfaces under exposures of low oxygen pressure studied by in situ reflection electron microscopy, *Surf. Sci.* **191**, 28-44.
16. Memmert, U. and Yu, M. L. (1991) Comparison between Si(100) and Si(111) in the reaction with oxygen at high temperatures, *Surf. Sci. Lett.* **245**, L185-L189.
17. Watanabe, H., Kato, K., Uda, T., Fujita, K., Ichikawa, M., Kawamura, T., and Terakura, K. (1997) Kinetics of initial layer-by-layer oxidation of Si(001) surfaces, submitted to *Phys. Rev. B*.
18. Goletti, C., Chiaradia, P., Moretti, L., Yian, W., Chiarotti, G., and Selci, S., (1996) Activated chemisorption of oxygen on Si(111)-2×1, *Surf. Sci.* **356**, 68-74.
19. Shklyaev, A. A., Aono, M., and Suzuki, T., (1996) Influence of growth conditions on subsequent submonolayer oxide decomposition on Si(111), *Phys. Rev.* B **54**, 10890-10895.
20. Zangwill, A. (1993) Scaling description of sub-monolayer epitaxial growth, in H. A. Atwater, E. Chason, M. Grabow, and M. Lagally (eds.), *Evolution of Surface and Thin Film Microstructure*, MRS Proceedings Vol. 280, Materials Research Society, Pittsburgh, pp.121-130.
21. Frantsuzov, A. A. and Makrushin, N. I. (1976) Growth of an oxide film on a clean silicon surface and the kinetics of its evaporation, *Thin Solid Films* **32**, 247-249.
22. Shklyaev, A. A. and Suzuki, T. (1997) Temperature dependence of the critical oxide cluster size on Si(111), submitted to *Surf. Sci.*

SCANNING TUNNELING MICROSCOPY ON OXIDE AND OXYNITRIDE FORMATION, GROWTH AND ETCHING OF Si SURFACES

H. Neddermeyer, T. Doege, E. Harazim, R. Kliese, A. Kraus, R. Kulla, M. Mitte and B. Röttger
Martin-Luther-Universität Halle-Wittenberg
Fachbereich Physik
D-06099 Halle, Germany

An overview is given on scanning tunneling microscopy (STM) studies of the interaction of O_2, NO and NO_2 with Si(111)7x7 and Si(100)2x1. The initial stages of adsorption and reaction have been measured in real-time which allows the observation of atomic processes in detail. The results show a large variety of individual effects which in some cases follow characteristic reaction paths. The assignment of the individual adsorption and reaction events to individual atomic effects, i.e., to adsorption of atomic or molecular species on specific adsorption sites is difficult, since the resulting local changes of the electronic structure are not known *a priori*. For O and O_2 a number of possible adsorption sites have been treated theoretically which may be used for their identification in the STM images. For Si(100)2x1 already from the beginning of the gas exposure different adsorption sites will be occupied. The N containing gases show similar effects as pure O_2 except for the reactivity and the distribution of specific adsorption events. Adsorption at higher temperatures gives rise to macroscopic changes of the surface which appear as Si etching and for longer exposure as oxide film formation. In contrast to results from the literature for Si(111) no indication for an island growth mode in the initial stage was found. In case of NO on Si(111)7x7 subsequent heating produces ordered Si nitride structures.

1. Introduction

Si oxide and nitride films are of great technological importance due to their dielectric and insulating properties. Unfortunately, thicker films which are of interest for applications cannot be used in scanning tunneling microscopy (STM) because of insufficient conductivity. Only very thin films with a thickness of one or at most a few monolayers (ML) may be imaged by STM when the additional potential barrier for tunneling is thin enough to allow tunneling. The principal advantage of STM compared to atomic force microscopy is the possibility to detect the local electronic structure which is particularly sensitive against adsorption, reaction and characterization of film formation. It is clear that reliable information on the adsorbate/substrate systems can only be obtained when the experiments are performed under ultrahigh vacuum (UHV) conditions. This means that the substrates have first to be prepared in UHV to obtain a well defined surface state and subsequently exposed to clean gases. This is particularly important for less reactive molecules such as NO where additional and more reactive species in the gas could produce misleading features in the STM images.

In spite of these difficulties the adsorption and reaction of Si surfaces with O and N containing gases have been studied by means of STM for some years by a number groups [1-10]. The first experiments have been performed at room-temperature and at very low O_2 pressure where adsorbate states have been identified [2-3]. Later, the studies have been extended to adsorption at higher temperatures where due to improvements of the instrumentation the effects could be monitored *in situ* and in real-time [5-6]. In addition to adsorption events one observes reactive etching due to formation of volatile SiO. One part of the observed local defects (which are also seen in the adsorption stage) have been assigned to the oxide nuclei. Similar experiments have been performed for N containing gases [1, 9, 10].

In the present work an overview is given on typical results in this field obtained by STM both at room and elevated temperature. The adsorption of O_2 on Si(111)7x7 at room-temperature as detected by STM in real-time is shown and explained in some more details. It will be demonstrated that in the initial stage the adsorption sites are located on or at least near the Si adatoms of the 7x7 unit cells. With increasing exposure deviations from the 7x7 symmetry are clearly detected which probably indicates a transition to Si oxide formation. The usefulness of measuring local *I/V* characteristics is demonstrated for the adsorption of NO_2 on Si(111)7x7. In a second step the same kind of measurements have been performed on a hot (770 K) Si(111)7x7 substrate. Initially, we see here the same adsorption phenomena as at room-temperature and in addition etching of step edges and on antiphase boundaries of 7x7 domains. With increasing exposure the formation of a thin oxide layer is observed on the entire surface which passivates the surface and prevents from further etching. As another example the exposure of Si(111)7x7 with NO at 860 K is shown which gives rise both to etching of step edges and formation of oxynitride islands. These islands can be transformed into a thin ordered Si nitride structure by further annealing at 1230 K. After these experimental results have been explained in some details an overview on other experiments is given (in particular O_2 on Si(100)) where only the main effects are summarized.

2. Experiment

The experiments have been performed in two different UHV chambers which are equipped with a "high-speed" STM working at room-temperature [4] and a high-temperature STM [11]. In both systems *in situ* measurements during adsorption have been made and conventional measurements after sample treatment in order check the influence of the scanning tip and the electrical fields during tunneling. Care has also been taken to exclude fragmentation or excitation of the molecules by ionization or thermal decomposition due to the instrumentation such as vacuum gauges, pumps or filaments. The sample treatment before the adsorption experiments consisted in slowly degassing at around 800 K and final flashing to above 1450 K where also Si nitrides desorb. This normally produces Si surfaces with a high degree of cleanliness and order. The local defect concentration possibly related to residual contamination or vacancies was in the order of less than 0.1 % and 1 % for Si(111) and Si(100), respectively. The exposure with the various gases is given in Langmuir (L) without correcting the reading of the ion gauges for the different gases (1 L is equivalent to an exposure of 1.33×10^{-6} mbar s). The images have been obtained in the constant current mode and the results are shown in form of constant current topographies (CCTs). Except for one example, not much effort has been made to measure local *I/V* curves. The measurement of such characteristics on O containing surfaces turned out to be rather difficult probably due to O takeover by the tip. However, already from the voltage dependence of the images very important qualitative information on the electronic states of the sample can be derived. For example, in case of NO/Si(111)7x7 one is able to distinguish between an etched part of the sample and a reacted part which both appear as depressions against the surrounding area. For the reacted part we needed an unusual high sample bias voltage (larger than 4 V) for optimum resolution. This was also found for O_2 adsorption and reaction on Si(111)7x7 where the adsorption states could well be imaged by using a sample bias $U = 2$ V, while the surface which we believe to be completely covered by a thin oxide layer could only be imaged with satisfactory resolution for $U > 4$ V.

3. Results and Discussion

3.1. Adsorption: O_2 on Si(111)7x7

The use of STM in adsorption experiments is motivated by the hope of obtaining information on the following structural and electronic details: (a) identification of adsorption sites and geometry, (b) reaction path and in particular presence of molecular precursor, (c) information on reaction products and (d) on structural and electronic changes of the substrate. In the present case, the transition to formation of an oxide film or to an oxide precursor is of particular importance and will be described in the subsequent section. Here only the changes of the surface due to adsorption of less than 0.1 ML of O are described.

One of the first examples for visualizing local chemical adsorption events in STM images has been described by Wolkow and Avouris for NH_3/Si(111)7x7 [1]. Not much later, the interaction of O_2 with Si(111)7x7 has been studied by a number of groups which demonstrated that the adsorption of O_2 directly affects the Si adatoms of the 7x7 unit cells. As a result of the O_2 exposure individual adatoms seem to have increased their height against the clean substrate (in a gray tone representation they appear as a bright state) and some seem to have disappeared (dark state). It was a straightforward idea to relate these contrast changes in the gray tone representations to the adsorption of O atoms in certain sites which then change their local electronic structure in a specific way. Since in the initial stage these changes are exactly found on the position of the adatoms, the incorporation of O in the dangling bond or the back-bond of the adatom may explain the contrast changes. By using a "high-speed" STM and imaging during O_2 adsorption a two-stage process has been identified. For this experiment an O_2 partial pressure in the STM chamber of 10^{-8} mbar has been employed which gave rise to only small number of atomic changes between subsequent images. The history of individual atomic positions with increasing exposure can then be followed on the time scale of the image repetition rate (5.8 s in this case) and precursor states or other dynamic or kinetic effects can be identified directly. This is demonstrated in Fig. 1 where the above mentioned typical changes between subsequent images of a series of CCT measurements are displayed. In Fig. 2 the above mentioned effects have been replotted from Fig. 1 in a magnified scale. In (a) the transition from a normal state to a bright state is shown, in (b) the "disappearance" of a bright state and in (c) the direct transition from a normal state to a dark state. These are the changes of the structure which become visible in the first part of the

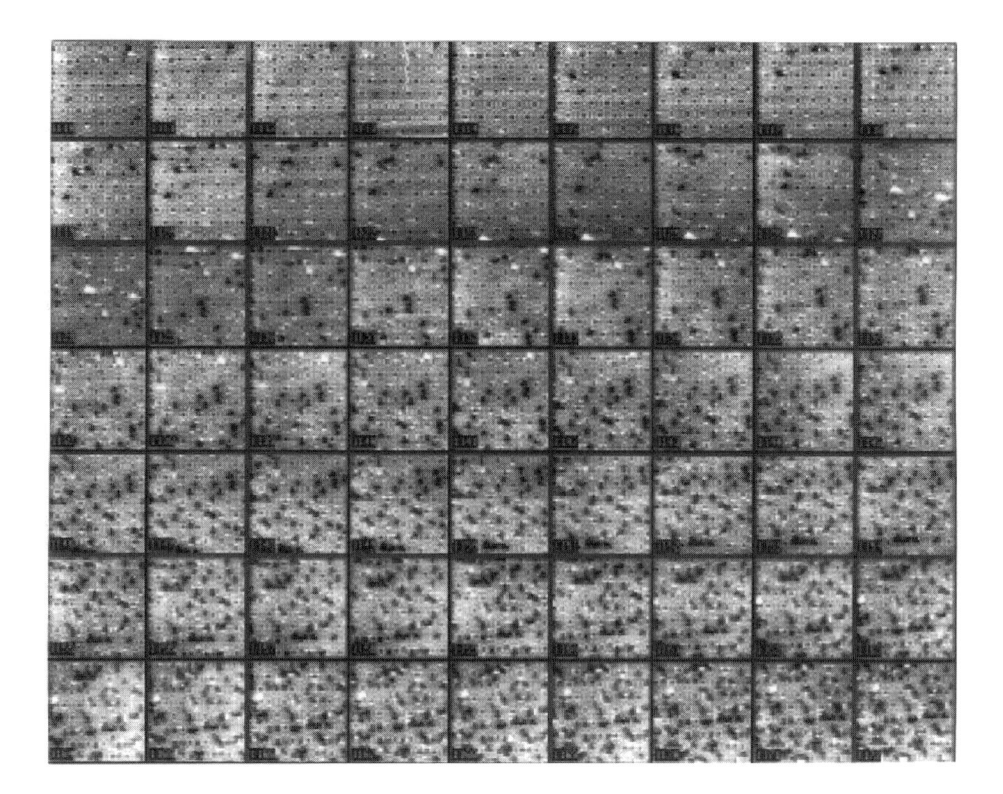

Figure 1. Sequence of CCTs obtained on Si(111)7x7 during O_2 adsorption. The frames have an area of 20x20 nm^2.

adsorption experiment where still all visible atomic structures are arranged in a perfect (though apparently incomplete) 7x7 arrangement. By following the individual history of the bright states we found that they were unstable turning then to dark which could be described by an exponential law. If we assume that the events leading to a dark state always go through the bright state (although not detectable by STM due to the limited time resolution) we obtained a lifetime of 100 s for the bright states. This corresponds to the time needed for adsorption of 1 ML in case of a sticking coefficient of one and means that the reactivity of the bright states to O_2 is in the order of one. It has to be mentioned that without O the bright states are stable. We did not find a clear indication for a tip-induced change from bright to dark. The observation of Avouris et al. that such changes did occur [3] might then be explained by a transfer of an O atom from the tip to Si surface.

For the interpretation of the bright and dark state one has to consider that they obviously constitute chemical and electronic changes of the Si adatoms of

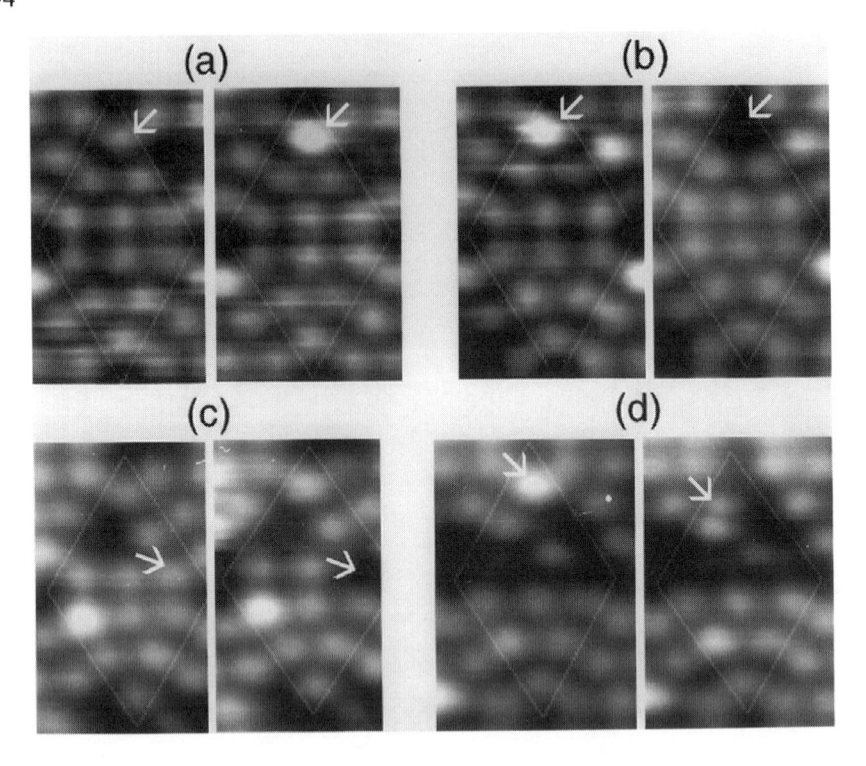

Figure 2. Individual processes during O$_2$ adsorption on Si(111)7x7.

the 7x7 structure. Since both states most likely correspond to those of a two-stage adsorption process, it is plausible to relate them to the adsorption of one and two O atoms, respectively. Intuitively, a bonding of an O atom to the dangling bond orbital is expected to reduce the density of states at the Fermi level because of the metallic character of the dangling bond surface states. Metallicity of these states will disappear upon bonding to an electronegative atom. This means that in the dark state one O atom is expected to be located on top of the Si adatom. Therefore, in the first step of the two-stage adsorption process the O atom is probably introduced in the back-bond position of the Si adatoms. This interpretation has already been given by Avouris et al [3] on the basis of tight-binding calculations and is thus supported by our more qualitative arguments. It should only be noted that Pelz and Koch came to a different conclusion on the nature of the dark state since they explained it by the incorporation of two O atoms in back-bond positions of the same adatom [2]. We note that in the course of the measurements we also observed distinct shifts of adatom structures due to beginning destruction of the 7x7 order (Fig. 2 (d)).

In case of high-temperature measurements we see the same kind of transitions (see below). Actually, the dark states have also been observed by Feltz et al. who have associated them with oxide clusters [6]. Also in the work of Ono et al. [7] this initial adsorption state is ascribed to oxidation. To our opinion, this terminology is somewhat misleading, since the initial adsorption state can at most be described as an oxide precursor. As will be discussed below, we are able to distinguish between the initial adsorption state of the sample as depicted in Figs. 1 and 2 and the formation of an oxide layer. At high temperature (770 K) and without the presence of O in the rest gas atmosphere the first one is unstable, it recovers to a clean 7x7 surface. On the contrary, the oxide film has been found unchanged after the same time and at annealing at the same temperature.

In Fig. 3 we show the result of a spectroscopic imaging of Si(111)7x7 which has been exposed to NO_2 and shows the same kind of features (dark and bright sites) as have been described for O on Si(111)7x7. The local I/V characteristics of the various kind of sites show systematic differences which may be attributed to differences in the local density of states. In the top part of the Fig. a CCT is reproduced showing the atomic sites of interest and in the bottom part the measured characteristics of these sites. It has to be noted that current characteristics may be influenced by electronic states of the tip which, unfortunately, is not always metallic in particular in case of measurements on O containing surfaces. However, differences of the curves when obtained with the same state of the tip may be attributed to the differences of the electronic states of the various atomic sites. According to the experimental results the unreacted adatom positions at the corner hole shows a maximum in the occupied density of states at -1.2 V (below the Fermi level) in agreement with the work of Avouris et al. and a very strong peak in the unoccupied density of states at 1.5 V which both have to be associated with the back-bonds (see curve 4). The initial adsorption state (bright site labeled 1) exhibits drastic shifts of the peak position both below and above the Fermi level which may qualitatively be attributed to the formation of new back-bonds by introduction of an O atom as has been proposed by Avouris et al. The dark state shows only some minor peaks on an increasing background towards lower and higher voltage, respectively. A qualitative interpretation is difficult in this case. Since the tunneling tip is penetrating more deeply into the surface, we believe that also neighboring sites my attribute by their electronic states to dI/dV. This could explain the general lack of structures in this curve (2). We note the characteristic peak of the density of states at -1.2 V is smeared out on site 3. We attribute this effect to the immediate neighborhood of reacted position 2.

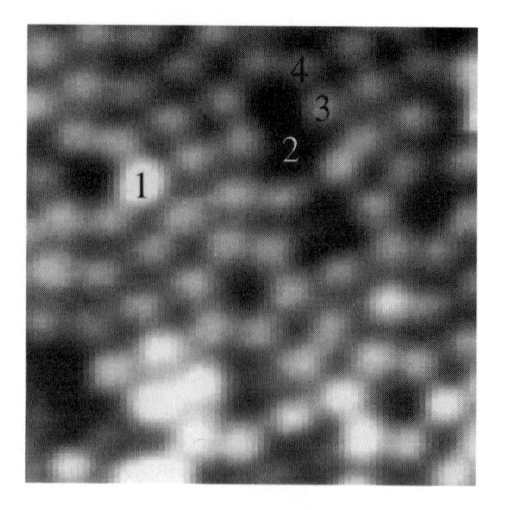

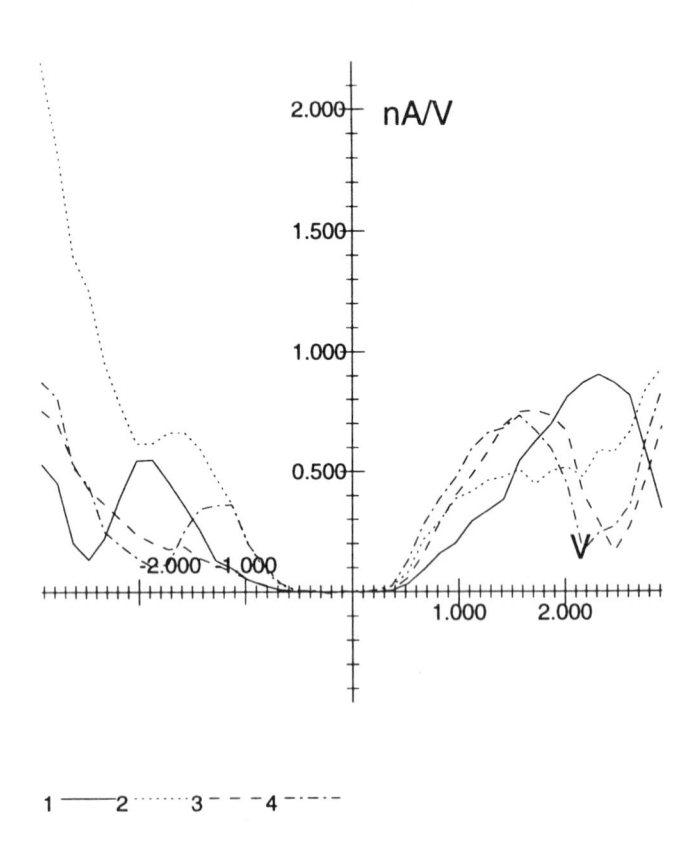

Figure 3. Spectroscopic measurements on NO$_2$ on Si(111)7x7. The local *I/V* curves have been differentiated numerically.

3.2. Reactive etching and oxide formation: O_2 on Si(111)7x7

For high-temperature adsorption an "active" and a "passive" oxidation stage is normally distinguished [12]. Below a critical O_2 pressure the adsorption leads to formation of volatile SiO and thus to etching of the Si surface and above the critical pressure a SiO_2 layer gradually develops. *In situ* STM measurements of the interaction of O_2 with Si(111)7x7 at high-temperatures by Feltz et al. have already shown that the mechanism of passive oxidation also includes an "active" part in the beginning of the gas exposure until one double layer of Si(111)7x7 has been removed [6]. In Ref. [6] a detailed description of the various processes leading to formation of a SiO_2 layer is also given. Our results, which on the first sight are qualitatively in agreement with the previous STM work show some additional details, however, with some interesting consequences for the mechanism of oxide formation in the range of the critical pressure. Some of the details concern the role of the adsorption states at high-temperature which we compare with our "fast" adsorption measurements at room-temperature. The second important aspect is that at higher exposure with O_2 we observe the formation of a passivating layer on the surface from where on the additional growth of the oxide film proceeds very slowly. To our opinion, the passivating layer corresponds to the formation of an oxide film with minimum thickness which would be important for further applications. The passivating layer has also been considered and described in a review by Engel devoted to surface physical studies of oxide formation on Si(111) and Si(100) [12].

Results of our real-time experiment are reproduced in Fig. 4. For these measurements we followed the changes of a Si(111)7x7 surface upon O_2 exposure by more than 3 hours. We have selected characteristic frames of the entire series showing the main stages of the O interaction. The O_2 exposure is given in the Fig.

The first frame (Fig. 4 (a)) shows the clean surface and a step edge which is stable at this temperature along [110]-like symmetry directions (left hand side)and displays a typical "frizziness" for other directions (right hand side). A domain boundary pins the end of the straight part of the step und runs to the right upper corner of the measured area (probably hardly visible on the reproduction of the Fig.). A number of defects are recognized (probably not in the reproduction but very clearly on the monitor screen), they appear as dark sites in the adatom pattern and have to be associated with some contamination. At this temperature they do not show diffusion. Exposing the surface to O_2 (from

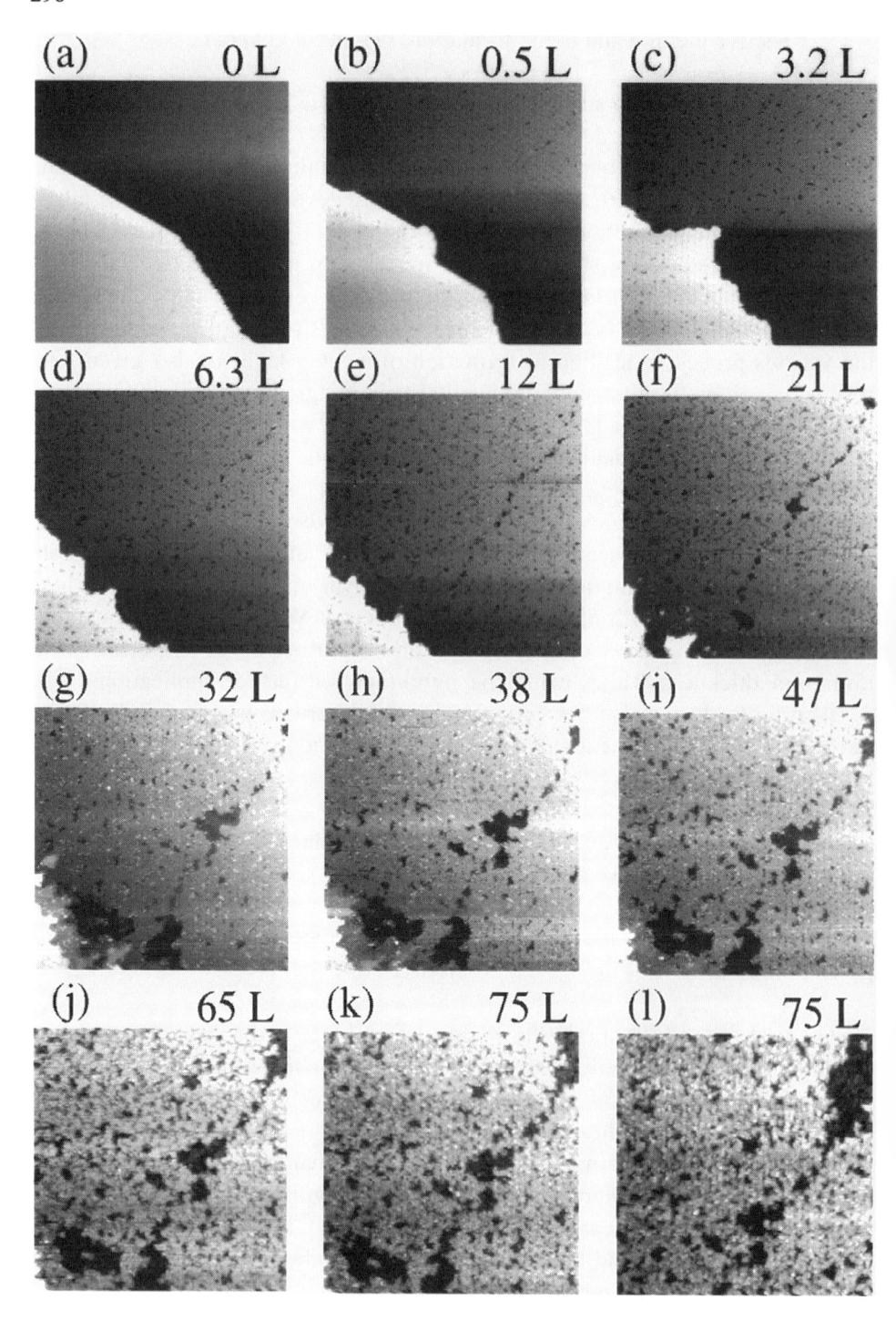

Figure 4. Sequence of CCTs obtained on Si(111)7x7 during O_2 adsorption at 770 K and an O_2 partial pressure of 10^{-8} mbar (10x10 nm^2).

(b) on) increases the number of these dark sites and produces bright sites in complete agreement with the room-temperature adsorption states. Most noticeable, however, is the reactive step etching resulting from SiO formation. The number of Si atoms removed from the step edge is by two orders of magnitude larger than the number of local atomic defects on the terraces arising from O_2 adsorption. Reactive step etching is explained by step-flow mode condensation of vacancies created by SiO desorption [6]. If we follow the time evolution at this stage we see that step edge etching approximately proceeds with a constant rate as well as that of local defect formation (Fig. 4 (c)). Not much later, however, the step etch rate slows down (from (f) and (g) on). At this stage some of the dark sites begin to condense into smaller clusters ((g) and (e)) which is particularly visible along the domain boundary. A more close inspection of the contrast in the images leads to the conclusion that only half of the local depressions correspond to the dark sites of the adsorption stage while the other half correspond to condensed vacencies (which may be partly reacted already). Obviously, the vacancies condensed in these smaller clusters are prevented from step-flow mode growth by a drastic decrease of the diffusion length. Since the undistorted 7x7 reconstruction is still present on most parts of the surface area, the production rate of volatile SiO should not be much different than in the beginning of the experiment. In the next growth stage many of these vacancy clusters show a tendency to grow laterally ((f) and (g)), some of them with exceptional large rate which means that the distribution of lateral size of the vacancy islands becomes non-uniform. There are also some indications for nuded zones of these smaller clusters around the bigger ones (Fig. 4(g)). In the basis of the latter ones the 7x7 reconstruction is sometimes well developed with a smaller density of defects than on the rest of the surface due to the native character of these bigger vacancy islands. At this stage of the adsorption experiment we made the surprising observation that without the presence of O_2 in the gas atmosphere and at a temperature of 770 K the surface completely recovers into a clean 7x7 reconstructed Si(111) surface, if only sufficient time is elapsed (10 to 20 hours in this case). This means that the observed features at this stage cannot yet correspond to SiO_2 species but merely represent adsorption stages (and etching) of the system O/Si(111). However, if we proceed with O_2 exposure the vacancy islands do not seem to grow any more ((h) and (i)). In addition, the 7x7 reconstruction becomes more and more distorted, also in the etched basis of the vacancy islands. At the end of the adsorption experiment no indications for 7x7 reconstruction or remainders of 7x7 unit cell halves are found ((j) and (k)). For optimum resolution the sample bias voltage had now to be increased to values up to 6 V. We explain this by the formation of a thin film of SiO_2 (or of an oxidic film with nearby stoichiometry) exhibiting already the wide band gap characteristics of SiO$_2$. This view is supported by the high thermal stability of the grown material. In contrast to the adsorption state described above, extended periods of annealing at the growth temperature did not change the surface structure. This means that a passivating layer has been produced which also prevents the surface from further etching attacks. Fig. 4 (l) has been measured after termination of the O$_2$ exposure in order to confirm that other parts of the surface unaffected by scanning do not look differently. Therefore, the measured area is slightly shifted.

In contrast to the previously described mechanism of oxide growth via a two-dimensional growth mode process we have to explain our data in a different way and therefore come to a different model for the initial oxide growth. Apparently, the full oxidation of the surface and the formation of the first oxide layer proceeds via a gradual reaction of the entire Si surface. To our opinion, we do not have indications for mobile oxide species on Si(111). This means that a two-dimensional island growth mode for the first oxide layer is not supported by our STM experiments. The impression of island-like features separated by deeper regions in previous experiments [7] was the basis for the conclusion that the protruding structures are the oxide islands implying that the deeper layer correspond to unreacted Si. Our data show that these deeper features have initially been formed by etching when vacancy diffusion was still possible and that they are later also oxidized continuously until the entire surface is covered with a SiO$_2$ layer of minimum thickness. The roughness of the surface exhibited by the oxide layer is a consequence of the initial quality of the surface and depends very much on the presence and distribution of defect structures on the clean surface such as steps and antiphase boundaries. These defect structures essentially determine the nucleation of the vacencies and the overall structure of the oxidized surface.

3.3. Reactive etching and oxynitride formation: NO on Si(111)7x7

Room-temperature exposure with on the order of 10 L NO at a partial pressure of 10^{-7} mbar leads to disordered adsorption structures (not shown here). On a hot (860 K) substrate the NO exposure leads to reactive etching at step edges and formation of oxynitride species in form of irregular islands. In Fig. 5 an overview of such a surface is reproduced. Two neighboring terraces are recognized. The step edge shows a meandering structure which is resulting from reactive etching of Si(111). The original position of the step (marked by "x") can still be recognized by the presence of a number of protruding irregular structures inside the area which before etching was part of the upper terrace.

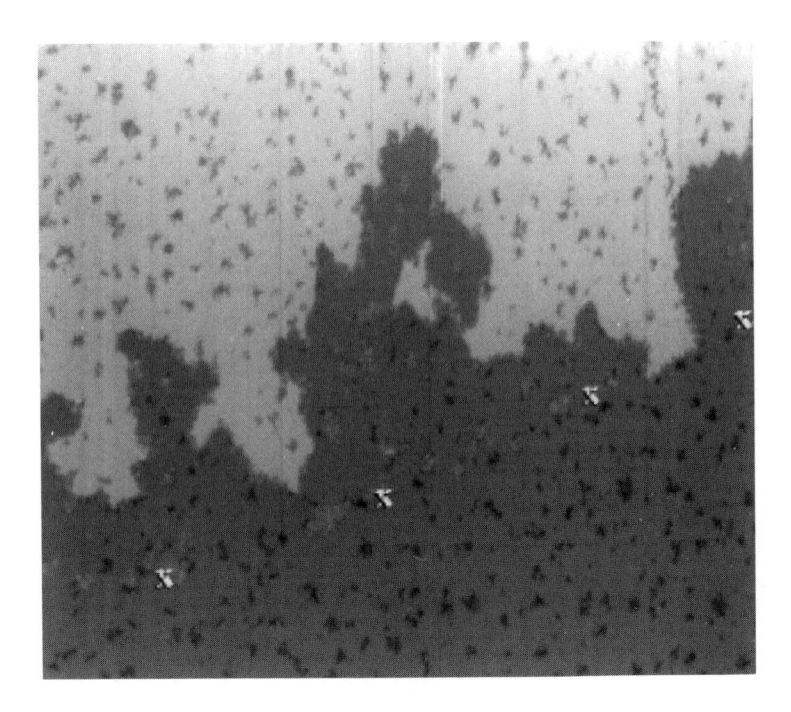

Figure 5. CCT from Si(111)7x7 which has been exposed to 7.5 L NO at 860 K (180x200 nm^2).

The clusters of protrusions inside the etched area can be assigned to small oxynitride islands which have been formed during NO adsorption prior to retraction of the step edge. They are actually measured at a deeper level than the unreacted terrace and correspond to the small depressions visible both on the upper and lower terrace. In Fig. 6 we have reproduced such a depression inside an area of 7x7 reconstruction with atomic resolution. There are no indications of residual parts of 7x7 unit cells in these depressions which rules out an explanation of these depressions as due to etching. Since, on the other hand, the 7x7 reconstructed part of the surface seems to be rather unperturbed and Auger electron spectroscopy (AES) of such a surface shows the presence of both O and N we conclude that O and N are contained in these depressions in form of an oxynitride. That the oxynitride species are seen as depression is probably of electronic origin. Due to a reduced density of states accessible for tunneling on these islands the tip approaches the surfaces more closely while scanning them.

3.4. Formation of Si nitride: NO on Si(111)7x7

To obtain a thin Si nitride structure on Si(111) either a surface as shown in Fig. 5 has to be subsequently annealed at fairly high temperatures (1230 K) or

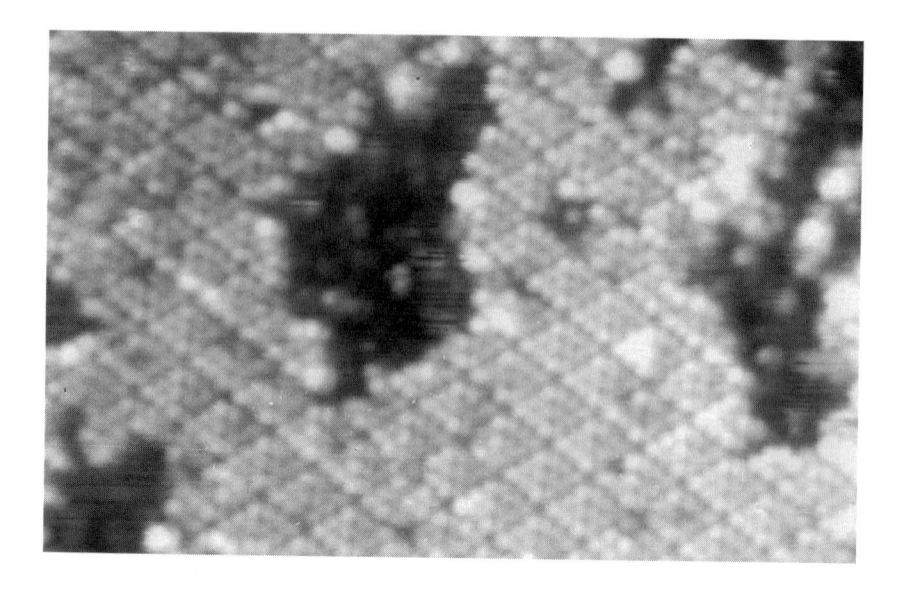

Figure 6. CCT from Si(111)7x7 exposed to 23 L NO at 925 K (25x40 nm^2).

Si(111)7x7 directly exposed to NO at this higher temperature. The surface then develops a thin ordered nitride structure as concluded from the appearance of sharp 8x8 spots in the LEED pattern which are also found in the two-dimensional Fourier transform of the STM images. Absence of O in this structure was inferred from no detectable O in the AES results. We note that the same 8x8 N/Si(111) structure has been generated by Bauer et al. after exposure of Si(111)7x7 to NH$_3$ molecules at high temperatures (see also Ref. [13]). The general behavior of the 8x8 nitride structure is quite interesting (Fig. 7). It obviously has the tendency to wet the surface and grows at the same time in an ideal two-dimensional layer as is demonstrated in Fig. 7. The nitride layer is marked by 8x8. The edges appear as a protruding wall which follows a smooth rounded curve instead of preferential growth along high-symmetry direction. About half of the surface shows the 7x7 structure (marked by 7x7) and inside the 7x7 regions some disordered parts with protruding features and 7x7 remnants are recognized. The concentration of N in the 8x8 structure is probably on the order of 1 ML as may be estimated from the initial coverage of the sample with oxynitride islands which is very similar and assuming that the oxynitride structures contain N in a concentration close to that of the Si atoms. We have not made attempts to derive an atomic model for the Si nitride structure (this might be also very difficult, see Ref. [9]). We note, however, that the general tunneling characteristics correspond to a wide band-gap insulator. The optimum resolution was only reached for U near 4 V and it is conceivable that

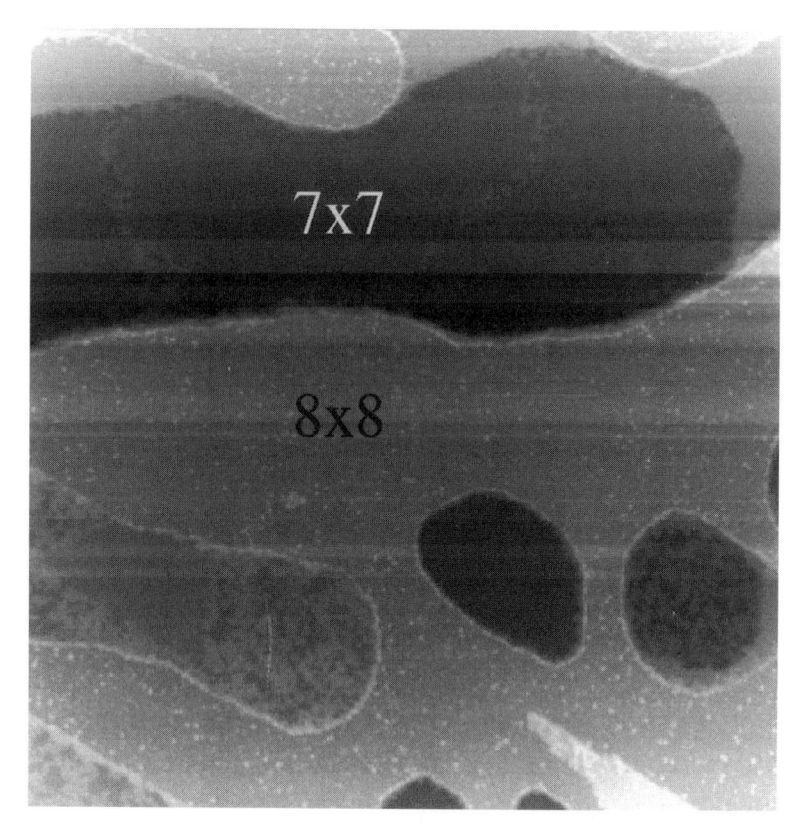

Figure 7. CCT from Si(111)7x7 exposed to 75 L NO at 860 K and subsequently annealed at 1230 K (250x250 nm^2).

electronic states of the two-dimensional nitride layer then contribute to tunneling.

3.5 O$_2$ on Si(100)2x1

Because of the technical importance of the oxidation of Si(100) some results on this system will also be described. Surface physical studies have mostly been performed on room-temperature adsorption of O$_2$. One can distinguish between a fast stage up to a coverage of 1 monolayer and a slow one, where an amorphous Si oxide layer is formed. Most STM experiments have been reported for the fast stage in the coverage regime up to a few 0.1 monolayer. It seems plausible that the most likely adsorption site of atomic O is a bridging position where it can replace a Si-Si bond or connect two dangling bonds of the 2x1 reconstructed Si(100) surface. A rather complex adsorption behavior

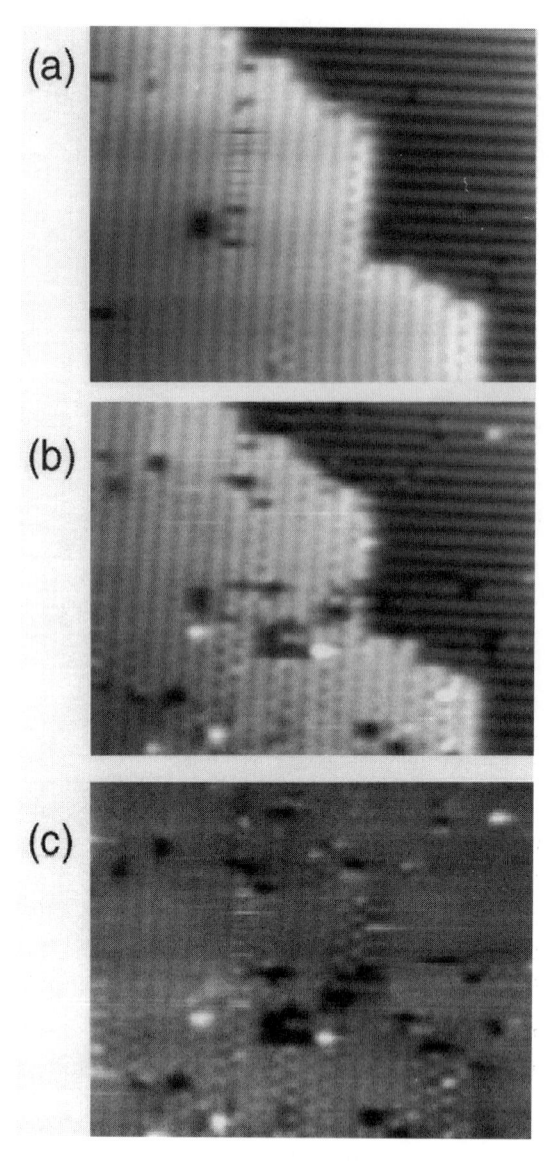

Figure 8. Difference image of Si(100)2x1 during O_2 adsorption (25x29 nm^2).

follows from theoretical work where in addition to the dimer bridge position a number of metastable adsorption sites have been proposed (see Ref. [11]). From experimental studies of the electronic structure it has been concluded that initial adsorption is in a bridge position between top- and second layer Si atoms which is followed by adsorption of O on the same Si atom. In the existing STM work a conclusive assignment of the observed phenomena to atomic adsorption or reaction events has not yet been reached (see Ref. [11]).

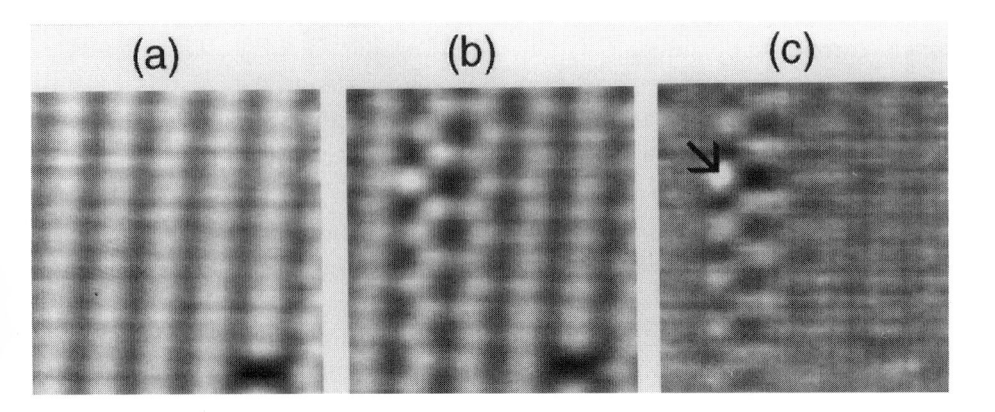

Figure 9. Difference image of Si(100)2x1 during O_2 adsorption showing an adsorption event which pins the flipping dimers and therefore gives rise to buckling in neighboring rows (4x4 nm^2).

An overview of the changes of the surface during O_2 adsorption may be obtained from Fig. 8 where a "difference" image from a series of measurements during O_2 take-up is reproduced. Frame (a) corresponds to the initial and (b) to the final state of the sample after it has been exposed to O_2 at a partial pressure of 1.5×10^{-8} mbar for nearly 3 min (corresponding to 2 L of O_2 exposure). On the clean parts of the surface the Si dimers are not resolved individually but rather appear as a line structure. The lower frame (c) represents the difference between the two upper measurements. The large number of effects which can be identified in the difference image is striking. They may be classified in the following scheme. (i) In the initial stage of O_2 adsorption the observed changes are found on individual dimers. Very often they appear as small protrusions in bridging positions between two neighboring rows of dimers. A local buckling in both rows of dimers is then observed. One example for this transition is shown in Fig. 9. This effect may be ascribed to chemisorption of an O atom in a non-dimer bridge position in the top Si layer. The buckling extends over a length scale of the order of 10 nm along the rows of dimers. It is caused by local pinning of the vibration of the asymmetric Si dimers at room-temperature which produces a symmetric appearance in the STM images (if not some other defects contribute to pinning and to buckling). (ii) The formation of dark sites in the surface is evident from Fig. 8. Some of them show the tendency to increase their size during O_2 adsorption. In this case the dimer structure of the next-lower terrace was found within the depression which obviously results from an O-induced etching process due to formation of volatile SiO species already at room-temperature. (iii) The removed Si atoms may also be found as protrusions on the surface. The formation of Si islands due to this kind of etching process is particularly obvious for O_2 adsorption at elevated temperatures. For room-temperature adsorption we have also attempted to obtain a fully oxidized surface and to measure it with STM. For longer exposure we have obtained images which look similar to those found for Si(111). However, we have not yet been able to clarify whether the surface was fully oxidized or only corresponded to a late adsorption state. Spectroscopic measurements and adsorption experiments at elevated temperatures are planned for the future in order to grow and characterize an oxide with minimum thickness also in this case.

4. CONCLUSION

STM is particularly useful for the analysis of the initial steps of oxide and nitride formation. On Si(111) it has been possible to grow a thin layer which most probably corresponds to an oxide layer of minimum thickness. On the basis of the electronic properties and of the achieved passivation of the surface this layer can no more be described as an adsorption state. Initially, on Si(111) and Si(100) adsorption steps can be followed on an atomic scale. However, the interpretation of the results has a rather qualitative nature so far, particularly for Si(100). Detailed calculations of the electronic structure for the possible adsorption sites should complement the experimental results in order to identify the various steps for formation of the oxide film.

ACKNOWLEDGMENTS

This work has been supported by the Volkswagen-Stiftung and through the Graduiertenkolleg "Dynamische Prozesse an Festkörperoberflächen" at the Ruhr-Universität Bochum.

REFERENCES

1. Wolkow, R. and Avouris, Ph. (1988) Atom-resolved surface chemistry using scanning tunneling microscopy, *Phys. Rev. Lett.* **60**, 1049-1052.

2. Pelz, J.P. and Koch, R.H. (1990) Successive oxidation stages of adatoms on the Si(111)7x7 surface, *Phys. Rev. B* **42**, 3761-3764.
3. Avouris, Ph., Lyo, I.-W. and Bozso, F. (1991) Atom-resolved surface chemistry: The early steps of Si(111)-7x7 oxidation, *J. Vac. Soc. Technol. B* **9**, 424-430.
4. Kliese, R., Röttger, B., Badt, D. and Neddermeyer, H. (1992) Real-time STM investigation of the initial stages of oxygen interaction with Si(100)2x1, *Ultramicroscopy* **42-44**, 824-831.
5. Feltz., A., Memmert, U. and Behm, R.J. (1992) In situ STM imaging of high temperature oxygen etching of Si(111)(7x7) surfaces, *Chem. Phys. Lett.* **192**, 271-176.
6. Feltz, A., Memmert U. and Behm, R.J. (1994) High temperature scanning tunneling microscopy studies on the interaction of O_2 with Si(111)-(7x7) surfaces, *Surf. Sci.* **314**, 34-56.
7. Ono, Y.,Tabe, M. and Kageshima H. (1993) Scanning-tunneling-microscopy observation of thermal oxide growth on Si(111)7x7 surfaces, *Phys. Rev. B* **48**, 14291-14300.
8. Johnson, K.E., Wu, P.K., Sander, M. and Engel T. (1993) The mesoscopic and microscopic structural consequences from decomposition and desorption of ultrathin oxide layers on Si(100) studied by scanning tunneling microscopy, *Surf. Sci.* **290**, 213-231.
9. Röttger, B., Kliese, R. and Neddermeyer, H. (1996) Adsorption and reaction of NO on Si(111) studied by scanning tunneling microscopy, *J. Vac. Soc. Technol. B* **14**, 1051-1054.
10. Bauer, E., Wei, Y., Müller, T., Pavlovska, A. and Tsong, I.S.T. (1995) Reactive crystal growth in two dimensions: Silicon nitride on Si(111), *Phys. Rev. B* **51**, 17891-17901.
11. Neddermeyer, H. (1996) Scanning tunelling microscopy of semiconductor surfaces, *Repts. Progr. Phys.* **59,** 701-769.
12 Engel, T. (1993) The interaction of molecular and atomic oxygen with Si(100) and Si(111), *Surf. Sci. Repts.* **1**, 91-144.
13. Khramtsova, E.A., Saranin A.A. and Lifshits V.G. (1993) Formation of the Si(111)8x8-N structure by reaction of NH_3 with a Si(111) surface, *Surf. Sci.* **280**, L259-L262.

THE INTERACTION OF OXYGEN WITH SI(100) IN THE VICINITY OF THE OXIDE NUCLEATION THRESHOLD

V.D. BORMAN, V.I. TROYAN and Yu.Yu. LEBEDINSKI
Moscow Engineering Physics Institute, Kashirskoe shosse 31, Moscow 115409, Russia.

The kinetics of silicon oxide formation on Si (100) at submonolayer coverages and surface roughening near the oxide nucleation threshold (NT) at temperatures T=915-940 K and oxygen pressure P=4×10^{-7} Torr are investigated by X-ray photoelectron spectroscopy and atomic force microscopy. Microscopic mechanisms of vacancies accumulation and roughening phase transition are proposed. We also discuss the reasons for the change of the oxidation mode.

In previous studies[1, 2], the oxide nucleation threshold (NT) for SiO_2 oxide on silicon surfaces was established as a dependence of oxygen pressure on temperature $P_C(T)$. The interaction of O_2 with silicon surfaces at low pressures and at elevated temperatures is characterised by two modes: oxide growth on the surface and surface etching via volatile SiO formation. These modes are referred to as "passive" and "active" oxidation, respectively.

The measurement of oxidation kinetics by X-ray photoelectron spectroscopy (XPS) was performed in the UHV analytical chamber (the base pressure 5×10^{-10} Torr) of XSAM-800 (Kratos) system in a real time regime. The energy resolution was 0.9 eV (Ag 5d$_{5/2}$) with spectra excitation by MgK$_\alpha$ source (1253.6 eV). The evolution of surface morphology after oxidation was monitored by atomic force microscopy (AFM). The samples (10x4 mm^2) were cut from n-type Si (100) wafers. The samples were cleaned in methanol and chemically etched in 5% HF, followed by heating to 1300 K for 10 min. in UHV.

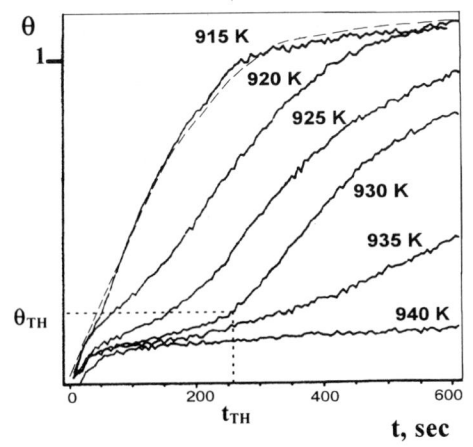

Figure 1. Dependence of oxygen coverage ($\theta=\theta_{CH}+\theta_{OX}$) on exposure time. The dashed line shows least-squares fitting of the curve at 915 K by formula (5).

The sample was heated by conducting electrical current. The temperature was monitored by measuring the resistance of the sample, calibrated by a thermocouple. In order to obtain the kinetics of initial oxidation, the intensity of the O1s peak was measured as a function of exposure time at oxygen pressure of $P=4\times10^{-7}$ Torr and several fixed temperatures from 915 to 940 K in the real time regime. For submonolayer coverages in question, we assumed a linear relationship between the intensity of the O1s peak and oxygen coverage θ. The oxygen coverage includes both chemisorbed and oxide oxygen ($\theta=\theta_{CH} + \theta_{OX}$). Several kinetics curves for different temperatures are shown on Fig. 1 (solid lines - experiment, dashed line - least-squares fitting by using formula (5)). The kinetic curve at low temperature (915 K) tends to saturate at t > 300 s with a saturation coverage of 1.2 ML. Consistent with previous XPS studies [3], silicon atoms in the surface layer(s) are in four different oxidation states Si^{n+} (n=1, 2, 3, 4) (see Fig. 2). In this region, one can observe some peculiarities of the kinetic curves. As seen in Fig. 1, the inhibition of the oxygen accumulation after the initial growth (chemisorption stage) occurs. We assume that the oxide formation in the vicinity of NT exhibits a threshold character: the oxidation starts only after achieving a critical values of oxygen coverage θ_C at the critical time t_C, in agreement with photoemission results [4]. At T=930 K, the critical coverage is equal to $\theta_C = 0.2$ ML. The dependence of the oxidation time τ_{OX} (the time necessary for the oxygen to cover up to $\theta=0.6$) on the temperature is presented on Fig. 3. One can see that τ_{OX} significantly increases while approaching temperature $T_C=905$ K. It is reasonable to suggest that this is the temperature of NT at the pressure of $P=4\times10^{-7}$ Torr.

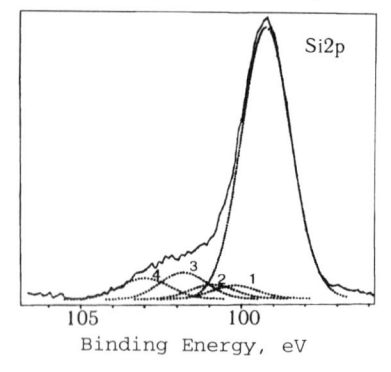

Figure 2. Si2p photoelectron spectra of the surface exposed at T=915 K and $P=4\cdot10^{-7}$ Torr

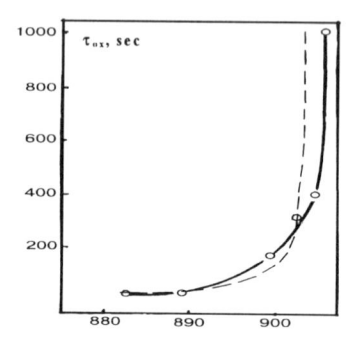

Figure 3. Dependence the oxidation time on the temperature (solid line). The dashed line shows the theoretical dependence of τ_{OX}, according to (2).

Recently, it has been shown that the surface becomes very rough in transition region [5,6,7]. An AFM image taken just after exposure to oxygen at 930 K is shown on Fig 4.

The silicon surface, originally being smooth on the atomic scale, becomes very rough. The deep (40 nm) quasi-periodic (300 nm in a surface plane) holes are observed on the background of low-dimension surface fluctuations. Thus, we can conclude that the oxide nucleation threshold lies in the temperature range in which the formation of silicon oxide and the roughening occur simultaneously and it is the transition from oxidation to roughening.

In accordance with the models developed in [8,9], the formation of submonolayer silicon oxide proceeds via nucleation and islands growth. The adsorbed oxygen atoms diffusing along the surface reach the oxide islands. They are trapped by the islands simultaneously with the atoms of the Si crystal which are built-in into the island. Therefore, the growth rate of the islands is determined by the difference $\theta_{CH} - \theta_{TH}$ (θ_{TH} - the threshold value of the surface coverage by the oxygen), i.e. by the supersaturation of the adsorbed oxygen.

The growth of the submonolayer SiO_2 oxide is described theoretically in [9] within the scope of the phenomenological Volmer-Weber-Zeldovich theory of the first order phase transitions. Taking into account diffusion interactions between islands and neglecting oxygen molecules direct trapping from the gas pahse on the oxide surface, one can obtain the following expression for the surface coverage by oxide, θ_{OX}, as a function of exposure time t:

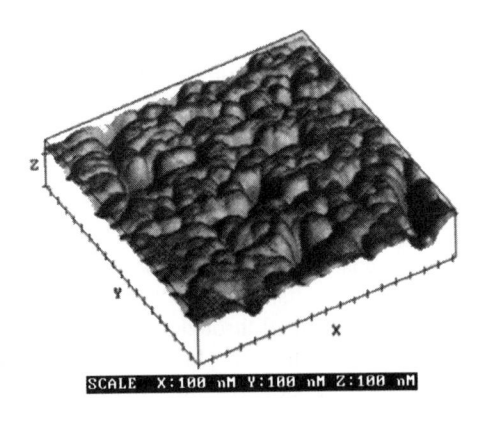

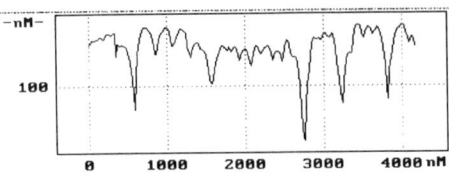

Figure 4. AFM image of the surface Si(100) and cross section profile after an exposition in oxygen at T=930 K and P=4·10^{-7} Torr

$$\theta_{ox} = (1 - \theta_{TH}/\theta_o) \cdot \tanh^2(t/\tau_{ox}), \quad \theta_o = Q \cdot s \cdot (\alpha + Q \cdot s \cdot \Omega)^{-1} \quad (1)$$
$$\tau_{ox} = A \cdot (T_{TH} - T)^{-1/2} \quad (2)$$

Here, Q is the oxygen molecules flux to the surface, s is the sticking coefficient, α is the frequency of oxygen escape from the surface, Ω is the area of oxygen atoms in adsorbate, τ_{OX} is a characteristic time of the oxide growth, the quantity does not depend on the difference $T_{TH} - T$; the temperature of NT at P=const, T_{TH}, is determined by the following expression:

$$P = P_0 e^{-E\alpha/T}, \quad (3)$$

where E_α is the activation energy of desorption of SiO molecules.

The formula (3) fits the experimental data from [1,2], obtained at higher pressures and temperatures compared to this work, with the value of E=3.5 eV close to that reported for the desorption of SiO, E=3.8±0.2 eV [2].

According to (1, 2), when approaching NT at P=const and T→ T_{TH}, $\tau_{OX} \to \infty$, while the value of $\theta_{OX} \to 0$. This is a general result for the first order phase transitions which does not depend upon the approximation made. The relationship (1) describes the experimental data at T=915K (Fig. 1). The experimental and theoretical curves $\tau_{OX}(t)$ are qualitatively consistent (Fig. 3). However, the value of the temperature of NT, T_{TH}, deduced from the experiment is larger than the theoretical value at P=4x10^{-7} Torr. The dependence θ_{OX} vs. t at T > 915 K does not fit even qualitatively. In this temperature range, the delay of oxidation onset is observed. The formation of the oxide begins only at t > t_{TH} and $\theta > \theta_{TH}$.

These peculiarities of oxidation can be explained by surface roughening at T > 915 K. Following [10], the roughening may be described as a vacancy "condensation" induced by oxygen adsorption at subsurface layers. During the exposure time t < t_{TH}, the adsorbed oxygen is accumulated on the surface. The atoms of the Si crystal escape from the surface as SiO molecules in the vicinity of NT which results in vacancy generation [11]. Pairs of "vacancy-adsorbed Si atom can" can also be formed through an attractive interaction of oxygen atoms with the vacancies [11]. According to [11], taking into account of such interactions results in the following expression for the equilibrium concentration of vacancies:

$$n_V = n_V^o \exp(-E_{ef}/T) \, , \quad E_{ef}=E_o-\theta_{CH}T \exp(U_{av}/T) \tag{4}$$

where U_{av} - the depth of the potential well of the oxygen adatoms - vacancies attraction.

According to (4), when the oxygen coverage is equal to

$$\theta_{CH}^V = \frac{E_o}{T} \exp\left(\frac{U_{av}}{T}\right) \tag{5}$$

the value of the effective activation energy of the vacancy formation becomes close to the temperature, and the number of vacancies can increase spontaneously ("vacancy instability").

Thus, in the vicinity of NT, under small exposure to oxygen, three kinds of species appear on the surface: oxygen adatoms, matrix (Si) adatoms, and vacancies. Their concentrations are different from the equilibrium values. Since the formation of the volatile molecules of SiO most likely occurs through the merging of oxygen and Si adatoms, exposure increase results in the growth of the concentration of vacancies and oxygen adatoms, while the concentration of Si adatoms decreases. The well known attractive interaction between vacancies at their sufficiently high concentrations can result in the formation of vacancy clusters, which were observed at silicon surfaces at small oxygen coverages in [12]. It has been proposed [10] that the roughness of the silicon surface during the adsorption of oxygen results from the vacancy condensation and the depth of the roughness "wells" is determined by the concentration of vacancies which have locally condensed in the thin subsurface layers.

Based on this, one can conclude that the observed evolution of the surface in the vicinity of NT can be interpreted in terms of a competition between two phase transitions, viz. oxidation phase transition and the phase transition which leads to the condensation of vacancies. If the time, t_{TH} for reaching the threshold overage, $\theta_{CH}(t_{TH})=\theta_{TH}$, is less than the time t_R of reaching the condition (5) of the spontaneous growth of the vacancy concentration, $\theta_{CH}(t_R)= \theta^v_{CH}$, then the oxide phase grows on the surface. This takes place at low temperatures in the vicinity of NT. The surface of the crystal remains atomically flat under these conditions. At higher temperatures, the reverse relation $t_R > t_{TH}$ is valid, and the rough morphology develops on the surface. We would like to note that the oxide islands are the drains for oxygen adatoms. Therefore, the value of θ_{CH} decreases during the formation of oxide, and the condition (5) can not be achieved even for longer oxygen exposures. As the rough surface develops, the concentration of atomic steps on the surface increases. These steps are the sources of Si adatoms, and thus this results in the SiO molecules flow increase. This is in agreement with the observed more than an order of magnitude higher flux of the evaporating SiO molecules for a rough surface compared to that for the flat one [2]. Finally, after the roughening of the surface the value θ_{CH} decreases and the condition of NT of the oxide phase can not be achieved.

At $t_{TH} \approx t_R$, two phase transitions should take place simultaneously, which corresponds to the observed oxidation of the rough surface.

The model of surface roughening induced by oxygen adsorption [10] yields a narrow distribution of the roughnesses as a function of their size. However, this does contradicts the observed (Fig. 3) AFM image. To explain the appearance of rare (quasi-periodic) deep wells on the background low scale surface fluctuations one should consider the dynamics of roughening and the accumulation of vacancies in the subsurface region of the crystal. Since vacancies are generated at the surface and diffuse into the crystal, the condition (5) is met first at the surface, and then in the bulk. As a result, the condition of the vacancy condensation is met earlier at a shallow (near-surface) depth, and open pores are formed at the surface. The process of vacancies accumulation also occurs during the development of the roughness up to the moment t_f, when due to the increase of the flux of desorbing SiO, the value θ_{CH} will become lower than θ^v_{CH} (5). The depth of the rare deep wells can be estimated as the maximal depth h, where the concentration of vacancies corresponding to the beginning of their condensation is achieved:

$$h \sim (t_f D_b)^{1/2} \qquad (6)$$

(D_b is the diffusion coefficient of vacancies in the bulk of the crystal). Since during the time t_f the surface roughness with the maximal size L is formed (L corresponds to the average distance between rare deep wells), then:

$$t_f \sim L^2/D_s \qquad (7)$$

(D_s is the diffusion coefficient of vacancies on the surface).

From (6), (7) it follows that the ratio of the deep well depth to the average distance between them is equal to:

$$h/L \sim (D_b/D_s)^{1/2} \qquad (8)$$

For silicon, the activation energy values are equal to $E_B=1.5eV$ [13] and $E_s=0.7eV$ [14]. Therefore, for the temperature T=900 K the ratio L/h ~ 30. This estimate is consistent with the observed surface morphology (see Fig. 3)

The work is partially supported by the program "Surface Atomic Structures" (funded by the Ministry of Science, Russian Federation).

References:

1. Lander, J.J., Morrison (1962) Low voltage electron diffraction study of the oxidation reduction of silicon, J. Appl. Phys. 33, 2089
2. Engel T. (1993) The interaction of molecular and atomic oxygen with Si(100) and Si(111), Surface Science Reports 18, 94-144.
3. Himpsel, F.J., McFeely, F.R., Tabeb-Ibrahimi, A., Yarmoff, J.A. (1988) Microscopic structure of the SiO_2/Si interface, Physical Review B38, 6084-6096.
4. Enta, Y., Takegawa, Y., Suemitsu., Miyamoto, N. (1996) Growth kinetics of thermal oxidation process on Si(100) by real time ultraviolet photoelectron spectroscopy, Applied Surface Science 100/101, 449 - 453.
5. Ross, F.M., Gibson, J.M. and Twesten, R.D. (1994) Dinamic observations of interface motion during the oxidation of silicon, Surface Science 310, 243 - 266.
6. Lu, H.C., Gusev, E.P., Garfunkel, E., Gustafsson, T. (1996) An ion scattering stady of the interaction of oxygen with Si(111): surface roughening and oxide growth, Surface Science 351, 111-128.
7. Feltz, A., Memmert, U. and Behm, R.J. (1994) High temperature scanning tunnelling microscopy studies on the interaction with Si(111)-(7×7) surface, Surface Science 314, 34-36.
8. Lutz, F., Kubler, L., Bischoff, J.L., Bolmont, D. (1989) Photoemission proof for a SiO_2 island growth mode initiated on the steps of Si(001) during thermal oxidation by O_2, Physical Review B 40, 11747-11750.
9. Borman, V.D., Gusev, E.P., Lebedinski, Yu., Yu., Troyan, V.I. (1994) Mechanism of submonolayer oxide formation on silicon surface upon thermal oxidation, Physical Reviev B 49, 5415-5423.
10. Borman, V.D., Tapinskaya, O.V., Tronin, V.N., Troyan, V.I. (1994) Dinamics of an adsorption - induced roughening transition as a phase transition in a vacancy subsystem, JETP Lett. 60, 718-724.
11. Borman, V.D., Gusev, E.P., Devyatko, Yu.N., Tronin, V. N., Troyan, V.I. (1994) On a mechanism of surface oxide formation near the nucleation thresshold, Surface Science 301, L239 - L244.
12. Devyatko, Yu.N., Tronin, V.N., (1990) Kinetic equation for a system of interaction point defects in irradiated metals, Physica Scripta 41, 355-364.
13. Vavilov, V.S., Kisevel, V.F., Mukashev, B.N.. (1990) Defects in silicon and at its surface, Moscow, Nauka.
14. Pimpinelli,A., and Villain, J. (1994) What does an evaporating surface look like? Physica A, 204, 521 - 542.

TUNNELING TRANSPORT AND RELIABILITY EVALUATION IN EXTREMELY THIN GATE OXIDES

M. HIROSE, W. MIZUBAYASHI, K. MORINO, M. FUKUDA and
S. MIYAZAKI
Department of Electrical Engineering, Hiroshima University, Higashi-Hiroshima 739, Japan

Direct tunnel current through n$^+$poly-Si /1.6 to 4.8nm thick SiO$_2$/p-Si(100) structures has been calculated on the basis of the WKB approximation. The measured current versus oxide voltage characteristics are well explained by the theory with the only one fitting parameter which is the tunneling electron effective mass m*$_{DT}$ = (0.29±0.02)m$_0$ independent of the oxide thickness. It is found that the quasi-breakdown of the oxides under constant current stressing in the direct tunneling regime is accompanied with multivalued gate-voltage fluctuations, indicating dynamic growth or shrinkage of a conducting filament near the SiO$_2$/Si interface. The areal size and length of the filament are evaluated from the analysis of the stress-induced leakage current. The charge to breakdown for oxides thinner than 3 nm exceeds 10^{4}C/cm^2 at an oxide field strength of 17 MV/cm.

1. INTRODUCTION

The gate oxide thickness of advanced MOSFETs with a feature size of 0.18μm is scaled down to less than 5nm. Even 2.0 to 1.5nm thick SiO$_2$ will be needed for future sub-0.1μm gate length transistors[1]. For such ultrathin gate oxides, the thickness uniformity is particularly important and the Si surface microroughness is to be controlled within an average microroughness of about 0.2 nm as measured by AFM. Since the Si oxidation appears to proceed through a layer-by-layer mechanism, the oxide thickness uniformity on the wafer is in general very good as confirmed by measured tunnel current through ultrathin gate oxides[2, 3]. In recent years it is shown that the measured tunnel current agrees fairly well with the result of model calculations[2-6]. For analyzing the tunnel current, the barrier height at the Si/SiO$_2$ interface and the effective mass of tunneling electron must be accurately given. It is interesting to note that most of reported values of the effective mass are close to 0.3m$_0$ as summarized in Table1[5-9], where the effective mass was obtained as a kind of fitting parameter to calculate the measured tunnel current based on the WKB approximation. The barrier height at the Si, Al or n$^+$poly-Si/ultrathin SiO$_2$ interface was assumed to be identical to a previously reported value for thick oxides. Recently the barrier height for the Si/1.6 to 3.5nm thick SiO$_2$ interface was directly determined and found to be almost identical to the value for thick oxides[10]. In the analysis of the tunnel current or the capacitance-voltage characteristic of n$^+$ or p$^+$ poly-Si gated ultrathin oxides, the poly-Si depletion effect must be taken into account when the poly-Si doping level is not sufficiently high or the oxide layer is extremely thin (1.5-2.5nm)[11].

Reliability of gate oxides above 3nm has been examined by Fowler-Nordheim tunnel current injection, where injected electrons travel through the oxide conduction band. In contrast to this, the leakage current under constant voltage or constant current stress for ultrathin (≤3nm) gate oxides is controlled by the direct tunneling at oxide voltages below 3.4eV which is close to the barrier height[6, 10]. It is shown that the time to dielectric breakdown is determined by the oxide electric field strength being similar to the case of the Fowler-Nordheim stress[12]. Dielectric wearout behavior of ultrathin oxides is different from the case of the thicker oxides. B mode stress-induced leakage current[13] or quasi-breakdown[14] is generally observed and the breakdown mechanism is explained by different models such as electron trap induced conducting path formation[13, 15] or local conducting filament formation[14, 16]. Under the constant current Fowler-Nordheim stress, multivalued gate-voltage fluctuations have been observed[17], being consistent with a model proposed by Ref. 15. The dielectric degradation is thought to be triggered with stress-induced electron traps[15] or formation of hypervalent Si in SiO_2 which results in dissociation of Si-O-Si bonds under electron injection[18].

In this paper the tunnel current through ultrathin (1.6-5nm) gate oxides is calculated on the basis of independently determined barrier height and oxide thickness. The dielectric degradation behavior is also quantitatively examined by analyzing the stress-induced leakage current, and a possible breakdown mechanism is discussed.

TABLE 1. Tunnel electron effective mass obtained by comparing measured current with WKB theory

Authors	Effective mass (m^*/m_0)	Barrier height (eV)	Gate electrode	Oxide thickness (nm)
Nagano[5]	0.36	3.34	n⁺poly-Si	3-5
Yoshida[6]	0.29±0.02	3.25	n⁺poly-Si	3-6
Depas[7]	0.30±0.02	3.2	Al	3-5
Hiroshima[8]	0.32±0.05	3.17	Al	3.5-5
Brar[9]	0.30±0.02	3.17	Al	1.7-3.5

2. EXPERIMENTAL

Gate oxides thicker than 3nm were grown on p-type, CZ Si(100) (8-10 Ωcm) in 2% dry O_2 diluted with N_2 at 1000°C or in dry O_2 at 900°C, or otherwise in wet atmosphere at 850 or 800°C. For the growth of SiO_2 thinner than 3nm dry oxidation in 2% O_2 diluted with N_2 at 850°C was employed. The detailed procedure of Si surface cleaning has been reported elsewhere[6]. The phosphorus diffusion for n⁺poly-Si gates was carried out in $POCl_3$ at 900°C for 10min followed by 10min annealing for oxides thicker than 3nm or at 850°C for 60min followed by 10min annealing for oxides thinner than 3nm. Constant current stress measurements by electron injection from n⁺poly-Si gates were performed at electric field strengths of $1.68 \times 10^7 \sim 4.4 \times 10^7$ V/cm which correspond to the current densities of -0.2 ~ -50A/cm².

3. RESULTS AND DISCUSSION

3.1. BARRIER HEIGHT AT THE SiO_2/Si INTERFACE

The valence band alignment at the ultrathin SiO_2/Si interface has been evaluated by the valence band spectra measured by x-ray photoelectron spectroscopy (XPS)[10, 19]. Figure 1 shows the valence band density of states (VBDOS) for thermally-grown 1.6-3.5nm-thick SiO_2 on Si(100) surfaces together with those of the hydrogen-terminated Si(100) surface and a 40nm-thick SiO_2. The top of the oxide VBDOS remains unchanged regardless oxide thicknesses. In order to determine the conduction band barrier height at the SiO_2/Si interface, the bandgap energy of ultrathin oxides has to be directly measured. It is recently shown that the inelastic energy loss of the O_{1s} core level photoelectron occurs through electron-hole excitation over the SiO_2 bandgap[20]. As shown in Fig. 2 rather broad feature of the plasmon loss signal peaked at ~22eV below the O1s core level peak is observed together with the photoelectron yield around 10eV which corresponds to the oxide bandgap excitation (see the inset). The energy bandgap for SiO_2 thicker than 1.7nm is clearly determined to be 8.95eV. From the results of Figs. 1 and 2, the barrier height is given as summarized in Table 2.

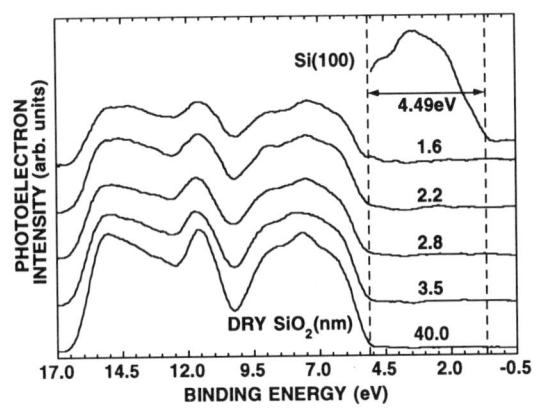

Figure 1. Valence band density of states for ultrathin dry oxides after having substracted the Si substrate component. The charging effect has been corrected for all spectra.

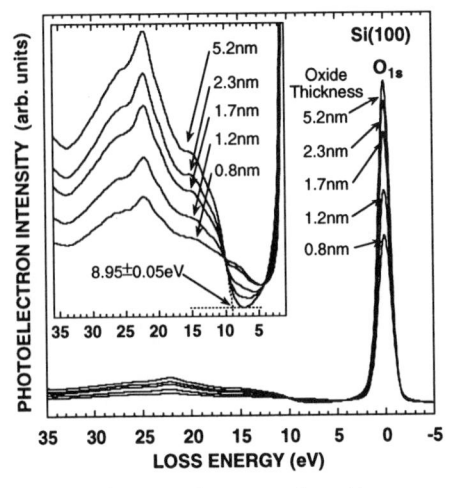

Figure 2. O_{1s} photoelectron and its energy loss spectra for oxides grown on Si(100) at 1000°C.

TABLE 2. Valence band alignment or hole barrier height and electron barrier height for the various ultrathin SiO$_2$/Si interfaces.

	Valence band alignment (eV)	Conduction band barrier height (eV)
1000°C dry SiO$_2$/Si(100)	4.49	3.34
850°C wet SiO$_2$/Si(100)	4.43	3.39
1000°C dry SiO$_2$/Si(111)	4.36	3.47

3.2. TUNNEL CURRENT ANALYSIS

Measured tunneling current for n$^+$poly-Si gate MOS capacitors is plotted as a function of oxide voltage Vox calculated from the capacitance-gate voltage characteristics as shown in Fig. 3. The calculated direct tunnel current J_{DT} in the figure is obtained from the following equation[6]:

$$J_{DT} = (q^2/2\pi h T_{ox}^2) \times (\phi_B - V_{ox}/2) \exp(-4\pi(2qm_{DT}^*)^{1/2} T_{ox} (\phi_B - V_{ox}/2)^{1/2}/h)$$
$$- (q^2/2\pi h T_{ox}^2) \times \phi_B \exp(-4\pi(2qm_{DT}^*)^{1/2} T_{ox} \phi_B^{1/2}/h). \quad (1)$$

Here, q is the electron charge, Tox the oxide thickness, ϕ_B the barrier height and m$^*_{DT}$ the direct-tunneling electron effective mass. In a previous work we employed m$^*_{DT}$ as the only parameter to fit the measured direct tunnel current to the calculated one for the gate oxide thickness range from 3 to 4.8nm, where m$^*_{DT}$ = (0.29±0.02)m$_0$ has been obtained for the given barrier height of 3.25eV[6]. It should be noted that phosphorus doping for poly-Si gates results in incorporation of phosphorus atoms in SiO$_2$ up to 4×10^{20} cm^{-3}[21]. This causes a little decrease (~0.1eV) of the barrier height as compared to the value in Table 2. As indicated in Table1, the obtained effective mass is quite consistent with other reported values. In addition to this, the oxide bandgap and barrier height remain unchanged over the SiO$_2$ layer thickness range 1.7 to 3.5nm as already shown in Fig. 2 and Table 2. Thus it is reasonable to assume that the value of m$^*_{DT}$ remains constant also for oxides thinner than 3nm. As a result, the measured tunnel current for 2.84 to 1.59nm thick oxides

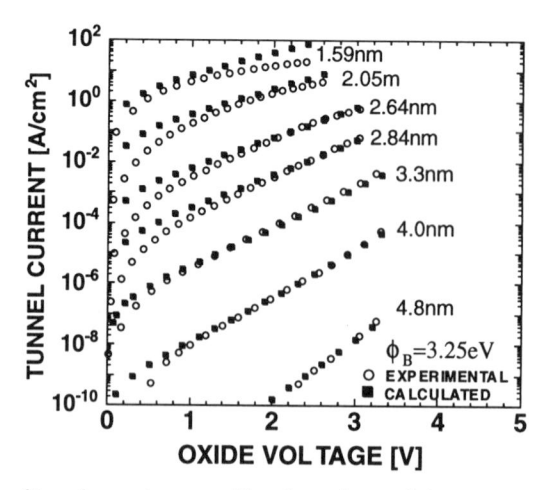

Figure 3. Measured tunnel current versus oxide voltage characteristics together with calculated ones.

is well fitted to eq.(1) as shown in Fig. 3. In the analysis of tunnel current we employed the oxide thickness measured by ellipsometry by using the index of refraction being equal to 1.460. The consistent thickness was obtained also by XPS. The oxide voltage in Fig.3 is derived by fitting the C-V characteristics measured at 100KHz to the theoretical one, in which the quantization effect in the p-Si surface accumulation layer is not taken into account and the n+poly-Si depletion effect at low positive gate voltages is neglected mainly because the phosphorus atom pile-up at the n+poly-Si/SiO$_2$ interface minimizes the poly-Si depletion[6, 11]. The result of Fig. 3 clearly indicates that the direct tunnel current calculated by a simple WKB approximation with a fitting parameter of the electron effective mass can well explain the observed current without taking into account atomic scale fluctuations of oxide thickness.

3.3. OXIDE RELIABILITY

A new failure mode appears in gate oxides thinner than 5nm. Under the constant current or voltage stress, a dramatic increase in the direct tunnel current component is observed prior to the dielectric breakdown[13, 14, 16]. Such oxide degradation mode is called B-mode stress induced leakage current (B-SILC)[13] or quasi-breakdown[14]. Two possible quasi-breakdown mechanisms have been proposed. One is formation of localized conducting paths via electron traps generated by electron injection[15], and the other is localized conducting-filament formation under the stress[14]. A typical behavior of the quasi-breakdown is illustrated in Fig. 4. The initial I-V curve is well fitted to the calculated direct tunnel current at oxide voltages below 3.4 V being close to the barrier height, above which the Fowler-Nordheim tunneling controls the current. Detailed procedure to analyze the tunnel current has been reported in Ref. 6. The charge injection over 1C/cm^2 from an n+poly-Si gate causes a significant increase in the direct tunnel current component although Q_{BD} of this capacitor is 5C/cm^2, whereas the substrate injection causes no significant degradation up to an injection level of 50C/cm^2.

For calculating the stress-induced direct tunnel current component, it is assumed that the oxide layer near the SiO$_2$/Si interface is locally deteriorated and becomes conductive as schematically illustrated in Fig. 4 (b) and (c). Since the slope of the direct tunnel current is sensitive to a change in local oxide thickness and the current level is determined by the area through which the quasi-breakdown current passes, the increased direct tunnel current component is well explained by this local degradation model without using any numerical fitting parameter, being different from the case of Ref.14. At Q_{inj} = 1C/cm^2 degradation of the oxide takes place by formation of multiple conducting regions in SiO$_2$ near the SiO$_2$/Si interface, and hence the effective oxide thickness is locally reduced to 3.5nm. The total degraded area is about 1% of the total 100×100μm^2 gate area. Further electron injection (2C/cm^2) results in formation of a conducting filament with a size of ~ 70×70nm^2 where the remaining oxide thickness is 1.6nm which controls the direct tunnel current component as indicated by the solid curve obtained from eq. (1). It should be noted that the resistivity of the conducting filament can be ignored because the stress-induced current is fully explained by the tunneling through the 1.6nm thick SiO$_2$[16]. This is a major difference of the present model from the previous one proposed in Ref. 14. The degradation behavior is quantitatively explained by the present model in which a localized conducting region is formed near the SiO$_2$/Si interface during electron injection from an n+poly-Si gate.

In order to get further insights in the origin of the conducting filament formation, we have changed the surface microroughness, the oxidation temperature and the oxidation

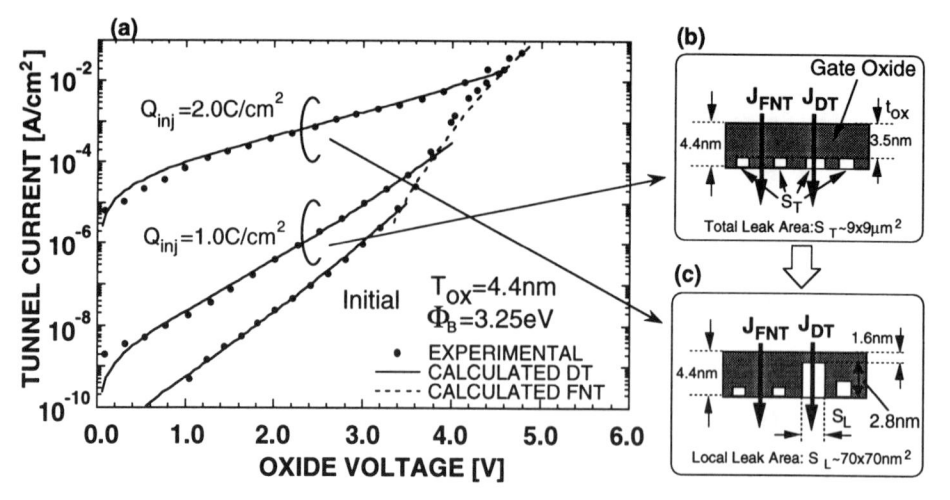

Figure 4. Tunnel current vs. oxide voltage or voltage drop in SiO$_2$ before and after charge injection for 4.4nm gate oxide grown at 1000°C (a). The solid lines correspond to calculated curves for the direct tunneling (DT) and the dashed line to Fowler-Nordheim tunneling (FNT) theory. A local degradation model for the oxide is schematically shown in (b) and (c).

atmosphere. However, the degradation behavior is independent of such changes. Next we have intentionally contaminated the Si wafer surface before the oxidation by exposing it to clean room air for 20min at which the native oxide layer growth is still negligible. The wafer surface is progressively contaminated with organic molecules by clean room air exposure exceeding 15min and the signal intensity saturates for exposure time above 45min. For tunnel MOS diodes fabricated on the wafer contaminated with organic molecules, the quasi-breakdown behavior is very similar to the result of Fig. 4. This implies that a low level of organic contamination is not necessarily a cause to trigger the quasi-breakdown in the oxide.

Next we have changed the gate area S_G and measured the areal size of the conducting filament S_L after the quasi-breakdown as shown in Fig. 5, where the areal size of the conducting filament normalized by the gate electrode area S_L/S_G is plotted against the quasi-breakdown charge Q_{qBD} for oxides with a thickness of 4.4nm. The result implies that S_L/S_G is approximately proportional to Q_{qBD}:

$$S_L \propto Q_{qBD} S_G. \qquad (2)$$

In addition to this the remaining oxide thickness tox above the filament remains almost unchanged, namely tox = 1.6nm for the gate oxide thickness Tox = 4.4nm even when the gate electrode area is changed. Thus the filament length Tox-tox is constant for a given oxide thickness, and the following empirical relationship is obtained between the volume of conducting filament S_L(Tox-tox) and the total electron number $Q_{qBD}S_G$ mainly injected to the filament region:

$$S_L(\text{Tox-tox}) = kQ_{qBD}S_G. \qquad (3)$$

Here, k is the proportional constant. Equation(3) suggests that total injected electrons

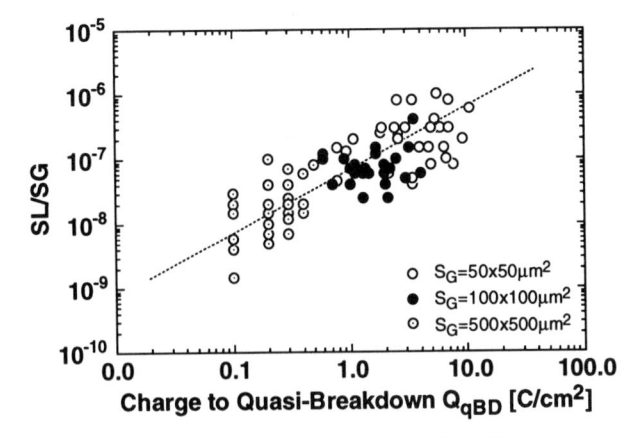

Figure 5. S_L/S_G versus Q_{qBD} for MOS capacitors with different gate areas.

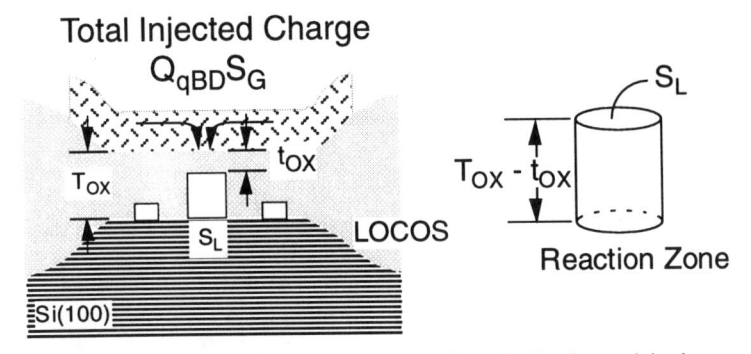

Figure 6. A schematic illustration of filament formation by electron injection.

from the n+poly-Si gate $Q_{qBD}S_G$ predominantly pass through the filament region as illustrated in Fig. 6 and determine the filament volume S_L(Tox-tox). A possible model to explain eq.(3) is that local dissociation of Si-O-Si bonds at hypervalent Si atom sites occurs through electron injection[18] and a chemical reaction might be promoted as follows:

$$e^- + Si\text{-}O\text{-}Si^* \rightarrow Si\text{-}Si + O. \quad (4)$$

Here Si* refers to hypervalent silicon. It is shown that the compressive stress in SiO_2 near the interface creates distorted $Si\text{-}O_4$ tetrahedra in SiO_2 and such strained bonds tend to form hypervalent or five-fold-coordinated Si atoms[18]. The electron injection into the strained region of SiO_2 promotes dissociation of Si-O-Si bonds at the hypervalent Si site, resulting in oxygen atom release and Si-Si bond formation which leads to a local conducting filament growth.

The in-depth distribution of the strained bonds in SiO_2 has been measured as a function of SiO_2 layer thickness by using an FT-IR-ATR (Attenuated Total Reflection) technique[22]. The result has shown that the compressively strained bonds are existing in SiO_2 within 2nm from the interface. The electron-induced dissociation of Si-O-Si bonds in SiO_2 creates atomic oxygen which diffuses in the oxide matrix to finally recombine with other oxygen

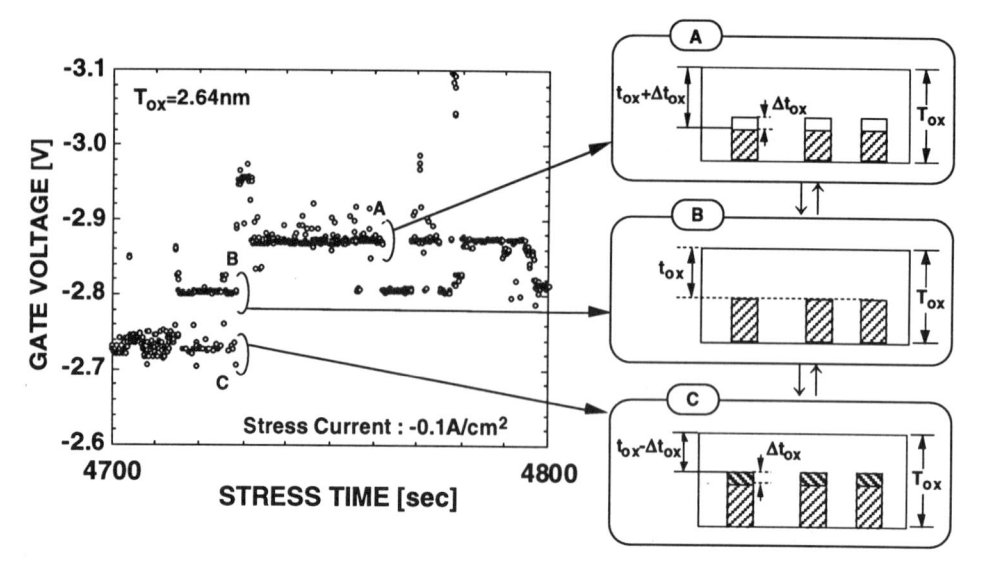

Figure 7. Multivalued gate-voltage fluctuations under the constant current stress for MOS capacitors with a oxide thickness of 2.64nm. The higher or lower gate voltage state corresponds to the state A or C, respectively.

atom to form the stable molecule or otherwise dissociated oxygen again reacts with a Si-Si bond to form a Si-O-Si bond. Therefore, the tunnel current passing through the conducting filament fluctuates as indicated in Fig. 7. This shows that the filament growth under the constant current stress is not smooth but the filament length dynamically changes, resulting in multivalued gate-voltage fluctuations. On the other hand such gate-voltage fluctuation was interpreted in terms of an electron-trap induced conducting path model[12, 15] or a variable range hopping model[17, 23]. The conducting filament model does not necessarily contradict with the electron-trap induced conducting path model although their formation mechanisms are apparently different at this stage.

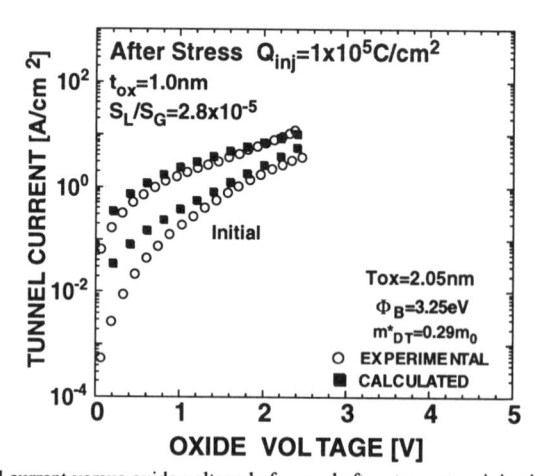

Figure 8. Tunnel current versus oxide voltage before and after stress at an injection level of 1×10^5 C/cm^2

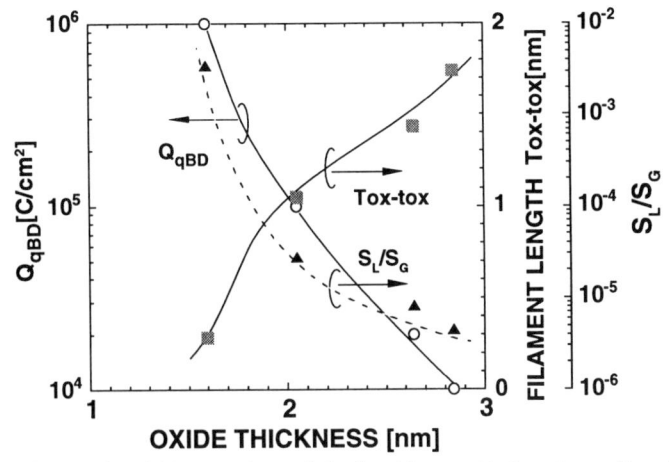

Figure 9. The filament length Tox-tox, the areal size S_L with respect to the gate area S_G and the injected charge Q_{qBD} to cause quasi- breakdown are plotted as a function of oxide thickness.

The filament formation model has been developed for oxides thinner than 5nm but thicker than 3nm. In order to examine whether the quasi-breakdown behavior for gate oxides thinner than 3nm is basically similar to the case of the thicker oxides the constant current stress has been applied to ultrathin gate oxides. As shown in Fig. 8 where the constant current stress induces a lager area of the conducting filament ($S_L/S_G = 2.8 \times 10^{-5}$) with a shorter filament length ($T_{ox}-t_{ox} = 1.05$nm) for 2.05nm-thick SiO_2 as compared to the case of the thicker oxides (Fig. 4). As indicated in Fig. 9 the filament length and the areal size smoothly change as the oxide thickness decreases, and the charge injection up to 10^4-$10^6 C/cm^2$ at oxide field strengths above 17MV/cm causes only quasi-breakdown. Thus the basic feature of the filament model appears to hold for oxides thinner than 3nm.

4. SUMMARY

It is shown that the WKB approximation to calculate the direct tunnel current well explains the measured current-voltage characteristics for ultrathin (1.6-4.8nm) gate oxides. The tunnel electron effective mass for the parabolic band is obtained to be $(0.29\pm0.02)m_0$ as a fitting parameter in the whole range of oxide thicknesses. This implies that the oxide thickness uniformity is extremely good and the electronic properties of gate oxides are kept unchanged regardless oxide thicknesses. Dielectric degradation behavior of ultrathin oxides under constant current stress is understood on the basis of a local wearout model where electron-induced conducting filament formation is thought to be a possible mechanism. Quantitative modeling of the quasi-breakdown yields the areal size and length of the localized filament formed in the oxide.

REFERENCES

[1] Momose, H. S., Nakamura, S., Ohguro, T., Yoshitomi, T., Morifuji, E., Morimoto, T., Katsumata, Y., and Iwai, H. (1997) Uniformity and reliability of 1.5nm direct tunneling gate oxide MOSFETs, 1997 Symposium

on VLSI Technol., Digest of Technical Papers, 15-16.
[2] Buchanan, D. A., and Lo, S.-H. (1996) Growth characterization and the limits of ultrathin SiO$_2$-based dielectrics for future CMOS applications, The Physics and Chemistry of SiO$_2$ and Si-SiO$_2$ Interface-3, editors Massoud, H. Z., Poindexter, E. H., and Helms, C. R., The Electrochmical Society, 3-14.
[3] Depas, M., Heyns, M. M., Nigam, T., Kenis, K., Sprey, H., Wilhelm, R., Crossy, A., Sofield, C. J., and Graef, D. (1996) Critical processes for ultrathin gate oxide integrity, The physics and Chemistry of SiO$_2$ and Si-SiO$_2$ Interface-3, editors Massoud, H. Z., Poindexter, E. H., and Helms, C. R., The Electrochmical Society, 352-366.
[4] Schuegraf, K. F., King, C. C., and Hu, C. (1992) Ultrathin silicon dioxide leakage current and scaling limit, 1992 Symposium on VLSI Technol., Digest of Technical Papers, 18-19.
[5] Nagano, S., Tsuji, M., Ando K., Hasegawa, E., and Ishitani, A. (1994) Mechanism of leakage current through nanoscale SiO$_2$ layer, J. Appl. Phys. 75, 3530-3535.
[6] Yoshida, T., Imafuku, D., Alay, J. L., Miyazaki, S., and Hirose, M. (1995) Quantitative analysis of tunneling current through ultrathin gate oxides, Jpn. J. Appl. Phys. 34, L903-L906.
[7] Depas, M., Van Mheirhaedhe, R. L., Laflere, W. H., and Cardon, F. (1993) Tunnel oxides grown by rapid thermal oxidation, Microelectronic Engineering, 22, 61-64.
[8] Hiroshima, M., Yasaka, T., Miyazaki, S., and Hirose, M. (1994) Electron tunneling through ultrathin gate oxide formed on hydrogen-terminated Si(100) surfaces, Jpn. J. Appl. Phys, 33, 395-398.
[9] Brar, B., Wilk, G. D., and Seabough, A. C. (1996) Direct extraction of the electron tunneling effective mass in ultrathin SiO$_2$, Appl. Phys. Lett. 69, 2728-2730
[10] Alay, J. L., and Hirose, M. (1997) The valence band alignment at ultrathin SiO$_2$/Si interfaces, J. Appl. Phys. 81, 1606-1608.
[11] Lo, S.-H., Buchanan, D. A., Taur, Y., Han, L.-K., and Wu, E. (1997) Modeling and characterization of n$^+$- and p$^+$-polysilicon-gated ultrathin oxides (21-26Å), 1997 Symposium on VLSI Technol., Digest of Technical Papers, 149-150.
[12] Depas, M., Degraeve, R., Nigam, T., Groeseneken, G., and Heyns, M. M.(1997) Reliability of ultra-thin gate oxide below 3nm in the direct tunneling regime, Jpn. J. Appl. Phys. 36, 1602-1608.
[13] Okada, K., and Kawasaki, S. (1995) New dielectric breakdown model of local wearout in thin silicon dioxides, Ext. Abstracts of 1995 Intern. Conf. on Solid Devices and Materials, 473-475.
[14] Lee, S. -H., Cho, B.-J., Lo, J.-C., and Choi, S.-H. (1994) Quasi-breakdown of ultrathin gate oxide under high field stress, Tech. Dig. IEDM (IEEE, Piscataway, 1994) 605-608.
[15] Degraeve, R., Groeseneken, G., Bellens, R., Depas, M., and Meas, H. (1995) A consistent model for the thickness dependence of intrinsic breakdown in ultrathin oxides, Tech. Dig. IEDM (IEEE, Piscataway, 1995) 863-866.
[16] Yoshida, T., Miyazaki, S., and Hirose, H. (1996) Analytical modeling of quasi-breakdown of ultrathin gate oxides under constant current stress, Ext. Abstracts of 1996 Intern. Conf. on Solid Devices and Materials, 539-541.
[17] Taniguchi, K., unpublished.
[18] Hasegawa, E., Akimoto, K., Tsukiji, M., Kubota, T., and Ishitan, A. (1993) Influence of the structural transition layer on the reliability of thin gate oxides, Ext. Abstracts of 1993 Intern. Conf. on Solid Devices and Materials, 86-88.
[19] Gruntharner, F. J., and Grunthaner, P. J.(1986) Chemical and electronic structure of the SiO$_2$/Si interface, Mat. Sci. Rep. 1, 148.
[20] Miyazaki, S., Nishimura, H., Fukuda, M., Ley, L., and Ristein, J. (1997) Structure and electronic states of ultrathin SiO$_2$ thermally grown on Si(100) and Si(111) surfaces, Appl. Sur. Sci., 113/114, 585-589.
[21] Morino, K., Miyazaki, S., and Hirose, M. (1997) Phosphorous incorporation in ultrathin gate oxides and its impact to the network structure, to be published in Ext. Abstracts of 1997 Intern. Conf. on Solid State Devices and Materials.
[22] Yamazaki, Y., Miyazaki, S., Bjorkman, C. H., Fukuda, M., and Hirose, M. (1994) Infrared spectra of ultrathin SiO$_2$ grown on Si surface, Mat. Res. Soc. Symp. Proc. 318, 418-424.
[23] Okada, K.(1997) A new dielectric breakdown mechanism in silicon dioxides, 1997 Symposium on VLSI Technol., Digest of Technical Papers, 143-144.

ELECTRICAL DEFECTS AT THE SiO$_2$/Si INTERFACE STUDIED BY EPR

JAMES H. STATHIS
IBM Research Division
T.J. Watson Research Center,
P.O. Box 218, Yorktown Heights NY 10598 USA

1. Introduction

Electron paramagnetic resonance (EPR) is one of the most powerful techniques for studying defects in electronic materials. It has been applied to the Si/SiO$_2$ system with considerable success over the past few decades. This paper will review recent work using this technique in conjunction with electrical characterization methods to study the physics and chemistry of electrically active defects in metal-oxide-semiconductor field effect transistors (MOSFETs). This work has revealed the crucial role played by atomic hydrogen (H^0) in the chemistry of defects in SiO$_2$. Experiments in which Si/SiO$_2$ structures are exposed to H^0 help to explain various phenomena resulting from electrical stress or radiation exposure in MOS structures. However, this work has also opened new questions about the nature of the dominant electrically active defects at or near the Si/SiO$_2$ interface.

2. Experimental Techniques

EPR is a standard analytical technique which is described in many texts, and its specific application to SiO$_2$ has been the subject of review articles.[1,2] Briefly, the sample is placed in a magnetic field and the unpaired spins (typically present on point defects) are detected by the resonant absorption of microwave radiation between the Zeeman levels. Spectra are usually displayed as the first derivative of the absorption signal. Conventional EPR uses a reflection bridge as shown in Figure 1a to detect the microwave loss. A related method is electrical detection of magnetic resonance, in which the EPR signal is a resonant change in some electrical property of the sample, as shown in Figure 1b. If the mobility or lifetime of free carriers is influenced by the presence of defects, then those defects may be detected with high sensitivity and selectivity using this method. Spin-dependent recombination (SDR) is used in a MOSFET or gate-controlled diode to detect generation-recombination centers at the Si/SiO$_2$ interface[3,4] and spin-dependent tunneling (SDT) is used to detect defects which cause trap-assisted tunneling across a thin SiO$_2$ film.[5]

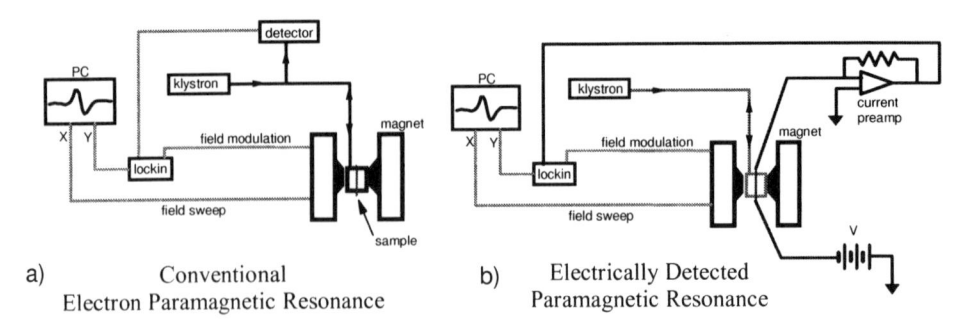

Figure 1. Diagram of magnetic resonance spectrometers for (a) conventional EPR and (b) electrically detected resonance.

The remote plasma system employed for atomic hydrogen exposure has been described in detail elsewhere.[6] Key features of the apparatus are optical baffles to prevent ultraviolet light from the plasma from reaching the sample, and a bolometric sensor to measure the atomic hydrogen concentration at the sample position.

Samples for these experiments include bare thermal oxides grown on (100) or (111) silicon wafers as well as fully-processed MOSFETs. The oxides were grown using standard thermal processing at 850-900°C, with thickness ranging from about 3 to 10 nm. Both *n* and *p* substrates with various doping levels were used. After exposing the bare oxide to the atomic hydrogen for various times, electrical characterization (capacitance-voltage (CV) or current-voltage (IV)) and EPR measurements were performed. Evaporated aluminum gates or a mercury probe were used for electrical measurement, and samples were ungated for EPR. SDR and SDT measurements were done on MOSFETs with polycrystalline Si gates. The devices were stressed using either Fowler-Nordheim tunneling stress at fields >8MV/cm, or using optically-induced electron injection at 2-4MV/cm.

3. The Chemistry of Hydrogen in SiO_2

The beneficial effects of hydrogen as a passivant for Si/SiO_2 interfaces have been recognized for decades, while at the same time[7] the detrimental effects of hydrogen-related defects have been known. While other impurities (e.g. alkali or other metals) have largely been eliminated in modern semiconductor fabrication, hydrogen remains a chronic problem because it is more difficult to eliminate entirely, nor would it necessarily be desirable to do so.[8]

The involvement of hydrogen in hot-electron-induced degradation and dielectric breakdown of SiO_2 films was suggested by a number of experiments, notably the observation of substrate dopant passivation and hydrogen redistribution during hot electron stress, and the enhanced degradation rate of hydrogen-soaked films.[9] The direct EPR observation of radiolytic *atomic* hydrogen (H⁰) in SiO$_2$ led Griscom[10] to describe interface state production in terms of the *depassivation* reaction

$$\text{Si-H} + \text{H}^0 \rightarrow \text{Si}\bullet + \text{H}_2 \tag{1}$$

This is consistent with the prevalent belief that the silicon dangling bond (Si•), known from EPR as the P$_b$ center, constitutes nearly all of the electrically active interface states.

Under continuous exposure to an atomic hydrogen flux, Reaction (1) will compete with the *passivation* reaction

$$\text{Si}\bullet + \text{H}^0 \rightarrow \text{Si-H} \tag{2}$$

P$_b$ centers will be passivated by hydrogen in samples with initial high P$_b$ density, and are generated if the starting P$_b$ density is low. The steady-state balance between passivation and depassivation determines the final P$_b$ density.[11,12] As shown in Figure 2, the results for P$_b$ on (111) are consistent with these two simple reactions. Whether starting from the fully depassivated (desiccated by vacuum anneal at 700-800°C) or from the fully passivated initial condition, the final P$_b$ density is essentially the same. For the

Figure 2. Passivation and de-passivation of P$_b$ centers on (111) silicon. (a) vacuum annealed; (b) vacuum annealed followed by H⁰ exposure; (c) as-grown sample after H⁰ exposure; (d) as-grown samples. From [11].

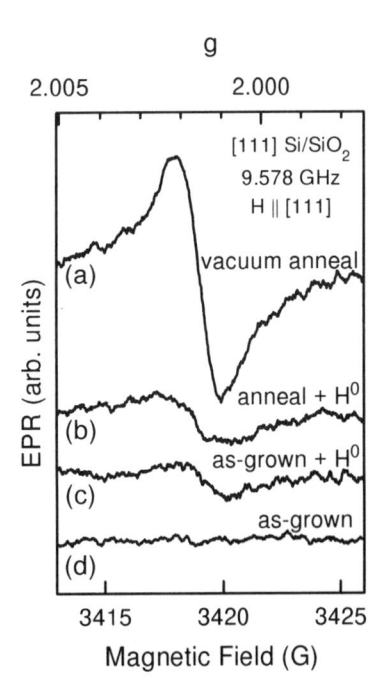

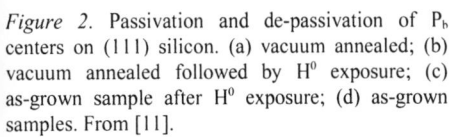

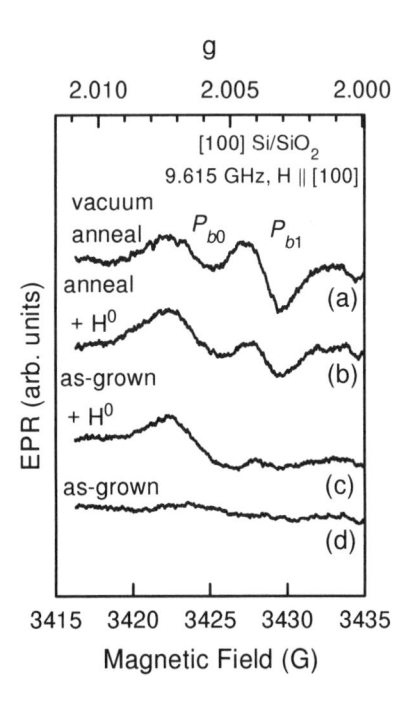

Figure 3. Passivation and de-passivation of P$_b$ centers on (100) silicon, as in Figure 2. From [12].

(100) interface there are two varieties of P_b center, called P_{b0} and P_{b1}, whose precise structures are still a matter of debate.[1,12,13] Figure 3 shows that the (100) centers exhibit quantitatively different behaviors. In the case of P_{b0}, room-temperature exposure to H^0 causes complete depassivation when starting from the fully-passivated initial condition (cf. Reaction (1)), and has no measurable passivating effect on the vacuum-annealed sample (cf. Reaction (2)). The P_{b1} center exhibits an intermediate and more complex behavior. Starting with the fully passivated interface, H^0 produces a small number of P_{b1} centers, similar to the case for P_b on (111). In contrast, when starting from the fully depassivated condition, only about 20 percent of the dangling orbitals can be passivated. This means that the starting condition is important for P_{b1}, suggesting that this defect has some dependence on the thermal history of the sample. For example, the vacuum depassivation step (heretofore envisioned simply as the thermally activated reverse of Reaction (2)) might create additional defect sites.

4. Atomic Hydrogen Effects Compared to Radiation or Electrical Stress

In this section the effects of H^0 on interface defects are compared to the effects caused by radiation and electrical stress. The evidence supports the idea that the release of atomic hydrogen is responsible for much of the hot-electron- and radiation-induced interface state generation in MOSFETs.[9] Atomic hydrogen exposure of (100)Si/SiO$_2$ interfaces produces electrically active interface states (D_{it}) with a broad range of time constants[12,14] and a distinct spectrum. Figure 4 compares the spectra of interface states produced by hot electrons and by H^0 exposure. The similarity is profound. Figure

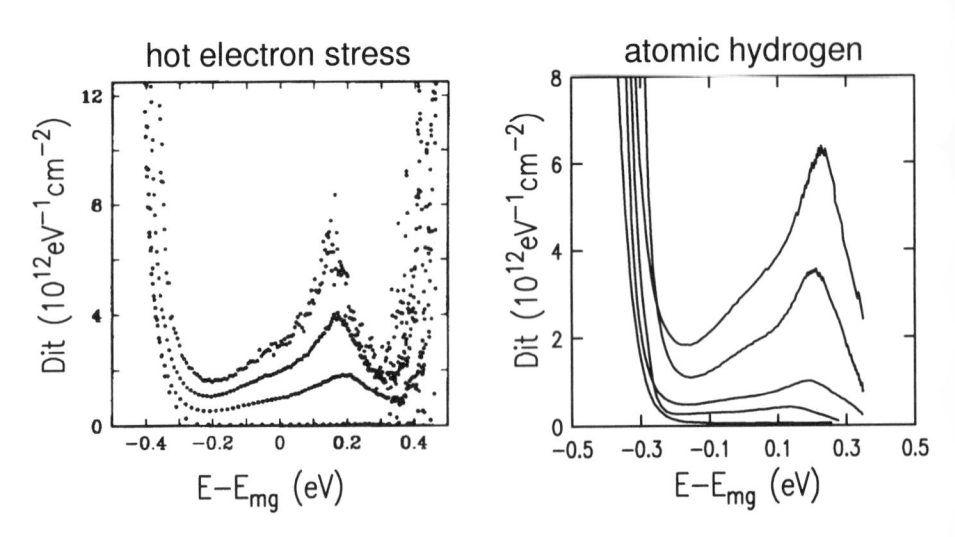

Figure 4. Comparison of interface state densities (D_{it}) for the (100) Si/SiO$_2$ interface, caused by hot electrons (left panel) and by atomic hydrogen (right panel). D_{it} has been measured on n-type samples using the high-low frequency capacitance method. (Data courtesy of D.A. Buchanan and E. Cartier.)

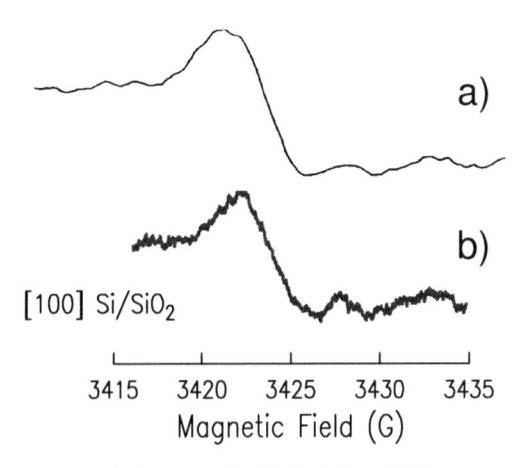

Figure 5. Comparison of atomic-hydrogen- and radiation-induced EPR spectra on (100) samples. a) gamma irradiation (from Kim and Lenahan, J. Appl.. Phys. **64**, 3551 (1988)); b) hydrogen-induced.

5 compares EPR spectra of hydrogen-exposed and γ-irradiated (100) interfaces. Again, the similarity is profound. The spectrum for both cases is dominated by the P_{b0} center, with only a small P_{b1} contribution. SDR measurements on electrically stressed MOSFETs likewise show a spectrum dominated by P_{b0} (see Figure 8 later). The fact that only (or mostly) P_{b0} is generated by radiation or hot electron stress can therefore be related, on a microscopic chemical level, to defect generation by atomic hydrogen which is released by the irradiation or hot electrons.

5. New Hydrogen-Related Defects

The data of the previous section show that H^0 produces both P_b centers and electrically active interface states in samples with initial low P_b and D_{it}. However, the D_{it} spectrum (Figure 4) is not consistent with the known P_b density of states.[15] Quantitative measurements reveal that the P_b center accounts for only a fraction of the interface states.[11] Figure 6 shows the hydrogen-induced growth of interface states (D_{it} at mid-gap) and P_b centers. It can be seen that P_b centers account for only ~5% of the fast interface states induced by hydrogen at the (111) interface. Similarly, for the (100) interface the total P_b density is only ≲15% of the hydrogen-induced D_{it}. These results are in marked contrast to previous studies which had found nearly one-to-one correlation between P_b and mid-gap D_{it} in irradiated or as-grown oxides, but are consistent with spin-dependent recombination measurements of P_b generation in electrically stressed n-MOSFETs.[4]

Many of the hydrogen-induced interface states behave as donor-like states with a broad distribution time constants, consistent with donor defects distributed throughout the near-interface region of the oxide within 20-30Å of the Si/SiO$_2$ interface.[14] These give rise to a characteristic hysteretic behavior in CV measurements, showing that the

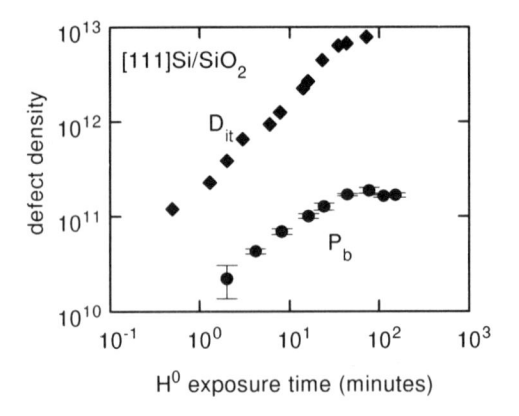

Figure 6. Mid-gap interface state density (D_{it}, eV^{-1}cm^{-2}) and interface dangling bond density (P_b, cm^{-2}) as a function of atomic hydrogen fluence.

defect can be slowly but reversibly neutralized and re-ionized by varying the surface potential. These defects are variously referred to as slow states (referring to their response time), border traps (referring to their purported physical location near the Si/SiO₂ interface), and anomalous positive charge (because they exhibit two charge states, positive or neutral, the positive state being the one noticed in CV measurements). Their behavior is to be contrasted with that of P_b centers, which exchange charge very rapidly with the Si substrate[16] and which are amphoteric, i.e. they can be positive, neutral, or negative depending on Fermi level position[1]. There may be more than one type of defect responsible for the slow states, but hydrogen is an integral part of the structure of at least some of them.[17] This again is in contrast to P_b centers, which are electrically active only when hydrogen is absent. EPR measurements have not found any resonance associated with these hydrogen generated donor-like defects.

6. Spin-Dependent Trap-Assisted Tunneling

As the SiO₂ layer thickness in MOSFETs is reduced below 5nm, an increasingly important problem is trap-assisted leakage currents through the oxide[18]. This leakage occurs even at low voltages in the direct tunneling range of ultra thin oxides. The defects causing the leakage can be generated by high-field stress or by direct H⁰ exposure. Little is known about the microscopic structure of these defects.

Spin-dependent transport measurements have been used to study the defect involved in stress-induced leakage currents.[5] The mechanism of spin-dependent trap-assisted tunneling is illustrated in Figure 7. In the absence of defects, the wave function of a free electron in the Si conduction band (left hand region of the diagrams) decays exponentially as it traverses the SiO₂ barrier. If a defect is present near the middle of the

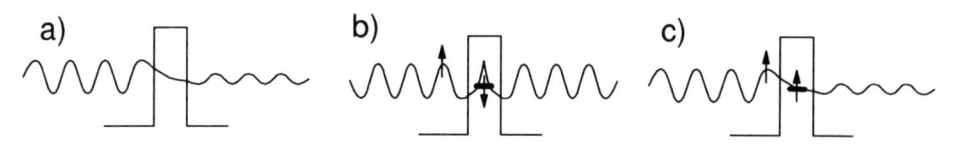

Figure 7. Mechanism of spin-dependent trap-assisted tunneling. (a) Tunneling though a barrier in the absence of defects. The transmission through the barrier is low. (b) Resonant tunneling with high transmission when the defect spin is orthogonal to the tunneling carrier. (c) Resonant tunneling suppressed when the spins are aligned in a magnetic field.

barrier then symmetry allows the electron to pass though with low attenuation, as in resonant tunneling[19]. However, if the defect is paramagnetic then the spin on the tunneling electron and that on the defect will be aligned in the magnetic field of the EPR spectrometer, and resonant tunneling will be suppressed by a factor equal to the product of the spin polarizations of the defect and the hopping electron. Flipping the spin by microwave absorption at the defect resonance allows the tunnel current to increase. The EPR spectrum of the defect responsible for the trap-assisted leakage is detected by the resonant change in tunnel current measured as a function of magnetic field.

The SDT spectrum obtained from a *n*-MOSFET with a 4.45nm SiO_2 gate oxide is shown in Figure 8. The absorption signal is shown. The oxide was stressed at 5.6V (positive gate) and the leakage current was then measured at 3V. At this voltage electrons from the MOSFET inversion layer enter the gate at an energy just below the oxide conduction band.

The SDT signal is anisotropic, as shown in the upper two curves of Figure 8, which were taken at two different orientations of the sample in the magnetic field. This at first

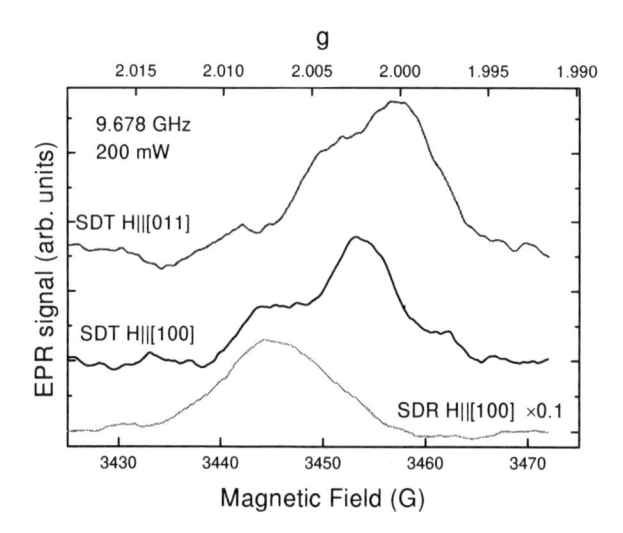

Figure 8. Spin-dependent tunneling (SDT) and spin-dependent recombination (SDR) signals in an electrically stressed 4.45nm MOSFET.

seems surprising for a defect in an amorphous SiO$_2$ film, since EPR signals in amorphous materials are usually powder patterns with no angular dependence. One possible explanation is that the defect is preferentially oriented along a substrate crystallographic axis. However, it is important to recognize that SDT, unlike conventional EPR, detects only those defects which participate in the transport process. Symmetry considerations suggest that for a *p*-like defect wave function, only defects with the orbital pointing along the direction of current flow can couple to the hopping electron, as illustrated in Figure 9. The SDT measurement will therefore be sensitive preferentially to those defects whose *p*-orbital is along the (100) direction normal to the substrate, and an angular dependence of the SDT signal will result.

The identity of this defect is not certain at this time, but the *g*-values are in the right range for oxygen dangling bonds. It is clearly not a P$_b$ center, even though P$_{b0}$ centers are detected in the same sample using the SDR technique (lower curve of Figure 8). Note that the SDT spectrum has a shoulder at the P$_{b0}$ position when the magnetic field is along (100), but the orientation dependence proves that this is not P$_{b0}$ since this shoulder disappears at the perpendicular orientation but the P$_{b0}$ signal shifts only slightly.[20]

The magnitude of the SDT signal, $\delta I/I = 2.4 \pm 0.3 \times 10^{-7}$ at the peak of resonance, is in very good agreement with expectation from the model outlined above. This is at least an order of magnitude smaller than typical SDR signals, which are enhanced by electron-hole correlation effects. The small size of the SDT signal makes more precise *g*-value measurements difficult.

7. Final Remarks

Experiments show that atomic hydrogen produces defects in SiO$_2$, very similar to the effects of hot electron stress or radiation. P$_b$ centers are generated by hydrogen,

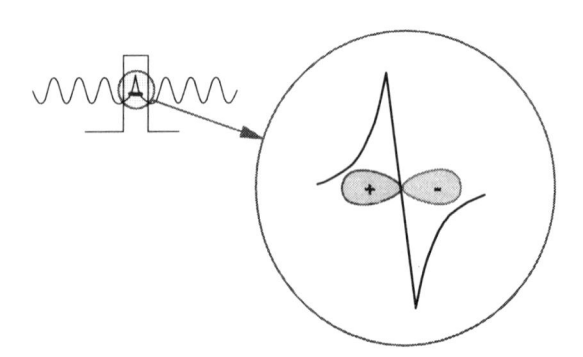

Figure 9. Explanation for the anisotropy of the SDT signal. For a defect *p*-orbital the tunneling wave function which was shown schematically in Figure 7 must actually be antisymmetric about the defect location as shown in the enlargement. Only orbitals oriented along the direction of current flow can couple in this way, as defects oriented in the perpendicular direction have zero overlap with the tunneling electron.

electrical stress, and radiation (mostly of the P_{b0} variety in (100) samples), but P_b centers cannot explain the majority of electrically active interface states, nor are they the main cause of leakage currents in thin SiO_2 films. It is clear that hydrogen is responsible for a suite of reactions beyond the simple Reactions (1) and (2). At the very least, any fully processed MOSFET contains various hydrogenous species within the oxide, such as water and hydroxyl groups, and atomic H released from the gate during electrical stress can react with these species in intermediate reactions, complicating the overall picture. Reactions (1) and (2) should be regarded as no more than schematic. The complete picture is not known at this time, and it seems clear that more research is needed to uncover the identity of the most important electrically active defect(s) which are created by atomic hydrogen or hot electron stress. Experiments with deuterated samples could be very helpful in further unraveling the role of hydrogen in defect generation in MOSFETs.

8. References

1. E.H. Poindexter and P.J. Caplan, Prog. Surf. Sci. **14**, 201 (1983).
2. D.L Griscom, in D.R. Uhlmann and N.J. Kreidl (eds.), *Glass Science and Technology, vol 4B* (Academic Press, New York, 1990) p. 151.
3. B. Henderson, M. Pepper, and R.L. Vranch, Semicond. Sci. Technol **4**, 1045 (1989).
4. J.H. Stathis and D.J. DiMaria, Appl. Phys. Lett. **61**, 2887 (1992).
5. J.H. Stathis, Appl. Phys. Lett. **68**, 1669 (1996).
6. E. Cartier and J.H. Stathis, Microelectron. Eng. **28**, 3 (1995).
7. A.G. Revesz, J. Electrochem. Soc. **126**, 122 (1979).
8. Unless all the hydrogen could be replace by deuterium; see J.W. Lyding, K. Hess, and I.C. Kizilyalli, Appl. Phys. Lett. **68**, 2526 (1996).
9. D.J. DiMaria and J.W. Stasiak, J. Appl. Phys. **65**, 2342 (1989), and references therein.
10. D.L. Griscom, J. Electron. Mater. **21**, 763 (1992).
11. E. Cartier, J.H. Stathis, and D.A. Buchanan, J. Appl. Phys. **63**, 1510 (1993).
12. J. H. Stathis and E. Cartier, Phys. Rev. Lett. **72**, 2745 (1994).
13. A.H. Edwards, in C.R. Helms and B.E. Deal (eds.), *The Physics and Chemistry of SiO_2 and the $Si-SiO_2$ Interface* (Plenum, New York, 1988), p. 271.
14. R.E. Stahlbush, E. Cartier, and D.A. Buchanan, Microelectron. Eng. **28**, 15 (1995).
15. E.H. Poindexter, in Z.C. Feng (ed.), *Semiconductor Interfaces, Microstructures, and Devices: Properties and Applications* (Insititute of Physics Publishing, Bristol, 1993) p. 229.
16. E. Cartier and J.H. Stathis, Appl. Phys. Lett. **69**, 103 (1996).
17. K.G. Druijf, J.M.M. deNijs, E. v.d. Drift, V.V. Afanas'ev, E.H.A. Granneman, and P. Balk, J. Non-Crystalline Sol. **187**, 206 (1995).
18. D.J. DiMaria and E. Cartier, J. Appl. Phys. **78**, 3883 (1995).
19. B. Ricco, M. Ya. Azbel, and M.H. Brodsky, Phys. Rev. Lett. **51**, 1795 (1983).
20. J.H. Stathis, Microelectron. Eng. **22**, 191 (1993).

TOWARDS AN ATOMIC SCALE UNDERSTANDING OF DEFECTS AND TRAPS IN OXIDE/NITRIDE/OXIDE AND OXYNITRIDE SYSTEMS

V.A. GRITSENKO*

Institute of Semiconductor Physics, Siberian Branch of Russian Academy of Sciences, Novosibirsk 630090, Russia

*Present address: Electronic Engineering Dept., The Chinese University of Hong Kong

1. Introduction

Amorphous dielectrics are key components of silicon devices. Silicon oxide, which has been used as the gate dielectric in MOS devices, is facing a number of challenges for deep sub-micron devices due to its low reliability because of electron and hole capturing, breakdown and boron penetration from poly-silicon. Silicon oxynitride (SiO_xN_y) will probably be used as the gate dielectric in the near future because of its boron blocking capabilities and higher reliability. Traps in SiO_xN_y determine the leakage current, breakdown and reliability. The nature of the traps still remains unclear.

Oxide/nitride/oxide (ONO) structures have been used as insulator in flat, trench, and stacked memory capacitors of DRAM's cell, as the dielectric between floating and top gate of floating gate EEPROM, and as the inter-polysilicon insulator and memory gate dielectric of radiation hard SONOS EEPROM devices. Anomalously large electron and hole capturing at the Si_3N_4-thermal SiO_2 interface however, has been observed [1-3], although the nature of these traps is not understood. The focus of this paper is to study the fundamental (atomic scale) properties of the SiO_xN_y and ONO structures, in particular, short-range order, electronic structure, the energy diagram, and the origin of the defects.

2. Short-range Order in Silicon Oxides of Different Composition

Short-range order in amorphous SiO_2, Si_3N_4 and SiO_xN_y is governed by the Mott rule which is: coordination number $CN = 8 - m$ (here m is the number of valence electrons). The electron distributions of Si, N, and O atoms are as follows: Si: $1s^2 2s^2 2p^6 3s^2 3p^2$; N: $1s^2 2s^2 2p^3$; O: $1s^2 2s^2 2p^4$. According to the electron distribution and the Mott rule in SiO_2, Si_3N_4 and SiO_xN_y the Si atom is coordinated to four O, N, or O and N atoms, the N atom in Si_3N_4 and SiO_xN_y is bonded to three Si atoms, and the O atom in SiO_2, and SiO_xN_y is bonded to the two Si atoms [4]. The same rule is observed in SiO_x (x<2) and SiN_x (x<4/3) [5]. The SiO_xN_y, SiO_x, and SiN_x consist of five sorts of tetrahedra SiO_vN_{4-v}, SiO_vSi_{4-v}, and SiN_vSi_{4-v}, respectively, were v = 0, 1, 2, 3, 4.

The structure of a-SiO$_x$ and a-SiN$_x$ is described by a random mixture model (RM) [6]. According to RM, a-SiO$_x$ consists of separate phases of Si, SiO$_2$, and suboxides, and a-SiN$_x$ of Si, Si$_3$N$_4$ and subnitrides (as can be seen from the XPS, Fig.1).

The structure of a-SiO$_x$N$_y$ can be described by a random bonding model (RB) [4]. In the RB model, the probability of tetrahedra of sort v in SiO$_x$N$_y$ is given by (1)

$$W(x,y) = \left(\frac{2x}{2x+3y}\right)^v \left(\frac{3y}{2x+3y}\right)^{4-v} \frac{4!}{v!(4-v)!} \quad (1)$$

For SiO$_x$N$_y$, the XPS (Fig.2) of the Si 2s core level can be deconvoluted as a sum of five peaks, corresponding to the tetrahedra with the probabilities given by (1). The SiO$_x$N$_y$ can be presented as a random mixture of Si, O, and N atoms, the coordination number of which follows from the Mott rule. This is a very useful property of SiO$_x$N$_y$ as a gate insulator. Due to the random distribution of the Si-O and Si-N bonds, the local fluctuations of dielectric permittivity (along the silicon-dielectric interface) are small. Therefore the fluctuations of the silicon surface potential in MOS devices with SiO$_x$N$_y$ are also small.

Fig. 1. Si 2s XPS of CVD SiO$_x$ of different composition [6].

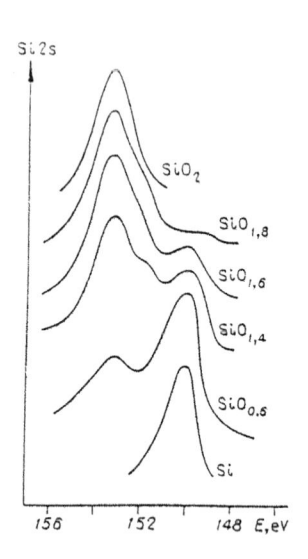

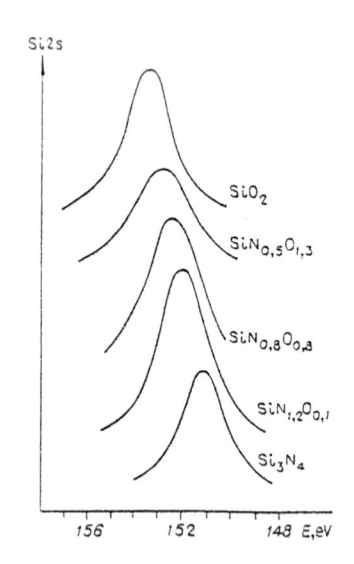

Fig. 2. Si 2s XPS of CVD SiO$_x$N$_y$ of different composition [4].

3. Electronic Structure of the a-SiO$_x$N$_y$

The Si-O bond in SiO$_x$N$_y$ is formed by Si 3s, 3p, 3d and O 2p, 2s bonding states, and the Si-N bond is formed by Si 3s, 3p, 3d and N 2p, 2s bonding states (Fig. 3). [4]. Near the top of the valence band E$_v$ there are nonbonding N 2p$_\pi$ and O 2p$_\pi$ orbitals. The N 2p$_\pi$ and O 2p$_\pi$ orbitals in SiO$_x$N$_y$ do not result in hole capturing in SiO$_2$, unlike early suggestions for SiO$_2$ [5].

The energy diagram of the Si-SiO$_x$N$_y$-Al structure, obtained from photoemission and fundamental absorption, shows that the electron barrier at the Si-SiO$_x$N$_y$ interface is linearly dependent on the refractive index (Fig.4) [7]. For a large nitrogen concentration, the Si- SiO$_x$N$_y$ hole barrier is small. The Si-SiO$_x$N$_y$ hole barrier value weakly depends on SiO$_x$N$_y$ composition. In order to block boron penetration, it is useful to increase the nitrogen concentration at the top interface of SiO$_x$N$_y$. However, in this case the hole barrier height can be small enough to allow hole injection from the positively biased poly-Si gate, which in turn may result in degradation. The resistivity against boron penetration and blocking hole injection from poly-silicon are the two opposite requirements to SiO$_x$N$_y$ composition. To achieve both of the above properties it is necessary to optimize the composition of SiO$_x$N$_y$.

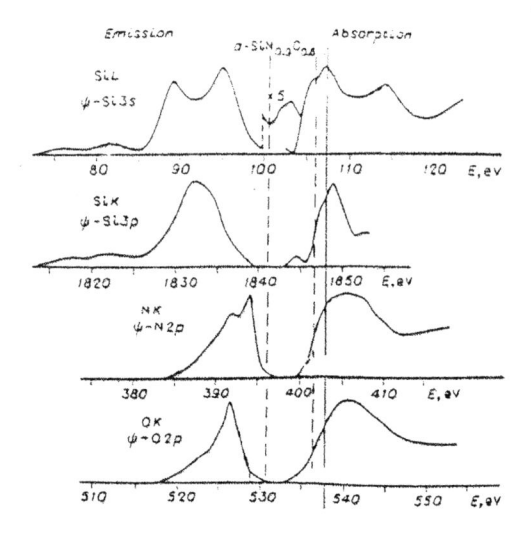

Fig. 3. X-ray emission and absorption spectra of SiO$_x$N$_y$, the symmetry of wave function in valence and conduction bands is marked [4].

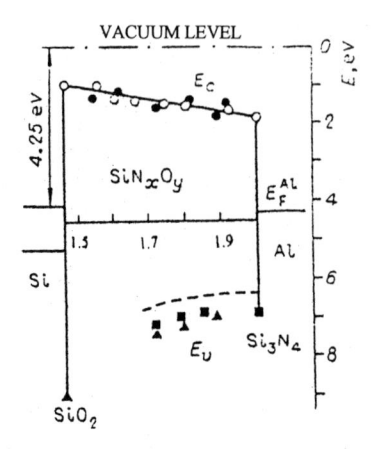

Fig. 4. Energy diagram of Si-SiO$_x$N$_y$-Al structures with different refractive index [7].

4. Energy Diagram of the ONO Structures

The electron barrier at the Si- Si$_3$N$_4$ interface is equal to 2.0 eV, and the hole barrier - 1.5 eV (Fig.5). This diagram is obtained from high field electron and hole injection measurements[8,9] and XPS studies. The hole barrier at the Si-Si$_3$N$_4$ interface is 0.6 eV lower than that obtained from photoemission [10]. The low hole barrier value explains the higher hole level injection from positively biased Si more than electron injection from negatively biased Si. In SONOS memory structures with a thin (≈20 Å) SiO$_2$ at both polarities on the metal at Si-dielectric interface, the major carriers are injected from silicon [11]. In SONOS structures with thick (>30 Å) SiO$_2$ layers, the dominant carriers are electrons [12]. The low leakage current, high breakdown field, high reliability and high yield of ONO structures (with respect to SiO$_2$) are related to electron capturing on traps in Si$_3$N$_4$ and the lower electric field at the injecting cathode. Electron capturing in Si$_3$N$_4$ is used in the ONO structures as a beneficial effect.

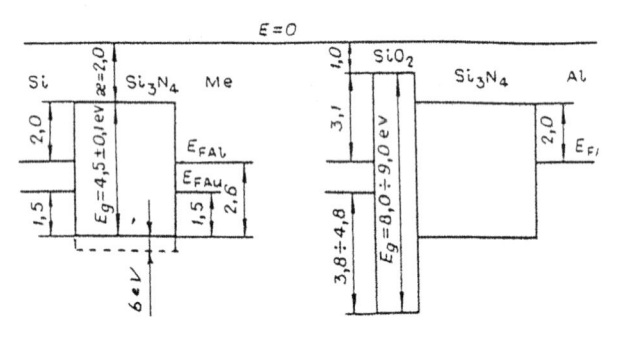

Fig. 5. Energy diagram of Si-SiO$_2$-Si$_3$N$_4$ -Me structures [8]. Dot line is the valence band position obtained in [10].

5. Chemistry of the Si$_3$N$_4$ - Thermal SiO$_2$ Interface

The top oxide in the ONO structures is usually produced by thermal oxidation of Si$_3$N$_4$. Anomalous large electron and hole capturing was observed at the Si$_3$N$_4$/SiO$_2$ interface in [1-3]. EELS plasmon spectroscopy of Si$_3$N$_4$/SiO$_2$ interface shows that a decrease in the electron beam energy from 3000 eV to 100 eV is accompanied by a decrease in the plasmon energy from 24.0 eV (Si$_3$N$_4$ bulk plasmon) to 20.0 eV (Fig.6) [13]. This value is lower than the bulk plasmon energy in SiO$_2$ (22.0 eV) and closer to the plasmon energy in Si (17.0 eV). This result can be interpreted by the presence of SiN$_x$ (x<4/3) or/and SiO$_x$ (x<2) at Si$_3$N$_4$-SiO$_2$ interface. This conclusion was confirmed by ellipsometric measurements. A very large refractive index of n=2.1 (n=1.96 for Si$_3$N$_4$ and n=1.46 for SiO$_2$) was observed at the Si$_3$N$_4$-SiO$_2$ interface (Fig. 7) [13]. The value of n=2.1 is typical of a silicon nitride enriched by excess silicon, i.e. SiN$_{x<4/3}$ [14].

The creation of the Si-Si bond during Si$_3$N$_4$ oxidation can be explained by the Mott rule, namely by the lower coordination number of oxygen in comparison with nitrogen. Replacing N-atoms for O-atoms may be accompanied by the creation of the Si-Si bonds according to (2)

$$2 \equiv Si_3N + 2 O \rightarrow 2 \equiv Si_2O + 2 \equiv Si^* \qquad \equiv Si^* + ^*Si\equiv \rightarrow \equiv Si-Si\equiv \quad (2)$$

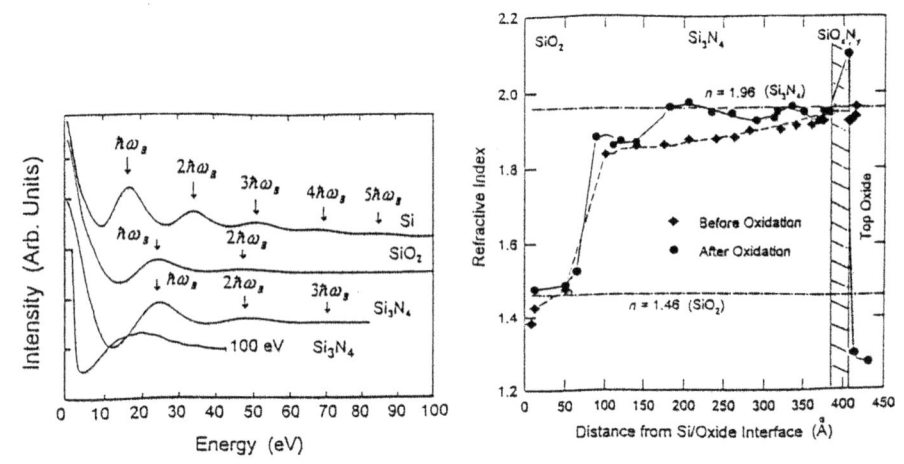

Fig. 6. EELS of Si, SiO$_2$ and Si$_3$N$_4$-SiO$_2$ interface   Fig. 7. The refractive index of Si-SiO$_2$-Si$_3$N$_4$-SiO$_2$ structure at 3000 eV and 100 eV [13]

A polaron model of multi-phonon capturing of electrons and holes on the Si-Si bond in Si$_3$N$_4$ was proposed and developed in [5, 9, 13, 15-18]. Similar model was proposed in [19]. Numerical simulations by MINDO/3 show that the Si-Si bond in Si$_3$N$_4$ and SiO$_2$ can capture both electrons and holes (Fig. 8.) [20]. According to these results, the Si-Si bonds act as electron and hole traps at the Si$_3$N$_4$-SiO$_2$ interface (Fig.9). The obtained result explains the anomalous large electron and hole capturing in the ONO structures [1-3].

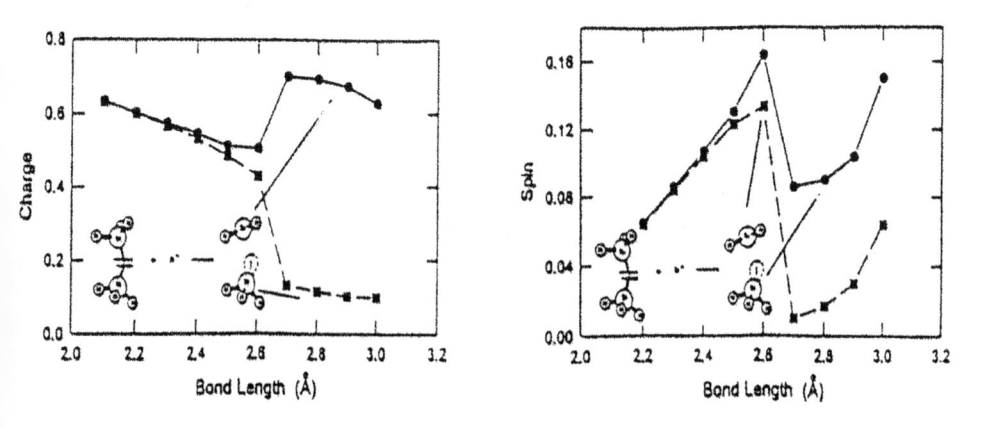

Fig. 8. Charge (left) and spin distributions (right) on the silicon atoms of Si-Si bond with captured hole in Si$_3$N$_4$ at different Si-Si distances [20].

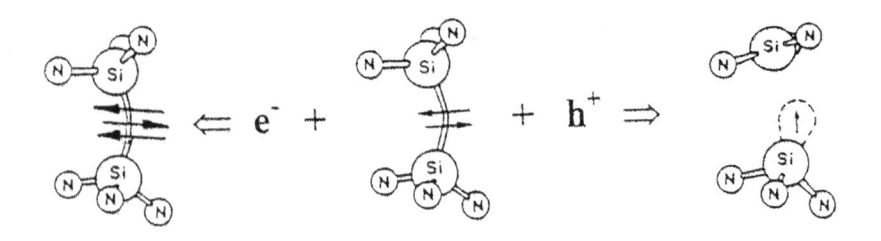

Fig. 9. The Si-Si bond as electron and hole trap in Si$_3$N$_4$.

6. Chemistry of the SiO$_x$N$_y$ - SiO$_2$ Interface Produced by SiO$_2$ Nitridation/Reoxidation

The ≡Si$_3$N species at the top surface of SiO$_x$N$_y$ are observed after SiO$_2$ nitridation [21,22]. The nitridation of SiO$_2$ in NH$_3$ results in the creation of a high density of electrons traps in SiO$_x$N$_y$. In order to reduce the trap density the reoxidation of SiO$_x$N$_y$ is performed after nitridation. However, at the top SiO$_x$N$_y$-SiO$_2$ interface, a high density of hole traps is observed [23]. The high density of E' centers is also observed at the top surface of reoxidized SiO$_x$N$_y$ [24]. The E' center is created after hole capturing on the Si-Si bond in SiO$_2$ according to the reaction ≡Si-Si≡ + h → ≡Si* + Si≡. Thus ESR data show that at the SiO$_x$N$_y$ - thermal oxide interface the Si-Si bonds exist. It is natural to propose that the Si-Si bonds are created as a result of nitrogen replacement at the SiO$_x$N$_y$ - thermal oxide interface according to the reaction (2). According to this hypothesis, the hole traps at the top surface of reoxidized nitrided oxide are the Si-Si bonds which are created by N atom replacing for O atoms.

7. A Hypothesis to Explain Why Nitridation Results in Hole Trap Removal from the Si-SiO$_2$ Interface

Hole traps (i.e. Si-Si bonds) always exist at the Si -thermal oxide interface[25]. The nitridation of SiO$_2$ results in removal of traps from the interface [26]. The nature of this phenomenon was unclear. Let's consider the interaction of nitrogen with the Si-Si bond at the Si-SiO$_2$ interface. The coordination number of N atom is three, so during the interaction with the Si-Si bond a nitrogen bridge ≡Si-N*-Si≡ between Si-atoms is formed according to the reaction (3)

$$2 \equiv Si\text{-}Si\equiv + 2 N \rightarrow 2 \equiv Si\text{-}N^*\text{-}Si\equiv \qquad (3)$$

Here -N*- is two coordinated nitrogen atom with a dangling electron. It was shown that the configuration of the nitrogen atom with the non-paired electron =N* is extremely unfavorable. High temperature annealing of Si$_3$N$_4$ with a high density of =NH bonds results in the diffusion of nitrogen into vacuum at the =NH bond breaking[27]. According to this observation the ≡Si-N*-Si≡ bridge will interact with other Si-Si bond as follows

$$2 \equiv\text{Si-N*-Si}\equiv + \equiv\text{Si-Si}\equiv \rightarrow 2 \equiv\text{Si}_3\text{N} \qquad (4)$$

The reactions (3) and (4) can be rewritten as

$$3 \equiv\text{Si-Si}\equiv + 2\text{ N} \rightarrow 2 \equiv\text{Si}_3\text{N} \qquad (5)$$

Hence according to (5) the reason why the Si-Si bonds are removed at nitridization from dielectric is the interaction of excess silicon with nitrogen. The proposed explanation is in qualitative agreement with the XPS study of Si-dielectric interface before and after nitridation in which a lower density of the Si-Si bonds at Si-dielectric interface was observed after nitridation[21]. An increase of the concentration of $\equiv\text{Si}_3\text{N}$ was observed at the same time.

Acknowledgement

The author thanks M.L Green, E.P. Gusev and E Garfunkel for their stimulating discussion on "How much nitrogen we need at the Si-SiO$_x$N$_y$ interface?" at the NATO conference in St.-Petersburg. This question led the author to the hypothesis discussed in the last section of this paper.

References:

1. Suzuki, E., Hayashi, Y. (1986). On the Oxide-Nitride Interface Traps by Thermal Oxidation of Thin Nitride in Metal-Oxide-Nitride-Oxide-Semiconductor Structures, IEEE Trans. Electron Devices, ED-33, 214-217.
2. Weinberg, Z.A., Stein, H.J., Nguen, T.N., and Sun J.Y. (1990) Ultrathin Oxide- Nitride-Oxide Films, Appl. Phys. Lett. 57, 1248-1251.
3. Ozawa, Y., Yamabe, K., Iwai, H., Sasaki, M., Toyota, M. (1995) An Improvement of Hot-Carrier Reliability in the Stacked Nitride-Oxide Gate n- and p-MISFET's, IEEE Transactions on Electron Devices, 42, 704-712.
4. Britov, I.A., Gritsenko, V.A., and Romaschenko, Yu.N. (1985) Short-Range Order and Electronic Structure of Amorphous SiN$_x$O$_y$, Letters to Journal of experimental and Theoretical Physics, 62, 321-327.
5. Gritsenko, V.A. (1993) Structure and Electronic Properties of Amorphous Dielectrics in Silicon MIS Structures, Ed. Science, Novosibirsk, Russia.
6. Gritsenko, V.A., Kostikov, Yu.P., and Romanov, N.A. (1981) SiO$_x$ as a Model Medium with Large-Scale Potential Fluctuation, Letters to Journal of Experimental and Theoretical Physics, 34, 3-6.
7. Gritsenko, V.A., Dikovskaja, N.D., and Mogilnikov, K.P. (1978) Band Diagram and Conductivity of Silicon Oxinitride Films, Thin Solid Films, 51, 353-357.
8. Gritsenko, V.A., Meerson, E.E. (1988) Injection of Electrons and Holes from Metal in MNOS Structures, Microelectronics (Sov), 17, 249-255.
9. Gritsenko, V.A. (1988) Electronic Structure and Optical Properties of Silicon Nitride, In "Silicon Nitride in Electronics", Elsevier, New York.
10. DiMaria, D.J., Arnett, P.C. (1975) Hole Injection into Silicon Nitride: Interface Barrier Energy by Internal Photoemission, Appl. Phys. Lett, 26, 711-714.
11. Ginovker, A.S., Gritsenko, V.A., Sinitsa, S.P. (1974) Two Band Conduction of Amorphous Silicon Nitride, Phys. Stat. Sol. B26, 489-495.
12. Gritsenko, V.A., Meerson, E.E., (1988) Monopolar and Bipolar Injection in MNOS Structures, Microelectronics (Sov), 17, 532-535.
13. Gritsenko, V.A., Morokov, Yu.N., Novikov, Yu.N., Petrenko, I.P. Svitasheva, S.N. (1997) Enriching of the Si$_3$N$_4$/ Thermal Oxide Interface by Excess Silicon in ONO structures, Microelectronic Engineering, 36, 123-124.

14. Bolotin, V.P., Britov, I.A, Gritsenko, V.A., Olshanezkii, B.Z., Popov, V.P., Romashenko, Yu. N., Serjapin, V.G., Tiis, S.A. (1990) Composition and Structure of Enriched by Silicon Silicon Nitride, Sov. Phys. Dokl., 310, 114-117.
15. Gritsenko, V.A., Pundur, P. (1986) Muliphonon Capturing and Radiative Transitions in a-Si$_3$N$_4$, Sov. Physics Solid State, 28, 1829-1830.
16. Gritsenko, V.A., Meerson, E.E., Travkov, I.V., Goltvjanskii, Yu.V. (1987) Nonstationary Electrons and Holes Transport by Depolarization of MNOS Structures: Experiment and Numerical Simulation, Microelectronics (Sov), 16, 42-50.
17. Gritsenko, V.A., Kostikov, Yu.P., Khramova, L.V. (1992) Electronic Structure of Si-H and N-H Bond in SiN$_x$:H, Sov. Phys. Solid State, 34, 1300-1303.
18. Gritsenko, V.A., Milov, A.D. (1996) Wigner Crystallization of Electrons and Holes in Amorphous Silicon Nitride, Antiferromagnetic Ordering of Localized Electrons and Holes as a Result of Resonance Exchange Interaction, Letters to Journal of Experimental and Theoretical Physics (Rus), 64, 531-536.
19. Kamigaki,Y, Minami, S., and Kato, H. (1990) A New Model of Electron and Hole Traps in Amorphous Silicon Nitride, J. Appl. Phys., 68, 2211-2220.
20. Gritsenko, V. A., Morokov, Yu.N.,. Novikov, Yu.N, Wong, H, Cheng, Y. C. (1997) Electronic Structure of Si-Si Bond in Si$_3$N$_4$ and SiO$_2$: Experiment and Simulation by MINDO/3, Pros. 1996' Material Research Conference, Boston, USA
21. Lu, Z.H., Tay, S.P., Cao, R., Pianetta, P. (1995) The Effect of Thermal N$_2$O Nitridation on the Oxide/Si(100) Interface Structure. Appl. Phys. Lett, 67, 2836-2838.
22. Lu, H.C., Gusev, E.P., Gustafsson, T., Garfunkel, E., Green, M.L., Brasen, D., and Feldman, L.C. (1996) High Resolution Ion Scattering Study of Silicon Oxyhitridation, Appl. Phys. Lett. 69, 2713-2715.
23. Malik, A., Vasi, J., and Chandorkar, A.N. (1993) The Nature of Hole Traps in Reoxidized Nitrided Oxide Gate Dielectrics. J. Appl. Phys. 74, 2665-2268.
24. Yount, J.T. and Lenahan, P.M. Dunn, G.J. (1992) Electron Spin Resonance Study of Radiation-Induced Point Defects in Irradiated and Reoxidized Nitrided Oxides, IEEE Trans. Nucl. Sci. 39, 2211-2219.
25. Himpsel, J., McFeely, F.R., Tabel-Ibragimi, A., Yarmoff, J.A., Hollinger, G. (1988) Microscopic Structure of the SiO$_2$/Si Interface, Phys Rev. B 38, 6084-6096.
26. Bhat, M., Yoon, G.V., Kim, J., Kwong, D.L., Arendt, A. White, J.M. (1994) Effect of NH$_3$ Nitridation on Oxide Grown in pure N$_2$O, Appl, Phys. Lett, 64, 2116-2118.
27. Chramova, L.V., Chusova, T.P., Gritsenko, V.A. Feofanov, G.N., Smirnova, T.P. (1987) Chemical Compositional Changing and Absorbtion Edge Red Shift in Annealed Silicon Nitride, Nonorganic Materials (Sov), 23, 73-76

A NEW MODEL OF PHOTOELECTRIC PHENOMENA IN MOS STRUCTURES

Outline and Applications

H.M. PRZEWLOCKI
Institute of Electron Technology
Al. Lotnikow 32/46, 02-668 Warsaw, Poland

1. Introduction

Photoelectric methods have been widely used to determine various parameters of the MOS system (for a summary see [1] and [2]). Until recently, all these methods were based on the physical model developed in the early seventies by Powell and Berglund [3-5], which allows prediction of the photocurrent vs. gate voltage, I(V_G) and photocurrent vs. wavelength, I(λ), characteristics of UV illuminated MOS structures.
The Powell-Berglund (PB) model applies, however, only when relatively high electric fields $\mathcal{E}$ exist in the dielectric (roughly when $|\mathcal{E}| > 10^5$ V/cm) and when the photocurrent is limited by the number of carriers that pass over the potential barrier at the emitter-dielectric interface.
Concerning the photocurrent-gate voltage I(V_G) characteristics, the PB model applies outside of a certain region lying close to the origin of the I-V_G coordinate system, as schematically illustrated by a rectangular box, shown in Fig. 1. Until recently, there was no physical model which would allow prediction of the I(V_G) characteristics inside of the box shown in Fig. 1, or in other words, in the region where low electric fields ($\mathcal{E} \cong 0$) are present in the dielectric. However, the I(V_G) characteristics of MOS structures can be easily and very accurately measured in this region, as illustrated in Fig. 2. These experimental characteristics have a number of interesting features, two of which are most important for this work:

1. The position of the zero-current point on the V_G axis (V_G^o) changes with changing the wavelength λ of the UV radiation illuminating the MOS structure.
2. The shapes of the I(V_G) characteristics change dramatically with changing λ.

These features are used as a basis of new measurement methods described briefly in chapter 5 of this paper.

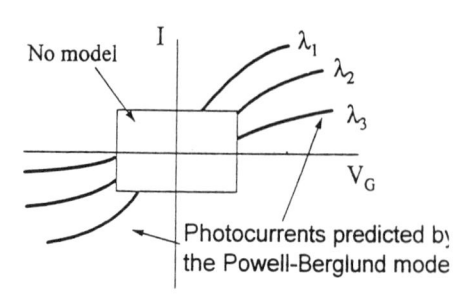

Figure 1. Schematic illustration of the regions in the photocurrent-gate voltage (I-V_G) plane, where the Powell-Berglund model applies

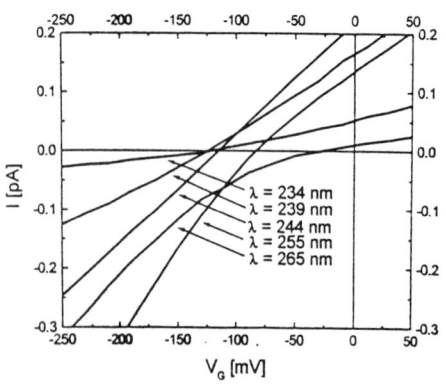

Figure 2. Experimental photocurrent vs. gate voltage ($I = f(V_G)$) characteristics in the region of low electric field in the dielectric, taken for an Al-SiO$_2$-Si(N$^+$) structure with SiO$_2$ layer thickness: $t_I = 147$ nm

However, to allow prediction of the I (V$_G$) and I (λ) characteristics in the region of low electric fields ($\mathcal{E} \cong 0$) in the dielectric, an appropriate physical model of internal photoemission had to be developed. Such a model has been developed, experimentally verified and used as a basis of new measurement methods.

This model and some of its applications are presented in the following chapters of this paper.

2. Formulation of the problem

An MOS structure with a semitransparent gate is considered which is illuminated by UV radiation, as shown in Fig. 3. The wavelength λ of the UV radiation is variable over the range of values at which photoinjection of electrons takes place, from both the gate and the substrate into the conduction band of the dielectric. The gate bias V$_G$ is variable over the range of values for which the voltage drop in the dielectric $|V_I|$ is small ($V_I \cong 0$).

The photocurrent I is measured in the external circuit, by a suitable electrometer. It is the main purpose of the following analysis to find the dependence of the photocurrent I, on the gate voltage V$_G$ and on the wavelength λ.

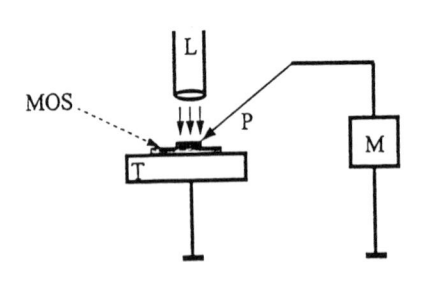

Figure 3. Schematic illustration of the setup considered in this work. MOS - the MOS structure with semitransparent gate, L - UV illumination system, P - contact probe, M - voltage source and current measurement unit, T - measurement stage.

In this analysis following assumptions are made:
- The photocurrent flowing in the external circuit of the setup, shown in Fig. 3, is due exclusively to the electrons photoinjected over the potential barriers from the

gate and from the substrate, into the conduction band of the dielectric (photoinjection of holes is assumed to be negligible).
- Due to large fluxes of electrons photoinjected into the dielectric, from the gate and from the substrate, a considerable space charge exists in the dielectric. This space charge consists of free electrons and of electrons which are trapped in the dielectric. It is assumed that at any position x in the dielectric, a fixed fraction θ of the total charge density resides in the conduction band (free electrons) while the remainder resides in traps (trapped electrons).
- The problem may be considered as one dimensional.

Considering the problem with these assumptions in mind, one should take into account that:
- At low electric fields in the dielectric ($\mathcal{E} \cong 0$) the diffusion component of the current I, plays a significant role.
- The current I flowing through the dielectric should be considered as a space charge limited current.

Hence, the problem is described by three equations:

The current flow equation:

$$j = q\mu n_c(x)\mathcal{E}(x) + \mu kT \frac{dn_c}{dx} \quad (1)$$

where j is current density, q electron charge, μ electron mobility in dielectric, n_c free-electron density in dielectric, $\mathcal{E}$ electric field in dielectric, k Boltzmann constant, T temperature and x is a coordinate perpendicular to the gate-dielectric and dielectric-substrate interfaces (x=0 at the gate-dielectric interface).

The Poisson equation:

$$\frac{\varepsilon}{q}\frac{d\mathcal{E}}{dx} = -n(x) \quad (2)$$

where ε is electric permittivity of dielectric and n is total density of electrons, i.e., the sum of free electron density n_C and trapped electron density n_T.

The relation between free and total electron densities:

$$n_c = \theta n \quad (3)$$

as previously assumed.

To solve the problem following dimensionless variables are introduced:

distance:
$$z = \frac{x}{t_I} \quad (4)$$

potential:
$$\phi = \frac{qV}{kT} \quad (5)$$

electric field:
$$E = \varepsilon \frac{qt_1}{kT} = -\frac{d\phi}{dz} \quad (6)$$

current density
$$J = \frac{jq^2 t_1^3}{\varepsilon \mu k^2 T^2 \theta} \quad (7)$$

electron density
$$N = \frac{nq^2 t_1^2}{\varepsilon kT} \quad (8)$$

where t_1 is thickness of the dielectric and $V = V(x)$ is potential in the dielectric.
Using these dimensionless variables and combining equations (1), (2) and (3), following relation is obtained:

$$J = \frac{d^3\phi}{dz^3} - \frac{d^2\phi}{dz^2} \cdot \frac{d\phi}{dz} \quad (9)$$

which should be solved, subject to boundary conditions, resulting from normalized Poisson equation (2):

$$\frac{d^2\phi}{dz^2}(z=0) = N(0) \quad (10a)$$

$$\frac{d^2\phi}{dz^2}(z=1) = N(1) \quad (10b)$$

where $N(0)$ and $N(1)$ are normalized electron densities at $z = 0$ and $z = 1$.
The formulation of this problem is described in more detail in [6].

3. Solution of the problem

Equation (9) can be integrated once, yielding:

$$Jz + C_1 = \frac{d^2\phi}{dz^2} - \frac{1}{2}\left(\frac{d\phi}{dz}\right)^2 \quad (11)$$

where C_1 is the first constant of integration.
Substitution of:

$$\phi = -2\ln y \quad (12)$$

transforms equation (11) into:

$$\frac{d^2 y}{dz^2} + \frac{1}{2}(Jz + C_1)y = 0 \quad (13)$$

Equation (13), which will be called the intermediate equation, will now be solved separately for J = 0 and for J ≠ 0.

3.1. SOLUTION FOR THE CASE OF J=0

For J=0, the solution of the intermediate equation (13) is given by:

$$y = A \cos \omega z + B \sin \omega z \qquad (14)$$

where A and B are constants of integration and:

$$\omega = \sqrt{C_1/2} \qquad (15)$$

Using eq. (12) and assuming that $\phi = 0$ for z = 0, one obtains A = 1 and:

$$\phi(z) = -2 \ln (\cos \omega z + B \sin \omega z) \qquad (16)$$

To find the value of B, eq. (16) is differentiated twice and use is made of the boundary conditions (10). As a result, one obtains [6]:

$$\phi(z) = -2 \ln \left\{ \frac{r \sin \omega z + \sin[\omega(1-z)]}{\sin \omega} \right\} \qquad (17)$$

where:

$$r = [N(0)/N(1)]^{1/2} \qquad (18)$$

and the value of ω is given by the equation:

$$N(0) \sin^2 \omega = 2\omega^2 \left(1 + r^2 - 2r \cos \omega \right) \qquad (19)$$

It results from eq. (17), that the dimensionless potential drop in the dielectric, $\Delta\phi$, is given by:

$$\Delta\phi = \phi(0) - \phi(1) = \ln \frac{N(0)}{N(1)} \qquad (20)$$

Hence, making use of eqs. (5) and (8), it can easily be shown, that for J=0, the total voltage drop in the dielectric V_I^o, is given by:

$$V_I^o = \frac{kT}{q} \ln \frac{n_G}{n_S} \qquad (21)$$

where n_G, n_S are electron densities in the dielectric, in the vicinities of its interfaces with the gate and the substrate.

Further, by considering the fractions of UV radiation absorbed by the gate and by the substrate and by making certain approximations, it can be shown [6], that the zero current gate voltage V_G^o, is given by:

$$V_G^o = \frac{kT}{q}\left(\ln\frac{A(\lambda)}{T(\lambda)}+\ln\frac{(h\nu-E_{BG})^{p_G}}{(h\nu-E_{BS})^{p_S}}\right)+C \qquad (22)$$

where $A(\lambda)$ is the fraction of UV radiation energy absorbed by the gate, $T(\lambda)$ is the fraction of UV radiation transmitted to (and absorbed by) the substrate, $h\nu$ is photon energy, E_{BG}, E_{BS} are the barrier heights for photoinjection from the gate and from the substrate, p_G, p_S are the exponential factors depending on the emitter type and C is given by:

$$C = \frac{kT}{q}\ln R + V_s^o + \phi_{MS} \cong \text{const} \qquad (23)$$

where R is a constant, V_s^o is the semiconductor surface potential, when J=0 and ϕ_{MS} is the effective contact potential difference in the MOS structure.

It has to be noticed here, that the voltage drop in the dielectric V_I^o, given by eq. (21), becomes equal zero, for $n_G = n_S$. (In fact, this is intuitively obvious, since for $n_G = n_S$ the diffusion component of the current disappears and since the total current is equal zero, it means that there is no voltage drop in the dielectric, i.e., $V_I^o = 0$). Due to the optical interference taking place in the MOS structure, the state of $n_G = n_S$ occurs only for certain, well defined wavelengths $\lambda = \lambda_O$ of the UV radiation. The values of λ_O depend primarily on the thickness of the dielectric layer t_I. For $\lambda = \lambda_O$, $V_I^o = 0$ and the zero current gate voltage V_G^o, becomes equal to the so called zero dielectric voltage, denoted V_{GO}.

It can be shown, that for $n_G = n_S$, the $I = f(V_I)$ characteristic and for $V_G = V_I + \text{const}$ also the $I = f(V_G)$ characteristic, become symmetrical with respect to the $I = 0$ point, i.e., in this case:

$$I(V_G - V_{GO}) = -I(V_{GO} - V_G) \qquad (24)$$

Eq. (24) is used in practice as an indication that the relation $\lambda = \lambda_O$ is fulfilled.

3.2. SOLUTION FOR THE CASE OF J ≠ 0

Considering the intermediate equation (13) for $J \neq 0$, the following substitutions are made:

$$y = s^{1/3}y_1 \qquad (25)$$

$$s = \frac{\sqrt{2}}{3} \frac{(Jz+C_1)^{3/2}}{J} \tag{26}$$

where the value of the variable s, given by eq. (26) can be either real (for $(Jz + C_1) \geq 0$) or imaginary (for $(Jz + C_1) < 0$). These two cases will now be considered separately.

3.2.1. Solution for real values of s
Substitutions given by eqs. (25) and (26), transform eq. (13) into a Bessel equation:

$$s^2 \frac{d^2 y_1}{ds^2} + s \frac{dy_1}{ds} + \left(s^2 - \frac{1}{9}\right) y_1 = 0 \tag{27}$$

with the solution:

$$y_1 = D[J_{1/3}(s) + C_2 J_{-1/3}(s)] \tag{28}$$

where $J_{1/3}(s)$ and $J_{-1/3}(s)$ are Bessel functions of the first kind, of the order 1/3 and -1/3 respectively, of the argument s and C_2 and D are the constants of integration. Using eqs. (12) and (25) to transform eq. (28), one gets:

$$\phi(s) = -2 \ln D s^{1/3} [J_{1/3}(s) + C_2 J_{-1/3}(s)] \tag{29}$$

In this equation, the independent variable z has been superceded by the variable s, defined by eq. (26).
For z = 0:

$$s = s_0 = \frac{\sqrt{2}}{3} \frac{C_1^{3/2}}{J} \tag{30}$$

while for z = 1:

$$s = s_1 = \frac{\sqrt{2}}{3} \frac{(J+C_1)^{3/2}}{J} \tag{31}$$

It results from eqs. (30) and (31), that:

$$J = \frac{9}{2} \left(s_1^{2/3} - s_0^{2/3}\right)^3 \tag{32}$$

and

$$C_1 = \frac{9}{2} s_0^{2/3} \left(s_1^{2/3} - s_0^{2/3}\right)^2 \tag{33}$$

From eq. (29) it follows, that the dimensionless potential drop $\Delta \phi$ in the dielectric, is given by:

$$\Delta \phi = \phi(s_0) - \phi(s_1) = 2 \ln \frac{s_1^{1/3} [J_{1/3}(s_1) + C_2 J_{-1/3}(s_1)]}{s_0^{1/3} [J_{1/3}(s_0) + C_2 J_{-1/3}(s_0)]} \tag{34}$$

To find the value of the integration constant C_2, eq. (29) is differentiated twice and use is made of the boundary conditions given by eqs. (10a) and (10b), yielding:

$$C_2 = \frac{J_{-2/3}(s_0) - p_0 J_{1/3}(s_0)}{J_{2/3}(s_0) + p_0 J_{-1/3}(s_0)} \tag{35}$$

and:

$$C_2 = \frac{J_{-2/3}(s_1) - p_1 J_{1/3}(s_1)}{J_{2/3}(s_1) + p_1 J_{-1/3}(s_1)} \tag{36}$$

where:

$$p_0 = \pm \sqrt{\frac{2}{9} N(0) \, s_0^{-2/3} \left(s_1^{2/3} - s_0^{2/3}\right)^{-2} - 1} \tag{37}$$

and:

$$p_1 = \pm \sqrt{\frac{2}{9} N(1) \, s_1^{-2/3} \left(s_1^{2/3} - s_0^{2/3}\right)^{-2} - 1} \tag{38}$$

Eliminating C_2 from eqs. (35) and (36), one obtains:

$J_{-2/3}(s_0)J_{2/3}(s_1) - p_0 J_{1/3}(s_0)J_{2/3}(s_1) + p_1 J_{-2/3}(s_0)J_{-1/3}(s_1) - p_0 p_1 J_{1/3}(s_0)J_{-1/3}(s_1) =$

$= J_{2/3}(s_0)J_{-2/3}(s_1) + p_0 J_{-1/3}(s_0)J_{-2/3}(s_1) - p_1 J_{2/3}(s_0)J_{1/3}(s_1) - p_0 p_1 J_{-1/3}(s_0)J_{1/3}(s_1);$ (39)

Equation (39) gives the exact relation, that exists for real values of s, between the s_0 and the s_1 values. It means that e.g. for a given (or assumed) value of s_0, a corresponding value of s_1 can be found by solving eq. (39). Once the corresponding pair of s_0 and s_1 values has been found, the values of J, C_1, C_2 and $\Delta\phi$ can be calculated using eqs. (32), (33), (35) or (36) and (34). This way a part of the J vs. $\Delta\phi$ characteristic, which corresponds with real values of s, can be calculated.

3.2.2. Solution for imaginary values of s

For imaginary values of s, one can use the t variable (instead of s variable defined by eq. (26)), which is defined by:

$$t = \frac{\sqrt{2}}{3} \frac{(-Jz - C_1)^{3/2}}{J} \tag{40}$$

The t variable is real, wherever the s variable is imaginary. In this case the use of eqs. (25) and (40), transforms the intermediate equation (13) into a modified Bessel equation:

$$t^2 \frac{d^2 y_1}{dt^2} + t \frac{dy_1}{dt} - \left(t^2 + \frac{1}{9}\right) y_1 = 0 \tag{41}$$

with the solution:

$$y_1 = D'[I_{1/3}(t) + C_2' I_{-1/3}(t)] \tag{42}$$

where $I_{1/3}(t)$ and $I_{-1/3}(t)$ are modified Bessel functions of the first kind, of the order of 1/3 and -1/3 respectively, of the argument t and D' and C_2' are the integration constants. Combining eqs. (42), (12), (25) and (40), one gets:

$$\phi(t) = -2\ln D'(it)^{1/3}[I_{1/3}(t) + C_2' I_{-1/3}(t)] \tag{43}$$

In eq. (43), the independent variable z has been superceded by the variable t, defined by eq. (40).
For z = 0:

$$t = t_0 = \frac{\sqrt{2}}{3}\frac{(-C_1)^{3/2}}{J} \tag{44}$$

while for z = 1:

$$t = t_1 = \frac{\sqrt{2}}{3}\frac{(-J-C_1)^{3/2}}{J} \tag{45}$$

It results from eqs. (44) and (45), that:

$$J = \frac{9}{2}\left(t_0^{2/3} - t_1^{2/3}\right)^3 \tag{46}$$

$$C_1 = -\frac{9}{2}t_0^{2/3}\left(t_0^{2/3} - t_1^{2/3}\right)^2 \tag{47}$$

From eq. (43) it follows, that the dimensionless potential drop $\Delta\phi$ in the dielectric, is given by:

$$\Delta\phi = \phi(t_0) - \phi(t_1) = 2\ln\frac{t_1^{1/3}[I_{1/3}(t_1) + C_2' I_{-1/3}(t_1)]}{t_0^{1/3}[I_{1/3}(t_0) + C_2' I_{-1/3}(t_0)]} \tag{48}$$

Similarly as in the previous section, the integration constant C2' is found by differentiating twice eq. (43) and making use of, boundary conditions, given by eqs. (10a) and (10b).

$$C_2' = -\frac{I_{-2/3}(t_0) - r_0 I_{1/3}(t_0)}{I_{2/3}(t_0) - r_0 I_{-1/3}(t_0)} \tag{49}$$

$$C_2' = -\frac{I_{-2/3}(t_1) - r_1 I_{1/3}(t_1)}{I_{2/3}(t_1) - r_1 I_{-1/3}(t_1)} \tag{50}$$

where:

$$r_0 = \pm \sqrt{\frac{2}{9} N(0) t_0^{-2/3} (t_1^{2/3} - t_0^{2/3})^{-2} + 1} \qquad (51)$$

$$r_1 = \pm \sqrt{\frac{2}{9} N(1) t_1^{-2/3} (t_1^{2/3} - t_0^{2/3})^{-2} + 1} \qquad (52)$$

Eliminating C_2' from eqs. (49) and (50), one gets:

$$I_{-2/3}(t_0)I_{2/3}(t_1) - r_0 I_{1/3}(t_0)I_{2/3}(t_1) - r_1 I_{-2/3}(t_0)I_{-1/3}(t_1) + r_0 r_1 I_{1/3}(t_0)I_{-1/3}(t_1) =$$
$$= I_{2/3}(t_0)I_{-2/3}(t_1) - r_0 I_{-1/3}(t_0)I_{-2/3}(t_1) - r_1 I_{2/3}(t_0)I_{1/3}(t_1) + r_0 r_1 I_{-1/3}(t_0)I_{1/3}(t_1) ; \qquad (53)$$

Equation (53) gives the exact relation that exists for real values of the variable t, between t_0 and t_1. For a given (or assumed) value of t_0, a corresponding t_1 value can be found by solving eq. (53). Once the corresponding pair of t_0, t_1 values has been found, the values of J, C_1, C_2 and $\Delta\phi$ can be calculated using eqs. (46), (47), (49) or (50) and (48). This way a part of the J vs. $\Delta\phi$ characteristic, which corresponds with real values of the t variable (or imaginary values of s variable), can be calculated.

It also has to be mentioned, that there exist situations in which the variable s is real in one part of the dielectric and becomes imaginary in the other part. In this case eq. (39) should be used in the region where s is real and eq. (53) should be applied in the other region. The interface between these two regions is at s=0.

4. Experimental verification

4.1. EXPERIMENTAL VERIFICATION FOR THE CASE J=0

To verify the theoretical considerations of section 3.1., a number of $V_G^o = f(\lambda)$ characteristics were taken for Al-SiO$_2$-Si (N$^+$) and Al-SiO$_2$-Si (P$^+$) structures, with different Al gate thicknesses t_{Al} (in the range of $t_{Al} = 8.0 \div 40.0$ nm) and with different thicknesses of the SiO$_2$ layer t_I (in the range of $t_I = 50 \div 400$ nm). Examples of such $V_G^o = f(\lambda)$ characteristics taken for structures with the same thicknesses of Al-gates, $t_{Al} \cong 15$ nm, but with different thicknesses t_I of the SiO$_2$ layers and with differently doped substrates, are shown in Fig. 4, in comparison with curves calculated using eq. (22).

The experimental characteristics shown in Fig. 4, were taken in the following way. For each λ value, a section of the photocurrent vs. gate voltage, $I = f(V_G)$, characteristic was taken, in the vicinity of the $I = 0$ point (usually from $V_G = V_G^o - 100$ mV, to $V_G = V_G^o + 100$ mV). The V_G^o values and the symmetry of each of these characteristics were monitored and recorded. Making use of the property expressed by eq. (24) and assuming again that $V_G^o \cong V_I^o + $ const it was concluded, that $\lambda = \lambda_0$ and $V_G^o(\lambda_0) = V_{GO}$, for each λ value, at which the I = f (V_G) characteristic becomes symmetrical in relation to the I = 0 point. The so determined V_{GO} values are marked in Fig. 4.

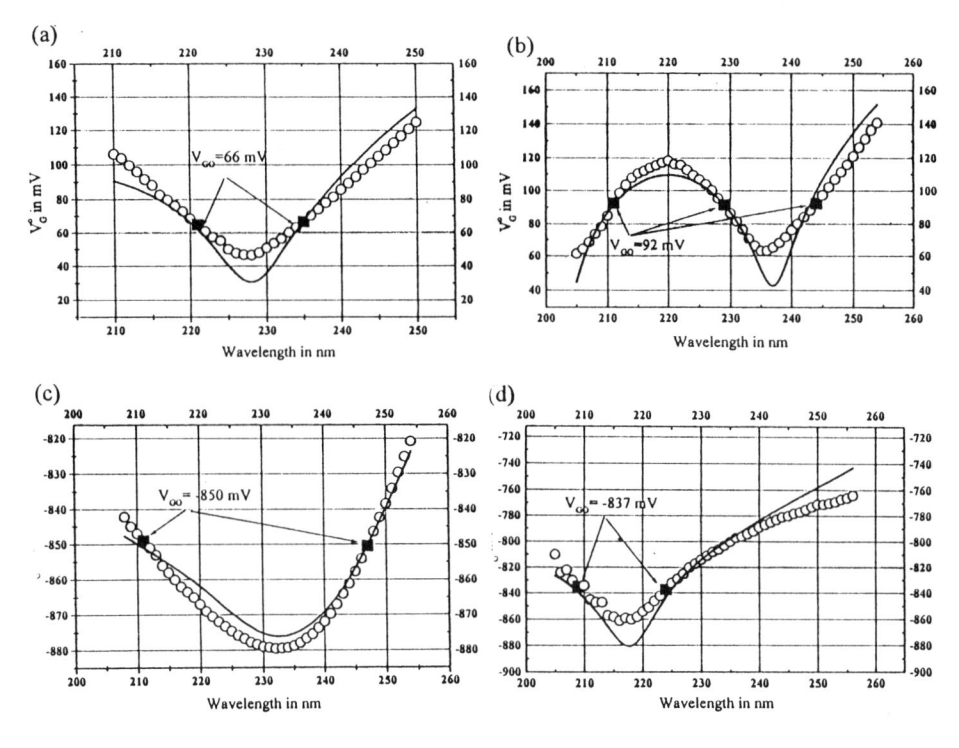

Figure 4. Comparison of $V_G^O = f(\lambda)$ curves calculated using eq. (22) (solid lines) with the experimental characteristics (circles), taken for Al-SiO$_2$-Si(N$^+$) and Al-SiO$_2$-Si(P$^+$) capacitors with different thicknesses t_I of SiO$_2$ layers. (a), (b) are for N$^+$ and (c), (d) are for P$^+$ substrate capacitors. Given below are t_I and C values used in calculations to obtain best fit between theoretical and experimental curves. Results of independent ellipsometric measurements of SiO$_2$ thickness t_I (ellips) are also given. The V$_{GO}$ values are marked in the figures by solid squares. (a) t_I = 209 nm, C = 50 mV, t_I (ellips) = 218 nm; (b) t_I = 375 nm, C = 50 mV, t_I (ellips) = 385,9 nm; (c) t_I = 62 nm, C = -872 mV, t_I (ellips) = 63.5 nm; (d) t_I = 196 nm, C = -855 mV, t_I (ellips) = 204.2 nm

The t_I and C values used for calculation of the theoretical V_G^O = f (λ) curves shown in Fig. 4, were chosen in such a way as to obtain the best fit between the theoretical and experimental curves. First, the t_I values assuring best fit were determined with high precision, from the position of the V_G^O = f (λ) curve minimum, which is very sensitive to the t_I value. Then, C values were chosen to obtain the appropriate vertical shift of the theoretical curve.

Concerning the parameters used to obtain the best fit between theoretical and experimental V_G^O = f (λ) curves, following remarks can be made.
- The t_I values used to obtain best fit were always (1.0 ÷ 4.5) % lower than the SiO$_2$ thickness t_I (ellips), determined by ellipsometry, as indicated in Fig. 4. This has been shown to be due to the oversimplified optical model, used to calculate the A(λ)

and T(λ) functions in eq. (22). Application of an improved optical model [7], makes the agreement between theory and experiment still better.
- The best fit C values for MOS structures with differently doped substrates, accurately reflect changes in ϕ_{MS} values, resulting from these differences in substrate doping [6], as predicted by eq. (23).

4.2. EXPERIMENTAL VERIFICATION FOR THE CASE J≠0

A series of measurements was made for a number of different structures, to verify the theory developed in this paper. For each of the MOS structures measured, a family of I=f(V_G) characteristics was taken, for different wavelengths λ and each of the characteristics of this family was compared with the corresponding theoretical characteristic, calculated using formulas derived in section 3.2.
An example of such an experimentally taken family of I = f(V_G) characteristics is shown in Fig. 2.
As can be seen in Fig. 2 and as already mentioned, the I = f(V_G) characteristics taken for different λ values, intersect the V_G axis at different points (have different V_G^o values) and are of different shapes.
To find the theoretical characteristic, which corresponds to a given λ value the following procedure was used. From the experimental characteristics the V_G^o and V_{GO} values are determined. Assuming that $V_G^o - V_{GO} = V_I^o$, the density ratio n_G/n_S is calculated using eq. (21). For the so determined n_G/n_S ratio the dimensionless photocurrent vs. voltage characteristics J = f(Δφ) can be calculated using the formulas derived in section 3.2.

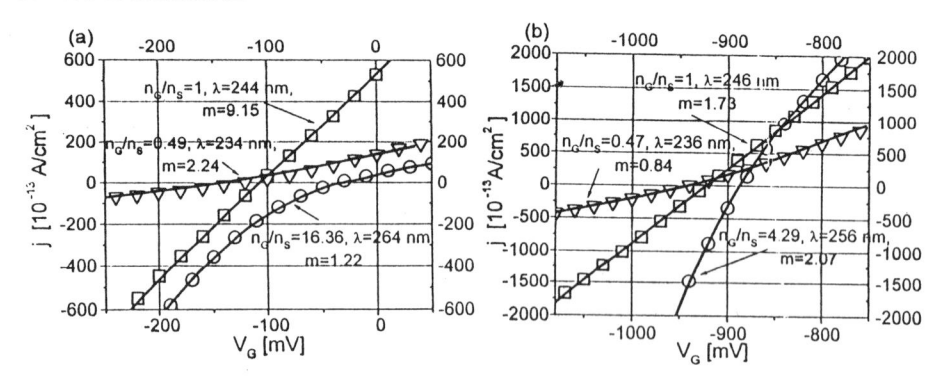

Figure 5. Comparison of j = f(V_G) curves calculated using equations (39) and (53) (solid lines), with the experimental characteristics (triangles, squares and circles) taken for (a) Al-SiO$_2$-Si(N$^+$) capacitor with oxide thickness t_I = 147 nm and (b) Al-SiO$_2$-Si (P$^+$) capacitor with t_I = 64 nm. Characteristics were taken for different wavelengths λ indicated in the figure. The n_G/n_S ratios and m coefficients used to obtain best fit are also indicated.

The calculated J = f(Δφ) characteristics should now be fitted to the corresponding experimental characteristics. This is done using the normalization eqs (5) and (7). Examination of these equations immediately reveals that for a given MOS structure they essentially contain only one unknown parameter θ, which then becomes the fitting factor (in the reasoning above, it is assumed that the electron mobility in SiO$_2$ is constant and is $\mu = 21$ cm^2/Vs [8]).

Selected experimental I = f(V$_G$) characteristics taken for two Al-SiO$_2$-Si structures are shown in Fig. 5, in comparison with the corresponding curves calculated using the formulas derived in section 3.2. Values of the best fit m parameter are indicated, where m = k · θ, with k being a constant. The values of m are given, instead of θ values, since there is still some doubt as to the exact value of k, which should be used in the relation given above.

As seen in Fig. 5, there is a remarkable agreement between the experimental characteristics and the curves calculated using the theory developed in this work.

5. Application of the model

The model presented above has already found several applications (and is expected to find more applications) in developing new measurement methods of the MOS system basic physical parameters. In this chapter one of these methods will be discussed, while other applications will only be mentioned. Following is the description of the photoelectric measurement method of the effective contact potential difference (ECPD) between the gate and the substrate of a MOS system, for which the ϕ_{MS} symbol is used.

The principles of the photoelectric ϕ_{MS} measurement were proposed already in the eighties [9-12], but development of the model discussed in this paper, allowed to make this method - by far - the most accurate of the existing methods of ϕ_{MS} determination [13, 14].

It is well known from theory of the MOS system [1, 2], that in general, the gate voltage V$_G$ is given, by the following sum:

$$V_G = V_I + V_S + \phi_{MS} \tag{54}$$

where V$_S$ is the semiconductor surface potential. For the gate voltage V$_G$ = V$_{GO}$, the voltage drop in the dielectric V$_I$ = 0, as discussed in section 3.1. and eq. (54) becomes:

$$V_{GO} = V_{SO} + \phi_{MS} \tag{55}$$

where V$_{SO}$ is the semiconductor surface potential at V$_G$ = V$_{GO}$. This situation is illustrated by the band diagram shown in Fig. 6. Using the method outlined in section 4.1. V$_{GO}$ can be determined with ± 1 mV accuracy. Hence, to accurately determine ϕ_{MS}, the value of V$_{SO}$ has to be found. There are two principal strategies of dealing with the V$_{SO}$ value. It can either be found by a separate measurement (e.g. from the C-V$_G$ characteristics), or it can be made negligible in comparison with ϕ_{MS}. The simplest way of making V$_{SO}$ negligible is to use MOS structures with

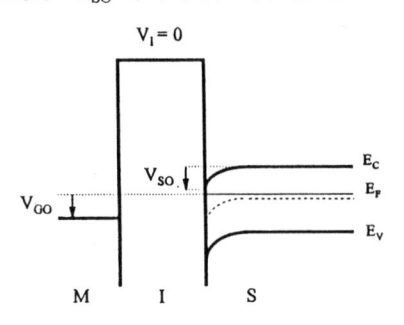

Figure 6. The energy band diagram of an MOS system biased with V$_{GO}$ gate voltage, at which V$_I$ = 0.

heavily doped substrates. In this case $V_{SO} < 5$ mV is readily achieved and $V_{GO} \cong \phi_{MS}$ results from eq. (55).

The method described above was shown [13] to be an order of magnitude more accurate than the classical method of ϕ_{MS} determination, used previously. In general, the accuracy of this method is estimated to be better than $\pm$ 10mV, while in case of optimized MOS samples, the accuracy of the order of $\pm$ 1mV can be achieved.

Several other useful applications of the model result from the fact that the accuracy of $\pm$ 1mV is readily achieved in measurement of the V_{GO} value (compare with the $\pm$ 100mV accuracy of the flat-band voltage, V_{FB}, determination from C-V_G characteristics). For instance, at $V_G = V_{GO}$, the effective charge in the dielectric Q_{eff} is equal: $- Q_S$, where Q_S is the semiconductor surface charge. Hence, if the surface potential V_{SO} is determined, Q_S can be calculated by standard methods and $Q_{eff} = - Q_S$, can be accurately determined.

Other measurement methods result from applications of the model for $J \neq 0$. The most obvious application in this case is due to the fact the θ factor defined by eq. (3) is an overall measure of trapping properties of the dielectric. At the same time, the m parameter which is proportional to θ, is the only parameter used to obtain fit between calculated and experimental $I = f(V_G)$ characteristics, as discussed in section 4.2. Hence, $I = f(V_G)$ characteristics of various MOS structures can be used to compare trapping properties of their dielectric layers.

6. Conclusions

A new model of photoelectric phenomena taking place in MOS structures at low electric fields in the dielectric has been developed and is presented in this paper. The model has been verified by calculating various characteristics of internal photoemission and comparing them with the experimental curves. As also shown in this article, the remarkable agreement between the calculated and the experimental characteristics supports the validity of the model. This model has found application in development of new measurement methods of the basic physical parameters of the MOS system. In particular, the photoelectric measurement method of the ϕ_{MS} factor, outlined in this paper is the most accurate of the existing methods of this parameter determination.

7. References

1. Nicollian, E.H. and Brews, J.R. (1982) *MOS Physics and Technology*, Wiley, New York.
2. Schroder, D.K. (1990) *Semiconductor Material and Device Characterization*, Wiley, New York.
3. Powell, R.J. (1970) Interface barrier energy determination from voltage dependence of photoinjected currents, *J. Appl. Phys.* **41**, 2424-2432.
4. Berglund, C.N. and Powell, R.J. (1971) Photoinjection into SiO$_2$: Electron scattering in the image force potential well, *J. Appl. Phys.* **42**, 573-579.
5. Powell, R.J. and Berglund, C.N. (1971) Photoinjection studies of charge distributions in oxides of MOS structures, *J. Appl. Phys.* **42**, 4390-4397.

6. Przewlocki, H.M. (1995) Photoelectric phenomena in metal-insulator-semiconductor structures at low electric fields in the insulator, *J. Appl. Phys.* **78**, 2550-2557.
7. Kudła, A., Brzezińska, D. and Wagner, T. (1997) Wyznaczanie modelu optycznego struktury MOS do badań fotoelektrycznych (in Polish). Determination of the MOS structure optical model for photoelectric investigations, *Elektronika* **38**, No. 4, 22-25.
8. Sah, C.T., et al (1988) Electron mobility in SiO_2 films on Si, in T.H.Ning (ed) *Properties of silicon*, INSPEC, The Institution of Electrical Engineers, London, 613-619.
9. Przewlocki, H.M., Krawczyk, S. and Jakubowski A. (1981) A simple technique of work function difference determination in MOS structures, *Phys. Stat. Sol. (a)* **65**, 253-257.
10. Przewlocki H.M. (1982) Work function difference in MOS structures; current understanding and new measurement methods, in: *Proc. Int. Workshop Physics Semicond. Devices.* Jain, S.C. and Radhakrishna, S. (Eds.), Wiley Eastern Ltd., New Delhi, 191-201.
11. Krawczyk, S., Przewlocki H.M. and Jakubowski, A. (1982) New ways to measure the work function difference in MOS structures, *Rev. Phys. Appl.* **17**, 473-480.
12. Przewlocki, H.M. (1987) On the properties of the contact potential difference in MOS structures, in: *Proc. V-th Int. School Physical Problems in Microelectronics.* J. Kassabov (Ed.) World Scientific, Singapore, 62-83.
13. Przewlocki, H.M. (1993) Comparison of methods for ϕ_{MS} factor determination in metal-oxide-semiconductor (MOS) structures, *Electron Technology* **26**, No. 4, 3-23.
14. Przewlocki, H.M. (1994) The importance, the nature and measurement methods of the ϕ_{MS} factor in MOS structures, *Electron Technology* **27**, No.1, 27-42.

POINT DEFECT GENERATION DURING Si OXIDATION AND OXYNITRIDATION

C. TSAMIS and D. TSOUKALAS
Institute of Microelectronics, NCSR 'Demokritos'
15310 Aghia Paraskevi, Greece

In this work we investigate the influence of interfacial nitrogen on the point defect injection kinetics during thermal oxidation of silicon in the high temperature regime (1050°-1150°C). Two different oxide growth techniques that introduce nitrogen at the interfacial region are investigated : a) N_2O oxidation and b) dry oxidation of N_2 implanted silicon. The interstitials that are injected during the oxidation process are monitored by the growth of pre-existing Oxidation Stacking Faults. We show that the existence of nitrogen at the interface can lead to an enhancement of the supersaturation of silicon interstitials at high temperatures (1050°-1150°C). The formation of a nitrogen rich layer at the interface alters the recombination processes for self-interstitials, and more interstitial atoms diffuse into the substrate. However, for N_2O oxidations and at lower temperatures this phenomenon is reversed and reduced supersaturation ratios are obtained.

1. INTRODUCTION

As the silicon device dimensions continue to shrink new techniques are being exploited for the development of very thin gate oxide films. Recent studies have shown that oxides grown in an N_2O ambient as well as dry oxides grown on N-implanted silicon substrates exhibit improved electrical properties, such as reduced interface trap densities, increased time to dielectric breakdown and improved hot electron endurance, due to the accumulation of nitrogen at the Si-SiO_2 interface [1,2]. Moreover, boron penetration from p+ gates through the gate oxide is suppressed, when nitrogen is incorporated within the oxide [3].

On the other hand, we know that thermal oxidation processes in general lead to an injection of silicon interstitials in the underlying silicon substrate as demonstrated by the Oxidation Enhanced Diffusion (OED) of B [4], P [5] and As [6], the Oxidation Retarded Diffusion (ORD) of Sb [7], and the growth of Oxidation Stacking Faults (OSF) [8] and Dislocation Loops [9]. Especially, the presence of nitrogen near the oxidizing interface can result in extremely high injection rates, i.e. the case of SiO_2 nitridation in NH_3 ambient [10,11]. During thermal nitridation the nitrogen species react quickly to form nitrogen rich-layers at the surface and at Si/SiO$_2$ interface. In the bulk of the oxide a much more slow replacement reaction takes place, displacing oxygen, some of which diffuses to the Si/SiO$_2$ interface growing a thin oxygen-rich layer and pushing the nitrogen-rich layer away from the interface [12]. Despite the very low oxidation rate, the diffusivity enhancement is quite large. Fahey et al [10] thermally nitrided 40nm of SiO$_2$ at 1100°C. They found that although only 2nm of oxide grew at the interface, after 1h at 1100°C, the diffusion of phosphorus was enhanced almost five times, more that twice the enhancement observed for normal wet oxidation, with an oxidation rate which is 100 times faster. Dunham [11] attributed this behavior to the existence of a nitrogen rich layer near the oxidizing interface, which alters the recombination processes of silicon interstitials by blocking their diffusion into the bulk of the oxide.

Quantitative estimates of the interstitial injection under various thermal processes are critical to correctly predict dopant diffusion and consequently the device operation. However no systematic work has been done in order to establish the influence of nitrogen on point defects concentration, for the case of N$_2$O oxides or for the dry oxides grown on N-implanted Si substrates. In this work we will investigate the influence of these growth techniques on point defect injection kinetics at high temperatures [13].

2. THERMAL OXIDATION IN N$_2$O AMBIENT

2.1 EXPERIMENTAL PROCEDURE

P type (100), CZ and FZ grown silicon wafers were used. At first, all the wafers were implanted with Boron (8×10^{13}cm^{-2}, 50 KeV) in order to increase the density of the nucleation sites where the Oxidation Stacking Faults (OSF) will form. Subsequently, all the wafers were oxidized in dry oxygen at 1100°C for 2h. This resulted in the formation of OSFs with a length of 11 μm. These defects were used to monitor the interstitial supersaturation during the N$_2$O oxidation. After the removal of the thermal oxide, half of the wafers were oxidized in dry oxygen at 900°C for 25 min and a thin oxide layer (about 10nm) was formed on the surface. All the wafers were then subjected to thermal oxidation in 100 % N$_2$O ambient with a flow rate of 4 standard liters per minute (slm), for various temperatures (1050°-1150°C) and times (5min up to 4h). The oxide film thickness was measured with an ellipsometer, assuming a bulk SiO$_2$ layer with refractive index of 1.46. The OSF length was measured with optical microscopy after revealing the defects with Wright etch.

2.2 RESULTS

Several discrepancies have been reported for the thickness of the oxides grown in an N$_2$O ambient. All these variations stem from the fact that the oxide thickness depends on the gas flow through the furnace [14]. In fig. 1 we report the oxide thicknesses that

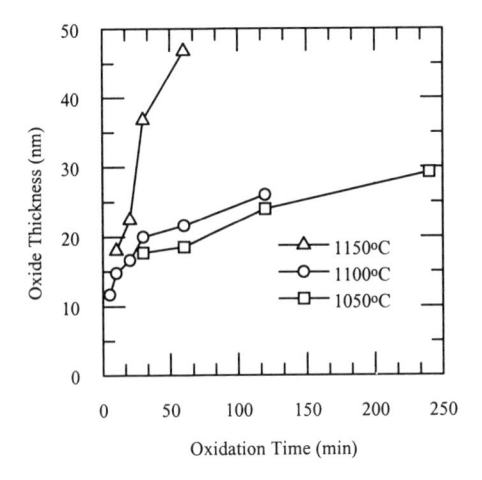

Figure 1. Thickness of N₂O oxide as a function of oxidation time for (a) 1050°C, (b) 1100°C and (c) 1150°C.

we obtained from our experiments, which are within the range reported by other researchers. It is interesting to note that there exists an important increase in the oxide thicknesses as the temperature is increased from 1100 to 1150°C. This high increase for temperatures above 1100°C has been reported in the literature [15].

Fig. 2 shows the increase ΔL of the OSF length due to oxidation in an N₂O ambient as a function of the oxidation time for different temperatures. In the same figure we have also included as a comparison the OSF length that would have resulted if the oxidations were carried out in 100% dry oxygen. These values were estimated using the model of Leroy [16], and are in excellent agreement with previously reported experimental data by Tsamis et al [17]. We notice that the length of the OSF grown in N₂O ambient is enhanced compared to those grown in 100% dry oxygen, despite that the oxidation rate in N₂O ambient is 3-4 times slower compared to dry oxidation [6].

We also observed that the OSF length does not depend on whether or not a thin thermal oxide (10nm) exists on top of the silicon surface prior to the N₂O oxidation. However, this observation holds for the conditions of the present experimental work and cannot be assumed as a general rule for all cases. It is probable that this oxide layer is thin enough not to influence the oxidation kinetics.

It is generally accepted that the OSF length dependence on oxidation time is given by

$$L_{SF} = \underbrace{a_1 \exp(-\frac{E_G}{kT}) t^n}_{A} - \underbrace{a_2 \exp(-\frac{E_{SHR}}{kT})}_{R_{SHR}} t \quad (1)$$

The first term of eq. 1 describes the increase of OSF length due to interstitial capture by the partial dislocation that bounds the fault while the second term describes the shrinkage of OSFs under inert conditions. E_G and E_{SHR} are the activation energies for the

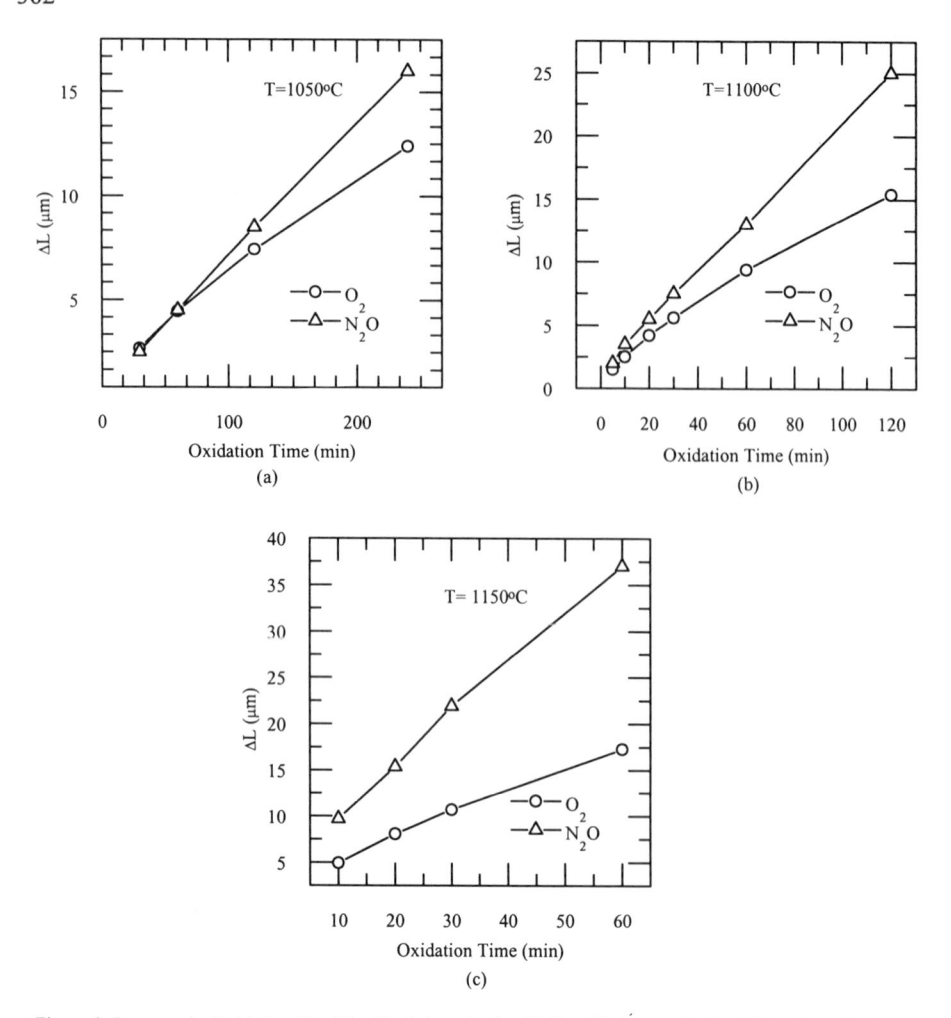

Figure 2. Increase in Oxidation Stacking Fault length after N_2O oxidation and after 100% O_2 oxidation as predicted by Leroy [4], as a function of oxidation time for (a) 1050°C, (b) 1100°C and (c) 1150°C.

growth and shrinkage respectively. Typical values are E_G=2.5eV and E_{SHR}=4-5eV.

With the aid of equations (1) and (3) we can calculate the values for the parameter A and the exponent n for OSF grown by oxidation in N_2O ambient. The shrinkage rate of OSF was assumed to be the same, both for dry and N_2O oxidation and was calculated by the model of Leroy [16]. The results are listed in Table 1.

We notice that the value of the exponent n is slightly higher for the case of N_2O oxidation.

Fig. 3 shows the temperature dependence of the growth term (parameter A) for OSFs growth in dry oxygen and N_2O. We find that the activation energy for OSFs grown in N_2O ambient is E_{GN} =3.6 eV, higher that the value E_{GD}=2.5eV for O_2

oxidation. Since the growth of OSF is driven by the generation of interstitials during the oxidation process, the difference observed in the activation energies is indicating that the presence of nitrogen at the Si/SiO$_2$ interface influences the mechanisms that lead to interstitial injection during thermal oxidation.

TABLE 1 Calculated values for parameter A and n from eq. 1

Temperature (°C)	N$_2$O		O$_2$	
	A(μm/min^n)	n	A(μm/min^n)	n
1050	.11	.90	.21	.75
1100	.4	.90	.48	.75
1150	1.62	.80	1.01	.75

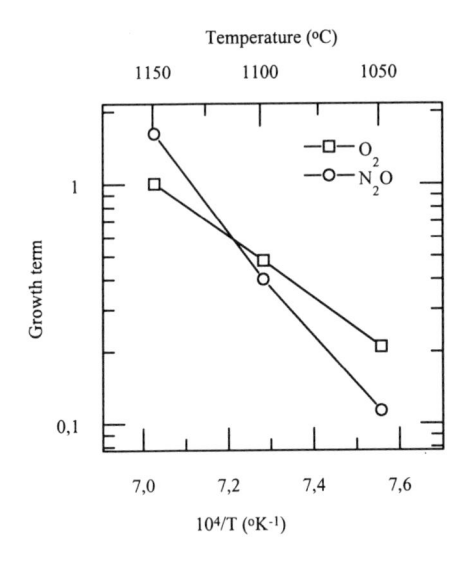

Figure 3. Temperature dependence of the Growth term (parameter A) of OSF in dry oxygen and N$_2$O.

2.3. DISCUSSION

It is well established that the OSF grow as silicon interstitials are captured by the Frank partial dislocation that surrounds the fault [16,18-19]. If C$_I$ is the silicon interstitial concentration in the vicinity of the partial dislocation under oxidizing conditions, then the growth rate of OSF can be expressed as :

$$\frac{dL}{dt} = K_{SF}(C_I - C_I^{SF}) \qquad (2)$$

or equivalently

$$\frac{dL}{dt} = K_{SF}(C_I - C_I^{eq}) + K_{SF}(C_I^{eq} - C_I^{SF}) \qquad (3)$$

where C_I^{eq} is the equilibrium concentration of self-interstitials and K_{SF} describes the interaction of interstitial with the partial dislocation.

C_I^{SF} is the silicon interstitial concentration that is in equilibrium with the fault and for the temperature range of interest can be approximated from the relation [17]

$$C_I^{SF} = C_I^{eq}(1 + \frac{\gamma}{kT}) \qquad (4)$$

where γ is the stacking fault energy (0.026 eV/atom).

The first term of eq.3 describes the growth of OSF due to intestitial capture while the second describes the shrinkage of OSF due to interstitial emission. Under inert condition the first term of eq.3 equals to zero and we can express the shrinkage rate R_{SHR} from the relation

$$R_{SHR} = K_{SF}(C_I^{eq} - C_I^{SF}) \qquad (5)$$

Placing eq. 5 and 6 to eq. 4 and integrating over time we can calculate the silicon interstitial average supersaturation ratio $<S_I(t)>$ [17]

$$<S_I(t)> \equiv \frac{1}{t}\int_0^t \frac{C_I - C_I^{eq}}{C_I^{eq}} = \left(-\frac{1}{R_{SHR}} \cdot \frac{L_{SF}}{t} + 1\right) \cdot \frac{\gamma}{kT} \qquad (6)$$

Using the experimental data for OSF growth (fig.2) and with the aid of eq. 6, we can calculate the supersaturation ratio of silicon interstitials for various temperatures and times in dry and N_2O ambient. Fig. 4 shows the time dependence of the average supersaturation ratio for various temperatures. Our results show clearly that the supersaturation ratio is higher in N_2O oxidation compared to dry one. But what is the cause of this enhanced interstitial supersaturation?

It is well established that both dry and wet oxidation of silicon leads to an injection of interstitials into the silicon substrate. An important concept that will help us

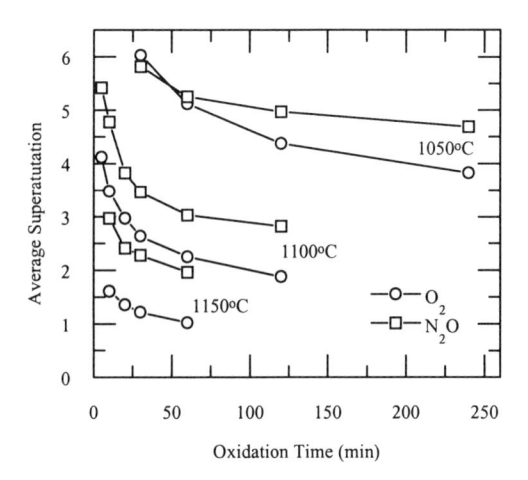

Figure 4. Average supersaturation ratio of silicon interstitial for dry and N$_2$O oxidation as a function of oxidation time for various temperatures.

to understand this behavior is the concept of "free volume", that was introduced by Tiller [20]. Due to the difference in the molecular volume between Si and SiO$_2$ there is a need for free volume in order for the larger SiO$_2$ network to fit in place of the silicon lattice. Otherwise large values of stress will appear at the interface and the oxidation process will be retarded. This constraint for free volume can be met by a net flow of interstitials away from the interface in the form of point defects. Lin et al [21] proposed that to satisfy the need for free volume approximately half of the silicon atoms are displaced to interstitial sites. These interstitial atoms can a) diffuse in the bulk of the silicon lattice, b) diffuse in the bulk of the oxide where they will be oxidized by the incoming oxygen atoms and c) undergo surface regrowth at surface kinks. It can be shown that the amount of interstitials diffusing into the silicon lattice is very low [22], and is usually ignored.

The interstitial supersaturation in the silicon lattice is determined by the balance between these two rates : the generation rate of silicon interstitials and the loss rate of interstitials, either by diffusing in the oxide and/or by undergoing surface regrowth. The various models proposed to explain the dependence of interstitial supersaturation on the oxidation rate differ on the mechanism they propose to dominate the loss rate of interstitials. Hu [18] assumes that the excess interstitials undergo surface regrowth at active sites on the silicon surface. On the other hand, Lin et al [21] and Dunham et al [22] favor the diffusion of silicon interstitial in the oxide as the dominating loss mechanisms. In their model, Taniguchi et al [23], supported the idea that a combination of both mechanisms could better explain the experimental results. It is obvious that any disturbance of the balance at the interface, resulting from variation of the interface or oxide properties, will influence point defect kinetics.

It has been demonstrated that N$_2$O decomposes at high temperatures in O$_2$, N$_2$ and NO [14] according to the reactions :

$$2N_2O \rightarrow 2NO + N_2 \qquad (7a)$$

$$2N_2O \rightarrow 2N_2 + O_2 \qquad (7b)$$

The composition of the dissociated species is dependent on the temperature and the pressure. At 1050°C in atmospheric pressure the proportions of NO, O_2 and N_2 are about 7.5, 29.6, and 62.9%, respectively [24]. The products of NO, O_2, and N_2 play important roles in the oxidation process. Since N_2 molecules are relatively inert, they simply diffuse out of the oxide film without any significant impact on the film chemistry. Oxygen is the main component responsible for the formation of the Si-O bonds. Nitric oxide (NO) is responsible for the incorporation of nitrogen in the dielectric film [14].

If we assume total decomposition of N_2O as it enters the furnace and ignore for the moment the formation of NO, then oxidation in N_2O ambient would be equivalent to thermal oxidation in N_2 and O_2 (33%) ambient. In this case we would expect the length of OSF to be much less than for oxidation in 100% O_2. However, this is not in accordance with our experimental results. It is the existence of the NO molecules that dramatically influences the silicon interstitial kinetics. NO is stable for sufficient time to allow its transport through the oxide during the oxidation and to react at the SiO_2-Si interface according to the reaction [25]

$$3Si + 4NO \rightarrow Si_3N_4 + 2O_2 \qquad (8)$$

The N in the nitride layer lattice becomes almost immediately displaced by O during continuing oxidation. This results in the formation of a nitrogen rich layer located near the Si/SiO_2 interface. The maximum nitrogen concentration is not located at the interface, but within the oxide at some distance from the interface. For the case of thermal oxidation in a conventional furnace the distance was calculated to be in the range of 5-12 and 3-16 Å for temperatures of 1050 and 1100°C respectively [25].

The formation of this nitrogen-rich layer near the oxidizing interface will have a significant impact on the point defect loss mechanisms. It will act as a barrier to the diffusion of silicon interstitials that are generated during the oxidation process back into the oxide. It is also possible that nitrogen will neutralize some of the interface sites where silicon interstitials can undergo surface regrowth. Independent of the mechanism that dominates the recombination kinetics, the interstitials will have no alternative but to diffuse into the silicon lattice. This will result in higher supersaturation ratios in the substrate, even though the oxidation rate, and consequently the generation rate of interstitials, is considerably lower than in dry oxidation.

Fig. 5 shows the temperature dependence of the average supersaturation ratio after oxidation in N_2O and O_2 for 60 min. As the oxidation temperature increases up to 1150°C, more nitrogen will be incorporated at the interface region and more pronounced will be the enhancement of the supersaturation ratio, as can be seen from fig. 2. If we extrapolate our results to temperatures below 1050°C we expect that the

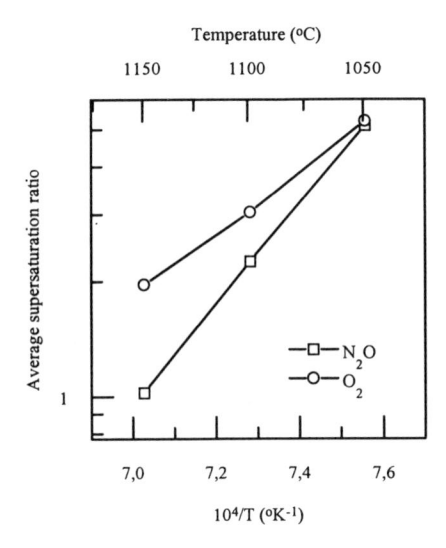

Figure 5. Average supersaturation ratio of silicon interstitials as a function of inverse temperature for dry and N$_2$O oxidation for oxidation time of 60 min.

influence of N$_2$O oxidation on point defect concentration will become less compared to that of 100% dry oxidation. As the temperature decreases less nitrogen is incorporated within the oxynitride [26]. Under these circumstances the increase in the fraction of interstitials injected into the silicon effected by the nitrogen rich interface layer, for the case of N$_2$O oxidation, is overwhelmed by the higher number of generated interstitials in dry oxidation. This result confirms the remark of Tsui et al [27] for a decreased silicon interstitial supersaturation, who observed a reduction of the Reverse Short Channel Effect when re-oxidation of the Source/Drain regions of a transistor is performed in N$_2$O ambient for the low temperature of 900°C, compared to re-oxidation in dry O$_2$.

We have to emphasize, however, that there exists also a time dependence on this phenomenon. It has been reported by Tang et al [26] that the total amount of nitrogen in the oxynitride increases with increasing temperature and growth time. During the initial stage of oxidation, the amount of nitrogen that is incorporated into the oxynitride will be not capable to alter significantly the silicon interstitials recombination processes and the supersaturation of interstitials for N$_2$O oxidation will be less that in dry oxygen, due to the lower oxidation (generation) rate in an N$_2$O ambient. However, as the amount of nitrogen in the oxynitride increases, it will have a significant impact on the loss mechanisms of silicon interstitials and will lead to higher supersaturation ratios. This time dependence can explain, within experimental errors, the cross over observed in fig. 4 for the supersaturation ratios in N$_2$O and dry oxygen, at 1050°C and for times less that 60 min.

3. THERMAL OXIDATION OF N$_2$-IMPLANTED SILICON

3.1 EXPERIMENTAL PROCEDURE

P type (100), CZ and FZ grown silicon wafers were used. As previously all the wafers were implanted with boron and subsequently oxidized at 1100 °C for 1h. This resulted in the formation of OSFs with a length of 5.5 µm. After the removal of the thermal oxide, half of the wafers were oxidized in dry oxygen at 950°C for 15 min and a thin oxide layer (about 15nm) was formed on the surface. All the wafers were implanted with N$_2$ at 25KeV with doses ranging from 5×10^{14} to 4×10^{15} cm^{-2}. Dry oxidations were carried out for different temperatures (1050-1100°C) and times (30 min up to 4h). The oxide film thickness was measured with an ellipsometer. Alternatively, mechanical measurement of the oxide thickness was performed by creating a grating pattern on the surface with lithography and measuring with a taly step. No significant differences were observed between the two measurements. The OSF length was measured as previously.

3.2 RESULTS

Fig. 6(a-b) shows the oxide thickness as a function of oxidation time for 1050 and 1100°C. These data correspond to the case where the implantation sacrificial oxide is not removed from the surface. If the sacrificial oxide is removed prior to dry oxidation then we observed that even at the highest implant dose the oxide growth is not suppressed, which indicates that during the ramp up the total amount of nitrogen in the wafer is reduced by outdiffusion.

Fig 7(a-b) shows the OSF length as function of the oxidation time for 1050 and 1100°C. We notice that the in this case the growth of the OSF is not enhanced by the presence of

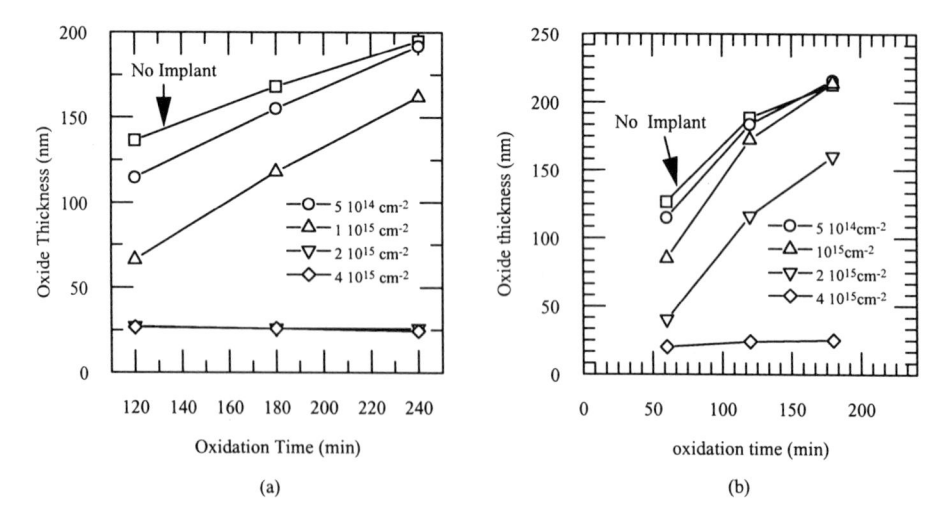

Figure 6. Oxide thickness as a function of oxidation time for (a) 1050 and (b)1100°C

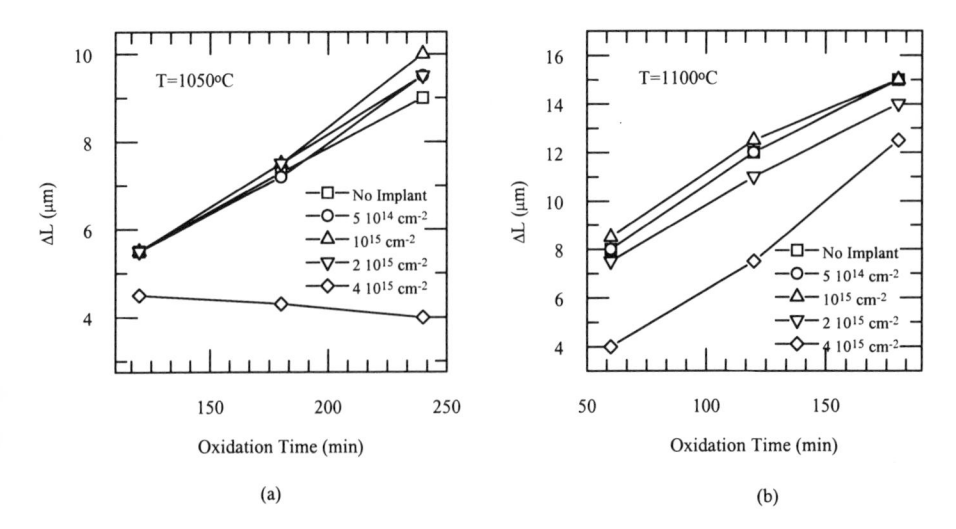

Figure 7. Increase of OSF Length as a function of oxidation time for (a) 1050 and (b)1100°C.

nitrogen. On the contrary, we observe that as the dose of the N_2 implantation increases the length of OSF obtained for the same oxidation conditions decreases. This is more clearly demonstrated in fig. 8 which shows the dose dependence of the OSF's length for 1100°C.

3.3 DISCUSSION

As we have shown previously, the length of the OSFs that are grown during

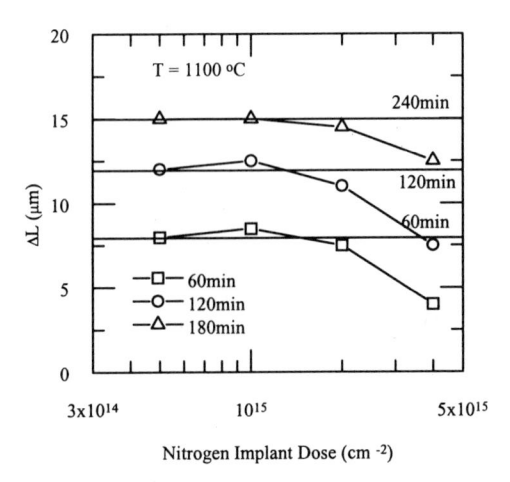

Figure 8. Increase of OSF Length as a function of N_2 implantation dose at 1100°C for various time. The straight lines correspond to the non-implanted silicon substrate

thermal oxidation in an N₂O ambient is enhanced considerably, compared to dry oxidation. However this does not seem to happen for OSF grown on N-implanted silicon. Although there exists an enhancement, especially during short times, it is not adequate to lead to larger OSFs. In fig. 9 we show the OSF length as a function of the oxide thickness for the three different cases. We notice that for the same amount of consumed silicon, the number of interstitials that are injected in the silicon substrate is higher for N₂O oxidation, with dry oxidation of N-implanted silicon following and the oxidation of non-implanted silicon injecting the lowest number.

To understand this different behavior of the N-implanted silicon, we notice a substantial difference between the two cases. During N₂O oxidation there is a continuous flux of NO towards the Si/SiO₂ interface. As we have already mentioned, the total amount of nitrogen within the oxide as well as the maximum concentration of nitrogen depends on the temperature and time of oxidation and increases for higher temperatures and longer times.

On the contrary, when oxidation of N-implanted silicon is performed, the total amount of nitrogen which will participate in the oxidation process is fixed, ie. the implantation dose. There are two experimental observations which help us to clarify the kinetics of nitrogen atoms in this case. Doyle et al [28] observed that during inert annealing at high temperatures (975°C) the nitrogen from the substrate moves towards the surface (nitrogen up-hill diffusion) and a nitrogen pile up appears at the interface even for very short anneal times. More recently, Liu et al [29] studied the boron penetration through dry oxides grown on N-implanted silicon. They observed that a N peak was formed within the thin oxide while no N was left in the silicon substrate beyond the oxide layer.

Further more, Hori et al [30] and Green et al [31] observed the displacement of nitrogen atoms from the interface and the reduction of nitrogen content due to external

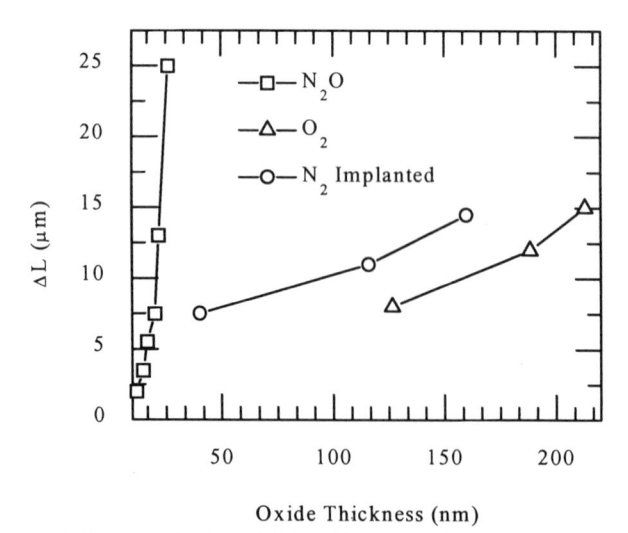

Figure 9. Oxidation Stacking Fault Length as a function of the oxide thickness.

O_2 reoxidation of dielectric films grown in N_2O ambient. This indicates that if there is no incoming NO species the nitrogen-rich oxynitride layer is pushed away from the interface by the new oxide growth at the interface.

Based on these two aspects we described, it is reasonable to speculate the following scenario. During temperature ramp up there is a flux of nitrogen atoms towards the interface. These atoms diffuse within the sacrificial oxide and a N peak appears in the oxide. At the same time, the total amount of nitrogen is reduced, since nitrogen atoms are outdiffused to the ambient. During the oxidation process, oxygen atoms diffuse through the oxide and react at the Si/SiO_2 interface. By this way a new oxide is formed at the interface and silicon interstitial are injected in to the substrate. Due to the existence of the N peak near the interface, the supersaturation of silicon interstitials at the initial stage of oxide growth will be high since the presence of N peak near the interface reduces the loss mechanisms of interstitials as we have discussed previously. This explains the experimental results of fig. 9. However, as the oxidation process continues the N peak moves away from the interface, and the recombination processes approach their normal conditions.

However, it is not possible from these experimental results to determine the exact physical processes that influence the supersaturation of silicon interstitials. More detailed information about the evolution of the nitrogen profile during high temperature annealing is necessary, in order to describe the mechanisms that determine silicon interstitial supersaturation in this case.

4. CONCLUSIONS

In conclusion we have studied the influence of interfacial nitrogen on point defect injection kinetics during a) N_2O oxidation and b) dry oxidation of N-implanted silicon. It was shown that oxidation in N_2O ambient at high temperatures leads to enhanced supersaturation of silicon interstitials. This behavior is attributed to the existence of a nitrogen-rich layer near the $Si-SiO_2$ interface; this obstructs the diffusion of silicon interstitials into the oxide and reduces reaction sites where they can be incorporated in the oxide, thus greatly enhancing the fraction of interstitials that are injected into the silicon. However, for lower temperatures this phenomenon is reversed, as less nitrogen is incorporated at the interface and oxidation in N_2O ambient leads to reduced interstitial supersaturation compared to dry one. In contrast, dry oxidation of N_2 implanted silicon substrates does not lead to significant enhancement of interstitial supersaturation, at least for all times. This was attributed to the different kinetics of nitrogen in this case.

5. REFERENCES

[1] Momose, H. H., Morimoto, T., Ozawa, Y., Yamabe, K., and Iwai, H. (1994) Electrical characteristics of rapid thermal nitrided-oxide gate n- and p- MOSFET's with less that 1 atom% nitrogen concentration,

IEEE Trans. Electron Devices, **41**, 546-552

[2] Liu C. T., Ma, Y., Becerro, J., Nakahara, S., Eaglesham, D. J., and Hillenius S. J. (1997) Light nitrogen implant for preparing thin-gate oxides, IEEE Electron Device Lett., **18**, 105-107

[3] Ma, Z. S., Chen, J. C., Liu, Z. H., Krick, J. T., Cheng, Y. C., Hu, C., and Ko, P. K. (1994) Suppression of Boron penetration in P+ polysilicon gate P-MOSFET's using Low-Temperature Gate-Oxide N_2O Anneal, IEEE Electron Device Lett., **15**, 109-111

[4] Antoniadis, D. A., Gonzalez, A. G., and Dutton, R. W. (1978) Boron in Near -Intrinsic <100> and <111> Silicon under Inert and Oxidizing Ambients-Diffusion and Segregation, J. Electrochem. Soc., **125**, 813-817

[5] Hu, S. M., Fahey, P., and Dutton R. W. (1983), On models of phosphorus diffusion in silicon, J. Appl. Phys. Lett., **54**, 6912-6918

[6] Antoniadis D. A., Lin A. M., and Dutton R. W. (1978) Oxidation-enhanced diffusion of arsenic and phosphorus in near-intrinsic <100> silicon, Appl. Phys. Lett., **33**, 1030-1032

[7] Mizuo, S., Kusaka, T., Shintani, A., Nanba, M., and Higushi, H. (1982) Effect of Si and SiO_2 thermal mitridation on impurity diffusion and oxidation induced stacking fault size in Si, J. Appl. Phys.,54, 3860-3866

[8] Murarka, S. P., and Quintana, G. (1977) Oxidation induced stacking faults in n- and p-type (100) silicon, J. Appl. Phys., **48**, 46-53

[9] Huang R., Y. S., and Dutton, R. W. (1993) Experimental investigation and modeling of the role of extended defects during thermal oxidation, J. Appl. Phys, **74**, 5821-5826

[10] Fahey, P., Barbuscia, G., Moslehi, M., and Dutton, R. W. (1985) Kinetics of thermal nitridation processes in the study of dopant diffusion mechanisms in silicon, Appl. Phys. Lett., **46**, 784-786

[11] Dunham, S. T. (1987) Analysis of the effect of thermal nitridation of silicon dioxide on silicon interstitial concentration, J. Appl. Phys., **62**, 1195-1199

[12] Vasquez, R. P., Hetch, H. H., Grunthaner, F. J., and Naiman, M. L. (1984) X-ray protoelectron spectroscopic study of the chemical structure of thermally nitrided SiO_2, Appl. Phys. Lett., **44**, 969-971

[13] Tsamis, C., Kouvatsos, D. N., and Tsoukalas D. (1997) Influence of N_2O oxidation of silicon on point defect injection kinetics in the high temperature regime, J. Appl. Phys., **69**, 2725-2727

[14] Tobin, P. J., Okada, Y., Ajuria, S. A., Lakhotia, V., Feil, W. A., and Hedge, R. I. (1994) Furnace formation of silicon oxynitride thin dielectric in nitrous oxide (N_2O): The role of nitric oxide (NO), J. Appl. Phys., **75**, 1811-1817

[15] Koyama, N., Endoh, T., Fukuda, H., and Nomura, S. (1996) Growth kinetics of ultrathin silicon dioxide films formed by oxidation in an N_2O ambient, J. Appl. Phys., **79**, 1464-1469

[16] Leroy, B. (1979) Kinetics of growth of the oxidation stacking faults, J. Appl. Phys., **50**, 7996-8005

[17] Tsamis, C., Tsoukalas, D., and Stoemenos, J. (1993) Comparison between the growth and shrinkage of oxidation stacking faults in silicon and silicon on insulator , J. Appl. Phys., **73**, 3246-3251

[18] Hu, S. M. (1981) Oxygen, Oxidation Stacking Faults, and related phenomena in Silicon", in J. Narayan and T. Y. Tan (eds) "Defects in Semiconductors", North-Holland, New York., 333-353

[19] Antoniadis, D. A. (1982) Oxidation Induced Point Defects in silicon, J. Electrochem. Soc., **129**, 1093-1001

[20] Tiller W. A. (1980) On the Kinetics of the Thermal oxidation of Silicon. I. A Theoretical Perspective, J. Electrochem. Soc., **127**, 619-629

[21] Lin, A. M., Dutton, R. W., Antoniadis, D. A., and Tiller, W. A. (1981) The growth of Oxidation Stacking Faults and the Point Defect Generation at the Si-SiO_2 Interface during Thermal Oxidation of Silicon, J. Electrochem. Soc., **128**,1121-1131

[22] Dunham, S. T., and Plummer, J. D. (1986) Point-defect generation during oxidation of silicon in dry

oxygen. I. Theory, J Appl. Phys., **59**, 2541-2555

[23] Taniguchi, K., Shibata Y., and Hamagushi, C. (1989) Theoretical model for self-interstitial generation at the Si/SiO2 interface during thermal oxidation of silicon, J. Appl. Phys., **65**, 2723-2727

[24] Kim, K., Lee, Y. H., Suh, M-S., Youn, C-J., Lee K-B., and Lee, H. J. (1996) Thermal Oxynitridation of silicon in N_2O Ambients, J. Electrochem. Soc., **143**, 3372-3376

[25] Hussey, R. J., Hoffman,T. L., Tao, Y., and Graham, M. J. (1996) A study of nitrogen incorporation during the oxidation of Si(100) in N_2O at High Temperatures , J. Electrochem. Soc., **143**, 221-228

[26] Tang, H. T., Lennard, W. N., Zinke-Allmang, M., Mitcell, I. V., Feldman, L. C., Green, M. L., and Brasen D. (1994) Nitrogen content of oxynitride films on Si(100), Appl. Phys. Lett., **64**, 3473-3475

[27] Tsui, P. G. Y. , Tseng, H., Orlowski, M., Sun, A. W., Tobin, P. J. , Reid, K., and Taylor, W. (1994) Suppression of MOSFET Reverse Short Channel Effect by N_2O Gate Poly Reoxidation Proces, IEDM Tech. Digest, **94,** 501-504

[28] Doyle, B., Soleimani, H., and Philipossian A. (1995) Simultaneous growth of different thickness gate oxides in silicon CMOS processing, IEEE Electon Device Lett., **16**, 303-305

[29] Liu, C. T., Ma, Y., Luftman H., and Hilenius S. J. (1997) Preventing Boron penetration through 25-A gate oxides with nitrogen implant in Si substrates, IEEE Electon Device Lett., **18**, 212-214, 1997

[30] Hori, T., Iwasaki H., and Tsuji, K. (1989) Electrical and physical properties of ultrathin reoxidized nitrided oxides preared by rapid thermal processing, IEEE Trans. Electron Dev., **36**, 340-349

[31] Green M. L., Brasen, D., Evans-Lutterodt K. W., Feldman. L. C., Krisch, K., Lennard W., Tang H.-T, Machanda L., and Tang, M.-T (1994) Rapid thermal oxidation of silicon in N_2O between 800 and 1200°C : incorporated nitrogen and interfacial roughness, Appl. Phys. Lett., **65**, 848-850

OPTICALLY INDUCED SWITCHING IN BISTABLE STRUCTURES: HEAVILY DOPED n⁺- POLYSILICON - TUNNEL OXIDE LAYER - n - SILICON

V.Yu. OSIPOV
*A.F.Ioffe Physicotechnical Institute,
194021, Polytechnicheskaya 26, St.-Petersburg, Russia.*

1. Introduction

The interest in metal-insulator-semiconductor (MIS) diode structures with the insulator layer thinner than 50 Å, when the conductivity of the layer can no longer be ignored, has arisen over 25 years ago and has been initially related to the mechanism of the current flow, particularly by tunneling. In recent years the interest in such structures with a partly conducting insulator, i.e., in metal- tunnel insulator- semiconductor (MTIS) structures has grown due to several factors. These factors include the following: the need to reduce the thickness of the gate insulator in silicon field-effect transistors so as to reduce the length of the channel to submicron dimensions in modern integrated circuits, the development of MIS transistors with a tunnel emitter, the feasibility of constructing highly efficient solar cells and photodetectors from MIS diodes, and the development of MIS switches and oscillators operating on the basis of the tunnel surface-barrier instability effect.

The observation of a switching effect in aluminum- tunnel-thin SiO_2 - n-Si structures was first reported in 1981. Switching from a high- to a low-resistance state was produced by optical radiation, and the latter state was maintained after the external irradiation was removed. Furthermore, the physical nature of this effect was discussed in detail in [1].

Investigations of switching effects in structures with a tunnel insulator were later conducted mainly on structures with a metal upper electrode [2]. At the same time, it is more important to study switching effects in STIS (semiconductor- tunnel insulator- semiconductor) structures with a polycrystalline silicon gate electrode than it is to study these effects in MTIS structures with a metal (for example, aluminum) upper electrode. This is because structures with a polycrystalline silicon electrode have, as a rule, a much lower resulting density of defects and are more reliable than structures with a metal electrode. The latter circumstance is due to the fact that in the first former case the properties of ultrathin oxide layers grown are more fully preserved with deposition of the gate electrode because the electrode is made of a material which interacts weakly with the surface of the insulator, whereas in the second case, since the gate metal (for example, aluminum) interacts chemically with the oxygen in the oxide, the structure of the oxide changes in the region next to the gate electrode. As a result, in practice it is important to use a polycrystalline silicon electrode, since the main problem in the MTIS structure technology is the production of ultrathin silicon dioxide insulator layers with stable and reproducible parameters.

Device structures with a polycrystalline silicon electrode are also of interest for a number of other reasons: The transparency of the polysilicon electrode in the corresponding spectral range makes it more convenient than a metal electrode for studying photoelectric effects and for applications, and the technology for fabricating polycrystalline gates and interconnections is now widely used in microelectronics.

2. Experimental Results

This paper reports the observation of optically induced switching in heavily-doped n^+-polysilicon (Si*) - tunnel SiO_2 - n-Si structures. A 23 Å thick tunnel-transparent oxide layer was grown on KEF-25 n-Si substrates with a resistivity of 25 Ohm*cm. The oxide was grown by thermal oxidation of the (100) silicon surface in a dry oxygen flow with a dew point of -65°C at a temperature of 700°C and pressure 760 torr. The wafer was exposed to the oxygen atmosphere for 5 min. (The uppermost silicon layer disturbed during polishing was removed from the surface by thermal oxidation of the silicon surface until a SiO_2 film of thickness of about 5000 Å was formed; this was removed when the standard operations of chemical cleaning, washing, and drying were applied. All this was done before growing the tunnel oxide film). The thicknesses of the SiO_2 layers were monitored by ellipsometry. Next, a 0.5 µm-thick polycrystalline layer was deposited on the oxide surface by decomposition of monosilane at 630°C. The polycrystalline silicon was doped with phosphorus up to the level $(2-4)*10^{19} cm^{-3}$. The structure was 6 mm in diameter. A ring-shaped aluminum electrode served as a contact to the polycrystalline silicon layer. The technology employed for growing defect-free tunnel-thin 17-50 Å oxide layers and the method for preparing high-quality Si*-SiO_2- Si tunnel structures were described previously in [3-5].

The fixed positive charge in the oxide was mainly due to an excess of trivalent silicon, Si^{3+}, and depended strongly on the technological conditions during oxide growth. The localization of the bulk of the fixed charge at a distance of 20 Å from the Si-SiO_2 interface was confirmed by the observation that flat-band voltage, deduced from the capacitance- voltage characteristics obtained under a reverse bias, changed slightly under the increasing of the oxide thickness in the range from 19 to 34 Å.

The current-voltage characteristic (IVC) of n^+ -Si* -SiO_2- n-Si structure in the region from -50 to +3.5 V (no illumination) is shown in Fig.1. One can see that with reverse applied bias the current flowing is much lower than with a forward bias. We shall now analyze the section of the IVC obtained under reverse bias.

As the reverse bias varies from 0 to 50 V, the dark current flowing through the structure increases up to ≈ 7.0 mA (curve 2 in Fig.2). Under the action of light in the wavelength range of the fundamental absorption of n-Si and with a reverse bias on the

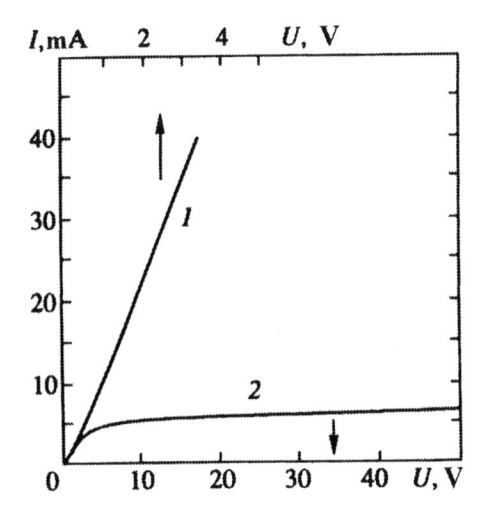

Figure 1. IVC of a structure in the region of forward (up to 3.6 V) and reverse (up to 50 V) applied biases. T = 293 K. 1 - forward bias, 2 - reverse bias.

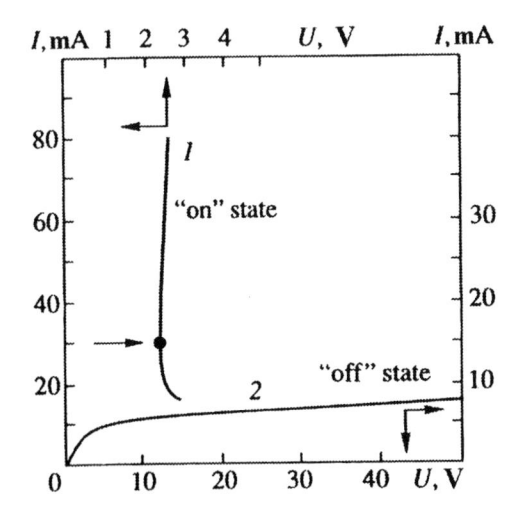

Figure 2. Switching curves for a structure in the region of reverse applied biases. Switching was achieved with a "white" light source with a bias of 50 V. T = 293 K. 1 - IVC in the "on" state with a high current flowing, 2 - IVC in a steady-state non-equilibrium depletion state. The open circle marks the point on the IVC in the "on" state corresponding to the minimum holding voltage.

sample of 50 V, the structure switches from a low-current state (I < 10 mA) into a high-current state (I > 20 mA). A voltage drop of not more than 2.9-3.0 V exists across the structure in the low-resistance state (20 < I < 100 mA). The low-resistance state also persists after the illumination pulse. The IVC of the structure in the low-resistance state is shown in Fig.2 (curve 1).

We shall now elucidate the physics of the observed switching process. In the high-resistance state, most of the reverse bias applied to the structure falls across the depletion layer of n-Si and only a very small part falls across the insulator (not more than 0.5 V). In this case, a so-called stationary non-equilibrium situation exists, where a depletion layer is formed in the silicon n-region, with a thickness W that increases with the voltage U on the structure as $W \sim (U + U_c)^{1/2}$, while a hole inversion layer at the SiO_2-n-Si boundary is not formed because of the high rate of tunneling loss of holes into the n^+ - polysilicon (Fig.3a). Here U_c is the flat-band voltage of the structure. The rates of thermal generation of minority carriers in the depletion region of n-Si are too low in this case for an inversion layer to be maintained.

Photogenerated minority carriers (holes) accumulate at the n-Si-SiO_2 boundary under the action of light. In the process, a non-equilibrium hole inversion layer forms at the interface, and the thickness of the depletion layer strongly decreases in the process (Fig.3b). The voltage applied to the structure is redistributed between the depletion region of n-Si and the insulator. Most of the voltage falls across the insulator and the inversion hole layer in the new situation, and a smaller part of the voltage falls across the depletion layer. In a specific experimental situation, a substantial part of the voltage (about 47 V) also falls across the load resistance connected in series with the structure. After the illumination pulse ends, the "on" low-resistance state is maintained because the Auger carrier generation process in the surface region of n-Si now acts as an internal source of minority carriers (holes), which are required to maintain the inversion layer at the n-Si- SiO_2 boundary (Fig.3b).

To be fair, we note that the Auger carrier-generation process was invoked to explain switching and persistence of the "on" state in aluminum-tunnel oxide- n-Si structures in early publications [1,2]. In turn, a detailed theoretical analysis of the Auger minority-carrier generation process in Al-SiO_2- n-Si structures was given in [6]. We shall explain the physics of this process.

In the "on" state a voltage (U_{ox}) exceeding 1.2 V falls across the insulator. Electrons tunneling from the conduction band of n^+-polysilicon become "hot" near the n-Si-SiO_2 boundary. Some of these "hot" carriers produce electron-hole pairs by impact ionization, which are separated by the electric field in the space-charge region in the n-Si (Fig.3b). In the process, holes accumulate near the n-Si-SiO_2 boundary. Therefore, the Auger generation rate is sufficient to compensate for tunneling leakage of holes into the valence band of n^+ - polysilicon and to maintain an inversion layer. The magnitude of the current flowing in a such state is determined mainly by electron tunneling from the conduction band of n^+- polysilicon into the conduction band of n-Si (the current I_n). In the process, electrons are transferred from polysilicon into silicon from the energy range below the Fermi level in n^+ - polysilicon and above the edge of the conduction band of polysilicon. For a voltage drop exceeding 1.2 V across the insulator, electrons also tunnel from the valence band of n^+-polysilicon into the conduction band of n-Si, but the contribution of this process to the current is very small to the extent that the probability of tunneling through the oxide is small for electrons from the low-lying

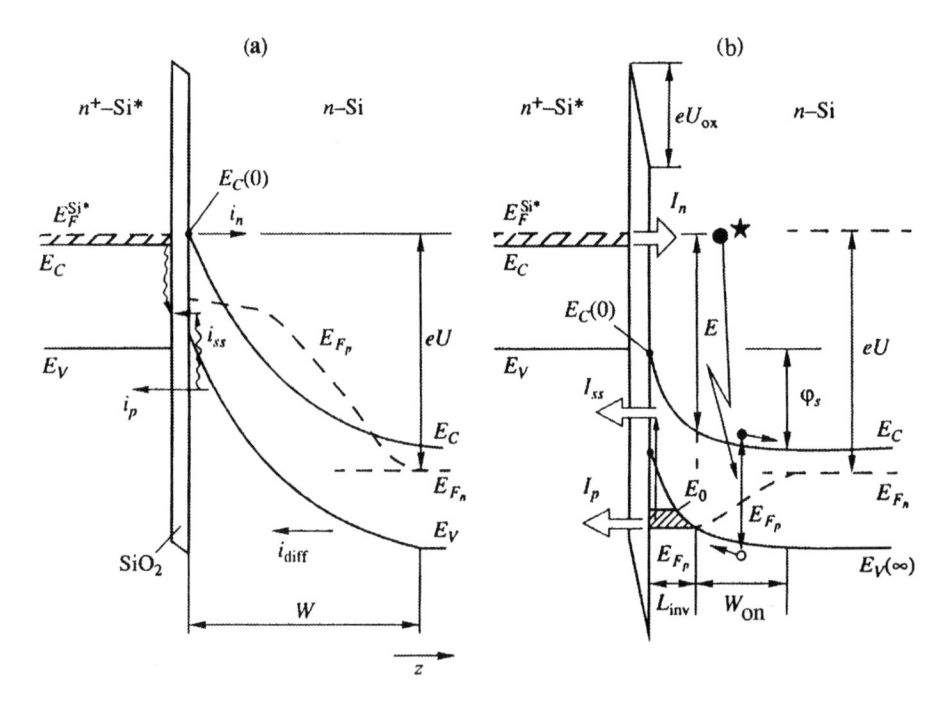

Figure 3. Energy diagram of an n⁺ -Si*- SiO₂ -n-Si structure in a steady-state non-equilibrium depletion state (a) and in the "on" state (b). a) E_{Fn} - Fermi level for majority carriers in n-Si, E_{Fp} - variation of the quasi-Fermi level for holes in the region of the depletion layer of thickness W, i_{diff} - hole diffusion current flowing from the n- substrate into the n-Si-SiO₂ interface, i_n - tunneling electron current from the conduction band of n⁺ - polysilicon into the conduction band of n-silicon, i_p, i_{ss} - hole currents from n - Si into n⁺ - polysilicon, which are associated with, respectively, the direct tunneling and tunneling-recombination processes at the interfaces; E_F^{Si*} -Fermi level of the majority carriers in n⁺ - polysilicon, E_c (0) - edge of the conduction band in silicon at the n-Si-SiO₂ boundary. The diagram was constructed for a bias U = 2.5 V; the difference E_F^{Si*} - $E_c(0) \approx 0$. For a high reverse bias this difference will now be E_F^{Si*} - $E_c(0) > 0$. b) L_{inv} - thickness of the hole inversion layer, W_{on} - thickness of the depletion region in n-Si in the "on" state, E_0 - position of the size-quantization ground state for holes in a potential well formed by inversion layers, E_{Fp} - variation of the quasi-Fermi level for holes in the region of the depletion layer and the inversion layer, φ_s - magnitude of the surface band bending in n-Si, E - energy of a "hot" electron tunnel-injected from polysilicon into n-Si up to the moment the electron traverses the inversion layer. The diagram was constructed for the case of external bias U =2.5 V and voltage drop on the oxide U_{ox} = 1.2 V. In the "on" state in the region of the inversion layer $E_{Fn} \approx$ const and E_{Fn} - E_{Fp} > 0. The hole diffusion current flowing from the inversion layer into the bulk of n-Si is not shown.

states on the energy scale. It can be assumed that mainly "hot" electrons, whose energies equal E_F^{Si*} - $E_c(z = 0)$, are injected from the n⁺-polysilicon through the oxide into the n-Si, i.e. the current is determined by the voltage drop U_{ox} on the insulator. Here E_F^{Si*} is the position of the Fermi level in n⁺- polysilicon, E_c (z = 0) is the edge of the conduction band in n-Si at the n-Si- SiO₂ interface at z = 0, and z is the longitudinal coordinate, measured from the n-Si-SiO₂ interface in a direction perpendicular to the plane of the interface. As a rule, only a few percent (2-5%) of the total number of carriers contributing to the current I_n give rise to Auger electron-hole pair production [2].

In the "on" state, as a result of the accumulation of minority carriers at the n-Si-SiO$_2$ interface and the appearance of a non-equilibrium hole inversion layer, the position of the quasi-Fermi level E_{Fp} for holes in the region of the inversion layer is lower than the Fermi level in the bulk of the n- semiconductor. As a result, diffusion flow of holes from the surface into the semiconductor (the current I_{diff}) always occurs in the "on" state. The current I_{diff} is opposite to the hole tunneling leakage current I_p from the inversion layer into n$^+$ - polysilicon (Fig.3b). We note additionally that there is present another channel for tunneling leakage of holes from the inversion layer into n$^+$ - polysilicon - through the surface states in the band gap in n-Si (the current I_{ss}). In the absence of illumination, the condition for the existence of the "on" state has the form

$$P > (I_p + I_{ss}) / I_n$$

Here P is the fraction of the carriers (of the total number of carriers contributing to the current I_n) which produce impact ionization and contribute to the Auger-generation current. In other words, this condition means that the Auger generation current (PI_n), which acts as an internal source of holes reaching the n- Si-SiO$_2$ boundary, should be greater than the hole tunneling leakage current in polysilicon ($I_p + I_{ss}$). The positive difference of these currents $PI_n - (I_p + I_{ss})$ determines the diffusion current I_{diff} of holes away from the interface. We note that P depends on the energy of the injected "hot" carriers and increases with the voltage on the structure in the "on" state.

It is interesting to note that the minimum voltage at which the "on" state is still maintained is 2.5 V. It exceeds the value E_{thr} = 1.4 - 1.5 eV, which determines the threshold kinetic energy of a "hot" carrier required for impact ionization in silicon. An examination of the energy diagram of the structure in the "on" state (Fig.3b) shows that the minimum holding voltage is determined by the sum of the voltage drop U_{ox} on the insulator and the surface band bending (φ_s) in n-Si at the holding threshold point. The sum of the voltage drops on the hole inversion layer and the depletion layer in n-Si determines the surface band bending φ_s. The injected "hot" carrier at the holding threshold of the "on" state can have at the SiO$_2$-n-Si interface a kinetic energy that is still insufficient for impact ionization, since it can happen that $E_F^{Si*} - E_c(0) < E_{thr}$ at z = 0. A hot carrier acquires additional energy, equal to the band bending in the inversion layer, and becomes capable of impact ionization only after traversing the entire inversion layer, whose thickness does not exceed 100 Å. It follows from this discussion that in the experimental situation some of the voltage (up to 0.7 V) in the "on" state also falls across the depletion layer, whose thickness W_{on} is substantially smaller than in the situation of the high-resistance state ($W_{on} \ll W$). The electric field of the depletion region in the "on" state separates the Auger-generated electron-hole pairs.

We note that the structure can be converted into the "on" state under dark conditions as well by further increasing the reverse bias up to 250-300 V (Fig.4). As

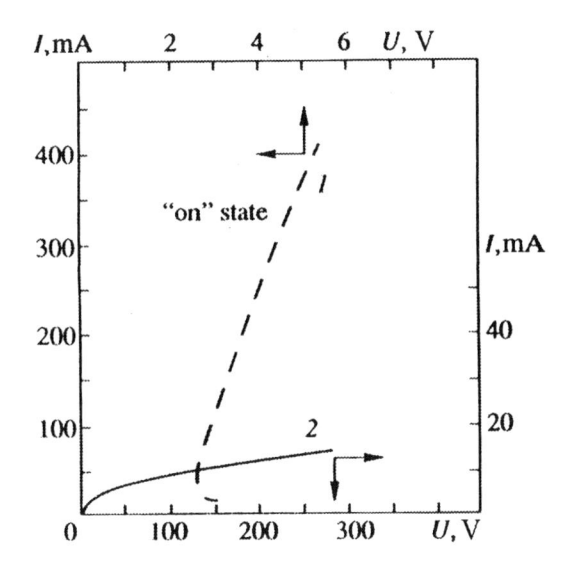

Figure 4. IVC of a structure with switching as a result of application of a high reverse bias. T = 293 K. 1 - IVC in the "on" state in the voltage range from 2.5 to 5 V, 2 - IVC in the state of steady-state non-equilibrium depletion in the voltage interval from 0 to 270 V.

the reverse bias increases, the thickness of the depletion layer in n-Si and the effective electric field at the SiO$_2$-n-Si interface increase. This electric field accelerates in the n-Si region the electrons which have tunneled from the conduction band of n$^+$ - polysilicon straight through the SiO$_2$ into the conduction band of n-Si. The electrons which have tunneled through the SiO$_2$ are accelerated because over the distance of emission of an optical phonon in silicon they acquire in the electric field of the space-charge region a greater energy than the energy of an optical phonon that they lose (E$_{ph}$ = 0.063 eV at T= 300K). At voltages of about 250 V, conditions for generation of the required number of minority carriers by Auger ionization are produced in the structure. This voltage is much lower than the voltage at which avalanche breakdown of n-Si with resistivity 25 Ohm*cm occurs in a sharp p$^+$-n junction. We note that at the same time that as the thickness W of the space-charge region increases, the thermal hole generation current (I$_{rg}$) in the space-charge region, whose magnitude is proportional to the thickness W and the carrier density in the semiconductor itself, increases.

As the voltage on the structure in the "on" state increases from 2.5 to 5 V, the positive charge due to the holes in the inversion layer increases and the voltage across the oxide increases too. In consequence, the energy of the electrons injected into n-Si and the efficacy of the Auger generation process increase. As one can see from Fig.4 (curve 1), the current depends virtually linearly on the voltage increment in the interval from 3 to 5 V.

The experimentally observed switching effect in n$^+$-Si -SiO$_2$- n-Si structures with minimum holding voltage of the "on" state of about 2.5 V attests to the high quality of the fabricated structures and the tunnel oxide. In some structures with d = 23 Å, grown on KEF-25 substrates, higher values of the minimum holding voltages (up to 4-5 V) were observed, and in a number of cases lower voltages (1.6-2.2 V) were observed. The latter fact merits closer study. For example, the presence of switching in structures with a minimum holding voltage of about 1.6 V signifies, in all probability, that in these structures in the "on" state not more than 0.15 eV falls across the depletion layer, the thickness of this layer decreases substantially, and the quasi- Fermi level for holes E_{Fp} in the region of the inversion layer lies near the edge E_v ($z \Rightarrow \infty$) of the valence band in n-Si.

3. Conclusions

1. The presence of a n^+-polysilicon electrode instead of a metal electrode gives additional specific features. For example, on the steady-state non-equilibrium depletion regime the bandgap in the gate electrode material blocks virtually completely the current transport channel associated with transfer of minority carriers (holes) from the valence band in n-Si into the gate material via surface states on the n-Si-SiO$_2$ interface (Fig.3a). For this reason, in the case of a polycrystalline silicon electrode the condition $E_{Fm} - E_c (0) > 0$ is more easily satisfied in the steady-state non-equilibrium depletion regime than in the case of a metal electrode [7]. A similar situation obtains in the "on" state: The lower the hole tunneling leakage current (I_{ss}) through the surface states at the n-Si-SiO$_2$ boundary, the greater the charge accumulated in the quasi-equilibrium hole layer and therefore the higher the voltage across the insulator.

2. We note that switching effects are not observed in heavily- doped p^+-polysilicon - tunnel-transparent oxide - p- Si structures [8].

3. It is timely matter to use n^+ -Si -SiO$_2$ - n- Si structures in the production of threshold optical-radiation detectors.

References

1. Lai, S.K., Dressendorfer, P.V., Ma, T.P. and Barker, R.C. (1981) Optically induced bistable states in metal/ tunnel oxide/ semiconductor (MTOS) junctions, *Appl. Phys. Lett.* **38**, 41-44.
2. Fossum, E.R., and Barker, R.C. (1984) Measurement of hole leakage and impact ionization currents in bistable metal- tunnel oxide-semiconductor junctions, *IEEE Trans. Electron. Devices* **ED-31**, 1168-1174.
3. Vul', A.Ya., Makarova, T.L., Osipov, V.Yu., Zinchik, Yu.S., and Boitsov, S.K. (1992) Kinetics of silicon oxidation and structure of oxide films of thickness less than 50 Å, *Sov. Phys. Semicond.* **26**, 62-67.
4. Boitsov, S.K, Vul', A.Ya., Osipov, V.Yu., Dideikin, A.T., Zinchik, Yu.S., Makarova, T.L. (1991) Passage of current through tunnel-transparent insulator in STIS structures, *Sov. Phys. Solid State* **33**, 1003- 1007.
5. Boitsov, S.K., Makarova, T.L., and Osipov, V.Yu. (1992) Problem of tunneling across an intermediate (19- 50 Å) oxide layer in semiconductor- tunnel transparent insulator- semiconductor structure, *Sov. Phys. Solid State* **34**, 777-781.
6. Ostroumova, E.V., and Rogachev, A.A. (1994) A simple model for the Auger transistor, *Semiconductors* **28**, 793-797.
7. Vul', A.Ya., Dideikin, A.T., Boitsov, S.K., Zinchik, Yu.S., and Sachenko, A.V. (1992) Photodetectors based on metal- tunnel insulator- semiconductor structures, *Sov. Phys. Semicond.* **26**, 166-170.
8. Vul', A.Ya., Dideikin, A.T., Osipov, V.Yu., Boitsov, S.K., Zinchik, Yu.S., and Makarova, T.L. (1992) Current-voltage and capacitance-voltage characteristics of silicon semiconductor- insulator- semiconductor structures with the insulator layer less than 50 Å thick, *Sov. Phys. Semicond.* **26**, 81-83.

HETEROJUNCTION Al/SiO$_2$/n-Si DEVICE AS AN AUGER TRANSISTOR

E.V.OSTROUMOVA and A.A.ROGACHEV
A.F.Ioffe Physical-Technical Institute,
Russian Academy of Sciences,
Politekhnicheskaja, 26, St.-Petersburg, Russia, 194021

Abstract

The paper is devoted to the investigation of current instabilities in the *Al-SiO$_2$- n-Si* Auger transistor. We succeeded for the first time in creating of the Auger transistor, in which in particular we used a metal-insulator heterojunction instead of a widegap semiconductor. The Auger transistor base is created by the holes, which are induced on silicon surface by electric field that exists in the thin oxide layer and is formed as a self-consistent quantum well near the n-silicon surface. The base width is about 10 Å and the well depth is equal up to 0.7 eV or even higher. The generation of electron-hole pairs by impact ionization (Auger generation) is the fastest physical process in semiconductors, which can be used for amplification and generation of electric signals. The impact ionization and drift regions are practically separated in the Auger transistor. The electron-hole pairs are generated in the transistor base and partly in the collector. The *S*- and *N*-type instabilities of the collector current in the Auger transistor in the case of circuit with a common emitter were investigated.

1. Introduction

A new type of semiconductor transistor, which may be conceived as a hybrid of two devices, namely TETRAN (Tunnel Emitter TRANsistor) with the base induced by an electric field [1] and a transistor with a widegap emitter known as the Auger transistor [2], is considered in the paper. This new Auger transistor was fabricated, like TETRAN was, on the *Al-SiO$_2$-n-Si* structure [3,4], but it has a much faster response than that of TETRAN. As a matter of fact this construction of the Auger transistor is the first operating one, since the Auger transistor proposed by H.Kroemer [2] was not realised, as we know, up to now, in spite of the heterojunction pair was theoretically chosen [5]. The Auger transistor under consideration (figure 1) has an emitter of hot electrons whose energy is so high that during loosing kinetic energy an electron generates one or more electron-hole pairs. In the given construction the emitter consists in the metal-silicon oxide heterojunction. It is important that the silicon oxide layer is tunnel-transparent thin. The current gain coefficient α (in a circuit with the common base) can be higher than one ($\alpha > 1$). The amplification is determined by the fastest available physical process, namely, the impact ionization or, in other words, the Auger generation [4]. In ordinary bipolar transistors the current amplification is associated with the diffusion or drift of charge carriers through the transistor base. The highest possible operating frequency of the transistors is determined by the characteristic times of diffusion and drift. In our transistor the base has a width of the order of 10^{-7}cm, which allows electrons to fly over the base ballisticaly (see figure 1). For this reason the electron movement through the base region is not the process limiting the highest possible operating frequency of the transistor. Thus, the highest possible frequency of the MIS Auger transistor is determined only by the time for charging the emitter-base capacitance. This time, however, is not very short, but at high operating current density up to 10^4A/cm^2 the highest possible operating frequency ranges is up to 10^{12} Hz [6]. It is approaching to the limiting operating frequency of the Read diode [7]. There exists a physical meaning in such a coincidence. Actually, the Read diode is an avalanche breakdown one, created by Read, who, apparently, had an idea to separate spatially the Auger ionization and drift processes. In the case of the two-contact diode it was achieved only partly. The spatial separation of the Auger generation and drift regions in the Auger transistor is practically complete. An electron gains kinetic energy and than looses it during generation of electron-hole pairs and phonons in the region of the order of 10^{-6} cm or less, whereas the drift region, i.e. the width of the electric field region in the collector, is 10^{-5}cm or larger.

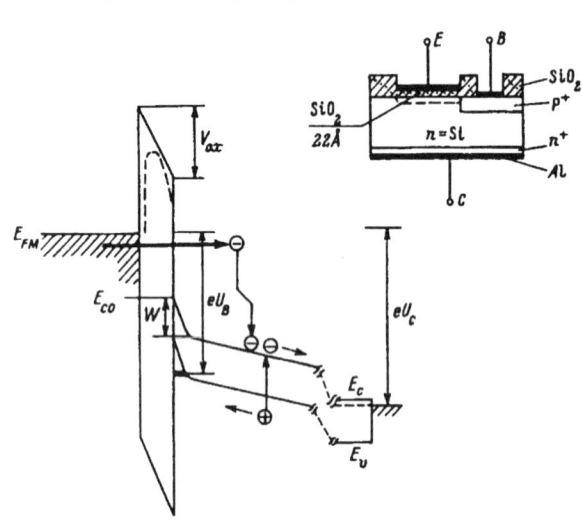

Figure 1. Energy diagram of the *Al-SiO₂ -n-Si* Auger transistor with a tunnel-thin oxide layer. The metal is biased negatively with respect to the semiconductor. The dotted line represents the potential barrier for electrons, with allowance for the image potential and penetration depth of the electron wave function into oxide.

Also of interest is the physics of noise in the Auger transistor. The nature of noise in it differs remarkably from that in the avalanche breakdown diodes. Electron-hole pairs in the Read diode are generated not only by fast electrons, but also by fast holes. The latter is important, since the noise of the electron-hole pair current is anomalously great. It seems one could reduce the capability of fast holes to create electron-hole pairs by replacing silicon in the construction by other semiconductors. However, in semiconductors such as *Si, Ge, GaAs, InP* and many others an electron and a hole give approximately equal contribution to the impact ionisation. This means that the excess noise is high. In the Auger transistor either only electrons or only holes (depending on the base being n-type or p-type) participate in the impact ionisation and, correspondingly, the excess noise is absent.

In this paper we discuss a model of the Auger transistor produced on the MIS Al-SiO_2-n-Si structure with the tunnel-thin oxide layer. For the first time the influence of surface quantum effects arising at superhigh electric fields on I-V characteristics of the Auger transistor has been studied [4,6,8,9]. The collector voltage determines the energy of the electrons in the vicinity of the SiO_2-Si heterojunction. First of all this leads to the formation of a self-consistent quantum well for holes (on the n-type silicon surface) induced by a strong electric field. The Auger process probability depends on the quantum well depth, since electrons flying ballisticaly over the well for holes increase their energy by the value of this well depth [4]. An existence of the self-consistent quantum well on the silicon surface determines the appearance of the N- and S-type of current-voltage characteristics of the Auger transistor.

2. Experimental Results

The Auger transistor is based on Al-SiO_2--n-Si structure with a tunnel-thin transparent layer of a thermal oxide. The oxide was thermally grown in emitter windows on a monocrystaline silicon n-type substrate with a (111) surface and the donor concentration of 10^{16} cm^{-3} (phosphor) at 700° C in the atmosphere of dry oxygen. According to ellipsometrical measurements the thickness of the oxide layer was 20-22 Å. The p^+ contacts to the base layer, which was induced by the electric field, were obtained by boron diffusion into silicon base windows ($p = 10^{18}$ cm^{-3}) and the n^+ contact to the collector was obtained by a diffusion of phosphor ($n = 10^{20}$ cm^{-3}) into the n-type substrate. The metallization of the structure was obtained by Al-sputtering with the silicon substrate temperature at 200° C (see insert at figure 1). The area of the emitter windows was 20 and 40 μm^2. The I-V characteristics of the transistor in a scheme with a common emitter are the typical I-V characteristics of a bipolar transistor at a low base current. An increase of current gain coefficient β with an increase of the base and collector currents was a characteristic feature of the collector characteristic of the Auger transistor. This I-V characteristic shows the S-type non-linearity under the increase of the collector current (figure 2). An appearance of current instability is

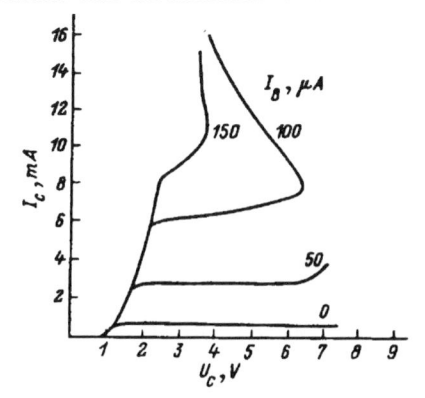

Figure 2. Experimental S-type I-V collector characteristics of the Auger transistor for common emitter configuration. T=300 K

typical for such a shape of the collector current. In this case the current density achieved up to 10^3 A/cm^2. Measurements of current density were seriously complicated by an inhomogeneous of current surface distribution. We believe that it is connected with an inhomogenity in the distribution of the thickness and composition of the tunnel oxide layer. The transistors operating at a high density of the collector current (10^4 A/cm^2 and higher) exhibit the *N*-shaped I-V characteristics of the collector current in the region of low voltage on the collector (figure 3). The non-linearity also corresponds to the instability of the collector current. In some cases the *S*-type shape characteristic changes to the *N*-type shape at low voltage (see figure 4).

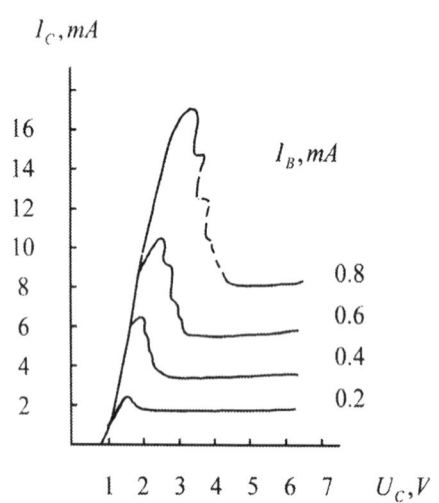

Figure 3. Experimental N-type I-V collector characteristics of the Auger transistor for common emitter configuration. T=300 K

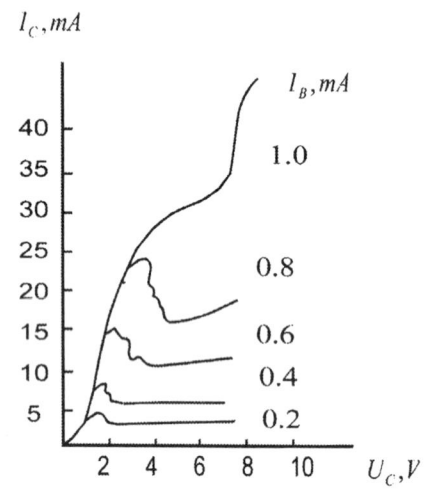

Figure 4. Experimental I-V collector characteristics of the Auger transistor for common emitter configuration. At low collector voltage the S-type characteristics change to the N-type. T=300 K.

3. Tunnel Currents of the Auger Transistor

The theoretical model of the MIS Auger transistor is based on the Hartree approximation to calculated the holes energy in a self-consistent quantum well on the silicon surface [10,11] and based on the quasi-classical approximation of current calculations. Figure 1 shows the energy diagram for an Auger transistor formed on the *Al-SiO₂ –n-Si* structure with a tunnel oxide layer. The base of the transistor is the hole charge layer that is induced by the electrical field of the oxide. The bigger the voltage drop across the oxide $V_{ox} = E_{FM} - E_{CO}$, the higher is the energy of injected electrons into silicon volume. When $E_{FM} - E_{CO} > E_g$ (see figure 1), it is possible to produce impact ionization of the electron-hole pairs in the silicon volume in vicinity of the collector. In the *Al-SiO₂–n-Si* structure under investigation the work function of electron from aluminium to silicon oxide and the affinity of electron for silicon crystal to silicon oxide are about the same and equal to 3.1-3.2 eV. This means that the barrier may be treated as a symmetrical one. As a result of such simplification the energy of an electron entering the semiconductor is

$$E_e^o = V_{ox} = E_{FM} - E_{CO} \tag{1}$$

The energy band near the Fermi level of the metal from which the electrons mainly tunnel into the semiconductor, has a width of about several hundredths of eV. The probability for the tunneling of electrons, which occupy the deep states under the Fermi level, is small because of the higher potential barrier for them.

An increase in the potential drop across the oxide V_{ox}/e is accompanied by an increase in the kinetic electron energy W and as a result the fast-electron-induced impact ionization is possible. The Auger current that appears as a result of the impact ionization begins to play an important role in the transistor current balance. The following currents (see figure 5) determine the operation of an Auger transistor:

1. The electron tunnel current, which is directed from the metal to the conduction band of the semiconductor, I_{mc};
2. The current due to the tunneling of holes from the interface states in the silicon forbidden gap to the metal, I_{ms};
3. The current of holes, which are attracted to the Si-SiO_2 interface by the oxide electric field, and tunneling to the metal through the oxide, I_{mv};
4. The Auger current, which increases both currents I_{ms} and I_{mv} and also collector current I_C.

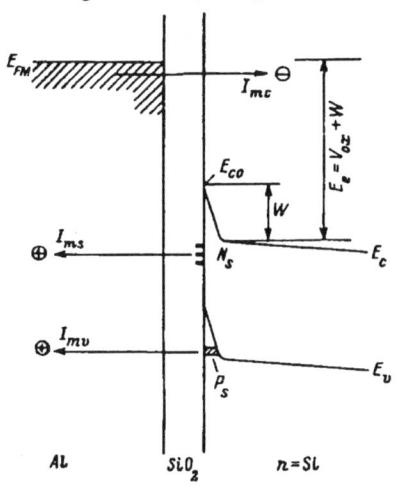

Figure 5. Tunnel currents diagram for the Auger transistor.

The tunneling through the oxide layer may be treated as current through the conduction band or valence band of oxide. The current connected with the holes tunneling into the metal is, nevertheless, always fewer than the electron tunnel current flowing from the metal into the conduction band of silicon, since the tunneling of holes from the silicon valence band to the metal via the oxide valence band is not likely to occur, because of the large effective mass of the hole in SiO_2: $m_h > 5m_e$. At the same time, the tunneling of electrons from the metal to the silicon valence band through the oxide conduction band takes place through a much higher barrier than that of electron tunneling to the silicon conduction band.

In describing the tunnel currents of electrons and holes usually we use the quasi-classical approximation. In this approximation a tunnel current through a potential barrier can be written as

$$I = \frac{eN}{\tau} \exp\left[-(2/\hbar)\int p\,dx\right], \tag{2}$$

where the integral $(2/\hbar)\int p\,dx$, which determines the carrier tunneling probability, is taken along the range, which is forbidden according to classical mechanics; N is the two-dimensional density of electrons that can take part in the tunneling, and τ is the time between collisions of charge carrier with the barrier wall. For an example, the tunnel current flowing from the metal to the semiconductor conduction band can be written in such a way

$$I_{mc} = \frac{eV}{16\pi^2 \hbar d^2} \exp\left\{-\frac{4\sqrt{2m^*}d}{3\hbar V_{ox}}[V^{3/2} - (V - V_{ox})^{3/2}]\right\}, \quad (3)$$

where $V = \varphi_m$ is the work function of an electron which escapes from the metal to the oxide conduction band; $m^* = 0.29 m_0$ is the tunneling effective mass of electrons in silicon oxide; d is the oxide thickness, and $V_{ox} = E_{FM} - E_{CO}$. Details of the calculations will be published elsewhere [12].

The base current $I_{ms} + I_{mv} = I_B$ and the Auger current flow in the direction opposite to the leakage currents I_{ms} and I_{mv}:

$$I_B + I_{Auger} - I_{ms} - I_{mv} = 0. \quad (4)$$

This equation in fact determines the balance of currents in the transistor. The Auger current is not involved in the emitter current, $I_E = I_C + I_{ms} + I_{mv}$, but proportional to the electron tunnel current I_{mc} and depends on the impact ionization factor A: $I_{Auger} = A I_{mc}$. The coefficient A reflects the dependence of the Auger current on the energy of injected electrons. The electron energy is determined by the potential drop across the oxide layer and the potential drop on the base layer (see figure 1).

In calculating the current-voltage characteristics of the Auger transistor we took the impact ionization coefficient A from an experimental study and approximated it to

$$A = k\left(\sqrt{1 + \frac{[(V_{ox} + W) - 1.5]^2}{1.5}} - 1\right) \quad (5)$$

in accordance with the experimental data of Ref.13.

Figure 6 shows the calculated characteristics of the collector current of an Auger transistor based on Al-SiO_2-n-Si structure. The S-type shape of the collector characteristics is a result of the ambiguity of the base current characteristic [4]. When the impact ionization appears; i.e. $A > 1$ the current gain coefficient $\beta = dI_C / dI_B$ is

$$\beta = \frac{I_C + AI_C}{I_E - (I_C + AI_C)} = \frac{\alpha}{1-\alpha} = \frac{\alpha_0 + A\alpha_0}{1-(\alpha_0 + A\alpha_0)} \quad (6)$$

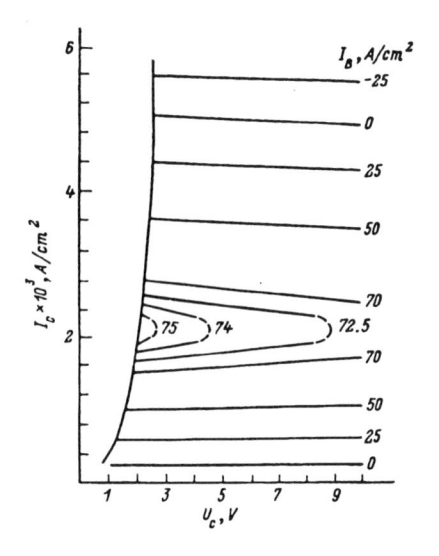

Figure 6. Calculated collector characteristics of an MIS Auger transistor; $n_d = 10^{16}$ cm^{-3}, $N_S = 10^{11}$ cm^{-2}, $d = 12$ Å

Here α_0 is the transistor gain in the absence of the impact ionization, $\alpha_0 < 1$. The coefficient α can be higher than one ($\alpha > 1$) if originally $\alpha_0 > 0.5$; i.e. the emitter injection coefficient must be quite high $\gamma > 0.5$. This result suggests that the higher the emitter efficiency, the smaller the impact ionization factor A is needed to obtain a value $\alpha > 1$. The tunnel currents of 10^3 A/cm and higher were observed experimentally in the Auger transistor, in which the oxide layer was 20-30 Å, as measured by means of an ellipsometer. To correctly estimate the barrier width in the quasi-classical calculations of the current-voltage characteristics, it should be taken into account that the electron density in the oxide layer in the direction normal to the surface is still very high up to 10Å penetration of the metal electron to the oxide. On this reason, the oxide thickness for our calculation was chosen to be of 10-15 Å.

4. Time Constants of the Auger Transistor

In the Auger transistor there are two important time constants. The slowest process in this Al-SiO_2-n-Si transistor is a charging of emitter-base capacity. This process strongly depends on current density and gives frequencies up to 10^{11} Hz if the collector current density is 10^4 A/cm^2. The highest frequency of oscillations (which for example fill radio pulse) may be up to 10^{11}-10^{12} Hz. This frequency is determined by the time needed for the electron to drift through the base-collector junction. The former time constant is important for S-characteristics. The latter is important for N-characteristics. It is possible to envisage such an application of the Auger transistor. In semiconductors such as *GaAs* or *InP* these frequencies are about an order of magnitude higher.

5. Conclusion

We investigated the S- and N- type instabilities in the collector current of the Auger transistors in a circuit with a common emitter. The physical reason of instabilities is obvious. The electric current depends not only on the value of the base current, but also on *the time*. It is mainly the dissolving time of the electron-hole clouds, which are created by high-energy electrons near the surface. There are two ways for the e-h cloud to be dissolved. First, dissolving it by the electric field of collector and second by dissolving it by the hole tunnel current to the metal. The former process leads to the N-type current-voltage collector characteristics and the latter one leads to S-type collector characteristics.

Acknowledgements

The work is supported by GKNT RF 213/68/4-2, FSNS 96-1011, RFBR 97-02-18106 and 97-02-18358 grants.

6. References

1. Simmons, J.G. and Taylor, G.W. (1986) Concepts of gain at an oxide-semiconductor interface and their application to the TETRAN and to the MIS switching device, *Solid-State Electronics*, **29**, 287-303.
2. Kroemer, H. (1972) *in the book*: A.G. Milns and J.J. Feucht, *Heterojunctions and Metal-Semiconductor Junctions*, New York, NY: Academic Press.
3. Grekhov, I.V., Ostroumova, E.V., Rogachev, A.A. and Shulekin, A.F. (1991) A silicon Auger transistor with a tunnel MIS emitter, *Pis'ma v Zh. Tekh. Fiz.*, **17**, 44-48, [*Sov. Tech. Phys. Lett.*, (1991), **17 (7)**, 476 –477,].
4. Ostroumova, E.V. and Rogachev, A.A. (1994) A simple model of the Auger transistor, *Fiz. Tech. Poluprovodn.*, **28**, 1411-1423 [*Semiconductors*, **28 (8)**, 793-799 (1994)].
5. Tiwary, S., Wang, W.I. and East, J.R. (1990) An analytic theory of the Auger transistor: a hot electron bipolar transistor, *IEEE Trans. Electron Devices*, **37**, 1121-1131.
6. Ostroumova, E.V. and Rogachev, A.A. (1997) Current instabilities in the Auger transistor, *Proc. of Intern. Symp. "Nanostructures: Physics and Technology"*, St.-Peterburg, Russia, 456-459.
7. Read, W.T. (1958) A proposed high-frequency, negative-resistance diode, *Bell. Syst. Techn. J.*, **37**, 401-446.
8. Ostroumova, E.V. and Rogachev, A.A. (1994) Quantum effects in MIS Auger-transistor, *Abstr. of II Intern. Symp. "Nanostructures: Physics and Technology"*, St.-Petersburg, Russia, 209-211.
9. Ostroumova, E.V. and Rogachev, A.A. (1996) MIS Auger transistor, *Proc. of the 26th European Solid State Device Research Conference ESSDERC'96*, Edit. by G.Baccarani & M.Rudan, Edition FRONTIERES, France, 245-248.
10. Ando, T., Fowler, A.B. and Stern, F. (1982) Electronic properties of two-dimensional system, *Reviews of Modern Physics* **54**, 438-672.
11. Rogachev, A.A. (1992) Electron-Hole Liquids in Semiconductors', *Handbook on Semiconductors*, Completely Revised Edition, Edited by T.S. Moss, **Vol.1** - *Basic Properties of Semiconductors*, Ed. by P.T.Landsberg, Elsevier Science Publiishers B.V., Chapter **9**, p.449-487.
12. Ostroumova, E.V. and Rogachev, A.A. (1997) Current instabilities in the Auger transistor, *Microelectronics and Reliabilities*, to be published.
13. Chang, Chi., Hu, Ch. and Brodersen, R.W. (1985) Quantum yield of electron impact ionization in silicon, *J. Appl. Phys.* **57(20)**, 302-309.

RADIATION INDUCED BEHAVIOR IN MOS DEVICES

V.V. EMELIANOV, G.I.ZEBREV, O.V.MESHUROV, A.V.SOGOYAN,
R.G.USEINOV
Research Institute of Scientific Instruments
RISI, Turaevo, Lytkarino-1, Moscow reg., 140061, Russia

1. Introduction

Despite many years of research, the nature of radiation-induced charge sources and basic mechanisms of charge relaxation in irradiated MOS devices remains controversial. Radiation-induced positive charge has been shown to be due to the oxygen vacancies having captured holes ($E`_\gamma$-centres) [1,2]. However the problem resides in the energetic and spatial distribution of the trapped holes and their electron states. Energy levels of near-interfacial radiation-induced defects of the oxide may be located above $E_C(Si)$, under $E_V(Si)$ or within Si forbidden gap [3 and ref. in 3]. At the same time, relaxation process dynamics should depend significantly on energy location of the trap electron states.

In this work, the original experimental results and the model of a postradiation response of radiation-induced damage in gate oxide of MOS structure versus temperature and oxide field is presented. We develop the previous works assuming that energy spectrum of defect's states is located within Si forbidden gap. The model deals with the defects in SiO_2 regardless of their microstructure. A novel conception of thermoactivated nature of tunneling exchange between defects and Si substrate is proposed. The predictions of the model suggested are verified in terms of thermal and field annealing effects.

2. Experimental procedure

The n-channel MOSFETs and p-substrate MOS capacitors with 58 nm gate oxide thickness were used in the experiments. These devices were made on the same test chip. The gate oxide was grown in wet O_2 containing HCl with the subsequent annealing in He.

The irradiations were performed by Co(60) γ - and Sr(90)-Y(90) electron sources. All devices were irradiated at the room temperature and the gate bias of +5V. The postradiation behavior of the MOSFETs and MOS capacitors was investigated by current-voltage (I-V) and high-frequency capacitance-voltage (C-V) measurements, respectively. The voltage ramp rate was 1 V/s. The values of the radiation-induced oxide charge density (Q_{ox}) and radiation-induced interface states density (N_{it}) were determined for the MOSFETs according to $N_{it}=\Delta S C_{ox}\varphi_B /kT\ln(10)$, $Q_{ox}=C_{ox}(\Delta V_t +q\Delta S\varphi_B /kT\ln(10))$ (where ΔS is the subthreshold voltage change of I-V characteristics, C_{ox} is the oxide capacitance, φ_B is the bulk Fermi potential, k is the Boltzmans factor, T is the absolute temperature, q is electron charge, ΔV_t is the threshold voltage change) and for the MOS capacitors according to $N_{it}=(\Delta V_{inv} -\Delta V_{mg}) C_{ox}/q$, $Q_{ox}= -\Delta V_{mg} C_{ox}$ (where V_{inv}, V_{mg} is the inversion voltage and the midgap voltage of MOS structure) [4].

In the first experiment the response of irradiated MOSFETs was tested during cyclical thermoelectric stresses. The devices were irradiated by the electrons to dose of 5 Mrad. Immediately after irradiation the device was thermoelectrically treated in accordance with a scheme in Table 1. All measurements were performed at room temperature.

TABLE 1. Postirradiation treatment sequence (experiment 1)

Time, hour	22	4	18	22	4	18	22	4	18	22	4	18
Temp. C	20	heating	120	20	heating	120	20	heating	120	20	heating	120
Bias, V	+5	+5	+5	-5	-5	-5	+5	+5	+5	-5	-5	-5

In the second experiment, the responses of MOSFETs and MOS capacitors were investigated as a function of measurement temperature. The devices were irradiated by the γ-source to dose of 2 Mrad. After irradiation the devices were annealed for 24 hours at the gate bias of +5V and temperature of 120C. Then the I-V and C-V characteristics were measured at the different temperatures in the range 20-120C. The measurements were performed at the voltage ramp from the inversion mode to the accumulation mode of the MOS structure, i.e. at the change of the gate bias from positive voltage to the negative gate bias.

The object of third experiment group was the recovery dynamics of radiation-induced damage in MOS oxides vs temperature and gate bias. The devices were irradiated by electrons to the dose of 5 Mrad and immediately after irradiation were annealed at temperatures of 20C, 67C, 120C (gate bias of +5V) and the gate bias of -5V,-1V,+1V,+5V (temperature of 120C). All measurements were performed at room temperature.

3. Experimental results

Figure 1 shows the values of Q_{ox}/q and N_{it} for n-channel MOSFETs as a function of time for the first experiment (see Table 1). The main general results are: first, radiation-induced positive charge annealing is reversible, and the charge can be restored under negative gate bias, second, both forward and reverse processes are accelerated by increasing the temperature. Note, for the investigated devices, time-depending value N_{it} and amplitude of Q_{ox}/q variations in two completed cycles of thermoelectric treatment don't change. Therefore, the reversible relaxation process is the only process of postirradiation charge relaxation in the gate oxide.

Dependencies of Q_{ox}/q and N_{it} on measurement temperature (see Fig.2) show that the values of Q_{ox}/q and N_{it} are increased with the rise of temperature. It should be emphasized that in this case there arises not annealing but a reversible and reproducible measurement temperature dependence.

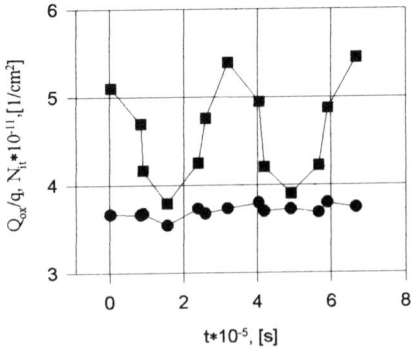

Figure 1. Gate voltage, temperature, radiation-induced oxide charge density (Q_{ox}/q - *square*) and interface states density (N_{it} - *circle*) vs. time after irradiation.

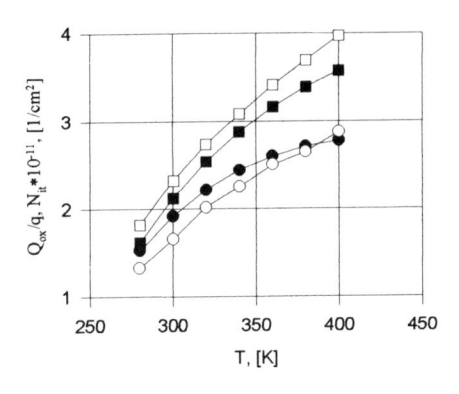

Figure 2. Radiation-induced oxide charge density (Q_{ox}/q - *square*) and interface states density (N_{it} - *circle*) for n-channel MOSFETs (*shade*) and p-substrate MOS capacitors (*clear*) vs. measurement temperature.

A normalised components of Q_{ox} vs. time of the positive gate bias annealing process at temperature 20C, 67C, 120C are presented in Fig. 3. Note, that long-time dependence of Q_{ox} vs. time is close to logarithmic. The slope of this function weakly depends on temperature. The value of activation energy was estimated by slope of the dependence of $t_{20\% \text{ annealing}}$ vs $1/T$. This value is 0.9eV.

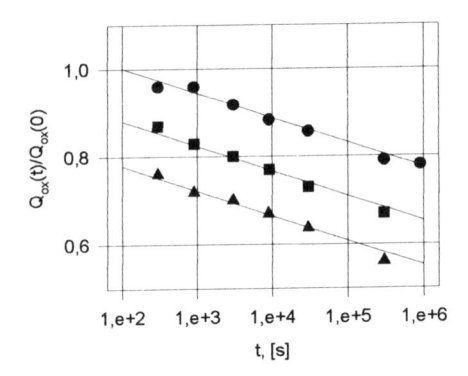

Figure 3. Normalised component of Q_{ox} vs. time after irradiation at different temperatures: 20C (circle), 67C (square), 120C (triangle).

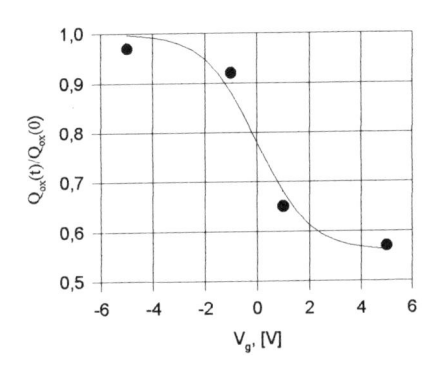

Figure 4. Normalised component of Q_{ox} vs. gate voltage (after irradiation V_t=1.46V).

A normalised component of Q_{ox} for the n-channel MOSFETs vs. gate voltage for the time of 3×10^5s is presented in Fig.4. Dependencies of the Q_{ox} vs. gate bias are saturated in the regions of the large values of the positive and negative gate bias. Note, that the most changes range of the Q_{ox} vs. V_g corresponds to the most changes range of surface potential on the surface Si/SiO$_2$ vs. V_g. So, the conclusion can be made that the dependence of the Q_{ox} on Vg is mainly caused by the electrical potential at the surface Si/SiO$_2$.

4. Physical model

A necessary (not sufficient) condition for existence of reversible rechargeable oxide traps is a location of their energy levels within the variation range of Fermi level in the Si vs. the gate voltage. According to the general thermodynamics the cause and direction of defect's charge state change are determined by trap energy position relative to the Fermi level. The carrier exchange between the Si and traps tends to restore the equilibrium at a given Fermi level position or, the same, at a given gate voltage V_g.

According to a single-particle approximation, direct tunneling of the Si-substrate electron is allowable for only the defect with energy level equal to the electron energy. The electron energies in initial and final states are substantially different when electron in valence or conduction bands tunnels to the oxide trap with energy level within Si forbidden gap. In this case, the electron tunneling without any phonon assistance is prohibited due to the energy and momentum conservation principle. Thermally exited vibration of the lattice leads to spontaneous deviations of the defect local displacement and fluctuations of its energy level. Energy level of the defect under a certain displacement can reach the silicon conduction or valence band edge. Hence, the tunnel communication between the Si-substrate and the defect becomes possible. As a result, if the electron energy in initial and final states is different the electron exchange process between silicon and oxide trap is of thermally activated nature. In the proposed model, the defect center rather than the substrate electron is thermally activated. An activation energy is the difference between the energy of a "defect + electron" system when the tunneling transition becomes possible and its energy in the equilibrium state H_{tun}-H_{ini}. The activation energy E_A of the electron transition from an energy level E to the defect level E_0 can be obtained using model [5,6]:

$$E_A(E) = H_{tun} - H_{ini} = \frac{1}{2}\frac{C}{\gamma^2}(E_0 - E)^2 \qquad (1)$$

where C is the elastic constant (about 30-35 eV/A^2); γ is the electron-phonon interaction constant in SiO$_2$ (2-4 eV/A).

In general case electron exchange between traps and conductive and valence band of semiconductor is possible. Positive charge relaxation kinetics due to the electron exchange between the oxide traps and the silicon substrate can be written as:

$$\frac{\partial \delta N^+(E_t,x)}{\partial t} = -\frac{\delta N^+(E_t,x)}{\tau_C^{cup}} - \frac{\delta N^+(E_t,x)}{\tau_V^{cup}} + \frac{\delta N - \delta N^+(E_t,x)}{\tau_C^{em}} + \frac{\delta N - \delta N^+(E_t,x)}{\tau_V^{em}} \qquad (2)$$

where δN is the concentration of traps with energies in the range E_t-E_t+δE_t, δN^+ is the concentration of positive charged traps with energies in the range E_t-E_t+δE_t, τ_C^{cup}, τ_C^{em}, τ_V^{cup}, τ_V^{em} are the times of the capture and emission of electrons in the exchange with conductive and valence band, respectively.

According to the thermoactivated tunneling mechanism of exchange and the principal of detailed equilibrium:

$$\tau_C^{cup} = \frac{1}{\sigma n v_T}\exp(E_{AC}/kT)\exp(x/\lambda), \quad \tau_C^{em} = \tau_C^{cup}\exp\left(\frac{E_F - E_t}{kT}\right)$$

$$\tau_V^{em} = \frac{1}{\sigma p v_T}\exp(E_{AV}/kT)\exp(x/\lambda), \quad \tau_V^{cup} = \tau_V^{em}\exp\left(\frac{E_t - E_F}{kT}\right) \quad (3)$$

where σ is the capture cross section of electron by traps, n,p are the concentrations of electrons and holes in conductive and valence bands of Si, v_T is the thermal velocity of electrons and holes in Si, E_{AC}, E_{AV} are the activation energies of exchange with conductive and valence bands of Si, λ is the tunneling length (0.06 nm), x is the trap distance from the interface, E_F is the Fermi level of Si.

For the simplicity, let us consider further only the exchange with conductive band of Si. Then under initial condition $\delta N(0) = \delta N^+(0)$ the solving of (4) for integrated positive charge density can be written as:

$$Q(t) = q\int_0^{d_{max}}\int_{E_V}^{E_F} D(E_t,x)\exp(-t/\tau_C^{cup})\partial E_t\partial x + q\int_0^{d_{max}}\int_{E_F}^{E_C} D(E_t,x)\partial E_t\partial x \quad (4)$$

where d_{max} is the maximum distance of a trap location from the interface, $D(E_t,x)$ is the energy-spatial density of traps. Using the tunneling front approximation [7] from (4) can be obtained:

$$\Delta Q(t) = Q(t) - Q(0) \approx \lambda q\left[\ln(\sigma n v_T t) - \frac{\overline{E_A}}{kT}\right]\int_{E_V}^{E_F} D(E_t)\partial E_t \quad (5)$$

where $\overline{E_A}$ is the average activation energy of the electron capture.

For the normalized component of the Q_{ox} in the case of the large positive gate bias, i.e. $E_F \approx E_C$ the expression (5) can be written as:

$$\eta = \frac{Q(t)}{Q(0)} \approx 1 - \frac{\lambda}{d_{max}}\left[\ln(\sigma n v_T t) - \frac{\overline{E_A}}{kT}\right] \quad (6)$$

5. Estimates and Fits

Simplified theoretical estimates (5), (6) predict that the dependence $Q_{ox}(t)$ on time is logarithmic. Slope of this logarithmic function does not depend on temperature. In general the dependence on temperature is due to the coefficient at activation energy. The field dependence is due to the values of n and E_F.

The maximal distance of spatial traps distribution d_{max}=2.4nm, the cross section of electron capture $\sigma \approx 10^{-12}$sm^2, the activation energy $\overline{E_A}$ =0.9eV are determined directly from the experimental data (see Fig.3). The obtained values of $\overline{E_A}$ and σ are satisfactory. At the same time the propagation depth of the tunneling front within the time scale of our experiments is usually estimated as 4-5 nm. However the exchange probability in the model presented is roughly proportional to the thermoactivative and tunneling components: $p \propto \exp(-E_A/kT) \cdot \exp(-x/\lambda)$. Due to this fact, the typical propagation depth of the spatial-energy front is less significant than the earlier estimations of simple spatial front propagation depth. From this point of view, the presented value of 2.4 nm seems to be acceptable. The results of simulation together with the experimental data are shown in Fig.3, Fig.4.

6. Conclusion

The postradiation response of the investigated MOS structures consists of the relaxation process due to "slow" interface states of donor type. The appearance of these states is caused by the capture of radiation-induced holes nearby the oxide-semiconductor interface. A mathematical model of reversible charge relaxation phenomenon is offered. Model predictions fits well to the obtained thermal-field relaxation dependences.

7. References

1. Conley, J.F., Jr., Lenahan, P.M., Lelis, A.J., Olham, T.R. (1995) Electron Spin Resonance Evidence that E'$_\gamma$ Centers Can Be Behave as Switching Oxide Traps, IEEE Transactions on Nuclear Science 42 (6), 1744-1749.
2. Witham, H.S., Lenahan, P.M. (1987) The Nature of The Deep Hole Trap in MOS Oxides, IEEE Transactions on Nuclear Science 34 (6), 1147-1151.
3. Lelis, A.J., Olham, T.R., Boesh, H.E., Jr., McLean, F.B. (1989) The Nature of The Trapped Hole Annealing Process, IEEE Transactions on Nuclear Science 36 (6), 1808-1815.
4. Benedetto, J.M., Boesh, H.E.,Jr. (1984) MOSFET and MOS Capacitor Separating Interface and Oxide Charge Effects in MOS Device Characteristics, IEEE Transactions on Nuclear Science 31 (6), 1461-1466.
5. Ngai, K.L., White, C.T. (1981) A Model of Interface States and Charges at Si-SiO$_2$ Interface: Its Predictions and Comparisons with Experiments, Journal Applied Physics 52 (1), 320-337.
6. Reilly, E.P., Robertson, J. (1983) Theory of Defects in Vitreous Silicon Dioxide, Physical Review B 26(6), 3780-3791.
7. McWhorter, P.J., Miller, S.L., Miller, W.M. (1990) Modeling the Anneal of Radiation-Induced Trapped Holes in a Varying Thermal Environmental, IEEE Transactions on Nuclear Science 37 (6), 1682-1689.

HYDROGENOUS SPECIES AND CHARGE DEFECTS IN THE Si-SiO$_2$ SYSTEM

EDWARD H. POINDEXTER and CHRISTOPHER F. YOUNG
Army Research Laboratory
Adelphi, Maryland 20783, U. S. A.

GARY J. GERARDI
William Paterson University
Wayne, New Jersey 07470, U. S. A.

1. Introduction

Hydrogen in its several speciations plays major roles in the Si-SiO$_2$ system. Some roles are beneficial, some are harmful, and some are unclear in their effects. There has been, nonetheless, little systematic research on H and its compounds *per se* in the Si-SiO$_2$ system. Much knowledge has been accumulated; but it is almost all focused on explaining one or another specific device phenomenon, rather than on illuminating of the behavior of hydrogen in itself. A number of these are listed below:

Negative-bias-temperature instability (NBTI)
Radiation/hot-electron generated defects and charges
Passivation/depassivation of interface trapped charge
Deal oxidation triangle
Oxidation-induced "fixed" oxide charge
Atomic-H paradox
H *vs* D effect in generation of charge defects
Diffusion, solubility of H$_2$O
Electron trapping in SiO$_2$
Anomalous positive charge

Here we seek scientific commonality in some of these H-related issues, and try to lay foundations for an overall physical chemistry of H in the metal-oxide-semiconductor (MOS) structure. It is usually necessary to consider H-related features of several phenomena in order to develop the best model for any one phenomenon. For compactness, our interpretation of any particular H-related problem presumes this wider view, and thus may seem in itself out-of-sequence or logically incomplete. It is, therefore, recommended that one read through the entire paper before studying any one section in detail.

In association with the subject hydrogenous species, the electrical defects of interest are oxidation-induced positive oxide charge Q_f (= qN_f, where q is electron charge);

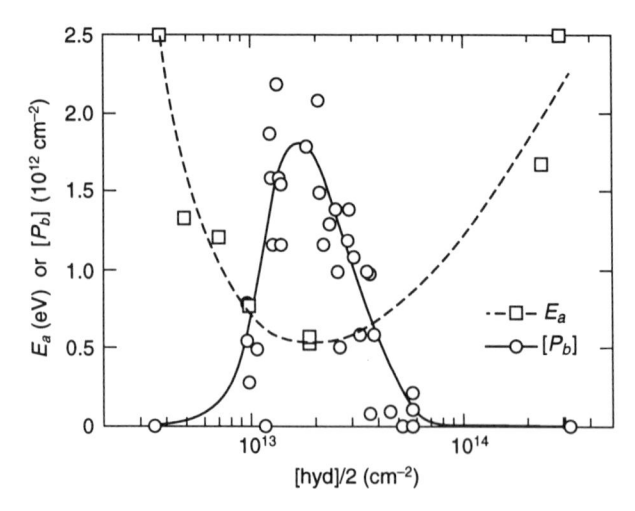

Figure 1. Generation of P_b centers by electric field at 300 K; E_a for thermal generation without field. The concentration of all H species present is [hyd].

otherwise-induced oxide charge Q_{ox}; interface trapped charge Q_{it} (= $qN_{it,}$ nee "surface states"); dangling Si bonds at the Si surface, called P_b centers (·Si≡Si$_3$, source of much of Q_{it}); and dangling Si bonds in the oxide, E' centers (O$_3$≡Si·....$^+$Si≡O$_3$), source of some kinds of Q_{ox}.

2. Hydrogen-Related Phenomena in the Si-SiO$_2$ System

2.1 NEGATIVE-BIAS-TEMPERATURE INSTABILITY

The negative-bias-temperature instability (NBTI)[1] causes an increase Q_{it} and Q_{ox} under negative gate stress of order MV/cm, and/or thermal stress up to 700°K. Each agent alone can cause the effect, but a larger/faster effect is produced by both acting together. Charge defects Q_{ox} and Q_{it} are created in roughly equal amounts. The NBTI in metal-oxide-semiconductor (MOS) devices is eliminated in manufacturing, largely by an empirical approach which in effect reduces the presence of water.

It has been observed[2-4] that dry-oxidized samples exposed to atomic H above 700°K, or steam-oxidized samples show the NBTI; dry samples exposed to molecular H$_2$ do not. A series of samples with graded H content were stressed with fields (5–10 MV cm^{-1}) or heat (to 700°K), applied both separately and together. The [P_b] generated by field alone at room temperature is shown in Fig. 1. It is maximized in a sample oxidized with 0.1 percent H$_2$O in O$_2$. Samples were also stressed at 650°K, without field. The heat-alone result was like the field-alone result. Activation energy E_a for P_b generation without field was determined, and is also shown in Fig. 1. The unstable mid-range of [H-species] is clearly shown for both heat and field stress.

The final value of Q_f is linearly proportional to interface hole concentration [h$^+$]. The hole concentration, variable over several orders of magnitude by applied fields, can thus have a very strong effect effect on interface chemistry. A reaction which fits qualitatively the several features noted thus far is

$$H-Si\equiv Si_3 + H_2O + h^+ \rightarrow {}^\bullet Si\equiv Si_3 + H_3O^+. \quad (1)$$

Reactions with a minimum in E_a are known in electrochemistry. One case was described by Marcus,[5] shown in the reaction coordinate diagrams of Fig. 2. The left-hand parabola of each pair encompasses the energy levels of hydrogen bound in the reactant molecule, and the right-hand, in the product. In the situation at (a), a transfer of H from left to right parabola is endothermic with a finite activation energy E_a. In (b), the reaction is slightly exothermic, and E_a is zero. At last, in (c), the reaction is very exothermic, and E_a is again finite; so the reaction is hindered in spite of its exothermicity.

Molecular orbital calculations and electrochemical observations[6] have shown that the partial reaction

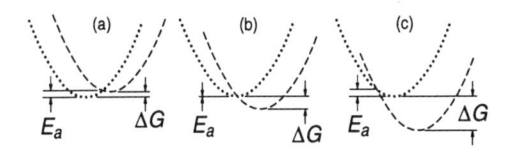

Figure 2. Reaction coordinates with variable free energy ΔG. As exothermicity increases, E_a is at first (a) finite, then (b) zero, then again (c) finite.

$$H^+ + nH_2O \rightarrow H_{2n+1}O_n^+. \quad (2)$$

is exothermic with $\Delta G = 8$ eV for $n = 1$, but ΔG increases to 13 eV for $n = 4$. Thus, this partial reaction becomes more exothermic by increase in availability of H_2O. The increase in the ΔG of the partial may well drive the reaction parabolas through the minimum E_a configuration. Further, when the corresponding kinetic equation for Eq. 1 is solved, it is found that several major quantitative aspects are in good accord with experimental measurements on the growth of Q_{ox} and Q_{it}, constraints imposed by non-varying [H] and [H–P_b], and the domination of the reaction by [h$^+$]. Thus, Eq. 1 is in overall good accord with numerous features of the NBTI.

2.2 RADIATION-GENERATED INTERFACE TRAPS AND CHARGES

Over the years, one general type of physico-chemical model has come to be accepted for the generation of interface traps and near-interface positive charges developed after a radiation exposure. Following an initial collision event in the oxide, secondary entities migrate from the site, and cause most of the ultimate damage at or near the Si-SiO$_2$ interface. One popular variant of the general model involves hydrogen transport; H is coming to be accepted as the dominant factor in regard to the portion of Q_{it} which develops slowly following the initial event. Radiation-generation of interface traps is augmented in a wet oxide. It was early proposed that radiation-generated holes in the oxide could promote dislodgement of hydrogen from bound sites in the oxide.[7] The hydrogen then migrates and generates interface traps:[8]

$$H-Si\equiv Si_3 + H \rightarrow {}^\bullet Si\equiv Si_3 + H_2. \quad (3)$$

Generation of Q_{it} by radiolytic H is supported by the correlated growth and decline of Q_{ox} upon post-irradiation infusion of molecular H_2, together with the subsequent growth of Q_{it}, Fig. 3.[8] The first stage is the annealing of positively-charged E' damage centers near the Si–SiO$_2$ interface, which releases hydrogen:

$$H_2 + E' \rightarrow H\text{-}E' + H^+. \qquad (4)$$

Molecular orbital calculations predict that this passivation of E' would be an easy reaction;[9] early experiments have shown that molecular H_2 passivates E' centers at room temperature with E_a of only about 0.3 eV.[10]

In the next phase of the damage process, protons are driven to the interface by the positive bias. Thermal passivation/depassivation studies (next section), however, do not favor direct generation of P_b centers by radiolytic atomic H, Eq. 3, since the same equilibrium condition (revealed in Q_{it}) could not occur for arbitrary, imposed [H] vs. [H$_2$], [product] vs. [reactant]. Objections to atomic-H depassivation models might be partially answered by a model in which the proton, bias-attracted electron e$^-$, and P_b site are incorporated into one reaction, instead of the two proposed earlier:

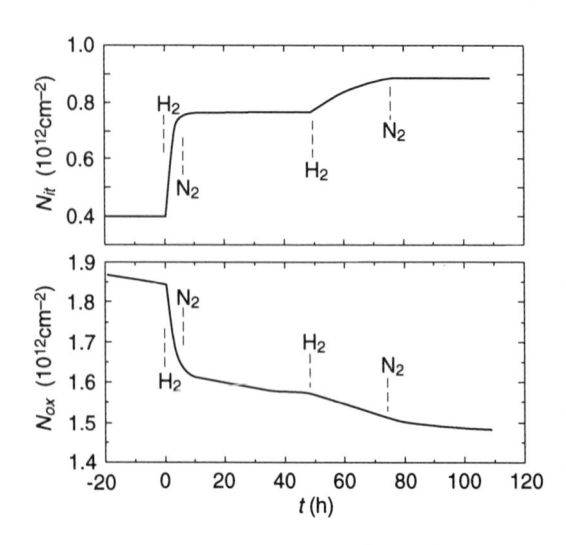

Figure 3. Concerted evolution of oxide charge N_{ox} and interface traps N_{it} upon introduction of molecular hydrogen and nitrogen. Radiation occurred at –65 h; gas introduced (300°K) as indicated. Adapted from Stahlbush et al [8].

$$H^+ + H\text{-Si}\equiv\text{Si}_3 + e^- \rightarrow \,\cdot\text{Si}\equiv\text{Si}_3 + H_2. \qquad (5)$$

A difficulty with any model proposing proton transport lies in the extreme attraction by H_2O (binding energy a minimum of 8 eV). It might be plausible that an H_3O^+ reaction occur, but proceed very slowly at 300°K.

$$H_3O^+ + H\text{-Si}\equiv\text{Si}_3 + e^- \rightarrow \,\cdot\text{Si}\equiv\text{Si}_3 + H_2O + H_2. \qquad (6)$$

The NBTI reaction coordinates, Fig. 2, and tests with high positive bias suggest that this reaction might be energetically possible only in a "dry" oxide, with perhaps only a single H_3O^+ involved. However, in general, radiation damage mechanisms remain unsettled; essential reactant concentrations and other controlling parameters are presently unknown. Concentration-dependent activation energy or a field-dependent concentration factor may be dominating influences.

2.3 PASSIVATION AND DEPASSIVATION OF P_b CENTERS

2.3.1 Unlimited passivant

In dry-oxidized Si-SiO$_2$ interfaces treated with a limitless supply of passivant, either H$_2$ or H, the same value of Q_{it} is finally approached, depending on the anneal temperature,[11] Fig. 4. Higher temperature causes a greater reduction in Q_{it}, and a sample previously annealed at a higher temperature depassivates at a lower temperature, tending toward the Q_{it} corresponding to the lower temperature. Yet, though pertinent and suggestive, these experiments are not ideal for developing reaction models; [H] is uncertain, and Q_{it} comprises more types of defects than the well-defined P_b center.

The P_b centers produced by expo-

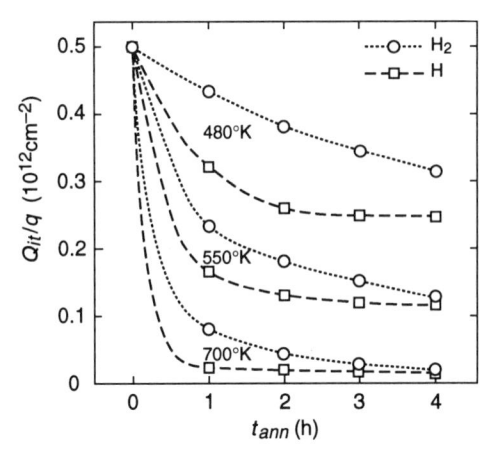

Figure 4. Passivation of interface traps Q_{it} by atomic H or molecular H$_2$. After DoThanh and Balk [11].

sure to atomic H at 300°K have recently been directly detected by EPR.[12] Oxides were grown on (111)Si in nominally dry O$_2$ at 1200°K. As grown, they showed no detectable P_b. (As this result is inconsistent with the Deal triangle, perhaps a little moisture was present.) Then, after vacuum anneal, a significant P_b signal developed, $[P_b] > 2 \times 10^{12}$ cm^{-2}. Both wafers were then subjected to extended anneal in atomic H at 300°K. Interestingly, $[P_b]$ in the vacuum-annealed wafer declined markedly; while in the as-oxidized wafer, P_b developed, with both wafers tending toward the same $[P_b]$, about 0.4×10^{12} cm^{-2}. These results are shown in Fig. 5.

2.3.2 Limited-passivant ambients

In one type of limited-passivant study, a previously H$_2$-passivated sample is exposed to nitrogen, argon, or air at 800°K or above.[13] Released hydrogenous species diffuse and escape from the outer surface. Thus P_b centers are depassivated. In another study, an Al film is deposited on an unpassivated oxide, and

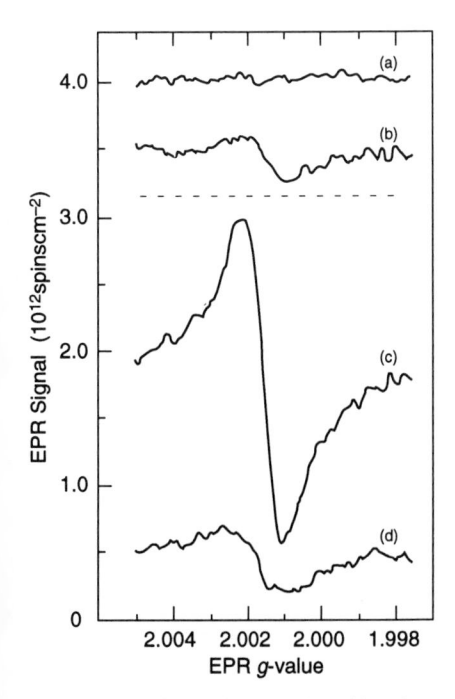

Figure 5. Passivation/depassivation of interface traps by atomic H, 300 K (a) P_b in as-oxidized wafer. (b) After exposure to H. (c) As-oxidized wafer, dessicated in vacuum. (d) Dessicated wafer exposed to H. After Cartier et al [12].

then heated.[14] The passivant is the limited amount of atomic H from reaction of the Al with outer surface silanols. Here, Q_{it} at first declines, but then increases again as time is prolonged, shown in Fig. 6. In both these experiments, depassivation kinetics were very well fit by simple dissociation of H–P_b. It was implied that depassivation by atomic-H attack might occur under special conditions; however, the constraints imposed by the kinetic data make it difficult to infer just what these special conditions might be.

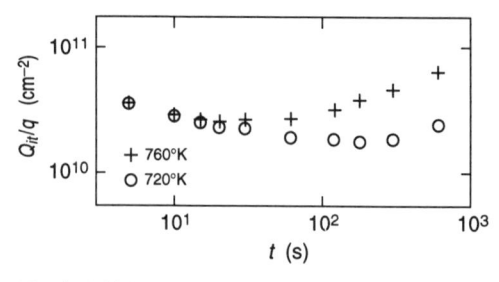

Fig. 6. Initial passivation of interface traps by atomic H from Al/silanol reaction, and eventual depassivation. After Reed and Plummer [17].

2.3.3 Proposed Passivation/Depassivation Models

Since the reaction of Eq.3 is not favored by kinetic depassivation studies, there may be other hydrogenous species which intervene and then control the ultimate Q_{it} equilibrium in either H or H_2 ambient and with either high or low initial Q_{it}. The data available do not allow detailed explanations of each study mentioned above, but some working hypotheses for pertinent broad situations can be conceived.

We consider first some schematic chemical reactions chosen to describe reactions which are expected to occur during a damp-to-wet oxidation. In addition to the oxidized interface $O_3{\equiv}Si{-}O{-}Si{\equiv}Si_3$, a number of other species should be produced:

$$Si_3Si{-\!\!-}Si{\equiv}Si_3 + O_2 + H_2O + h^+ \rightarrow$$

$$O_3{\equiv}Si{-}O{-}Si{\equiv}Si_3 + H{-}Si{\equiv}Si_3 + H{-}O{-}Si{\equiv}O_3 + H_3O^+ . \qquad (7)$$

During post-oxidation H anneal we might expect that residual O will be scavenged:

$$H + O_2 \rightarrow H_2O . \qquad (8)$$

If the H anneal is above 500°K, H_2O will be released from paired silanols in the SiO_2:

$$H{-}O{-}Si{\equiv}O_3 \rightarrow SiO_2 + H_2O \qquad (9)$$

The usual passivation reaction will occur in parallel, of course:

$$H + {\cdot}Si{\equiv}Si_3 \rightarrow H{-}Si{\equiv}Si_3. \qquad (10)$$

The water will depassivate P_b centers per the NBTI reaction of Eq. 1; if the system is presumed to fall into the midrange of $[H_2O]$, then the depassivation reaction may occur in the range of E_a which enables it to overcome the direct passivation of P_b by H.

If a damp/wet oxidized sample is annealed in vacuum or Ar (or in N_2, with some proviso that ancillary N compounds may be produced), residual H_2O, O_2, and H_3O^+ will be driven out; but while this is going on, the passivated P_b centers will be depassivated, perhaps by two reactions. The first is the simple dissociation

$$H-Si\equiv Si_3 \to {}^{\bullet}Si\equiv Si_3 + H . \tag{11}$$

This direct release of H from H–P_b was consistent with the kinetics in the studies above.[13,14] In addition, the NBTI reaction may well occur before all the H_2O has diffused away.

Extended exposure to H must then establish the final steady-state $[P_b]$ observed in the above studies. Atomic H is not a passive player in the chemical scene; it does not merely roam around or attach gently; rather, it attacks. It is a powerful reducing agent, and given time, might commence to reduce Si–O bonds, most likely those in the interface plane, in preference to the those within the SiO_2 phase.

$$H + O_3\equiv Si-O-Si\equiv Si_3 \to H_2O + O_3\equiv Si-H + H-Si\equiv Si_3. \tag{12}$$

Along with this, the passivation reaction of Eq. 10 and the depassivation reaction of Eq. 1 occur in parallel. A steady state balance develops, and the resultant $[P_b]$ depends only on temperature which sets the rates of the several reactions in accord with activation energies; the $[P_b]$ is ultimately independent of the initial conditions or oxidation procedures. If H_2 is substituted for H, the ultimate balance will be reached more slowly; but the same general behavior should follow. Moreover, oxide initially grown nominally dry oxygen will sooner or later reach the same steady state under sufficiently prolonged hydrogen exposure.

Again, these are hypotheses, kinetic differential equations with proper reactant/product concentrations are needed to test the proposals realistically.

2.4 OXIDATION TRIANGLE

If an oxide is grown in oxygen, and rapidly cooled in O_2, then the resultant interface dangling bond centers $[P_b]$ or interface traps (Q_{it}/q), and near-interface oxide positive charges (Q_f/q) decline linearly with the oxidation temperature. They define one leg of the well-known Deal triangle, Fig. 7.[1] Initial $[P_b]$ can be as high as $N_{it}/2$ after low-T oxidation; properly measured, $[P_b] \approx N_f$.

The effect of nitrogen anneals on Q_f and

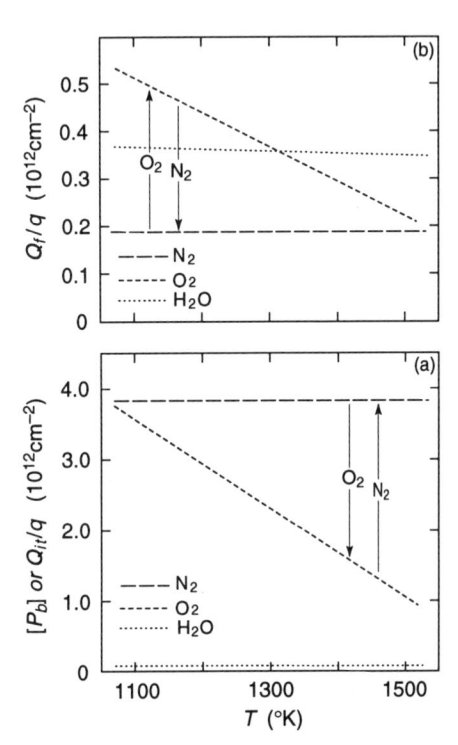

Figure 7. Deal triangle for oxidation of (111)Si in O_2 or H_2O, and anneal in N_2. (a) Oxide charge after oxidation, fast pull, followed by N_2 anneal, slow pull. (b) Corresponding behavior of interface traps. When correctly measured on the same unannealed, O_2-oxidized wafer, $Q_f/q \approx Q_{it}/q$.

Q_{it} defines the horizontal baseline of each triangle. Nitrogen anneal raises $[P_b]$ and N_{it} to a uniform high value, but reduces N_f. Re-exposure to O_2 restores all quantities to the value characteristic of the temperature. The triangle has sometimes been interpreted to arise from a variable degree of lattice mismatch, depending on the different thermal expansions of Si and SiO_2. However, this idea does not explain the nitrogen anneal, and has other faults.

It is suggested here that the triangle, supposedly applicable to perfectly dry oxides, is a result of an inevitable small amount of water. At high temperature, low solubility of H_2O favors a lower initial $[P_b]$ or Q_{it}, and Q_f. A nitrogen anneal at high temperature further dries the oxide, reducing Q_f by reduction in $[H_3O^+]$; and it also allows interface H–Si≡Si_3 centers to dissociate, thus increasing $[P_b]$.

The significance of H_3O^+ is supported by the oxide charge condition after steam oxidation. The oxide charge is very nearly independent of temperature, Q_f/q being about 0.3×10^{12} cm^{-2} over the range of Fig. 7.[15] The constant value in the face of a large $[H_2O]$ shows the importance of water in establishing a buffered interface condition. Further, higher Q_f is generally observed for p-doped Si, consistent with the idea that oxide H_2O could more easily abstract a hole from the Si surface necessary to form H_3O^+.

2.5 OXIDATION-ASSOCIATED OXIDE CHARGE

The oxidation-associated charge Q_f is near the Si interface, but the defect center responsible has not been identified. The discussion of the Deal oxidation triangle tacitly offers H_3O^+ as the source of Q_f in the as-oxidized wafer. There has never been an accepted model for the oxidation-induced positive charge; EPR has shown that it is definitely not due to (positively-charged) E' centers. The observed and extrapolated behavior of H_3O^+ in SiO_2 is in good accord with several aspects of the oxide charge. The reaction of Eq. 1 provides a plausible explanation for the initial, as-oxidized equality of Q_{it} and Q_f. The NBTI studies showed that the H_3O^+ ion is not dischargeable by electric field alone. A dry nitrogen anneal would easily annihilate H_3O^+ via reaction or diffusion away from the stabilizing presence of the Si interface. The H_3O^+ ion is invisible to EPR, and its dissociation products are invisible. Altogether, H_3O^+, strongly implicated in the NBTI, is a reasonable source for the Q_f resulting from the thermal oxidation.

2.6 ATOMIC HYDROGEN

The behavior of purported atomic H in SiO_2 is ambiguous. Atomic H is readily detected by EPR in radiation-damaged thermal films[16] (and fused silica[17] and quartz[18]) below 100°K. The EPR disappears above 100°K, indicating a great decrease in [H]. Nevertheless, the existence of atomic H in some

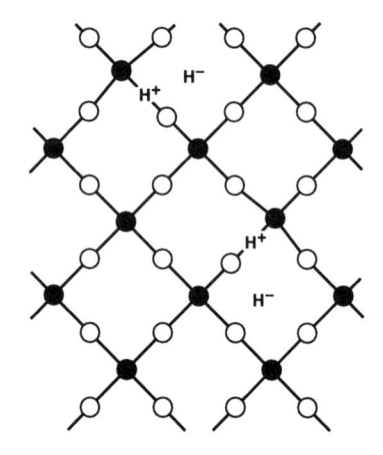

Figure 8. Proposed model of atomic hydrogen disproportionation products in silica. The proton enters Si–H–O structure; negatively charged H$^-$ atom occupies negative-U interstitial site.

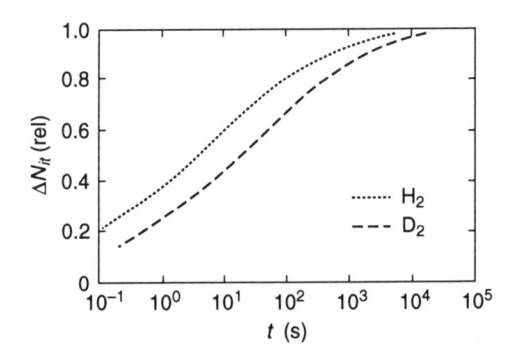

Figure 9. Time-dependent growth of interface traps in radiation-damaged H$_2$- or D$_2$-annealed Si–SiO$_2$ structures. Adapted from Saks and Brown [24].

guise in thermal SiO$_2$ at 300°K and above is indicated by the easier passivation of P_b centers[19,20] or interface traps Q_{it}[11] as compared to molecular H$_2$, Fig. 4.[11] The E_a is found to lie in the range 0.3 to 0.6 eV for passivation by atomic H,[14] compared to 2.6 eV for molecular H$_2$.[13]

The H problem in Si-SiO$_2$ may be related to the proposed behavior of atomic H in Si. Its disposition and transport have been found to be explainable by a dual-moiety model.[22-23] Atomic H seems to exist in (at least) two states in silicon: as a proton, H$^+$, and as a negative ion, H$^-$. The proton finds an energetically stable site in the bond-centered location, symbolized as ≡S$\underline{\text{H}^+}$Si≡. The negative ion is judged to be stable in a negative-U interstitial position. There is some debate about whether the H$^-$ site is further stabilized (The ion is weakly stable in isolation.) by negative electron-electron correlation involving the surrounding lattice, or by some possibly intermittent interaction with the proton. Perhaps a similar disproportionation of atomic H into H$^+$ and H$^-$ occurs in silica, as shown in Fig. 8.

2.7 DEUTERIUM VS. HYDROGEN EFFECTS

It has been observed that a Si-SiO$_2$ interface passivated with deuterium instead of hydrogen is more resistant to radiation damage, Fig. 9.[24] No detailed model was proposed. A stronger deuterium advantage has now been shown for the case of hot-electron damage.[25] No "chemistry-based" model has been offered; but a plausible "physics-based" explanation has appeared.[26] It is proposed with well-reasoned arguments that the heavier D–Si≡ moiety in its bending or "wagging" mode is coupled less effectively to lattice phonons generated by the impinging hot electrons. Hence the bond does not suffer from such a large build-up of motional energy as does the lighter H–Si≡ moiety, Fig. 10, and, therefore, does not break so readily. This argument is in good accord with the observation that H bonded to a bare Si surface is more easily dislodged by the beam in an electron microscopy exposure than is D.[25]

Chemical models, however, deserve some consideration. The H or D might not be dislodged directly by impact-stimulated lattice modes, but rather, by some sequential

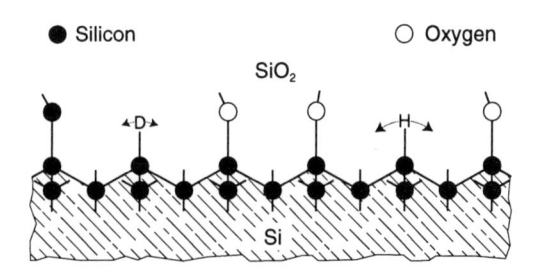

Figure 10. Proposed model for reduced generation of deuterium-passivated interface traps by hot-electron injection compared to hydrogen-passivated. Note the reduced amplitude of bending or "wagging" motion for D versus H.

chemical mechanism like that for radiation damage. The converse of an NBTI-like reaction, Eq. 1, might apply, even though there is no D effect on the NBTI itself. A tunneling step could make the deuteron much harder to dislodge than the proton; a diffusion-controlled step would be less dramatic. The limited data do not allow any firm conclusion on these possibilities, or on whether "physics" or "chemistry" approaches are more pertinent.

3. Hydrogen Disposition in Other Materials

3.1 BULK SILICA

Some relevant studies on quartz and fused silica well deserve mention here. Hydrogen in bulk silica[27] eventually exists only in a molecular or other bound form, mostly as silanol (OH) or adsorbed H_2O. Free molecular H_2O exists in bulk SiO_2 if hydrogenous species are present in quite large concentration, or if there are pore spaces larger than the usual interatomic voids, and with suitable pretreatment. Single H_2O molecules isolated in interstitial sites are not seen by IR spectroscopy, and must be rare. Most of the water is normally bound as silanol or OH groups on Si atoms which are not 4-fold coordinated to oxygens. The silanols may be either isolated—standing alone, or vicinal—situated adjacent to another silanol. Vicinal silanols comprise chemisorbed H_2O in silica. Heating between 400°K and 800°K liberates H_2O from two OH, and the remaining oxygen forms a bonding siloxane (≡Si–O–Si≡) bridge. In contrast, isolated silanols are very difficult to dislodge, requiring temperatures of 1200°K; and then they liberate protons or H atoms, not the OH group. Several dispositions of H-species in silica are shown in Fig. 11.

Excess molecular H_2O has been found to attach to isolated or vicinal silanols, forming physisorbed H_2O; such water is removed at temperatures below 400°K, with E_a of 0.5 eV.[27] A second layer of water can attach strongly to this first layer, with a higher activation

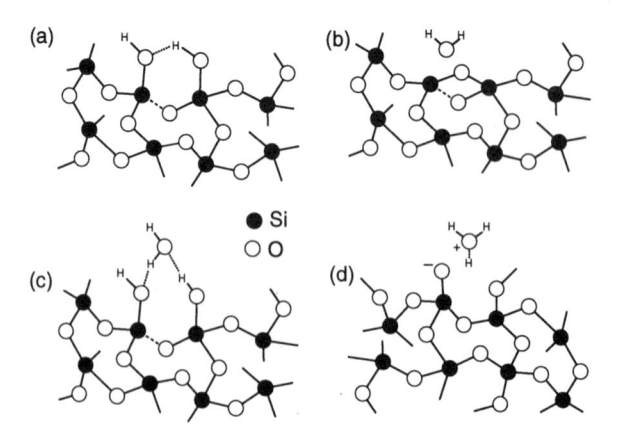

Figure 11. Hydrogenous species in silica. (a) Hydrogen-bonded vicinal silanols. (b) H_2O released from vicinal pair; formation of siloxane linkage, ≡Si–O–Si≡. (c) H_2O H-bonded to silanols = physisorbed H_2O. (d) Abstraction of silanol H⁺ by H_2O to form hydronium ion H_3O^+.

energy for release, reflecting a tendency to form attached or independent clusters of H_2O molecules. Resultant clusters and even microdroplets are seen in quartz by IR spectroscopy.[28] The spectral lines are very broad, due to the dynamically changing O–H bond lengths and strengths in liquid water. The tendency to cluster portends the greater exothermicity of chemical reactions whose products can include such clusters, as the NBTI.

On quartz surfaces (internal and external) in wet ambients, some silanol protons are dislodged and join H_2O molecules to form H_3O^+. The resultant acidic fraction in turn gives rise to self-destructive chemical weakening of the SiO_2 network (and eventual local dissolution). Molecular orbital study indicates that the H_3O^+ attaches to a Si–O–Si bridge oxygen, causing a lengthening and weakening of the Si–O bonds.[29] Any preexisting bond strain of only a few percent renders the Si–O bond more susceptible to such chemical attack.[30] These chemical features corroborate the observed susceptibility of the usually H-rich oxide near the Si-SiO_2 interface to radiation-induced E'-center formation, and perhaps also P_b center formation.

In regard to the atomic H problem, there is no direct spectroscopic evidence on possible adsorption of a bare proton to a bridge O between two Si atoms in quartz or fused silica.[28] Further, molecular orbital theory noted above indicates that H^+ can attach to the bridge oxygen only if bonded in a hydronium ion, or with a water molecule closely adjacent. In another study,[31] of a particular type of E' center in quartz, atomic H is found situated between two Si atoms, $O_3\equiv$Si–H....Si$\equiv O_3$. The structure is bistable, with H transferring from one Si to the other. None of these quartz findings suggests a viable harbor in thermal silica for non-paramagnetic but chemically vigorous atomic H in some guise.

3.2 A HYDROGEN-FREE SYSTEM

There is presently very high interest in integrated circuits which are built upon thin Si layers atop buried oxides. One such approach is that of "separation by implantation of oxygen," or SIMOX. The SIMOX oxide, produced in a truly water-free milieu, is perhaps the driest SiO_2 to be found anywhere; it offers a possible totally dry baseline for study of the very first stages of H or H_2O infusion and reaction in thin film silica.

4. Concluding Remarks

Despite the many H-related Si-SiO_2 studies, two crucial factors—reactant concentrations and the influence of the electrified interface—have not been give enough attention. Reaction E_a in water clusters is strongly controlled by the number of H_2O molecules present, and charged reactant concentration at the interface is drastically varied by electric field. Further, there has been too little consideration of H phenomena from other disciplinary areas, such as acid-base theory, electrochemistry, glass chemistry, and quartz mineralogy. As a result, many significant electrical and chemical effects in the Si-SiO_2 system have been but vaguely modeled. From consideration of neglected aspects, new working hypotheses for some H-related phenomena are offered here. In most cases, additional new H species seem necessary, beyond the usual H^+, H, H_2, OH, and H_2O.

Whether or not the hypotheses suggested are valid, it is hoped that they will provide a framework useful in tracking the ultimate explanations.

5. References

1. Deal, B. E., Sklar, M., Grove, A. S., and Snow, E. H. (1967) Characteristics of the surface-state charge (Q_{ss}) of thermally oxidized silicon, *J. Electrochem. Soc.* **114**, 266-274.
2. Gerardi, G. J., Poindexter, E. H., Caplan, P. J., Harmatz, M., Buchwald, W. R., and Johnson, N. M. (1989) Generation of P_b centers by high electric fields: Thermochemical effects, *J. Electrochem. Soc.* **136**, 2609-2614.
3. Blat, C. E., Nicollian, E. H., and Poindexter, E. H. (1991) Mechanism of negative-bias-temperature instability, *J. Appl. Phys.* **69**, 1712-1720.
4. Gerardi, G. J., Poindexter, E. H., Harmatz, M., Warren, W. L., Nicollian, E. H., and Edwards, A. H. (1991) Depassivation of damp-oxide P_b centers by thermal and electric field stress, *J. Electrochem. Soc.* **138**, 3765-3770.
5. Marcus, R. A. (1960) Exchange reactions and electron transfer reactions including isotopic exchange, *Discuss. Faraday Soc.* **29**, 21-31.
6. Conway, B. E. (1981) *Ionic Hydration in Chemistry and Biophysics*, Elsevier, Amsterdam.
7. Revesz, A. G. (1977) Chemical and structural aspects of the irradiation behavior of SiO_2 films on silicon, *IEEE Trans. Nucl. Sci.* **NS-24**, 2102-2107.
8. Stahlbush, R. E., Edwards, A. H., Griscom, D. L., and Mrstik, B. J. (1993) Post-irradiation cracking of H_2 and formation of interface states in irradiated metal-oxide-semiconductor field-effect transistors, *J. Appl. Phys.* **73**, 658-667.
9. Edwards, A. H., unpublished results.
10. Li, Z., Fonash, S. J., Poindexter, E. H., Harmatz, M., Rong, F., and Buchwald, W. R. (1990) Hydrogen anneal of E' centers in thermal SiO_2 on Si, *J. Noncryst. Solids* **126**, 173-176.
11. DoThanh, L., and Balk, P. (1988) Elimination and generation of Si-SiO_2 interface traps by low-temperature hydrogen annealing, *J. Electrochem. Soc.* **135**, 1797-1801.
12. Cartier, E., Stathis, J. H., and Buchanan, D. A. (1993) Passivation and depassivation of silicon dangling bonds at the Si/SiO_2 interface by atomic hydrogen, *Appl. Phys. Lett.* **63**, 1510-1512.
13. Brower, K. L. (1990) Dissociation kinetics of hydrogen-passivated (111)Si-SiO_2 interface defects, *Phys. Rev. B* **42**, 3444-3453.
14. Reed, M. L., and Plummer, J. D. (1988) Chemistry of Si-SiO_2 interface trap annealing, *J. Appl. Phys.* **63**, 5776-5793.
15. Razouk, R. R., Lie, L. N., and Deal, B. E. (1981) Kinetics of high pressure oxidation of silicon in pyrogenic steam, *J. Electrochem. Soc.* **128**, 2214-2220.
16. Brower, K. L., Lenahan, P. M., and Dressendorfer, P. V. (1982) Defects and impurities in thermal oxides on silicon, *Appl. Phys. Lett.* **41**, 251-253.
17. Tsai, T. E., Griscom, D. L., and Friebele, E. J. (1989-II) Medium-range structural order and fractal annealing kinetics of radiolytic atomic hydrogen in high-purity silica, *Phys. Rev. B* **40**, 6374-6380.
18. Markes, M. E., and Halliburton, L. E. (1979) Defects in synthetic quartz: Radiation-induced mobility of interstitial ions, *J. Appl. Phys.* **50**, 8172-8180.
19. Johnson, N. M., Biegelsen, D. K., and Moyer, M. D. (1981) Low-temperature annealing and hydrogenation of defects at the Si-SiO_2 interface, *J. Vac. Sci. Technol.* **19**, 390-394.
20. Caplan, P. J., and Johnson, N. M., unpublished results.

21. Estreicher, S. K., and Maric, D. M. (1993) What is so strange about hydrogen interactions in germanium? *Phys. Rev. Lett.* **70**, 3963-3966.
22. Van de Walle, C. G., Denteener, P. J. H., Bar-Yam, Y., and Pantelides, S. T. (1989-II) Theory of hydrogen diffusion and reactions in crystalline silicon, *Phys. Rev B* **39**, 10791-10808.
23. Johnson, N. M., Herring, C., and Van de Walle, C. G. (1994) Inverted order of acceptor and donor levels of monatomic hydrogen in silicon, *Phys. Rev. Lett.* **73**, 130-133.
24. Saks, N. S., and Brown, D. B. (1993) The role of hydrogen in interface trap creation by radiation in MOS devices—a review, in C. R. Helms and B. E. Deal (eds.), *The Physics and Chemistry of SiO_2 and the Si-SiO_2 Interface*, Plenum Press, New York, p. 455-463.
25. Lyding, J. W., Hess, K., and Kizilyalli, I. C. (1996) Reduction of hot electron degradation in metal oxide semiconductor transistors by deuterium processing, *Appl. Phys. Lett.* **68**, 2526-2528.
26. Van de Walle, C. G. (1996) Comment on "Reduction of hot electron degradation in metal oxide semiconductor transistors by deuterium processing," *Appl. Phys. Lett.* **69**, 2441.
27. Iler, R. K. (1979) *The Chemistry of Silica*, John Wiley and Sons, New York.
28. Kronenberg, A. K. (1994) Hydrogen speciation and chemical weakening of quartz, in P. J. Heaney, C. T. Prewitt, and G. V. Gibbs (eds.) *Silica: Physical Behavior, Geochemistry, and Materials Applications*, Mineralogical Society of America, Washington, p. 123-176.
29. Xiao, Y., and Lasaga, A. C. (1994) *Ab initio* quantum mechanical studies of the kinetics and mechanisms of silicate dissolution: $H^+(H_3O^+)$ catalysis, *Geochim. et Cosmochim. Acta* **58**, 5379-5400.
30. Heggie, M., and Jones, R. (1987) Density functional analysis of the hydrolysis of Si–O bonds in disiloxane. Application to hydrolytic weakening in quartz, *Phil. Mag. Lett.* **55**, 47-51.
31. Isoya, J., Weil, J. A., and Halliburton, L. E. (1981) EPR and *ab initio* SCF-MO studies of the Si·H–Si system in the E_4' center of α-quartz, *J. Chem. Phys.* **74**, 5436-5448.

THE ROLE OF HYDROGEN IN THE FORMATION, REACTIVITY AND STABILITY OF SILICON (OXY)NITRIDE FILMS

F.H.P.M. HABRAKEN, E.H.C. ULLERSMA, W.M. ARNOLDBIK
Atomic and Interface Physics, Debye Institute,
University of Utrecht
P.O. Box 80.000, 3508 TA Utrecht, The Netherlands
and
A.E.T. KUIPER
Philips Research Laboratories
Prof. Holstlaan 4
5656 AA Eindhoven, The Netherlands

ABSTRACT.
In the last decennium it has become clear how hydrogen and hydrogenated gases are involved in the formation of thin dielectric nitride and oxynitride films. As a consequence hydrogen is incorporated in the deposited or grown films, where it plays a role in their physical, chemical and electrical reactivity and stability. Hydrogen is able to migrate in and desorb from the films via several mechanisms. These mechanisms are concisely reviewed. We consider the processes of wet oxidation and nitridation in the Si-O-N system as two manifestations of a single chemical reaction system. In this system hydrogen stabilizes intermediate reaction products, allowing multi-step reactions to proceed. Interruption of the process or, more specifically, isolation of the intermediate species from the reactants results in incorporation of these hydrogenated intermediates in the material. It appears that the reactivity of oxynitrides strongly increases for increasing O/N concentration ratio of the material. The use of isotope-sensitive high-energy ion beam methods is emphasized.

1. Introduction

Deposited silicon nitride and oxynitride films find their most important applications as electrical and chemical isolation layers in silicon-based very-large-scale integrated (VLSI) technology. Hydrogenated amorphous insulator and semiconductor films are applied in other type of devices like solar cells and thin film transistors. The interest in the role of hydrogen in these silicon-related amorphous semiconductor and insulator films stems from three interconnected considerations. First, the films are very often manufactured in a process where hydrogen containing gases are used and this mere fact causes hydrogen to be incorporated in the material. Second, the presence of hydrogen in these materials is usually associated with deviations from the "ideal" amorphous network structure, and consequently, a relation between defects and hydrogen is often emphasized. Third, and may be much more important, the presence of hydrogen in these materials is a major cause of their instability, because hydrogen can migrate in and out the films at relatively low temperatures, which may result in detrimental macroscopic and microscopic defects. These considerations have stimulated numerous studies of the preparation of these films and of the physical and chemical behaviour of hydrogen therein, and its relation with electrical and other properties of the material. It is the main objective of this paper to present an illustration of the first and the third phenomenon mentioned. For a more detailed recent review the reader is referred to [1].

In the overview we make a distinction between different kinds of layer manufacture. First we consider material which is deposited from the gas phase in a chemical vapour process (CVD) at high temperature or at low temperature, where a plasma is used to activate the reactant gas mixture (plasma enhanced CVD (PECVD)). Second we discuss thermally grown material. In that case layer growth occurs by diffusion of one or more of the reactants through the solid. Here we will emphasize the thermal nitridation of SiO_2 films on silicon and the process of thermal oxidation of silicon nitride and (oxy)nitride films, denoted by SiO_xN_y (or $SiO_xN_yH_z$, if a substantial amount of hydrogen is present in the material). In this class of materials one can span the entire range from "pure" silicon nitride (Si_3N_4) to silicon dioxide (SiO_2). The oxynitride composition is characterized by their oxygen-to-nitrogen (O/N or O/(O+N)) concentration ratio.

The paper is organized as follows. At first we briefly mention the high-energy ion-beam analysis techniques to determine quantitatively layer thicknesses and compositions with an emphasis on hydrogen and deuterium depth profiling. Subsequently, we disuss the composition of low and high temperature CVD oxynitride films. We devote a longer section to the behaviour of hydrogen in the material during annealing at temperatures above the temperature of deposition. Here we mention several mechanisms of hydrogen desorption from the material and discuss the experiments from which the existence of these mechanisms has been inferred. Finally, we discuss the role of hydrogen in the conversion of thermally grown oxide to oynitrides and the oxidation of CVD (oxy)nitrides.

2. Layer characterization and hydrogen depth profiling.

High-energy (MeV) ion beam techniques are especially useful because of their ability to reveal reliably absolute concentration depth profiles and abundancies of elements on a substrate surface. Among these techniques Rutherford Backscattering Spectrometry is the best known and most widely applied technique. However, its poor sensitivity for light elements on a substrate, which contains a significant concentration of heavier elements, and its inherent impossibility to detect hydrogen on a surface makes this method less suitable for our purposes. In contrast the complementary technique of Elastic Recoil Detection (ERD) is very well suited to determine hydrogen, deuterium, oxygen and nitrogen depth profiles These elements and isotopes are relevant for the work described here.

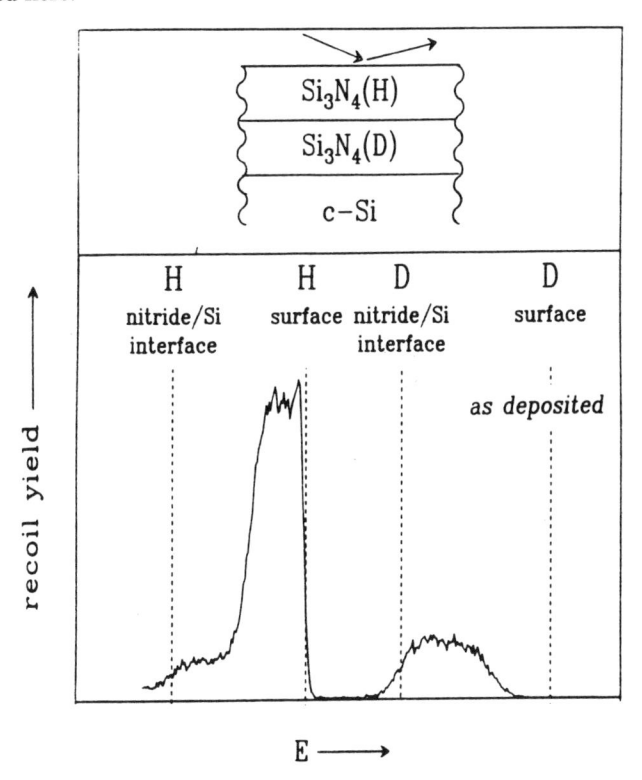

Figure 1. ERD spetrum of an as deposited LPCVD silicon nitride double layer structure. The top layer contains hydrogen and the bottom layer contains both hydrogen and deuterium. The ERD conditions were: primary beam: 10 MeV Si, angle of incidence with surface plane: 28°, recoil angle 33°, Mylar foil thichness: 6 μm. The surface and interface positions of H and D are indicated.

In brief: a parallel beam of MeV (preferably heavy) ions hit the material to be investigated. A very few of the ions recoil a nucleus residing in the sample surface region in a binary elastic scattering process in the direction of a detector. This detector counts the recoil particles and measures their energy. The quantitativity of the method stems from the fact that the cross section of the (Rutherford) scattering process is precisely known and not influenced by chemical environment and other matrix effects. Depth resolving power is due to the (reasonably well) known stopping power of ions in solids. [2]

In the most simple set up, the scattered primary particles are prevented from reaching a (silicon surface barrier) detector by an absorber foil of several μm thickness in front of it. In this case identification of the various recoils is done by considering their energy. This procedure does sometimes lead to ambiguous results. In more advanced detection systems the recoil particles are identified explicitly in a transmission telescope, magnetic spectrometer, or a time-of-flight device. In this way one ideally obtains background free separate depth profiles of the various isotopes and elements. For a review on ERD, see [3].

If one is interested in H and D profiles only, the simple set up satisfies. Figure 1 shows that relevant H and D profiles are obtained using a beam of 10 MeV ^{28}Si projectiles, a recoil angle of 33° and a Mylar absorber foil thickness of 6 μm thickness. In these conditions the probe depth amounts to 150-200 nm and the depth resolution at the surface is 5 nm. At a depth of 100 nm it is estimated to amount to 10 nm.

Excellent hydrogen profiles are also obtained by performing Nuclear Reaction Analysis (NRA) using the nuclear reaction $^1H(^{15}N,\alpha\gamma)^{12}C$ [4]. However, the use of this technique is impeded when also deuterium is present in the surface region of a material because of interfering nuclear reactions of D with the ^{15}N projectile.

In the nitrides and oxynitrides hydrogen is mostly bonded to Si or N. Information about these bonds is gathered using (Fourier Transform) Infrared absorption Spectroscopy.

3. Low temperature plasma CVD silicon oxynitrides

Low temperature silicon nitride films are obtained in a plasma of SiH_4 and NH_3, and sometimes N_2 at temperatures between 150-500°C. The amount of incorporated hydrogen strongly depends on the deposition temperature. It can be as high as 39 at.% at a deposition temperature of 150°C [5], but is lower at higher deposition temperature [6,7]. The hydrogen content may be further lowered by excitation of only one of the reactants e.g. NH_3 or N_2/He [8]. Hydrogen is bonded to N and Si. The concentration ratio Si-H/N-H has been deduced to be a unique function of the Si/N ration in the nitride films: for nitrogen-rich films, most of the hydrogen is bonded to nitrogen, whereas for nitrogen-poor films most of the incorporated H is found in Si-H configurations [6].

Low temperature silicon oxynitrides are usually grown by mixing N_2O into the PECVD nitride reactant mixture at the expense of the ammonia input [9]. The reactant gas flow ratio $R_2=(N_2O+NH_3)/SiH_4$ appears to control the Si-H concentration in the grown films: if this ratio is large (i.e. ≥40) the Si-H bond concentration is below 1 at.%. However, the decrease of the Si-H bond concentration with increasing R_2 is accompanied by an increase of the N-H concentration, such that the total H concentration remains approximately constant.

The ratio Si-H/N-H is important for the thermal stability of oxynitrides. It has been shown that oxynitrides, which contain only a small concentration of Si-H bonds (i.e. ≤1at.%) remain stable during annealing at temperatures up to 850°C, whereas samples which contain hydrogen in both Si-H and N-H configurations in significant amounts already loose a part of their hydrogen during annealing at 500°C [10]. This difference was ascribed to a cross linking effect: the mutual presence of Si-H and N-H bonds in significant concentrations makes the following reactions possible at already low temperatures:

$$\text{Si-H} + \text{N-H} \Rightarrow \text{Si-N} + \text{H}_2 \qquad (1a)$$
$$\text{Si-H} + \text{Si-H} \Rightarrow \text{Si-Si} + \text{H}_2 \qquad (1b)$$

Evidence for these cross linking reactions has been found in the increase of the Infrared absorbance in the Si-N region as a result of the annealing [10]. It is not yet completely clear what the fate is of the H_2 molecules formed in the cross linking reaction. These films blistered and/or cracked during the anneals, so it is plausible that the H_2 molecules as such desorb from the film.

This mechanism of hydrogen loss as a result of cross linking and molecular outdiffusion of hydrogen without retrapping has recently been evidenced to occur in plasma deposited hydrogenated a-SiC films [11]. The basis of the experiments was that H and D were spatially separated in as deposited films. The evidence was obtained by performing thermal desorption spectroscopy (TDS) and ERD on double layer structures of hydrogenated and deuterated silicon carbide films (fig. 2). In the TDS spectra at low temperature H_2 and D_2 were simultaneously detected in the gas phase but the amount of HD remained low initially. The ERD spectra indicated that the extent of interdiffusion f H and D at the considered temperature was negligible. The conclusion therefore is that H_2 (and D_2) molecules are formed from 2 H (D) atoms, both bonded in the same small volume, and desorb without being retrapped, irrespective of the depth in the film from which they originate. This type of experiments has not yet been performed (to our knowledge) in low temperature (oxy)nitride films, but they certainly will elucidate this matter.

In the oxynitrides, which contain only N-H bonds, hydrogen apparently is not able to disappear from the film as a result of cross linking. This is plausible since the reaction:

$$\text{Si-N-H} + \text{Si-N-H} \Rightarrow \text{Si-N-N-Si} + \text{H}_2 \qquad (1c)$$

is probably energetically not favourable. In these films another mechanism of hydrogen loss is operative, notably at higher temperatures (see section 4). From experiments in which [12] the oxynitride films were annealed in a D_2/N_2 mixture at various temperatures it has been deduced that the activation energy for the exchange reaction

$$D_2 + =\text{N-H} \Rightarrow =\text{N-D} + \text{HD} \qquad (2)$$

amounts to about 1.5 eV, irrespective of the O/N ratio of the PECVD oxynitride. The rate of this exchange reaction increases with increasing O/N concentration ratio for O/N>0.5. This is the first example of the increased reactivity of oxynitrides for O/N>0.5. From the same experiments it was deduced that the hydrogen diffusion rate strongly increases with increasing O/N ratio. This represents the second example of the increased reactivity with increasing O/N for O/N>0.5 (fig. 3). After a pre-anneal at 1000 °C, when most of the hydrogen was escaped, the diffusion rate appeared reduced, but the O/N dependence was preserved.

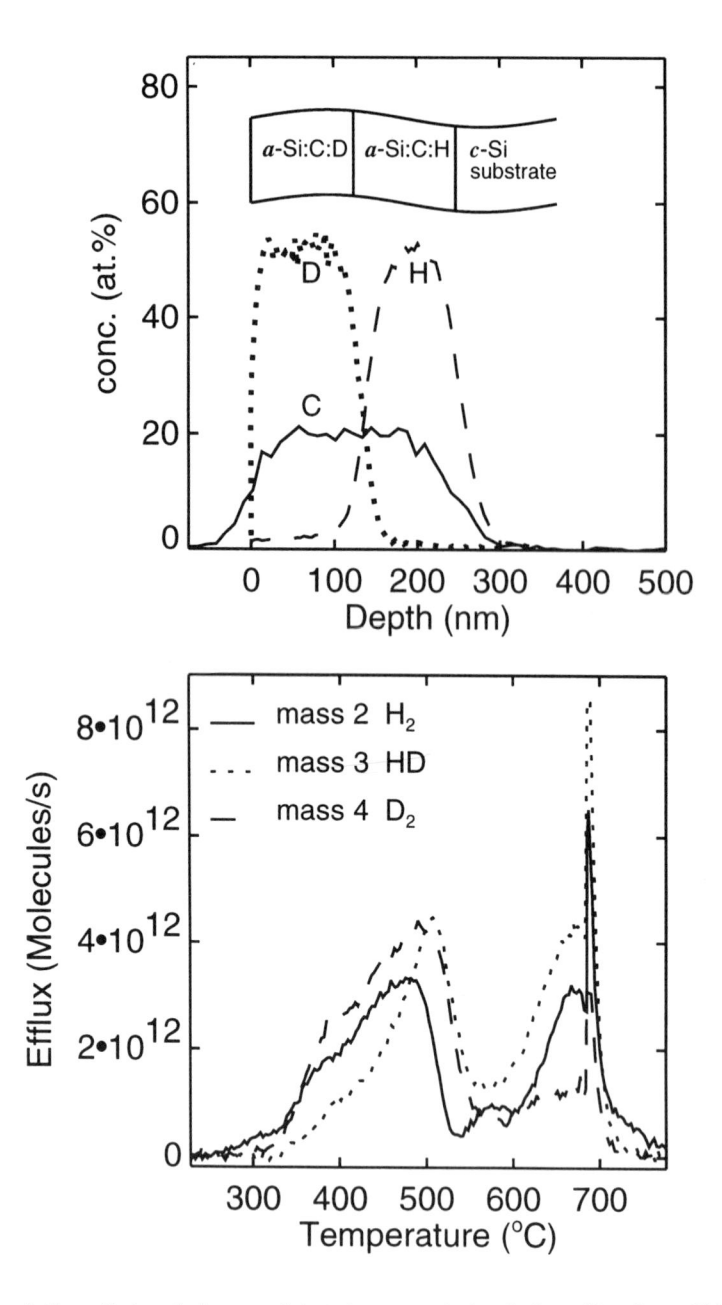

Figure 2. Top: Carbon, hydrogen and deuterium concentration depth profiles of an a-Si:C:H/a-Si:C:D/ c-Si layer structure. Down: Thermal desorption spectrum of the layer structure. The temperature ramp amounted to 0.33°C/sec. From [11].

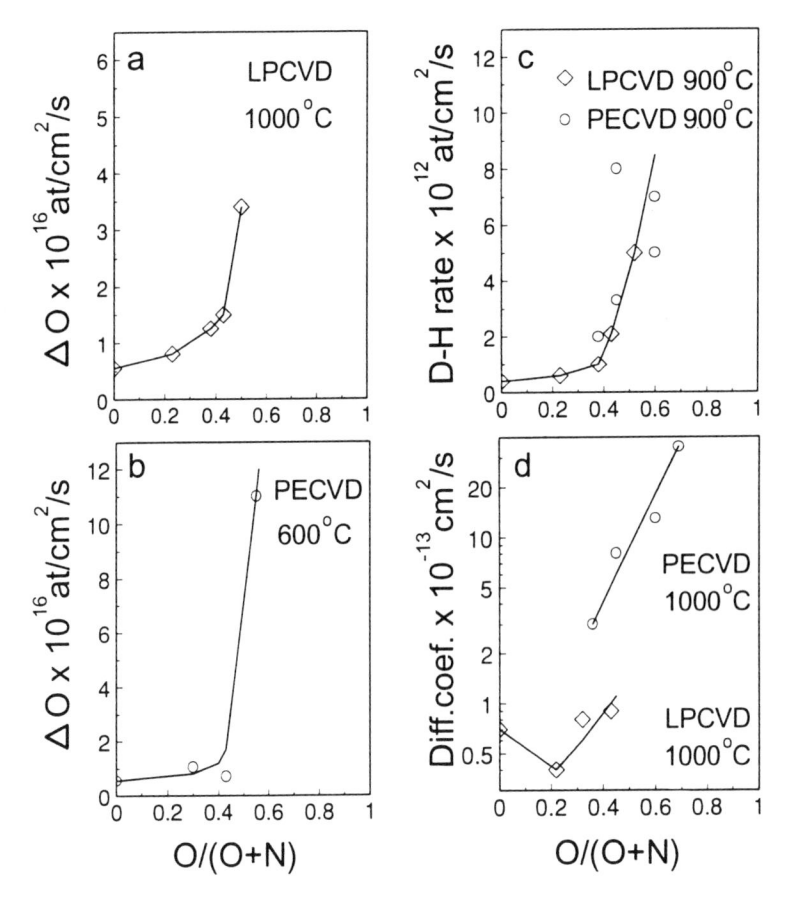

Figure 3. Oxidation rate of (a) LPCVD and (b) PECVD oxynitrides, (c) D_2/N-H exchange rates, normalized to correct for the difference in D concentration for the various samples and (d) hydrogen diffusion coefficients for LPCVD and PECVD oxynitrides with =N-H bonds only. Reprinted from [1].

4. High temperature CVD silicon oxynitrides

High temperature oxynitride films are usually prepared in a Low pressure CVD (LPCVD) process from SiH_2Cl_2, NH_3 and N_2O at temperatures above 750°C. They contain 1-5 at% hydrogen, depending on the O/N ratio and on the temperature. It is mainly bonded to N and it originates from the NH_3 [13]. When these layers are annealed above the deposition temperature hydrogen is released from these films. The mechanism is however different from the cross linking and molecular diffusion process, as described in section 3.

We have deposited double layers of (oxy)nitrides in a LPCVD process at 820°C. The double layers were grown subsequently from NH_3 and ND_3. Also the order of growth was reversed [14]. Figure 4 shows the total amount of D in the layer structure after annealing at 1000°C for various times. We note that deuterium, when it is present in the lower layer, needs some anneal period, before its amount starts to decrease. This is in contrast to the situation that D is in the top layer. What happens becomes clear if one considers the ERD H and D depth profiles (fig.5). D and H diffuse across the interface in the double layer structure and D starts to disappear as soon as it reaches the surface. The absolute value of the diffusion coefficient does depend on the O/N ratio. It is in the order of 10^{-13} cm^2/sec at 1000°C [14]. The interdiffusion process is characterized by an activation energy of about 3 eV. This value of 3 eV is close to the bond energy of N-H bonds, which has led to the conclusion that N-H bond decomposition is the rate limiting step in the diffusion process.

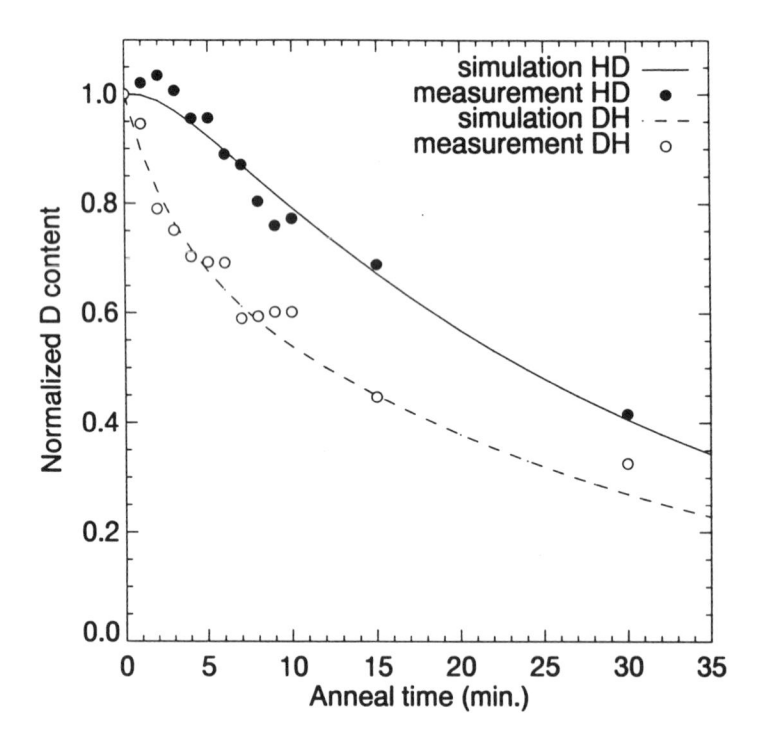

Figure 4. Total deuterium content of LPCVD Si$_3$N$_4$ H and D double layer structures. Solid dots: ND$_3$ grown layer between substrate and NH$_3$ grown layer. Open dots: NH$_3$ grown layer between substrate and ND$_3$ grown layer. Reprinted from [12].

Subsequent to the bond breaking, the hydrogen (deuterium) atom becomes trapped in a nitrogen-related trapping site or exchanges with another nitrogen bonded hydrogen or deuterium atom. It has been suggested that if the bond breaking occurs within a distance of bout 10 nm from the immediate surface, the hydrogen atom is able to desorb

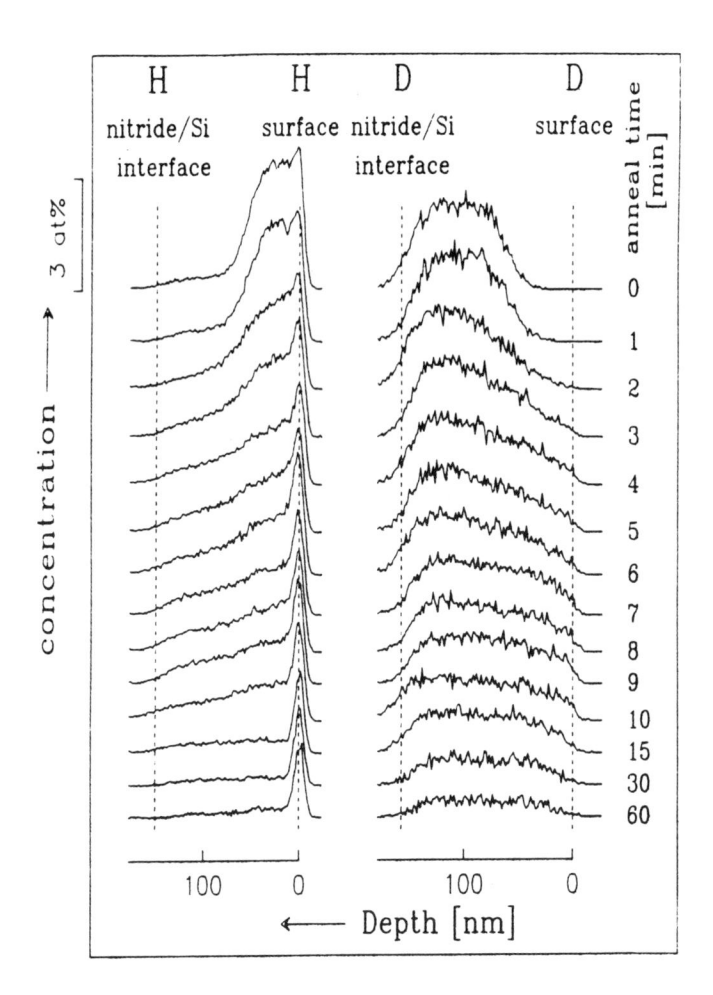

Figure 5. Hydrogen and deuterium concentration depth profiles of the LPCVD silicon nitride sample having the D layer below the H layer, after annealing for the indicated period at 1000°C.

into the gas phase. A SiO$_2$ capping layer is not able to prevent the desorption [14]. The observed diffusion rates are such that at the deposition temperature of 820°C some diffusion must have occurred. This then may be a cause of the defective nature of the LPCVD silicon (oxy)nitride films, since it is expected that as a result of the H desorption N (and some Si) dangling bonds arise.

The diffusion coefficient exhibits a minimum for O/N values around 0.4; it strongly increases for O/N > 0.4-0.5. This is a manifestation of the second example of an increased reactivity for O/N>0.4-0.5 (fig. 3). Annealing in D$_2$ gas of oxynitrides containing hydrogen results in exchange of D and H. The extent of exchange parallels the diffusion coefficient: it also has a minimum at O/N = 0.4 and increases strongly for O/N>0.4 (see also section 3). Consistently, in an earlier study it has been established that around O/N=0.4-0.5 the hydrogen is the most stable in LPCVD silicon oxynitride films [13].

5. Thermal nitridation of SiO$_2$ in NH$_3$.

The formation of thermal (oxy)nitride layers in the temperature range 800-1200°C has been studied in the hope that one could combine the beneficial properties of thermally grown SiO$_2$ (low interface state density) and of Si$_3$N$_4$ layers (high dielectric constant, chemical integrity, diffusion barrier characteristics). Nitrogen is incorporated in SiO$_2$ films on Si via annealing in NH$_3$ [15,16]. In this process N is accumulated predominantly in the regions at the surface and at the interface. Along with N, hydrogen is incorporated in concentrations of a few at.%, depending a.o. on the temperature and duration of nitridation. It is more or less homogeneously distributed over the dielectric film. The presence of H, in the form of NH$_3$, is essential for N incorporation in SiO$_2$ [1]. The reaction mechanism of the NH$_3$ process can be represented as follows [1]:

$$\text{Si-O-Si} + \text{NH}_3 \Rightarrow \text{Si-OH} + \text{Si-NH}_2 \Rightarrow \text{Si-(NH)-Si} + \text{H}_2\text{O} \qquad (3)$$

The main point here is that N is incorporated first as NH$_x$ and that H$_2$O or OH is a reaction product, which may escape in the NH$_3$ ambient, or may oxidize the Si substrate. Indeed, some authors have demonstrated the presence of N-H groups in nitrided oxides [17]. In this view, the presence of H in the entire film indicates that the reaction volume spans the entire film. The bonding of O and N to one or two H atoms keeps both elements mobile in the network, which is essential in the mechanism of nitridation beyond a surface layer [18,19]. As evidence for this mobility, Serrari *et al* [20] have elegantly shown that, during nitridation of SiO$_2$ in NH$_3$, oxygen species also diffuses through the oxide layer. Nitrogen atoms become immobilized when they are two- or three-fold coordinated with Si. Similar deductions have been made on the basis of isotopic tracer experiments in the recent work of Baumvol *et al* [21,22,23].

The (oxy)nitride structures, that develop at the outer interfaces of the oxide during nitridation are constituting diffusion barriers for NH$_3$ and H$_2$O, thereby slowing down the rate of uptake of N. It has been argued that the most nitrogen-rich regions in the dielectric have the stoichiometry of about Si$_2$N$_2$O [24]. When this concentration has been reached the nitrogen uptake virtually stops. We conclude that the reactivity of oxynitrides towards NH$_3$ is appreciable for O/N>0.5, but small for O/N<0.5.

6. Thermal oxidation of LPCVD silicon nitride and oxynitride.

The study of the oxidation of LPCVD (oxy)nitrides has been motivated by its occurrence during the LOCOS isolation processing scheme in VLSI technology [25]. A report of an extensive study of this process for the temperature range 800-1000°C is found in [26]. In the process of oxidation of silicon (oxy)nitrides the SiO$_2$ layer grows in a layer-by-layer fashion on top of the (oxy)nitride. The importance of the role of hydrogen becomes immediately clear if one considers the rate of oxidation as a function of the H_2O/O_2 gas ratio. If there is no water in the gas phase, virtually no oxide formation occurs, in the considered temperature range. The oxidation rate increases approximately linearly with the water content of the oxidizing atmosphere [26]. A further important observation concerns the hydrogen pile up at the SiO_2/oxynitride interface (see figure 6). This peak is ascribed to intermediates in the reaction sequence:

$$Si_3\text{-}N + H_2O \Rightarrow Si\text{-}OH + Si\text{-}(NH)\text{-}Si \qquad (4a)$$
$$Si\text{-}(NH)\text{-}Si + H_2O \Rightarrow Si\text{-}NH_2 + Si\text{-}OH \qquad (4b)$$
$$Si\text{-}NH_2 + H_2O \Rightarrow NH_3 + Si\text{-}OH \qquad (4c)$$
$$Si\text{-}OH + Si\text{-}OH \Rightarrow Si\text{-}O\text{-}Si + H_2O \qquad (4d)$$

Thus NH_3 is an end product of the oxidation reaction. Experimental evidence for this deduction is found in [1].

The rate of oxidation increases with O/N ratio of the oxynitride, especially for O/N>0.5 (see figure 3), concomitant with a decrease of the apparent activation energy from 2 eV for O/N<0.5 to 1.2 eV at O/N=0.7. This is our final example of the increase of reactivity of oxynitrides for O/N>0.5 (fig. 3).

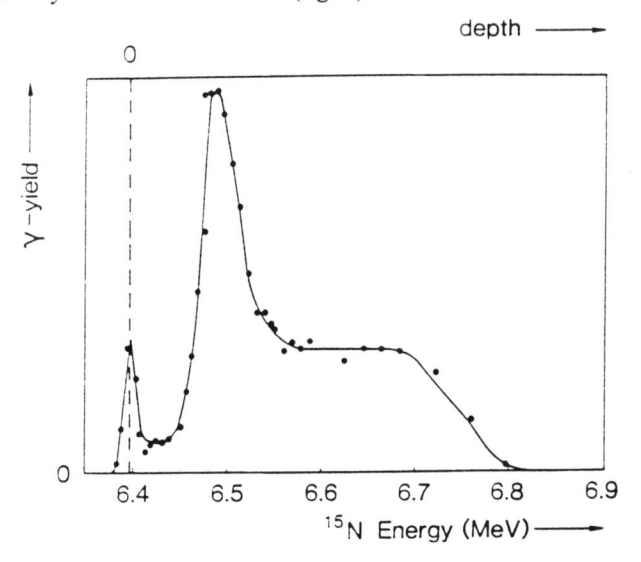

Figure 6. NRA obtained hydrogen concentration depth profile in a 150 nm thick oxynitridide film (O/N=0.7) after wet oxidation at 950°C for 9 hrs. The small peak near 6.4 MeV corresponds to the sample surface, the larger one around 6.5 MeV to the oxide/oxynitride interface. Reprinted from [26]

Since hydrogen apparently is indispensable for the oxidation to occur it is interesting to study the oxidation of low temperature oxynitrides. It appears [27] that oxynitrides which contain mainly H and N in N-H groups, oxidize at a high rate, with an estimated low activation energy of 0.4 eV. This is explained by noting that a =N-H group is linked in the amorphous network to two neighbouring Si atoms, a situation that is structurally comparable to an -O- atom in SiO_2. If eq. 4a indeed represents the first step in the oxidation of nitride, then this step can be omitted at the location of a =N-H group, which then forms a weak spot in the network for oxidation. Support for this model comes from the work of Chiang *et al* [28]. They have found that the rate of hydrolysis correlates with the =N-H concentration in PECVD nitride (They have also found that oxygen proved ineffective as oxidizing agent and that NH_3 is a product of the hydrolysis reaction). So we infer that eq. 4a represents very probably the rate limiting step in the wet oxidation of high temperature (oxy)nitride films.

7. Epilogue

We have indicated that several mechanisms are possible for the outdiffusion of hydrogen. It has been shown that the reactivity (including the diffusion rate) in the various types of (hydrogenated) oxynitrides shows a strong increase for the O/N concentration ratio larger than 0.5. A clear explanation for this effect is not yet available. In the nitridation and oxidation reactions the central role of Si-(NH)-Si groups has been identified. This specific group may be considered as an iso-structural analogue of the Si-O-Si group of SiO_2. This consideration may be a clue for the understanding of the O/N dependence of the nitridation and oxidation reactions. In both versions of the same reaction system hydrogen stabilizes intermediate reaction products, allowing multi-step reactions to proceed. In this sense hydrogen profiling reveals where reactions take place: in the nitridation of SiO_2 the entire film is involved, whereas in the oxidation of (oxy)nitrides the reactions take place at the oxide/oxynitride interface, which gives rise to an H interface pile up. This on its turn reflects again the structural aspects in the Si-O-N system: the dense structure of Si_3N_4 poses a diffusion barrier for oxidizing species, which causes reactions to take place at the interface and provokes a layer-by-layer type of oxide growth. In contrast, SiO_2 has an open structure, which apparently is rather reactive, such that fast reactive diffusion of NH_3 in the film is readily possible. The oxynitride composition close to Si_2N_2O (O/N=0.5) marks the transition from the nitride to the oxide structure [1].

8. References

1. Habraken, F.H.P.M. and Kuiper, A.E.T. (1994) Silicon nitride and oxynitride films, *Materials Science and Engineering* **R12**, 123-175.
2. Feldman, L.C. and Mayer, J.W. (1986) *Fundamentals of surface and thin film analysis*, Elsevier Science Publishers, Amsterdam
3. Arnold Bik, W.M. and Habraken, F.H.P.M. (1993) Elastic recoil detection, *Rep. Prog. in Phys.* **56**, 859-902.
   See also: Arnoldbik, W.M. and Habraken, F.H.P.M. (1995) Applications of elastic recoil detection in materials analysis, in *Application of particle and laser beams in materials technology*, edited by P.

Misaelides, NATO ASI series E, vol **233**, Kluwer Academic Publishers, Dordrecht.
4. Lanford, W.A., Trautvetter, H.P., Ziegler, J.F., and Keller, J., (1976) New precision technique for measuring the concentration versus depth of hydrogen in solids, *Applied Phys. Lett.* **28**, 566-568.
5. Chow, R., Lanford, W.A., Ke-Ming, W. and Rosler, R.S. (1982) Hydrogen content of a variety of plasma-deposited silicon nitrides, *J. Appl. Phys.* **53**, 5630-5633.
6. Claassen, W.A.P., Valkenburg, W.G.J.N. , Willemsen, M.F.C. and van de Wijgert, W.M. (1985) Influence of Deposition Temperature, Gas Pressure, Gas Phase Composition, and RF Frequency on Composition and Mechanical Stress of Plasma Silicon Nitride Layers, *J. Electrochem. Soc.* **132**, 893-898.
7. Cotler, T.J. and Chapple-Sokol, J. (1993) High quality plasma-enhanced chemical vapor deposited silicon nitride, *J. Electrochem. Soc.* **140**, 2071-2075.
8. Lucovsky, G., Richard, P.D., Tsu, T.V., Lin, S.Y. and Markunas, R.J. (1986) Deposition of silicon dioxide and silicon nitride by remote plasma enhanced chemical vapor deposition, *J. Vac. Sci. & Technol. A* **4**, 681-694
9. Denisse, C.M.M., Troost, K.Z., Oude Elferink, J.B., Habraken, F.H.P.M ., van der Weg, W.F. and Hendriks, M. (1986) Plasma enhanced growth and composition of silicon oxynitride films, *J. Appl. Phys.* **60**, 2536-2542.
10. Denisse, C.M.M., Troost, K.Z., Habraken, F.H.P.M., van der Weg, W.F. and Hendriks, M. (1986) Annealing of plasma silicon oxynitride films, *J. Appl. Phys.* **60**, 2543-2547.
See also: Schliwinski, H.J., Schnakenberg, U., Windbracke, W., Neff, H. and Lange, P. (1992) Thermal Annealing effects on the mechanical properties of plasma-enhanced chemical vapor deposited silicon oxide films, *J. Electrochem. Soc.* **139**, 1730
11. Ullersma, E.H.C., Inia, D.K.,van Sark, W.G.J.H.M., Habraken, F.H.P.M ., van der Weg, W.F., Westerduin, K.T. and van Veen, A. (1997) Evidence for molecular diffusion of hydrogen in a-Si:C:H films, to be published.
12. Arnoldbik, W.M., Marée, C.H.M., Habraken, F.H.P.M. (1994) Deuterium diffusion into plasma-deposited silicon oxynitride films, *Applied Surface Science* **74**, 103-113.
13. Habraken, F.H.P.M., Tijhaar, R.H.G., van der Weg, W.F., Kuiper, A.E.T. and Willemsen, M.F.C. (1986) Hydrogen in low-pressure chemical-vapor-deposited silicon (oxy)nitride films, *J. Appl. Phys.* **59**, 447-453
14. Arnoldbik, W.M., Marée, C.H.M., Maas, A.J.H., van den Boogaard, M.J., Habraken, F.H.P.M. and Kuiper, A.E.T. (1993) Dynamic behaviour of hydrogen in silicon nitride and oxynitride films made by low-pressure chemical vapor deposition, *Phys. Rev. B.* **48**, 5444-5456.
15. Ito, T., Nozaki, T. and Ishikawa, H. (1980) Direct Thermal Nitridation of Silicon Dioxide Films in Anhydrous Ammonia Gas, *J. Electrochem. Soc.* **127**, 2053-2057.
16. Habraken, F.H.P.M., Kuiper, A.E.T., Tamminga, Y. and Theeten, J.B. (1982) Thermal nitridation of silicon dioxide films, *J. Appl. Phys.* **53**, 6996-7002.
17. Koba, R. and Tressler, R.E. (1988) Thermal nitridation of SiO_2 Thin Films on Si at 1150°C , *J. Electrochem. Soc.* **135**, 144-150.
18. Vasquez, R.P. and Madhukar, A. (1986) A kinetic model for the thermal nitridation of SiO_2 /Si, *J. Appl. Phys.* **60**, 234-242.
19. Habraken, F.H.P.M. and Kuiper, A.E.T. (1990) Oxidation of silicon (oxy)nitride and nitridation of silicon dioxide: Manifestations of the same chemical reaction system?, *Thin Solid Films* **193/194**, 665-774.
20. Serrari, A., Chartier, J.L., le Bihan, R., Rigo, S. And Dupuy, J.C. (1991) Study of Atomic transport mechanism of oxygen during thermal nitridation of silicon dioxide, *Applied Surf. Sci.* **51**, 133- 138.
21. Baumvol, I.J.R., Stedile, F.C., Ganem, J.-J., Tremaille, I. and Rigo, S. (1996) Thermal nitridation of SiO_2 films in ammonia: the role of hydrogen. *J. Electrochem. Soc.* **143**, 1426-1434.
22. Baumvol, I.J.R., Stedile, F.C., Ganem, J.-J., Tremaille, I. and Rigo, S. (1996) Thermal nitridation of SiO_2 films in ammonia: isotopic tracing of nitrogen and oxygen in the initial stages. *J. Electrochem. Soc.* **143**, 2938-2945.
23. Baumvol, I.J.R., Stedile, F.C., Ganem, J.-J., Tremaille, I. and Rigo, S. (1996) Thermal nitridation of SiO_2 films in ammonia: isotopic tracing of nitrogen and oxygen in further stages and in reoxidation. *J. Electrochem. Soc.* **143**, 2946-2952.
24. Vasquez, R.P., Madhukar, A., Grunthaner, F.J. and Naiman, M.L. (1986) An x-ray photoelectron spectroscopy study of the thermal nitridation of SiO_2 /Si , *J. Appl. Phys.* **60**, 226-233.

25. Kooi, E., van Lierop, J.G. and Appels, J.A. (1976) Formation of Silicon Nitride at a Si/SiO$_2$ Interface during local Oxidation of Silicon and during Heat-Treatment of Oxidized Silicon in NH$_3$ Gas, *J. Electrochem. Soc.* **123**, 1117-1120.
26. Kuiper, A.E.T., Willemsen, M.F.C., Mulder, J.M.L., Oude Elferink, J.B., Habraken, F.H.P.M. and van der Weg, W.F. (1989) Thermal oxidation of silicon nitride and oxynitride films, *J. Vac. Sci. & Technol.* **B7**, 455-465.
27. Denisse, C.M.M., Smulders, H.E., Habraken, F.H.P.M. and van der Weg, W.F. (1989) Oxidation of plasma enhanced chemical vapour deposited silicon nitride and oxynitride films, *Applied Surf. Sci.* 39, 25-32.
28. Chiang, J.N., Ghanayem, S.G., and Hess, D.W. (1989) *Chem. Mater.* **1**, 194

HYDROGEN-INDUCED DONOR STATES IN THE MOS SYSTEM:

Hole Traps, Slow States and Interface States.

J.M.M. DE NIJS, K.G. DRUIJF[a] and V.V. AFANAS'EV[b]
DIMES
Delft University of Technology
P.O. Box 5053
2600GB Delft
The Netherlands

We propose that atomic hydrogen, trapped at regular network oxygen atoms, produces a hole trap, a slow state and an interface state. The overcoordinated oxygen configuration resulting from this interaction is associated with an electronic donor state.

Extensive studies of the metal-oxide-silicon (MOS) system have led to a broad consensus that atomic hydrogen (H^0) plays a key role in the degradation of the system.[1] Unfortunately, our understanding of H^0 in the MOS system still is rather limited. Cartier et al. have observed that exposure of the Si/SiO$_2$ system to H^0 indeed produces large numbers of interface states. However, only a fraction of these are silicon dangling bonds centers; the majority are defects with unknown microscopic structure.[2] H^0 is also directly involved in the generation of slow states, also known as "anomalous positive charge (APC) centers" or "border traps".[1,3] These slow states are related to defects in the near-interfacial region of the oxide, but like hydrogen-induced interface states their microscopic nature has not been established. McLean postulated that H^0 also exists as a proton in the oxide layer.[4] If so, H^0 must be capable of trapping a hole. Apparently, H^0 is associated with an interface state, a slow state, and with a hole trap.

In this paper we will discuss a unified model to explain this behavior of H^0. The basic concept of the model is that H^0 interacts with a network oxygen atom thus forming an overcoordinated configuration which constitutes an electrical donor state.[5-8] H^0 is not permanently trapped; its retention time depends on the charge state and on the strain in the Si-O-Si bond. H^0 is not trapped in the neutral, relaxed configuration, in agreement with the condition that it easily diffuses through the (unstrained) bulk of the oxide layer. However, since the network is strained near the Si/SiO$_2$ interface, the model implies trapping in this region. In fact, it postulates a distribution of

---

a) present address: Philips Semiconductors, 6534AE Nijmegen, The Netherlands
b) present address: Depart. of Physics, University of Leuven, B-3001 Leuven, Belgium.

retention times associated with the strain distribution. Charging the configuration positively increases the bonding energy of H, thus enhancing the stability.

For our studies we used Si(100) samples oxidized in dry O_2 at 1000°C followed by a 20min. post oxidation anneal in N_2 at the same temperature. Transparent capacitors were fabricated by resistive evaporation of a thin Al layer. The samples had 30-35nm oxides and were not subjected to a post metallization anneal. Vacuum ultraviolet (VUV) radiation (hv≈10eV) from a Kr lamp was used to generate charge carriers and H^0 in the oxide layer. After VUV exposure, the oxide was neutralized by injection of ≈10^{14} electrons cm^{-2} at 0V bias using a Hg lamp (hv<6eV).

The first observation we made was that the exposure produces unstable donor-type interface states.[5] This point is illustrated in Fig. 1A, where we show a series of CV curves recorded after exposure and neutralization of a capacitor. During the time between the CV measurements the capacitor was shorted. Immediately after the exposure, the system contains ≈4×10^{12} cm^{-2} interface states above midgap; however, this number gradually decreases. Since the oxide layer had been neutralized, the stretch-out in negative direction demonstrates that the defects are donors. Fig. 1A also reveals that the annealing rate decreases with time, which suggests that the states do not have equal lifetimes but that some are more stable than others.

Applying 0V bias causes the donor centers to be neutral. We have also studied the annealing process when the states are positively charged by applying -7V gate bias.[6,7] As seen from Fig. 1B, in this case the states are conserved. The donors anneal when neutral but not when charged.

Our model postulates that the interface states are caused by trapping of H^0 near the Si/SiO$_2$ interface. When neutral this bond is rather weak, but sufficiently strong to retain the atom for some time. Upon escape, it can be trapped by another oxygen; however, it can also dimerize or diffuse into the silicon. These latter two cases result in a net reduction of the amount of H^0, which explains the annealing effect. When positively charged, the atom

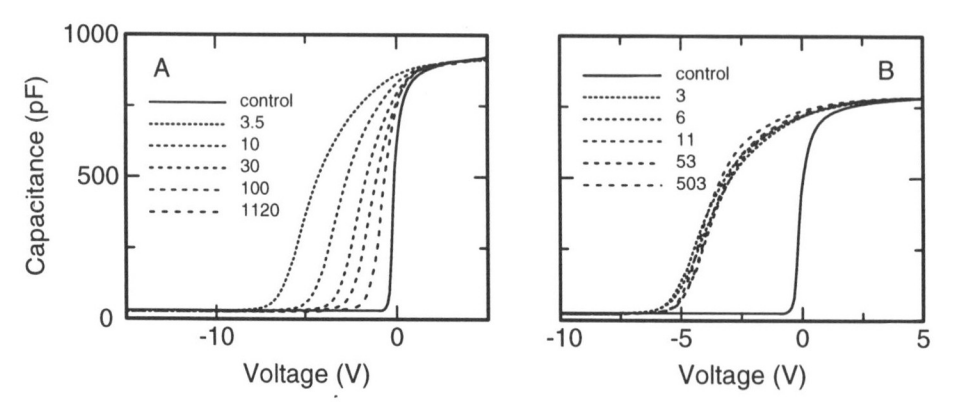

Fig.1 HF CV curves for a n-type capacitor after VUV exposure and neutralization. The curves were recorded after certain time intervals, as indicated in min., for 0V (A) and -7V (B) bias during storage. The solid lines indicate the CV curve before irradiation.

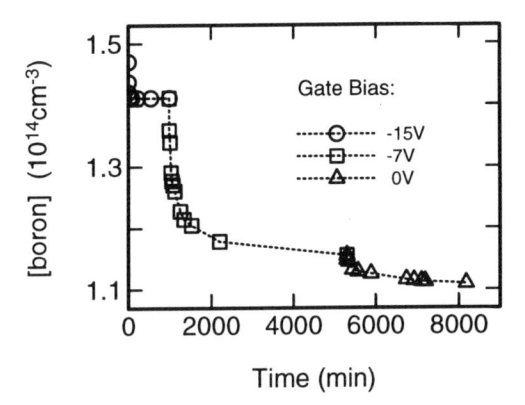

Fig. 2. Evolution of activated substrate doping level ([boron]) for a p-type capacitor after VUV exposure and neutralization. During the first 1000 min, -15V bias was applied, followed by a period at -7V after which the capacitor was shorted.

would be trapped more strongly, inhibiting its escape.

The feature of H^0 escape into the silicon during annealing of the interface states can be used to check the model.[6,7] To this end, the deactivation of the B dopant was used as a probe for H^0. After generation of interface states we subjected the capacitor for a period of 1000min. to -15V bias, thus charging all defects positive. As shown in Fig. 2, apart from some initial B deactivation, there is no significant effect. This shows that the supply of H^0 has ceased. After 1000min. the bias is changed to -7V. The states above midgap now are positively charged, but those below midgap are neutral so that the hydrogen can escape. The B data clearly reveal a large H^0 effusion. Finally after ≈5200min., we applied 0V bias, thus destabilizing the interface states above midgap. Again, this change is immediately followed by H^0 emission. This experiment demonstrates that the annealing of the interface states is accompanied by release of H^0.

Concerning slow states in stressed Al-gated capacitors there is a consensus that these are unstable donors centers.[1] Furthermore, Stahlbush et al. have demonstrated that they are produced by H^0.[3] Below we will show that also VUV exposure produces such unstable slow donors.[9]

Fig. 3A shows the result of a switching experiment for a sample subjected to an exposure of ≈3x10^{16} photons cm^{-2} followed by a neutralization step. The specimen clearly reveals a reversible V_{mg} behavior, the fingerprint of the slow state. The reversible V_{mg} shift is superimposed on an irreversible recovery process associated with annealing of fast interface states.

In a different experiment (Fig. 3B), we could show that the slow states have donor character. After irradiation we applied a constant negative bias which would make acceptor-type defects neutral and donor-type defects positive. Next, the charge state was assessed by photoinjection of ≈2x10^{13} electrons cm^{-2}: if positively charged, the states will have a large capture cross section (>10^{-13}cm^2), whereas being neutral they will have a small cross section (<10^{-15}cm^2). The data show that the states are easily discharged which proves that they are indeed donor-type defects. From this result we

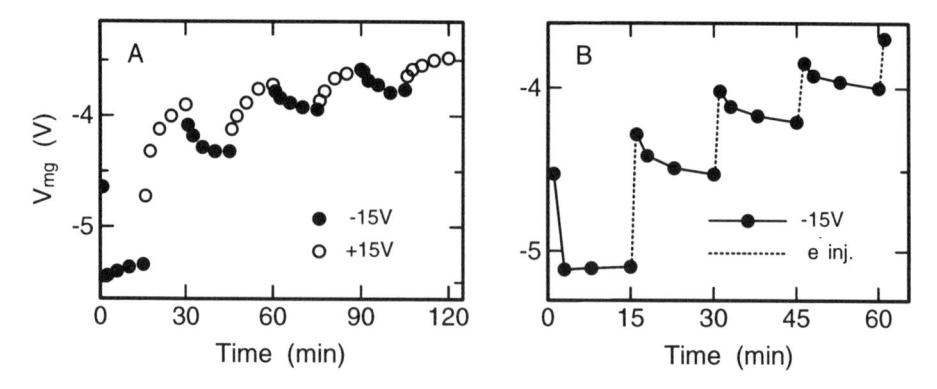

Fig.3 Slow state response of VUV-irradiated and neutralized samples (A) and capture of photoinjected electrons by positively charged slow states centers (B). In 'B' the sample is biased at -15V, but this bias was periodically interrupted for UV electron injection at 0V.

conclude that the slow states and interface states are of a similar nature.

In 1980 McLean postulated drift of protons through the oxide layer to explain post irradiation build up of interface states.[4] Further support for this assumption was obtained from studies of Saks and Rendell and of Edwards.[10,11] Below we will provide proof for the generation of protons in the oxide.[12] For these experiments we used capacitors with a 66nm oxide. Hole injection was accomplished by VUV exposure at +1MV/cm.

As shown in Fig. 4, injection of $\approx 10^{16}$ holes cm^{-2} induces a 40V midgap shift. After injection, a large recovery is observed, associated with discharging of the oxide layer and with annealing of interface states. An estimation of the latter contribution was made by neutralizing the protons immediately after hole injection, which showed that the annealing only accounts for a smaller part, i.e. the recovery is predominantly caused by dis-

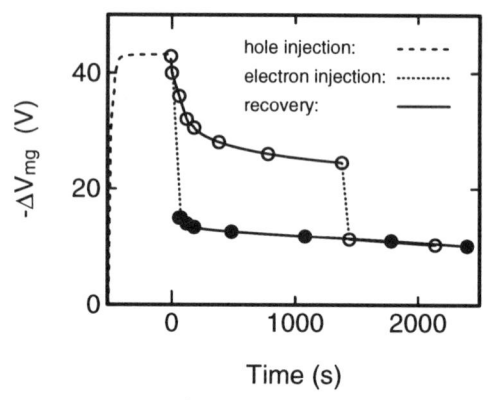

Fig. 4. ΔV_{mg} upon injection of 10^{16} holes cm^{-2}. The dashed line (t<0) shows the charge accumulation during VUV injection. For t>0 the figure shows the recovery effect of V_{mg} for a VUV-irradiated capacitors (open symbols) and the annealing of donor-type interface states for an equivalent samples that was neutralized after hole injection (solid symbols). The first set was also neutralized, but only when the recovery process had gone to completion.

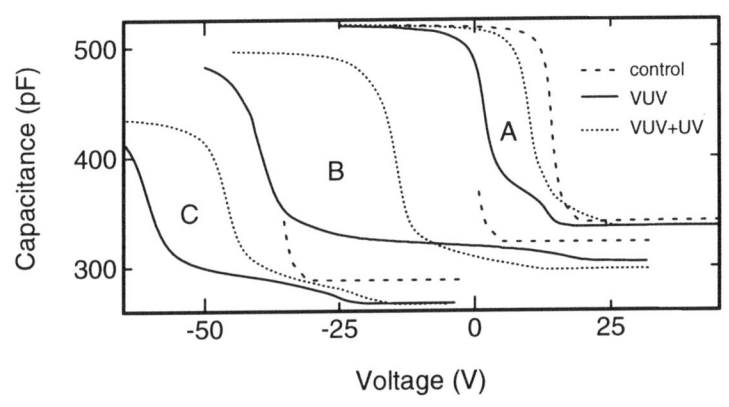

Fig. 5. HF CV curves of capacitors subjected to various doses of VUV hole injection, before (solid lines) and after (dotted lines) neutralization by UV electron injection. The dashed lines indicate the control curves before starting the procedure. Sets A and B correspond to capacitors that received a total fluence of respectively 2×10^{14} and 1×10^{16} holes cm^{-2}. Neutralization and measurement of the CV curves (solid and dotted) were performed immediately after hole injection. In case C, the capacitor was injected with 1×10^{16} holes cm^{-2}; however, the procedure was continued only after a delay of 2000s.

charging. After ≈1000s the effect has slowed down, suggesting that the discharging has ceased. Apparently the discharging takes place in the first minutes, in agreement with McLean's data.[4]

To verify whether the fast oxide discharging phenomenon is caused by proton drift, we have checked if neutralization of the oxide is accompanied by H^0 release or not. The result is shown in Fig. 5. Each of the three cases (A, B and C) concerns a capacitor subjected to hole injection followed by neutralization. The CV curves were recorded before and after hole injection, and after the neutralization step. In case A, the sample was subjected to a small hole fluence so that the O vacancies are charged whereas only minor amounts of H^0 are released. As shown by the unchanged inversion capaci-

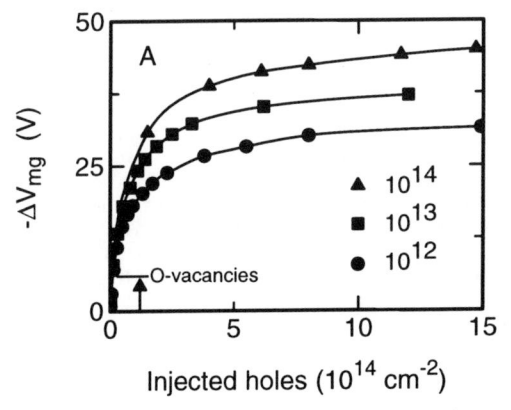

Fig. 6. ΔV_{mg} during hole injection in the oxide of MOS capacitors at different injection rates (indicated in holes cm^{-2}s^{-1}). ΔV_{mg} caused by hole trapping at O vacancies is also indicated.

tance, neutralization it is not accompanied by H^0 release. Capacitor B was subjected to a large hole fluence and all steps were performed after each other without delay. The data demonstrate the release of substantial amounts of H^0 during hole injection and upon neutralization. In case C, we also applied a large hole fluence, but now we waited for 2000s before continuation of the procedure to allow the oxide discharging process to complete. The data show that in this case the neutralization was not accompanied by H^0 release. This result clearly proves that the oxide discharging process is caused by mobile protons.

The generation of protons in the oxide implies that H^0 constitutes a hole trap; albeit with anomalous behavior. The number of hole traps is equal to the H^0 concentration which is governed by the rates of H^0 generation and elimination. Because of this, the number of hole traps should depend on the hole injection rate. The data in Fig. 6 show indeed enhanced trapping for larger injection rates. This hole trap with small cross section ($<10^{-15} cm^2$) is apparently the same one as reported by Stivers and Sah.[13]

The authors would like to thank prof. dr. P. Balk for a critical reading of the manuscript.

References

1. Stahlbush, R.E., (1996) Slow and fast state formation caused by hydrogen, in: Massoud, H.Z., Pointdexter, E.H. and Helms, C.R. (eds.), *The Physics and chemistry of SiO₂ and the Si/SiO₂ interface*, Vol **96-1**, The Electrochemical Soc., Pennington, NY, 525-37.
2. Cartier, E., Stathis, J.H. and Buchanan, D.A. (1993) Passivation and depassivation of Si dangling bonds at the Si/SiO₂ interface by atomic hydrogen, *Appl. Phys. Lett.* **63**, 1510-12.
3. Stahlbush, R.E., Cartier, and E., Buchanan, D.A., (1995) Anomalous Positive Charge formation by atomic hydrogen exposuree, *Microelectron. Eng.* **28**, 15-18.
4. McLean, F.B. (1980) A framework for understanding radiation-induced interface states in MOS structures, *IEEE Trans. Nucl. Sci.*, **NS-27**, 1651-63.
5. Druijf, K.G., de Nijs, J.M.M., Drift, E. van der, Granneman, E.H.A. and Balk, P., (1994) The nature of defects in the Si-SiO₂ System generated by VUV Irradiation, *Appl. Phys. Lett.*, **65**, 347-49.
6. Nijs, J.M.M. de, Druijf, K.G., Afanas'ev, V.V., Drift, E. van der and Balk, P. (1994) Hydrogen-induced donor-type Si/SiO₂ interface states, *Appl. Phys. Lett.* **65**, 2428-30.
7. Druijf, K.G., Nijs, J.M.M. de, Afanas'ev, V.V., Drift, E. van der and Balk, P., (1995) On the microscopic nature of donor-type Si/SiO₂ interface states. *J. Non-Crystal. Solids*, **187**, 206-209.
8. Edwards, A.H., (1991) Interaction of H and H₂ with the silicon dangling orbital at the (111) Si/SiO₂ interface. *Phys. Rev. B*, **44**, 1832-38.
9. Druijf, K.G., de Nijs, J.M.M., Drift, E. van der, Granneman, E.H.A. and Balk, P., (1995) Recovery of VUV-irradiated MOS systems, *J. Appl. Phys.*, **77**, 3657-67.
10. Saks, N.S. and Rendell, R.W. (1992) The time-dependence of post-irradiation interface state build-up in deuterium annealed oxides, *IEEE Trans. Nucl. Sci.*, **NS-39**, 2220-29.
11. Edwards, A.H. and Germann, G., (1988) Interaction of hydrogen molecules with intrinsic defects in a- SiO₂, *Nucl. Instr. Meth. Phys. Res.* **B32**, 238-47.
12. Afanas'ev, V.V., Nijs, J.M.M. de, and Balk, P. SiO₂ hole traps with small cross section (1995) *Appl. Phys. Lett.* **66**, 1738-40.
13. Stiver, A.R. and Sah, C.T. (1981) A study of oxide traps and interface states of the Si/SiO₂ interface, *J. Appl. Phys.* **51**, 6292-04

FUTURE TRENDS IN SiC-BASED MICROELECTRONIC DEVICES

A.A. LEBEDEV and V.E. CHELNOKOV

A.F.Ioffe Physico-Technical Institute Russian Academy of Science.
Politechnicheskaya 26, St.Petersburg 194021, Russia

1. Introduction

Because of the large bandgap (3.0 eV for 6H SiC and 3.2 eV for 4H SiC) and radiation hardness, silicon carbide is prospective material for high temperature microelectronic devices. In SiC power devices unit switching power and speeds exceeding those in silicon by an order of magnitude or more are obtainable, with operating temperatures up to $1000^0 C$.

In the present work, a theoretical analysis of SiC parameters important for electronic devices producing will be made. Also a comparative description of different technological methods used for SiC growth will be done: for substrates growth - Lely method and modified Lely method; for n-type epilayer growth - sublimation epitaxy (SE), container free liquid phase epitaxy (CFLPE), and chemical vapor deposition (CVD); for p-n junction production - SE, CFLPE, CVD, Al implantation (ID) and boron diffusion; for mesa structures formation - plasma-ion etching.

Currently SiC diodes with breakdown voltage about 4.5 kV [1] have been realised. However, such SiC p-n junctions usually have very small working area $\leq 10^{-3}$ cm^2. Therefore to decrease defect density in SiC epilayers or to increase working area of the devices is one of the main problems in developing SiC power electronics.

Another problem is surface protection of manufactured devices. On the one hand, it is necessary to decrease surface breakdown, and on the other - to cover mesa-structure by a dielectric, with the critical electric field strength more than that for air. In this paper, we will try to generalize experience, which has been obtained in our laboratory during investigation of different SiC diodes, prepared by different technologies.

2. Theoretical estimations

In conventional semiconductor electronics, the working temperature T_w of a p-n junction structure is defined as the temperature at which the reverse current, at fixed reverse voltage close to the breakdown voltage, does not exceed an experimentally chosen value. This criterion was used to calculate theoretical parameters of SiC, as the first widegap material for high-temperature electronics.

Figure 1 shows the p-n junction T_w for several widegap materials (and also Ge and Si) plotted against the bandgap. It seems that T_w grows with E^g nearly linearly. Electrophysical parameters are given in Table 1. For BN, the admissible working temperature of a p-n junction is equal to the melting temperature. In diamond, a noticeable vacuum graphitization is observed at 1700 K, i.e. at a temperature lower than $T_w = 2100$ K. The latter circumstance suggests that the classical definition of admissible T_w for a p-n junction seems to be insufficient for the widegap materials. It is probably necessary to introduce another temperature - the temperature of the thermal stability of the material itself. The maximum value of this temperature is the melting temperature and the minimum value is Debye temperature T_D.

Figure 2 gives a relation between admissible T_w determined in a way mentioned above and T_D for some semiconductors which have been conveniently subdivided into 4 groups:

(i) Ge, Si, BP, β-SiC with $T_D = 1,5 T_W$

(ii) α-SiC, C (diamond) with $T_D = 1,5 T_W$

(iii) GaAs, GaP, GaN, AlN, BN with $T_w = 1,5 T_D$

(iv) GaAs$_{1-x}$P$_x$, CdS, InS, with $T_w = (2,2 - 4) T_D$

The group (iv) materials can hardly be successful for semiconductor device technology, and they also have low thermal conductivity. The group (ii) materials have a higher thermal conductivity than that of group (i) or group (iii)

TABLE 1. Comparison of selected semiconductors room temperature properties.

Parameter	Si	GaAs	3C-SiC	6H-SiC	4H-SiC
Band gap (eV)	1.1	1.42	2.30	3.0	3.2
Breakdown field for 10^{17} cm^{-3} (MV/cm)	0.6	0.6	>1.5	2-3	2-3
Saturated Electron Drift (cm/s)	10^7	8×10^6	2.5×10^7	2×10^7	$2. \times 10^7$
Electron mobility (cm^2/Vs)	1100	6000	750	200-300	800
Hole mobility (cm^2/Vs)	420	320	40	60	115
Thermal Conductivity (W/cm K)	1.5	0.5	5.0	4.9	4.9
Commercial wafers	12"	6"	none	1,375"	1,375"

Another parameter essential in a p-n junction material is the critical strength of avalanche breakdown E_{cr}. Figure 3 shows the calculated plots of $E_{cr} = f(N_d)$ for widegap materials for abrupt p-n junction[2]. The inequality of the impact ionization coefficient was also taken into account. It seems that α-SiC (6H) posses the highest electrical strength. Its properties are presented along the

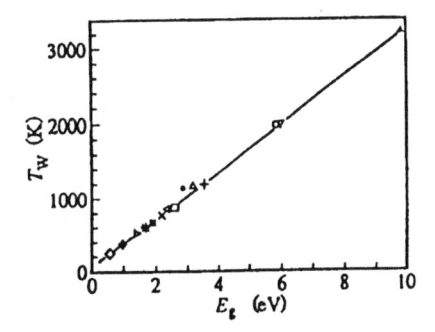

FIGURE 1. Operating temperature of the p-n junction versus the bandgap for some semiconductors. (▲) BN, (∨) AlN, (o) C (diamond), (+) ZnS, () GaN, (●) α - SiC, ( ) CdS, (<) β - SiC, (×) GaP, (■) BP, (∗) GaAs$_{1-X}$P$_X$, (>) GaAs, (◆) Si, (◊) Ge.

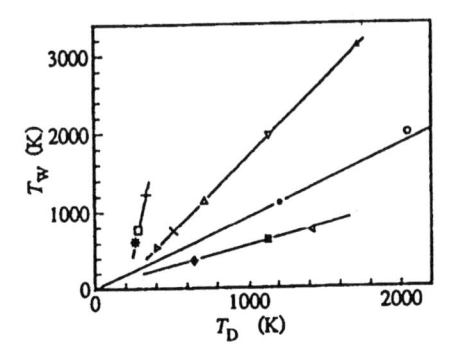

FIGURE 2 Comparison of operating and Debye temperature for some semiconductors (notations as in Fig.1).

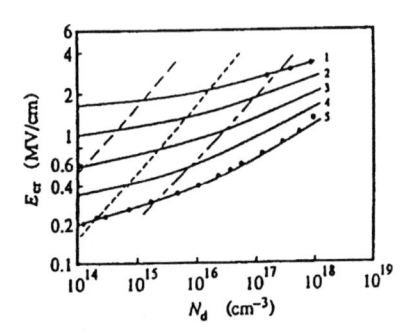

FIGURE 3. Critical strength of avalanche breakdown field E_{cr} versus impurity concentration in the slightly doped p-n region for some wideband materials. SiC (6H) (curve 1), GaN (curve 2), C (Diamond) (curve 3), BP (curve 4), Si (curve 5). $U_{br} = 10^4$ V (———), $U_{br} = 10^3$ V (– – –), $U_{br} = 10^2$ V (— - —).

crystallographic C-axis (this polytype is anisotropic; the p-n structures is made in direction normal to this axis). The points in Figure 3 correspond to experimental data for Si and SiC, and the dashed lines indicate isovolts, i.e. these are the lines for the equal reverse voltage of the pn junction.

In Figure 4 (a, b) we present the breakdown voltage U_{br} as a function of the space charge region thickness W and of impurity concentration in a slightly doped p-n junction for several widegap materials. A breakdown voltage of 10 kV can be achieved for BN at $N_d-N_a = 10^{16}$ cm^{-3}. For Si this value can be achieved at $N_d-N_a = 8 \times 10^{12}$ cm^{-3}. The space charge region thickness for Si is then 1.2 mm and for BN is 4 $\times 10^{-3}$ cm. At U_{br} = const and $N_d \sim E_{cr}^2$, $W \sim E_{cr}^{-1}$ follows from a simultaneous solution of Poissons's eqution and the breakdown voltage equation for an abrupt p-n junction. When selecting a material for semiconductor electronic devices, one should take into account the power losses in a forward switched pn junction.

Figure 5 shows the relative values of admissible forward current for some widegap materials. Forward current of a Si junction was taken as unit scale. We see that only diamond, α-SiC, BN and AlN are superior to Si, while $A^{II}B^{VI}$ materials, $A^{III}B^V$ solid solution and GaAs do not have advantages due to low thermal conductivity.

The above considerations together with elaborate p-n junction technology have determined the choice of SiC as the first material for high-temperature electronics. In papers [3-5], it was studied in detail the application of silicon carbide for producing different types of semiconductor devices. It has been shown that SiC increases the power efficiency, the RF limit and reliability of basic types of RF avalanche-fligth diodes, various transistors, commutative diodes, varactors etc.

As a conclusion of this part of the paper, one can summarize that

(i) Any type of semiconductor devices can be based on SiC.

(ii) Value of (Power) X (Frequency) for SiC will be on 3-5 orders more than for existing power devices.

(iii) Using of the SiC can increase temperature working range of power devices by one order of magnitude or more.

(iv) The range of the reverse voltages can be increased by 3 orders of magnitude.

3. Technology

3.1. SUBSTRATES

At the present time there are two main methods for producing SiC substrates: Lely method [6] (Lely-substrates) and modified Lely method [7] (LM-substrates). Both type of substrates were studied by X-ray topography (using Bragg reflection and Lang's method) and X-ray diffractometry. The studies have revealed the following advantages of Lely substrates:

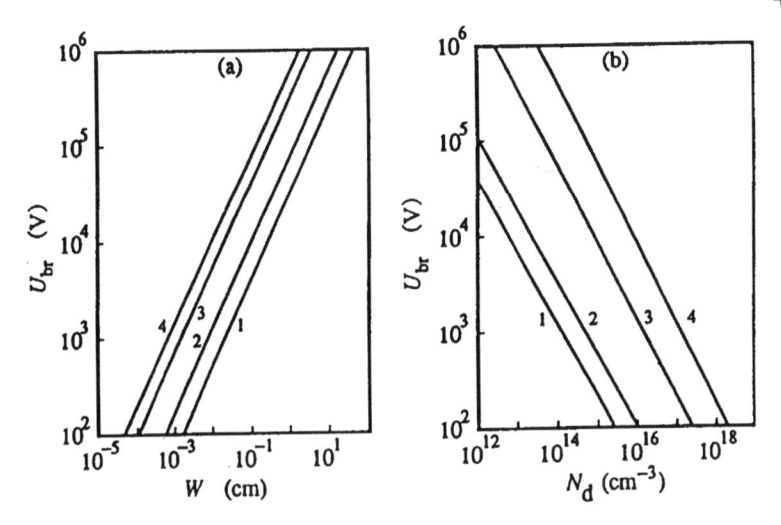

FIGURE 4. Breakdown voltage dependence on (a) the space charge region thickness and (b) impurity concentration in slightly doped p-n region. Si (curve 1), BP (curve 2), α-SiC (curve 3), BN (curve 4).

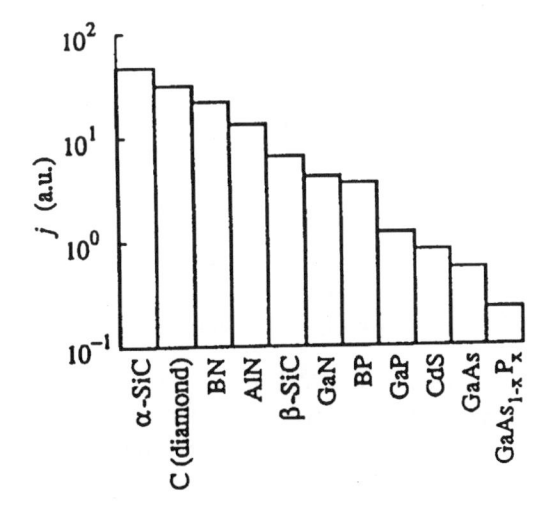

FIGURE 5. Diagram of admissible forward currents for some wideband semiconductors.

(a) The middle value of dislocations density was found 10^2- 10^5 cm^{-2} for LM substrates (Fig.6) and 10^1 -10^3 cm^{-2} for Lely substrates (with the absence of pinholes)(Fig.7);

(b) For LM substrates it is usual uniform distribution of dislocation on the substrate area; in Lely substrates the density of dislocations is nonuniform, they are concentrated especially near the "growth leg" point at which the crystal growth begins on graphite members of the growth chamber. Thus, the area having dislocation density $N_d \leq 10^1$ cm^{-2}, or practically dislocations free, increases with increasing dimensions of the crystal.

(c) For Lely substrates are prefer dislocations lying in the basal plane of the crystal. Such dislocations create no etching pits on the (0001) Si surface and exert no influence on the perfection of epitaxial layers grown on Lely substrates. In LM substrates there are many dislocation and different size inclusions which are oriented perpendicular to the basal plane of crystal.

So structural perfection of the Lely substrates is essentially higher than LM substrates. But LM substrates has one important advantage - they have considerably more area. However this advantage is very important for mass production devices, like blue LED's. For SiC power devices, which can be used in space technical or in nuclear station equipment, where quality and reliability of the devices are more important than its price, this advantage is not so important. Beside this, presents of the Lely substrate with area 3-5 cm^2 [8] give hopes that after optimization of the Lely growth technology it will be possible to growth substrates with area equal to LM substrates 1 inch. in diameter.

3.2 EPILAYERS.

3.2.1. Epilayers growth.

The epilayers and p-n structures studied in this work were grown on the (0001) Si face of single crystal silicon carbide substrates of polytype 6H and 4H. structures. In all cases the dopants were Al (p-type) and N (n-type). The thickness of epilayers was 1-2 µm for p-type and 5-10 µm for n-type. The value of electron mobility and hole lifetime is practically the same for all type of epilayers :100-300 cm^2/ Vs and 10^{-8} - 10^{-7} s, respectively.

During our experiments on CVD growth on Lely 6H SiC substrate we observed inversion of the conductivity type of growing epilayers, depending on the C/Si ratio in the gas phase. Increasing this ratio from 1 to 5-8 reduced the concentration of uncompensated donor levels in the films grown from 1-3 10^{18} cm^{-3} to 2-4 10^{16} cm^{-3}. At higher C/Si ratios the resulting layers were of p-type with the N_a - N_d concentrations varying from 1 10^{16} (C/Si ~ 8) to 6-8 10^{17} cm^{-3} (C/Si =10-15). Investigation of the p-type layers grown in this work has shown that their conductivity for all value of N_a-N_d is governed by the presence of deep acceptor centers lying in the energy interval Ev + (0.15 - 0.25) eV, which parameters is close to parameters of structural defects (L-centre, Ev + 0,24 eV) always present in sublimation grown 6H SiC.

437

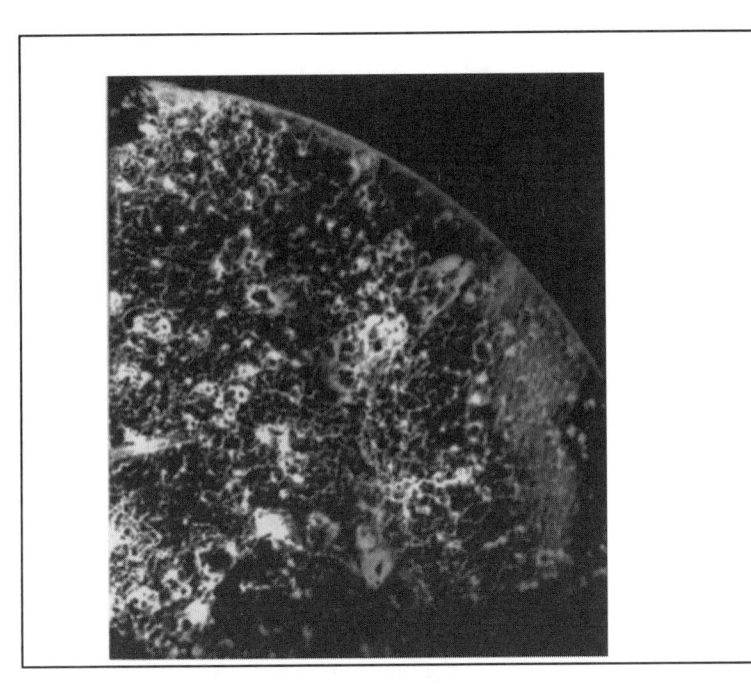

FIGURE 6. X-ray topograph of LM substrate
(Lang's method, 1120 reflection, $Mo_{K\alpha1}$-radiation)

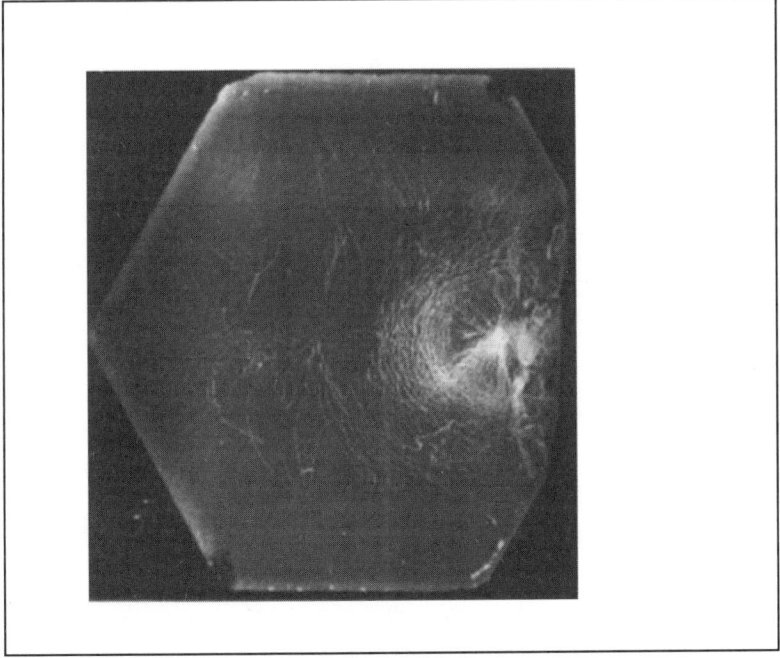

FIGURE 7. X-ray topograph of Lely substrate
(Lang's method, 1120 reflection, $Mo_{K\alpha1}$-radiation)

3.2.2. Deep levels parameters.

It was found that the same deep levels which had been observed in SE grown samples are present in CVD p-n structures [L-center (E_v + 0,24 eV),"shallow boron" (E_v+0.35 eV), i-center (E_v+0.52 eV), D-center (E_v + 0.58 eV), S-center (E_c - 0.35 eV), R - center (E_c - 1.27 eV)]. Concentration of D-centers in the initial SE epitaxial layers was found to increase upon additional diffusion of boron, and that of i-centers to become higher after fabricating a p-n structure by ion-beam doping with Al.

It has been found that i and D centers in 6H connected with yellow and green("boron" and "defect" electroluminescence bands, correspondingly [9]. Also it were founded that R and S centers could govern the recombination processes in the 6H SiC p-n structures [10]. In 4H SiC, acceptor centers were found having approximately the same ionization energies and a similar ratio of hole and electron capture cross-sections σ_p and σ_n ($\sigma_p/\sigma_n \gg 1$); these centers initiated analogous electroluminescence processes [20,21] (see table 2)

3.2.3. Self compensation of the SiC epilayers

Previously, the net concentration of deep acceptor centers (E_{na}) in SE p-n structures has been found to be invariably higher by an order of magnitude than the net concentration of donor-like deep centers [9]. The E_{na} values in SE structures with $N_d - N_a = 10^{17}$ cm^{-3} amounted to 10-50% of N_d-N_a. The E_{na} value was little affected by variation of the concentration of shallow donor levels. So, at N_d-$N_a = 10^{18}$ cm^{-3} the degree of compensation was as low as 3%, and at N_d-$N_a = 10^{16}$ cm^{-3} overcompensation of epitaxial layers occurred. We observed in CVD p-n structures the same dependence of E_{na} on N_d-N_a (figure 8), but with net concentration of deep acceptor centers in CVD p-n structures being invariably smaller by 2-3 orders of magnitude than in SE p-n structures with the same value of $N_d - N_a$. So, no overcompensation of CVD epilayers occurred down to N_d-$N_a \sim 10^{14}$ cm^{-3}.

In CFLPE and CVD n-type epilayers net concentration of the deep acceptor levels were on 1-2 orders less. But minimum value of N_d-N_a, which we have in CFLPE epilayers was about 5×10^{16} cm^{-3} [10], and in CVD epilayers this value can achieve 10^{14} cm^{-3} [11].

TABLE 2. Parameters of the main detected Deep Centers.

Polytype	Type of center	Energy position, eV	Electron cross-section, cm^2 (σ_n)	Hole cross-section, cm^2 (σ_n)	Participation in recombination processes
6H SiC	S	Ec - 0.35	10^{-15}	1-3 x10^{-13}	radiation-free
	R	Ec - 1.27	10^{-14}	1-2 x10^{-15}	
	L	Ev + 0.24	10^{-18}	2-4 x10^{-15}	radiation-free
	I	Ev + 0.52	2-6x10^{-21}	1-3 x10^{-17}	
	D	Ev + 0.58	1-3x10^{-20}	1-3 x10^{-16}	? DEL "boron"
4H SiC	analoge L	Ev + 0.27	< 10^{-17}	5-6x10^{-15}	?
	analoge i	Ev + 0.5	< 10^{-21}	5x10^{-17}	DEL
	analoge D	Ev + 0.54	1-3x10^{-21}	1-3x10^{-16}	"boron"

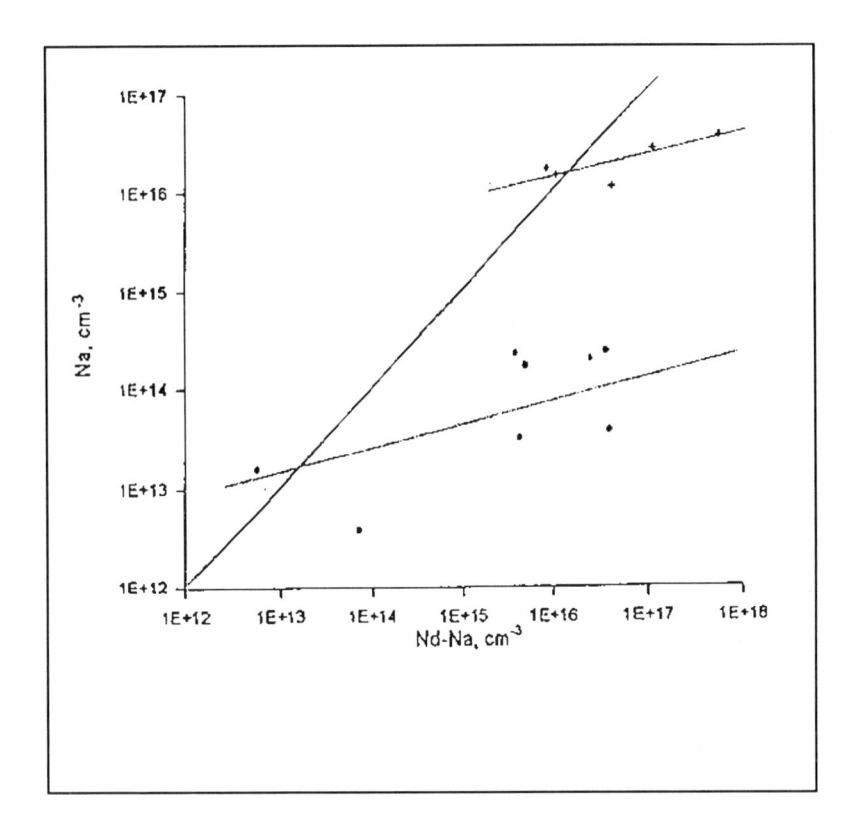

FIGURE 8. Dependence of the net concentration of deep acceptor centers on N_d-N_a concentration in SE (+) and CVD (•) p-n structures. Straight line corresponds to E_{na} equal to 0.5 (N_d-N_a).

3.3. P-N JUNCTION

In p-n structures produced by SE, CFLPE and CVD epitaxy plots of $1/C^2$ - U (where C is p-n junction capacitance and U is applied voltage) were straight lines in the entire voltage range and yielded the same N_d-N_a value as were obtained for the Shotky-barrier structured formed on the surface of the n-type epilayers before formation of the p-n junction. Thus the junctions in this structures were abrupt and asymmetric.

In junctions produced with ion-implantation, capacitance-voltage characteristics of the barrier plotted in $1/C^2$ - U coordinates were slightly deviating from a straight line showing a kink in a region of low reverse voltage (fig.9)[12]. DLTS measurements of the such structures shows that concentration of deep acceptor levels (i-center, E_v + 0.52 eV) increased by a factor of 3-5 near the p-n junction compared with its level through the rest of the base (Fig.10) . So after preparation of p-n junction by ion implantation of Al on base of low doped n-type epilayers ($< 10^{16}$ cm^{3}) it was arisen overcompensation of the n-type base near metallurgical boundary of p-n junction and "S" type current voltage characteristics can be observed [13,20].

But from the other hand p-emitter produced by ion implantation has more higher concentration of the electrical activity impurity, than emitter produced by another technological methods. This is one of advantages of such p-n junction formation technology, which make more easy preparation of the low resistance ohmic contact. In case of optimization of implantation and subsequent annealing it is possible to decrease concentration of the acceptor levels, arising in the n-base after implantation and obtain diode on base of low doped n-SiC epilayers .

It were investigated diodes fabricated on CREE substrates with epitaxial layers of p and n type for and the following techniques were used for preparing ohmic contacts:

- evaporation of Al or Ni without additional firing-in, T_{sample} ~200°C;
- firing-in an ohmic contact (Ni), T ~ 1000°C;
- firing-in a Ti (Al) contact to the p type, T ~1400°C;
- firing-in an Al (Au W) contact to the p type, T ~1900°C.

It was found that, when the temperature of thermal treatment of 130 μm-diameter diodes was raised, the voltage of the fabricated diodes dropped from 250 - 150 V to 10 - 100 V (Measurements were made on air without any special surface protection.)(Table 3). In this case the breakdown voltage was approximately 1.5 - 2 times lower for regions with high dislocation density.

Then a p-type layer was formed by the sublimation method on two substrates having no p-type CVD layer. Capacitance-voltage measurements shows practically the same value of the N_a-N_d for both type of epilayers 10^{19} cm^{-3} (CVD) and 0.7 10^{19} cm^{-3} (SE). From X-ray diffractometry investigation it was founded that for CVD p-type changes of the lattice parameter Δd/d is about 3,5 10^{-4} which is correlates to N_a-N_d concentration ~ 10^{21} cm^{-3} . So concentration of the doping

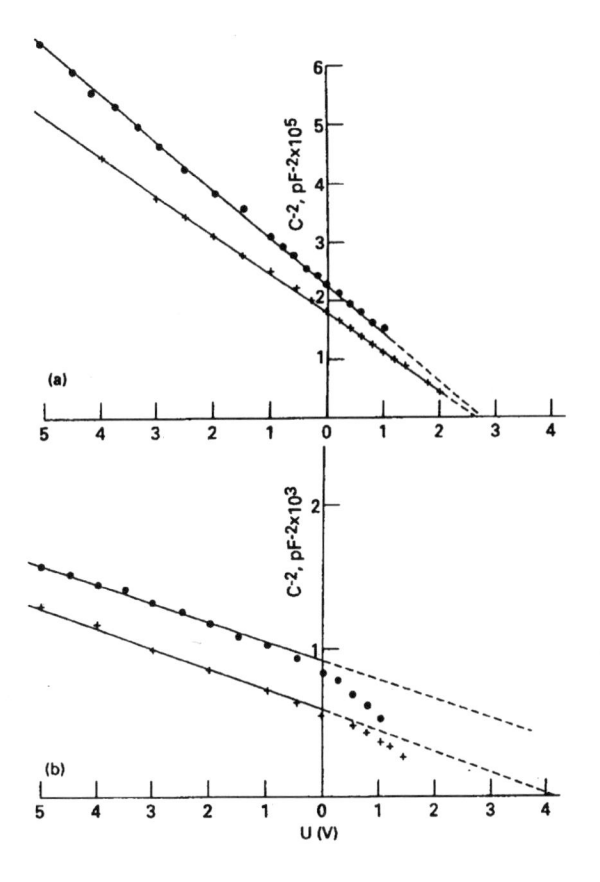

FIGURE 9. Capacitance-voltage characteristics of (a) SE stuctures and (b) ID structures. (+) 6H SiC, (•) 4H SiC.

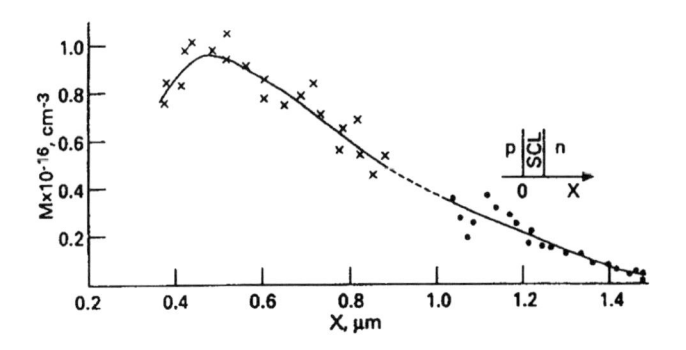

FIGURE 10. Distribution of i-centers in p-n junction regions of ID structures. SCL is the space charge layer. (×) SiC 6H, (•) SiC 4H.

impurity is significantly higher than we can see from electrical measurements. At the same time for the diodes with p-type produced by SE, value of $\Delta d/d \sim 0$, so all impurities are in electrically active position.

Based on these samples, diodes with ohmic Al (Au W) contacts were formed. The breakdown voltages of these diodes were 240 - 300 V (d ~130 μm) or ~200 V (d ~600 μm). The investigations we performed indicate that breakdown voltage is strongly affected, apart from structural defects, by the metallurgical boundary between n- and p-type layers whose quality is much higher in the case of high-temperature sublimation growth.

TABLE 3. Dependence of the diode breakdown voltage on diode thermal treatment.

	Cree Inc.	substrates	numbers		
	BO853-6 p-type CVD	E-0437-11 p-type CVD	E-0464-14 p-type CVD	E-0464-11 p-type SE	B 0853-7 p-type SE
N_d-N_a concentration in n-type layers, cm^{-3}	8 10^{15}	1-3 10^{16}	3-4.2 10^{16}	1-3 10^{16}	0.3-1 10^{15}
Temperature treatment C°			U_m, V		
		For	130 μm	diameter	diodes
200	245,7	112,2	206		
1000	211,5	102,5	156		
1400	115	21,3	136		
1900	70,7	21,2	107,5	237,5	300
		For	600 μm	diameter	diodes
1400	60				
1900	50	25,6	3.3	199	208

3.4. MESA-STRUCTURE FORMATION AND CONTACTS

Because of the high chemical stability of SiC it is possible to make etching of its (0001) facet only by plasma-ion methods [14]. After such etching the surface of the semiconductor was strongly damaged and was not electrically neutral. It is a reason for arising breakdown at considerably lower voltages than the values implied by the impurity concentration in the base. In this case the breakdown was surface-like in character, starting at currents of 10^{-6} to 10^{-8} A and was irreversible. For decreasing probability of the surface-like breakdown it is necessary to make surface electrical field strength less than the same in volume.

This can be achieved by profiling of the mesa-structure during etching [15]or and by diffusion of the acceptor impurities into the surface region of the n-base [16]. Also electrical activity of the surface can be decreased by special treatment, for ex. oxidation.

Another problem is the fact that for more higher reverse voltage take place breakdown through the air along the surface of the mesa-structure. One of the maximum values (about 2200 V) of breakdown voltage was obtained when the samples were placed in a dielectric liquid (Fluorient[TM] FC-77) with a high value of electric breakdown field strength [17]. So for obtained high temperature and high power devices it is necessary to cover mesa-structure by the thermostable dielectric with critical electric field strength more than same for air.

Optimum ohmic contacts for SiC diodes may be contacts which simultaneously has low resistance, can be used as a mask during plasma-ion etching and can be easy to solder to package. At the present time contact, which can satisfied all this requests, is Ni-based contacts [18].

4. Devices

At the present time possibility of formation practically all type of SiC-based semiconductor devices are already shown. Among them are: rectified diodes [22], stabilitrons [23], Schottky diodes [24], FET's [25-27], UV photodetectors [28], light emitting diodes for different spectrum range [27,29], bipolar junction transistor [27], thyristors [30] and first integrated circuts [31]. Some of this devices are now present at the marketplace. Blue light emitting diodes were the first SiC based devices which reach significant volume sales. Working parameters of other devices are close to previous made theoretical estimation, but its small size and big price did not permits wide industrial production. Now it is understandable that for receiving best parameters of each kind of devices it is necessary to use optimum technological combination, which will be include different type of growth technology, surface protection, metallization, oxidation etc.

5. Conclusions

In [19] was drawn optimum, from authors point of view, technological scheme for SiC high power diodes producing. substrates (grown by Lely method) + n-type epilayres with low doped region on the top (CVD) + p-type (SE or implantation) + low resistance ohmic contacts(on base of the metals with high melting temperature, which can be simultaneously as a mask for plasma-ion etching) + mesa formation with profiling (plasma-ion etching) + surface protection (dielectric with critical electric field strength more than same for air) + packaging in metal-ceramic packages.

Results, described in this paper allowed to draw such scheme for producing some other SiC devices (Table 4).

Acknowledgments

This work is supported in part by the U.S. Ministry of Defense.

TABLE 4. Optimum technological combination for producing diferent SiC devices.

Technology									Devices
Substrate		N -type layer			P-type layr				
Lely	ML	SE	LPCFE	CVD	SE	LPCFE	CVD	Ion im.	
+		+			+				UV
	+		+			+			photodet
				+			+		ector
								-	
+		-			-				Blue
	+		+			+			light
				+			+		emitted
								+*	diodes
+		+			-				Green
	+		+			-			light
				+			-		emitted
								+	diodes
+		+			+				FET's
	+		-			-			
				+			+		
								-	
+		+			+				Middle
	+		+			-			power
				+			+		diodes
								+	
+		-			+				High
-		-			-				power
				+			-		diodes
								-	

*) for 4H SiC only.

References

1. Kordina. O., Bergman J.P, Henry A., Jansen. E., Savage S., Andre J., Ramberg L.P., Lindefelt U.,. Hermansson W and Bergman K.(1995) A 4,5 kV 6H silicon carbide rectifier,.*Appl.Phys.Lett. J* (**67**) 11:1561-1563.
2. Otblesk A.E. and Chelnokov V.E. (1980) Wide bandgap semiconductors - perspective materials for high power semiconductor electronic, *Proc. III All-Union Conf on Wide-gap Semiconductors*, LETI, Leningrad, 1979, pp 197-211.
3. Chelnokov V.E. (1992) SiC bipolar devices, *Mat Science and Engineering*, **B11** pp 103-111.
4. Chow T.P. and Tyagi R(1994) Wide Bandgap Semiconductors for Superior High-Voltage Unipolar Power Devices *IEEE transition on electron devices*, **41**, pp1481-1483.
5. Neudeck P.G. (1995) Progress towards high temperature, high power SiC devices, *Inst.Phys.Conf.Ser* 141,Chapter1, ,pp 1-6
6. Lely J.A..(1955) Darstellung von Einkristallen von Siliziumcarbid und Beherrrshing von Art und Mende der eingebeunten verunreininungen *Ber.Dt.Keram.Ges.* **55**, : 229-231.
7. Tairov, Yu.M. and Tsvetkov V.F. (1978) Investigation of growth processes of ingots of silicon carbide single crystals *J. of Crystal Growth* **43**, 209-212.
8. Lebedev, A.A., Tregubova, A.S. Chelnokov V.E., Scheglov M.P., Glagovskii A.A. (1997) Growth and investigation of the big area Lely-grown substrates, *Mat Science and Engineering* **B 46**, 291-295.
9. Lebedev,A.A. and Chelnokov V.E.. (1994) Measurement of electrophysical properties of silicon carbide epitaxial films, *Diamond and Related Materials* 3, 1393-1397.
10. Anikin M.M., Zubrilov A.S., Lebedev A.A., .Strel'chuk A.M and .Cherenkov A.E(1991) "Recombination processes in 6H-SiC p-n structures and the influence of deep centers" *Sov.Phys.Semicond* **25**,: 289-293.

11. Larkin D.J., Neudeck P.G., Powell J.A., and Matus L.G. (1993).Site competition epitaxy for ontrolled doping of CVD silicon carbide *Proc. of the 5th conference Silicon Carbide and related materials, 1-3 Nov 1993,Washington,DC,USA*.p.51-54.
12. Anikin M.M., Lebedev A.A., Syrkin A.L., and Suvorov A.V. (1985) Investigation of the deep levels in SiC by capacitance spectroscopy methods *Sov.Phys.Semicond* **19**,:69-71.
13. Anikin M.M., Lebedev A.A., Popov I.V., Strel'chuk A.M., Suvorov A.V., Syrkin A.L., and Chelnokov V.E.(1986) Structure with an ion-implanted p-n junction in epitaxial 4H-SiC with an S-type current-voltage characteristic" *Sov.Phys.Semicond* **20**, 1036-1037.
14.. Syrkin A.L., Popov I.V., and Chelnokov V.E.(1986) Reactive plasma-ion etching of Silicon Carbide, *Sov.Tech.Phys.Lett* **12**, 99-101.
15. Lanous F,.Planson. D, Locatelli M.L, Chante J.P. (1996) Angle etch control for silicon carbide power devices to be published in *Appl.Phys.Lett.*
16. Lebedev A.A., Andreev A.N., Mal'tsev A.A., Rastegaeva M.G., Savkina N.S. and Chelnokov V.E. (1995) "Fabrication and investiigation of 6H-SiC epitaxial-diffused p-n structures" *Semiconductors* **29**, 850-853.
17. Neudeck P.G., Larkin D.J., Powell J.A. , Matus L.G., and Salupo C.S. (1994),2000 V 6h-SiC p-n junction diodes grown by chemical vapor deposition. *Appl.Phys.Lett.***64**,: 1384.
18. Rastegaeva M.G., Andreev A.N., .Zelenin V.V, Babanin A.I., and Chelnokov V.E.. (1995) "Complex use of nickel based metalization in process of the SiC-6H device formation: ohmic contacts, masking and packaging"*Technical Digest of Int. Conf. on SiC and Related Materials, Kyoto,Japan* 18-21 Sept.1995, 152-153.
19. Lebedev A.A., Rastegaeva M.G., Savkina N.S., Tregubova A.S., Chelnokov V.E. and Scheglov M.P.(1997) High Temperature SiC based rectified diodes: new results and prospects ,*Inst.Phys.Conf Ser* **155**,Chapter 8, 605-608.
20. Anikin M.M., Lebedev A.A., Poletaev N.K., Strel'chuk A.M., Syrkin A.L.., Chelnokov V.E.(1994), Deep centers and blue-green electroluminescence of 4H SiC. *Semiconductors* **28**, 288 .
21. Lebedev A.A., and. Poletaev N.K (1996), Deep centers and Electroluminescence of the boron doped 4H-SiC p-n structures. *Semiconductors* **30**, 238.
22. Anikin M.M, Lebedev A.A., Popov I.V., Sevast'yanov V.E., Syrkin A.L., Suvorov A.V., Chelnokov V.E. and Shpynev G.P. (1984) Silicon carbide rectified diode *Sov.Tech.Phys.Lett.* **10**, 444.
23. Andreev A.N., Anikin M.M., Zelenin V.V., Ivanov P.A., Lebedev A.A., Rastegaeva M.G., Savkina N.S., Strel'chuk A.M,Syrkin A.L. and Chelnokov V.E. (1995) High temperature silicon carbide stabilitrons for the voltage range from 4 to 50 V,*Mat.Science & Engineering* **B29**, 190-193.
24. Anikin M.M., Andreev A.N., Lebedev A..A., Pyatko S.N., Rastegaeva M.G., Savkina S.N., Strel'chuk A.M., Syrkin A.L. and Chelnokov V.E. (1991), *Sov.Phys.Semicond.* **25**, 198-201.
25. Diamon H., Yamanaka M., Shinohara M., Sakuma E., Misawa S., Endo K., and Yoshida S.(1987) Operation of Schottky-barrier field-effect transistors of 3C-SiC up to 400^0C, *Appl.Phys.Lett.* **51**, 2106-2108.
26. Sheppard S.T., Melloch M.R. and Cooper J.A.,(1994) Characteristics of Inversion-Channel and Buried-Channel MOS Devices in 6H-SiC, *IEEE transition on electron devices* **41**, 1257-1263.
27 .Palmour J.W., Edmond J.A. Kong H.S. and Carter C.H. Jr (1993) 6H- silicon carbide devices and applications, *Physica B* **185** , 461-465.
28. Anikin M.M Andreev A.N., Pyatko S.N., Savkina N.S., Strel'chuk A.M., Syrkin A.L. and Chelnokov V.E. (1992) UV photodectecor in 6H-SiC *Sensors and actuators A* **33**, 91-93.
29. Ziegler G and Theis D.(1981) A New Degradation Phenomenon in Blue light Emitting Silicon Carbide Diodes, *IEEE transactions on electron devices*, **ED-28**, 425-427.
30. Palmor J.W., and Lipkin L (1994) High temperature power devices in silicon carbide, *Transactions of secound Inter. High Temp. Electronic Convference*, **1**, Charlotte, NC, USA , XI 3-XI 8.
31. Brown D.M., Chezzo M., Kretchmer J., Krishnamurthy V, Michon G., and Gati G, (1994) High temperature silicon carbide planar IC technology and first monolithic SiC operational amplifier IC, *Transactions of secound Inter. High Temp. Electronic Convference,* **1** Charlotte, NC, USA , XI 17-XI 22.

THE INITIAL PHASES OF SiC-SiO$_2$ INTERFACE FORMATION BY LOW-TEMPERATURE (300 °C) REMOTE PLASMA-ASSISTED OXIDATION OF Si AND C FACES ON FLAT AND VICINAL 6H SiC

G. LUCOVSKY AND H. NIIMI
Departments of Physics, Materials Science and Engineering, and Electrical and Computer Engineering, North Carolina State University, Raleigh, NC 27695-8202, USA; e-mail: gerry_lucovsky@ncsu.edu

The initial stages of SiC-SiO$_2$ interface formation by low temperature (300 °C) remote plasma assisted oxidation (RPAO) have been studied by on-line Auger electron spectroscopy (AES) for flat and vicinal 6H SiC(0001) wafers with Si(0001) and C faces (000$\bar{1}$). The paper focuses on i) interfacial bonding and ii) oxidation rates for thickness to about 2 nm. Plasma-assisted oxidation of 6H SiC is compared with i) thermal oxidation of SiC and ii) plasma-assisted oxidation of flat and vicinal Si(111).

1. Introduction

There is considerable interest in SiC as a semiconductor material for high temperature devices, including power transistors. To fabricate these devices, it will be necessary to gain an increased understanding of the chemical bonding at the SiC-SiO$_2$ interface, and in particular to develop processing that minimizes interfacial defects and transition regions, thereby maximizing device performance and reliability. The oxide that forms on SiC during high temperature oxidation in O$_2$ ambients is SiO$_2$, and this oxidation process yields SiC-SiO$_2$ interfaces and gate dielectrics with sufficiently low defect densities for field effect transistor (FET) operation [1]. However, channel mobilities in depletion mode FETs are significantly lower than expected from bulk SiC properties and much research has been focused on explaining these differences and on improving channel mobilities. Since SiO$_2$ is the dominant solid state oxidation product, oxidation products involving carbon atoms must include gaseous molecules such as CO. As oxide growth progresses these oxidation by-products must be transported from the growth interface where they are generated, through the oxide and out of the film. The formation of gaseous CO does not preclude the generation of additional interfacial carbon atom bonding arrangements, including Si-C and C-O as have been reported in silicon oxycarbide interfacial transition regions [2]. The molar volume mismatch between SiC and SiO$_2$ is greater than the corresponding mismatch between Si and SiO$_2$, so that intrinsic levels of strain at thermally-grown SiC-SiO$_2$ interfaces are anticipated to be at least as high as the values of approximately 5x10^9 dynes/cm^2 reported for thermally-grown Si-SiO$_2$ interfaces [3]. This paper deals with SiC-SiO$_2$ interface formation by an alternative low temperature method that has been successfully applied to crystalline Si device technology [4,5]. As applied to Si, this approach provides independent control of Si-SiO$_2$ interface formation and oxide film grown. The Si-SiO$_2$ interface and an ultra-thin oxide layer (~0.5nm) are formed by low temperature (300 °C) remote plasma-assisted oxidation (RPAO), and the remainder of the oxide, or composite dielectric layer is then deposited by low temperature (300 °C) remote plasma enhanced chemical-vapor deposition (RPECVD). Device quality interfaces are obtained by subjecting the plasma processed Si-SiO$_2$ heterostructure to a post-deposition rapid thermal anneal (RTA), e.g., for 30 s at 900 °C. The 900 °C anneal promotes both chemical and structural interface relaxations, including a minimization of interfacial sub-oxide bonding arrangements in transition regions between the Si crystal and the stoichiometric bulk oxide [6]. A similar plasma-assisted process has recently been applied to SiC, and has yielded SiC-SiO$_2$ interfaces with defect state densities comparable to what has been achieved on SiC by conventional high temperature thermal oxidation [2,7]. The potential advantages of this approach to SiC-SiO$_2$ interface formation are: i) it reduces the requirement for transporting gaseous reaction products such as CO molecules away from the interface as the oxide grows; and ii) it reduces strain induced interface defect generation, since only a small fraction of the total oxide layer thickness is formed by consumption of the SiC substrate. The focus of this paper is a study of SiC-SiO$_2$ interface formation by RPAO using on-line AES to monitor interfacial bonding and the initial oxide growth rate of flat and vicinal 6H SiC wafers with Si and C faces.

2. Experimental Results

The samples used in this study were n-type 6H SiC purchased from Cree Research, Inc. with Si and C faces. One set of wafers was oriented on a principal axis, in either the (0001) direction (Si face) or the (000$\underline{1}$) direction (C-face), and the other set of wafers were vicinal and off-cut at approximately 3.5 degrees in 1120 directions. The surfaces of the on-axis Si and C face sample were prepared from a bulk ingot, as was the surface of the off-axis sample with the C face. The off-axis sample with the Si faces was an epitaxially-grown film. The surfaces of these wafers were cleaned following a procedure suggested by Cree Research, Inc.: the ex-situ cleaning consisted of i) sequential rinses in tri-chloro-ethane (TCE), acetone and methanol, followed by ii) a conventional two bath RCA clean. The sacrificial oxide formed in the second step of the RCA clean was removed by rinse in dilute HF. Samples were then loaded into a multichamber system, which provides separate UHV compatible chambers for RPAO and AES. The experimental procedure was to alternate AES measurements with RPAO processing. A similar approach had been applied to RPAO of Si, and through analysis of the AES data, two kinds of information were obtained: i) the chemical bonding at the Si-SiO$_2$ interface, and ii) the oxide thickness as a function of oxidation time [8].

Figures 1(a) and (b), respectively, give differential AES spectra for flat and vicinal SiC with Si faces obtained by RPAO using an O$_2$ source gas; Figures 2(a) and 2(b) show similar spectra for flat and vicinal SiC with C faces. The experimental processing conditions are: i) a substrate temperature of 300 °C, ii) a process pressure of 300 mTorr, iii) a plasma power to the He/O$_2$ mixture of 30 W, and iv) flow rates 200 standard cubic centimeters per minute (sccm) of He, and 20 sccm of O$_2$. The as-loaded samples showed surface contamination by oxygen and nitrogen that was not removed by the ex-situ cleaning process. No additional in-situ cleaning was attempted. This was in part due to the limited temperature capabilities of the plasma-processing chamber; the substrate heater in this chamber can not heat a wafer to a temperature greater than about 500 °C. The nitrogen contamination problem must be overcome before reliable studies of interface bonding using an N$_2$O source gas can be analyzed to determine if preferential nitrogen atom incorporation takes place at the SiC-SiO$_2$ interface. The SiC samples used in this study were n-type with a N-atom doping level of ~1.4x10^{18} cm^{-3}, so that the N contamination may possibly have its origin in the doping of the SiC wafers;

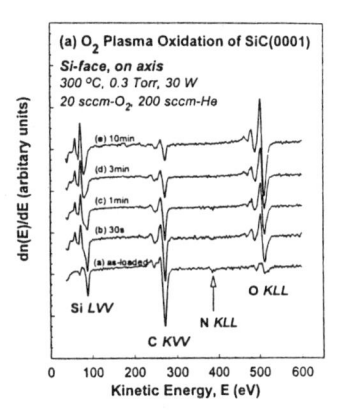

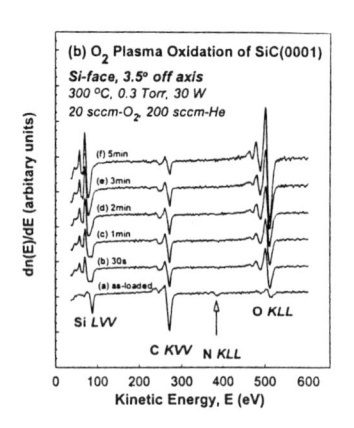

Figure 1. Derivative mode AES spectra for (a) flat and (b) vicinal 6H SiC(0001) with Si faces obtained by RPAO at 300 °C using O_2 as the oxygen atom source gas.

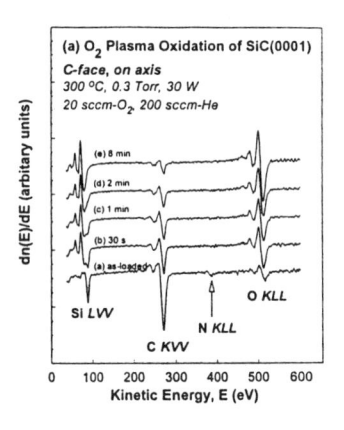

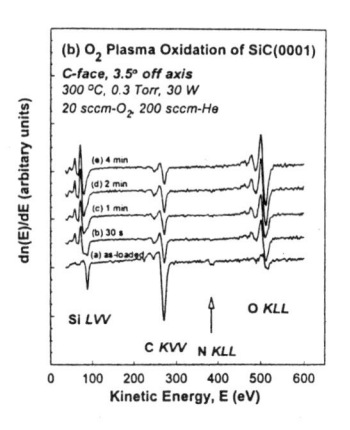

Figure 2. Derivative mode AES spectra for (a) flat and (b) vicinal 6H SiC(0001) with C faces obtained by RPAO at 300 °C using O_2 as the oxygen atom source gas.

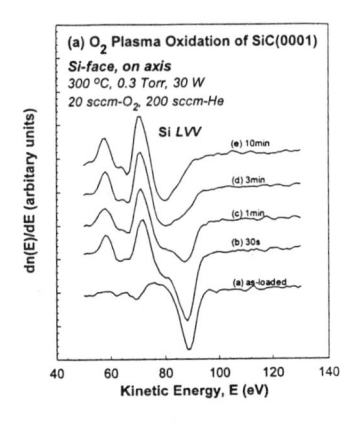

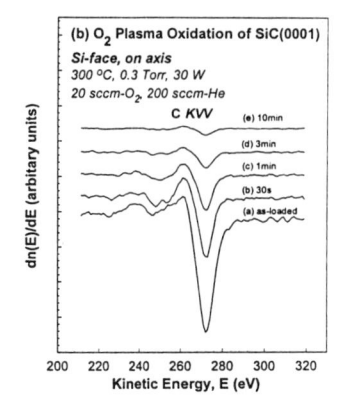

Figures 3. Changes in the (a) Si_{LVV} AES and (b) C_{KVV} AES features as a function of the oxidation time for flat 6H SiC wafers with Si faces for O_2 RPAO.

however, the surface N concentrations can not be attributed simply to the doping because the areal density of dopant atoms in a 2 nm thickness (~2 electron escape depths) is ~3x10^{11} cm^{-2}, which is considerably less than the minimum AES detection limit of ~10^{13} cm^{-2}. In the case of Si, as-loaded samples generally displayed some degree of sub-monolayer oxygen contamination, but no AES detectable nitrogen signal so that definitive studies of N interface bonding could be performed with both O$_2$ and N$_2$O source gases [7]. Prior to oxidation, the flat and vicinal SiC wafers displayed essentially the same (± 3 %) Si/C surface ratios so that changes in these ratios could be correlated with the effects of the oxidation processes. As the oxidation time for the SiC wafers was increased, there were four changes evident in the SiC spectra in Figs. 1(a) and (b) and 2(a) and (b): i) the line shapes and multiplicity of features within the Si$_{LVV}$ manifold changed, ii) the C$_{KVV}$ signal strength decreased, iii) the N$_{KLL}$ signal strength decreased, and iii) the O$_{KLL}$ signal strength increased. These spectral changes are consistent with the growth of an SiO$_2$ film on the SiC substrate, and are better displayed in the figures that follow.

Figures 3(a) and (b), and 4(a) and (b) display, respectively, changes in (a) the Si$_{LVV}$ and (b) the C$_{KVV}$ (AES) features as a function of the oxidation time for flat (Fig. 3) and vicinal (Fig. 4) SiC with Si faces. Similar spectra have been obtained for the samples with C faces. Consider first the Si$_{LVV}$ features in Figs. 3(a) and 4(a). The feature at ~88 eV is associated with Si-C bonds, in particular with a Si atom that is bonded to four C atoms. As the oxidation proceeds this feature decreases in strength as the Si-O feature at ~76 eV increases. These changes in relative intensity are due to an increase of the oxide thickness with time, and are consistent with the relative values of the electron escape depth, ~ 0.6 nm, and the oxide thickness (see below). There is also a shift of the Si-C feature to lower energy as the oxidation proceeds; this is more easily seen in the spectra for the off-axis sample. Figures 5(a) and (b) display the integrated Si$_{LVV}$ spectra for the flat and vicinal samples with Si faces. It is evident from Figs. 3(a) and 5(a) that as the intensity of the Si-C feature decreases, the position of this spectral feature moves to lower energy. In particular in Fig. 3(a), the Si-C feature is evident as a distinct spectral peak at ~88 eV prior to oxidation, after the oxidation has progressed for 3 minutes it has shifted to lower energy and appears as shoulder at ~84 eV on the 76 eV peak. In contrast, the C-Si feature in the C$_{KVV}$ spectra in Figs. 3(b) and 4(b) simply decrease in strength as the oxidation process proceeds. The absence of any significant spectral change in this region of the AES spectrum is indicative of the fact that C-O bonds are not formed to the limit of AES sensitivity, ~10^{13} cm^{-2}, during the RPAO process. Similar conclusions with regard to the lack of new features in the C$_{KVV}$ regime are also evident in the AES spectra of all samples used in this study, including flat and vicinal samples with C faces. The development of interfacial C-O bonds would produce a satellite peak at lower energy, which would increase in relative strength to the C-Si feature as the oxide growth proceeded and the AES signal became more sensitive to the SiC interface than the bulk.

The oxide thickness can be obtained to ±5 % from the relative intensity changes in the C$_{KVV}$ feature using the characteristic escape depth of 0.96 nm for 275 eV electrons. A similar approach has been applied for the determination of the oxide thickness for the RPAO process on Si, and has been validated by a direct measurement of oxide thickness by cross-sectional high resolution transmission electron microscopy. Figures 6(a) and (b) shows log-log plots, respectively of the oxide thickness versus the oxidation time for the RPAO of flat and vicinal (a) Si(111) and (b) 6H SiC(0001) with a Si face using the O$_2$ source gas. In each instance the data can be fit by a power law function, $t_{ox} = \alpha t^\beta$, where t_{ox} is the oxide thickness in nm, t is the oxidation time in minutes, and α

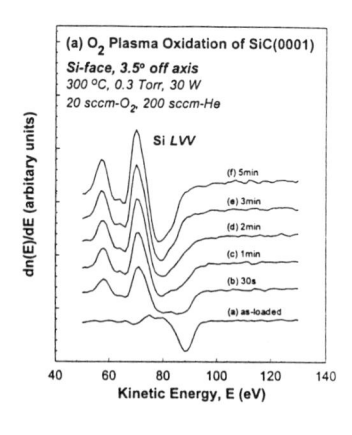

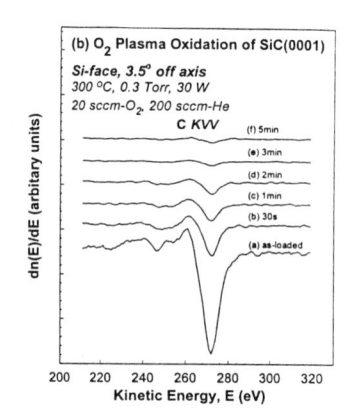

Figure 4. Changes in the (a) Si$_{LVV}$ AES and (b) C$_{KVV}$ AES features as a function of the oxidation time for vicinal 6H SiC wafers with Si faces for O$_2$ RPAO.

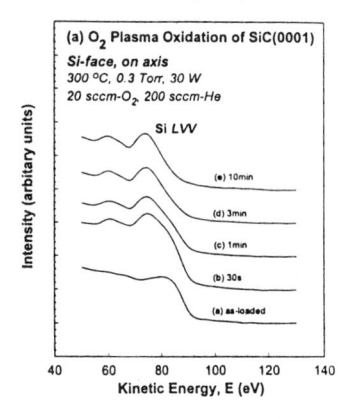

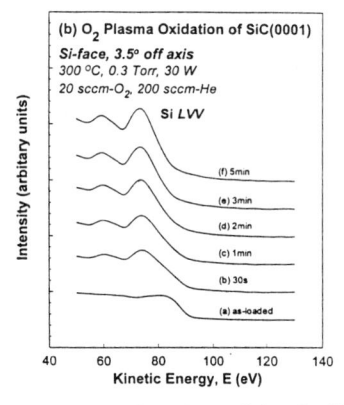

Figures 5. Integrated, n(E), Si$_{LVV}$ AES spectra as a function of the O$_2$ RPAO oxidation time for (a) flat and (b) vicinal 6H SiC(0001) with Si faces.

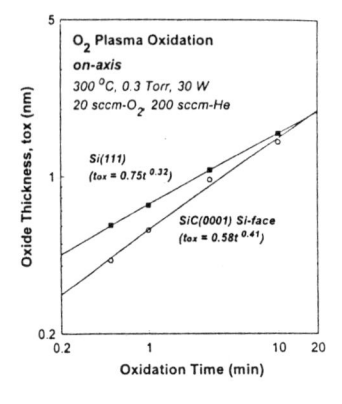

Figure 6. Log-log plots of the oxide thickness versus the oxidation time for the O$_2$ RPAO of flat and vicinal (a) Si and (b) 6H SiC. The data are fit to a power law dependence, $t_{ox} = \alpha t^\beta$, as discussed in the text.

and β are fit parameters [8]. For the all of the SiC samples of this study the power law factors are essentially the same, 0.40 ± 0.01 (see Fig. 7), whereas for the Si(111) surfaces they are smaller, ~ 0.31±0.01. The value of β for flat Si(100) surfaces is less than for Si(111), ~ 0.28 ± 0.01. Since $t_{ox} = \alpha t^\beta$, the oxidation rate, R, is given by:

$$R = d(t_{ox})/dt = \alpha\beta t^{(1-\beta)}. \quad (1)$$

When comparing oxidation rates for surfaces for which the values of β are the same, as for the different SiC surfaces, the relative values of R are independent of time, and are simply given by the ratios of the respective values of α. On the other hand, if the values of β are different as they are for comparisons between SiC and Si, then

$$(R_{SiC}/R_{Si})_{ij} = [(\alpha_{ij}\beta)_{SiC} / (\alpha_{ij}\beta)_{Si}] \, t^{(\beta(SiC)-\beta(Si))}, \quad (2)$$

and the relative rates are time dependent. Table I includes three sets of oxidation rate comparisons: i) for oxidation using for the oxidation of flat and vicinal Si(111): ii) for oxidation of flat and vicinal 6H SiC(0001) with Si and C faces; and iii) comparisons between SiC and Si(111), wherein the Si(111) data are scaled for a 3.5 degree off cut angle. In all instances the rates of oxidation for vicinal wafers are faster than for flat wafers with the same surface termination; however, there are significant differences in these relative rates. For the Si(111) wafers the oxidation rate for the vicinal samples off-cut ~5 degrees in the 11$\underline{2}$ direction is 1.6 (see Fig. 6(a)). If we assume that increases in oxidation are directly proportional to the number surface steps/cm, which for small angles, < 10 degrees, is also proportional to the off cut angle, then the increased rate, scaled down for a 3.5 degree off cut angle is approximately 1.4. For the 6H SiC(0001) wafers; the increased rates of oxidation for vicinal as compared to flat wafers are 1.6 for Si faces, and 1.1 for C faces. The oxidation rate of SiC with a C face is about 1.2 times larger than for SiC with a Si; however the oxidation for vicinal SiC with a C face is slower than that of SiC with a Si by a factor of 0.82. Finally, the comparison between the oxidation rates of Si(111) and SiC(0001) is not as straightforward as those cited above because of the different β factors in the empirical growth rate laws. Table I includes comparative oxidation rates at three different times: after 30 seconds, 60 seconds and 3 minutes. From these comparisons, the following result is obtained: i) for the SiC with Si faces, (a) for flat faces, the oxidation rate for SiC is faster for times greater than about 60 seconds, and (b) for vicinal surfaces, the oxidation rate for SiC is faster for times of at least 30 seconds or more; and ii) for SiC with C faces, (a) for flat faces, the oxidation rate for SiC is faster for times greater than at least 30 seconds, and (b) for vicinal wafers, the oxidation rate is slower for times less than 3 minutes or 180 s, but is approximately equal at 3 minutes. Since the both prefactor terms, α, and β values differ, another way to compare the relative oxidation processes for Si(111) and SiC(0001) is to compare the oxide thickness directly. These comparisons have not been scaled to take into account the differences in off cut angles between the Si and SiC samples and therefore give qualitative trends to be expected for other off cut angles. Comparisons between Si(111) and SiC(0001) with Si faces give the following results: $t_{ox})Si > t_{ox})SiC$ for $t_{ox} < 2$ nm for flat surfaces, and < 3 nm for vicinal surfaces; for Si(111) and SiC(0001) with C faces: $t_{ox})Si > t_{ox})SiC$, for $t_{ox} < 1$ nm for flat surfaces, and < 3 nm for vicinal surfaces. In general, as seen from these comparisons, Table I and Fig. 6, the oxidation rate ratio (R_{SiC}/R_{Si}) increases with increasing time.

Some preliminary studies of the oxidation rate of SiC using plasma-excited N_2O/He have been performed on flat and vicinal SiC with Si faces (see Fig. 8). For the same flow rates of same flow rates of N_2O and O_2, the N_2O process is faster than the O_2

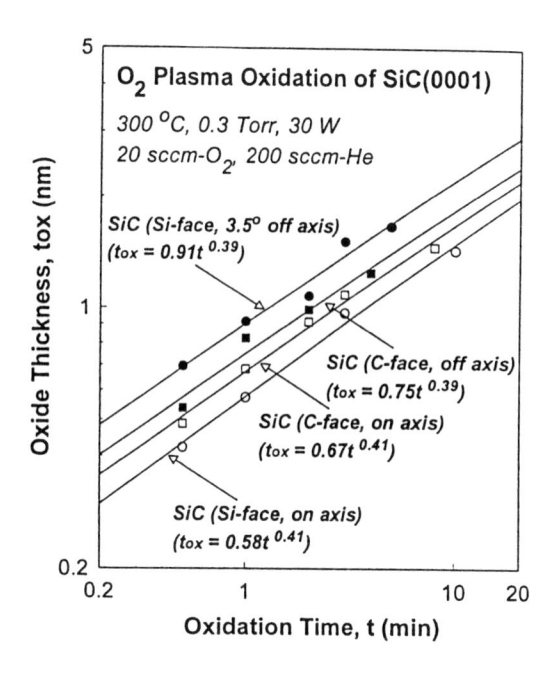

Figure 7. Log-log plots of the oxide thickness versus the oxidation time for the O_2 RPAO of flat and vicinal 6H SiC with Si and C faces (four plots are shown). The data are fit to a power law dependence, $t_{ox} = \alpha t^\beta$.

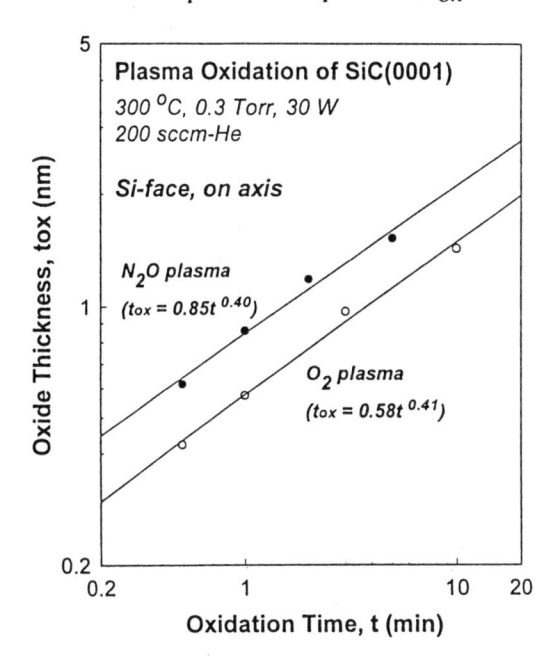

Figure 8. Log-log plots comparing the oxide thickness versus the oxidation time for the RPAO of flat SiC using O_2 and N_2O as the oxygen atom source cases. The data are fit to a power law dependence, $t_{ox} = \alpha t^\beta$.

process on SiC, but slower on Si by a factor of 1.2 [8]. However, when the flow rates of N$_2$O and O$_2$ are adjusted to make the same amount of O available, the N$_2$O and O$_2$ processes give essentially the same results for Si(100). For SiC, the following results are found: i) for the flat wafers the O$_2$ process is slower by a factor of 1.5, and ii) for the vicinal wafers by a factor of 1.1.

Table I Oxidation Rates for Flat and Vicinal Si and 6H SiC with Si and C Faces

(a) Oxidation of Si(111)

Wafer Surface	Comparative Oxidation Rate ($R_{vicinal}/R_{flat}$)
Si(111)	1.56 (5 degree off set)
	1.39 (scaled to 3.5 degrees)

(b) Oxidation of 6H SiC(0001)

Wafer Surface	Comparative Oxidation Rate ($R_{vicinal}/R_{flat}$)
6H SiC Si Face	1.57 (3.5 degree off set)
6H SiC C Face	1.12 (3.5 degree off set)

Wafer Surface	Comparative Oxidation Rate (R_C/R_{Si})
Flat Faces	1.16
Vicinal Faces	0.82

(c) Comparisons Between SiC and Si(111)* (R_{SiC}/R_{Si})

(i) SiC with Si Faces

Wafer Surface	Comparative Oxidation Rate		
	30 s	60 s	180 s
Flat	0.94	0.99	1.08
Vicinal	1.13	1.19	1.30

(ii) SiC with C Faces

Wafer Surface	Comparative Oxidation Rate		
	30 s	60 s	180 s
Flat	1.08	1.15	1.26
Vicinal	0.85	0.91	0.99

* α values for Si(111) vicinal surfaces have been normalized to 3.5 degree off cut angle comparative rates calculated from: $(R_{SiC}/R_{Si})_{ij} = [(\alpha_{ij}\beta)_{SiC} / (\alpha_{ij}\beta)_{Si}]\, t^{(\beta(SiC)-\beta(Si))}$

3. Discussion

3.1 STEP EDGE BONDING

In order to gain insight into the origins of the differences in oxidation rates between flat and vicinal wafers, it is first necessary to discuss the step edge bonding of the vicinal wafers. The dangling bond geometry for Si(111) wafers off cut in the 11$\bar{2}$ direction has been discussed in Ref. 9 and references therein. The steps are effectively two atom layers high, and the step edge atoms have a single dangling bond. The steps on 6H SiC with Si terminated surfaces are more complex due to the larger unit cell. The repeat pattern of 6H SiC extends over twelve atomic planes of alternating Si and C atoms as shown in Fig. 9. For off cut angles in the 11$\bar{2}$0 direction, the vicinal surfaces can either include two different six atom steps that are separated by Si atom terminated terrace regions, or alternatively consist of six two atom layer steps/6H repeat pattern. Figure 9 indicates the bonding arrangements for the two different six atoms steps. The first set of steps are in an ABC sequence with one carbon atom dangling bond per double layer Si-C

455

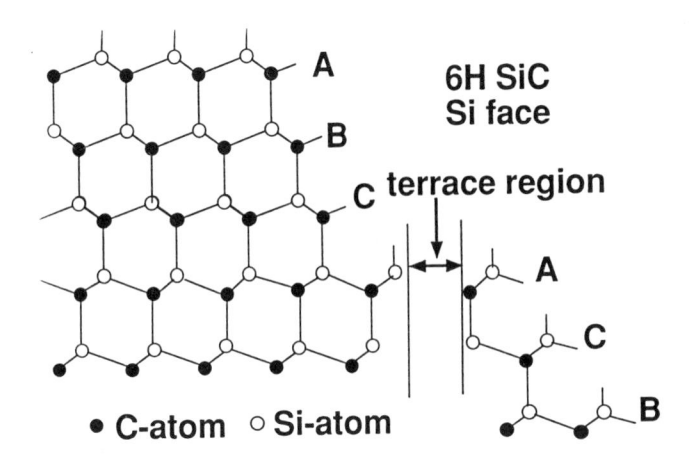

Figure 9. Schematic representation of surface and step bonding for 6H SiC vicinal wafers. The C-atom dangling bonds for the first 6-atom, 3-layer sequence are one per step edge C-atom; for the second 6-atom, 3-layer sequence, there are two dangling bonds per step edge C-atom. Steps are separated by terrace regions with Si-dangling bonds.

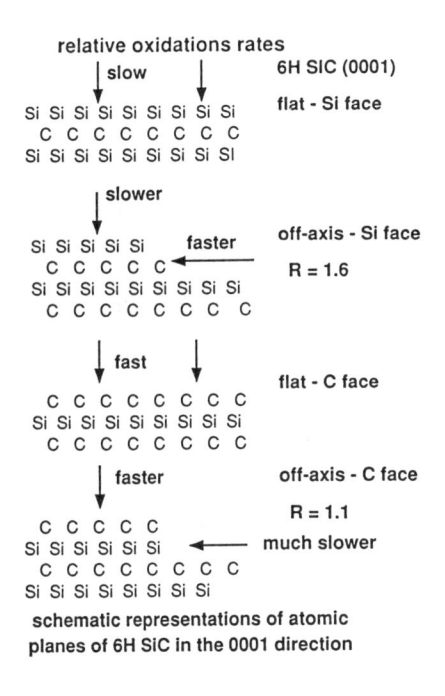

Figure 10. Schematic representations of surface and step edge oxidation processes for flat and vicinal 6H SiC with Si and C faces.

component, and the second set of steps are an ACB sequence with two carbon atom dangling bonds per double layer Si-C component. The average number of dangling bonds per step in 6H SiC(0001) is 1.5 dangling bonds per step edge atom, whereas in Si(111) off cut in the 11$\underline{2}$ direction it is only 1 dangling bond per step atom. The spacing between single steps on Si(111) off cut in the 11$\underline{2}$ direction is approximately 5 nm for a 3.5 degree off cut angle, so that the ratio of step edge to terrace dangling bonds is approximately 0.06. For a 3.5 off cut angle on SiC, the dangling bond ratio is higher by a factor of 1.5 due to the different step edge terminations of the ABC and ACB sequences. If the 6H SiC surface has two atom steps, as is likely for the wafers used in this study, the dangling bond considerations are the same. Starting with the same ABC sequence as in Fig. 9, the first three two atom steps are terminated by C atoms with one dangling, and the second three two atom steps by C atoms with two dangling bonds. This means that the number of dangling bonds per step on the average is still 1.5 times greater than for off cut Si wafers.

3.2 AES RESULTS

The analysis of the AES spectra for the flat and vicinal SiC surface RPAO processes using the O_2 source gas indicates: i) the solid state oxidation product is SiO_2; ii) there are no interfacial C-O bonds; and iii) oxidation proceeds initially more rapidly on vicinal surfaces than on flat surfaces, paralleling what has been found for Si. The primary solid state oxidation product for thermal oxidation of SiC is also SiO_2; however, there have been reports of the observation of C-O at or near the $SiC-SiO_2$ interface [2]. This means that there may be significant differences in the oxide formation chemistries by low temperature plasma-assisted and high temperature thermal oxidation processes. For example, in the plasma assisted processes the oxidation species are typically long-lived molecular metastables such as O_2^* or positive molecular ions such as O_2^+, whereas in the thermal oxidation process the oxidation species are typically O_2 molecules, and at higher temperatures, O atoms as well. Since the thermal and plasma-oxidation processes are qualitatively different with respect to the formation of C-O bonds, then the differences between the two types oxidation processes could be expected to require different post-oxidation and post-deposition procedures for forming low defect density $SiC-SiO_2$ interfaces. However, this is not the case, as is discussed in 3.3 of this section of the paper. Before discussing this aspect of the experimental results, additional aspects of the power laws fits are discussed below.

Differences between the oxidation rates of flat and vicinal wafers of 6H SiC are included in Table I. Significant differences between oxidation rates between flat and vicinal wafers indicate different modes of oxidation. For example if the oxidation rate for vicinal wafers is greater than for flat wafers, as for 6H SiC with a Si face and for Si(111), then the modes of oxidation of the flat and vicinal wafers are different (see Fig. 10). When this difference is large, it can be concluded that the oxidation of the flat wafers proceeds in a direction normal to the wafer surface, whereas the oxidation of the vicinal wafers proceeds in a direction parallel to the step surfaces. On a microscopic scale, this means that the step edge atoms of the vicinal surfaces of Si(111), and 6H SiC with a Si, have a higher reaction rate with the plasma activated oxygen species than the surface atoms. For oxidation of Si(111), the step edge atoms are Si atoms with a single dangling bond per step edge atom, whereas for the 6H SiC with the Si face, the are C atoms with two different dangling bond configurations: half of the step edge atoms have a single dangling bond, and the other half have two. These comparisons between Si(111) and SiC indicate that C step edge atoms of vicinal SiC are oxidized faster than the Si surface atoms by about a factor of about 50%, the same factor for Si step edge atoms of vicinal Si(111). However, when the difference in dangling bond factors are taken into account, the relative increase, normalized to the step edge dangling bond density, is smaller for the vicinal SiC. The oxidation rates of flat and vicinal 6H SiC with C faces are not significantly different (see Table I). This means that surface C atoms are oxidized more effectively that step edge Si atoms. Comparing Si(111) and SiC with C faces, the following observations can be made: i) surface C atoms of flat 6H SiC are oxidized about 20 % faster than surface Si atoms of Si(111); and ii) step edge Si atoms of Si(111) vicinal wafers are oxidized faster than surface Si atoms of Si(111), but ii) step edge Si atoms of 6H SiC with C faces, are not oxidized significantly faster than the surface C atoms. The power laws fits to RPAO data show that the oxidation process proceeds very rapidly initially and then slows down considerably with increasing oxidation time. This means chemical reaction rates between excited oxygen species and Si- and Si- and C-atom dangling bonds on Si and SiC, respectively, are decreased after Si-O bonds and a superficial layer of SiO_2 are formed at the onset of RPAO [10].

For the case of RPAO of Si the oxidation rate for flat surfaces is faster using O_2 than N_2O for the same relative flow rates of He and O_2 or N_2O [8]. For the RPAO of SiC the situation is reversed for flat surfaces and the oxidation rate is faster using N_2O. Studies are currently underway to address this difference. A model has been proposed for RPAO of Si that explains the differences in the oxidation rates using O_2 or N_2O source gases [11]. This model was developed primarily to explain the retention of nitrogen atoms at the Si-SiO_2 interface during oxide growth using the N_2O source gas. The surface contamination of SiC by nitrogen makes it impossible to determine whether there is nitrogen retention at the SiC-SiO_2 interface during oxide growth in N_2O. This issue is presently under study, and will require the development of procedures for ex-situ surface and or in-situ preparation that remove residual nitrogen atom contamination from the SiC surface.

3.3 ELECTRICAL PROPERTIES OF SiC-SiO_2 INTERFACES

This section summarizes results of recent electrical measurements made on MOS capacitors, in particular it compares values of D_{it} as extracted from capacitance-voltage measurements on SiC interfaces prepared on p-type substrates. The experimental studies of E. Stein von Kaminski et al. [12] have demonstrated mid-gap D_{it} values in the low 10^{11} cm^{-2}-eV^{-1} range for interfaces prepared by high temperature thermal oxidation. These device structures used metal electrodes. The attainment of low defect densities required two annealing steps: i) a post-oxidation an Ar/ H_2 mixture at 1150 °C, and ii) a conventional PMA in an H_2 containing ambient at 400 °C after metallization. When annealing procedures that did not include H_2 in the 1150 °C anneal were applied to the RPAO formed interfaces, the D_{it} values were significantly higher, in the 10^{13} cm^{-2}-eV^{-1} range [7]. However as shown in Ref. 7, these D_{it} values were significantly reduced for the RPAO interfaces following a high temperature (1150 °C) anneal in an Ar/H_2 mixture. It is interesting to note that the difference in D_{it} values between RPAO devices subjected to the Ar and Ar/H_2 anneals is very nearly equal to the density of carbon atom dangling bonds at the step edges. This suggests the possibility that carbon atom dangling bonds are not terminated by during the oxidation or oxide deposition steps, or during the conventional PMA.

Consider first the RPAO interfaces. Since our studies show no C-O bonding in interfacial regions for either the flat or vicinal wafers, it is suggested that the step edge carbon atom dangling bonds may remain unterminated after the oxidation process. The high defect density after the conventional 400 °C PMA in a hydrogen containing ambient further suggests that the step edge carbon atom dangling bonds are not hydrogen terminated by this process. If this is the case then hydrogen atom production during the 1150 °C anneal in Ar/H$_2$ is sufficiently high to produce C-H bonding at the step edges. Since thermal oxidation processes show evidence for C-O bonding in interfacial transition regions [2] and the RPAO process does not, this suggests that there might be differences related to the way step carbon atom dangling bonds are terminated, either during thermal and plasma oxidations, or in post oxidation annealing. However, this appears not to be case, since interfaces produced by high temperature thermal oxidation, also require a high temperature anneal in an H$_2$ containing ambient. This means that even though C-O bonds can be found in silicon oxycarbide transitions regions after high temperature thermal oxidation [2] they may not be formed in sufficient numbers at carbon atom step edges to neutralize dangling bond defects. Alternatively, the local bonding environment of carbon atoms in interfacial regions and at the Si-C step edges is different and may be a contributing factor to oxygen atom termination.

Finally, it is also interesting to note that fabrication of low D$_{it}$ interfaces on n-type SiC by RPAO and thermal oxidation does not require a high temperature anneal in an H$_2$ containing ambient [7]. There are two possible explanations for not requiring such an anneal for the RPAO formed interfaces: i) the active interfacial defects are in the lower half of the SiC bandgap and hence are more active in p-type material, or ii) the formation of C-O bonds at an SiC interface is Fermi level dependent favoring C-O bond formation in n-type material. The wafers that we studied were nitrogen doped n-type wafers, and there was no evidence of interfacial C-O bond formation. However, the concentration of step edge C atom dangling bonds that are available for termination by C-O bonding is below the AES detection limit. The studies on flat and vicinal 6H SiC(0001) wafers with C faces also indicate no detectable C-O bonding. It is clear that these many of the questions identified with respect to electrical results on SiC devices need additional experimental and theoretical studies before they can be satisfactorily resolved. In particular, electron spin resonance experiments, that can readily distinguish between Si and C atom dangling bonds may be helpful in resolving the issues discussed in the last two paragraphs.

4. Future Directions

There are still many unresolved issues relative to the formation and properties of SiC-SiO$_2$ interfaces that impact of device technology. One interesting issue deals the differences between oxidation process on flat and vicinal SiC substrates. This has been probed in a limited way in this paper; however, there are a number of important issues to study. First as a point of reference, RPAO should be studied on Si(111) as a function of the off-cut angle, and the off-cut direction. Note that there is one-dangling bond per step edge atom for Si(111) off cut in a 11$\underline{2}$ direction, but two dangling bonds per step edge atom if the off-cut is in the $\underline{1}$12 direction. Since SiC occurs in several different polytypes, it would interesting to do a more comprehensive study of plasma-assisted oxidation where the effects of step edge dangling bonds can be studied for geometries where there is either one or two dangling bonds per step edge atom, and not a mixture of both types of dangling bond configurations as on the vicinal 6H Si(0001) faces. Studies of cubic β-SiC with C and Si faces, off cut in 11$\underline{2}$ and $\underline{1}$12 directions would also help considerably in quantifying oxidation rate issues.

With respect to interface formation and characterization, it will be interesting to compare the initial stages of rapid thermal oxidation (RTO) of Si and C faces of 6H SiC with the

RPAO studies presented in this paper. Of interest are the oxidation kinetics, e.g., will the RTO process also be characterized by a power law time dependence?, and if so, will the power law exponent, β, be the same as for RPAO? In a similar context, it will be interesting to compare RTO and RPAO on Si(111) as well. Much of the recent research on gate dielectrics for Si technology has focused on ultra-thin dielectrics wherein direct tunneling is the dominant transport mechanism through the oxide. It would be interesting to initiate similar studies on SiC. The plasma processing approach of this paper is particularly well suited to comparative studies of both Fowler-Nordheim and direct tunneling in Si(111) and SiC since the effects of the interfaces can be separated from bulk properties of the plasma-deposited dielectric thin films. It will also be interesting to compare the properties of FETs, in particular channel mobilities, between devices fabricated by conventional thermal processing and devices in which the oxide films are prepared by the two-step plasma-assisted oxidation/deposition processes of this paper. Finally, it will interesting to study interface nitridation. This study could be done initially on p-type wafers in order to avoid the possibility of N-dopant atoms from be a source of surface contamination.

Acknowledgments

Supported the Office of Naval Research. The authors wish to acknowledge a collaboration with A. Gölz and Professor H. Kurz of RWTH-Aachen, Germany.

References

1. L.A. Lipkin and J.W. Palmour, J. Electronic Materials 25, 909 (1996), and references therein.
2. B. Hornetz, H-J, Michel and J. Halbritter, J. Mater. Res. 9, 3088 (1994).
3. J.T. Fitch, E. Kobeda, G. Lucovsky and E.A. Irene, J. Vac. Sci. Technol. B 7, 153 (1989).
4. T. Yasuda, Y. Ma, S. Habermehl and G. Lucovsky, Appl. Phys. Lett. 60, 434 (1992).
5. G. Lucovsky, Yi Ma, S.V. Hattangady, D.R. Lee, Z. Lu, V. Misra, J.J. Wortman, and J.L. Whitten, Jpn. J. Appl. Phys. 33, 7061 (1994).
6. G. Lucovsky, A. Banerjee, B. Hinds, B. Claflin, K. Koh and H. Yang, J. Vac. Sci. Technol. B 15 (1997), in press.
7. A. Gölz, R. Janssen, E. Stein von Kamienski and H. Kurz, in The Physics and Chemistry of SiO2 and the Si-SiO2 Interface, Ed. by H.Z. Massoud, E.H. Poindexter and C.R. Helms (Electrochemical Soc., Pennington, 1996), p. 753.
8. G. Lucovsky, H. Niimi, K. Koh, D.R. Lee and Z. Jing, in Ref. 7, p. 441.
9. C.H. Bjorkman, C.E. Shearon, Jr., Y. Ma, T. Yasuda, G. Lucovsky, U. Emmerichs, C. Meyer, K. Leo and H. Kurz, J. Vac. Sci. Technol. B 11, 964 (1993).
10. 14. P. Thanikasalam, T.K. Whidden and D.K. Ferry, J. Vac. Sci. Technol. B 14, 2840 (1996).
11. K. Koh, H. Niimi and G. Lucovsky, MRS Symp. Proc. (1997), in press.
12. E. Stein von Kamienski et al., Microelectronic Engineering 28, 201 (1995).

CHALLENGES IN THE OXIDATION OF STRAINED SiGe LAYERS

VALENTIN CRACIUN*, JUN-YING ZHANG AND IAN W. BOYD
Electronic and Electrical Engineering,
Torrington Place, University College London,
London WC1E 7JE,
United Kingdom
*permanent address: Laser Department
National Institute for Laser, Plasma and Radiation Physics
Bucharest V, RO-76900
Romania

Abstract

The use of ultraviolet (UV) and vacuum ultraviolet (VUV) photons generated from low pressure Hg lamps and excimer lamps, respectively, to enhance the growth of ultrathin dielectrics films on SiGe strained layers and on Si at temperatures below 550 °C is described in this paper. The thickness, structure (composition) and electrical properties of the grown oxide layers were investigated by ellipsometry, Raman and electron spectroscopy, scanning and transmission electron microscopy, Rutherford backscattering, Fourier-transform IR spectroscopy, capacitance-voltage and current-voltage measurements in order to characterise the oxidation process. The use of the UV and VUV radiation during the oxidation resulted in significant enhancements of the growth rate when compared with thermal oxidation, especially for the novel excimer lamp sources, which are more powerful light sources than Hg lamps. Using this technique, high quality, stoichiometric SiO_2 layers were grown on Si or Si capped SiGe strained layers, without any measurable relaxation of the strained substrate. New effects such as the formation during the UV-assisted oxidation of SiGe samples of nanocrystalline Ge particles inside the grown oxide layer which exhibit visible photoluminescence and stress effects induced by the oxidation process are also presented.

1. Introduction

The use of high quality strained SiGe layers grown by molecular beam epitaxy (MBE) or chemical vapour deposition (CVD) on Si substrates holds great promise to enhance some of the Si properties like hole mobility or open exciting new applications in the area of heterojunction bipolar transistors, quantum wells or superlattices [1, 2]. Oxidation processes performed at high temperatures in various ambients are a key ingredient of any Si based integrated circuit (IC) manufacturing process. The grown oxide layers perform a multitude of tasks on the IC chip. Therefore, the study of the SiGe oxidation has many motivations, such as the growth of high quality oxide layers for MOS applications with a SiGe channel, the passivation of the structures, or simply to investigate the modifications of the Si thermal oxidation process induced by the presence of Ge atoms.

The need to minimise the thermal budget of VLSI and ULSI electronic devices is a necessity, more so for structures incorporating strained SiGe layers. It has been shown that such layers begin to relax if processed at temperatures in excess of 800 °C [3]. On the other hand, the thermal oxidation of Si or SiGe at temperatures below 800 °C proceeds at very low growth rates. Consequently, various low temperature techniques are being explored that involve substitute sources of energy for the enhancement of the growth rate. Over the years, photo-induced processing has in particular received considerable attention [4, 5]. One advantage of photoprocessing is that the surface is not subjected to damaging ionic bombardment which can be the case in plasma assisted systems [6]. Depending upon the wavelength used, the energy supplied can excite the surface through bond-weakening or bond-breaking, lead to structural modification by exciting charge-carriers and induce the formation of charged species, or generate new reactive species. So, lasers and lamps can be used to produce silicon (Si) dielectrics. Experiments on the use of Hg lamps to stimulate SiGe and Si oxidation at temperatures where thermal oxidation would be considered negligible are discussed here. Recently, new types of VUV pseudo-continuous high power (up to 10 W) excimer lamps have been developed [7, 8]. Their use to grow oxides at even lower temperatures than those obtained with Hg lamps are also presented.

2. Thermal Oxidation Of Silicon Germanium Layers

It was found almost one decade ago that the presence of Ge atoms at the interface between the Si substrate and the grown oxide layer strongly influence the classical Deal-Grove oxidation kinetics of Si [9]. Since then, many other investigations have tried both to quantify this effect in terms of Ge concentration, temperature and ambient atmosphere (wet versus dry) and to use it to enhance the growth rate of thin oxides [10-12]. There is now a reasonable understanding of the kinetics of thermal oxidation of SiGe layers and the multilayer structure of the grown oxides.

For low to medium Ge content (<60 %) it has been found that the oxide grows faster on SiGe samples than on Si. The structure of these oxides usually begins with a thin, Ge free, stoichiometric SiO_2 layer. All the Ge atoms initially present in this region were snow-ploughed to the SiO_2/Si interface, where they accumulated in a layer with a higher concentration than that of the as-grown SiGe layer (it is well known that because of the large difference between the Gibbs free energies of formation of SiO_2 and GeO_2, Si atoms are always preferentially oxidised [13]). As shown by Frey et al. [14], during the oxidation of SiGe layers, the beginning of the Ge entrapping into the grown oxide depends on the actual segregated Ge concentration at the interface. When the Ge concentration in this accumulation layer reaches values around $N_{th}=2 \times 10^{15}$ atoms/cm^2, some of Ge atoms start to be incorporated into the advancing SiO_2 front, resulting in a mixed SiGe oxide layer. This process will continue as long as the Ge concentration at the interface remains higher than N_{th}. If the oxidation process is carried out long enough, then when most of the Ge atoms have been incorporated into the growing oxide and the interface concentration is smaller than N_{th}, the Ge atoms are again rejected by the advancing oxide layer which thereafter will consist of pure SiO_2. Because of these atomic rearrangements at the interface between the oxide and the substrate, the electrical properties of the grown layers, although dramatically improved in recent years, still lag far behind the best reported values for the SiO$_2$/Si structures [11, 15].

To suppress this atomic rearrangement and avoid the SiGe layer relaxation, alternative low-temperature oxidation processes such as plasma (microwave or ECR) assisted oxidation [16, 17] or ion beam induced oxidation [18] which are performed at temperatures low enough such that Ge atoms are practically immobile have been employed. It was thus possible to obtain a congruent (Si$_{1-x}$Ge$_x$)O$_2$ layer, whose stoichiometry reflects the initial Si$_{1-x}$Ge$_x$ strained layer stoichiometry.

However, an interesting process associated with this (SiGe)O$_2$ formed layer was the discovery that Ge atoms trapped inside could be precipitated during an annealing treatment at temperatures above 500 °C into nanocrystalline (nc) particles [19]. These oxide layers containing the nc-Ge particles were further shown to exhibit visible photoluminescence (PL) [20, 21]. As the annealing process produces particles having a broad distribution of sizes [21, 22], it has not been possible to clearly identify the particle sizes responsible for the PL effect. As there is growing evidence that very small clusters, which have more than 50 % of their constituent atoms on the outer shell, tend to crystallise in a differently ordered structure to that found for bulk materials [23], there have been suggestions that such new crystalline structures could be responsible for the observed PL properties of nc-Ge particles [24]. The UV-assisted oxidation process discribed here has allowed us to clearly show that the nc-Ge particles responsible for visible PL retained the diamond-type crystalline structure.

3. UV-assisted oxidation

3.1. HG LAMP OXIDATION SYSTEM

The Hg lamp-based UV oxidation system used (see figure 1) consisted of a low pressure Hg lamp contained within a stainless steel chamber which can be evacuated by a turbomolecular pump down to 10^{-6} torr and then filled with high purity VLSI grade oxygen (O$_2$) [25, 26]. The lamp was water-cooled, approximately 20x20 cm in size, and emitted strongly in the UV at 254 nm (hv=4.98 eV) with a weaker output in the VUV at 185 nm (hv=6.7 eV). For simplicity, the term UV is used to describe the collective output from this lamp. The substrates used were either p-type (100) oriented single crystal Si with a resistivity of 2-10 Ωcm, or MBE grown Si$_{0.8}$Ge$_{0.2}$ strained layers measured by Rutherford Back-Scattering (RBS) to be 150-160 Å thick on 1000 Ωcm p-type (100) Si substrates. Some of these SiGe layers had a 15 nm thick Si capping layer also MBE grown. The samples were placed on a heater at a distance of some 4.5 cm below the lamp and could be resistively heated to 550 °C. Prior to UV exposure in 1 atmosphere of O$_2$, the samples were cleaned in hydrofluoric (HF) acid (1 %) and rinsed in deionised water.

After UV exposure for various times, the thickness of each oxide grown was measured by single wavelength or spectroscopic ellipsometry. RBS and Auger X-ray Electron Spectroscopy (AES) were used to determine the depth distribution and chemical state of Si, Ge and O species in the oxidised layers as a function of depth. RBS (random and channelling) measurements were performed using 2 MeV He$^+$ ions and a scattering angle of 170° and the spectra were interpreted using the RUMP

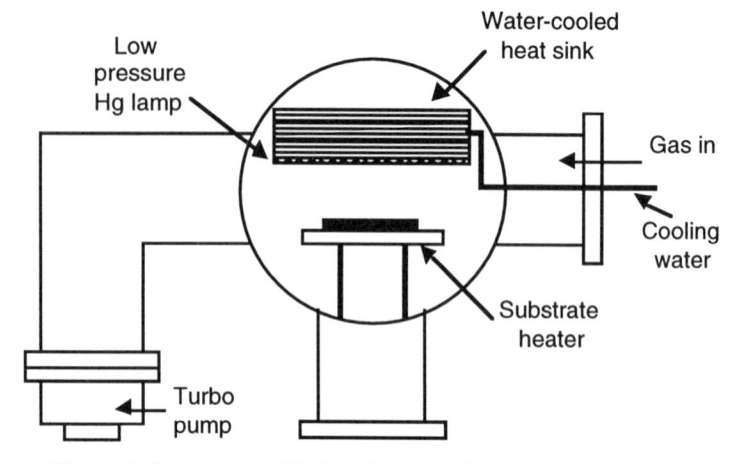

Figure 1. Low pressure Hg lamp-based oxidation system [25, 26].

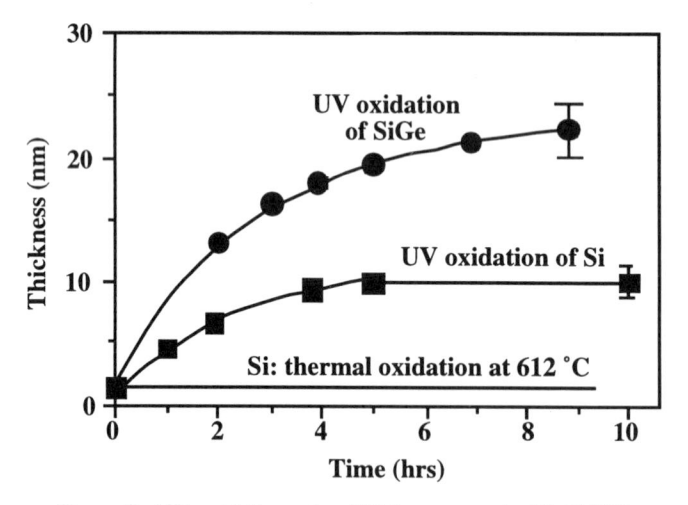

Figure 2. UV oxidation rate of SiGe compared with Si [28].

simulation program [27]. Raman and transmission electron microscopy were performed to identify the crystalline structure of the layers formed. Fourier transform IR spectroscopy (FTIR) was use to check the molecular structure of the layers whereas capacitance voltage (C-V) and current-voltage (I-V) measurements were performed in order to assess their electrical properties.

3.2. SiGe OXIDATION

The oxide layer thicknesses grown at 550 °C on the SiGe and on the Si samples are shown on Fig. 2, together with data for Si thermal oxidation at 612 °C. From the plot of the square of the oxide thickness as a function of the UV exposure time resulted that the UV-assisted oxidation process broadly follows the usual parabolic law found for Si thermal oxidation. The striking feature of the results is the further enhanced oxidation rate of SiGe over that of Si (roughly a factor of two) under identical irradiation conditions [28]. Although an enhancement of the Si oxidation rate in the presence of Ge atoms has been observed for wet oxidation conditions [29], these present observations are the first yet reported showing such an effect during dry conditions.

RBS measurements confirmed continued oxide growth in the SiGe for longer irradiation times. Although not shown here, RBS spectra also indicated that most of the Ge atoms were continuously rejected by the growing oxide layer with only a small fraction being trapped inside the grown layer. As the leading edge of the Ge RBS profiles acquired from samples oxidised for different irradiation times does not change significantly this implies that the trapped Ge atoms in the SiO_2 layer are completely immobile during prolonged oxidation.

The structure of the oxidised samples, as revealed by simulations of the RBS spectra and confirmed by AES depth profiling measurements, consists of three regions [28]. The outermost layer of 4-5 nm thickness is always pure SiO_2. This covers a region of SiO_2 containing from 5-7% of trapped Ge. The remaining unoxidised SiGe lies beneath these SiO_2 layers. Channelled RBS spectra revealed that the SiGe layer retained some epitaxial order [30]. The Ge concentration at the interface just prior to the beginning of the entrapping was calculated from the thickness of the formed pure SiO_2 layer and the initial Ge concentration at 2×10^{15} atoms/cm^2, in perfect agreement with the estimation of the model of Frey et al. [14] for thermal SiGe oxidation.

Nevertheless, it remains intriguing as to why so few Ge atoms are trapped inside the growing SiO_2 layer when thermodynamic considerations and other experimental data [31] suggest that at the relatively low temperature used, almost all the Ge should have been incorporated into the growing oxide as GeO_2. From the RBS simulation of the oxidised sample structure, an interdiffusion coefficient of Ge in Si of about 10^{-18}cm^2/s is estimated. When extrapolating the diffusion coefficients measured between 850 - 1100 °C [32] to our oxidation temperature of 550 °C, however, one obtains a maximum possible diffusion coefficient of 10^{-20} - 10^{-21} cm^2/s. It is thus very clear that not only during the usual thermal oxidation [33] but also now during UV assisted oxidation, Ge and Si atoms attain much larger than expected diffusion coefficients.

FTIR spectra obtained from the processed samples and displayed on Fig. 3 showed only the absorption bands corresponding to SiO_2, i.e. with peaks at 1075 cm^{-1} (asymmetric stretching vibration), 800 cm^{-1} (bending vibration), and 450 cm^{-1} (rocking mode) implying that the entrapped Ge was in a different chemical form than that of GeO_2 found during the thermal oxidation. Valuable detail of the entrapped Ge has recently been provided by TEM, Raman and XPS studies [34]. Cross-sectional high resolution TEM micrographs showed the presence of well crystallised material inside the grown oxide layer. The size of these nanoparticles varied between 2 and 8 nm, and the average d_{111} interplanar distance was found to be around 0.33 ± nm, corresponding to either pure Ge or very Ge-rich SiGe. TEM and Raman studies have also clearly indicated that the nc-Ge particles formed during low temperature UV-assisted oxidation of SiGe preserve the diamond-type crystalline structure even when their dimension was less than 3 nm. Such layers containing the nc-Ge particles were found to exhibit visible photo-luminescence in the 550-800 nm range, with a peak at 2.2 eV [35].

To further check the origin of the PL effect, part of a 6 hrs oxidised SiGe sample was dipped in diluted HF in order to remove the grown oxide layer but, as checked by Raman and XPS analysis, not the remaining SiGe layer. The total visible light

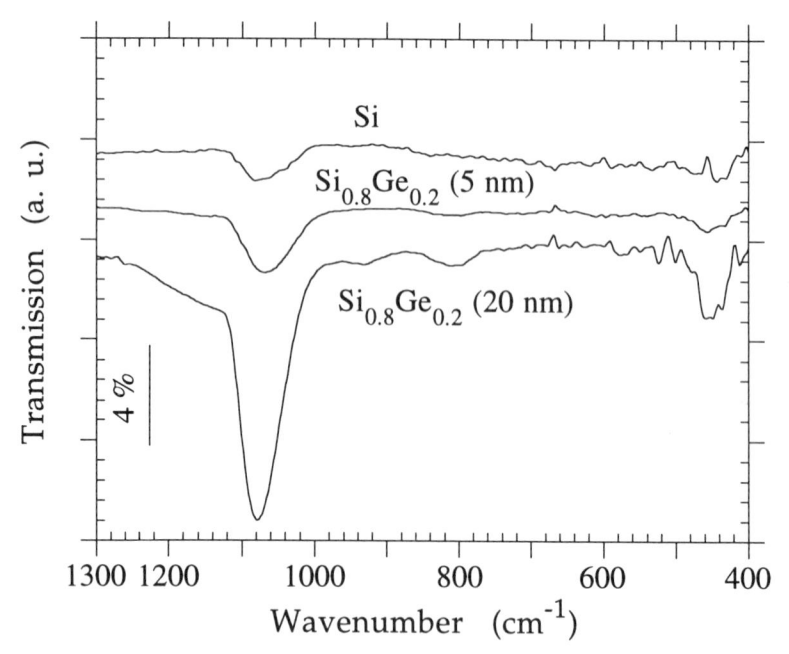

Figure 3. FTIR spectra of Si and SiGe samples oxidised at 400 °C for 4 hours.

emission in the 550-800 nm range resulting from excitation with an Ar$^+$ laser at 488 nm was then measured. While the as-grown part showed a bright emission, the HF-dipped part did not show any at all. This confirmed that the nc-Ge particles, having the diamond-type crystalline structure are responsible for the visible photoluminescence exhibited by SiGe UV-oxidised samples.

Another interesting question raised by this study is the survival of the nc-Ge particles formed during prolonged oxidation under conditions when Si atoms situated at greater depths below the SiO$_2$ layer are oxidised. Grazing incidence X-ray diffraction (GIXD) spectra were recorded for the HF-dipped and as-grown regions. Both spectra exhibit a very clear (311) SiGe diffraction peak. This is rather unexpected, because the samples position was fixed during the measurements. This peak certainly comes from the SiGe layer still present on both samples whose presence has been clearly indicated by both XPS and Raman measurements. Its presence implies that the SiGe layer suffered fragmentation during its continuous rejection by the growing oxide layer into small randomly oriented regions. Similar fragmentation of a buried SiGe layer has been recently reported by Fatemi et al. [36], and showed to be caused by the thermal stress produced during a rapid cool-down step from 800 °C. However, as the temperature during the UV-assisted oxidation process was only 550 °C and the cooling-down process was slow, the thermal stress could not be responsible for the fragmentation of the buried SiGe layer in our case. Another observation from the GIXD spectra was the displacement of the SiGe (311) peak position for the as-grown sample with respect to its position for the sample which had the SiO$_2$ layer removed. This is a clear indication of a stress level within the oxidised sample which can therefore account for the fragmentation of the buried SiGe layer.

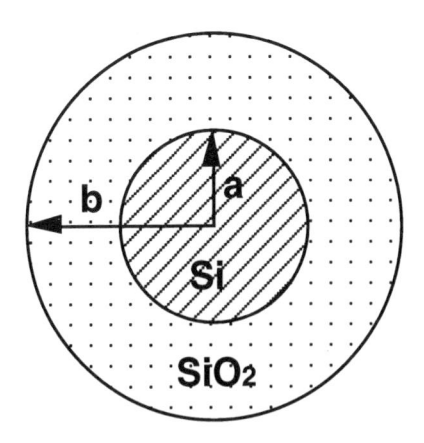

Figure 4. Oxidation of a small spherical particle

The retardation of the oxidation in particles exhibiting small radii (usually less than 50 nm) is a well documented effect for Si [37-38] and has been explained by the presence of a high stress level induced by the formation of new oxide layers. This can be understood by extending the usual Deal-Grove model [39] to the oxidation of a small spherical particle as shown in Fig. 4 (the notations are similar to that of Ref. [37] in order to allow easy comparison with the oxidation of cylindrical structures to be made).

The oxidant concentration in the oxide layer is obtained by solving the Laplace equation in polar co-ordinates together with the usual boundary conditions at the O_2 source and oxide-substrate interface, respectively. For the oxidant concentration at the interface, $C_i(a)$, one obtains:

$$C_i(a) = \frac{1}{k_s} \frac{C^*}{\frac{1}{k_s} + \frac{1}{h}\frac{a^2}{b^2} + \frac{1}{D}\frac{a}{b}(b-a)} \qquad (1)$$

where C, C*, h, k_s, D, a, and b are the oxidant concentration in the SiO_2, the oxidant solubility in the SiO_2, the surface mass transfer of the oxidant, the surface reaction rate constant, the radius of the crystalline core, and the radius of the oxide surface, respectively (see Fig. 4). The oxidant concentration at the interface for planar oxidation is already well-known [Deal-Grove, Ref. 39]:

$$C_i(x_0) = \frac{1}{k_s} \frac{C^*}{\frac{1}{k_s} + \frac{1}{h} + \frac{1}{D}(x_0)} \qquad (2)$$

Since a<b, the oxidant concentration at the interface is higher for a spherical particle as compared to a planar surface, and the smaller the radius the larger the difference, similar to the results of Kao et al. [37] obtained for cylindrical convex structures. As the oxidation rate at the oxide-semiconductor interface is:

$$\frac{dx_0}{dt} = \frac{1}{N} k_s C_i \qquad (3)$$

where N is the number of oxidants required to form a cubic unit of oxide, a small spherical particle should always oxidise faster than a planar surface. However, the experimental data presented by Okada and Iijima [38] concerning the oxidation of small Si particles showed quite the contrary. For temperatures less than 950 °C, the oxidation rate decreases with reduced temperature and particle radius, compared to that of planar Si, becoming vanishingly small for particle radii smaller than a certain value. This self-limiting oxidation effect has been explained by the stress effect due to the increase of the volume associated with the transformation of the semiconductor material into oxide. In the case of Si it has been shown [40] that D is the term which is mostly affected by the stress, the oxidation rate on small radii structures being diffusion-limited.

Using the resolvable oxidation data collected from oxidised cylindrical Si structures of various sizes and at several temperatures ranging from 850 to 950 °C, Liu et al. [40] have calculated the dependence of the activation energy of the diffusivity D as a function of the ratio of the outer radius to the inner radius of the structure, $\gamma=a/b$. Our measurements have indicated that the average dimension of the nc-Ge particle is around 5 nm [34, 35]. The Ge concentration of the starting material was 20 %. This implies that a remaining 5 nm radius Ge particle would have come from a 7.5 nm radius SiGe particle which was selectively oxidised. The oxidation of 2.5 nm of Si results in a 5.7 nm thick oxide layer. Such a particle thus has $\gamma =2.1$ and, if it were made of Si, the diffusion activation energy would increase from 2.10 eV for planar oxidation [41] to 2.22 eV [40]. This corresponds to a decrease of the oxygen diffusion coefficient at 550 °C from 1.52×10^{-13} cm^2/s to 2.8×10^{-14} cm^2/s, and then one obtains from (1) and (3) a decrease of the oxidation rate by almost an order of magnitude.

Unfortunately, it is not yet possible to estimate for the Ge case the oxidation retardation because of lack of experimental data. Nonetheless, these simple numerical estimations give a plausible explanation for the survival of the nc-Ge particles during prolonged oxidation of SiGe samples.

Hg LAMP OXIDATION OF Si

As the electrical measurements performed on the SiGe oxidised samples have shown a very poor oxide, with high leakage currents and trapped charge, we have investigated the oxidation of Si capped SiGe strained layers and pure Si wafers [30]. We did not find any difference in stoichiometry, growth rate, or electrical parameters between these two sample lots, so both will be called Si in the following.

The oxide thicknesses grown on the Si as a function of UV exposure time for substrate temperatures from 400 °C to 550 °C are shown in Fig. 5. Oxidation below 400 °C was particularly slow and therefore not studied. As can be seen from the figure, several distinct growth regions can be defined. At each temperature used, an initially rapid near-linear growth regime (as best can be determined from the data) is followed by a transition region. After some characteristic time, a very small growth rate occurs

Several factors may contribute to the observed reaction enhancement when compared to that of thermal oxidation. Some of the UV light (particularly that whose photon energy is greater than 5.1 eV) will be absorbed by the O_2 in the gas phase, according to the reaction paths:

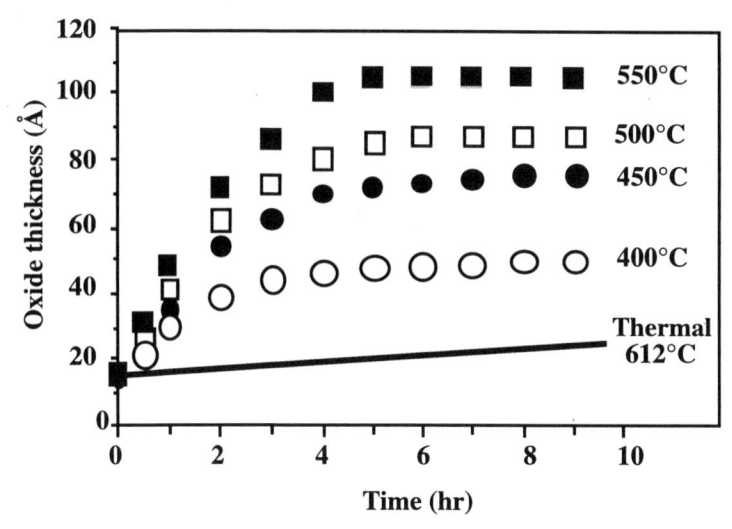

Figure 5. Oxide thickness on Si against UV exposure time for a range of temperatures [26].

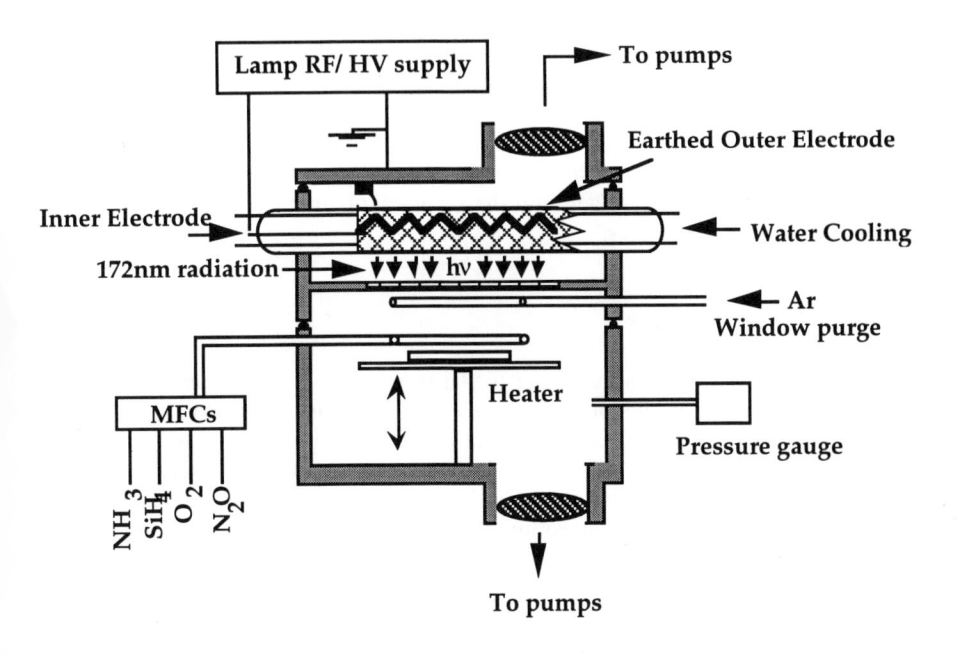

Figure 6. General experimental arrangement for oxidation/photodeposition [43].

$$O_2 + h\nu\ (\lambda=185\ nm) \longrightarrow 2O \qquad \sigma_1=1.6\times10^{-20}\ cm^2 \qquad (4)$$
$$O + O_2 + M \longrightarrow O_3 + M \qquad k_2=2.8\times10^{-12}\ cm^3molec^{-1}s^{-1} \quad (5)$$
$$O_3 + O \longrightarrow 2O_2 \qquad k_3=2.8\times10^{-15}\ cm^3molec^{-1}s^{-1} \quad (6)$$
$$O_3 + O_2 \longrightarrow O_2 + O_2 + O \quad k_4=7.7\times10^{-16}\ cm^3molec^{-1}s^{-1} \quad (7)$$

Since radiation at $\lambda = 254$nm is also emitted by the lamp, the following reaction also occurs [42]:

$$O_3 + h\nu\ (\lambda=254\ nm) \longrightarrow O_2 + O \qquad \sigma_5 = 1.1\times10^{-17}\ cm^2 \qquad (8)$$

leading to photodissociation, the release of oxygen atoms, and ultimately the formation of ozone, which is known to be a stronger oxidising agent than its allotropic cousin, molecular oxygen [43].

Therefore during UV exposure, the SiO_2 layer is supplied with molecular and atomic oxygen as well as ozone from the surrounding gas. As it is well known that O_3 decomposes extremely rapidly at temperatures above 100 °C [42] we proceed by assuming that in these experiments the role of ozone is predominantly to act as a supplier of O and O_2 species to the oxide surface. Thus, as a consequence of the UV radiation, O atoms and O^- and O_2^- ions can be formed in addition to the usual species present during thermal oxidation [43]. Any radiation reaching the surface will also induce photoelectron ejection from the Si into the SiO_2. The relative importance of these mechanisms is difficult to assess because of the lack of knowledge concerning even the thermal Si-O_2 reaction. Nevertheless, a tentative model to explain these enhancement and saturation features has been developed and is described in Ref [43].

FTIR measurements of the oxides indicated, from the position of the Si-O stretching vibration at 1075 cm^{-1} and the full width at half maximum (FWHM) value of 74 cm^{-1}, that the layers were essentially identical to those conventionally prepared by thermal methods. Interface roughness, a parameter which can affect both the stoichiometry of the first oxide layers and thickness measurements, was studied using cross-sectional high resolution TEM. The average roughness was found to be the order of 1 nm, a value corresponding to approximately 3 layers of SiO_2. Such roughness is not unusual for Si-SiO_2 interfaces on oxidised samples prepared using the RCA cleaning method [44] and can in fact limit the accuracy of any thickness measurements performed on oxidised Si samples.

4. Excimer Lamp Sources

Whilst the advantages of lamps over laser for large area processing is quite apparent it nevertheless remains that the photon fluxes and the wavelength availability are somewhat restricted. This has lead to interest in the development of alternative lamp sources, one of which is the excimer lamp whose operating principle is based on the so-called dielectric barrier discharge (silent discharge). Light emission of these excimer lamps relies on the radiative decomposition of excimer states created in a gas by a silent discharge in a high pressure (few hundred mbar) gas column [45]. In a dielectric barrier discharge lamp, one or both electrodes is electrically insulated. A high voltage (7-10 kV) and high frequency (100-500kHz) supply is applied, causing an arc discharge to occur randomly across the surface of the dielectric. The charge build-up on the dielectric surface immediately decreases the field in the discharge gap and extinguishes the arc. The duration of an arc is $\approx$2-5ns and several are formed quasi-simultaneously at a frequency of twice the driving frequency. Each individual current filament is known as a microdischarge because of the short time duration and low electrical energy involved. The self-extinguishing feature of this type of discharge enables the use of high gas pressure ($\approx$500mbar) without causing electrode sputtering and associated contamination problems common with traditional arcs. In the case of rare gas halide mixtures, and in particular for the mixtures of ArF, KrF and XeCl, the common laser frequencies of 192, 248 or 308nm are obtained. For excited molecular complexes in pure rare gases, lower wavelength continua are generated at 126, 146 and 172nm for Ar, Kr, and Xe. The well-known pumping mechanism [46] reveals the lack of any bound ground level indicating the splitting of the dimer formed when emission occurs, and therefore the absence of self-absorption of the radiation by the gas phase [47].

Examples of application towards dielectric formation are given here, were (100) oriented n-type Si samples were oxidised under the 172 nm radiation produced by a Xe_2^* lamp [48]. The distance between the sample and lamp was 1 cm with the power density on the sample being about 100 mW/cm^2 (see Fig. 6).

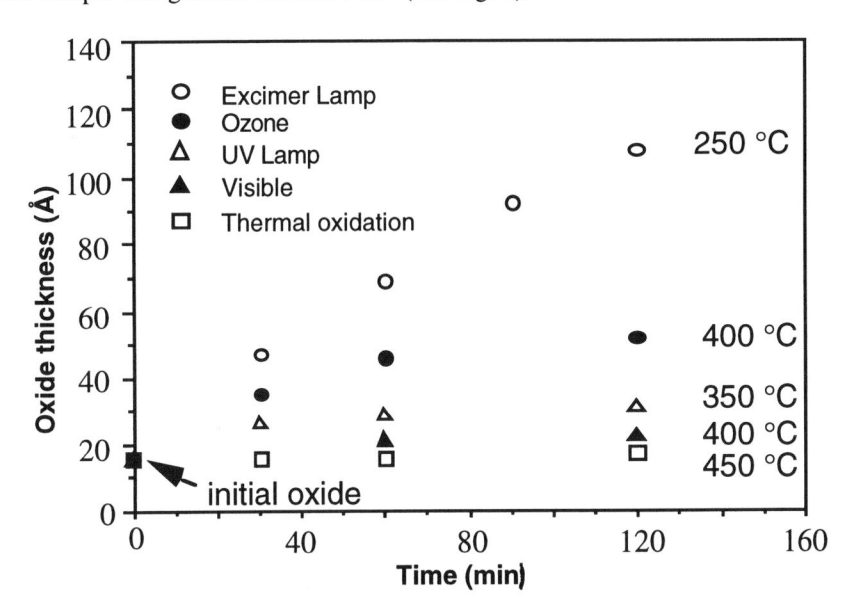

Figure 7. Oxide thickness as a function of exposure time for different photon and thermal sources [48].

Figure 7 shows the oxide thickness of the films grown at 5 mbar as a function of exposure time compared with our previously obtained data for layers grown using a low pressure mercury lamp, visible radiation, ozone and conventional furnace oxidation. As can be seen, the oxidation rate using the excimer lamp is by far the highest. The excimer lamp induced oxidation rate at 250 °C is more than three times greater than obtained using a low pressure mercury lamp at 350 °C.

The FTIR spectra obtained for these films were similar to those recorded for thermally grown oxides [49], with peaks at 1075 cm^{-1}, 800 cm^{-1} and 450 cm^{-1}. The full width half-maximum of the stretching peak at 1075 cm^{-1} was 75 cm^{-1}, much smaller than that reported for excimer laser-induced oxides (133 cm^{-1}) [50].

Simple MOS capacitors have been fabricated using excimer lamp grown oxides 11 nm thick, with an evaporated Al top contact of area 3.2×10^{-3} cm^2. Table 1 shows the surface state charge density and the leakage current density of as-grown films to be 2.4×10^{11} cm^{-2} and 10^{-3} A/cm^2 at 1.2V, respectively.

$C_{ox} = 48.2$ pF
$C_{fb} = 8.3$ pF
$V_{FB} = -0.77$ V
$Q_{ss}/q = 4.5 \times 10^{10}$ /cm^2
Area $= 3.2 \times 10^{-3}$ cm^2

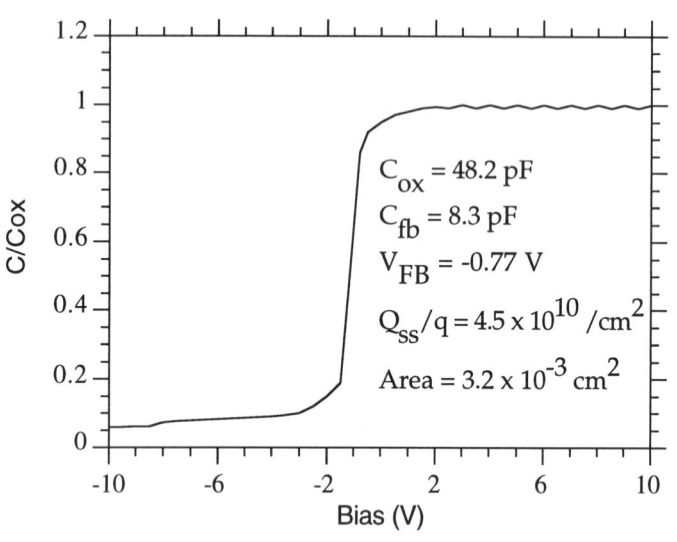

Figure 8. A typical high frequency C-V characteristic of an MOS capacitor incorporating a VUV grown oxide annealed for 2 hrs [48].

The electrical properties of these films are not yet comparable with the best high temperature thermal oxide films grown under strict clean room conditions. It has been recently found that UV-O$_3$ annealing of Ta$_2$O$_5$ films improves their electrical properties significantly [51]. This technique has been applied to our films, which were subsequently annealed at 400 °C in 1 atmosphere of O$_2$ and irradiated by a 172 nm excimer lamp. A comparison of the electrical properties of the films after different annealing times is shown in Table 1. After 2 hrs annealing, the fixed oxide charge density changed from 2.4×10^{11} cm^{-2} for the as-grown material, to 4.5×10^{10} cm^{-2}. The I-V characteristics of the MOS capacitors also show that after the VUV annealing, the leakage current reduced dramatically. Figure 8 shows a typical high frequency C-V trace of one MOS capacitor incorporating a VUV grown oxide after 2 hrs VUV annealing. The flat-band voltage V_{FB} is slightly negative, which indicates the presence of positive fixed charges near the SiO$_2$/Si interface. The positive fixed oxide charge number density (Q$_f$/q) calculated from the flat-band voltage is found to be 4.5×10^{10} cm^{-2}. Clearly, very good electrical properties can be obtained for the SiO$_2$ layers by this VUV-induced low temperature oxidation and annealing technique. It is important to note at this stage that the optimum growth and annealing steps may not yet have been achieved nor understood.

TABLE 1. A comparison of the electrical properties of the oxides after different annealing times (172 nm lamp, 1000 mbar O_2, 400 °C) [48].

Annealing time (hrs)	Charge on surface (cm^{-2})	Leakage current density at 1.2 V (A/cm^2)
0	2.4×10^{11}	1×10^{-3}
1	7.5×10^{10}	8×10^{-4}
2	4.5×10^{10}	4×10^{-4}

5. Conclusions

The microstructure of dielectric layers formed by low temperature UV assisted dry oxidation of SiGe strained layers has been studied. The initial oxidation rate is higher than for pure Si under identical conditions. The role of the UV-generated ozone is to supply oxygen atoms to the oxide surface. The UV radiation provides another mechanism for enhancement, although its precise action role needs further studies.

A 3-layered structure, similar to that found for thermal oxidation is formed after prolonged oxidation of SiGe consisting of pure SiO_2 on top of a SiO_2 layer which contains nc-Ge particles, under which lies unoxidised SiGe. The diffusion coefficients of Si and Ge extrapolated to the low temperatures used in our investigations cannot account for this Ge enrichment behaviour. The nc-Ge particles formed during the UV-assisted oxidation process always retain the diamond type crystalline structure and exhibit visible photoluminescence. The electrical properties of the grown dielectrics are quite poor and for MOS applications the best solution so far seems to be the oxidation of a Si layer grown on top of the SiGe layer. Although the temperature used for this process is low enough that there is no thermal relaxation of the strained layers, the high stress level caused by the oxidation process itself can produce defects in the SiGe layer.

These experiments provide exciting evidence for the possibility of realistic applications of UV radiation for producing rapid low temperature oxidation of Si and Si alloys. This UV-induced low temperature oxidation process appears to be very attractive for future ULSI technology, especially when followed by a UV/O_3 annealing step, as the lamp technology used can be readily extended to large area (e.g. 12 inch) wafers.

Acknowledgements: The authors wish to thank Alec Reader for TEM and RBS analysis and P. Andreazza for GIXD analysis.

References

1. People, R. (1986) Physics and applications of Ge_xSi_{1-x}/Si strained layer heterostructures *J. Quant. Electron.* **QE22**, 1696-1711.
2. Fukatsu, S. (1995) Optoelectronic aspects of strained $Si_{1-x}Ge_x/Si$ quantum wells, *J. Mater. Sci.: Mat. in Electronics* **6**, 341-349.
3. Zaumseil, P., Fischer, G. G., Brunner, K., and Eberl, K (1997) Comparison of the thermal stability of $Si_{0.603}Ge_{0.397}/Si$ and $Si_{0.597}Ge_{0.391}C_{0.012}/Si$ superlattices structures, *Appl. Phys. Lett.* **81**, 6134-6136.
4. Boyd, I.W. (1987) *Laser Processing of Thin Films and Microstructures: Oxidation, Deposition and Etching of Insulators*, Springer, New York.
5. Eden, J.G. (1992) *Photochemical Vapour Deposition*, Vol. 122 of Chemical Analysis, J. Wiley & Sons Inc., Canada.
6. Nishino, S., Honda, H., Matsunami, H. (1986) SiO_2 film deposition by KrF excimer laser irradiation *Jpn. J. Appl. Phys.* **25**, L87-L89.
7. Kogelschatz, U. (1992) Silent-discharge driven excimer UV sources and their application, *Appl. Surf. Sci.* **54**, 410-423.
8. Boyd, I. W., and Zhang, J. Y. (1997) New large area ultraviolet lamp sources and their applications, *Nucl. Instr. Meth.* **B121**, 349-356.
9. Holland, O. W., White, C. W., and Fathy, D. (1987) Novel oxidation proces in Ge^+-implanted Si and its effect on oxidation kinetics *Appl. Phys. Lett.* **51**, 520-522.
10. LeGoues, F.K., Rosenberg, R., Nguyen, T., Himpsel, F., and Meyerson, B.S. (1989) Oxidation studies of SiGe, *J. Appl. Phys.* **65**, 1724-1729.
11. Nayak, D.K., Kamjoo, J, Park, K. S., Woo, J. C. S., Wang, K. L. (1992) Rapid isothermal processing of strained GeSi layers *IEEE Trans. Electron Dev.* **39**, 56-63.
12. Eugene, J., LeGoues, F.K., Kesan, V. P., Iyer, S. S., and d'Heurle, F. M. (1991) Diffusion versus oxidation rates in silicon-germanium alloys *Appl. Phys. Lett.* **59**, 78-80.
13. Castle, J. E., Liu, H. D., Watts, J. F., Zhang, J. P., Hemment, P. L. F., Bussmann, U., Robinson, A. K., Newstead, S . M., Powell, A. R., Whall, T. E., Parker, E. H. C. (1991) An investigation of $Si_{0.5}Ge_{0.5}$ alloy oxidation by high dose oxygen implantation, *Nucl. Instr. Meth.* **B55**, 697-700.
14. Frey, E. C., Yu, N., Parikh, N. R., Swanson, M. L., and Chu, W. K. (1993) Transition between Ge segregation and trapping during high-pressure oxidation of Ge_xSi_{1-x}/Si, *J. Appl. Phys.* **74**, 4750-4755.
15. Mukhopadhyay, M., Ray, S. K., Nayak, D. K., Maiti, C. K. and Shiraki, Y. (1995) Properties of SiGe oxides grown in a microwave oxygen plasma *J. Appl. Phys.* **78**, 6135-6140.
16. Mukhopadhyay, M., Ray, S. K. , Nayak, D. K., and Maiti, C. K. (1996) Ultrathin oxides using N_2O on strained $Si_{1-x}Ge_x$ layers, *Appl. Phys. Lett.* **68**, 1262-1264.
17. Tchikatilov, D., Yang, Y. F. , and Yang, E. S. (1996) Improvement of SiGe oxide grown by electron cyclotron resonance using H_2O vapor annealing, *Appl. Phys. Lett.* **69**, 2578-2580.
18. Vancauwenberghe, O., Hellman, O. C., Herbots, N., and Tan, W. J., (1991) New SiGe dielectrics grown at room temperature by low-energy ion beam oxidation and nitridation, *Appl. Phys. Lett.* **59**, 2031-2033.
19. Paine, D. C., Caragianis, C., Kim, T. Y., and Shigesato, Y. (1993) Visible photoluminescence from nanocrystalline Ge formed by H_2 reduction of $Si_{0.6}Ge_{0.4}O_2$, *Appl. Phys. Lett.* **62**, 2842-2844.
20. Liu, V. S., Chen, M. J. S., Nicolet, M.-A., Engels, V. A., and Wang, K. L. (1993) Nanocrystalline Ge in SiO_2 by annealing of $Ge_xSi_{1-x}O_2$ in hydrogen *Appl. Phys. Lett.* **62**, 3321-3323.
21. Zhu, J. G., White, C. W., Budai, J. D., Withrow, S. P., Chen, Y. (1995) Growth of Ge, Si, and SiGe nanocrystals in SiO_2 matrices, *J. Appl. Phys.* **77**, 4386-4389.
22. Maeda, Y. (1995) Visible photoluminescence from nanocrystallite Ge embedde in a glassy SiO_2 matrix: Evidence in support of the quantum-confinement mechanism *Phys. Rev.* **B51**, 1658-1670.
23. Saito, Y. (1979) Crystal structure and habit of silicon and germanium particles grown in argon gas *J. Cryst. Growth* **47**, 61-72.
24. Kanemitsu, Y., Uto, H., Matsumoto, Y., and Maeda, Y. (1992) On the origin of visible photoluminescence in nanometer-size Ge crystallites, *Appl. Phys. Lett.* **61**, 2187-2189.
25. Nayar, V., Patel, P., and Boyd, I.W. (1990) Atmospheric pressure, low temperature (<500 °C) UV/ozone oxidation of silicon *Electronics Letters* **26**, 205-206.

26. Kazor, A., and Boyd, I.W. (1991) UV-assisted growth of 100A thick SiO_2 at 550 °C *Electronics Letters* **27**, 909 -911.
27. Doolittle, L.R. (1986) A semiautomatic algorithm for rutherford backscattering analysis *Nucl. Instrum. Meth. Phys. Res.*, **B15**, 227-231.
28. Craciun, V., Boyd, I. W., Kersten, W. J., Reader, A. H., Hakkens, F. J. G., Oosting, P. H., and Vandenhoudt, J. D. E. W. (1994). Microstructure of oxidised layers formed by the low-temperature ultraviolet-assisted oxidation of strained $Si_{0.8}Ge_{0.2}$ layer on Si .*J.Appl. Phys.* **75**, 1972-1974.
29. Paine, D.C., Caragianis, C., and Schwartzman, A.F. (1991) Oxidation of $Si_{1-x}Ge_x$ alloys at atmospheric and elevated pressure *J. Appl. Phys.* **70**, 5076-5084.
30. Craciun, V. and Boyd, I. W. (1993) Low temperature UV induced oxidation of silicon and silicon-germanium strained layers, in R. B. Fair and B. Lojek (eds.), First International Conference on Rapid Thermal Annealing, Scottsdale, Arizona, pp. 363-368.
31. Walle, G.F.A. van de, Ijzendoorn, L.J. van, Gorkum, A.A. van, Heuvel, R.A. van den, and Theunissen, A.M.L. (1990) Thermal stability of strained $Si/Si_{1-x}Ge_x/Si$ structures *Semicond. Sci.Technol.* **5**, 345-347.
32. Paine, D. C., Caragianis, C., and Shigesato, Y. (1992) Nanocrystalline germanium synthesis from hydrothermally oxidized $Si_{1-x}Ge_x$ alloys*Appl. Phys. Lett.* **60**, 2886-2888.
33. Rai, A.K., and Prokes, S.M. (1992) Wet oxidation of amorphous SiGe layer deposited on Si(001) at 800 and 900 °C *J. Appl. Phys.* **72**, 4020-4025.
34. Craciun, V., Reader, A.H., Vandenhoudt, D.E.W., Best, S.P., Hutton, R.S., Andrei, A., Boyd, I.W. (1995) Low temperature UV oxidation of SiGe for preparation of Ge nanocrystals in SiO_2 *Thin Solid Films* **255**, 290-294.
35. Craciun, V., Boulmer-Leborgne, C., Nicholls, E. J., and Boyd, I. W. (1996) Light emission from germanium nanoparticles formed by ultraviolet assisted oxidation of silicon-germanium *Appl. Phys. Lett.* **69**, 1506-1508.
36. Fatemi, M., Thomson, P. E., and Twigg, M. E. (1995) Thermal stress-induced, high-strain fragmentation of buried SiGe layers grown on Si, *Appl. Phys. Lett.* **67**, 2678-2680.
37. Kao, D.-B., McVittie, J. P., Nix, W. D., Saraswat, K. C. (1988) Two-dimensional thermal oxidation of silicon-II. Modeling stress effects in wet oxides *IEEE Trans. Electron. Dev.* **ED-35**, 25-37.
38. Okada, R., and Iijima, S. (1991) Oxidation property of silicon small particles *Appl. Phys. Lett.* **58**, 1662-1663.
39. Deal, B. E. and Grove, A. S. (1965) General relationship for the thermal oxidation of silicon *J. Appl. Phys.* **36**, 3770-3778.
40. Liu, H. I., Biegelsen, D. K., Ponce, F. A., Johnson, N. M., and Pease, R. F. W. (1994) Self-limiting oxidation for fabricating sub-5 nm Si nanowires *Appl. Phys. Lett.* **64**, 1383-1385.
41. Massoud, H. Z., Plummer, J. D., and Irene, E. A. (1985) Thermal oxidation of silicon in dry oxygen *J. Electrochem. Soc.* **132**, 1745-1753.
42. Blaluch, D. L., Cox, R. A., Hampson, R. F., Herr, J., Tore, A., and Watson, R. T. (1980) Evaluated kinetic and photochemical data for atmospheric chemistry *Phys. Chem. Ref. Data* **9**, 295-471.
43. Boyd, I. W. (1996) Dielectric photoformation on Si and SiGe, in F. Roozeboom (ed.), Advances in Rapid Thermal and Integrated Processing, Kluwer Academic Publisher, Dordrecht, pp. 235-264.
44. Meuris, M., Verhaverbeke, S., Mertens, P.W., Heyns, M.M., Hellemans, L., Bruynseraede, Y., and Philipossian, A. (1992) The relationship of the silicon surface roughness and gate oxide integrity in NH_4OH/H_2O_2 mixtures*Jpn. J. Appl. Phys.* **31**, L1514-L1517.
45. Elliasson, B. and Kogelschatz, U. (1991) Modeling and applications of silent discharge plasmas *IEEE Trans. Plasma Sci.* **19**, 309-323.
46. Rhodes, Ch. K. (1984) *Excimer Lasers*, Topics in Appl. Phys., Vol. 30, Springer, Berlin.
47. Gilbert, J, and Baggot, A. (1991) *Essentials of Molecular Photochemistry* Blackwell, Oxford.
48. Zhang, J. Y., Boyd, I. W. (1996) Low temperature photo-oxidation of silicon using deep UV radiation, *Electron. Lett.* **32**, 2097-2098.
49. Boyd, I.W., and Wilson, J.I.B. (1982) A study of thin silicon dioxide films using infrared absorption techniques *J. Appl. Phys.* **53**, 4166-4172.
50. Richter, H., and Orlowski, T.E. (1984) Ultrafast UV-laser-induced oxidation of silicon: control and characterization of the $Si-SiO_2$ interface *J. Appl. Phys.* **56**, 2351-2355.
51. Shinriki, H. and Nakata, M. (1991) $UV-O_3$ and dry-O_2: Two-step annealed chemical vapor-deposition Ta_2O_5 films for storage dielectrics of 64-Mb DRAM's *IEEE Tansactions on Electron Devices* **38**, 455-462.

THE CURRENT STATUS AND FUTURE TRENDS OF SIMOX/SOI, NEW TECHNOLOGICAL APPLICATIONS OF THE SIC/SOI SYSTEM.

J. STOEMENOS
*Aristotle University of Thessaloniki Physics Department,
54006 Thessaloniki, Greece.*

Abstract

The state of the art of the Silicon On Insulator (SOI) technology is presented The significant difference in the formation of thermally grown oxide and the buried oxide (BOX) produced by high dose oxygen implantation in silicon (SIMOX) will be highlighted. The different sources of the defects in the Si-overlayer and the SiO_2 buried layer produced during implantation and annealing treatment including Si-islands formation and strained Si-Si bonds in the BOX are discussed. A comparative study of the two most successful technologies SIMOX and wafer bonding is included. The feasibility to extend the SOI structures in the SiC is shown.

1. Introduction

One of the most promising structures, which are suitable for the deep-sub-0.1µm devices, is Silicon On Insulator (SOI) which is especially useful for the low-voltage, low-power, CMOS technology [1,2]. MOSFET devices fabricated on SOI have several advantages such as reduced parasitic capacitance, latch-up elimination, soft-error-rate reduction and simple device isolation [3].

SOI permits the realization of fully depleted (FD) MOSFET which exhibit sharper subthreshold slope, reduced short-channel effects and are free of kink phenomena, even without body contacts. Moreover, they are more dynamically stable than partially depleted devices in terms of dynamic floating body effects [4]. The SOI structures are also of substantial importance for the future Si nanodevices operating on the basis of quantum mechanical phenomena [5].

From all the different approaches for SOI technologies only SIMOX and wafer bonding (WB) are seriously considered for production in an industrial scale [6]. In the present paper the main characteristics of the SIMOX wafers will be discussed. SIMOX permits an excellent thickness control of the Si-overlayer even for thickness below 50nm, which is especially useful for FD MOSFETs.

Three are the main methods for realization of wafer bonded SOI structures, namely by plasma-assisted chemical etching (PACE) [7], by bond and etch-back SOI (BESOI) [8] and by the new Smart-Cut® technology [9]. The advantages and the disadvantages of SIMOX and the wafer bonded SOI methods regarding the quality of the Si-overlayer and the buried oxide will be discussed.

The wafer bonding technique is also applicable in other semiconductors, which can be bonded with SiO_2. The formation of such structures with the SiC (SiCOIN) seems to be very promising for SiC devices working at high temperature [10]. Also the advantages of the 3C-SiC epitaxially grown on the Si-overlayer of SOI wafers will be presented [11].

2. SIMOX technology

Silicon separation by implanted oxygen (SIMOX) in order to form a buried SiO_2 layer is one of the best examples of the successful co-operation of recent technological advancements in the field of ion-implantation and materials science [12].

2.1 General characteristics of standard SIMOX

Commercially available SIMOX wafers are produced in the Eaton NV-200 implanters at energy 190 keV with a dose $1.8 \times 10^{18} O^+ cm^{-2}$ at implantation temperature 600 to 650° C, subsequently annealed for 6h at 1320° C under Ar + 1% O_2 atmosphere. In standard SIMOX the Si-overlayer and the buried oxide (BOX) layer are about 220nm and 390nm thick, respectively [13]. The dislocation density in the Si-overlayer of the standard material is of the order $1-5 \times 10^5$ cm^{-2}. These are threading edge type dislocations with Burger vectors of 1/2<110> types. They appear in pairs with distance usually less than 0.1 µm, denoted by the letter T in Fig.1 However a multistep implantation and annealing process can reduce the dislocation density to 10^4 cm^{-2} [14].

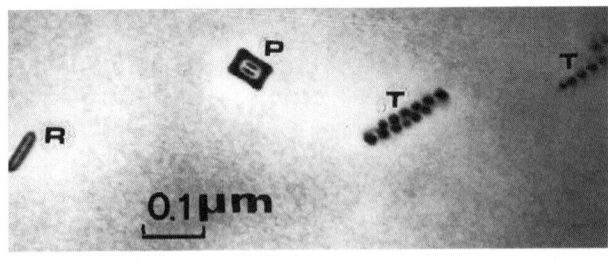

Fig.1. Defects in the Si-overlayer of the standard SIMOX, pair dislocations are denoted by the letter T, stacking fault complex in the form of orthogonal pyramid is denoted by the letter P, a shallow prismatic stacking fault is denoted by the letter R. The fault is located on the Si/SiO$_2$ interface.

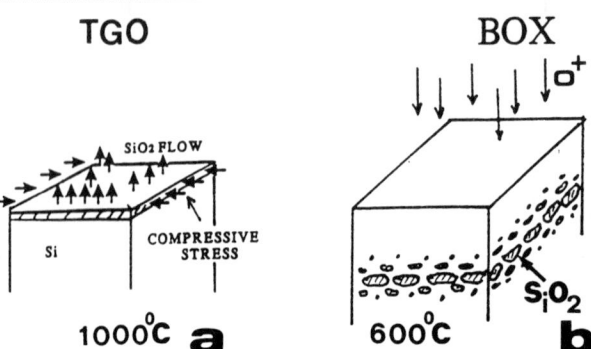

Fig.2. Schematic showing the differences in the formation of Si/SiO$_2$ interfaces in a) thermally grown oxide b) buried oxide.

At the back side of the Si-overlayer, orthogonal shaped stacking faults complexes and prismatic SF are observed, denoted by the letter P and R respectively in Fig.1. Their density is about 10^6 cm^{-2}. The prismatic SFs are very shallow defects, which consist from opposite facing SFs terminated in a Lomer Cottrell sessile dislocation.

2.2 Formation of the buried oxide layer.

The formation of the BOX in SIMOX is significantly different from the thermally grown oxide of the bulk silicon. In thermally grown oxide the conversion of Si to SiO$_2$ involves a 2.2 fold increase in molar volume. In the case of thermally grown oxide an additional 1.2 molar volume, per unit volume of oxidized Si, must be obtained in the direction normal to the Si surface. This extra volume is mainly accommodated by viscous flow of the oxide at temperatures above 950° C, as schematically depicted in Fig.2a. Below this temperature viscous flow is not observed. In this case intrinsic stress is developed [15].

Volume accommodation is also possible by emission of Si atoms, which become self-interstitial (Si$_I$) according to the form:

$$xSi + O_2 \rightarrow SiO_2 + (x-1) Si_I \qquad (1)$$

A completely free of strain SiO$_2$ formation requires x=2.2 which implies an excess of Si interstitial in the silicon matrix. Theoretical and experimental works on the silicon self-diffusion reveal, that Si$_I$ have high formation energy and very low migration energy. The sum of these two activation energies is about 5eV [16-19]. Therefore, under normal oxidation conditions the formation of SiO$_2$ by emission of Si$_I$ is insignificant because of the high activation energy of the Si$_I$ formation. The ratio of the Si interstitials to oxidized silicon atoms is less than 10^{-3} [20,21].

In SIMOX where the BOX is formed at 600° C the oxidation by emission of Si$_I$, as it is described by equation (1), dominates because during ion implantation the energy to break the silicon bonds is provided by the ion beam that creates silicon vacancies and self-interstitials. The former participate in the formation of the SiO$_2$ while the later migrate almost athermally by virtue of their high diffusivity towards the surface [22], as shown in Fig.2b.

According to equation (1) for a free of strain oxidation, a flux 0.63F$_o$(cm^{-2}s^{-1}) of Si$_I$ out of the BOX is required, where F$_o$ is the flux of the implanted oxygen. Therefore in order to avoid the development of defects due to self-interstitial supersaturation a fast migration of the Si$_I$ to the surface, which is the natural sink for Si interstitial, must be maintained during implantation. Moreover self-interstitial supersaturation can build up a chemical potential which opposes the oxidation reaction. Hu [16] estimated the Si$_I$ supersaturation considering the flux of the Si$_I$ interstitials migration to the surface in the case of SiO$_2$ precipitate formation in Czochralsky Si. In a first approximation the same equation is applicable in the case of SIMOX. Thus the

Si$_I$ supersaturation C_I/C_I^* is related with the oxygen flux as follows

$$\frac{C_I}{C_I^*} = 1 + \frac{0.63 \, F_o w^2}{8 \, D_s C_s} \qquad (2)$$

Where w is the distance from the free surface, D_s is the Si$_I$ self-diffusivity and C_s the Si concentration equal to 5.5×10^{22} atoms cm^{-3}, C_I^* is the equilibrium Si$_I$ concentration. The supersaturation C_I/C_I^* can be reduced by reducing the oxygen flux F_o, the implantation depth w or by increasing the Si self-interstitial migration diffusivity D_S [16]. Molecular dynamic simulations in Si have been used to follow low-energy ion/surface [22]. The simulation shows that the Si$_I$ diffusivity has a maximum along [100] direction and a minimum along <111> directions in (001) wafers. In contrast the migration energy for vacancies is higher than Si$_I$ and there is no tendency for vacancies diffusion to proceed preferentially toward the surface. The differences in migration energy of the Si$_I$ and vacancies as well as the preferential migration of the former toward the surface readily explains the formation of voids which were observed in the top of the Si-overlayer [23].

2.3 Defects due to Si$_I$ supersuturation

Most of the defects in SIMOX are related with Si$_I$ supersaturation and the difficulty of the Si$_I$ to be incorporated in low energy lattice sites at the surface. Significant Si growth occurs at the surface due to the Si$_I$ migration there. This is a homoepitaxial growth where the BOX is the source of the Si atoms. The quality of the epitaxial layer is strongly depended from the quality of the initial surface, the growth temperature, the vacuum and the deposition rate, which in SIMOX is the flux of the Si$_I$ towards the surface. Since the implantation temperature is around 600°C, the conditions for a perfect Si epitaxial growth are similar with these of MBE, conditions, which are not satisfied in the conventional implanters. The growth defects at the uppermost part of the Si-overlayer are mainly small dislocation loops of extrinsic type, semiloops and segments of dislocations as it is shown by the cross-section and plane-view micrographs in Figs.3a. Most of these semiloops escape to the surface during the subsequent high temperature annealing (HTA). However some of them are extended downwards to the Si-overlayer in the form of semiloops. These loops are pinned at the Si/BOX interface resulting in the formation of pairs of threading dislocations denoted by the letter T in Fig.1. The important role of the surface to be the sink of Si$_I$ becomes evident by implanting through a very thin screen oxides 40 to 50 nm. In this case the dislocation density increases to about three orders of magnitude [24].

Up to now we considered the problem of the easy incorporation of the Si$_I$ on the surface by assuming free migration of Si$_I$ from the BOX to the surface. The requirement for this process is the existence of easy paths for the Si$_I$ migration to the surface. This is not always ease because of the SiO$_2$ precipitates in the Si-overlayer.

481

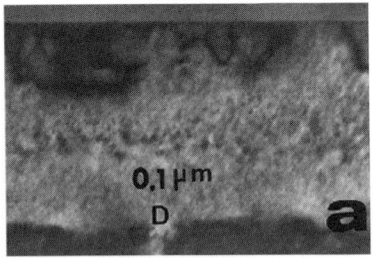

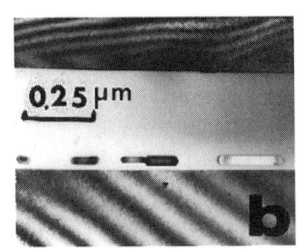

Fig.3.a) Defects at the uppermost part of the Si-overlayer after low dose implantation, $0.15 \times 10^{18} O^+ cm^{-2}$. Segments of dislocations and semiloops are evident at the uppermost part. A very high density of dislocation is also observed in the implanted zone. b) Cross-section micrograph from the standard SIMOX, small Si islands were formed at the backside of the BOX.

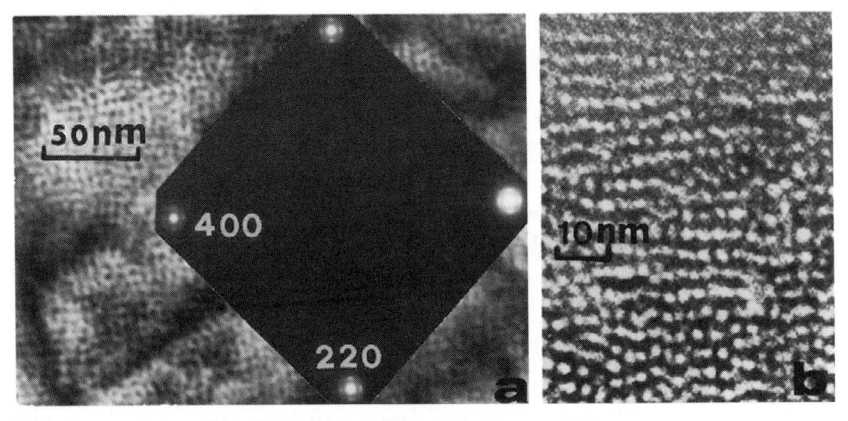

Fig.4 Periodic arrangement of the SiO_2 precipitates in the Si-overlayer of the as implanted SIMOX. The precipitates form a 3D structure along the three-<100> directions. a) Plane view micrograph, the diffraction in the inset reveals satellite spots around the main spots due to the periodic arrangement of the SiO_2 precipitates b) Periodic precipitates in cross-section.

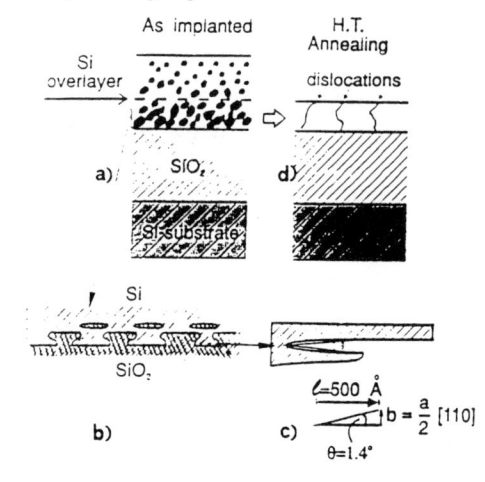

Fig.5. Schematic showing the generation of dislocations in the Si-overlayer during the HTA due to the small Si matrix misorientation near the Si/SiO_2 interface. a) SIMOX formation, the black dots represent precipitates. b) The structure of the Si-overlayer near the Si/SiO_2 interface. c) The generation of a perfected dislocation due to misorientation by an angle θ of the agglomerated Si areas during HTA. d) Generation of threading dislocation in the Si-overlayer.

Oxygen implantion has a Gaussian distribution, which results in the formation of SiO_2 precipitates in the Si near the BOX, as shown in Fig.2b. The SiO_2 precipitates inhibit the Si_I diffusion because the diffusivity of Si in SiO_2 is extremely low [25]. The inability of the Si to migrate as self-interstitial during oxidation giving the necessary space for the formation of relaxed SiO_2 results in the formation of densified SiO_2. It must be pointed out that oxidation occurs during oxygen implantation even at the liquid nitrogen temperature [26]. The oxidation is preferentially shifted at the front side of the BOX and leaves Si islands at the backside of the BOX as shown in Fig.3b, also densifies the BOX. Due to the stressed SiO_2 multiple stacking faults are formed in the Si-overlayer near the Si/BOX interface [27]. The development of stress also results in a slight misorientation of the Si islands inside the BOX which are responsible for the formation of defect during the HTA as will be shown in paragraph 2.4.

A three dimensional periodic arrangement of SiO_2 precipitates in the Si-overlayer could facilitate the Si_I migration to the surface. A periodic arrangement of SiO_2 precipitates along the three equivalent <100> crystallographic directions has been already observed, as shown in Fig.4a and 4b, when this periodic arrangement appears the final dislocation density is very low [28]. The periodicity of the SiO_2 precipitates is about 5nm and the possibility to use this natural periodic structure in the future nano-devices is attractive.

Defects due to the radiation damage are formed near the Si-overlayer/BOX interface These defects are evident during the very early stage of implantation before the SiO_2 precipitates to be visible, they are small segments of dislocations, denoted by the letter D, as shown in Fig.3a. These defects are eliminated during the HTA because they are pinned by the growing SiO_2 precipitates and finally are annihilated [29].

Point defects are produced during implantation in the BOX by the trapping of Si atoms forming strained bonds like $O_3 \equiv Si \cdots Si \equiv O_3$ [30,31]. The strained Si-Si bonds can relax by capturing a hole and forming an E_1' center. Other point defects are attributed to the strained Si-O-Si bonds formed due to the densification of the buried oxide [32].

SiO_2 can be densified by a 30% reduction in volume suggesting a significant modification of the structure by producing smaller ring networks of SiO_2 tetrahedral [32]. These bonds are more sensitive to the irradiation and can be viewed according to the reaction:

$$O_3 \equiv Si\text{-}O\text{-}Si \equiv O_3 + En \rightarrow O_3 \equiv Si\text{-}O^{\cdot} + \cdot Si \equiv O_3 \qquad (3)$$

Where *En* symbolizes a source of energy, which may be either particle or ionizing radiation. The oxide network is relaxed to a new equilibrium state, which does not necessarily involve recombination of the broken bonds [32,33]. The densified BOX is partially relaxed during the high temperature annealing, as is reveals the elimination of the multiple stacking faults in the Si-overlayer near the Si/BOX interface [27].

2.4 Defect formation in the Si-overlayer close to the BOX during HTA.

Significant reconstruction occurs in the Si-overlayer at the zone near the

Si/SiO$_2$ interface during the HTA. The reverse process of equation (1) describes the dissolution of the SiO$_2$ precipitates. Therefore a perfect balance between the absorbed and emitted Si$_I$ is satisfied. The dissolution of the SiO$_2$ precipitates and the absorption of Si$_I$ is a three dimensional coalescence process that implies dislocation formation for accommodating the translation and rotation displacements between the agglomerating silicon areas, as diagrammatically shown in Fig.5, [13]. The prismatic defects that were observed at the back of the Si-overlayer in SIMOX are similar with the small prismatic faults which were observed in homo-epitaxially grown Si layers [34], suggesting that are growth defects produced during the Si-overlayer reconstruction near the Si/SiO$_2$ interface.

2.5 Point defects in BOX due to HTA

A long annealing of the thermally grown oxide above 1000° C in N$_2$ or Ar ambient results in a redox reaction of the form

$$O_3 \equiv Si - O - Si \equiv O_3 \rightarrow O_3 \equiv Si\cdot \cdot Si \equiv O_3 + O \qquad (4)$$

The free oxygen is diffused to the SiO$_2$/Si interface where it reacts forming SiO$_2$ there [35 and 37]. The reaction occurs even in oxidizing atmosphere if the thermal oxide is sealed by a poly-Si film schematically depicted in Fig.6. It was suggested that single $O_3 \equiv Si\cdot$ centers could be formed upon separation by diffusion of the two parts of the $O_3 \equiv Si\cdot \cdot Si \equiv O_3$ structure. The separated trivalent center could relax upon capture of an electron, therefore acting as electron trap [35,38]. In the same way trivalent centers can be created in the BOX during the HTA in SIMOX. After HTA only crystalline Si-islands exist in the BOX, as TEM observations reveal [13]. Spectroscopic ellipsometry, Raman spectroscopy, infrared transmission and x-ray photoelectron spectroscopy measurements performed in standard SIMOX [39], agree with TEM observations and electron spin resonance measurements (ESR) [40]. The BOX is similar to Si-rich SiO$_2$, otherwise characterized as SIPOS. In SIPOS amorphous-Si precipitates with a mean size of 10 nm exist, if the annealing temperature is lower than 1150° C. Above this temperature the a-Si is completely crystallized [41]

The coexistence of amorphous and crystalline Si precipitates in the same system after a HTA at 1300°C, as has been proposed for the BOX in Ref. [42], is thermodynamically improbable. Recent experiments in SIMOX and bonded SOI reveal that both exhibit similar radiation induced trapping properties in the BOX [43, 44]. This is unexpected for the bonded material because the SiO$_2$ is formed by thermal oxidation. The similarity in the behavior is attributed to the redox reaction, which occurs above 1000°C.

2.6. Low dose SIMOX

Low dose SIMOX has a low density of dislocation and a high throughput of SIMOX wafer production. The BOX layer is about 80nm thick resulting to better heat dissipation. Typical implantation conditions are oxygen dose $4 \times 10^{17} O^+ cm^{-2}$ and energy 120keV, subsequently subjected to HTA. This process gives a Si-overlayer about 200nm

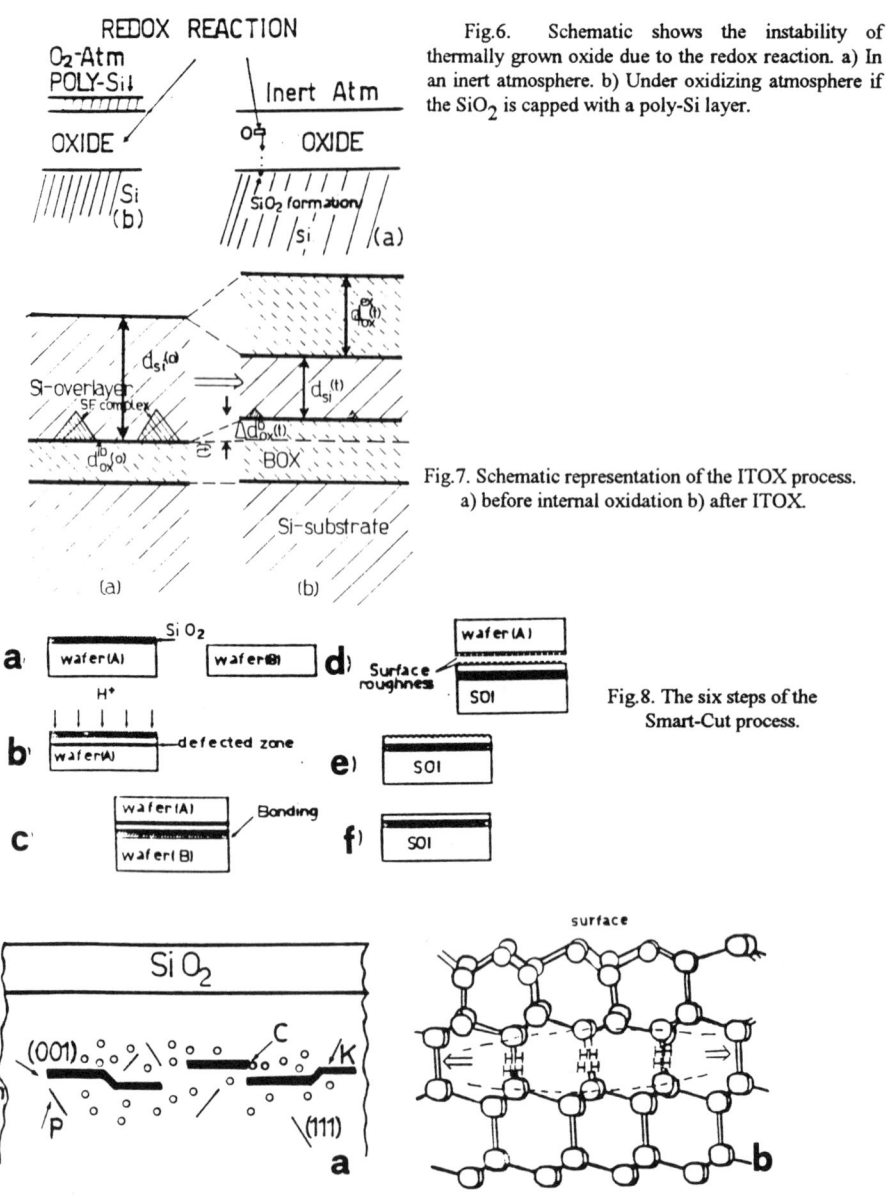

Fig.6. Schematic shows the instability of thermally grown oxide due to the redox reaction. a) In an inert atmosphere. b) Under oxidizing atmosphere if the SiO$_2$ is capped with a poly-Si layer.

Fig.7. Schematic representation of the ITOX process. a) before internal oxidation b) after ITOX.

Fig.8. The six steps of the Smart-Cut process.

Fig.9. Schematic shows the defects after hydrogen implantation. a) Small spherical vacancies, {111} plate like defects denoted by the letter P, (001) plate like defects denoted by the letter C and combination of {111} and (001) plate like defects denoted by the letter K. b) Broken Si-Si bonds parallel to the surface of The crack tip propagates on the (001) plain near the surface.

thick with a dislocation density 10^4cm^{-2}, however the density of the small SF complexes in the Si-overlayer near the BOX is 10^7 cm^{-2}, higher than in standard SIMOX [45]. This is expected, because the small SF complexes are formed during the reconstruction of the Si/SiO$_2$ interface during HTA, this reconstruction is more intensive for lower doses.

The BOX is of high quality [46], because the density of the strained -Si-Si- bonds is lower. The lower thickness of the BOX permits an easy migration of the Si_I. Unfortunately the density of conducting defects is in the range 2-7 cm^{-2} [47]. The small Si islands in the BOX, with a density 10^7 cm^{-2}, are responsible for the low breakdown electric field.

The relatively inferior electrical characteristics of the BOX can be substantially improved by internal thermal oxidation (ITOX). High temperature oxidation of the Si-overlayer in SIMOX increases the thickness of the BOX [48], as shown in Fig.7. The oxidation of the BOX is attributed to the increase of the oxygen interstitial concentration in the Si-overlayer beyond its normal solubility [49]. The increase is assumed to be proportional to the oxidation rate of the Si wafer. The internal oxidation also eliminates the small SF complexes at the backside of the Si-overlayer, as shown in Fig.7.

High oxidation growth rates are essential, for an efficient ITOX process. This occurs only at very high oxidation temperatures, oxidation at 1350°C gives an internal to external oxidation ration of the order 1/14 [49]. ITOX is a high temperature oxidation process and therefore is not possible to eliminate completely the strained Si-Si point defect due to the redox reaction. Conditions of annealing in an optimum oxidizing ambient can be realized in SIMOX, if a low dose oxygen implantation, say 10^{15} cm^{-2}, is performed at 1000° C after the HTA at 1300° C. Today brief implantation at 1000°C is feasible from the technological point of view [50].

3. Bonded Silicon on Insulator (BSOI)

In BSOI technology two thermally oxidized silicon wafers are brought to contact at room temperature. Before contacting the wafers were flushed by deionised water, drying by spinning under an infrared lamp. The bonding kinetics includes hydrogen bonds at low temperatures up to 250°C and Si-O-Si bond formation at higher temperatures up to 1100°C. The next step is the controlled thinning of one of the bonded wafers for the formation of the device film, which involves a chemo-mechanical thinning in one of the bonded wafers down to 1µm with thickness uniformity 0.3µm. The bonded SOI wafers have a very low defect density suitable for power electronics and bipolar applications [8].

3.1 Bond and Etch-back SOI (BESOI)

Bond and Etch-back SOI involves an etch-stop built in one of the bonded wafers. This is a double epi approach where the first, a heavily boron doped epi layer, serves as etch-stop and the second as device layer. The high boron concentration creates stress and consequently generates misfit dislocations. Germanium of a suitable concentration is added during the growth of the B doped epi layer to compensate the stress caused by the boron, [8]. Silicon films about 100nm thick and thickness variation of 10nm can be produced by this method.

Since boron diffuses rapidly at temperatures as low as 900°C the epitaxial growth of the Si-B-Ge etch-stop layer and the undoped device layer have to be performed well below this temperature. The annealing process to strength the wafer bonding is also limited at this temperature [6].

3.2 Plasma-Assisted Chemical Etching (PACE)

Chemo-mechanically thinned bonded wafers, with device film thickness in the range of the μm are locally etched by a small plasma tool. High-pressure plasma, confined in a 3 to 30mm diameter cup, locally etches the Si film of the wafer that is scanned under it. A computer utilizing the film thickness data controls the etching. High pressure and low energy plasma <<1eV, contribute to the final surface quality, which is comparable to that of bulk silicon.

The PACE process is capable to produce bonded wafers having device film thickness down to 50nm with 10% uniformity [7].

Ultrathin wafer bonded layers, less than 200nm thick, etched by SECCO reveal disk like defects in the order of $10^4 cm^{-2}$. The exact origin of these defects is not known. It is believed that these are local areas of enhanced stress, due to the development of interface gas bubbles [6].

3.3 Smart-Cut® technique

The Smart-Cut® technique is a combination of hydrogen ion implantation, followed by wafer bonding which permits the transfer of thin device film on the bonded wafer [9]. The process can be descried in six steps, as shown in Fig.8.

1) Wafer (A) is thermally oxidized, the thickness of the oxide corresponds to the BOX layer after bonding Fig.8a.

2) Hydrogen implantation of Wafer (A) through the oxide to a depth at which the Si will split in the 4th step, Fig.8b. Defects are produced in the implanted zone these are microcavities denoted by the letter C, {111} platelets denoted by the letter P and more extended (001) platelets parallel to the surface denoted by the letter K, as schematically shown in Fig.9a.

3) Wafers A and B, the second covered only by the native oxide, are bonded at room temperature, after a cleaning procedure similar to the standard BSOI process, Fig.8c.

4) A low temperature annealing (400 to 600°C) results in the split of wafer A in to two parts. The split corresponds in the middle of the implanted zone. A thin Si film and the oxide layer remains bonded on the wafer B. The splitted surface is rough, as shown in Fig.8d.

5) A second annealing at 1100°C stabilizes the bond, also eliminates the microdefects at the surface, as shown in Fig.8e.

6) Finally a touch-polishing step is applied in order to eliminate the roughness at the surface, this is the Unibond® wafer, Fig.8f. The wafer A can be used again as the handle wafer for a new Unibond® wafer reducing production cost.

The reduction of the fracture resistance after hydrogen implantation and the formation of blister and crater at dose above $10^{17} H^+ cm^{-2}$ have been observed by several authors [51-55]. The smart-cut® technique has been developed on the bases of these properties by Bruel [56].

The split of the (001) Si wafers after H-implantation and subsequent annealing at 600°C is attributed to the small {111} and (001) plate like defects, especially to the second. These are microcracks [54], which are formed by hydrogen passivation of the broken Si-Si bonds produced during H-implantation between two adjacent Si planes. The resulting Si-H bonds form planes of hydrogen atoms as schematically shown in Fig.9b. This configuration results in a small amount of bowing of the neighboring Si lattice planes parallel to the hydrogen planes. Thus strain is created facilitating the propagation of the crack tip by disrupting the Si lattice along the hydrogen plane. The platelets are places for concentration of molecular hydrogen, which plays a very important role for the split of the wafer during the low temperature annealing, around 600°C.

Cleavage planes in Si are the (111) planes; therefore microcracks are expected to have lower energy of formation and consequently the higher concentration on these planes. However, observations reveal that the (001) microcracks are the largest and the most numerous. From all the equivalent {001} platelets only the (001) parallel to the surface are formed.

The {111} platelets appear in all the four equivalent <111> orientations. [57]. The difference in the distribution of the plate like defects can be explained by the lower formation energy requirement in the (001) oriented platelets near the surface, because these can be easily deformed. It is noticed that H-implantation in (111) wafers results in the formation of (111) plate like defects parallel to the surface. In this case, none of the three other equivalent (111) platelets appear, as shown in Ref. [58], (figure II-18). The absence of the equivalent {111} platelets, which form 70° angle with the (111) surface, is related with the difficulty to be deformed due to their high angle with the surface.

It is estimated that only the one eighth of the implanted hydrogen is accommodated in the platelets [52], other trapping sites for hydrogen, are the small vacancy clusters and other damage defects. During the low temperature annealing hydrogen from the small vacancy clusters migrates to the larger (001) platelets which grow resulting in a crack propagation parallel to the surface.

The Smart-Cut® process permits the transfer of single crystalline Si films in other substrates which can be bonded with silicon like glass or quartz The electrical characteristics of the transferred Si film are comparable to that of the bulk Si [59].

4. SiC on Insulator (SiCOIN)

The Smart-Cut® technique is also applicable in SiC. Direct wafer bonding of an oxidized 6H or 4H-SiC films on Si substrate in order to form SiC on insulator (SiCOIN) is very attractive. Wafer bonding of 6H or 4H-SiC on insulator by the Smart-Cut® technique results in high quality low cost SiCOIN structures, because the same 6H-SiC wafer can be use for several times producing many SiCOIN wafers [10]. A 6H-SiC film 0.2µm thick can be transferred to the SiO_2/Si wafer.

For the realization of the SiCOIN structure the process described in Fig.9, paragraph 4.3 was applied. The wafer (A) is now a 6H or 4H-SiC wafer which is covered by a 1µm thick deposited SiO_2 which will be the future buried oxide. Deposited oxide is preferable because the oxidation kinetics in SiC is very low. The optimum hydrogen dose is $8 \times 10^{16} H^+ cm^{-2}$ and the temperature of transfer is 900°C, finally a heat treatment above 1100°C is applied in order to strengthen the chemical bonds [60].

Defects due to non-perfect bonding are observed having a density of $10^3 cm^{-2}$, these are holes with a mean size 50μm and blisters, and both are attributed to small particles trapped at the bonded interface. The electrical properties of the SiCOIN structure are not known yet. However a hydrogen concentration 8×10^{19} at/cm^3 remains in the SiC overlayer [60].

5. Epitaxial deposition of 3C-SiC on SOI.

Cubic SiC (3C-SiC) epitaxially grown on the Si-overlayer of SOI wafers has many advantages compared with the standard 3C-SiC epitaxially grown on Si wafers, because eliminates the effect of the Si substrate to the electrical properties of the 3C-SiC overgrown especially at elevated temperatures. The separation of the 3C-SiC from the Si substrate by the buried oxide layer results in lower leakage current improving of the electrical characteristics of the 3C-SiC [61]. Also the softening of the SiO$_2$ buried layer at the growth temperature of 3C-SiC, allows the relieve of stresses associated with the differential thermal expansion coefficient between the 3C-SiC over-layer and the Si wafer [11, 62].

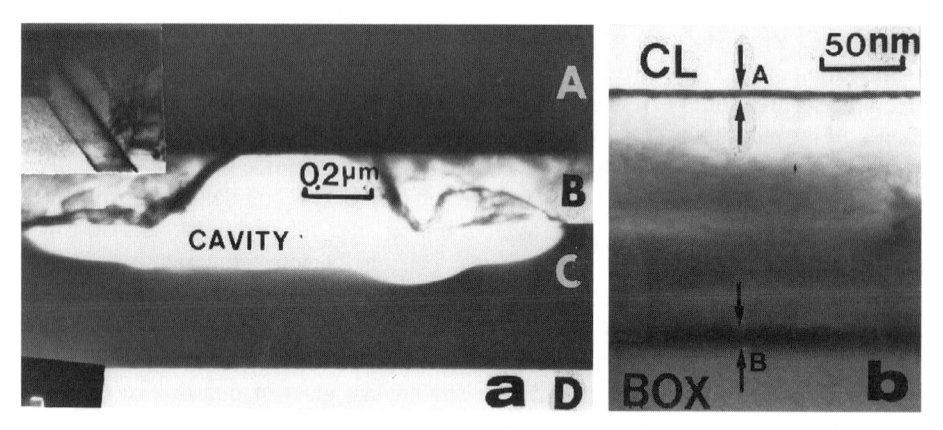

Fig.10. a) Formation of cavities in the Si-overlayer and the BOX. The letters A, B, C and D denote 3C-SiC, Si-overlayer, BOX and Si-substrate, respectively. The SiC near the surface is shown in the inset in the upper left side corner b) Cross section from a SIMOX specimen implanted by carbon through a SiO$_2$ capping layer 130nm thick. The specimen was implanted at 1030°C with a dose 0.5×10^{17} C$^+$cm^{-2} at energy 100keV. CL denotes the SiO2 capping layer. Two very homogeneous SiC layers were formed, in both the SiO$_2$/Si interfaces denoted by arrows.

The structure of epitaxially grown 3C-SiC on SIMOX is shown in Fig.10a. The following layers are observed, the 3C-SiC layer denoted by the letter A, the Si-overlayer denoted by the letter B, and the BOX denoted by the letter C, Si-islands are evident near the back interface which are denoted by the letter D. The structure of the 3C-SiC film is similar with this grown on the standard Si-wafers. Stacking Faults (SF) at the upper most part of the film having a density 3×10^9 cm^{-2}, are evident in the inset in the upper left corner of Fig.10a. No defects were formed in Si-overlayer. However very often this layer was interrupted by large cavities as it is shown in the right side of Fig.10a. These cavities include the Si-overlayer and part of the BOX revealing that during the 3C-SiC growth a huge mass transport of Si and O_2 occurred. The formation of these cavities does not disturb the 3C-SiC film revealing that the mass transport of at the Si-overlayer and the BOX occurs after the formation of the first 3C-SiC layer in the early stage of growth.

The cavities in the BOX always have as starting point one of the reversed empty pyramids bounded by {111} planes, which are formed in the Si-overlayer at the SiC/Si, during the carbonization process [63]. As the pyramid touches the Si/SiO$_2$ interface a preferential dissolution of the SiO$_2$ in BOX layer occurs. A different contrast is observed around the cavities, which is attributed to a very thin layer of polycrystalline SiC. Very often the large cavities in the BOX produce depressions at the surface. The density of the cavities is 4×10^7 cm^{-2} and their mean size 0.5µm [64]. The size and density of the cavities is reduced as the deposition temperature is reduced. The possible mechanisms of the formation of the cavities in the BOX are under investigation. A way to avoid the formation of cavities in the Si-overlayer is the formation of the carbonized SiC layer on the Si-overlayer in SOI, by low dose high temperature carbon implantation through a SiO$_2$ capping layer into the near surface region of the underling silicon. Such implantation, results in the formation of a continuous 3C-SiC epitaxial layer at the SiO$_2$ / Si interface without the formation of cavities at the Si side, as shown in Fig.10b [65]. It is believed that the chemical driving force for the preferential nucleation of 3C-SiC at the SiO$_2$/Si interface during high temperature implantation, is associated with the high density of Si dangling bonds and strained Si-O bonds in the vicinity of the interface [66]. The growth of the 3C-SiC layer is controlled by the current density of the carbon ions permitting a very slow growth rate. Therefore giving to the arriving atoms the chance to be accommodated in low energy lattice sites resulting in a better SiC film

Conclusions

It is widely accepted that silicon will continue to dominate other semiconductors. It will cover more than 98% of the semiconductor market for the next ten years [67]. Thanks to the excellent flexibility of the Si-SiO$_2$ system. The SOI structures will dominate, due to their advantages, in the new generation of the ULSI circuits.

Acknowledgment.

The author is indebted to Dr. C.Jaussaud, B. Aspar and Dr. J. Margail LETI / Grenoble and Dr. P.L.F. Hemment at the University of Surrey for their collaboration in performing this work and for the helpful discussions.

References:
1) H. Iwai, H. S. Momose, M. Saito, M. Ono, Y. Katsumata, Microelect. Eng. **28**, 145 (1995)
2) J. P. Colinge, "Silicon-on-Insulator Technology: Materials to VLSI" Kluwer Ac. Pub. London 1991.
3) T. Tsuhiaki , Microelect. Eng. **28**, 371 (1995)
4) M. A. Guerra, Solid State Technol. 75, (1990)
5) J. P. Colinge, Microelect. Eng. **28**, 423 (1995)
6) U. Gosele, M. Reiche and Q.-Y. Tong , Microelect. Eng. **28**, 391 (1995)
7) G.J. Gardopee, P.B. Mumola, P. J. Clapis, C.B. Zarowin, L.D. Bollinger and A.M. Ledger, Microelect. Eng. **22**, 347 (1993)
8) W. P. Maszara, Microelect. Eng. **22**, 299 (1993)

9) B. Aspar, M. Bruel, H. Moriceau, C. Maleville, T. Poumeyrol, A.M. Papon, A. Claverie, G. Benassayag, A.J. Auberton- Herve, T. Barge, Microelect. Eng. **36**, 233 (1997)
10) L. Di Cioccio, Y. Le Tiec, F. Letertre, C. Jaussaud, M. Bruel, Electronics Let. **32**, 1144 (1996)
11) F. Namavar, P. Colter, A. Cremins-Costa, C-H. Wu, E. Gagnon, D. Perry and P. Pirouz. Mat. Res. Soc. Symp. Proc. Vol. 423 (1996) p409
12) S. Cristoloveanu, J. Elect. Soc. **138**, 3131 (1991).
13) J. Stoemenos, A. Garcia, B. Aspar, and J. Margail, J. Electrochem. Soc. **142**, 1248   (1995)
14) J. Margail and J. Stoemenos Patent No B9390LC (24/04/1987).
15) E. A. Lewis and E. A. Irene, J. Vac. Sci. Technol. **A4**, 916 (1986).
16) S. M. Hu in "Oxygen, Carbon, Hydrogen and Nitrogen in Crystalline Silicon", MRS Vol 59 (1985) p.223
17) G. D. Watkins Ist. Phys. Conf. Ser. No. 23, 1 (1975).
18) K. Car, P. J. Kelly, A. Oshiyama and S. T. Pantelides, Phys. Rev. Lett. **52**, 1814 (1984); **54**, 360 (1985).
19) Y. Bar-yan and J. D. Joannopoulos, Phys. Rev. Lett. **52**, 1129 (1984).
20) S. M. Hu, J. Appl. Phys. **45**, 1567 (1974).
21) S. T. Dunham and J. D. Plummer, J. Appl. Phys. **59**, 2551 (1986).
22) M. Kitabake and J. E. Greene, Mat. Res. Soc. Symp. Proc. Vol 223, (1991) p9
23) W. P. Maszara, J. Appl. Phys. **64**, 123 (1988).
24) J. Margail, J. M. Lamure and A. M. Papon, Mat. Sc. Eng. **B12**, 27 (1992).
25) G. Brebec, R. Seguin, C. Sella, J. Bevenot and C. Martin. Acta Met. **28**, 327 (1980).
26) F. Namavar, T. I. Budnick, F. H. Sanchez and H. C. Hayden Proc. Mat. Res. 233 (1986).
27) S. Visitserngtrakul, C. O. Jung, T. S. Ravi, B. Cordts, D. E. Burke and S.J. Krause, Inst. Phys. Conf. Ser. No 100 Microsc. Semicond. Mater. Conf. Oxford 10-13 April 1989, Ed. A. G. Gullis and P. D. Augustus (Inst. of Physics, Bristol 1989) p557.
28) J. Stoemenos, J. Margail, M.Dupuy and C. Jaussaud. Phys. Scripta **35**,42 (1987).
29) J. Stoemenos, K. J. Reeson, A. K. Robinson and P. L. F. Hemmment, J. Appl. Phys. **69**, 793 (1991).
30) T. Makino and J. Takahashi Appl. Phys. Let. **50**, 267 (1987).
31) R.C. Barklie, A. Hobbs, P.L.F. Hemment and K. Reeson J. Phys. C. Solid State Phys. **19**, 6417 (1986).
32) R. A. B. Devine, J. Arndt. Phys. Rev. **B39**, 6132 (1989).
33) R.A.B. Devine, RADEX Conference Saint-Malo, France (1993).
34) R. K. Lawrence, H. L. Hughes and A. G. Revesz, IEEE Int. SOI Conf. Proc. Marriott-Florida Oct. 6-8 (1992) p106.
35) P. Balk, M. Aslam and D. R. Young Solid-State Electronics **27** 8/9, 709 (1984).
37) M. Aslam, R. Singh and P. Balk, Phys. Stat. Solid (a) **84**, 659 (1984).
38) R.A.B. Devine, D. Mathiot, W.L. Warren, D.M. Fleetwood, B. Aspar Appl. Phys. Lett. **63**, 2926 (1993).
39) P. J. McMarr, B. J. Mrstik, M. S. Barger, G. Bowden, J. R. Blanco, J. Appl. Phys. **67**, 7211 (1990).
40) J. F. Conley, Jr. and P. M. Lenahan, P. Roitman, Appl. Phys. Lett. **60**, 2889 (1992).
41) A. Hartstein, J. C. Tsang, D. J.DiMaria and D. W. Dong Appl.Phys Lett. **36**. 836 (1980)
42) A.G. Revesz and H.L. Hughes, Microelronic Eng. **36**, 343 (1997)]
43) K. Vanheusden and A. Stesmans."Silicon-On-Insulator Thechnology and Devices" Ed. by S. Cristoloveanu, Electoch. Soc. Vol.94-11, (1994) p.197.
44) G. Gruber, P. Pailler, J.L. Autran, A.J. Auberton- Herve, T. Barge, Microelect. Eng. **36**, 387 (1997)
45) B. Aspar, C. Guilahalmenc, P. Pudda, A. Garcia., A.M. Papon, A.J. Auberton- Herve, Microelect. Eng. **28**, 411 (1995)
46) Y. Omura, S. Nakashima, K. Izumi and T. Ishii, IEEE Trans., ED-40, 5 (1993)
47) V.V. Afanas'ev, B.Aspar, A. Auberton-Herve, G. Brown, W. Jenkins, H.L.Hughes, and A.G. Revesz, Proc. IEEE In. Con. SOI (1996), p56
48) S. Nakashima, T. Katayama, Y.Miyamura, A. Matsuzaki, M. Kataoka, D. Ebi, M. Imai,K. Izumi and N. Ohwada, J. Electrochem. Soc. **143**, 244 (1995)
49) U. Gosele, E. Schroer and J.-Y. Huh, Appl. Phys. Lett. **67**, 241 (1995)
50) R. Schork, P. Pichler, A. Kluge and H. Ryssel. Nuc. Instr. Meth. in Phys. Res. B59/60, 499 (1991)
51) W.K. Chu, R.H. Kastl,R.F. Lever, S Mad, Masters, Phys.Rev. **B16**, 3851 (1977)
52) S. Romani and J.H. Evans Nucl. Instrum. Meth B **44**, 313 (1987).
53) M.F. Beaufort,H. Garem,J.Lepinoux and J.C.Desoyer, Scripta Met. et Materialia **25**,1187(1991)
54) N.M. Johson, F.A. Ponce, R.A.Stree and R.J. Nemanich, Phys. Rev.**B35**, 4166 (1987)
55) M.F. Beaufort,H. Garem and J.Lepinoux Phil. Mag. **69**, 881 (1994)
56) M. Bruel Electron. Lett. **31**, 1201(1995)
57) B. Aspar, M. Bruel, H. Moriceau, C. Maleville, T. Poumeyrol, A.M. Papon, A. Claverie, G. Benassayag, A.J. Auberton- Herve, T. Barge, Microlelectronic Eng. **36**, 233 (1997)
58) T. Poumeyrol, These " Etude du mecanisme de transfert dans le procede Smart-Cut: Application a

l'elaboration d'une structure SOI (UNIBOND)", L'Institut National Polytechnique de Grenoble, Sept.25 (1996)
59) D. Munteanu, C. Maleville, S. Cristoloveanu, H. Moriceau, B. Aspar, C. Raynaud., O. Faynot, J.-L. Pelloie, A.J. Auberton- Herve, T. Barge, Microlelectronic Eng. **36**, 395 (1997)
60) L. Di Cioccio, F. Letertre, Y. Le Tiec, A. M. Papon, C. Jaussaud, M. Bruel, Mat. Sci. Eng. **B46**, 349 (1997)
61) W. Reichert, R. Lossy, J.M. Gonzalez Sirgo, E. Obermeier and J. Stoemenos In. Con. on SiC and Related Materials (ICSRM) Kyoto, Japan, Sep. 1995, Eds S. Nakashima, H. Matsunami, S. Yoshida and H. Harima. Inst. Phys. Con. Ser. No142,(1995) p 129
62) C. Dezauzier, J.M. Bluet,S. Contreras, J. Camassel, J. Pascual, J.L. Robert, L.DiCioccio, J. Stoemenos. ibid. p 453
63) J.P. Li, A.J. Steckl, I. Golekci, F. Reidinger, L. Wang, X.J. Ning, and P. Pirows, Appl. Phys. Let.**62** 3135 (1993)
64) W. Reichert, E. Obermeier and J. Stoemenos, European-CSRM, Heraclion-Crete, Greece Oct. 7-9 (1996) P2-64.
65) A. Nejim, P.L. Hemment, J. Stoemenos E-MRS Spring Meeting ,June 4-7,1996, Strasbourg,France Symposium I, I-VI.4, also in Nucl. Instr. Meth. Phys. Res. B **120**, 120 (1996)
66) C. R. Helms *"The Si-SiO₂ System"* Materials Science Monographs Vol. 32, Chapter 3, Ed. P. Bulk, Elsevier N.Y. (1988)
67) G. Kamarinos, Mat. Scien. Eng. A**199**, 45 (1995)

LOCAL TUNNEL EMISSION ASSISTED BY INCLUSIONS CONTAINED IN BURIED OXIDES

L. MEDA,
via Meda 35, 20141 Milano, Italy

G. F. CEROFOLINI
ST Microelectronics, 20041 Agrate, Milano, Italy

1. Introduction

Silicon-on-Insulator (SOI) structures are interesting candidates for dielectric insulation of future mos devices because, in principle, they allow larger and larger integration. In spite of the existence of different reliable technologies to obtain buried oxides (BOXes), their use is still limited because of their high production cost.

For BOXes prepared via separation by implanted oxygen (SIMOX) [1], a way to reduce the production cost is to decrease the implantation time, by lowering the oxygen dose. BOX thickness reduction from 400 nm to 100 nm has been demonstrated to be feasible, with the advantages of minor cost, better heat dissipation, and minor defectiveness in the top silicon layer [2].

The dynamics of BOX formation is complex and involves different phenomena: oxide nucleation and precipitation (occurring during oxygen implantation) and coalescence and Ostwald's ripening (occurring during the annealing) [3]. After these processes, peculiar defects are observed in BOX layers, in addition to dislocations in the top silicon due to the relief of local stress. While dislocations could be avoided by lowering the oxygen dose, BOX defects are persistent and this explains the interest in analyzing how they affect electrical performances.

The major defects in BOXes are pinholes and silicon inclusions. Pinholes interrupt the dielectric insulation by connecting top silicon with the substrate; they are more abundant in thinner oxide layers and their origin could be due to dust particulates on the silicon surface, behaving as a mask during oxygen implantation [4]. While pinholes are possibly caused by environmental factors, silicon inclusions are intrinsically related to the BOX formation, and they are present in both thin and thick layers [5-6]. Both these kinds of defects are harmful: pinholes affect device yield, while inclusions impact mainly on electrical performances. Due to their intrinsic nature, accounting for inclusion effects is an interesting challenge. Their structure can be observed by TEM and their electrical behaviour can be studied by preparing capacitors, built using BOX as dielectric layer.

To characterize BOX layers containing inclusions, scaled-area capacitors have been prepared and their current-voltage (*I-V*) characteristics have been determined. The effects of inclusions are manifold: they reduce the breakdown voltage because of the reduction of the effective BOX thickness, and sustain a field assisted tunneling current (Fowler-Nordheim current) enhancing the surface electric field. In the following we shall show that the overall behavior of the *I-V* characteristics is simply accounted for by considering only these effects. The analysis of the experimental characteristics will allow us to establish a correlation between structural parameters of silicon inclusions and electrical behavior.

2. Experimental

Oxygen ions were implanted into (100) oriented, p-type silicon wafers of 4 in. diameter, in the dose range 4.6×10^{17} - 1.2×10^{18} cm^{-2}. The ion implanter was an Eaton NV-200, operating at 200 keV and the sample temperature during the implantation was 620 °C. Subsequently, the implanted wafers were annealed at 1350 °C for 6 h in Ar + 1% O$_2$ in a furnace, to form SOI structure with sharp interfaces. The BOX thickness, depending on the oxygen implanted dose, ranged from 90 nm to more than 300 nm, underneath a silicon layer of about 200-50 nm.

Capacitors with scaled areas (1×10^{-4} cm^2, 9×10^{-4} cm^2, 2.4×10^{-3} cm^2, 1.7×10^{-2} cm^2) were built on samples of about 100 nm BOX thickness, by sputtering aluminum on SOI samples, and defining the areas by plasma etching, keeping the BOX layer as dielectric. Current-voltage characteristics were obtained using a HP4145 parameter analyzer, by ramping the voltage from 0 to 40 V, with steps of 0.5 V and duration of 0.3 s, in the temperature range 20-300 °C. The measurement conditions were destructive for the capacitor oxide as the breakdown voltage was always reached. A total area of about 300 cm^2 was investigated.

3. Results of I-V Characteristics

Typical *I(V)* curves of the described devices are shown in fig. 1; at a rough inspection their behaviour is similar to capacitors with thin thermal oxides, combined with an apparent Fowler-Nordheim behavior at higher fields. The nominal electric field, V/t_{BOX}, is by itself insufficient to account either for the apparent tunneling current, or for the relatively low breakdown voltage, so that we are forced to admit a role of silicon inclusions.

A resistive behavior was observed only for largest area capacitors, thus suggesting pinholes through the oxide. Assuming a Poisson statistics, the yield of good capacitors as a function of the area, determines the pinhole density: around 10 cm^{-2}. This value depends on the cleanliness of the environment.

I-V characteristics measured at different substrate temperature showed that the conduction process is not thermally activated, thus suggesting that the current through the thin oxide, in the pre-breakdown regime, is due to a tunnel effect.

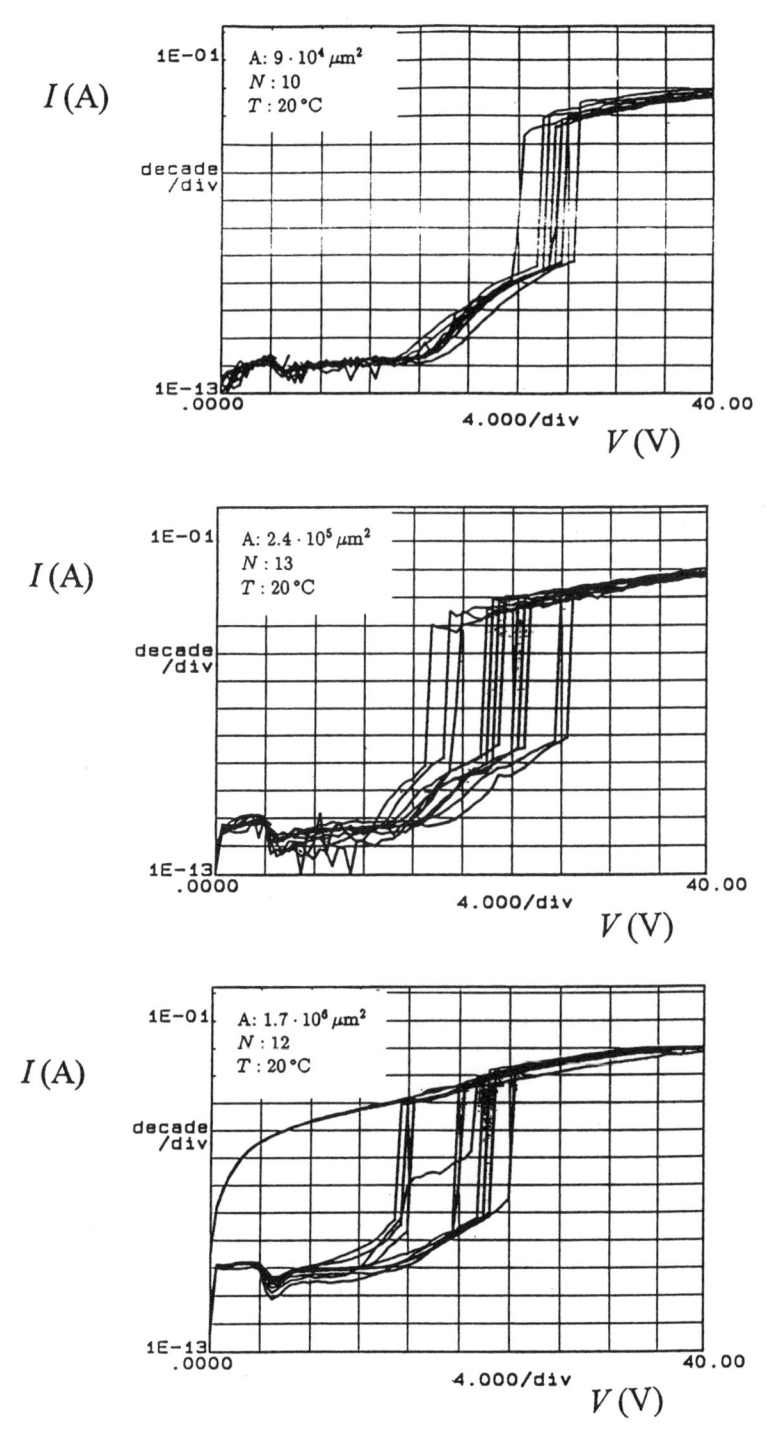

Fig. 1 - *I-V* characteristics measured on scaled area capacitors (9 10^4 μm^2; 2.4 10^5 μm^2 ; 1.7 10^6 μm^2)

4. Model

In a preliminary analysis we observed a rough correlation between the current measured immediately before the breakdown and the breakdown voltage itself [6]. Assuming that the breakdown occurs in the region hosting the biggest inclusion, the observed correlation suggests a fundamental role of inclusions in determing the tunneling current.

The present model tries to fit almost the whole characteristics, by assuming that the current density J, emitted from a surface under a strong electric field E, is given by the Fowler-Nordheim (FN) equation [7]

$$J(E) = A\, E^2 \exp\left(-\frac{B}{E}\right) \qquad (1)$$

where A and B are assumed to be constant for assigned materials.

Equation (1) holds true up to the breakdown voltage V_{BD}, where carrier multiplication produces a divergence of J at $V=V_{BD}$. This is mathematically described limiting $J(E)$ to the domain $\{E: 0 \leq E \leq E_{BD}\}$. In turn, $E_{BD}(t)$ is an increasing function of the oxide thickness t. For sufficiently thick oxides one has $E_{BD} = E_{BD}(\infty) = 8.3 \times 10^6$ V/cm. When the dielectric layer is reduced to zero the breakdown field is described as

$$E_{BD}(t) = E_{BD}(\infty) + \Delta E_{BD}\, \frac{\lambda}{\lambda + t} \qquad (2)$$

where λ is the mean free path of the electrons in the oxide, and ΔE_{BD} is the enhancement $\Delta E_{BD} = E_{BD}(0) - E_{BD}(\infty)$. However, geometric inhomogeneities in the BOX (inclusions, roughness, etc...) produce a distribution of the electric field, so that we cannot assume $E=V/t_{BOX}$, but rather we must describe the emitting surface by means of a distribution $\phi(E)$ of the electric field. By definition $\phi(E)\, dE$ is the fraction of surface with electric field between E and $E + dE$.

If different emitting regions are not correlated, the current flowing through the capacitor is given by

$$I(V) = \int_{-E_{min}}^{E_{max}} J(E)\, \alpha(E)\, dE \qquad (3)$$

where $\alpha(E) = \phi(E) a_{tot}$, and a_{tot} is the emitting surface. Of course, the distribution of the electric field $\phi(E)$ depends on the applied voltage V, and we have inserted this dependence in $I(V)$. In so doing, we have shifted the problem of descibing the actual capacitor to the problem of solving the Laplace equation with Neumann boundary conditions for the real capacitor geometry (interface topography). In most cases, such a knowledge is not available and at most one knows a distribution of inclusion thickness, and of the distances between interface and inclusions. To account for the effect of inclusions on the electric field, we consider the following elementary model.

The inclusion, of thickness x, intercepts $V x / t_{BOX}$ equipotential surfaces, characteristic of the BOX without inclusions, but all of them are expelled into the remaining BOX because of the high dielectric constant of silicon: a half are ejected in the lower part of the BOX, and the other half in the upper part. Stipulating that the lower and upper electric fields are uniform, in the lower part of the BOX the electric field is given by

$$E_{lo} = \frac{V}{t_{BOX}} \left(1 + \frac{x}{2 t_{lo}} \right)$$

where t_{lo} is the distance of the inclusion from the lower SiO$_2$-Si interface. The average electric field in the oxide containing inclusions is given by an average between the lower and the upper fields

$$E_{av} = \frac{V}{t_{BOX}} \left(1 + \frac{x}{t_{BOX} - x} \right)$$

The ratio $k = E_{lo}/ E_{av}$ defines the electric field enhancement due to the vicinity of the inclusion to the emitting surface. This quantity depends on geometric factors only

$$k(x, t_{lo}) = \left(1 + \frac{x}{2 t_{lo}} \right) \left(1 - \frac{x}{t_{BOX}} \right)$$

The electric field, responsible for the tunneling current, is hence enhanced, due both to the reduced oxide thickness, and to the vicinity of inclusions to the injecting interface. The smaller is this distance, the denser the equipotential surfaces expelled by the intercepting inclusions.

The model stands on the hypothesis that the flowing current is determined only by the injecting surface, and depends on E_{lo} only, while the BV is determined by the average field E_{av}. As a consequence, the *I-V* characteristics measured by injecting current into one and the opposite interface of a BOX, having a non-symmetric distribution of inclusions, must be different.

Given a certain distribution function of inclusion areas, the tunneling current flowing through the BOX containing silicon inclusions is believed to be given by many contributions: the leakage current I_0; the sum of currents resulting from electric fields enhanced, because of the presence of inclusions with different size and position; and the FN current tunneling through the BOX in correspondence to the areas without inclusions. Defining $J_i^*(V) = J(E_{lo}(V, x_i, t_{lo,i}))$, where E_{lo} depends on V, x_i, and $t_{lo,i}$ as described above, the current flowing through the BOX is given by

$$I(V) = I_0 + \Sigma_i \, a_i \, J_i^*(V) + (a_{tot} - \Sigma_i \, a_i) J(V/t_{BOX})$$

where a_i is the area of the i-esima inclusion of thickness x_i.

4.1. ACCOUNTING FOR THE EXPERIMENTAL CURVES

The calculation of *I(V)* requires the knowledge of the distribution function of inclusion thickness. The BV data of ref. [5] suggests that in the interval (x_{min}, x_{max}), the distribution function of inclusions depends on thickness x as exp($-x/s$), where s is a suitable length.

Taking into account that these distributions form a discrete family, we can write that the total area of inclusions with thickness x_i is given by

$$\frac{a_i}{a_{tot}} = c \, \exp\left(-\frac{x_i}{s}\right)$$

where: a_{tot} is the capacitor area, the sum being extended to all the inclusions, and c is a normalization constant.

In the following calculations we have chosen a simple distribution of four size inclusions, typically going from 25 nm to 55 nm with step $s = 10$ nm; in some cases the largest size x_{max} has been increased up to 70 nm. Other parameters have been fixed: the FN equation constants for thermal silicon dioxide-silicon interface ($A=1.3 \times 10^{-6}$ A/V^2, $B=2.32 \times 10^8$ cm/V); the BOX thickness ($t_{BOX}=100$ nm); the capacitor area ($a_{tot}=10^{-3}$ cm^2);

the breakdown field ($E_{BD}(\infty) = 8 \cdot 10^6$ V/cm); the enhancement of the breakdown electric field for thin oxides ($\Delta E_{BD}=10^7$ V/cm); the mean free path of electrons in the oxide ($\lambda=5$ nm); the leakage current ($I_0=10^{-12}$ A). In the simulations, performed with MATHEMATICA 2.2 [8], we have considered the effect of the variation of several parameters:

- the dimension of the biggest inclusion;
- the position of inclusions with respect to the interface;
- the comparison of different inclusion distributions;
- the weight of the biggest inclusion compared with all the others;
- the inclusion asymmetry seen for opposite injections.

It can be observed that:

1 - the biggest inclusion determines the BV, that is also related to the maximum current just before BV (fig. 2);

2 - a key role is played by the distance of the inclusion from the injecting interface, because this distance enhances the emitting field and hence the tunneling current (fig. 3);

3 - the inclusion distribution has also an important role in affecting the tunneling current; in fact, fixing the biggest dimension and considering two different distributions one steeper than the other one, a higher tunneling current corresponds to the steeper distribution (fig. 4);

4 - a single big inclusion carries a current comparable with the contribution of all the others, as is seen for the curves of fig. 5;

5 - when inclusions are arranged in an asymmetric way in the BOX, the tunneling current is different by changing the direction of injection (fig. 6).

Revesz noticed this fact and proposed a similar explaination of conduction via tunneling of electrons between silicon small clusters into the BOX [9].

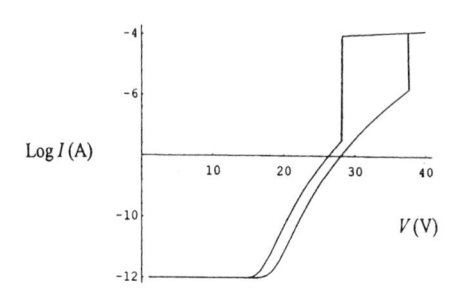

Fig. 2 I-V characteristics calculated for two inclusion distributions differing only for the largest inclusion (70 nm and 60 nm); the calculations have been performed with fixed $t_{lo} = 7$ nm.

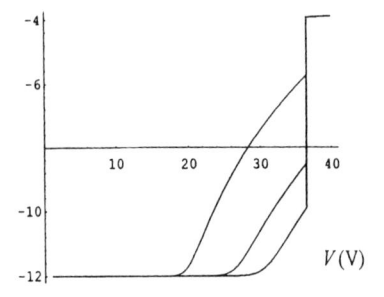

Fig. 3 Comparison of the I-V characteristics calculated for different t_{lo} = 8 nm, 12 nm, and 15 nm; for fixed inclusion distribution (from 60 nm to 40 nm)

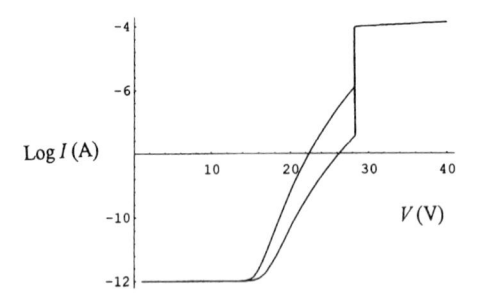

Fig. 4 Comparison of characteristics for two inclusion distributions; the upper curve refers to the steeper one (from 70 nm to 50 nm), the lower to the smoother one (from 70 nm to 25 nm); $t_{io} = 7$ nm.

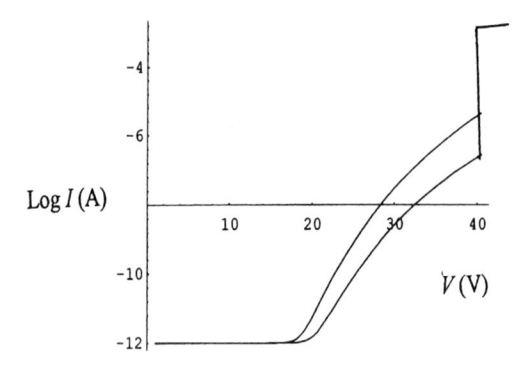

Fig. 5 Current flowing through the largest inclusion (55 nm) is comparable with the total current for an inclusion distribution (from 55 nm to 40 nm); $t_{io} = 7$ nm.

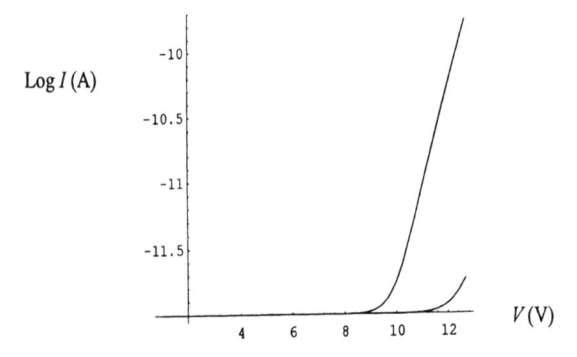

Fig. 6 Polarity dependence calculated for a given inclusion distribution (from 88.5 nm to 40 nm), and for fixed $t_{io} = 5$ nm at the lower side. The upper curve, corresponding to the sie with smaller t_{io} (direct current)

5. Discussion and Conclusions

The major attractive of the SOI structure is its ability to provide dielectric insulation of the active zone from bulk silicon. Any leakage current is therefore detrimental and should be avoided. The comprehension of causes responsible for leakage is expected to be able to provide clues for their elimination; hence the interest toward a detailed understanding of mechanisms of conduction through the BOX.

Our analysis has demonstrated that for all voltages the current can be described as the sum of a nearly constant leakage current I_0 and a Fowler-Nordheim current. The reasons for almost constant current at lower fields are not clear: capacitor charging, trapping-detrapping at the interface or bulk states, as well as low field conductive paths (e.g. silicon molecular chains through the BOX) might play a role. From the quantitative point of view, this leakage current is however small and affects device performances only at a minor degree. The FN current at higher fields, instead, may become quite high and, moreover, the destructive breakdown occurs at voltages appreciably lower than the one expected by the BOX thickness. In the literature, at least three models have been proposed to explain this behavior.

Silicon Clusters. This hypothesis was advanced by Revesz et al. in 1993 [9], who assumed that the oxygen implantation was unable to produce fully stoichiometric SiO_2. After annealing the BOX would result in silica containing small silicon clusters. The anomalously high conduction, therefore, would be a FN current assisted by a dramatic reduction of the barrier height, resulting by the formation of an impurity band in the oxide. Though we cannot rule out this hypothesis for all SOI structures, this explaination deos not account for the reduced BV observed for BOXes.

Geometric Disuniformities. Any disuniformity in BOX thickness is responsible for an increase of the electric field. Since the measured capacitance is consistent with the one expected for a BOX of thickness t_{BOX}, only a small fraction of the total area is flawn by disuniformity. About the nature of disuniformity we mention interfacial asperities due to silicon crystallization at the Si-SiO_2 interface, or silicon inclusions in the BOX. The former defect was advocated by Hall and Wainwright in 1996 [10], who overlooked the structural distribution of inclusions.

Silicon inclusions close to the injecting interface was the defect proposed by Meda et al. in 1994 [6]; this explaination takes into account both the FN current and the reduced BV. The very facts that there exists a distribution of inclusions which account for both the experimental *I-V* characteristics and BV, and also that electron micrographies gave evidence for inclusions with roughly the right distribution (see fig. 7), makes us confident that the major mechanism for the anomalous characteristics measured in BOXes of SOI structures is Fowler-Nordheim emission, sustained by the electric field enhanced by the presence of silicon inclusions close to the silicon -oxide interface.

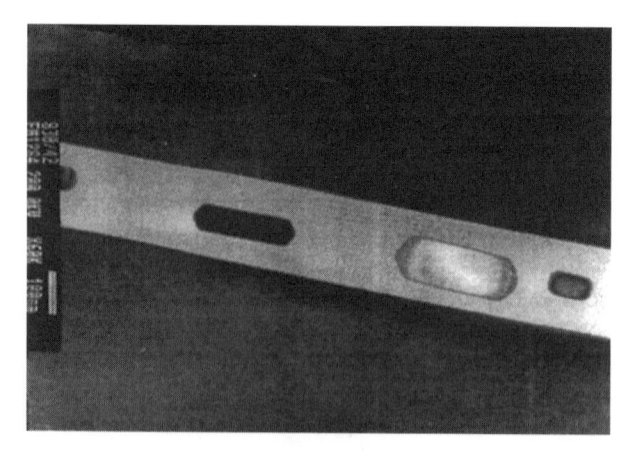

Fig. 7 TEM cross-section showing a burried oxide layer of about 180 nm containing silicon inclusions of different size.

6. References

1. Izumi K., (1990) *Vacuum* 42, 333
2. L. Meda L., S. Bertoni S., Cerofolini G.F., Spaggiari C. , (1993)*Nucl. Instr. and Methods* B80/81, 813
3. Cerofolini G.F., Bertoni S., Meda L., Spaggiari C., (1994)*Mater. Sci. Eng.* B22, 172
4. Meda L., Bertoni S., Cerofolini G.F., Pagliano P., Pandini G., Spaggiari C., (1995) *Ion Implan. Tech. 94*, Elsevier Sci., 706
5. Meda L., Bertoni S., Cerofolini G.F., Gassel H., (1994) *Nucl. Instr. and Methods* B84, 270
6. Meda L., Bertoni S., Cerofolini G.F., Spaggiari C., Gassel H., (1994) *Proc. 6th Intl. Symp. SOI Technoogy and De*vices, The Electrochem. Soc., Pennington NJ, vol. 94-11, 224
7. Lenzlinger M. and Snow E.H., (1969) *J. Appl. Phys.* 40, 278
8. Wolfram S., (1991) *Mathematica*, Addison-Wesley Publishing Company second edition.
9. Revesz A. G., Brown A. G., Hughes H. L., (1993) *J. Electrochem. Soc.*, 140, 3222
10. Hall S., Wainwright S.P., (1996) *J. Electrochem. Soc.*, 143, 3354

Authors Index

V. Afanas'ev	425	G. Lucovsky	147, 447
W. Arnoldbik	411	A. Markovits	131
I. Baumvol	165, 227	H. Z. Massoud	103
V. Borman	309	A. Mattheus	217
I. W. Boyd	461	L. Meda	493
D. Brasen	181	O. V. Meshurov	391
R. Car	89	C. Minot	131
G. Cerofolini	117, 493	M. Mitte	289
V. Chelnokov	431	S. Miyazaki	315
M. Clement	25	Y. Mizubayashi	315
V. Craciun	461	K. Morino	315
J. J. M. de Nijs	25, 425	H. Neddermeyer	289
S. Dimitrijev	191	H. Niimi	447
T. Doege	289	V. Osipov	375
D. Druijf	425	E. V. Ostroumova	383
G. Dufour	165	A. Pasquarello	89
V. Emelianov	391	E. Poindexter	397
L. C. Feldman	1, 181	H. M. Przewlocki	343
M. Fukuda	315	C. Radtke	227
J. Ganem	165	N. Re	117
E. Garfunkel	1, 39, 181	G. M. Rignanese	89
G. J. Gerardi	397	S. Rigo	165
L. Gosset	165	F. Rochet	165
A. Grassl	217	A. A. Rogachev	383
M. L. Green	39, 181	B. Rottger	289
V. Gritsenko	335	T. Salgado	227
E. Gusev	1, 39, 181	A. Shklyaev	277
T. Gustafsson	39, 181	C. J. Sofield	79
A. Gschwandtner	217	A. V. Sogoyan	391
F. H. P. Habraken	411	T. Sorsch	181
H. B. Harrison	191	P. Soukiassian	257
E. Harazim	289	J. H. Stathis	325
T. Hattori	241	F. Stedile	165, 227
M. Hirose	315	J. Stoemenos	477
M. S. Hybertsen	89	A. M. Stoneham	79
G. Innertsberger	217	J. Stoemenos	477
F. Jolly	165	A. Szekeres	65
R. Kliese	289	A. Talg	217
A. Kraun	289	P. Tanner	191
A. Kuiper	411	I. Trimaille	165
R. Kulla	289	V. Troyan	309
C. Krug	227	C. Tsamis	359
Yu.Yu. Lebedinski	309	D. Tsoukalas	359
A. Lebedev	431	E. Ullersma	411
W. N. Lennard	181	R. G. Useinov	391
H.-F. Li	191	C. F. Young	397
H. C. Lu	39, 181	G. I. Zebrev	391
Z.-H. Lu	49	J.-Y. Zhang	461

List of Workshop Participants

1. **Dr. B. H. Bairamov**
Ioffe Physico-Technical Institute
Russian Academy of Sciences
26 Polytechnicheskaya St,
194021 St Petersburg
Russia

2. **Prof. Israel Baumvol**
Instituto de Fisica - UFRGS
Av. Bento Goncalves, 9500,
91540-000 Porto Alegre, Brazil
and
Universite Paris 7, 2 place Jussieu, 75251
Paris Cedex 05, France
e-mail: baumvol@gps.jussieu.fr

3. **Dr. V. F. Borodzula**
St.-Petersburg Techical Univ.
St. Petersburg,
Russia
e-mail: anny@nest.neva.ru

4. **Dr. Gianfranco Cerofolini**
SGS-THOMSON Microelectronics
20041 Agrate MI
Italy
e-mail: Gianfranco.cerofolini@st.com

5. **Dr. Valentin Craciun**
Institute of Atomic Physics
Lasers Department
Bucharest V, Magurele
RO-76900
Romania
e-mail: craciv@roifa.ifa.ro

6. **Dr. Jan J. M. de Nijs**
DIMES/TU Delft
PO Box 5053
NL-2600
Delft
The Netherlands
e-mail: denijs@dimes.tudelft.nl

7. **Prof. Yuri N. Devyatko**
Dept. of Theoretical Physics
Moscow Enginer. Physics Inst.
Moscow 115409
Russia
e-mail: devyatko@veshn.mephi.msk.su

8. **Dr. Alexey Efremov**
Inst. of Semicond. Physics
Nat. Academy of Sci. of Ukraine
Prospect Nauki, 45, Kiev, 252028
Ukraine
e-mail: lvg@div9.semicond.kiev.ua

9. **Vladimir Emelianov**
Research Institute of Scientific Inst
Minatomenergo Russia
Lutkarino-1, Turaeva
Moscow Region 140061
Russia
e-mail: avsog@spels.msk.ru

10. **Dr. Anatoli Evtukh**
Inst. of Microelectronics
Kiev
Ukraine
e-mail:lvg@div9.semicond.kiev.ua

11. **Prof. Leonard C. Feldman**
Dept. of Physics and Astronomy
Vanderbilt University
Nashville, TN 37235
USA
e-mail: feldman@ctrvax.vanderbilt.edu

12. **Prof. Eric Garfunkel**
Department of Chemistry
Rutgers University
Piscataway, NJ 08854
USA
e-mail: garf@rutchem.rutgers.edu

13. **Dr. Martin Green**
Bell Laboratories
Lucent Technologies
Murray Hill, NJ 07974
USA
e-mail: marty_green@bl11.lucent.com

14. **Dr. Vladimir Gritsenko**
Institute of Semicond. Physics
Russian Academy of Science
Siberian Branch, 630090 Novisibirsk
Russia
e-mail: vlad@isph.nsk.su

15. **Prof. Evgeni Gusev**
Depts. of Physics and Chemistry
Lab. for Surface Modification
Rutgers University
Piscataway, NJ 08854
USA
e-mail: gusev@physics.rutgers.edu

16. **Prof. Torgny Gustafsson**
Dept. of Physics and Astronomy
136 Frelinghuysen Road
Piscataway, NJ 08854-8019
USA
e-mail: gustaf@physics.rutgers.edu

17. **Prof. Frans H. P. Habraken**
University of Utrecht
Utrecht
The Netherlands
e-mail: F.H.P.M.Habraken@fys.ruu.nl

18. **Prof. H. Barry Harrison**
School of Microelectronic Engin.
Faculty of Science and Technol.
Griffith University
Queensland, 4111
Australia
e-mail: B.Harrison@me.gu.edu.au

19. **Prof. Takeo Hattori**
Dept. of Electrical & Electronic
Engineering
Musashi Institute of Technology
1-28-1 Tamazutsumi
Setagaya-ku, Tokyo 158
Japan
e-mail: hattori@ipc.musashi-tech.ac.jp

20. **Prof. Masataka Hirose**
Research Center for Nanodevices and
Systems
Hiroshima University
1-4-1 Kagamiyama
Higashi-Hiroshima 739
Japan
e-mail: hirose@sxsys.hiroshima-u.ac.jp

21. **Gudrun Innertsberger**
Univ. of Munchen and Siemens
Gruenwalderstr. 46/4
D-81547 Munchen
Germany
e-mail: gudrun.innertsberger@hlistc.siem

22. Dr. G. G. Kareva
Institute of Physics
St. Petersburg University Peterhof
St. Petersburg, 198904
Russia
e-mail: apver@onti.niif.spb.su

23. Dr. V. T. Khartsiev
Ioffe Phys.-Technical Institute Russian
Academy of Sciences
26 Polytechnicheskaya St.
194021 St. Petersburg
Russia
e-mail: kharv@les.ioffe.rssi.ru

24. Dr. Sergey V. Kidalov
Ioffe Phys.-Technical Institute
Russian Academy of Sciences
26 Polytechnicheskaya St.
194021 St. Petersburg
Russia
e-mail: kid@vul.ioffe.rssi.ru

25. Dr. S. Kojkov
St.-Petersburg Technical Univ.
St. Petersburg
Russia

26. Dr. A. Krylov
St. Petersburg Electrotechnical State University
St. Petersburg
Russia
e-mail: vald@vald.usr.etu.spb.ru

27. Dr. Alexander Lebedev
Ioffe Phys.-Technical Institute
Russian Academy of Sciences
26 Polytechnicheskaya St.
194021 St. Petersburg
Russia
e-mail: shura@lebedev.ioffe.rssi.ru

28. Prof. Vladimir Litovchenko
Inst. of Semicond. Physics,
National Academy of Sci. of Ukraine,
Prospect Nauki
45, Kiev, 252028
Ukraine
e-mail: lvg@div9.semicond.kiev.ua

29. Dr. Zheng-Hong Lu
Inst. for Microstructural Sciences
National Research Council of Canada
Ottawa, Canada K1A OR6
e-mail: Zheng-Hong.Lu@nrc.ca

30. Prof. Gerald Lucovsky
Department of Physics
North Carolina State University
Raleigh, NC 27595-8202
USA
e-mail: Gerry_Lucovsky@ncsu.edu

31. Prof. Gunther Luepke
Dept. of Physics and Astronomy
Vanderbilt University
Nashville, TN 37235-0291
USA
e-mail: luepkeg@ctrvax.vanderbilt.edu

32. Prof. Hisham Z. Massoud
Dept. of Electrical and Computer Engineering
Duke University
Durham, NC 27708-0291
USA
e-mail: massoud@ee.duke.edu

33. Alexander Mattheus
Siemens AG
Semiconductors Center
Otto-Kahn-Ring 6
D-81739 Munich
Germany
e-mail: Mattheus@msgw.hlistc.siemens.de

34. Dr. Read McFeely
IBM T. J. Watson Research Center
Yorktown Heights, NY 10598
USA
e-mail: mcfeely@watson.ibm.com

35. Dr. Laura Meda
Enichem
via Fauser 4
28100 Novara
Italy
e-mail: meda@enichem.geis.com

36. Prof. Christian Minot
Lab. Chimie Theorique,
Universite P et M Curie,
4 place jussieu 75252
Paris cedex 05
France
e-mail: christian.minot@lct.jussieu.fr

37. Steven Moscowits
Department of Chemistry
University of Washington
Seattle, WA 98195
USA
e-mail: mosk@u.washington.edu

38. Prof. Hennig Neddermeyer
Martin-Luther-University
Halle-Wittenberg, FB Physik
D-06099 Halle
Germany
e-mail: neddermeyer@ep3.uni-halle.de

39. Dr. Vladimir Osipov
Ioffe Physico-Technical Institute F
Academy of Sciences
26 Polytechnicheskaya St.
194021 St. Petersburg
Russia
e-mail: osipov@fic.ioffe.rssi.ru

40. Dr. E. V. Ostroumova
Ioffe Physico-Technical Inst.
Russian Academy of Sciences
26 Polytechnicheskaya St.
194021 St. Petersburg
Russia
e-mail: trirog@mail.wplus.net

41. Dr. Alfredo Pasquarello
IRRMA
Ecublens
CH-1015 Lausanne
Switzerland
e-mail: pasquarello@irrmasg17.epfl.ch

42. Dr. Edward Poindexter
Army Research Laboratory
AMSRL-SE-EI
2800 Powder Mill Road
Adelphi, MD 20783
USA
e-mail: edward_poindexter@emh3.arl.mil

43. Prof. Henryk M. Przewlocki
Institute of Electron Technology
Al. Lotnikow 32/46
PL 02 668 Warszawa
Poland
e-mail: hmp@ite.waw.pl

44. Prof. Serge Rigo
Universite Denis Diderot - Paris 7
Groupe de Physique des Solides
Tour 23 2 place Jussieu
75251 Paris Cedex 05, and
Ministere de L'Education Nat.
France
e-mail: Serge.Rigo@dgrt.mesr.fr

45. Dr. A. A. Rogachev
Ioffe Physico-Technical Institute R
Academy of Sciences
26 Polytechnicheskaya St.
194021 St. Petersburg
Russia
e-mail: trirog@mail.wplus.net

46. Dr. Sergey N. Rogozhkin
Dept. of Theoretical Physics
Moscow Eng. Physics Inst.
Moscow 115409
Russia
e-mail: h@rsv.mephi.msk.su

47. Dr. Alexander Shklyaev
Inst. of Semicond. Physics
Russian Academy of Science
Siberian Branch, 630090 Novisibirsk
Russia
e-mail: shklyaev@jrcat.or.jp

48. Dr. V. I. Sokolov
Ioffe Physico-Techical Institute Ru
Academy of Sciences
26 Polytechnicheskaya St.
194021 St. Petersburg
Russia
e-mail: zam@zam.spb.ru

49. Prof. Patrick Soukiassian
Commissariat a l'Energie Atomique,
Saclay,
DSM-DRECAN-SRSIM
Batiment 462, 91191
Gif sur Yvette Cedex
France
e-mail: psoukiassian@cea.fr

50. Dr. James H. Stathis
IBM TJ Watson Research Center
Yorktown Heights, NY 10598
USA
e-mail: stathis@watson.ibm.com

51. Prof. Fernanda Stedile
Instituto de Quimica - UFRGS
Av. Bento Goncalves, 9500
915400-000 Alegre
Brazil
e-mail: stedile@if.ufrgs.br

52. Prof. A. Marshall Stoneham
Dept. of Physics and Astronomy
University College London
London WC1E 6BT
UK
e-mail: ucapams@ucl.ac.uk

53. Prof. John Stoemenos
Aristotle University of Thessaloniki
Department of Physics
54006 Thessaloniki
Greece
e-mail: stoimeno@ccf.auth.gr

54. Dr. Anna Szekeres
Institute of Solid State Physics
Tzarigradsko Chaussee 72
Sofia 1784
Bulgaria
e-mail: aszeki@center.phys.acad.bg

55. Prof. Norman Tolk
Dept. of Physics & Astronomy
Vanderbilt University
Nashville, TN 37235
USA
e-mail: tolk@macpost.vanderbilt.edu

56. Dr. Victor Torres
Departamento de Fisica
Universidade de Aveiro
3810 Aveiro
Portugal
e-mail: vtorres@ideiafix.fis.ua.pt

57. Dr. Isabelle Trimaille
Universite Paris 7
2 place Jussieu
75271 Paris Cedex 05
France
e-mail: trimaill@gps.jussieu.fr

58. Prof. Victor Troyan
Interdepartmental Analytical Lab.,
Moscow Eng. Physics Institute
Moscow 115409
Russia
e-mail: mal@park.mephi.ru

59. Dr. Christos Tsamis
NCSR "Demokritos"
Institute of Microelectronics
15310 Aghia Paraskevi
Athens
Greece
e-mail: ctsamis@cyclades.nrcps.ariadnet.gr

60. Dr. V. Ya. Uritsky
St. Petersburg Electrotechnical
State University
St. Petersburg
Russia
e-mail: vald@vald.usr.etu.spb.ru

61. Dr. Oleg A. Usov
Ioffe Physico-Technical Institute
Russian Academy of Sciences
26 Polytechnicheskaya St.
194021 St. Petersburg
Russia
e-mail: usov@exd.ioffe.rssi.ru

62. Prof. Alexander Ya. Vul'
Ioffe Physico-Technical Inst. Russian
Academy of Sci.
26 Polytechnicheskaya St.
194021 St. Petersburg
Russia
e-mail: vul@vul.ioffe.rssi.ru

63. Dr. M. V. Zamoryanskaya
Ioffe Physico-Technical Institute
Russian Academy of Sciences
26 Polytechnicheskaya St.
194021 St. Petersburg
Russia
e-mail: zam@zam.spb.ru